D0079015

Basic Calculus

Springer

New York
Berlin
Heidelberg
Barcelona
Budapest
Hong Kong
London
Milan
Paris
Singapore
Tokyo

Basic Calculus

From Archimedes to Newton to its Role in Science

Alexander J. Hahn

Department of Mathematics
University of Notre Dame

Springer

Textbooks in Mathematical Sciences

Series Editors

Thomas F. Banchoff
Brown University

Jerrold Marsden
California Institute of Technology

Keith Devlin
St. Mary's College

Stan Wagon
Macalester College

Gaston Gonnet
ETH Zentrum, Zürich

Cover Illustration: A photo adaptation of the Earth and Moon as viewed from the Apollo 8 spacecraft. Original photo courtesy of the National Space Science Data Center and Dr. Richard J. Allenby, Jr., Principal Investigator (68-118A-01A).

Library of Congress Cataloging-in-Publication Data
Hahn, Alexander, 1943–
    Basic calculus : from Archimedes to Newton to its role in
science / Alexander J. Hahn.
      p. cm. — (Textbooks in mathematical sciences)
    Includes bibliographical references and index.
    ISBN 0-387-94606-3 (hardcover : alk. paper)
    1. Calculus—History.   I. Title   II. Series.
  QA303.H154   1997
  515′.09—dc21   97-997

Printed on acid-free paper.

Production managed by Timothy Taylor; manufacturing supervised by Joe Quatela.
Photocomposed copy prepared from LaTeX files by The Bartlett Press, Marietta, GA.
Printed and bound by Edwards Brothers, Inc., Ann Arbor, MI.
Printed in the United States of America.

9 8 7 6 5 4 3 2 1

ISBN 0-387-94606-3 Springer-Verlag New York Berlin Heidelberg   SPIN 10516485

For the Instructor

This text could as well have the title *The Story of Calculus*. It weaves a tale of mathematics from the trigonometry of the Greeks, the analytic geometry of Descartes, the calculus of Leibniz and Newton, to the role of calculus in a number of disciplines of basic science and modern economics. This calculus is elementary and current, but follows the subject's historical flow as it was driven by the important problems of the day.

The Purpose: Mathematics texts usually present their subject with a focus on a mastery of subroutines, leaving interesting connections with other disciplines to footnotes and exercises, and failing to illuminate the impact of mathematics in a persuasive manner. It is the twofold purpose of this text to develop calculus from within its rich historical context and to demonstrate its broad and formidable informative power. The subject is painted with a broader brush that includes basic elements of science, but the emphasis is always on the careful development of the mathematics and the information that it provides.

Experience at Notre Dame: My colleagues and I use this text in two different first-year calculus courses at the University of Notre Dame. One is a two-semester calculus sequence for arts and letters honors students. The other is a one-semester course of elementary applications of calculus for regular arts and letters students and architecture majors. The overall response from these students has been very positive, even though they have brought very different backgrounds and abilities to the classroom. This has been evident from the high formal course evaluations by the students as well as their supportive glowing informal comments. Indeed, I was awarded the university-wide Madden award for teaching excellence in 1995 in large measure because of their "applause." I should hasten to point out, however, that this text is not "mathematics as usual." Some students take to it easily, but others find it difficult to leave behind the "plug and chug" mentality that they bring from high school. It goes without saying that the instructor needs to be very sensitive to the nature and level of the audience. The pace of the course, the amount of reviewing, the selection of materials, the exposition of the elementary aspects of science, and the choice of homework exercises (from routine, to demanding, to horizon-expanding), are *all* factors that must be given very careful attention, because all will have considerable impact on the scope and substance of the course.

Description of the Contents: Chapter 1 is an exposition of basic trigonometry and its applications (to the concerns of the Greek thinkers). Chapter 2 has an elementary look at the mathematics and astronomy of the "Ptolemaic system." Chapter 3 introduces conic sections and area, and presents Archimedes' analysis of the area of a parabolic section. Chapter 4 develops analytic geometry and the astronomy of Galileo and Kepler. Chapters 5 and 6 together constitute a modern course in essential calculus based on the work of Leibniz and Newton. Chapter 7 focuses on the heart of Newton's *Principia Mathematica*. Chapters 8, 10, and 13 develop calculus in a more complete and definitive way, in part as a response to fundamental questions that arise in the arguments of Leibniz and Newton. Chapter 9 consists of four mathematical studies: a pulley problem of De L'Hospital, the suspension bridge, an experiment of Galileo with incline planes, and geometric optics. Chapter 11 analyzes radioactivity, three nuclear clocks and what they tell us, and the dynamics of populations. Chapter 12 presents the basics of business calculus. Chapter 14, finally, develops fundamental concepts of physics (work, energy, impulse, and momentum) and applies calculus to a variety of topics ranging from the motion of a bullet in the barrel of a gun, to the propulsion of rockets, to the expanding universe.

Logical Interdependence: This text contains several components that are essentially self-contained. One such component is the calculus in Chapters 5 and 6 together with some analytic geometry (from Chapter 4). Chapter 7 and fundmental facts about conic sections (from Chapter 3) and the calculus of velocity (from Chapter 6) is another. Chapter 9 along with a some basic calculus (the chain rule and the max/min strategy from Chapter 8) is a third component. A fourth is Chapter 11 and some elementary properties of exponential and logarithm functions (from Chapter 10). A fifth component consists of the business calculus of Chapter 12 together with very basic differential and integral calculus (as covered in Chapters 5 and 8). Finally, Chapters 13 and 14 together comprise a challenging component of integral calculus and physics.

Examples of Courses from This Text: A number of different first-year calculus courses can be taught from this text to audiences ranging from arts and letters to science students. Some specific suggestions follow:

Course A: This fast-paced, one-semester course proceeds linearly from Chapter 1 through Chapter 7. It covers basic college mathematics from geometry to trigonometry and from analytic geometry to calculus. It does so from "Archimedes to Newton," in other words, from within the context of the scientific questions that helped drive these disciplines. This outline has been followed in the first semester of the arts and letters honors course at Notre Dame. A colleague has taught this course with Chapter 8 in place of Chapter 7. The contrast between the calculus of Leibniz/Newton and the more definitive approach of Chapter 8 makes it possible to discuss the rationale for rigor in mathematics.

A less demanding version of the outline proceeds as follows:

Course B: The approaches of Leibniz and Newton in Chapters 5 and 6 together provide a modern course of differential and integral calculus. This course is intuitive, devoid of excess baggage, concentrates on the essence of the matter, and includes important applications. If analytic geometry needs attention, add Section 4.3. This core material can be supplemented in a number of ways. The origins of integral calculus can be traced by studying Archimedes' analysis in Chapter 3. After a review of the conic sections, also in Chapter 3, the students are ready for the challenging look at the heart of Newton's *Principia* that Chapter 7 undertakes.

A more traditional two-semester calculus course for science students can be taught using the following model:

Course C: Start with a review of the conic sections from Chapter 3 and continue with the analytic geometry (and its applications to astronomy) of Chapter 4. Then cover Chapters 8 to 11 and Chapters 13 and 14. These chapters provide a traditional calculus course with a thorough look at applications. A number of different topics in science are linked together by the common thread of calculus that gives essential information to them all. The many exercises can be used to expand the scope of topics.

A toned-down version of this outline is the basis of a second-semester course of elementary applications of calculus taught to arts and letters students and architecture majors at Notre Dame:

Course D: This course has as its single goal the illustration of the proposition that calculus informs a variety of different disciplines with absolutely crucial information. The course has any basic calculus course as prerequisite and (with insertions of appropriate reviews) focuses on a selection of applications that can include the pulley problem of L'Hospital, radioactivity and nuclear clocks, models of populations, bullets in rifles, and Hubble's law and Einstein's universe.

Course E: An essential one-semester "business calculus" can be taught from Chapters 5, 6, 8, and 12.

Because there is such a rich historical discussion throughout the text, judicious selections of chapters can provide the basis for a variety of specialized courses for specific programs of study. For example, astronomy is a recurring theme in this book, and the materials in Chapters 1, 2, 4, 7, and 14 can serve as the basis of a course "Elements of mathematics in astronomy." Chapters 4, 5, 6, and 7 along with appropriate reading materials, such as those authored by Herbert Butterfield and Thomas Kuhn, form the substance of a course "revolutions in science and mathematics." A slower and more elementary "mathematics and science of the Greeks" could flow from Chapters 1, 2, and 3 and Section 4.1 and selected reading from Aristotle and Plato.

All of the courses just discussed can be supplemented with the web site

`http://www.nd.edu:80/-hahn/`

Acknowledgments

 big hug of thanks to my wife Marianne and daughter Anneliese. To the former for her unflagging encouragement and support, to the latter for assisting with the preparation of the manuscript, and to both of them for hanging in there with a guy writing a book. A warm word of thanks also to my parents for their affection and for being impressed by all the activities of their son regardless of their merit.

A word of gratitude to Timothy O'Meara, who led the way long ago when he enlivened the calculus courses that many of us taught with him with compelling historical elements.

A heartfelt thank-you to Eileen Kolman for supporting me and my efforts at a difficult time, and another to Bart Braden for his very generous assistance.

I am grateful to my colleague Bruce Williams for the effective and energized way in which he taught from this text and for the many clarifications and corrections that he provided.

Thanks to Serafim Arzoumanidis, of Steinman Company, in Manhattan, for enlightening me about suspension bridges, and to Laura and Maria Eidietis for spending many hours fishing for facts in the university library.

Special thanks go to my honors students of the academic year 1994/95: Chrissy Chiacco, Ed Dawson, Jeremy Hutton, Tica Lee, Sean Macmanus, Molly McCracken, David McGarry, Anne McNicholas, Amber Neely, Kristen O'Connor, Eileen Scully, Rae Sikula, Jerry Steinhofer, and Ashleigh Thompson.

And thanks also to those of the academic year 1995/96: Claire Boyle, Dina Brick, Melanie Caesar, Johnathan D'Amore, Maria Eidietis, David Freddoso, Adam Haubenreich, Christina Hilpipre, Kristine Howard, Fred LaBrecque, Shane Lampman, Mark Leen, Nora Mahoney, Therese Mitros, Josh Nemeth, Jeffrey Nichols, K.C. Nocero, Andrew Nutting, Yvette Piggush, John Schuessler, Allison Sinoski, Sarah Taylor, and Alice Zachlin.

The interest and enthusiasm with which they, Ryan Carlin, Laura Colleton, Phillip Sicuso, and many other students engaged the topics of this course sustained my own interest and enthusiasm throughout my long "journey of calculus."

I would also like to express my gratitude to many of my colleagues and former colleagues at the University of Notre Dame for providing many facts and insights, especially Kevin Barry, Steve Battile, Ralph Chami, Norm Crowe, Cornelius Delaney, Barry Keating, David Kirkner, Charles Kulpa, Jerry Jones, Gary Lamberti, David Lodge, Michael Lykoudis, Ernan McMullin, Gerard Misiolek, Ken Olson, Joseph Powers, Terry Rettig, Fred Rickey, Rich Sheehan, Stephan Stolz, and Eric York.

To the women in the office of the mathematics department, Patti Strauch, Joan Hoechstman, Cheryl Huff, Judy Hygema, and Carole Martin, thanks for the many tasks, both large and small, along the way.

A warm word of thanks to the staff at Springer-Verlag, especially to Tim Taylor for his unflagging and diligent shepherding of the production process.

Finally, a word of deep appreciation to all my friends and colleagues who help make the University of Notre Dame such a wonderful scholarly environment.

Alexander J. Hahn
Notre Dame, 1998

Contents

For the Instructor v

Part I. From Archimedes to Newton 1

1. The Greeks Measure the Universe 3

    1.1. The Pythagoreans Measure Length 4
    1.2. The Measure of Angles 6
    1.3. Eratosthenes Measures the Earth 9
    1.4. Right Triangles 10
    1.5. Aristarchus Sizes Up the Universe 13
    1.6. *The Sandreckoner* 17
    1.7. Postscript 21
    Exercises 23

2. Ptolemy and the Dynamics of the Universe 29

    2.1. A Geometry of Shadows and the Motion of the Sun 30
    2.2. Geometry in the *Almagest* 32
    2.3. The Solar Model 35
    2.4. The Modern Perspective 39
    2.5. Another Look at the Solar Model 41
    2.6. Epicycles 43
    2.7. Postscript 45
    Exercises 46

3. Archimedes Computes Areas 51

    3.1. The Conic Sections 52
    3.2. The Question of Area 56
    3.3. Playing with Squares 58
    3.4. The Area of a Parabolic Section 60
    3.5. The *Method* 63
    3.6. Postscript 65
    Exercises 65

4. A New Astronomy and a New Geometry...................... 71

4.1. A New Astronomy... 72
4.2. The Studies of Galileo... 76
4.3. The Geometry of Descartes..................................... 81
4.4. Circles and Trigonometry...................................... 86
4.5. The Ellipse.. 90
4.6. Cavalieri's Principle... 92
4.7. Kepler's Analysis of the Orbits................................ 94
4.8. The Method of Successive Approximations.................... 99
4.9. Computing Orbital Information................................ 101
4.10. Postscript... 103
     Exercises.. 103

5. The Calculus of Leibniz.................................... 109

5.1. Straight Lines... 111
5.2. Tangent Lines to Curves... 117
5.3. Areas and Differentials... 120
5.4. The Fundamental Theorem of Calculus........................ 125
5.5. Functions.. 128
     A. The Derivative... 132
     B. Antiderivatives.. 134
5.6. Some Applications.. 135
     A. Finding Maximum and Minimum Values..................... 135
     B. Volumes... 136
     C. Lengths of Curves... 138
5.7. Postscript... 140
     Exercises.. 141

6. The Calculus of Newton.................................... 147

6.1. Areas Under Simple Curves..................................... 148
6.2. The Fundamental Theorem of Calculus (Again)............... 153
6.3. Computing Definite Integrals.................................. 156
6.4. Moving Points... 160
6.5. The Trajectory of a Projectile................................. 166
6.6. Applications to Ballistics?...................................... 170
6.7. Postscript... 173
     Exercises.. 174

7. The *Principia*.. 179

7.1. Equal Areas in Equal Times..................................... 181
7.2. Analyzing Centripetal Force.................................... 185

7.3.   The Inverse Square Law 187
7.4.   Test Case: The Orbit of the Moon 191
7.5.   The Law of Universal Gravitation 193
7.6.   Incredible Consequences 195
7.7.   Postscript 197
       Exercises 198

Part II. Calculus and the Sciences 205

8. Analysis of Functions 207

8.1.   Putting a Limit to the Test 209
8.2.   Continuous Functions 212
8.3.   Differentiability 217
8.4.   Derivatives as Rates of Change 220
8.5.   About Derivatives 222
       A. Computing Derivatives 222
       B. Some Theoretical Concerns 226
8.6.   Derivatives of Trigonometric Functions 228
8.7.   Increase and Decrease of Functions 230
8.8.   Maximum and Minimum Values 236
8.9.   Postscript 240
       Exercises 240

9. Connections with Statics, Dynamics, and Optics 249

9.1.   The Pulley Problem of De L'Hospital 251
       A. The Solution Using Calculus 251
       B. The Solution by Balancing Forces 253
9.2.   The Suspension Bridge 257
9.3.   An Experiment of Galileo 265
       A. Sliding Ice Cubes and Spinning Wheels 265
       B. Moments of Force and Inertia 267
       C. The Mathematics for Galileo's Experiment 269
9.4.   From Fermat's Principle to the Basic Telescope 272
       A. Fermat's Principle and the Path of a Light Ray 272
       B. Basic Properties of Lenses 277
       C. Quantitative Analysis of Lenses 280
9.5.   Postscript 285
       Exercises 286

10. Basic Functions and Their Graphs 299

10.1. Exponential Functions 300
10.2. Inverse Functions 302

10.3.	Logarithms	306
10.4.	Returning to a Problem of Leibniz	309
10.5.	Inverse Trigonometric Functions	311
10.6.	Concavity	314
10.7.	Asymptotes	318
10.8.	Graphing	320
10.9.	Postscript	328
	Exercises	328

11. The Exponential Function and the Measurement of Age and Growth ... 333

11.1.	Nuclear Activity	335
	A. The Atom and its Nucleus	335
	B. The Discoveries of Rutherford	336
11.2.	The Earth's Geologic History	342
	A. Reading Nuclear Clocks	344
	B. The Potassium–Argon Clock	347
11.3.	Life Evolves: A Timeline	348
11.4.	Dating the More Recent Past	352
11.5.	The World's Population	356
	A. Statistics and Trends	356
	B. The Logistics Model	358
	C. Applying the Logistics Model	362
11.6.	The Growth of Microorganisms	364
	A. The Exponential Phase	365
	B. Experimenting with *E. coli*	368
	C. Fermentation Processes	370
11.7.	Postscript	372
	Exercises	373

12. The Calculus of Economics ... 385

12.1.	Basics of Banking	386
	A. Interest	386
	B. Investment Plans	389
	C. Annuities and Bonds	390
12.2.	Inflation and the Consumer Price Index	395
12.3.	Supply and Demand in a Market	397
	A. Supply and Demand Functions	398
	B. OPEC and the Price of Oil	401
12.4.	A Firm's Cost of Production	403
	A. Cost Functions	403
	B. The Method of Least Squares	408

      C. Cost Analysis for Electric Companies 410
12.5. Price, Revenue, and Profit 412
      A. Revenue Functions 413
      B. Maximizing Profit 414
      C. Profit-Maximizing for a Refinery 416
12.6. Consumer Surplus 419
12.7. Postscript 422
      Exercises 423

13. Integral Calculus: Meaning and Method 435

13.1. Riemann Sums and the Definite Integral 436
      A. A Return to the Approach of Leibniz 436
      B. A Return to the Approach of Newton 439
      C. The Fundamental Equality 442
13.2. The Definite Integral as Surface Area 443
13.3. Methods of Integration 448
      A. The Substitution Method 448
      B. Trigonometric Substitutions 450
      C. Integration by Parts 451
13.4. Polar Coordinates 453
      A. Graphing Polar Equations 453
      B. Areas in Polar Coordinates 460
13.5. Differential Equations 462
13.6. Postscript 466
      Exercises 467

14. Integral Calculus and the Action of Forces 475

14.1. Work and Energy 476
      A. Variable Force and Work 476
      B. Kinetic and Potential Energy 482
14.2. Interior Ballistics 485
      A. Analysis of Springs 485
      B. The Force in a Gun Barrel 488
      C. The Springfield Rifle 490
14.3. Momentum 492
      A. Force as a Function of Time 493
      B. Conservation of Momentum 495
14.4. The Calculus of Rocket Propulsion 498
      A. Thrust 499
      B. The Rocket Equation 501
14.5. Surprises about Gravity? 505
      A. Gravity and Shape 506

B. The Case of the Sphere 507

14.6. Returning to Newton's *Principia* 511

A. Forces and Polar Coordinates 511

B. The Inverse Square Law 513

14.7. Hubble's Law and Einstein's Universe 515

14.8. Postscript 522

Exercises 522

Index **535**

Part I

From Archimedes to Newton

The Greeks Measure the Universe

This chapter is an exposition of fundamental matters that will be in use throughout this book. It introduces basic concepts of geometry and trigonometry and shows how the Greeks—who developed these disciplines into usable theories—put them to use.

1.1 The Pythagoreans Measure Length

The primary purpose of a number system is to count and measure things. This is where we will begin. The Pythagorean theorem[1] asserts that in a right triangle (see Figure 1.1), the equality $a^2 + b^2 = c^2$ relates the lengths of the sides. We are assuming that some unit of length (nowadays it would be a foot, yard, mile, inch, or meter, for example) is given. The Pythagorean theorem works "in reverse" also. Namely, if in some triangle the lengths of the sides satisfy the relationship $a^2 + b^2 = c^2$, then the triangle is a right triangle with c the length of the hypotenuse and a and b the lengths of the other two sides.

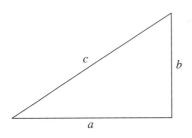

Figure 1.1

It appears that "rope stretchers" used this fact in early construction. Take, for example, a rope that has a length of 12 units (Figure 1.2): If the rope is stretched out horizontally on the ground in the triangular pattern indicated in Figure 1.3, then, since the lengths of the sides satisfy the relationship $4^2 + 3^2 = 16 + 9 = 25 = 5^2$, the Pythagorean theorem asserts

that the angle at 4 is a right angle. The construction of two perpendicular walls can begin!

Figure 1.2

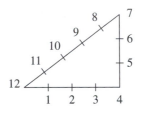

Figure 1.3

Assume that μ is some unit of length. We will say that a straight segment is *measurable* in μ if its length in units of μ is $\frac{m}{n}$, where m and n are both positive integers. So the segment is measurable if its length can be expressed precisely in whole or fractional units of μ. For instance, if the unit μ is the inch, then the segments of lengths $26\frac{4}{7} = \frac{186}{7}$ inches, as well as $53\frac{14}{29} = \frac{1551}{29}$ inches and $1726\frac{951}{3657} = \frac{6312933}{3657}$ inches are measurable.

The question presents itself as to whether all straight segments are measurable in the unit μ. Assume that the segment pictured directly below is measurable and that it has length $\frac{m}{n}$.

Take another segment of the same length, place them together at right angles, and form the right triangle pictured in Figure 1.4. Is the resulting hypotenuse h measurable? We will see (surprise?) that it is not.

Assume that it is. Then the length of h is some fraction $\frac{r}{s}$ of positive integers. By Pythagoras's theorem, $\left(\frac{r}{s}\right)^2 = \left(\frac{m}{n}\right)^2 + \left(\frac{m}{n}\right)^2 = 2\left(\frac{m}{n}\right)^2$. By clearing denominators, $r^2 n^2 = 2m^2 s^2$. Now set $x = rn$ and $y = ms$, and

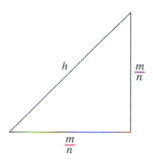

Figure 1.4

observe that $x^2 = 2y^2$. Factoring 2 out of x as many times as possible gives $x = 2^a x_1$ with x_1 odd. In the same way, $y = 2^b y_1$ with y_1 odd. (If, for instance, x were equal to 172, then $172 = 2 \cdot 86 = 2 \cdot 2 \cdot 43 = 2^2 \cdot 43$. So $a = 2$ and $x_1 = 43$.) By a substitution we get $(2^a x_1)^2 = 2(2^b y_1)^2$. So by the rules for exponentiating,

$$2^{2a} x_1^2 = 2^{2b+1} y_1^2.$$

Notice that $2a \neq 2b + 1$, since $2a$ is even and $2b + 1$ is odd. Therefore, either $2a > 2b + 1$ or $2a < 2b + 1$.

1. Assume that $2a > 2b + 1$. So $2^{2a-(2b+1)} x_1^2 = y_1^2$. It follows that y_1^2 is even. Since y_1 is odd, it has the form $y_1 = 2k + 1$ for some integer k. So $y_1^2 = 4k^2 + 4k + 1 = 4(k^2 + k) + 1$, which is the sum of an even integer and 1. But this means that y_1^2 is odd. Thus y_1^2 is both even and odd. This is impossible.
2. Assume next that $2a < 2b + 1$. This implies that $x_1^2 = 2^{(2b+1)-2a} y_1^2$. By the argument in Step 1, it is now x_1^2 that is both even and odd. This too is impossible.

Reflect over what has been done. The discussion started with the supposition that h is measurable in μ. Then the argument moved in a strictly logical way to impossible consequences. The inescapable conclusion is that h cannot be measurable in μ.

The preceding proof was not taken from an old Greek text. It is, however, from the point of view of the precision of the logic and the flow of its argument, very much in the spirit of Greek mathematics. The point is this: While the hypotenuse h is a perfectly valid

geometric construction, it cannot be measured with the numbers of the Pythagoreans. In particular, their numerical considerations ran into limitations that geometrical ones did not. The suggestion presents itself that this was an important reason why Greek geometry and trigonometry flourished in a way that numerical analyses and algebra did not.

Pythagoras and his followers had founded a Greek colony in today's southern Italy in the 6th century B.C. They formed a cult based on the philosophical principle that mathematics is the underlying explanation of all things from the relationship between musical notes to the movements of the planets of the solar system. Indeed, they held that all reality finds its ultimate explanation in numbers and mathematics. As we have just seen, however, the numbers of the Pythagoreans were unable to come to grips with the very basic matter of measuring length. It is believed that when the Pythagoreans realized this, a crisis ensued within their ranks that contributed to their downfall.[2] In any case, it seems somewhat ironic that the Pythagorean theorem—an assertion about the lengths of certain segments—derives its name from a school (or person) that did not possess a number system with which length could always be measured.

Today we can put it this way: The Pythagorean number system did not have enough numbers. It consisted only of the numbers of the form $\frac{m}{n}$, i.e., the rational numbers, and it did not include other real numbers. In fact, the preceding demonstration of the nonmeasurability of h shows that $\sqrt{2}$ is a number that is not rational. (Take $m = n = 1$.) It is, in other words, irrational. On the other hand, we know that

$$\sqrt{2} = 1.4142\ldots$$

$$= 1 + 4\frac{1}{10} + 1\frac{1}{100} + 4\frac{1}{1000} + 2\frac{1}{10000} + \cdots,$$

and hence that $\sqrt{2}$ can be constructed in terms of a decimal expansion. This infinite process gives rise to the number line. Fix a unit of length and take a straight line that runs infinitely in both directions.

Fix a point and label it 0. Mark off a point one unit to the right of 0 and label it 1. Continue in this way to get 2, 3, ... Do a similar thing on the left of 0, but use −1, −2, . . . to label the points. Continue in this way with tenths of units, hundredths of units, and so on, to achieve the following relationship: Every point on the line corresponds to a real number, in other words, a number given by a decimal expansion, and every such number corresponds to a point on the line. See the illustration for $\sqrt{2} = 1.4142\ldots$ in Figure 1.5.

The Greeks shied away from infinite processes and did not hit upon the construction of the real numbers. (This did not occur until the 16th century!) This means that they did not have the coordinate— also called Cartesian—plane. Therefore, they could not graph algebraic equations and make use of the interplay between algebra and geometry that lies at the core of modern mathematical analysis. In this course we will not shy away from real numbers: we will make use of them throughout.

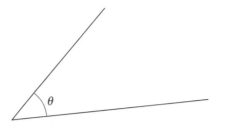

Figure 1.6

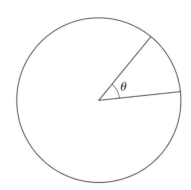

Figure 1.7

1.2 The Measure of Angles

We turn next to the study of angles. Take two line segments with a common endpoint; they form a wedge. The issue is to devise a numerical way of measuring the angle, i.e., the amount θ of the opening of the wedge. See Figure 1.6.

Place the wedge in a circle with the common point at the center. See Figure 1.7. One way of measuring θ is with degrees. This unit is designated by ° and is a legacy of the base 60 number system of the Babylonians. Declare the entire circle to have 360 degrees and then apportion degrees proportionally. So a wedge consisting of half the circle will have 180°,

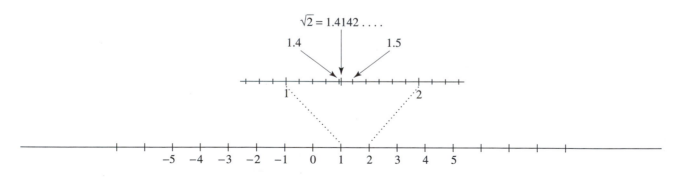

Figure 1.5

one consisting of a quarter circle will measure 90°, and so on. In this way, any angle can be assigned a certain number of degrees. It is known that the angles of any triangle add up to 180°.

A more useful numerical measure of an angle is the radian measure. It is based on the measurement of length. Let r be the radius of the circle in Figure 1.8 and let s be the length of the arc that is cut out of the circumference by the wedge. The radian measure of the angle θ is defined to be the ratio $\frac{s}{r}$. We will write $\theta = \frac{s}{r}$. An immediate question that arises is this: Does the radian measure of the angle depend on the size of the circle into which it is placed? If this measure is to be a meaningful concept, it should not.

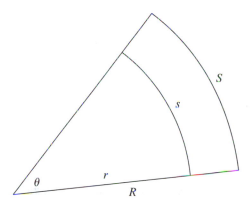

Figure 1.9

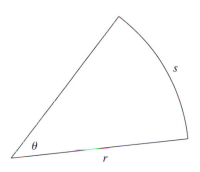

Figure 1.8

Place θ into another circle. Let R be its radius and S the length of the arc that θ cuts out. See Figure 1.9. Is $\frac{s}{r} = \frac{S}{R}$? This is the question that must be addressed. Consider what follows to be a "thought experiment," rather than something that is carried out in practice.

Let n be a positive integer (n can be, say, 5 or 500 or 40,000), and partition the wedge into n equal pieces. Connect the intersection points with straight line segments and let their respective lengths be d_n and D_n. (Note that these lengths depend on the n that you have picked.) The case $n = 4$ is shown in Figure 1.10. Observe that each of the smaller triangular wedges (those with sides of length r) is similar to each of the

larger ones (with sides of length R). This is because the corresponding angles are equal. It follows that $\frac{d_4}{r} = \frac{D_4}{R}$. In the same way, $\frac{d_n}{r} = \frac{D_n}{R}$ for any n. Observe next that if n is taken to be large, then the lengths of the n segments each of length d_n add up to approximately s. So the number nd_n is nearly equal to s, and therefore $\frac{nd_n}{r}$ is nearly equal to $\frac{s}{r}$. If n is taken sequentially larger and larger, the numbers $\frac{nd_n}{r}$ close in on $\frac{s}{r}$. We abbreviate this by writing

$$\lim_{n \to \infty} \frac{nd_n}{r} = \frac{s}{r}.$$

The symbol lim is short for limit and refers to the closing-in process. The symbol ∞ represents "infinity"

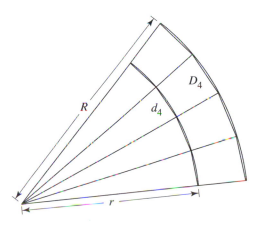

Figure 1.10

and $\lim_{n\to\infty}$ means that n, by being taken larger, is being "pushed to infinity." Proceeding in exactly the same way with the larger arc, we get that

$$\lim_{n\to\infty} \frac{nD_n}{R} = \frac{S}{R}.$$

Since $\frac{d_n}{r} = \frac{D_n}{R}$ for any n, we see that

$$\frac{s}{r} = \lim_{n\to\infty} \frac{nd_n}{r} = \lim_{n\to\infty} \frac{nD_n}{R} = \frac{S}{R}.$$

So our thought experiment has established the required equality $\frac{s}{r} = \frac{S}{R}$. Observe that the radian measure of an angle is a ratio of lengths. It is therefore a real number.

We will learn later that "limit" procedures such as the one we just considered are the cornerstone of calculus.

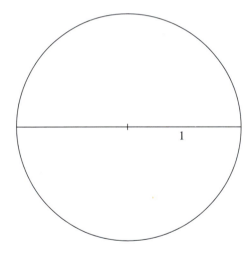

Figure 1.11

Now take a circle of radius 1 (Figure 1.11). The length of one-half of its circumference is a number that is denoted by π. Consider the angle $180°$ and observe that its radian measure is $\frac{\pi}{1} = \pi$. Now take a circle of radius r and let c be its circumference (Figure 1.12). Using this circle, we get that the radian measure of $180°$ is equal to $\frac{c/2}{r}$. Since the radian measure of an angle is the same regardless of which circle is used, it follows that $\frac{c/2}{r} = \pi$. This gives the well-known formula

$$c = 2\pi r$$

for the circumference of a circle of radius r.

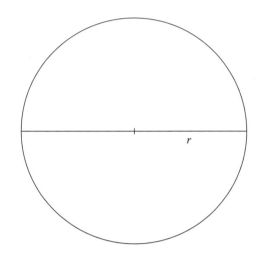

Figure 1.12

Consider the circle of Figure 1.11 and check that degrees and radians correspond as shown in Table 1.1.

Table 1.1	
Degrees	**Radians**
$360°$	2π
$180°$	π
$90°$	$\frac{\pi}{2}$
$60°$	$\frac{\pi}{3}$
$45°$	$\frac{\pi}{4}$
$30°$	$\frac{\pi}{6}$
$1°$	$\frac{\pi}{180}$

Since π represents length, it is some real number. What is its decimal expansion?[3] Take a circle of radius 1 and center C and divide the upper semicircle into three equal parts, as shown in Figure 1.13. Each of the angles is equal to $\frac{\pi}{3}$. Concentrate on the wedge on the right. Let A and B be the indicated points of intersection, and observe that the isosceles triangle $\triangle CAB$ is equilateral. (Why?) Therefore, AB has length 1. Choose the point P on the circle such that the segment CP bisects the angle $\angle ACB$ and draw the tangent $A'B'$ to the circle at P. Since the bisector CP and the tangent $A'B'$ are perpendicular (by a basic property of the circle), the angles at A' and B' are both equal to $\frac{\pi}{3}$. It follows that the triangles $\triangle CA'P$ and $\triangle CB'P$ are equal. Put[4] $A'P = PB' = t$. Since the three angles of $\triangle CA'B'$ are equal, it follows that $\triangle CA'B'$ is equilateral. Therefore, CA' has length $2t$, and by the Pythagorean theorem applied to $\triangle CPA'$, it follows that $1 + t^2 = (2t)^2$. Thus $3t^2 = 1$, and hence $t = \frac{1}{\sqrt{3}}$. So $A'B'$ has length $2t = \frac{2}{\sqrt{3}}$. Since the radian measure of the angle $\frac{\pi}{3}$ is the length of the arc AB, it seems plausible (we will be content with plausibility in this discussion—see also Exercise 15) from Figure 1.13, that

$$1 = AB < \frac{\pi}{3} = \text{length arc } AB < A'B' = \frac{2}{\sqrt{3}}.$$

Now multiply through by 3 to get

$$3 < \pi < \frac{2 \cdot 3}{\sqrt{3}} = \frac{2\sqrt{3}\sqrt{3}}{\sqrt{3}} = 2\sqrt{3}.$$

This corresponds to $3 < \pi < 3.47$.

Archimedes (287–212 B.C.)—we will encounter him again soon—used an argument, similar in principle but much more involved (it uses many more triangles instead of our three), to show that

$$3\frac{10}{71} < \pi < 3\frac{1}{7}.$$

This gives $3.1408 < \pi < 3.1429$. The correct expansion begins $\pi = 3.14159\ldots$. It turns out (and this is difficult to establish) that π is an irrational number.

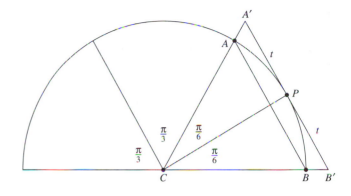

Figure 1.13

1.3 Eratosthenes Measures the Earth

Eratosthenes (276–194 B.C.) was director of the Museum of Alexandria, as Euclid had been before him. Under the sponsorship of Egypt's Ptolemaic rulers, this was an early version of a government funded research institute.

It was the commonly accepted view among Greek philosophers that the Earth is round. Eratosthenes was able to measure its size. He knew that at noon

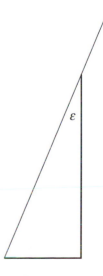

Figure 1.14

at the summer solstice of every year (this occurs on or about June 21 in the Northern Hemisphere) in the town of Syene (near today's Aswan in Egypt), the Sun shines into the very bottom of a deep well. This means that at precisely that time of the year the Sun is directly overhead at this location. In Alexandria, again precisely at noon at the summer solstice, Eratosthenes measured the length of the shadow of a gnomon[5] in vertical position (using a plumbline) and determined the angle ε in Figure 1.14 to be 7.5°. Eratosthenes knew also that Syene was (roughly) due south of Alexandria at a distance of about 500 miles.[6]

He next let r be the radius of the Earth and noticed that the basic situation is given by Figure 1.15 with $\varepsilon = 7.5°$ and $s = 500$. The rest was easy. On the one hand,

$$\varepsilon = \left(7\frac{1}{2}\right)\frac{\pi}{180} = \frac{15(3.14)}{360} = 0.13 \text{ radians.}$$

But on the other hand, $\varepsilon = \frac{s}{r} = \frac{500}{r}$. Therefore,

$$r = \frac{500}{\varepsilon} = \frac{500}{0.13} = 3850 \text{ miles.}$$

So Eratosthenes—quite literally with a stick, some observations, and geometry, cemented together by pure thought—had measured the size of the Earth! Of course, he had only an approximation. The correct value of the radius of the Earth is 3950 miles.

1.4 **Right Triangles**

Trigonometry is the study of the right triangle and the applications of this study. Indeed, the word trigonometry is the Greek rendition of "measuring the triangle." This section recalls some ba-

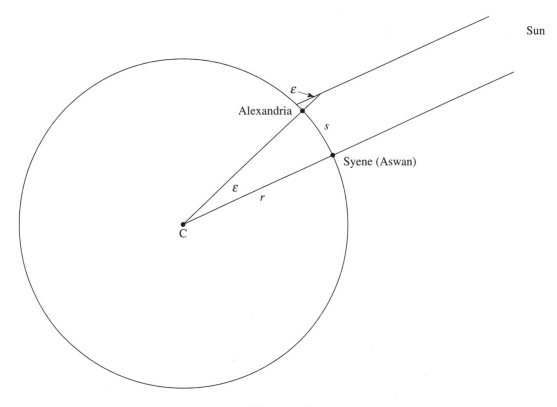

Figure 1.15

sic facts, all known—with different terminology and notation—to the Greeks.

Consider the right triangle in Figure 1.16. For the given angle θ, define the "sine," "cosine," and "tangent" to be the following ratios of lengths:

$$\sin \theta = \frac{a}{h}, \quad \cos \theta = \frac{b}{h},$$

and

$$\tan \theta = \frac{\sin \theta}{\cos \theta} = \frac{\frac{a}{h}}{\frac{b}{h}} = \frac{a}{b}.$$

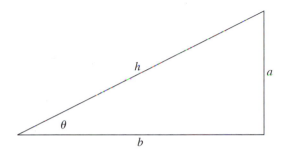

Figure 1.16

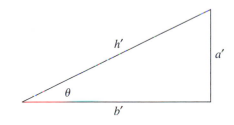

Figure 1.17

Only the size of the angle θ matters in these definitions and not the right triangle into which it is placed. This is easily seen as follows. Suppose another right triangle is used; see Figure 1.17. Superimposing the two triangles shows that they are similar. (See Figure 1.18.) Therefore, $\frac{a'}{h'} = \frac{a}{h}$, $\frac{b'}{h'} = \frac{b}{h}$, and $\frac{a'}{b'} = \frac{a}{b}$, so it doesn't matter which right triangle is used to compute $\sin \theta$, $\cos \theta$, and $\tan \theta$.

There are many identities that relate $\sin \theta$, $\cos \theta$, and $\tan \theta$. For example, by Pythagoras's theorem, $a^2 + b^2 = h^2$. So $\frac{a^2}{h^2} + \frac{b^2}{h^2} = 1$, and therefore,

$$\sin^2 \theta + \cos^2 \theta = 1.$$

It is customary to write $\sin^2 \theta$ instead of the less efficient $(\sin \theta)^2$.

We will now compute $\sin \theta$, $\cos \theta$, and $\tan \theta$ for some standard values of θ. Consider the right triangle in Figure 1.19. Since it is isosceles, the acute angles are each $45°$ or $\frac{\pi}{4}$ radians. Therefore,

$$\sin \frac{\pi}{4} = \frac{1}{\sqrt{2}}, \quad \cos \frac{\pi}{4} = \frac{1}{\sqrt{2}}, \quad \text{and } \tan \frac{\pi}{4} = \frac{1}{1} = 1.$$

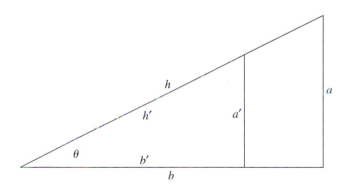

Figure 1.18

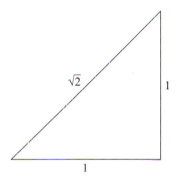

Figure 1.19

Next, take the equilateral triangle that has sides of length 2 and height h. By Pythagoras's theorem, $h^2 + 1^2 = 2^2$. So $h^2 = 3$ and $h = \sqrt{3}$. See Figure 1.20. Recalling that the angles of an equilateral triangle are each equal to 60° or $\frac{\pi}{3}$ radians, we see that

$$\sin\frac{\pi}{3} = \frac{\sqrt{3}}{2}, \ \cos\frac{\pi}{3} = \frac{1}{2}, \text{ and } \tan\frac{\pi}{3} = \frac{\sqrt{3}}{1} = \sqrt{3}.$$

Since each of the smaller angles at the top is $\frac{\pi}{6}$, it follows that

$$\sin\frac{\pi}{6} = \frac{1}{2}, \ \cos\frac{\pi}{6} = \frac{\sqrt{3}}{2}, \text{ and } \tan\frac{\pi}{6} = \frac{1}{\sqrt{3}}.$$

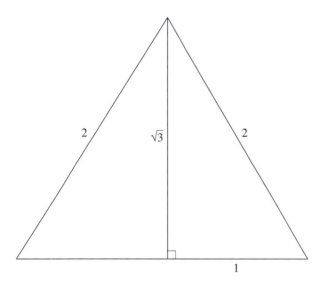

Figure 1.20

Observe that $\sin 0$, $\cos 0$, and $\tan 0$ do not make sense, for the simple reason that there is no right triangle with an angle of 0° or, equivalently, 0 radians.

Suppose, however, that θ is very small. A look at the right triangle with hypotenuse 1 in Figure 1.21 shows that $\sin\theta = \frac{a}{1} = a$ is also very small. So if θ is close to 0, then $\sin\theta$ is also close to zero. Take θ sequentially smaller and smaller. This rotates the hypotenuse downwards and pushes a to zero. Therefore, $\sin\theta = a$ is pushed to zero. We summarize this by writing

$$\lim_{\theta\to 0} \sin\theta = 0.$$

It now makes sense to set $\sin 0 = 0$. For entirely similar reasons, $\sin\frac{\pi}{2}$ does not make sense either. This time focus on the angle ϕ in Figure 1.21. As the hypotenuse rotates, ϕ is pushed to $\frac{\pi}{2}$, and in the process, $\sin\phi = \frac{b}{1} = b$ is pushed to 1. In limit notation,

$$\lim_{\phi\to\frac{\pi}{2}} \sin\phi = 1.$$

So it makes sense to set $\sin\frac{\pi}{2} = 1$. A similar analysis for the cosine shows that $\lim_{\theta\to 0}\cos\theta = 1$ and $\lim_{\theta\to\frac{\pi}{2}}\cos\theta = 0$; so we set $\cos 0 = 1$ and $\cos\frac{\pi}{2} = 0$.

Table 1.2			
θ	$\sin\theta$	$\cos\theta$	$\tan\theta$
0	0	1	0
$\frac{\pi}{6}$	$\frac{1}{2}$	$\frac{\sqrt{3}}{2}$	$\frac{1}{\sqrt{3}}$
$\frac{\pi}{4}$	$\frac{1}{\sqrt{2}}$	$\frac{1}{\sqrt{2}}$	1
$\frac{\pi}{3}$	$\frac{\sqrt{3}}{2}$	$\frac{1}{2}$	$\sqrt{3}$
$\frac{\pi}{2}$	1	0	?

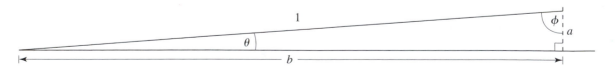

Figure 1.21

Table 1.2 summarizes this information. The reader is asked to fill in the question mark.

We conclude with some basic properties of the sine. Consider the circle of radius 1 and the right triangle shown in Figure 1.22. Observe that $\sin\theta = \frac{a}{1} = a$ and $\theta = \frac{s}{1} = s$ radians. Since $s > a$, it follows that

$$\theta > \sin\theta.$$

If θ is small, note that the lengths s and a will be close to each other. We symbolize this by writing $s \approx a$. Therefore, for small θ,

$$\sin\theta \approx \theta.$$

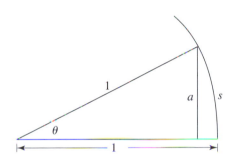

Figure 1.22

Table 1.3		
θ **in degrees**	θ **in radians**	$\sin\theta$
30°	$\frac{\pi}{6} = 0.5237$	0.5000
15°	0.2618	0.2588
10°	0.1746	0.1737
3°	0.0524	0.0524
2°	0.0349	0.0349
1°	0.0175	0.0175

Take $\pi = 3.142$ and use your calculator to verify the data in Table 1.3. Observe that there is close agreement between $\sin\theta$ and θ, but only if θ is taken in radians. So when computing $\sin\theta$, make sure that your calculator is in "radian mode" and not "degree mode."

1.5 Aristarchus Sizes Up the Universe

Not much is known about Aristarchus of Samos (310–230 B.C.). He received his education directly or indirectly from Aristotle's Lyceum (the institute in Athens). The most important fact about him is this: He believed that the universe is Sun centered (or heliocentric), that the Sun is fixed, and that the Earth revolves around the Sun and rotates about its own axis in the process.

What will interest us is Aristarchus's use of "cosmic" trigonometry in his treatise *On the Magnitudes and Distances of the Sun and Moon*. His analysis rests on the following hypotheses and observations[7]:

A. The Moon receives its light from the Sun.
B. The Moon revolves in a circle about the Earth with the Earth at the center.
C. When an observer on Earth looks out at a precise half moon, the angle $\angle EMS$ in Figure 1.23 is 90°. At that moment the angle $\angle MES$ can be measured to equal 87°.
D. At the instant of a total eclipse of the Sun, the Moon and the Sun (as viewed from the Earth) subtend the same angle, and this angle can be measured to be 2°. Refer to Figure 1.24.
E. During a lunar eclipse, the shadow indicated in Figure 1.25 has width 4 times the radius r_M of the Moon. (This was based on how long the Moon was observed to be in the Earth's shadow.)

What did Aristarchus deduce from these observations? Let

$$r_E = \text{the radius of the Earth}$$

$$r_M = \text{the radius of the Moon}$$

$$r_S = \text{the radius of the Sun}$$

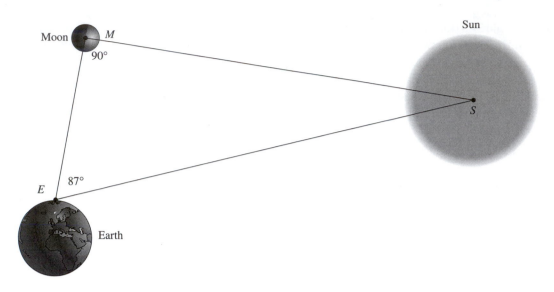

Figure 1.23

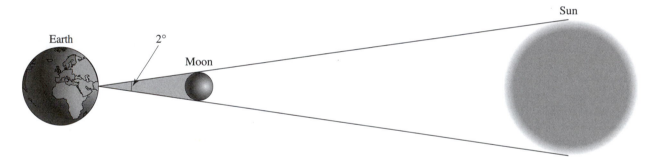

Figure 1.24

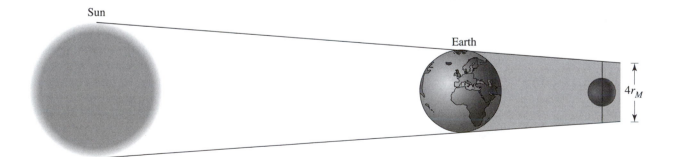

Figure 1.25

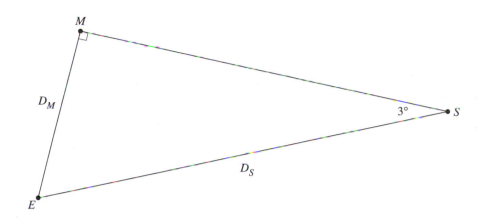

Figure 1.26

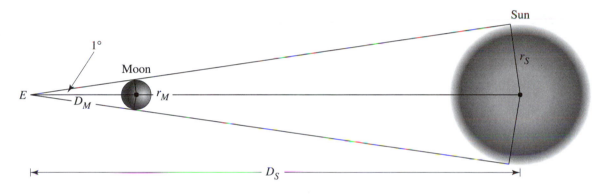

Figure 1.27

D_M = the distance from the Earth to the Moon

D_S = the distance from the Earth to the Sun

Figure 1.26 comes directly from observation **C**. Observe that $3° = \frac{\pi}{60}$. Therefore, $\frac{D_M}{D_S} = \sin 3° = \sin \frac{\pi}{60}$. Since $3° = \frac{\pi}{60}$ radians is a small angle, $\sin \frac{\pi}{60}$ is approximately equal to $\frac{\pi}{60}$. (See Table 1.3.) In view of Archimedes's estimate, we take $\pi = 3.14$. (Aristarchus was a contemporary of Archimedes and would certainly have been aware of such estimates.) This gives $\frac{\pi}{60} = 0.052$. To make things simple, we will take $\sin \frac{\pi}{60} = \frac{\pi}{60} = 0.05 = \frac{1}{20}$. So $\frac{D_M}{D_S} = \sin \frac{\pi}{60} = \frac{1}{20}$. In this way Aristarchus arrived at the estimate

$$\frac{D_S}{D_M} = 20.$$

Refer next to observation **D** and Figure 1.24. This is the situation of the solar eclipse. Figure 1.27 is an elaboration of Figure 1.24. Radii of the Moon and the Sun have been inserted and some distances have been labeled. The angle at E indicated as being equal to $1°$ is obtained by bisecting the $2°$ angle of Figure 1.24. Notice that the Sun is larger than the Moon. By similar triangles, $\frac{r_S}{r_M} = \frac{D_S}{D_M}$, and therefore

$$\frac{r_S}{r_M} = 20.$$

Observe also that $\frac{r_M}{D_M} = \sin 1°$. Since $1° = \frac{\pi}{180}$ radians, $\frac{r_M}{D_M} = \sin \frac{\pi}{180} \approx \frac{\pi}{180}$. Taking $\pi = 3.14$, it follows that $\frac{\pi}{180} = 0.017$. Note that $\frac{1}{60}$ is very close to 0.017. So to make things simple, we will take $\frac{r_M}{D_M} = \sin \frac{\pi}{180} = \frac{1}{60}$.

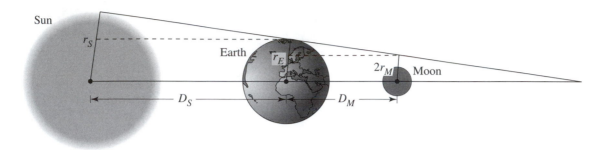

Figure 1.28

So Aristarchus obtained the approximation

$$\frac{D_M}{r_M} = 60.$$

From observation **E** and Figure 1.25, he obtained Figure 1.28. This figure shows a light ray that is tangent to both the Sun and the Earth. The radii of the Sun and the Earth drawn into the figure are both perpendicular to this light ray. The extension of the radius of the Moon indicated in the figure is perpendicular to this light ray as well. Because the two triangles with the "dotted" bases are similar, it follows that

$$\frac{r_E - 2r_M}{r_S - r_E} = \frac{D_M}{D_S}.$$

Recalling that

$$\frac{D_M}{D_S} = \frac{r_M}{r_S}$$

we get

$$\frac{r_E - 2r_M}{r_S - r_E} = \frac{r_M}{r_S}.$$

After cross-multiplying, $r_S r_E - 2r_S r_M = r_M r_S - r_M r_E$. So $r_S r_E + r_M r_E = 3 r_S r_M$. Dividing this last equation through by $r_S r_M$ gives us $\frac{r_E}{r_M} + \frac{r_E}{r_S} = 3$. Since $\frac{r_S}{r_M} = 20$, we have $r_S = 20 r_M$. By substitution, we get

$$3 = \frac{r_E}{r_M} + \frac{r_E}{20 r_M} = \frac{20 r_E + r_E}{20 r_M} = \frac{21 r_E}{20 r_M}.$$

Therefore, $\frac{r_E}{r_M} = \frac{60}{21} = \frac{20}{7}$, and

$$r_M = \frac{7}{20} r_E.$$

Since $\frac{r_S}{r_E} = \frac{r_S}{r_M} \cdot \frac{r_M}{r_E} = 20 \cdot \frac{7}{20} = 7$, it follows that

$$r_S = 7 r_E.$$

With Eratosthenes's value of $r_E = 3850$ miles, Aristarchus got the approximations

$$r_M = 1350 \text{ miles} \quad \text{and} \quad r_S = 27{,}000 \text{ miles}.$$

Since $\frac{D_M}{r_M} = 60$, he estimated that $D_M = 60 r_M = 80{,}000$ miles; and inserting this value into $\frac{D_S}{D_M} = 20$, he got

$$D_S = 1{,}600{,}000 \text{ miles}.$$

Aristarchus's argument was actually more elaborate and complicated than this. He used different approximations and got slightly different answers. For example, instead of $\frac{r_M}{r_S} = \frac{1}{20}$, he obtained $\frac{1}{20} < \frac{r_M}{r_S} < \frac{1}{18}$; and instead of $\frac{r_M}{D_M} = \frac{1}{60}$, he had $\frac{1}{60} < \frac{r_M}{D_M} < \frac{1}{45}$. However, the essence of his analysis has been retained.

Table 1.4 compares Aristarchus's estimates to modern values. Notice that Aristarchus's value for the radius of the Moon is rather accurate. However, his estimates for the distance to the Moon, the radius of the Sun, and its distance from the Earth are off by factors of 3, 16, and 50, respectively.

Table 1.4

	Aristarchus	Actual
r_E radius of Earth	3850 miles[a]	3950 miles
r_M radius of Moon	1350 miles	1080 miles
r_S radius of Sun	27,000 miles	432,000 miles[b]
D_M Earth to Moon	80,000 miles	238,868 miles[c]
D_S Earth to Sun	1,600,000 miles	93×10^6 miles[c]

[a]As already pointed out, this is taken from Eratosthenes. While Eratosthenes lived about 40 years after Aristarchus, similar estimates had been made earlier.

[b]The Sun consists of gas. The radius given is that of the photosphere, the illuminated part. The part from the center to 30% of its radius has in essence all the shining power and 60% of the mass. The part from the center to 60% of its radius has 95% of the mass.

[c]Radar measurements.

In any case, using only some basic observations and pure thought (i.e., mathematics), Aristarchus provided at least some idea of the magnitude of the distances in the solar system and began to unravel some of its mystery. In fact, Aristarchus's strategy is correct in principle. With more accurate measurements of the angles involved, he would have done much better; see Exercise 26. More serious is the problem of refraction. A ray of light bends as it moves from one medium to another of different density, and in particular, it bends when it moves through the atmosphere. The distance D_S was not computed with suitable accuracy until the 17th century, when the Italian astronomer Casini came within $7\frac{1}{2}$% of the correct value. More accurate values were not obtained until the latter part of the 18th. The calculations for these estimates used information about the observed path of Venus across the Sun.

1.6 **The Sandreckoner**

Archimedes—the most famous scientist of antiquity—was born around 285 B.C. in the Greek city state of Syracuse, a port on the Mediterranean island of Sicily. There is historical reference to the fact that he spent considerable time in Alexandria, and there seems little doubt that he studied with the successors of Euclid. After his studies, he returned to Syracuse and lived there in complete absorption with his mathematical investigations.

Late in the 3rd century B.C., Syracuse became embroiled in the struggle between Rome and Carthage for control of the western Mediterranean. In his *Parallel Lives*, the historian Plutarch (about 46–126 A.D.) recounts Archimedes's efforts in the defense of the city against the Romans:

When, therefore, the Romans assaulted the walls in two places at once, fear and consternation stupefied the Syracusans, believing that nothing was able to resist that violence and those forces. But when Archimedes began to ply his engines, he at once shot against the land forces all sorts of missile weapons, and immense masses of stone that came down with incredible noise and violence, against which no man could stand; for they knocked down those upon whom they fell, in heaps, breaking all their ranks and files. In the meantime huge poles thrust out from the walls over the ships, sunk some by the great weights which they let down from on high upon them; others they lifted up into the air by an iron hand [and soon] such terror had seized upon the Romans, that, if they did but see a little rope or a piece of wood from the wall, instantly crying out, that there it was again, Archimedes was about to let fly some engine at them, they turned their backs and fled.

The Roman attack on Syracuse was repelled. A lengthy siege was later successful and Syracuse was conquered and destroyed in 212 B.C. Archimedes perished during the destruction. Plutarch relates several versions of his death. The one most widely cited finds Archimedes, oblivious to the city's capture, absorbed in the study of a particular diagram that he had sketched in the sand. When a Roman soldier confronted him, Archimedes requested time to complete his deliberations. The impatient soldier, however, ran him through with his sword.

The work of Archimedes is impressive—as we shall soon see. Plutarch speaks of Archimedes's

> purer speculations [and] studies, the superiority of which to all others is unquestioned, and in which the only doubt can be, whether the beauty and grandeur of the subjects examined, or the precision and cogency of the methods and means of proof, most deserve our admiration. It is not possible to find in all geometry more difficult and intricate questions, or more simple and lucid explanation. No amount of investigation of yours would succeed in attaining the proof, and yet once seen, you immediately believe you would have discovered it; by so smooth and so rapid a path he leads you to the conclusion required.

Archimedes was also the quintessential eccentric scientist. His deep absorption in thought

> made him forget his food and neglect his person, to that degree that when he was occasionally carried by absolute violence to bathe, or have his body anointed, he used to trace geometrical figures in the ashes of the fire, and diagrams in the oil on his body.

A famous episode recounts how, after a particularly satisfying discovery, Archimedes ran through the streets of Syracuse in naked celebration shouting "Eureka, Eureka!" (Eureka, or $\varepsilon\upsilon\rho\varepsilon\kappa\alpha$, is Greek for "I have found it.")

It is, of course, difficult to separate fact from fiction and reality from legend in Plutarch's account of Archimedes's remarkable talents as inventor of machines of war. The ingenuity of Archimedes's speculations about geometry and physics, however, can be corroborated. Much of his work has survived in transmitted form.

We turn now to Archimedes's scheme for writing large numbers. You will encounter the basic Greek number system in the exercises. It allows for numbers as large as

$$99,999,999 = 9999(10,000) + 9999 = {}_{,}\delta\lambda\varsigma\delta M,\overline{\delta\lambda\varsigma\delta}$$

but not larger. Archimedes enlarged the basic Greek number system into an incredible scheme. He introduced the number

$$MM = (10,000)(10,000) = 100,000,000 = 10^8.$$

and referred to numbers up to MM as numbers of the *first order*. He then built the following tower of "orders" and "periods" of numbers. In modern notation:

First order: The numbers N with
$$1 \leq N < 10^8 = 10^{8 \cdot 1}$$

Second order: The numbers N with
$$10^{8 \cdot 1} = 10^8 \leq N < 10^{16} = 10^{8 \cdot 2}$$

Third order: The numbers N with
$$10^{8 \cdot 2} = 10^{16} \leq N < 10^{24} = 10^{8 \cdot 3}$$

Fourth order: The numbers N with
$$10^{8 \cdot 3} = 10^{24} \leq N < 10^{32} = 10^{8 \cdot 4}$$

$$\vdots$$

10^8th order: The numbers N with
$$10^{8(10^8 - 1)} \leq N < 10^{8 \cdot 10^8}.$$

These are the numbers of the **first period**. This is only the beginning. The numbers of the **second period** are also partitioned into orders:

First order: The numbers N with
$$10^{8 \cdot 10^8} \leq N < 10^{8(10^8 + 1)}$$

Second order: The numbers N with
$$10^{8(10^8 + 1)} \leq N < 10^{8(10^8 + 2)}$$

Third order: The numbers N with
$$10^{8(10^8 + 2)} \leq N < 10^{8(10^8 + 3)}$$

Fourth order: The numbers N with
$$10^{8(10^8+3)} \le N < 10^{8(10^8+4)}$$

$\vdots$

10^8th order: The numbers N with
$$10^{8(10^8+10^8-1)\cdot 8} \le N < 10^{8(10^8+10^8)}.$$

The **second period** ends with (one less than) the number $10^{8(10^8+10^8)} = (10^{8\cdot10^8})^2$. The **third, fourth, fifth,** etc.... **periods** follow. Finally, with the **10^8th period** and the number $(10^{8(10^8)})^{10^8} = 10^{8\cdot10^{16}}$, the array stops.

To put all of this into perspective, note that the visible universe is thought to have on the order of 10^{80} atoms (about 99.9% of them hydrogen and helium). This is Eddington's number, after the British astrophysicist Arthur Stanley Eddington. Since $10^{80} = 10^{8\cdot10}$, this is the first number of the 11th order of the first period. The point is not so much the usefulness of Archimedes's scheme, but rather the grandiose nature of his speculations. Archimedes, in other words, thought big!

Archimedes looked for a context in which to illustrate the utility of his cosmic array of numbers. Having found it, he was evidently very pleased to address his manuscript *The Sandreckoner* to his benefactor the king of Syracuse. He began by giving the king an astronomy lesson:

Aristarchus brought out a book consisting of some of the hypotheses, in which the premises lead to the result that the universe is many times greater than that now so called. His hypotheses are that the fixed stars and the Sun remain unmoved, that the Earth revolves about the Sun in the circumference of a circle, the Sun lying in the middle of the orbit, and that the sphere of the fixed stars, situated about the same center as the Sun, is so great ...

Then he stated his purpose:

I say then that, even if a sphere were made up of sand, as great as Aristarchus supposes the sphere of the fixed stars to be, I shall still prove that, of the numbers named by me, some exceed in multitude the number of grains of sand in a mass which is equal in magnitude to the sphere referred to, provided that the following assumptions are made ...

In other words, Archimedes imagined the entire cosmos to be packed with sand and proposed to count the number of grains of sand in question!

Our description of Archimedes's discussion will use earlier notation: r_E for the radius of the Earth, r_M for the radius of the Moon, r_S for the radius of the Sun, and D_S for the distance between the Sun and the Earth.

What assumptions did Archimedes make? The first concerned r_E:
$$r_E \le 47,500 \text{ miles.}$$

We have seen that Eratosthenes's rather accurate estimate was 3850 miles. So here Archimedes thought too big. Remember, however, that it was his purpose to display the vastness of his number scheme. Following Aristarchus, Archimedes supposed that
$$r_M < r_E.$$

Recalling that Aristarchus had shown that r_S is about 20 times greater than r_M, he next assumed that
$$r_S \le 30 r_M.$$

This is in fact too small. A look at Table 1.4 shows, after a quick calculation, that $r_S = 400 r_M$.

Because $r_S \le 30 r_M < 30 r_E \le (30)(47,500)$, Archimedes got
$$r_S \le 1,425,000 \text{ miles.}$$

While these inequalities are based on a combination of earlier estimates and pure speculation, Archimedes next turned to careful observation and delicate geometrical arguments. He used a long rod

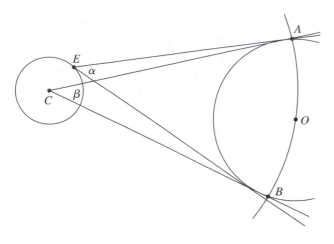

Figure 1.29

with a small disc fastened at its end. He pointed it in the direction of the Sun just after sunrise and carefully measured the angle α that the Sun subtends in the sky. He determined that $(\frac{90}{200})^\circ < \alpha < (\frac{90}{164})^\circ$ or, in radian measure,

$$\frac{1}{200}\frac{\pi}{2} < \alpha < \frac{1}{164}\frac{\pi}{2}.$$

The implied estimate $\alpha \approx \frac{1}{2}^\circ$ for the so-called angular diameter of the Sun is accurate. (Aristarchus took it to be 2°. Refer to Figure 1.24.)

Archimedes imagined an observer on the Earth looking out at the Sun just after sunrise. He constructed the diagram of Figure 1.29 and positioned the observer at the point E. Two tangent lines are drawn from E to the Sun. Note that α is the angle that these tangent lines determine at E. From C, the center of the Earth, he placed two more tangent lines to the Sun and let β be the angle that they determine. Taking the center of the Sun to be O, he obtained a circular arc by rotating CO, and he let A and B be the two points of intersection of this arc with the tangents from C. Since $\beta = \frac{\text{arc } AB}{D_S}$,

$$D_S = \frac{\text{arc } AB}{\beta} \approx \frac{2AO}{\beta} = \frac{2r_S}{\beta}.$$

Since the Sun is far away and E and C relatively close, observe that $\alpha \approx \beta$. (Archimedes showed much

more. By a very careful argument that used delicate geometry and trigonometry, for example, a formula equivalent to $\frac{\sin \varepsilon}{\sin \gamma} < \frac{\varepsilon}{\gamma} < \frac{\tan \varepsilon}{\tan \gamma}$ for $0 < \gamma < \varepsilon < \frac{\pi}{2}$, he verified that $\beta < \alpha < \frac{100}{99}\beta$.) So Archimedes obtained the approximations

$$D_S \approx \frac{\text{arc } AB}{\alpha} \approx \frac{2r_S}{\alpha}.$$

Inserting the inequalities $r_S \le (30)(47{,}500)$ miles and $\frac{1}{200}\frac{\pi}{2} < \alpha$, he found that $D_S < \frac{(60)(47{,}500)(400)}{\pi} < 160 \times 10^6$ miles. He therefore obtained

$$D_S < 160 \times 10^6 \text{ miles.}$$

Let D_* be the distance to the stars. Speculating about another of Aristarchus's assertions, Archimedes took

$$D_* < 1.6 \times 10^{12} \text{ miles.}$$

It is a strange curiosity that these last two inequalities correspond in a sense to correct values (Table 1.5).

Archimedes could now turn to the computation of the number of grains of sand needed to fill the sphere of the universe, i.e., the sphere of radius D_*. See Figure 1.30. He began with a poppy seed and assumed that a sphere the size of a poppy seed could hold 10,000 grains of sand. (The sand that Archimedes has in mind is evidently very finely grained.) He next estimated that the diameters of 40 poppy seeds added to one "fingerbreadth." Now, the volume of a sphere of radius r is

Table 1.5		
	Archimedes[a]	**Actual**
D_S Earth to Sun	160×10^6 miles	93×10^6 miles
D_* Earth to Stars	1.6×10^{12} miles	24×10^{12} miles[b]

[a]The numbers in the column are to be understood to be bigger than the distances to which they refer. For example, D_S is less than the indicated number. Note also that these numbers were based mostly on speculation. These distances were not determined with any finality until more than 2000 years later.

[b]This is the approximate distance to the nearest stars; see the discussion in Section 1.7 (Postscript).

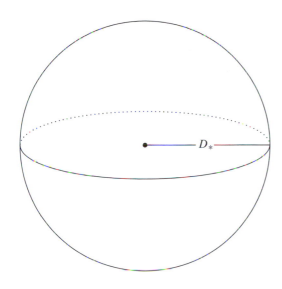

Figure 1.30

$\frac{4}{3}\pi r^3$, or $\frac{4}{3}\pi(\frac{d}{2})^3 = \frac{\pi}{6}d^3$ in terms of its diameter d. It follows that if the diameter of a sphere is increased by a factor of $2, 3, 4, \ldots, k, \ldots$, then its volume is increased by the factor 2^3, 3^3, 4^3, ..., $k^3,\ldots$. So a sphere of diameter one finger-breadth has a volume of 40^3 times greater than that of a poppy seed. Therefore, it can be filled with $10{,}000 \times 40^3 = 4^3 \times 10^3 \times 10^4 \approx 10^9$ grains of sand. Continuing in this way, Archimedes concluded that 10^{63} grains of sand will fill a sphere of radius 1.6×10^{12} miles. Since $D_* < 1.6 \times 10^{12}$ miles, this is more than enough to fill the sphere of the universe. Because Archimedes's number system handles 10^{63} with ease, his goal was accomplished!

By a remarkable coincidence, the 10^{63} grains of sand correspond to about 10^{80} atoms. So Archimedes arrived at a total amount of cosmic matter that isn't far from Eddington's 20th century estimate.

1.7 Postscript

Trigonometry—on a bigger scale yet—is used (*is used today!*) to determine the distances to the near stars. The principle is simple. Imagine yourself in a moving car looking out at the scenery. As the car moves, your perspective is constantly changing. The objects that are near will zoom past, those that are farther will move past more slowly, and the distant ones, say, the Moon, will hardly move at all. The Earth, too, is moving. As it does, it affords different perspectives on the heavens. Plotting the positions of stars carefully and regularly month after month reveals that some stars change their position in a detectable and measurable way and that others remain fixed. The greater the change in a star's position, the closer it is. The smaller the change in position the farther away it is.

To measure the distance to a near star A, proceed as follows. Consider the Earth in the two opposed positions E and E' in its orbit about the Sun. (See Figure 1.31.) When the Earth is at E, make note of the position C_1 of A against the fixed pattern of distant stars in a constellation with a telescope. Since A is near, its position in the constellation is observed to have shifted to, say, C_2 when the Earth is at E'. When the Earth is back at E, measure the angle $\theta = \angle C_1 E C_2$

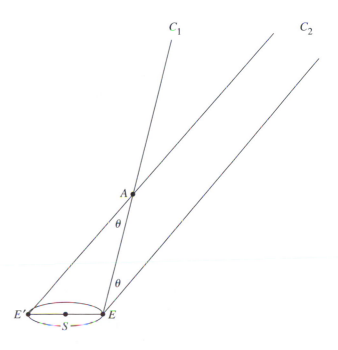

Figure 1.31

between the positions C_1 and C_2. Since the stars in the constellation are very far, the lines $E'C_2$ and EC_2 are parallel. So $\angle EAE' = \theta$.

The angle $p = \frac{1}{2}\theta$ is called the *stellar parallax* of the star A. Let D_A be the distance from the Earth to the star A. This distance is in essence the same as that from the Sun S to A. The distance between the Earth and the Sun is some 93 million miles. See Table 1.5. Astronomers refer to this distance as the *astronomical unit*. So 1 AU = 93×10^6 miles. So the triangle $\triangle ASE$ can be approximated by the circular sector shown in Figure 1.32. Letting p_{rad} be the stellar parallax of A in radian measure, it follows that

$$p_{\mathrm{rad}} = \frac{1}{D_A},$$

where D_A is given in AUs.

Since stellar parallax is always extremely small, it is customary to measure it in seconds, where 1 second is $\frac{1}{60}$ of 1 minute, which is $\frac{1}{60}$ of 1 degree. So $1° = 3600$ seconds. Therefore,

$$1 \text{ radian} = \left(\frac{180}{\pi}\right)^{\circ}$$

$$= \frac{3600 \cdot 180}{\pi} \text{ seconds} = 2 \times 10^5 \text{ seconds}.$$

So the parallax p is converted from radians to seconds by $p_{\mathrm{sec}} = 2 \times 10^5\, p_{\mathrm{rad}}$.

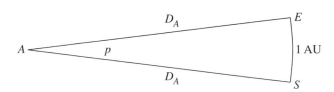

Figure 1.32

The distances to the stars are so vast that they are not measured in AUs but in *light years*. One light year, 1 LY for short, is the distance that light travels in one year. Since light travels 186,000 miles in one second, it travels approximately 6×10^{12} miles in one year. So

we take 1 LY = 6×10^{12} miles. Since 1 AU = 93×10^6 miles, it follows that 1 LY = $\frac{6 \times 10^{12}}{93 \times 10^6}$ AU. Therefore, 1 AU = $\frac{93 \times 10^6}{6 \times 10^{12}}$ LY = 1.5×10^{-5} LY. The equality $p_{\mathrm{rad}} = \frac{1}{D_A}$ can now be converted as follows:

$$p_{\mathrm{sec}} = 2 \times 10^5\, p_{\mathrm{rad}} = 2 \times 10^5 \frac{1}{D_A} \text{ AU}$$

$$= (2 \times 10^5)(1.5 \times 10^{-5})\frac{1}{D_A}\mathrm{LY} = \frac{3}{D_A}\mathrm{LY}.$$

Therefore, the distance D_A from the Earth to the star A in light years is

$$D_A = \frac{3}{p_{\mathrm{sec}}},$$

where p_{sec} is the parallax of A measured in seconds.

The parallax of the faint star Proxima Centauri is about 0.76 seconds, or $\frac{1}{5000}°$. Since this is the nearest star, its parallax is the largest. Inserting $p_{\mathrm{sec}} = 0.76$ into the preceding equation shows that the distance to Proxima Centauri is approximately $\frac{3}{0.76} = 4$ LY. Since 1 LY = 6×10^{12} miles, Proxima Centauri is about 24×10^{12} miles distant (see Table 1.5). Stellar parallaxes were measured with good accuracy in the 19th century. For example, Wilhelm Bessel, the German mathematician and astronomer, measured the stellar parallax of 61 Cygni in 1838 and came within 10% of the modern value of 0.27 seconds. Popping $p_{\mathrm{sec}} = 0.27$ into the equation just derived, we find that 61 Cygni is 10.9 LY away. Sirius, the brightest star, has a parallax of about 0.38 seconds. This yields a distance of about 8 LY.

Modern methods for measuring distances to the more distant stars and galaxies include analyzing the convergence or divergence of the motion of individual stars in a cluster to determine the distance of the cluster; and analyzing the light of a particular star to determine its intrinsic luminosity, which, in combination with its brightness as measured from Earth, gives an indication of its distance. The size of the universe is, of course, considerably bigger than even Archimedes speculated. Some galaxies are millions of LYs distant and some have a size of hundreds of thousands of LYs.

Exercises
1A. The Greek Number System

One version (there are variations) of the traditional number system of the Greeks is the following:

Units	Tens	Hundreds
$\alpha = 1$ (alpha)	$\iota = 10$ (iota)	$\rho = 100$ (rho)
$\beta = 2$ (beta)	$\kappa = 20$ (kappa)	$\sigma = 200$ (sigma)
$\gamma = 3$ (gamma)	$\lambda = 30$ (lambda)	$\tau = 300$ (tau)
$\delta = 4$ (delta)	$\mu = 40$ (mu)	$\upsilon = 400$ (upsilon)
$\varepsilon = 5$ (epsilon)	$\nu = 50$ (nu)	$\phi = 500$ (phi)
$\varsigma = 6$ (digamma)	$\xi = 60$ (xi)	$\chi = 600$ (chi)
$\zeta = 7$ (zeta)	$o = 70$ (omicron)	$\psi = 700$ (psi)
$\eta = 8$ (eta)	$\pi = 80$ (pi)	$\omega = 800$ (omega)
$\theta = 9$ (theta)	$\varsigma = 90$ (koppa)	$\lambda = 900$ (sampi)

(The digamma, koppa, and sampi were taken from an older alphabet of the Phoenicians.)

Other numbers are formed by juxtaposition, using the rule that larger numbers go on the left and smaller ones on the right. For example: $\kappa\varepsilon = 25$, $\lambda\zeta = 37$, $\upsilon\pi\eta = 488$. To designate thousands, the units symbols were used with a stroke before the letter to avoid confusion. For example: $,\gamma = 3000$ and $,\beta\tau\pi\delta = 2384$. (To distinguish between numerals and letters, the Greeks sometimes put a bar over the numerals: $\overline{,\varepsilon\chi o} = 5670$.) For 10,000 an M (for $\mu\nu\rho\iota\alpha o = $ myriad) was used. This was combined with other symbols as follows:

$$\beta M = 20{,}000,$$

$$\iota\delta M_{,}\eta\phi\xi\zeta = 14(10{,}000) + 8567 = 148{,}567$$

$$\upsilon M = (400)(10{,}000) = 4{,}000{,}000$$

$$\omega\mu\varepsilon M_{,}\beta\tau\pi\delta = 845(10{,}000) + 2384 = 8{,}452{,}384$$

Addition and multiplication were cumbersome; this is probably one reason why Greek algebra lagged behind Greek geometry.

1. Write the numbers 85; 842; 34,547; 2,875,739 using the Greek system.

1B. Greek Algebra

Problems (2)-(5) are taken from *The Greek Anthology*.[8]

2. The Muses stole and divided among themselves, in different proportions, the apples I was bringing from Helicon. Clio got the fifth part, and Euterpe the twelfth, but divine Thalia the eighth. Melpomene carried off the twentieth part, and Terpsichore the fourth, and Erato the seventh; Polyhymnia robbed me of thirty apples, and Urania of a hundred and twenty, and Calliope went off with a load of three hundred apples. So I come to thee with lighter hands, bringing these fifty apples that the goddesses left me. How many apples did I bring from Helicon?

3. Make me a crown weighing sixty minae, mixing gold and brass, and with them tin and much wrought iron. Let the gold and brass together form two-thirds, the gold and tin together three-fourths, and the gold and iron three-fifths. Tell me how much gold you must put in, how much brass, how much tin, and how much iron, so as to make the whole crown weigh sixty minae. [A number of references describe the mina as a unit of weight roughly equal to one pound. So it seems that this crown was intended for no ordinary mortal.]

4. Throw me in, silversmith, besides the bowl itself, the third of its weight, and the fourth, and the twelfth; and casting them into the furnace stir them, and mixing them all up take out, please, the mass, and let it weigh one mina. [The first thing is to decide what the question is.]

5. Brick-maker, I am in a great hurry to erect this house. Today is cloudless, and I do not require many more bricks, but I have all I want but three hundred. Thou alone in one day couldst make as many, but thy son left off working when he had finished two hundred, and thy son-in-law when he had made two hundred and fifty. Working all together, in how many hours can you make these? [Hint: If there is not enough information, supply it.]

1C. The Quadratic Formula

6. Consider the quadratic polynomial $x^2 - 5x + 4$. Take the x coefficient -5; divide it by 2 to get $-\frac{5}{2}$; squaring gives $\left(-\frac{5}{2}\right)^2 = \left(\frac{5}{2}\right)^2$. Now rewrite $x^2 - 5x + 4$ as follows:

$$x^2 - 5x + 4 = x^2 - 5x + \left(\frac{5}{2}\right)^2 - \left(\frac{5}{2}\right)^2 + 4.$$

Check that $x^2 - 5x + \left(\frac{5}{2}\right)^2 = \left(x - \frac{5}{2}\right)^2$. Therefore

$$x^2 - 5x + 4 = \left(x - \frac{5}{2}\right)^2 + 4 - \left(\frac{5}{2}\right)^2$$

$$= \left(x - \frac{5}{2}\right)^2 + \frac{16}{4} - \frac{25}{4}$$

$$= \left(x - \frac{5}{2}\right)^2 - \frac{9}{4}.$$

We have done what is called "completing the square" for the polynomial $x^2 - 5x + 4$. Now answer the following:

   i. For which values of x is $x^2 - 5x + 4 = 0$?
   ii. What is the least value that $x^2 - 5x + 4$ can have?

7. Solve the equation $3x^2 + 21x + 12 = 0$ by completing the square for the polynomial $x^2 + 7x + 4$.

8. Let a, b, and c be constants and consider the equation $ax^2 + bx + c = 0$. What is x equal to if $a = 0$? If $a \neq 0$, use the strategy of Problems 6 and 7 to show that the solutions are given by the *quadratic formula*:

$$x = \frac{-b \pm \sqrt{b^2 - 4ac}}{2a}.$$

1D. Rational and Real Numbers

9. The matter of measurability is perhaps most concretely illustrated by the consideration of the numbers that underlie our monetary system. All dollar amounts are expressed in the form: $\$x\frac{yz}{100}$. So only rational numbers, indeed only certain rational numbers, are allowed. In a supermarket one will occasionally find, say, 3 items for a dollar. A single item is not measurable within the system. Why not?

10. Consider the numbers $1.333333\ldots$, $2.676767\ldots$, and $4.728728728\ldots$. Show that they are rational numbers. [Hint: Let r be the number. In the first case, consider $r - 10r$.]

11. What are the decimal expansions of the rational numbers $\frac{5}{4}$ and $\frac{468}{198}$? [Hint: Carry out the divisions.]

   Note: It turns out that a real number r is rational precisely if its decimal expansion has a repetitive pattern after some point. For example,

$$234.5\underline{9}999999999\ldots = 234.6 = 234\frac{6}{10}$$

and

$$52.\underline{36}36\ldots = 52\frac{468}{198}$$

are rational numbers. So is

$$35.34672\underline{638}638638\ldots.$$

12. Recall that a prime number is a positive integer $p > 1$ that has no divisors except 1 and p. Euclid pursued the study of prime numbers in Book 9 of the *Elements*. The

following fact about prime numbers is known as the Fundamental Theorem of Arithmetic: Every positive integer n is a product $n = p_1^{k_1} \cdots p_i^{k_i}$ of powers of distinct prime numbers $p_1, \ldots, p_i$; and there is only one way of doing this (aside from rearranging the order of the factors). For example, $54 = 2 \cdot 3 \cdot 3 \cdot 3 = 2^1 \cdot 3^3 = 2 \cdot 3^3$ is the unique factorization of 54 into prime powers.

   i. Determine the factorizations of 28, 192, and 143 into powers of distinct primes.
   ii. Use the Fundamental Theorem of Arithmetic to show that $\sqrt{3}$ is irrational.
   iii. Show that a positive integer n is a square precisely when all the exponents of its factorization into primes are even integers.
   iv. Let n be a positive integer. Use the Fundamental Theorem of Arithmetic to show that $\sqrt{n}$ is rational only if n is square.

1E. Angles and Circular Arcs

13. Fill in the following blanks.

   i. 1 radian = _____ degrees
   ii. 1 degree = _____ radians
   iii. $78.5° = $ _____ radians
   iv. 1.238 radians = _____ degrees

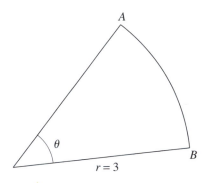

Figure 1.33

14. In the circular sector of Figure 1.33, $\theta = 57.3°$. What is the length of the arc AB?

15. Start with Figure 1.13 of the text. Take the tangent to the circle at A and let P' be its point of intersection with $A'P$. By Figure 1.34, it is plausible that arc $AP \leq AP' + P'P$. Show that $AP' < A'P'$. Conclude that the

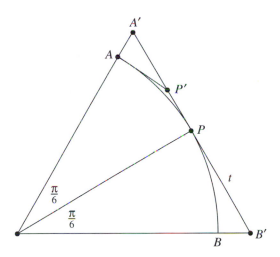

Figure 1.34

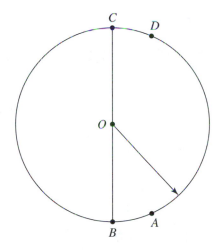

Figure 1.35

inequalities arc $AP < A'P = t$ and arc $AB < A'B' = 2t$ are plausible.

16. A rock is hurled with a sling. Just before it flies off, it is twirled in a circular arc of radius 3 feet at 4 revolutions per second. With what speed (in feet per second) does the rock fly off?

17. The circle in Figure 1.35 has center O and a radius of 2 feet. The arrow is rotating clockwise at a rate of one revolution in twelve hours (so think of it as the hour hand of a clock). The points A and D are positioned in such a way that the segment connecting them is parallel to the diameter BC of the circle. Determine the length of the arc CD, given that the arrow requires 7.5 hours to rotate from A to D.

1F. Basic Trigonometry

18. Use the appropriate triangles to fill in the values:

    i. $\cos \frac{\pi}{6} =$ _____   **ii.** $\cos \frac{\pi}{4} =$ _____

  iii. $\cos \frac{\pi}{3} =$ _____   **iv.** $\tan \frac{\pi}{6} =$ _____

    v. $\tan \frac{\pi}{4} =$ _____   **vi.** $\tan \frac{\pi}{3} =$ _____

19. Use Figure 1.21 of the text to determine the following limits:

    i. $\lim\limits_{\theta \to 0} \cos \theta =$ _____

   ii. $\lim\limits_{\phi \to \frac{\pi}{2}} \cos \theta =$ _____

  iii. $\lim\limits_{\theta \to 0} \tan \theta =$ _____

  iv. $\lim\limits_{\phi \to \frac{\pi}{2}} \tan \theta =$ _____

20. Illustrate with a diagram that if $\theta' > \theta > 0$, then $\cos \theta' < \cos \theta$ and $\tan \theta < \tan \theta'$.

21. The secant of θ is defined by $\sec \theta = \frac{1}{\cos \theta}$. Verify the identity $\sec^2 \theta = \tan^2 \theta + 1$.

22. Compare (use a calculator) the values of α, $\sin \alpha$, $\tan \alpha$ for

    i. $\alpha = 0.1$ radians:
       $\sin \alpha =$ _____, $\tan \alpha =$ _____.

   ii. $\alpha = 0.01$ radians:
       $\sin \alpha =$ _____, $\tan \alpha =$ _____.

  iii. $\alpha = 0.001$ radians:
       $\sin \alpha =$ _____, $\tan \alpha =$ _____.

1G. Distances and Sizes in the Universe

23. Compute r_M, r_S, D_M, and D_S using Aristarchus's argument and

    a. Keep $r_E = 3850$ miles.
    b. In hypothesis **C**, take $89°50'$ instead of $87°$. (The angle measure $'$ is called minute; $1'$ is equal to $\frac{1}{60}^{\circ}$.)
    c. In hypothesis **D**, take $\frac{1}{2}^{\circ}$ instead of $2°$. (So the angle in Figure 1.27 is $\frac{1}{4}^{\circ}$ instead of $1°$.)
    d. In hypothesis **E**, take $5r_M$ instead of $4r_M$.

    Round off to get an accuracy up to 4 decimal places. Compare your conclusions with the modern values from Table 1.4.

24. Both Aristarchus and Archimedes assumed that the Earth is a sphere. Is this reasonable in view of the mountain ranges on its surface? You are given that the radius r_E of the Earth is 3950 miles, that the height of Mount Everest is 29,028 feet, that 1 mile = 5280 feet,[9] and that the radius of a basketball is 4.7 inches. If the Earth were shrunk to the size of a basketball, how high would Mount Everest be? Is this higher than one of the little mounds—officially called a pebble—on a basketball? These have a height of about 0.02 inches.[10]

25. Since 3950 miles = 20,856,000 feet and 4.7 inches = 0.39 feet, the shrinking factor in Problem 24 is $\frac{0.39}{20,856,000}$, or about $\frac{1}{50,000,000}$. Show that if the radius r_M of the Moon, the distance D_M to the Moon, the radius r_S of the Sun, the distance D_S to the Sun, and the distance D_* to the nearest star were shrunk by this factor, we would have (approximately):

$r_E = 4.7$ inches (the radius of a basketball, as we have already seen)

$r_M = 1.37$ inches (about the radius of a baseball which is 1.43 inches)

$D_M = 25$ feet

$r_S = 45$ feet

$D_S = 1.86$ miles

$D_* = \frac{1}{2}$ million miles

26. Shrink the Sun to the size of a basketball. What is the shrinking factor? Shrink the rest of the solar system by this factor and compute $r_E, r_M, D_M, r_S, D_S,$ and D_* in this case.

27. Reconstruct Archimedes's argument that 10^{63} grains of sand more than fill the sphere of the universe. Assume that 1 finger breadth is $\frac{2}{3}$ of an inch.

28. The near stars Barnard, G51-15, and Ross 780 have stellar parallaxes of 0.55, 0.27, and 0.21 seconds, respectively. Determine their distances in light years.

Notes

[1] This theorem was probably already known to the Babylonians in 1700 B.C., more than 1000 years before the time of Pythagoras (about 570 to 500 B.C.).

[2] The Pythagorean philosophy was not without later influence, however. The book by G.L. Hershey, *Pythagorean Palaces* (Cornell University Press, Ithaca N.Y. and London, 1976), was written to establish the fact that in "the Italian Rennaissance domestic architecture was largely ruled by Pythagorean principles."

[3] This is a question that occupied the mathematicians of antiquity and is still relevant today. Our discussion will give only a very elementary perspective.

[4] When a segment, such as $A'P$ or PB', appears in a mathematical expression, the reference here and elsewhere in this text will always be to its length.

[5] A gnomon is simply a straight stick or rod. The word comes from gno, "to know," in ancient Greek. Our words prognosis and physiognomy are derived from it.

[6] The Greeks used stadia, not miles. Ten stadia are the equivalent of about one mile.

[7] The detail on the Earth in the figures that follow should not mislead the reader into thinking that the Greeks of the third century B.C. understood the scope and shape of Africa, Northern Europe, and the Atlantic Ocean. The Greeks were very familiar with the territory near the Mediterranean Sea and the armies of Alexander the Great advanced all the way to today's India and Afghanistan. However, a true concept of the extent of the continental land masses and oceans began to develop only with the voyages of discovery in the 15th century.

[8] *The Greek Anthology* is a collection of Greek poems, songs, and riddles. Some of these were compiled as early as the 7th century B.C. and others as late as the 10th century A.D. Harvard University Press produced a new edition of *The Greek Anthology* in 1993.

[9] So Mount Everest is about 5.5 miles high. As an aside, note the Earth is actually not quite a sphere. The Earth's rotation has caused an "equatorial bulge," so that the Earth's diameter through the equator is about 26 miles more than that through the poles.

[10]According to Rawlings Sporting Goods, the official circumference of a basketball is from 29.5 to 30 inches. This translates to a radius r of 4.695 to 4.775 inches. A pebble has an official height from 0.013 to 0.025 inches.

2

Ptolemy and the Dynamics of the Universe

So far we have considered the "static" situation of the universe: the sizes of the Earth, Moon, and Sun, and the various distances between them. Next we turn to the question of its "dynamics": according to what schemes do the heavenly bodies move?

2.1 A Geometry of Shadows and the Motion of the Sun

The Greek philosophers perceived the basic structure of the universe to be this: The Earth is spherical in shape, it is fixed, and it lies at the center of the universe. The Moon, Mercury, Venus, Sun, Mars, Jupiter, and Saturn are the planets. (Neptune, Uranus, and Pluto were not yet known.) All is surrounded by the *celestial sphere* of fixed stars, which rotates about its axis once a day. The motion of the planets (including the Sun) has two components: they are carried along by the daily rotation of the sphere of stars, but they also have their own independent motion. The planets, in particular, are observed to change their position relative to the background of fixed stars. Each of the Greek thinkers provided additional details. For example, according to Aristotle, each of the planets and the Sun is attached to its own crystal sphere whose rotation moves it along; and according to Pythagoras, each of the heavenly bodies contributes a certain tone to an orchestrated "music of the spheres." The reasons given in support of this scheme included, "We must suppose the Earth, the Hearth of the House of the gods, to remain fixed and the planets with the whole embracing heavens to move" and "The motion must be circular because circles and spheres are the only perfect shapes." The Sun-centered model of Aristarchus was dismissed as being in obvious contradiction to the facts.

The essence of the structure of the Greek universe is illustrated in Figure 2.1. The axis is the continuation of the north-south axis of the Earth. The *celestial equator* is the projection of the Earth's equator against the celestial sphere. The *ecliptic* is the projection of the path of the Sun against the celestial sphere. Both the celestial equator and the ecliptic are great circles on the celestial sphere. The two points where these circles intersect are called the *vernal* and *autumnal equinox* positions. The angle ε between the planes determined by the celestial equator and the ecliptic is called the *obliquity of the ecliptic*. The point on the ecliptic where the Sun reaches its highest position above the plane of the celestial equator is the *summer solstice* position. The point of lowest position below this plane is the *winter solstice* position. Note that this terminology is still in use today.

Today we know that the Earth rotates once a day about its north-south axis and that it moves in a year-long orbit about the Sun. However, from the vantage point of an observer on the surface of the Earth—and this was the point of view of the Greek philosophers—the daily rotation of the Earth is seen as the daily rotation of the celestial sphere of stars. The motion of the Earth in its orbit around the Sun is seen as that of the Sun against the celestial sphere in a yearly cycle. Our attention turns to the observations by the Greeks of this apparent motion of the Sun.

The following ritual must have been performed by the Greek astronomers thousands of times. Plant a gnomon—this is the pole Eratosthenes used in his measurement of the Earth—at a point A in the middle of a flat horizontal stretch of ground. Use a plumb line to set it in vertical position GA. Refer to Figure 2.2. At some time in the morning, after the Sun has risen in the east, mark the endpoint W of the shadow (it lies west of A) and measure the length AW. As the Sun moves up in the sky, the shadow will shorten and rotate. When the Sun starts its descent, the shadow will begin to lengthen. At some time before the Sun sets in the west, the tip of the shadow will reach a point E (toward the east) on the ground where its length AE will equal AW. By the symmetry of the process, the bisector of the angle WAE lies on the north-south line through A. This line is called the *meridian* through A. At the moment in the day when the shadow falls on the meridian, it has reached its shortest length; the Sun is now at its highest point in the sky. This moment is *noon* on that day. The length of a day is defined to be

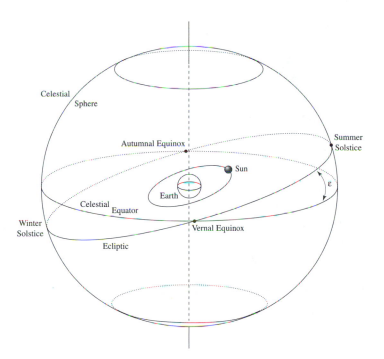

Figure 2.1

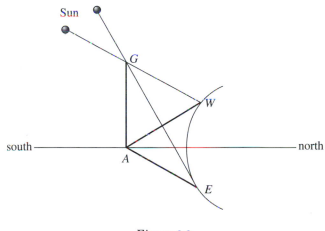

Figure 2.2

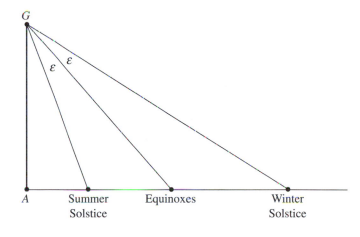

Figure 2.3

the duration of time between two successive noons. The division of a day into 24 equal units defines the hour. The hour, in turn, is divided into 60 equal units called minutes. (Note the influence of the Babylonian base 60 number system.)

Since the focus is on the movement of the Sun against the celestial sphere, the daily rotation of this sphere (in other words, the daily rotation of the Earth) must be discounted. To achieve this, gnomon measurements are taken exactly at noon each day. The day on which the noon shadow of the gnomon is shortest, in other words, the day of the year on which the Sun is at its very highest point in the sky, is summer solstice. Winter solstice is the day on which the noon shadow of the gnomon is longest. This is the day of the year on which the noon Sun is at its lowest point. Summer solstice is the day of longest daylight, and winter solstice is the day of shortest daylight. There are two days on which day and night have the same length. The one occurring before summer solstice is the vernal equinox, and the other, occurring after summer solstice, is the autumnal equinox. Some thought (and reference to Figure 2.1) shows that this occurs when the Sun is in the plane of the celestial equator. This corresponds to the shadow position determined by the bisection of the angle given by the shadow's summer and winter solstice positions: see Figure 2.3. The angle ε obtained by the bisection is the obliquity of the ecliptic.

The seasons of the year are determined as follows: *Spring* is the period from vernal equinox to summer solstice; *summer* runs from summer solstice to autumnal equinox; *autumn* from autumnal equinox to winter solstice; and *winter* from winter solstice to vernal equinox. The year is defined to be the time period between two successive summer solstices (or winter solstices, or autumnal equinoxes...).

The astronomer-mathematician Hipparchus lived and worked from about 190–125 B.C. in Nicaea (50 miles southeast of Constantinople) and also on the island of Rhodes. His shadow measurements gave the following results:

$$\text{Spring} \quad 94\tfrac{1}{2} \text{ days}$$
$$\text{Summer} \ 92\tfrac{1}{2} \text{ days}$$
$$\text{Autumn} \ 88\tfrac{1}{8} \text{ days}$$
$$\text{Winter} \ \ 90\tfrac{1}{8} \text{ days}$$

By addition, Hipparchus obtained that the year consists of $365\tfrac{1}{4}$ days. He measured ε to be about $24°$.

Refer back to Figure 2.1 and observe that the vernal equinox, summer solstice, autumnal equinox, and winter solstice positions divide the ecliptic into four

equal circular arcs. The fact that the lengths of the seasons are different means that the Sun traces out these four arcs in different lengths of time. This, however, stands in clear contradiction to the two fundamental principles of Greek philosophy:

(1) The Earth is fixed and positioned at the center of the universe (the geocentric hypothesis).
(2) All other bodies in the universe move with constant velocity along circular paths (which have the Earth at their center).

Another phenomenon stood in conflict with these principles: while the general pattern of motion of the planets (against the backdrop of the stars) was from west to east, they were observed to stop on occasion and reverse direction. This is the phenomenon of the planets' "retrograde motion." These difficulties appear to have moved Plato (in the *Dialogue Timaeus*) to exhort the philosophers of his day to do the apparently impossible: to develop a theory of motion for the universe that adhered to the fundamental principles (1) and (2) *and* was consistent with observed reality.

This is precisely what Claudius Ptolemy (about 100–165 A.D.) of Alexandria set out to do. His "Ptolemaic System of the Universe" is developed in his great work, the *Almagest*, one of the most influential texts in the history of science. It is an encyclopedic compilation—a grand synthesis—of Greek mathematics and astronomy. The title *Almagest* has a later origin. It comes from the Greek word *megiste* ($\mu\varepsilon\gamma\iota\sigma\tau\eta$), meaning "greatest" (probably from references to the work as the "Great Collection" or "Great Astronomer") to which Islamic astronomers added the Arabic prefix *al* (the).

2.2 Geometry in the *Almagest*

The *Almagest* contains the first surviving comprehensive exposition of Greek trigonometry (and is thus also of importance in the history of mathematics). Presupposing a thorough knowledge of the *Elements* of Euclid, Ptolemy includes careful expositions of specialized foundational results—for example, those of Apollonius (about 270–190 B.C.) of Perga (an ancient city that was located on the south coast of today's Turkey), those of Hipparchus, as well as his own. The trigonometry in the *Almagest* is primarily a science of numerical calculations—often very intricate—of triangles in the plane and on the sphere. This emphasis on numerical methods goes counter to the commonly held notion that Greek mathematics deliberately avoided such methods in favor of purely geometrical ones. Ptolemy, however, gives no indication of the rules and techniques underlying his calculations, and it is indeed true that the Greeks were reluctant to deal with such matters in the form of formal treatises.

From the large body of mathematics in the *Almagest*, we will consider only a few (very few) facts from plane geometry. This suffices for the purposes of this text.

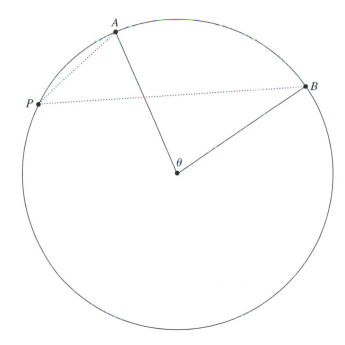

Figure 2.4

Proposition 2.1. Let A and B be any two points on the circle in Figure 2.4 and let θ be the angle that they determine along with the center. Let P be any other point on the circle. Then

$$\angle APB = \frac{\theta}{2}.$$

One way to verify this is as follows. Consider first the special case shown in Figure 2.5, where the segment PB is a diagonal. Let $\angle APB = \alpha$. Notice that the triangle formed by A, P, and the center is isosceles and, in particular, that the angle at A is also equal to α. Adding the three angles of this triangle shows that $2\alpha + (\pi - \theta) = \pi$. Therefore, $2\alpha = \theta$, and hence $\angle APB = \alpha = \frac{\theta}{2}$, as required.

Now let's return to the general case. Draw a diameter PC from P through the center and let φ be the indicated angle. See Figure 2.6. By the special case already verified,

$$\angle APC = \frac{1}{2}(\theta + \varphi) \quad \text{and} \quad \angle BPC = \frac{1}{2}\varphi.$$

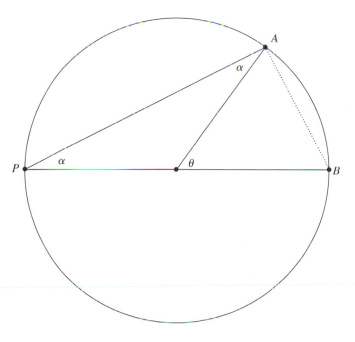

Figure 2.5

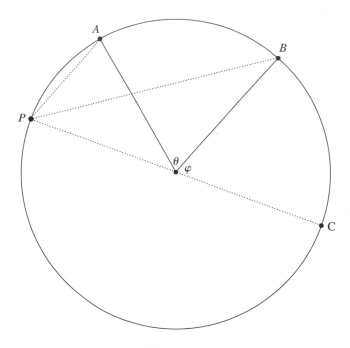

Figure 2.6

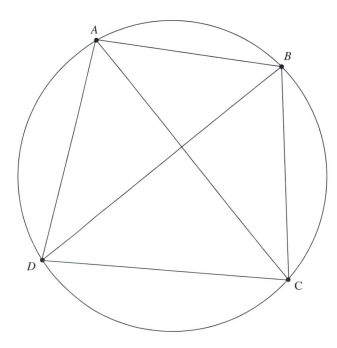

Figure 2.7

Since $\angle APB = \angle APC - \angle BPC = \frac{1}{2}(\theta + \varphi) - \frac{1}{2}\varphi = \frac{\theta}{2}$, the proof of Proposition 2.1 is complete.

Proposition 2.2. Fix A and B on the circle as in Figure 2.4. Then $\angle APB$ is the same, no matter what point P on the circle is taken.

This is so because for any point P, the angle $\angle APB$ is equal to one-half of the central angle determined by A and B.

It is now possible to derive the following famous theorem.

Ptolemy's Theorem. Let $A, B, C,$ and D be any four points on a circle. Then

$$(AC) \cdot (BD) = (AB) \cdot (CD) + (AD) \cdot (BC).$$

Figure 2.7 shows the typical situation. Pythagoras's theorem is a special case of Ptolemy's theorem. This can be seen as follows. Complete any right triangle to a rectangle and inscribe the rectangle into a circle. Ptolemy's theorem applied to the four vertices of the triangle provides Pythagoras's theorem. Exercise 1 asks you to provide the details. Ptolemy's theorem can also be used to prove some of the basic identities of trigonometry. These include $\sin(\alpha + \beta) = (\sin\alpha)(\cos\beta) + (\cos\alpha)(\sin\beta)$ and the analogous formula for the cosine. However, there are more efficient ways to verify these identities. See Exercise 6. The proof of Ptolemy's theorem follows below. It relies on Proposition 2.2 and basic properties of similar triangles. Since it will not play a role in this book, little is lost by skipping it.

Proof of Ptolemy's Theorem. Let E be the indicated point of intersection in Figure 2.8 and draw AF such that $\angle DAF = \angle BAE$. By Proposition 2.2, $\angle ADB = \angle ACB$. So $\angle DAF = \angle BAE = \angle BAC$ and $\angle ADF = \angle ADB = \angle ACB$. Now refer to the triangles $\triangle ADF$ and $\triangle ACB$ of Figure 2.9. Note that these triangles have two angles in common (labeled 1 and 2). Since the sum of the angles of any triangle is $180°$, the angles labeled 3 are also equal. It follows that

△*ADF* and △*ACB* are similar. Take one more look at Figure 2.9. Since *AD* and *AC*, and *DF* and *BC*, are corresponding pairs of sides,

$$\frac{AD}{AC} = \frac{DF}{BC}. \qquad (2.1)$$

By another application of Proposition 2.2 to Fig-

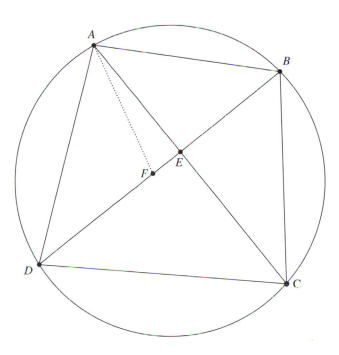

Figure 2.8

ure 2.8, ∠*ACD* = ∠*ABD* = ∠*ABF*. Since

∠*DAC* = ∠*DAF* + ∠*FAE* = ∠*BAE* + ∠*EAF* = ∠*BAF*

the triangles △*ACD* and △*ABF* have two angles in common. They are the angles labeled 1 and 2 in Figure 2.10. Therefore, they too are similar. Since ratios of corresponding sides are equal, we see that

$$\frac{AB}{AC} = \frac{BF}{CD}. \qquad (2.2)$$

By equation (2.1), $DF = \frac{AD}{AC}BC$, and by equation (2.2), $BF = \frac{AB}{AC}CD$. By a look at Figure 2.8,

$$BD = BF + DF = \frac{AB}{AC}CD + \frac{AD}{AC}BC.$$

Multiplying through by *AC* finally shows that

$$(AC) \cdot (BD) = (AB) \cdot (CD) + (AD) \cdot (BC).$$

The proof of Ptolemy's theorem is now complete.

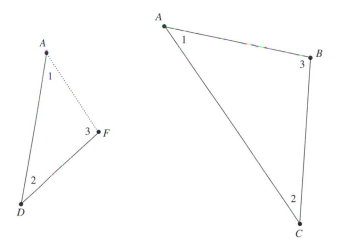

Figure 2.9

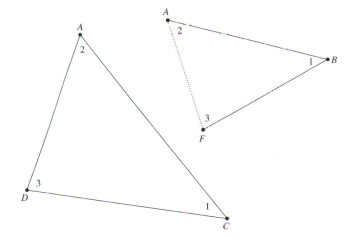

Figure 2.10

2.3 The Solar Model

Since the Sun affects everything else (consider, from the modern point of view, the fact that all the planets are in orbit around it), Ptolemy began his astronomy with it. As already pointed out, he was confronted with the following basic question: Is it possible to develop a theory that has the Sun moving with constant velocity in a circular path around the Earth and that at the same time explains the observed differences in the lengths of the seasons?

Ptolemy's solution, in principle, is this: retain the Earth at the center of the celestial sphere; retain the Sun's circular orbit and retain its constant orbital velocity; but move the center of this orbit away from the Earth! While this does compromise the fundamental principles of Greek philosophy in part, it does retain the essential elements.

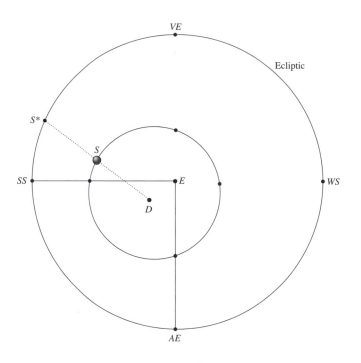

Figure 2.11

The details of the solution follow. This part of the *Almagest* is taken from the work of Hipparchus. Consider Figure 2.11. The smaller circle represents the path of the Sun; it is centered at the point D. The larger circle represents the ecliptic. Recall that

this is the projection of the Sun's path on the celestial sphere. The Earth E is at the center of the ecliptic. The vernal equinox, summer solstice, autumnal equinox, and winter solstice positions are indicated by VE, SS, AE, and WS. Refer back to Figure 2.1. The Sun S moves along its circular path at constant speed. Its projection, S^*, moves on the ecliptic. Compare the motion of the Sun S to that of S^* and notice that S^* will require more time to move from SS to AE than from AE to WS, or from WS to VE, or from VE to SS. In other words, according to this scheme, summer is the longest season!

According to Hipparchus's observations, however, spring—and not summer—is the longest season. This means that in the model of Figure 2.11, D is not in the correct position. But it should now be apparent that it will be possible to keep the Earth at the center of the universe, to retain uniform circular orbits, and at the same time to allow for the differences in the lengths of the seasons.

So where exactly does D have to be placed? The answer to this question is the essence of the solar model of the *Almagest*. Figure 2.12 shows the ecliptic and the path of the Sun. The points E, D, S, and S^* continue to have the same meaning. In order to make spring the longest season, D is now placed somewhere above and to the left of E. The rest of the notation is as follows:

J is the position of the Sun in its orbit at the moment of vernal equinox.

G is the position of the Sun at autumnal equinox.

C is the position of the Sun at summer solstice.

H is the position of the Sun at winter solstice.

A is the position of the Sun, called *aphelion*, when it is farthest from the Earth.

A^* is the projection of A against the ecliptic.

Finally, K and F, and B and I, are the points of intersection of the indicated vertical and horizontal segments with the circular path of the Sun, and L is the intersection of the indicated segments.

Denote the angle between the vernal equinox and aphelion positions, i.e., the angle JEA, by λ_A. Let r

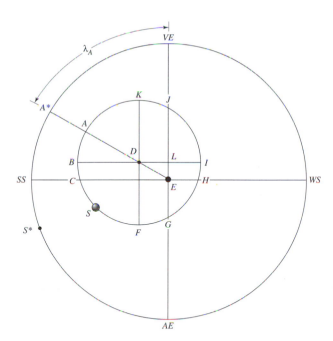

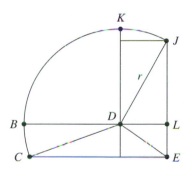

Figure 2.14

Figure 2.12

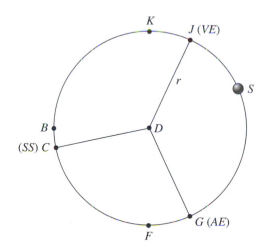

Figure 2.13

be the radius of the Sun's orbit, i.e., $r = DS$, and let e be the length of the segment ED. The solution of the problem of the precise placement of D relative to E is this: Use the information about the lengths of the seasons given by Hipparchus and determine the

parameters λ_A, r, and e, or at least the ratio $\frac{e}{r}$, in such a way that the resulting scheme is consistent with these observations. The steps in Ptolemy's analysis are these: Compute the lengths of the arcs JK and BC. This gives information about the triangle DLE. This triangle then determines the parameters. The details follow. When arc JK, arc BC, etc. appear in a mathematical expression, the reference will always be to the length of the particular arc.

Consider the motion of the Sun S on its circular path, and let ω be the angle traced out by the rotating segment DS in one day. Since the Sun completes the full circle in $365\frac{1}{4}$ days, it follows that $\omega = \frac{360°}{365\frac{1}{4}} = 0.986°$. Therefore,

$$\omega = (0.986°)\frac{\pi}{180°} = 0.0172 \text{ radians.}$$

Now refer to Figure 2.13. Because the Sun moves from J to C during spring, which is $94\frac{1}{2}$ days long, it follows that $\angle JDC = 94\frac{1}{2}\omega = 1.625$ radians.[1] Since the radian measure of this angle is also $\frac{\text{arc } JC}{r}$, it follows that arc $JC = 1.625r$. Because summer is $92\frac{1}{2}$ days long, the very same argument shows that arc $CG = (92\frac{1}{2}\omega)r = 1.591r$. Therefore,

$$\text{arc } JC + \text{arc } CG = 1.625r + 1.591r = 3.216r.$$

But also, referring to Figure 2.13 again,

$$\text{arc } JC + \text{arc } CG = \text{arc } JK + \text{arc } KF + \text{arc } FG$$

$$= \text{arc } JK + \pi r + \text{arc } FG.$$

Because arc $JK =$ arc FG (see Figure 2.12), we now get

$$2\,\text{arc}\,JK + \pi r = \text{arc}\,JC + \text{arc}\,CG = 3.216r.$$

Therefore, $2\,\text{arc}\,JK = 3.216r - 3.142r = 0.074r$. So arc $JK = 0.037r$. Observe, therefore, that $\angle KDJ = \frac{\text{arc}\,JK}{r} = 0.037$ radians. In degrees this is approximately $2°$, which is a small angle.

A look at Figure 2.14 (it has been extracted from Figure 2.12) shows that $DL \approx \text{arc}\,JK = 0.037r$. More precisely, since $\theta = \angle KDJ = 0.037$ is small, $\frac{DL}{r} = \sin\theta \approx \theta = \frac{\text{arc}\,JK}{r}$. (See Table 1.3 in Section 1.4.) We will therefore take

$$DL = 0.037r.$$

Ptolemy gave a similar approximation for EL. Refer back to Figure 2.13 and observe that

$$\text{arc}\,BC = \text{arc}\,JC - \text{arc}\,JB$$

$$= \text{arc}\,JC - (\text{arc}\,JK + \text{arc}\,KB)$$

$$= 1.625r - 0.037r - \frac{\pi}{2}r$$

$$= 1.625r - 0.037r - 1.571r$$

$$= 0.017r.$$

The angle $\angle BDC = \frac{\text{arc}\,BC}{r} = 0.017$ radians is a small angle; it is approximately $1°$. So by another look at Figure 2.14, $EL \approx \text{arc}\,BC = 0.017r$. We will therefore take

$$EL = 0.017r.$$

Figure 2.15 shows the right triangle DLE as extracted from Figure 2.12 with the insertion of the information just derived. Ptolemy could now compute the basic parameters e and λ_A. By the Pythagorean theorem,

$$e = \sqrt{DL^2 + EL^2} = \sqrt{(3.7 \times 10^{-2}r)^2 + (1.7 \times 10^{-2}r)^2}$$

$$= \sqrt{13.69(10^{-2}r)^2 + 2.89(10^{-2}r)^2}$$

$$= 10^{-2}r\sqrt{13.69 + 2.89}$$

$$= 10^{-2}r\sqrt{16.58} = 0.041r.$$

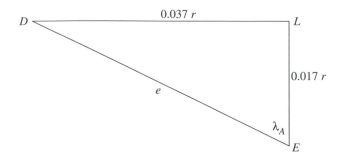

Figure 2.15

It follows that e is approximated by

$$e = \frac{1}{24}r.$$

Because $\sin\lambda_A = \frac{DL}{e} = \frac{0.037r}{0.041r} = \frac{0.037}{0.041} = 0.90$, and Ptolemy knew trigonometry, he concluded that

$$\lambda_A = 64°.$$

Ptolemy's determination of the position of the center of the Sun's circular orbit relative to the Earth was complete. The *Almagest* also contains an estimate for r (the distance from the Earth to the Sun), but it is no more accurate than that of Aristarchus. See the values for D_S in Table 1.4 of Section 1.5.

Once Ptolemy established his theory of the Sun's motion, he used it to confirm Hipparchus's observation that winter has $90\frac{1}{8}$ days. By Figure 2.12,

$$\text{arc}\,HJ = \text{arc}\,HI + \text{arc}\,IJ$$

$$= \text{arc}\,BC + (\text{arc}\,IK - \text{arc}\,JK)$$

$$= 0.017r + \left(\frac{\pi}{2}r - 0.037r\right)$$

$$= 0.017r + (1.571r - 0.037r)$$

$$= 1.551r.$$

Note that the angle that the Sun sweeps out during winter is $\angle HDJ$. Its radian measure is $\frac{\text{arc}\,HJ}{r} = 1.551$. Since it sweeps out $\omega = 0.0172$ radians per day, it follows that winter must be $\frac{1.551}{0.0172} = 90.17 \approx 90\frac{1}{6}$

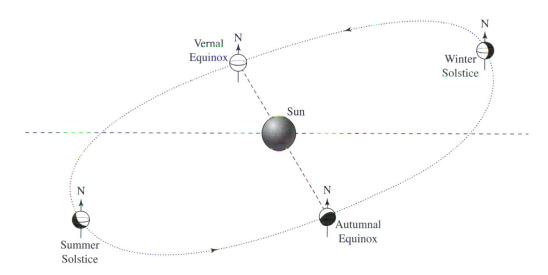

Figure 2.16

days long. This corresponds closely to the $90\frac{1}{8}$ days observed by Hipparchus.

2.4 The Modern Perspective

Let us now turn to today's heliocentric point of view. The motion of the Earth has two primary features. One is the circular orbit of the Earth with the Sun in the interior of the circle. (This orbit is in fact not quite circular. It is a circle that is "stretched" very slightly along a diameter. This subtlety will be taken up in Chapter 4.) The other is the Earth's daily rotation about the axis through the two poles. This north-south axis is perpendicular to the equator, but it is not perpendicular to the Earth's orbit. This is because the plane determined by the Earth's equator is not the same as that of the orbit. For purposes of the present discussion, think of what is going on in the following way. Consider the plane parallel to the Earth's equator and through the center of the Sun. We will call it the "plane of the Sun." See Figure 2.16. This figure show a perspective of the Sun from "above." The two dotted lines in the figure are perpendicular lines in this plane. The arrow N (for north) lies on the axis of the Earth's rotation and is, therefore, perpendicular to the plane of the Sun. This arrow always points in the same direction: north. The Earth passes alternately above and below the plane of the Sun. At summer solstice, the Sun shines down most directly on the Earth's northern hemisphere. From the perspective of Figure 2.16, the Earth is now at its lowest point below the plane of the Sun. On this day the Sun is above the horizon for the longest period of time in the northern hemisphere; it is the day of longest daylight. On the day of winter solstice, the smallest portion of the northern hemisphere is exposed to the Sun; it is the day on which the Sun is above the horizon for the shortest period of time. There are two instances where the center of the Earth lies in the plane of the Sun. Since the axis of rotation is perpendicular to this plane, it follows that on the two days when this occurs, every location on the Earth receives 12 hours of sunshine. These are the days of vernal and autumnal equinox. From the perspective of Figure 2.16, the Earth is above the plane of the Sun during its motion from autumnal equinox to winter solstice to vernal equinox (i.e., during autumn and winter); and the Earth is below the plane of the Sun as it moves from vernal equinox to summer solstice and back to

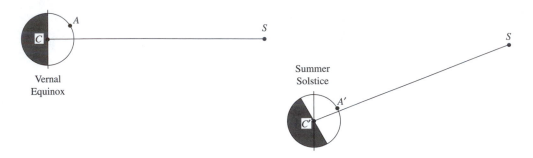

Figure 2.17

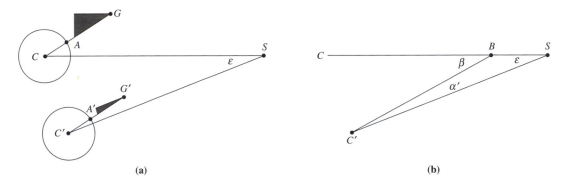

Figure 2.18

autumnal equinox (i.e., during spring and summer). The figure indicates the dark and sunlit regions of the Earth at each of the four relevant positions.

The question that we will turn to in this section as well as the next is this: What information do the methods of the Greeks provide when they are applied to the heliocentric point of view?

Consider the plane perpendicular to the plane of the Sun and through both the center of the Sun and the center of the Earth at summer solstice. (Use Figure 2.16 to visualize this plane.) Now rotate the vernal equinox position of the Earth into this plane as shown in Figure 2.17. The Earth's axis of rotation and the regions of sunlight and shadow on the Earth are indicated in the figure. In each case, it is noon in Alexandria. The points C and C' are the center of the Earth, A and A' refer to Alexandria, and S represents the center of the Sun. Now put the diagrams of Figure 2.17 together. This is done in Figure 2.18(a). The segments AG and $A'G'$ are gnomons planted in Alexandria at noon. (They are drawn larger than to scale in order to make them and the shadows that they cast visible in the diagram.) A moment's thought shows that ε is the angle between the plane of the Sun and the plane of the Earth's orbit. Denote the angle at G by α and that at G' by α'. Notice that the angle at C' is also α'.

Figure 2.18(b) is derived from Figure 2.18(a) as follows. Extend the segment $C'A'G'$ to a point B on CS and let β be the angle CBC'. Because the segments $C'B$ and CG are both determined by a gnomon placed in Alexandria, they are parallel. Therefore, $\beta = \alpha$. The angles ε, α', and $\pi - \beta$ belong to the same triangle. So

$$\pi = \varepsilon + \alpha' + (\pi - \beta) = \varepsilon + \alpha' + (\pi - \alpha)$$

and hence $\varepsilon + \alpha' - \alpha = 0$. It follows that $\varepsilon = \alpha - \alpha'$.

This discussion has put one of the shadow measurements of the Greeks (as discussed in Section 2.1) into the heliocentric context. By measuring the angles α and α' and computing the obliquity of the ecliptic $\varepsilon = \alpha - \alpha'$, the Greeks had in effect computed the angle between the plane of the Earth's equator and the plane of the Earth's orbit about the Sun. Recall that the Greeks measured ε to be $24°$. A more accurate value for this angle is $23\frac{1}{2}°$.

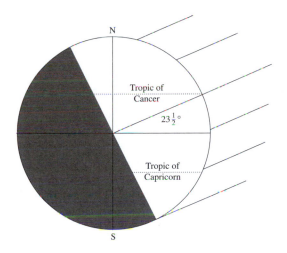

Figure 2.19

Figure 2.19 is a "close-up" of the summer solstice situation in Figure 2.17. A look at the geometry shows that the Sun will be directly overhead at noon at any point on the surface of the Earth that lies $23\frac{1}{2}°$ above the equator. The line that this angle determines is the tropic of Cancer. The tropic of Capricorn is the corresponding line for the Southern Hemisphere.

Consult an atlas and confirm that Aswan, Egypt, lies very near the tropic of Cancer. We now understand why the Sun at summer solstice shines into the very bottom of the well in Syene (the ancient city near today's Aswan). Recall that this fact was instrumental in Eratosthenes's computation of the radius of the Earth.

2.5 Another Look at the Solar Model

This section turns to the questions: What quantitative information do Ptolemy's mathematical methods provide about the circular orbit of the Earth around the Sun? What, in particular, do they tell us about the position of the Sun within the Earth's orbit?

We will make use of today's values for the lengths of the seasons. Spring: 92 days, 18 hours, 20 minutes, or 92.764 days; summer: 93 days, 15 hours, 31 minutes, or 93.647 days; autumn: 89 days, 20 hours, 4 minutes, or 89.836 days; and winter: 88 days, 23 hours, 56 minutes, or 88.997 days.[2] This is the information that will be fed into Ptolemy's model of Section 2.3.

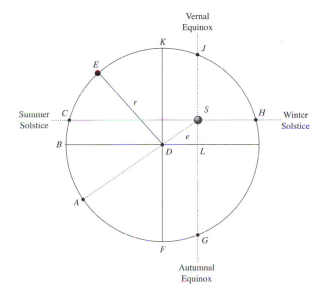

Figure 2.20

Figure 2.20 shows the Earth's circular orbit. The points S and E represent the Sun and the Earth, respectively. The point D is the center of the Earth's orbit. The radius of the circle is r, and e is the distance from S to D. The points J, C, G, and H are the positions—now of the Earth—at vernal equinox, summer solstice, autumnal equinox, and winter solstice. Since summer is the longest season, the center D of the circle must be positioned below and to the left in relation to the Sun. Why? Aphelion, the position of the Earth farthest from the Sun, is labeled A. We will suppose that the Earth moves with a constant angular velocity of $\omega = \frac{360°}{365\frac{1}{4}} = 0.986°/\text{day} = 0.0172$ radians/day.

During spring, the Earth moves from J to C. Looking at the radian measure of $\angle JDC$ in two ways, we get

$$\angle JDC = \frac{\text{arc } JC}{r} = 92.764\omega$$

$$= (92.764)(0.0172) = 1.596.$$

So arc $JC = 1.596r$. Doing the same for summer shows that

$$\angle CDG = \frac{\text{arc } CG}{r} = 93.647\omega$$

$$= (93.647)(0.0172) = 1.611.$$

So arc $CG = 1.611r$, and arc $JCG = 1.596r + 1.611r = 3.207r$. On the other hand, since arc $JK = \text{arc } FG$, we have arc $JCG = 2\,\text{arc } JK + \pi r$. Therefore, $2\,\text{arc } JK = 3.207r - 3.142r = 0.065r$. It follows that

$$DL \approx \text{arc } JK = 0.033r.$$

As before, we will regard this and subsequent approximations as equalities. Since

$$\text{arc } CG = \text{arc } CB + \frac{\pi}{2}r + \text{arc } FG$$

$$= \text{arc } CB + 1.571r + \text{arc } JK,$$

we get

$$1.611r = \text{arc } CB + 1.571r + 0.033r,$$

or arc $CB = 1.611r - 1.571r - 0.033r = 0.007r$. So

$$SL \approx \text{arc } CB = 0.007r.$$

Now take $DL = 0.033r$ and $SL = 0.007r$ and consider the right triangle SDL in Figure 2.21. By Pythagoras,

$$e = \sqrt{DL^2 + SL^2} = \sqrt{(33 \times 10^{-3}r)^2 + (7 \times 10^{-3}r)^2}$$

$$= \sqrt{1089(10^{-3}r)^2 + 49(10^{-3}r)^2} = 10^{-3}r\sqrt{1089 + 49}$$

$$= 10^{-3}r\sqrt{1138} = 0.034r.$$

Since $\sin\theta = \frac{DL}{e} = \frac{0.033r}{0.034r} = \frac{0.033}{0.034} = 0.970$, we can conclude (use the inverse sine button of a calculator) that $\theta = 76°$.

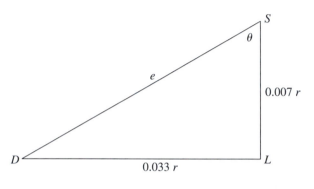

Figure 2.21

Ptolemy's mathematics has produced a model of the Earth's orbit around the Sun. This orbit is a circle that has the Sun at a distance e from the center. While we know today that the Earth's orbit is in fact an ellipse, we will see in Chapter 4 that this ellipse is very nearly a circle. We also know that the Sun is not at the center of the ellipse, but that it is off-center (at a focus). Ptolemy's model, in other words, agrees with the essence of the actual situation. We will now consider other elements that Ptolemy's model captures.

The distance from the Earth to the Sun is known to vary by about 1.7%. This means that one-half the difference between the maximum and minimum distances divided by the average distance is about 0.017. Let's now return to the model. Here, too, the distance from the Earth to the Sun varies. Refer back to Figure 2.20 and note that the maximum distance (attained in the summer) is $AD + DS = r + e$, and the minimum distance (attained in winter) is $r - e$. So one-half the

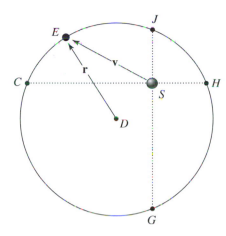

Figure 2.22

difference between the maximum and minimum distance is $\frac{1}{2}[(r+e)-(r-e)] = \frac{1}{2}(2e) = e$, and the average distance is $\frac{1}{2}[(r+e)+(r-e)] = r$. Therefore, the ratio $\frac{1}{2}(\text{max} - \text{min})/\text{average}$ is

$$\frac{e}{r} = 0.034,$$

or about 3.4%. This ratio is twice the actual value. Again, however, the model reflects the phenomenon.

Figure 2.22 shows the Earth E in its circular orbit around the Sun S with D the center of the orbit. It is extracted from Figure 2.20. Let $\mathbf{r}$ be the arrow from D to E and recall that $\mathbf{r}$ rotates with constant angular velocity. Let $\mathbf{v}$ be the arrow from S to E. During winter, E moves from H to J (winter solstice to vernal equinox). In the process, $\mathbf{v}$ rotates from H to J and hence through exactly 90°. At the same time, $\mathbf{r}$ also rotates from H to J, and hence through an angle less than 90°. Therefore, during winter $\mathbf{v}$ rotates, on average, faster than $\mathbf{r}$. During summer, both $\mathbf{r}$ and $\mathbf{v}$ rotate from C to G. Again $\mathbf{v}$ rotates through 90°, but $\mathbf{r}$ now rotates through more than 90°. So during summer $\mathbf{v}$ rotates more slowly than $\mathbf{r}$. Since $\mathbf{r}$ rotates at constant angular velocity, it follows that $\mathbf{v}$ has, on average, greater angular velocity during winter, when the Earth is closer to the Sun, than during summer, when it is farther away. This is in fact true, and

consistent (as we shall see in Chapter 4) with Kepler's second law of planetary motion!

2.6 Epicycles

Let's return to the *Almagest*. Having dealt with the Sun, Ptolemy next turned to the Moon. The theory of the Moon is the most challenging. Indeed, the *Almagest* presents a progression of three different lunar theories of increasing complexity, a complexity forced by persistent discrepancies in lunar observations. In retrospect, this is hardly surprsing. The distance from the Earth to the Moon varies by more than 6% over a *month* (the period of its orbit). The Moon can get as close as 220,000 miles to the Earth and as far as 250,000 miles away. While the orbit is in the general shape of an ellipse, it is not a closed loop. This complexity is explained by the fact that the Moon is subject to the gravitational forces of both the Earth and the Sun. The mathematics of the interplay of the gravitational forces of three masses is known as the three-body problem. It is very difficult and still not completely solved today.

Another phenomenon that Ptolemy's astronomy had to explain was the *retrograde* motion of the planets. The planets—as observed from the Earth against the background of the stars—move near the ecliptic. However, their motion is irregular. They are observed to intermittently slow down, stop, reverse direction for some time (this is the *retrograde* phase), only to stop once more and then continue in the original direction. For example, the apparent path of Saturn as Ptolemy observed it in the years 132–133 A.D. is pictured in Figure 2.23.

Given the Sun-centered perspective, the explanation of what Ptolemy observed is straightforward. The Earth completes one orbit around the Sun in a year, while Saturn takes about 30 Earth years to complete an orbit. Figure 2.24 shows the (dotted) lines of sight of someone on Earth looking out on Saturn against the background of the fixed stars, just as Ptolemy had done. Observe that the general pattern matches

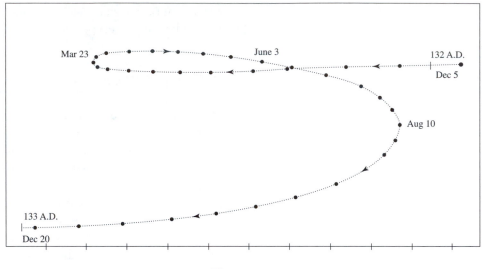

Figure 2.23

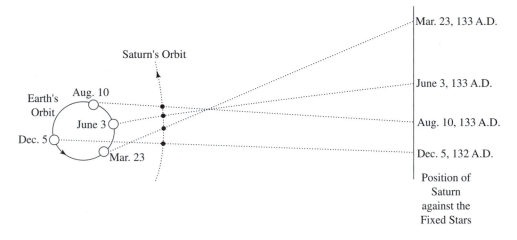

Figure 2.24

that of Figure 2.23. The observation that the planets move near the ecliptic is explained by the fact that the planes of their orbits are almost the same as that of the Earth's orbit. The planes are not exactly the same, however, and this explains the fact that the planets are observed to rise and fall. For instance, Figure 2.23 shows that Saturn is higher in the sky in December of 132 A.D. than in December of 133 A.D.

How does the *Almagest* deal with the irregularities of the Moon's orbit, and how does it—with its Earth-centered perspective—come to grips with the phenomenon of the retrograde motion of the planets? The answer, in short, is that it combines circular motions in a variety of ways. Let's focus on the retrograde motion of the planets. Figure 2.25 shows a planet P in a circular orbit with radius r and center the point B. In turn, B moves along another circle. It has the

Earth E as its center and radius R. The larger circle is called the *deferent* and the smaller one the *epicycle*. Both rotations occur at constant speeds. The arrows designate the directions. Suppose that P moves faster around B than B does around E. When P is in the position shown, the two motions—as viewed from E—are in opposite directions. But P will be seen to move from left to right due to the greater speed of P. However, as P continues around B, it will eventually move in the same direction as B. Now P will be seen to be moving from right to left. So this scheme of two rotating circles explains, at least in principle, the observed reversal of a planet's direction. This model is the basis of the *Almagest*'s lunar and planetary theories.

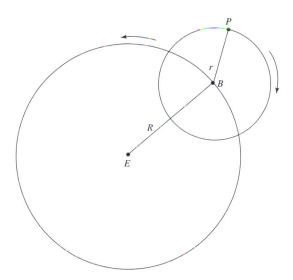

Figure 2.25

To make his theory stay in tune with his observations, Ptolemy was forced to make adjustments to the scheme of Figure 2.25. These are indicated in Figure 2.26. He modified the orbit of the planet P by moving the center of the deferent to a point C away from E. In addition, he corrected the speed of the point B by choosing yet another point A and stipulating that the segment from A to B rotates with constant speed.

The various parameters—the positions of C and A relative to E, the radii R and r, and the speeds of the rotations—were different for each planet. When even this scheme did not mesh with observations, he introduced additional circles. Different configurations of circles were required. Those for the inner planets (Mercury and Venus) were more complicated than those of the outer planets (Mars, Jupiter, and Saturn), and (as already remarked), the Moon's was the most complicated of all. In summary, Ptolemy's solar system was an extremely intricate clockwork of spokes and wheels.

2.7 Postscript

Our overview of the *Almagest* is complete. Given the parameters it starts with—an Earth-centered solar system in which celestial objects move in circular orbits at constant speeds—it succeeded in a remarkable way. We have seen how it combines mathematics and empirical data derived from careful observations to produce a mathematical model that simulates the Sun's orbit about the Earth. It is thus an early example of mathematical physics.

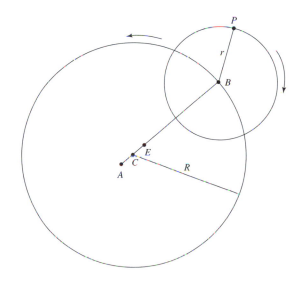

Figure 2.26

The *Almagest* is unrivaled in antiquity as an example of how a large and important class of natural phenomena could be described in mathematical terms. On the whole, it is to be viewed not as a description of the physical universe but rather as an abstract mathematical scheme designed to predict the positions of the objects moving within it. In this sense it agreed sufficiently with observations to be acceptable to practical astronomers. Observe that from the perspective of a fixed Earth, the planets do (in essence) move along epicycles carried by the deferent of the Sun's orbit about the Earth (see Figure 2.26). Therefore, from the perspective of the Earth, Ptolemy's is the correct geometry.

The *Almagest* contains a catalogue of stars (1022 of them). This was necessary because crucial observations (those of the planets for instance) are made in reference to them. Arab astronomers expanded this catalogue and compiled atlases of stars that proved invaluable to early Spanish and Portugese navigators. Ptolemy's *Almagest* remained the foundation of astronomy for fourteen centuries. In fact, the scheme of Figure 2.1 is still the most common coordinate system for listing planetary and stellar positions. An observer on the Earth can specify the location of a star *X* by measuring α (called the star's *right ascension*) and δ (called its *declination*) in terms of angles from the point of the observation; see Figure 2.27.

Ptolemy's clockwork did not become obsolete until the 16th and early 17th centuries, when Copernicus introduced the heliocentric universe (reintroduced is more appropriate, in view of Aristarchus), Galileo discovered the phases of Venus and the moons of Jupiter, and Kepler demonstrated that the orbits of the planets were ellipses.

Exercises
2A.   Some Geometry

1. Consider any rectangle and let *A*, *B*, *C*, and *D* be its vertices. Inscribe it into a circle with center at the intersection of the diagonals. Use Ptolemy's theorem to verify Pythagoras's theorem.

2. Place an angle θ at the center of a circle of radius 2. Let the endpoints of the arc that it cuts from the circle be *A* and *B*. What does θ equal (in degrees) if the length of the arc from *A* to *B* is $1\frac{1}{2}$?

3. Refer to Figure 2.28. The points *A*, *B*, and *D* are on a circle of radius 3. The length of the arc *AB* is 4. Determine the angle $\angle ADB$ first in radians and then in degrees.

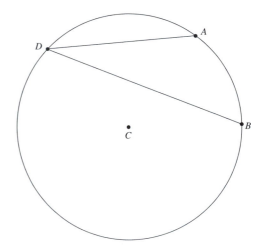

Figure 2.28

2B.   Some "Inverse" Trigonometry

4. Use a calculator to fill in the following blanks (use three decimal accuracy):

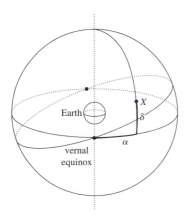

Figure 2.27

i. $\sin \underline{\quad}° = 0.534$
ii. $\sin \underline{\quad}° = 0.219$
iii. $\sin \underline{\quad}° = 0.834$
iv. $\sin \underline{\quad}$ (in radians) $= 0.002$
v. $\sin \underline{\quad}$ (in radians) $= 0.664$
vi. $\tan \underline{\quad}° = 0.774$
vii. $\tan \underline{\quad}° = 1.478$
viii. $\tan \underline{\quad}° = 5.000$
ix. $\tan \underline{\quad}$ (in radians) $= 10.473$
x. $\tan \underline{\quad}$ (in radians) $= 27.664$

5. What angle α (in degrees and radians) has $\sin\alpha = \frac{1}{2}$? What angle β (degrees and radians) has $\tan\beta = 1$? What angle φ has $\sin\varphi = \frac{\sqrt{3}}{2}$?

2C.   More Trigonometry

6. Let α and β be two angles, both less than $\frac{\pi}{2}$, and arrange them into a triangle as shown in Figure 2.29.

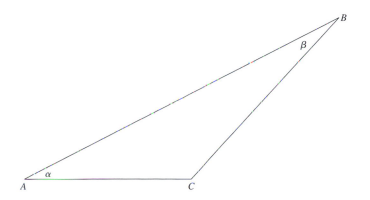

Figure 2.29

Complete this triangle to Figure 2.30 in such a way that the segment AD is perpendicular to BD; also draw the perpendicular EC to AB.

i.   Show that $\gamma = \alpha + \beta$.
ii.   Show that $\frac{BD}{AB} = \frac{EC}{AC}$ and hence that $BD = \frac{EC}{AC}(AE + EB)$.
iii.   Using (i) and (ii), show that $\sin(\alpha + \beta) = (\sin\alpha)(\cos\beta) + (\cos\alpha)(\sin\beta)$.

It can also be shown (we will forego the pleasure) that

iv. $\cos(\alpha + \beta) = (\cos\alpha)(\cos\beta) - (\sin\alpha)(\sin\beta)$.

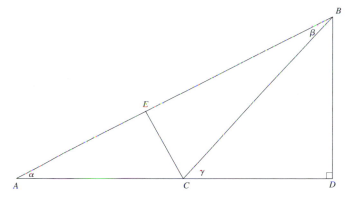

Figure 2.30

Note: The identities (iii) and (iv) are crucial for the development of the calculus of the trigonometric functions.

7. Let θ be any angle, and place it into a circle of radius 1, center C, and diameter DA, as shown in Figure 2.31. Draw the segments BD and BC, and draw a perpendicular BE down to DA.

i.   Show that $\angle BDC = \frac{\theta}{2}$.
ii.   Show that $\tan\frac{\theta}{2} = \frac{\sin\theta}{1+\cos\theta}$.
iii.   Deduce from (ii) that $\tan\frac{\theta}{2} = \frac{1-\cos\theta}{\sin\theta}$. [Hint: Multiply (ii) by $\frac{1-\cos\theta}{1-\cos\theta}$.]
iv.   Use (ii) and (iii) to show that $1 - \tan^2\frac{\theta}{2} = \frac{2\cos\theta}{1+\cos\theta}$.
v.   Use (ii) and (iv) to show that $\tan\theta = \frac{2\tan\frac{\theta}{2}}{1-\tan^2\frac{\theta}{2}}$.

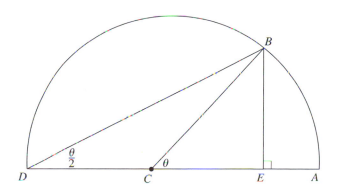

Figure 2.31

2D.   Ptolemy's Mathematics

8. Redo the calculations of Section 2.3 by starting with $\omega = 0.02$ and by working to an accuracy of two decimal places. What values for the ratio $\frac{e}{r}$ and the angle λ_A do you get? What does this tell us about Ptolemy's calculations?

9. Use the model of the Earth's orbit around the Sun in Section 2.5 to determine the days of the year on which the Earth is farthest (*aphelion*) and closest (*perihelion*) to the Sun. Compare this with the data given in the *World Almanac*. [Hint: Summer solstice occurs on June 21 or June 22, and the Earth is in position C in Figure 2.20 on that day. How long will it take for the Earth to get to position A?]

10. Ptolemy computed the distance D_M between the Earth and the Moon as follows. In Figure 2.32, E and M are the centers of the Earth and Moon, respectively, and r_E is the radius of the Earth. Take $r_E = 3,850$ miles from Eratosthenes. The point A represents Alexandria and AQ is perpendicular to EM. Ptolemy observes the angle α to be $50°55'$ and computes β (from lunar longitude and obliquity of ecliptic data) to be $49°48'$. Supply the details to the following outline of Ptolemy's argument:

    i.   Show that $\alpha = \beta + \mu$.
    ii. Show that $D_M = r_E \cdot \frac{\sin\alpha}{\sin\mu}$ using the addition formula for the sine.
    iii. Obtain $D_M = 150,000$ miles.
    iv. Compare this with the modern value.

2E.   Diophantus of Alexandria

Diophantus (of Alexandria, about 250 A.D.) was a mathematician who pursued arithmetic and algebra.

11. The following passage from *The Greek Anthology*[3] provides some information about Diophantus's life. Use it to determine his age. Diophantus passed one-sixth of his life in childhood, one-twelfth in youth, and one-seventh more as a bachelor. Five years after his marriage was born a son who died four years before his father, at half of his father's [final] age.

12. Solve the following problem from Diophantus's book *Arithmetica*: Find four numbers, the sum of every arrangement three at a time being 22, 24, 27, and 20.

13. Solve the following, also from the *Arithmetica*: In the right triangle ABC, right angled at C, AD bisects the angle at A. Find the set of smallest integers for AB, AD, AC, BD, DC, such that the ratios $DC:CA:AD$ are equal to 3:4:5. [Hint: Use of the formula from Exercise 7(v) and the fact that if a prime number divides a product of two integers, then it must divide one of the factors.]

Notes

[1]The computations that follow are carried out with three decimal accuracy, even though the data on which they are based (the length of the seasons and the year) do not meet this standard. These computations thus violate today's principles of reliability. However, they do reflect Ptolemy's own both in

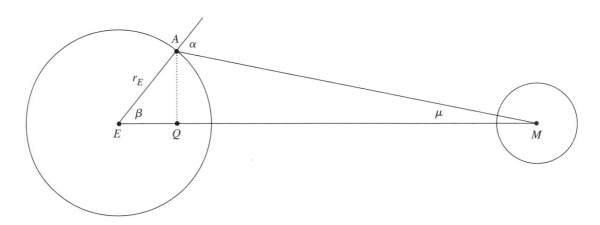

Figure 2.32

spirit and in terms of their numerical conclusions. The computations take $\pi = 3.142$.

[2]The reason for the difference between the lengths of the seasons computed from Ptolemy's time and now has to do with the fact that the year (the number of days between successive summer solstices, or vernal equinoxes) is slightly less than the $365\frac{1}{4}$ days measured by Hipparchus. It is actually 365 days, 5 hours, 48 minutes, and 46 seconds long (11 minutes and 14 seconds less than $365\frac{1}{4}$ days).

[3]See Note 8 of Chapter 1.

3

Archimedes
Computes Areas

Several centuries have passed since the achievements of Claudius Ptolemy. During this time, Christianity emerged and the Germanic invasions brought instability. Faith and survival, not mathematics and science, became important. Islam conquered North Africa, the Middle East, and parts of Europe. The Latin West was thus cut off from regular contact with the Greek East, and the mathematical and scientific discoveries that we have described were largely lost in Western Europe between 650 A.D. and 1150 A.D. It was not until the beginning of the 13th century that the works of the Greek scholars were rediscovered in Arabic manuscripts and translated into Latin. This stimulated the pursuit of scientific thought (much of it dominated by attempts to understand Aristotelian science in its relationship to Christian revelation) in the new universities that had sprung up by that time.

The Renaissance—the rebirth—began in Italy in the 14th century. Greek philosophy, architecture, art, science, and mathematics provided a crucial stimulus. The recovery of Ptolemy's maps[1] reintroduced the method of defining geographical position in terms of coordinates, that is to say, by the circles of latitude and the meridians of longitude on the globe. The Western world learned the mathematics and astronomy of Euclid, Aristarchus, Archimedes, Apollonius, and Ptolemy. The interaction and confrontation with Greek science generated the energies that would later help bring about the invention of calculus and the scientific revolution.

We will focus on what the Europeans learned from the Greeks about three remarkable families of curves: the parabola, the ellipse, and the hyperbola. These *conic sections*—obtained by intersecting a cone with a plane—were well known to Euclid. They were analyzed extensively in the encyclopedic treatise *Conics* of Apollonius of Perga. The parabola and the ellipse—as we will see in subsequent chapters—form the basis for our understanding of both the motion of projectiles and the orbits of the planets around the Sun. After a look at some basic properties of these curves, we will turn to the ingenious mathematical discoveries that Archimedes made about them.

3.1 **The Conic Sections**

We will recall—without proofs—a few of the basic properties of the conic sections that we will need later. A unit of length is given. As before, if, say, P and Q are two points in the plane, PQ or QP will be the segment between them; if PQ appears in an equation, it will be understood to be the length of the segment PQ.

In the plane, fix both a line D and a point F not on D. The collection of all points P such that the length

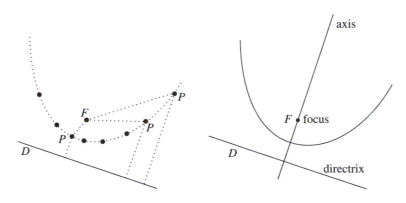

Figure 3.1

of PD is equal to that of PF is called a *parabola*. (See Figure 3.1.) The line D is called the *directrix* and the point F is called the *focus*. The line through the focus that is perpendicular to the directrix is the *axis* of the parabola.

Proposition 3.1. Let P be any point on a parabola and let the segment AB be tangent to the parabola at P. Consider the line from P that is parallel to the axis of the parabola and let Q be a point on this line, as shown in Figure 3.2. Then $\angle QPA = \angle FPB$.

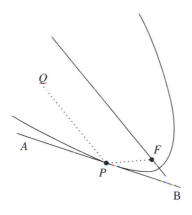

Figure 3.2

This is the "reflection property" of the parabola. We point out in passing that when light is reflected by a mirror, the angle of incidence is equal to the angle of reflection. It follows that if a mirror has the shape obtained by rotating a parabolic arc one revolution around its axis, then any light ray parallel to the axis will be reflected through the focus. This is the reason for the use of parabolic mirrors both in telescopes (for collecting light) and in the headlights of cars (for projecting light).

Again, consider any parabola. Cut it with any straight line, and let S and S' be the points of intersection. (See Figure 3.3(a).) For some point V on the parabola, the tangent line at V is parallel to the cut SS'. The parabolic region SVS' is called a *parabolic section*, and V is the *vertex* of the parabolic section. A *section*, generally speaking, is a region obtained from another by a cut. Let A be the midpoint of the segment SS'; connect A with V, and take E on the tangent such that ES is parallel to VA. (Refer to Figure 3.3(b).) Finally, let B be any point on VE and let C be the point on the parabola such that BC is parallel to VA. (See Figure 3.4.)

Proposition 3.2. Take B to be the midpoint of VE. Then C is the vertex of the parabolic section SVC. (See Figure 3.5.)

Proposition 3.3. For any point B on EV (see Figure 3.4),

$$\frac{BC}{ES} = \frac{VB^2}{VE^2}.$$

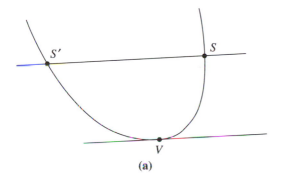

(a)

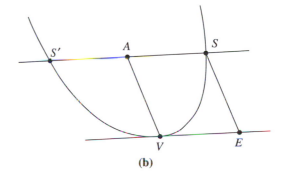

(b)

Figure 3.3

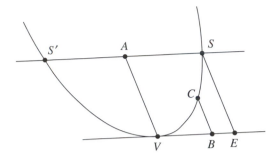

Figure 3.4

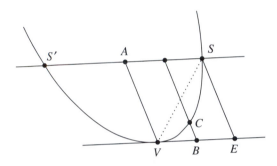

Figure 3.5

We will study the ellipse next. Fix any two points F_1 and F_2 in the plane and a constant k that is greater than or equal to the distance between F_1 and F_2. Consider the collection of all points P such that the distance from P to F_1 plus the distance from P to

F_2 is equal to k. Refer to Figure 3.6(a). Any such collection of points is an *ellipse*. The points F_1 and F_2 are the *focal points* of the ellipse. The midpoint C of the segment F_1F_2 is called the *center* of the ellipse. Any segment between two points on the ellipse and through the center is a *diameter* of the ellipse. The diameter through the two focal points is the *major axis* of the ellipse, and the diameter perpendicular to the major axis is the *minor axis*. See Figure 3.6(b). Consider a circle with center C and radius r. From the preceding point of view, it is an ellipse with focal points $C = F_1 = F_2$ and $k = 2r$.

Proposition 3.4. Let P be any point on an ellipse and let the segment AB be tangent to the ellipse at P, as shown in Figure 3.7. Then $\angle F_1PA = \angle F_2PB$.

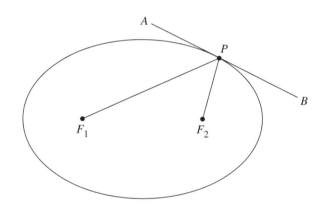

Figure 3.7

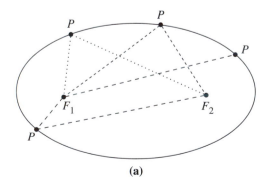

           (a)   **(b)**

Figure 3.6

This is the "reflection property" of the ellipse. We note in passing that it is used in a surgical procedure to remove kidney stones. A source of high intensity sound waves is located at a point, say F_1, and an elliptical reflector is placed in such a way that F_1 is one of the focal points. The patient is positioned so that the kidney stone is at the other focus, F_2. The sound waves emanating at F_1 scatter in all directions, but those that strike the reflector are reflected through F_2. The strong concentration of sound waves at F_2 shatters the kidney stone.

Proposition 3.5. Let PQ be any diameter of an ellipse and let B be any point on PQ. Choose A such that AB is parallel to the tangent at P. (See Figure 3.8.) Then the ratio $\frac{QB \cdot BP}{AB^2}$ is the same no matter where B is chosen on PQ.

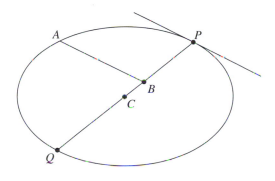

Figure 3.8

Example 3.1. Consider Proposition 3.5 in the case of a circle. See Figure 3.9. Because the focal points coincide and are located at the center C, it follows by Proposition 3.4 that the tangent at P is perpendicular to the diameter QP. Since AB is parallel to the tangent, AB and QP are perpendicular. By an application of Proposition 2.1 of Section 2.2, the triangle $\triangle PAQ$ has a right angle at A. (See Exercise 1.) Compare the triangles $\triangle PBA$ and $\triangle PAQ$. They both have right angles and they have $\angle APB$ in common; it follows that they are similar. By the same argument,

$\triangle QBA$ is also similar to $\triangle PAQ$. Therefore, $\triangle PBA$ is similar to $\triangle QBA$, and hence $\frac{PB}{AB} = \frac{AB}{QB}$. So, $\frac{QB \cdot PB}{AB^2} = 1$. We have shown that in the case of the circle, the ratio of Proposition 3.5 is always equal to 1.

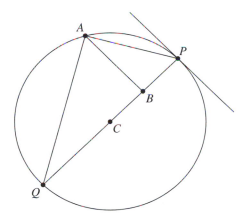

Figure 3.9

The hyperbola is defined next. Again, fix two points F_1 and F_2 in the plane and a constant $k \geq 0$. Suppose that k is less than or equal to the distance between F_1 and F_2. The collection of points P such that the absolute value of the difference between PF_1 and PF_2 is equal to k is a *hyperbola*. (Refer to Figure 3.10.) The points F_1 and F_2 are the *focal points* of the hyperbola. The line determined by the two focal points is called the *axis*. The midpoint C between the two focal points is the *center* of the hyperbola.

Hyperbolas will play only a minor role in this text. As an aside, note that space probes designed to investigate an outer planet are often redirected by allowing them to pass near an inner planet. The gravitational field of the inner planet will accelerate a probe and bend its path along a hyperbolic curve before allowing it to escape and continue.

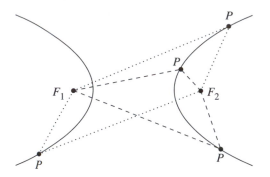
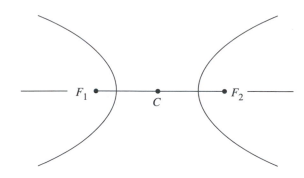

Figure 3.10

3.2 The Question of Area

W‌e now turn to the concept of area for regions in the plane. If b is the base of a rectangle and h the height (Figure 3.11), then its area A is simply defined to be $A = bh$. This is the basic situation, and the determination of the area of any other region is ultimately reduced to this case. Consider next a parallelogram with base b and height h. By cutting off the indicated triangle on the left and reattaching it on the right, as shown in Figure 3.12, the given parallelogram is transformed—without change in area—into a rectangle with the same base b and height h. It follows that the area A of the parallelogram is $A = bh$.

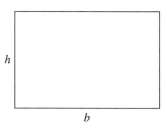

Figure 3.11

Turn next to the case of a triangle with base b and height h. Take a copy of the same triangle, flip it, and attach it to the original, as shown in Figure 3.13. The resulting larger figure is a parallelogram with

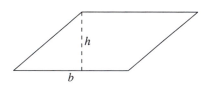
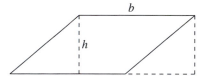

Figure 3.12

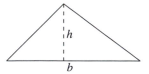
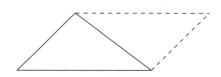

Figure 3.13

base b and height h. This larger area is equal to bh, as shown before. It follows that the area A of the original triangle is $A = \frac{1}{2}bh$.

We next consider the more subtle situation of the area of a sector of a circle with radius r. A *sector* is a pie-shaped region. Consider the sector determined by the center C and the segments CB and CD and let θ be the angle of the sector; see Figure 3.14. We will determine the area A of the sector by thinking of it as consisting of lots of very thin triangular wedges.

Partition the given sector into, say, n smaller sectors all of the same size. This determines $n - 1$ points between B and D. Consecutively connecting all of these points inscribes an isosceles triangle in each of the sectors. All of these triangles have the same shape and size. The base as well as the height of the isosceles triangles certainly depend on n, so we denote them by b_n and h_n respectively. The case $n = 5$ is shown in Figure 3.15. The bases of the triangles—taken together—approximate the arc BD. So $b_n n$ approximates the length of arc BD. The area of each inscribed triangle is $\frac{1}{2}h_n b_n$. Taken together, these n triangles approximate the sector BCD. Let $A_n = \left(\frac{1}{2}h_n b_n\right)n = \frac{1}{2}h_n(b_n n)$ be the sum of the areas of all n triangles. So A_n approximates the area of the sector BCD. When n is taken to be sequentially larger and larger, two things happen simultaneously:

(1) The numbers h_n close in on the radius r, and the numbers $b_n n$ on arc BD. As a result, the numbers $\frac{1}{2}h_n(b_n n)$ close in on $\frac{1}{2}r(\text{arc } BD)$.

(2) The numbers A_n close in on the area of the sector BCD.

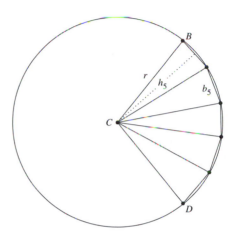

Figure 3.15

Because the numbers $A_n = \frac{1}{2}h_n(b_n n)$ can close in on one number only, it follows that the area of the sector BCD must be equal to $\frac{1}{2}r(\text{arc } BD)$. In the notation of limits, we can abbreviate what was just asserted by writing

$$\text{Area of sector } BCD = \lim_{n \to \infty} A_n = \lim_{n \to \infty} \frac{1}{2}h_n(b_n n)$$

$$= \frac{1}{2}r(\text{arc } BD).$$

Since we know that $\theta = \frac{\text{arc } BD}{r}$ radians, we have now established the fact that the area A of a sector of radius r and angle θ (given in radians) is equal to

$$A = \frac{1}{2}r^2\theta.$$

With $\theta = 2\pi$, this gives us the familiar expression πr^2 for the area of a circle of radius r.

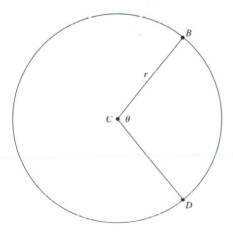

Figure 3.14

3.3 **Playing with Squares**

$\boxed{C}$ onsider a square with side of length a. Of course, the area A of the square is $A = a^2$. Nothing else needs to be said. Nevertheless, we will now consider a completely different approach to this area. It may seem completely idiotic, but it will pay dividends later.

Subdivide the square into four equal squares and color three of them black, as shown in Figure 3.16. Denote the area of the resulting black "L" by **B**. (Note that $\mathbf{B} = \frac{3}{4}A$, but this is irrelevent for now.) Do the same thing with the remaining white square in the upper right corner. Since each of these subdividing squares has area one-fourth that of the original white square, it follows that this smaller black L has area $\frac{1}{4}\mathbf{B}$; see Figure 3.17. Repeating this construction again produces a black L, this time of area $\frac{1}{4}\left(\frac{1}{4}\mathbf{B}\right) = \left(\frac{1}{4^2}\right)\mathbf{B}$, and so on.

Now consider the following steps. Step 0 is the creation of the black area **B**. Step 1 adds the black L of area $\frac{1}{4}\mathbf{B}$ to **B** and produces the black area $\mathbf{B} + \frac{1}{4}\mathbf{B}$. Step 2 adds the next black L and produces the black area $\mathbf{B} + \frac{1}{4}\mathbf{B} + \frac{1}{4^2}\mathbf{B}$. Step 3 adds the next black L of

area $\frac{1}{4}\left(\frac{1}{4^2}\mathbf{B}\right) = \frac{1}{4^3}\mathbf{B}$, and so on. The results of Steps 0 through 3 are illustrated in Figure 3.18. They produce the black areas

$$\mathbf{B}, \quad \mathbf{B} + \frac{1}{4}\mathbf{B}, \quad \mathbf{B} + \frac{1}{4}\mathbf{B} + \frac{1}{4^2}\mathbf{B}, \quad \mathbf{B} + \frac{1}{4}\mathbf{B} + \frac{1}{4^2}\mathbf{B} + \frac{1}{4^3}\mathbf{B}.$$

This construction can be continued indefinitely. Make note of the emerging pattern and observe that for any number n (possibly very large), Step n creates the black area

$$\mathbf{B} + \frac{1}{4}\mathbf{B} + \frac{1}{4^2}\mathbf{B} + \frac{1}{4^3}\mathbf{B} + \cdots + \frac{1}{4^n}\mathbf{B}.$$

Each step produces an approximation of the area A of the original square. The area of the small white square that remains in the upper right corner is the error of the approximation. Let's compute these errors. After Step 0, the area of the small white square is $\frac{1}{4}A$. After Step 1, it is one-fourth of this, or $\frac{1}{4^2}A$. After Step 2, it is one-fourth that of Step 1, or $\frac{1}{4^3}A$. And after Step 3, it is $\frac{1}{4^4}A$. Continue the pattern, and observe that the area of the white square after Step n is $\frac{1}{4^{n+1}}A$. For example, the error of Step 14 is $\frac{1}{4^{15}}$ (and hence less than 0.000000001) of the area A. Since $\frac{1}{4^{n+2}}$

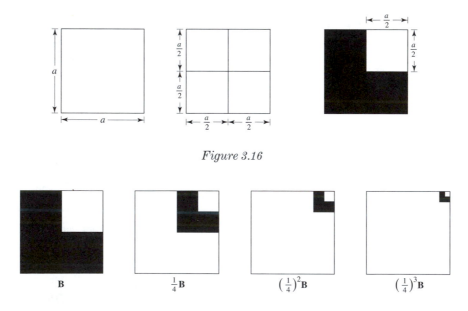

Figure 3.16

B $\frac{1}{4}\mathbf{B}$ $\left(\frac{1}{4}\right)^2\mathbf{B}$ $\left(\frac{1}{4}\right)^3\mathbf{B}$

Figure 3.17

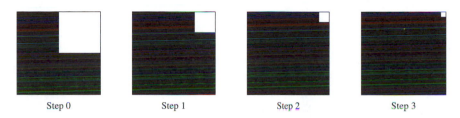

Step 0   Step 1   Step 2   Step 3

Figure 3.18

is less than $\frac{1}{4^{n+1}}$, it follows that the error of each step is less than the preceeding error.

In summary, for any n, Step n produces the partition

$$A = \left(\mathbf{B} + \frac{1}{4}\mathbf{B} + \frac{1}{4^2}\mathbf{B} + \cdots + \frac{1}{4^n}\mathbf{B}\right) + \frac{1}{4^{n+1}}A$$

of the area of the original square into a black area and a white area. The black area is an approximation of A, and the white area $\frac{1}{4^{n+1}}A$ is the error of the approximation. The approximation can be made as accurate as we need to have it. Simply choose n large enough so that the error $\frac{1}{4^{n+1}}A$ is small enough!

In the present case of the square, all this is rather superfluous because we already know that its area is $A = a^2$. The point is, however, that this strategy can be applied to compute the areas of regions that are not known beforehand. We will see shortly how Archimedes used it to compute the area of a parabolic section.

We conclude with a consequence of our analysis. By Figure 3.16, $\mathbf{B} = \frac{3}{4}A$; so $A = \frac{4}{3}\mathbf{B}$. Substituting this into the preceding equation for A gives

$$\frac{4}{3}\mathbf{B} = \left(\mathbf{B} + \frac{1}{4}\mathbf{B} + \frac{1}{4^2}\mathbf{B} + \cdots + \frac{1}{4^n}\mathbf{B}\right) + \left(\frac{1}{4^{n+1}} \cdot \frac{4}{3}\mathbf{B}\right).$$

After canceling $\mathbf{B}$, we get $\frac{4}{3} = 1 + \frac{1}{4} + \frac{1}{4^2} + \cdots + \frac{1}{4^n} + \frac{1}{4^n} \cdot \frac{1}{3}$. Thus, for any positive integer n,

$$1 + \frac{1}{4} + \left(\frac{1}{4}\right)^2 + \cdots + \left(\frac{1}{4}\right)^n + \frac{1}{3}\left(\frac{1}{4}\right)^n = \frac{4}{3}.$$

This is an identity established by Archimedes. Note that $1 + \frac{1}{4} + \left(\frac{1}{4}\right)^2 + \cdots + \left(\frac{1}{4}\right)^n = \frac{4}{3} - \frac{1}{3}\left(\frac{1}{4}\right)^n$. Note also that for larger and larger n, the term $\frac{1}{3}\left(\frac{1}{4}\right)^n$ gets closer

and closer to 0. Therefore, the numbers

$$1 + \frac{1}{4}$$

$$1 + \frac{1}{4} + \left(\frac{1}{4}\right)^2$$

$$1 + \frac{1}{4} + \left(\frac{1}{4}\right)^2 + \left(\frac{1}{4}\right)^3$$

$$\vdots$$

$$1 + \frac{1}{4} + \left(\frac{1}{4}\right)^2 + \cdots + \left(\frac{1}{4}\right)^n$$

$$\vdots$$

close in on $\frac{4}{3}$. We abbreviate this observation by writing

$$\lim_{n\to\infty} \left(1 + \frac{1}{4} + \left(\frac{1}{4}\right)^2 + \cdots + \left(\frac{1}{4}\right)^n\right) = \frac{4}{3}.$$

Let's rewrite the preceeding sum more compactly:

$$\sum_{i=0}^{n} \left(\frac{1}{4}\right)^i = 1 + \frac{1}{4} + \left(\frac{1}{4}\right)^2 + \cdots + \left(\frac{1}{4}\right)^n.$$

This "sigma notation" (Σ is the capital Greek sigma, i.e., "S," so it suggests "Sum") was not used by Archimedes. It was introduced only in the 18th century by the Swiss mathematician Leonhard Euler. The stategy implicit in it is this: The first i is $i = 0$. Plug it into $\left(\frac{1}{4}\right)^i$ to get $\left(\frac{1}{4}\right)^0 = 1$ and write $+$. Next, put in what you get by letting $i = 1$, namely $\left(\frac{1}{4}\right)^1$ and write $+$. Continue until the last i, namely $i = n$, is reached. In a similar way, when $n = 6$,

$$\sum_{i=3}^{6} 2^i = 2^3 + 2^4 + 2^5 + 2^6 = 8 + 16 + 32 + 64 = 120$$

With this shorthand notation the limit equation can be rewritten $\displaystyle \lim_{n \to \infty} \sum_{i=0}^{n} \left(\frac{1}{4}\right)^i = \frac{4}{3}$.

3.4 **The Area of a Parabolic Section**

W e now turn to Archimedes's work *Quadrature of the Parabola*, in which he computes the area of a section of a parabola. Quadrature refers to "squaring," and squaring a given region bounded by one or more curves refers to the classical problem of constructing a rectangular region with the same area as the given one.

Take any parabolic section SVS', where V is the vertex. (Refer back to Figure 3.3(a).) Inscribe the triangle $\triangle SVS'$, as shown in Figure 3.19. Archimedes's objective is the proof of the fact that the area of the parabolic section is equal to four-thirds the area of the inscribed triangle. Archimedes's argument consists of a number of steps, each carefully laid out and each rather routine. The argument as a whole, however, is formidable. Let's follow along.

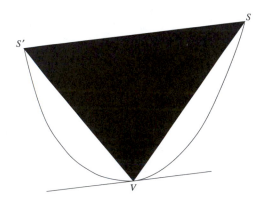

Figure 3.19

Refer to Figure 3.20. Let A be the midpoint of the segment $S'S$. Draw a line through S parallel to VA, and let its point of intersection with the tangent at V be E. Do a similar thing on the other side of VA.

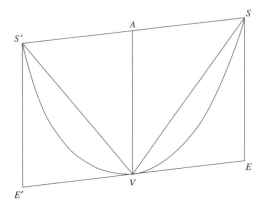

Figure 3.20

Next, Archimedes takes B to be the midpoint of the segment VE and draws the segment BD parallel to VA. (See Figure 3.21.) The point C is the intersection of BD with the parabola, and G is the intersection of BD and VS. He does the same thing on the other side.

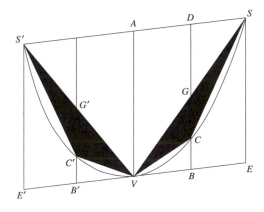

Figure 3.21

The key step in the argument—focus on the right side of the figure—is now this:

Proposition A. The area of the triangle $\triangle VAS$ is equal to four times the area of the triangle $\triangle VCS$.

Archimedes argues as follows. Because B is the midpoint of VE, it follows that $\frac{VB}{VE} = \frac{1}{2}$. So by Proposition 3.3 of Section 3.1, $\frac{BC}{ES} = \frac{VB^2}{VE^2} = \frac{1}{4}$. Since $ES = BD$, he gets

$$BC = \frac{1}{4}BD.$$

Next consider Figure 3.22, which is extracted from Figure 3.21. By similar triangles, $\frac{BG}{ES} = \frac{VB}{VE} = \frac{1}{2}$. Since $ES = BD$, it follows that $BG = \frac{1}{2}BD$. So G is the midpoint of BD, and hence $BG = GD$. Because $BG = \frac{1}{2}BD$, it follows that $\frac{1}{2}BG = \frac{1}{4}BD$. But $\frac{1}{4}BD = BC$, so $BC = \frac{1}{2}BG$. Therefore, C is the midpoint of BG, and $CG = BC$. By a substitution,

$$CG = \frac{1}{2}GD.$$

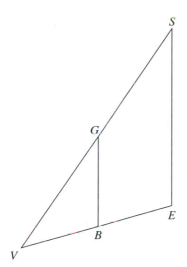

Figure 3.22

Figure 3.23 consists of two figures extracted from Figure 3.21. Concentrate on the two triangles at the top. Note that they have the same height. Since $CG = \frac{1}{2}GD$, it follows that

$$\text{Area } \triangle DGS = 2 \text{ Area } \triangle GCS.$$

Concentrating on the bottom figure, we get in the same way that Area $\triangle DGV = 2$ Area $\triangle GCV$. Adding

areas,

$$\text{Area } \triangle DGS + \text{Area } \triangle DGV$$
$$= 2(\text{Area } \triangle GCS + \text{Area } \triangle GCV).$$

Refer back to Figure 3.21 and notice that this equality says that Area $\triangle DVS = 2$ Area VCS. Refer to Figure 3.21 once more and observe that Area $\triangle DVA =$ Area $\triangle DVS$. (Why?) It follows that Area $\triangle VAS = 2$ Area $\triangle DVS$. Combining equalities gives

$$\text{Area } \triangle VAS = 4 \text{ Area } \triangle VCS$$

as required in Proposition A.

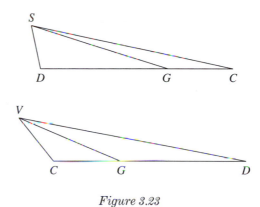

Figure 3.23

The same argument applied to the left side of Figure 3.21 shows that

$$\text{Area } \triangle VAS' = 4 \text{ Area } \triangle VC'S'.$$

Therefore,

$$\text{Area } \triangle SVS' = 4(\text{Area } \triangle VCS + \text{Area} \triangle VC'S').$$

So Archimedes has verified

Proposition B. In reference to Figure 3.21,

$$\text{Area } \triangle VCS + \text{Area } \triangle VC'S' = \frac{1}{4} \text{ Area } \triangle SVS'.$$

The brilliance of Archimedes's argument becomes apparent now: Since B and B' are the midpoints of VE and VE', respectively, Proposition 3.2 of Section

3.1 applies and asserts that C and C' are the respective vertices of the parabolic sections VCS and $VC'S'$. Consider now the two parabolic sections VCS and $VC'S'$ with their circumscribed parallelograms; see Figure 3.24. Applying Proposition B to each of them shows that the sum of the areas of the two black triangular regions on the right is $\frac{1}{4}$ Area $\triangle VCS$ and that of the two black triangles on the left is $\frac{1}{4}$ Area $\triangle VC'S'$. Since we have already seen that

$$\text{Area } \triangle VCS + \text{Area } \triangle VC'S' = \frac{1}{4} \text{Area } \triangle SVS',$$

the four black triangles together have area

$$\left(\frac{1}{4}\right)^2 \text{Area } \triangle SVS'.$$

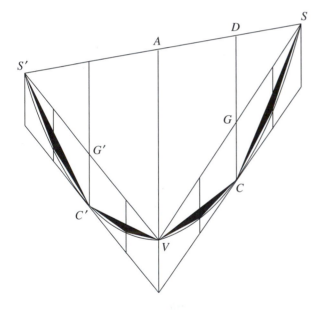

Figure 3.24

Let $\mathbf{B} = \text{Area } \triangle SVS'$. Recall that the black area of Figure 3.21 is $\frac{1}{4}\mathbf{B}$, and that of Figure 3.24 is $\left(\frac{1}{4}\right)^2 \mathbf{B}$. Archimedes's procedure of forming triangles can be repeated again and again. The principle is now the same as that in the approximation of the area of the square in Section 3.3. When the black areas produced

by each step are taken together, they provide an approximation of the area of the parabolic section. The first three approximations, obtained by adding the black regions already determined (those of Figures 3.19, 3.21, and 3.24), are shown in Figure 3.25. Observe the pattern, and notice that when the areas of the triangles produced after n steps are added together, an approximating inscribed region of area

$$A_n = \left(1 + \frac{1}{4} + \left(\frac{1}{4}\right)^2 + \left(\frac{1}{4}\right)^3 + \cdots + \left(\frac{1}{4}\right)^n\right)\mathbf{B}$$

is obtained. Now let n get larger and larger, and recall from Section 3.3 that

$$\lim_{n \to \infty} \left(1 + \frac{1}{4} + \left(\frac{1}{4}\right)^2 + \left(\frac{1}{4}\right)^3 + \cdots + \left(\frac{1}{4}\right)^n\right) = \frac{4}{3}.$$

Because the numbers A_n close in on

$$\frac{4}{3}\mathbf{B} = \frac{4}{3} \text{Area} \triangle SVS',$$

it follows that this is the area of the parabolic section SVS'. The proof of Archimedes's theorem is now complete:

Archimedes's Theorem. The area of the parabolic section SVS' is equal to four-thirds the area of the inscribed triangle SVS'.

We have witnessed a remarkable argument. Its mathematical precision is decidedly superior to most of what would follow over the next 1800 years! In its last step (the "limit" step), the proof you saw differs from Archimedes's original. The Greeks did not consider (or perhaps, more accurately, did not accept as legitimate) arguments that used "infinite" processes. Instead, Archimedes let A be the area of the parabolic section SVS' and $A' = \frac{4}{3}$ Area $\triangle SVS'$, and he demonstrated the equality $A = A'$ by showing that $A > A'$ and $A < A'$ are both impossible. About 2000 years later, arguments involving infinite processes become the heart and soul of integral calculus; however, mathematicians wrestled with their legitimacy until the 19th century.

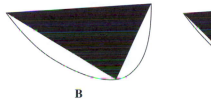
B

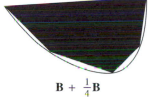
$\mathbf{B} + \frac{1}{4}\mathbf{B}$

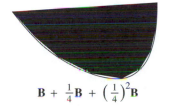
$\mathbf{B} + \frac{1}{4}\mathbf{B} + \left(\frac{1}{4}\right)^2\mathbf{B}$

Figure 3.25

3.5 The *Method*

There were references in the transmitted classical literature to a certain treatise of Archimedes called *Method*. The work itself, unfortunately, appeared to have been lost. However, in the early 1900s a Danish historian learned of a certain catalogue describing the holdings of the libraries of the Greek Orthodox Church. The catalogue listed a document that contained—in two superimposed layers—both a newer and an older manuscript, the latter of a mathematical nature. He traveled to Constantinople in the summer of 1906 and found a document consisting of about two hundred leaves of parchment. It contained Greek Orthodox prayers from the 13th and 14th centuries, and indeed, most of the leaves showed a more or less distinct underlayer: a mathematical manuscript in 10th century notation! Most of it was readable. A careful study revealed that it contained work of Archimedes. For the most part it was work already known, but among the new things there was the *Method*. This was an important find. It gave insights into the method that Archimedes had used to discover the connections between various areas and volumes and, in particular, between the areas of the parabolic section and the inscribed triangle.

The essence of his method is this: geometrical regions are compared by slicing them into thin strips and placing these (surprise!) at the two ends of a lever. To illustrate it, we return to the parabolic section.

Refer again to Figure 3.21 and extend the segment VC so that it intersects ES at F. (See Figure 3.26.) By Proposition A,

$$\text{Area } \triangle VAS = 4 \text{ Area } \triangle VCS.$$

Since the triangles VAS and VES are identical, their areas are the same. Therefore,

$$\text{Area } \triangle VES = 4 \text{ Area } \triangle VCS.$$

Recall from the proof of Archimedes's theorem that B is the midpoint of VE and C is the midpoint of BG. Since $\triangle VEF$ is similar to $\triangle VBC$, we have $\frac{BC}{EF} = \frac{1}{2}$, or $EF = 2BC$. Because C is the midpoint of BG, it follows that $2BC = BG$. Therefore, $BG = EF$. Now, $\triangle VES$ is similar to $\triangle VBG$ and B is the midpoint of VE, so $\frac{BG}{ES} = \frac{1}{2}$. Hence $ES = 2BG = 2EF$, and it follows that F is the midpoint of ES.

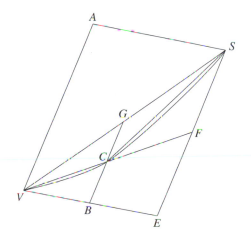

Figure 3.26

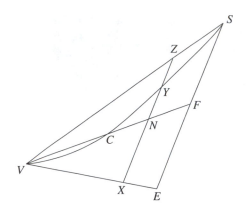

Figure 3.27

Now let X be an arbitrary point on VE and draw XZ parallel to ES. (See Figure 3.27.) Let N and Y be the indicated points of intersection. Archimedes next proceeded to the important step of verifying the equality

$$XZ \cdot NF = YZ \cdot VF.$$

By similar triangles, $\frac{VN}{VX} = \frac{VF}{VE}$, and therefore $VN = \frac{VX \cdot VF}{VE}$. Also by similar triangles, $\frac{XZ}{VX} = \frac{ES}{VE}$, and hence $XZ = \frac{VX \cdot ES}{VE}$. By an application of Proposition 3.3 of Section 3.1, $\frac{XY}{ES} = \frac{VX^2}{VE^2}$. By combining these equalities,

$$VN \cdot XZ = \left(\frac{VX \cdot VF}{VE} \right) \left(\frac{VX \cdot ES}{VE} \right)$$

$$= \frac{VX^2}{VE^2} \cdot VF \cdot ES = \frac{XY}{ES} \cdot VF \cdot ES$$

$$= XY \cdot VF.$$

So by Figure 3.27, $(VF - NF) \cdot XZ = VN \cdot XZ = XY \cdot VF = (XZ - YZ) \cdot VF$. After multiplying out and subtracting, $-XZ \cdot NF = -YZ \cdot VF$. The equality $XZ \cdot NF = YZ \cdot VF$ is now verified.

Now think of both the triangle $\triangle VES$ and the parabolic section VCS (Figure 3.26) to be made out of very thin, rigid (no bending), and homogeneous (no lumps) material. Imagine $\triangle VES$ being sliced up into very thin strips, as shown in Figure 3.28. To every strip XZ of $\triangle VES$ there corresponds—by Fig-

ure 3.27—the strip YZ of the parabolic section VCS. In this way, the parabolic section is sliced into strips also. Choose a unit of weight in such a way that a strip 1 unit long weighs 1 unit. So the weight of any strip is numerically equal to its length. Again refer to Figure 3.27 and extend the segment VF to a point K such that $FK = VF$; see Figure 3.29. Consider the segment VFK to be a lever with its fulcrum at F. Take any strip XZ of $\triangle VES$ and the corresponding strip YZ of the parabolic section. Refer to the Exercises 3C for the law of the lever, and notice that the equality $XZ \cdot NF = YZ \cdot VF$ means that each strip XZ balances the strip YZ placed at K. Doing this for each strip of $\triangle VES$ shows that $\triangle VES$ is balanced by the sum total of all of the strips of the parabolic section VCS placed at K.

Observe next that the product weight × distance for the right side of the lever is equal to

(Area of parabolic section VCS) $\cdot FK$

$$= \text{(Area of parabolic section } VCS) \cdot VF.$$

Now consider the left side of the lever. In his work *On the Equilibrium of Planes*, Archimedes showed that the *center of mass* (also called the *centroid*) of the triangle $\triangle VES$ is the point M on VF with $MF = \frac{1}{3}VF$. (See Exercise 13.) Considering the entire weight of $\triangle VES$ to be concentrated at M, the product weight × distance for the left side of the lever is

$$(\text{Area } \triangle VES) \cdot MF = (\text{Area } \triangle VES) \cdot \frac{1}{3}VF.$$

Since the system is in balance, Archimedes has now established that

(Area of parabolic section VCS) $\cdot VF$

$$= (\text{Area } \triangle VES) \cdot \frac{1}{3}VF.$$

Canceling VF and using the fact (noted earlier) that Area $\triangle VES = 4$ Area $\triangle VCS$ finally shows that

$$\text{Area of parabolic section } VCS = \frac{4}{3} \text{ Area } \triangle VCS.$$

Noting finally that C is the vertex of the parabolic section VCS (by Proposition 3.2 of Section 3.1), we

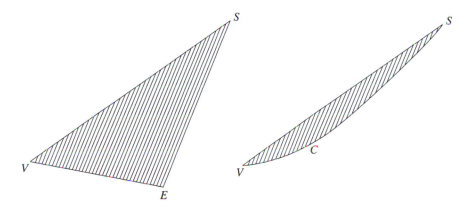

Figure 3.28

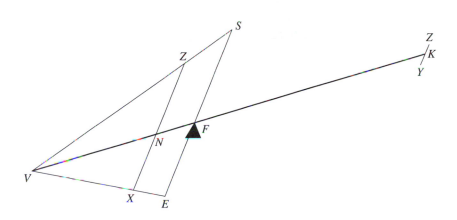

Figure 3.29

now see that Archimedes has established his theorem again, this time for the parabolic section *VCS*.

3.6 Postscript

W e have taken a glimpse at the mathematical genius of Archimedes. When Plutarch praised "the precision and cogency of the methods and means of proof" of Archimedes's studies, we can now understand what he meant. When Archimedes's investigations (and also those of Euclid, Aristarchus, Apollonius, Ptolemy, and the other classical scholars) were appreciated and understood, an important

source of light began to displace some of the darkness of the Middle Ages. In reference to mathematics, there can be little doubt that Archimedes's analysis of the parabola influenced the development of integral calculus in the 17th century. Of course, since it was rediscovered only in 1906, the influence of the *Method* could have been only indirect.

Exercises
3A.   The Circle and Related Areas

1.   Consider a triangle that is inscribed in a circle and has a diagonal of the circle as one of its sides. Use Proposition 2.1 of Section 2.2 to show that this triangle is a

right triangle. Show this again, by using Ptolemy's theorem and Pythagoras's theorem.

2. Compute the area of the circular sector of radius 7 and angle $\frac{\pi}{5}$. Then compute the area of the circular sector of radius 5 and an arc of length 8.

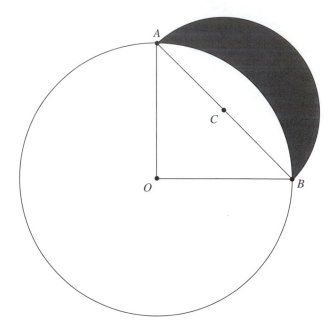

Figure 3.31

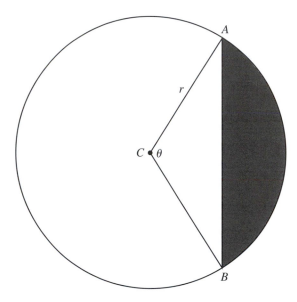

Figure 3.30

3. Show that the shaded area of the circle in Figure 3.30 is equal to $\frac{1}{2}\theta r^2 - r^2 \sin\frac{\theta}{2}\cos\frac{\theta}{2}$.

4. Compute the section of the circle in Exercise 3 if $r = 5$ and $\theta = 50°$.

5. In the diagram of Figure 3.31, O is the center of a circle and the radii AO and OB meet at O at right angles. The second arc connecting A and B is half of the circle with diameter AB and center C. Show that the area of the moon-shaped region—such regions are called *lunes*[2]—produced by the two circular arcs is equal to the area of the triangle AOB.

6. Suppose that the circle of Exercise 5 has radius r. Show that the area of the lune is $\frac{1}{2}r^2$ and that of the section of the circle between the lune and $\triangle AOB$ is equal to $\frac{\pi}{4}r^2 - \frac{1}{2}r^2$.

3B.   Sigma Notation and Areas

7. Write the sums $1 + 2 + 3 + 4$; $1 + 2^2 + 3^2 + 4^2 + 5^2$; and $1 + 2^2 + 3^3 + 4^4 + 5^5 + 6^6$ using sigma notation.

8. The squares in Figure 3.32 are each of side 1. Continue the indicated pattern to show that $\lim\limits_{n\to\infty}\sum\limits_{i=1}^{n}\left(\frac{1}{2}\right)^i = 1$.

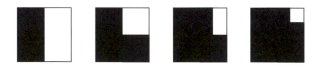

Figure 3.32

9. Consider a triangle with base b and height h. Refer to Figure 3.33. Inscribe into the triangle the rectangles R_1, R_2, and R_2', where the base of R_1 is $\frac{1}{2}b$, and the bases of R_2 and R_2' are $\frac{1}{2}\left(\frac{1}{2}b\right) = \frac{1}{4}b$; and continue the pattern indicated. Verify that the area of this triangle is $\frac{1}{2}bh$ by filling the triangle with this sequence of rectangles. In other words, show that

$$\frac{1}{2}bh = \text{Area } R_1 + (\text{Area } R_2 + \text{Area } R_2') + \cdots.$$

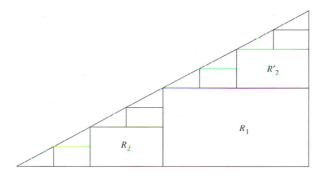

Figure 3.33

10. Consider a parabolic section where the cut is 7 units long and the perpendicular distance from the cut to the vertex is 4 units. What is the area of the section?

3C.   Archimedes's Law of the Lever

A lever consists of a rigid arm and a pivot called a *fulcrum*. The arm is considered to be weightless (in practical situations it would be of negligible weight). In Figure 3.34 the arm is represented by the segment through the points P_1 and P_2 and the fulcrum is positioned at F. If two point-masses of weights m_1 and m_2 are placed on the lever at points P_1 and P_2 in such a way that $m_1 d_1 = m_2 d_2$, where d_1 is the distance from P_1 to F and d_2 is the distance from P_2 to F, then the system balances, i.e., it is *in equilibrium*. This is Archimedes's law of the lever. Notice that a great weight m_1 can be balanced by a small weight m_2 provided that d_2 is large enough. The lever was most certainly used by the Egyptian pyramid builders. Archimedes is to have said (in reference to the lever), "Give me a place to stand on and I will move the Earth."

11. Suppose $d_1 = 1$ and $m_1 = 2000$. If $d_2 = 10$, what weight has to be placed at P_2 to raise m_1?

12. Determine the center of mass of the system consisting of the two masses shown in Figure 3.35. The bigger block has a mass of 80 kilograms and the smaller a mass of 15 kilograms. The distance between the centers of the blocks is 9 meters.

3D.   Facts Needed in the *Method*

An important step in Archimedes's argument concerns the center of mass (the centroid) of the triangle.

13. In the triangle $\triangle VES$ of Figure 3.36, F and F' are the midpoints of the segments ES and SV, respectively. Show that the areas of $\triangle VFE$ and $\triangle VFS$ are the same [Hint: they have equal bases; do they have equal heights?] Deduce that the centroid M must lie on the segment VF. By the same argument M must lie on EF'. So M is the intersection of VF and EF'.

14. Now extend the segment EF' of Figure 3.36 to a segment $EF'J$ in such a way that VJ is parallel to ES. Show that $\triangle FME$ and $\triangle VMJ$ are similar and that $\triangle ESF'$ and $\triangle JVF'$ are similar. Deduce that $FM = \frac{1}{2}VM$, and hence that $FM = \frac{1}{3}VF$, as proved by Archimedes.

Figure 3.35

15. Refer to the proof of Archimedes's theorem in Section 3.5. Show that the verification of the equality $XZ \cdot NF = YZ \cdot VF$ is unaffected when XZ is placed to the left of C, in other words, when Figure 3.27 is replaced by Figure 3.37.

Figure 3.34

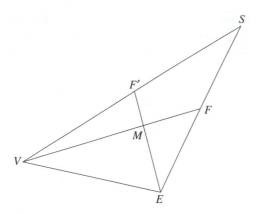

Figure 3.36

3E. Some Mathematics from the Middle Ages

16. The mathematician Gerbert (later Pope Sylvester II, 999–1003 A.D.) solved the following problem considered very difficult at the time. Let x and y be the legs of a right triangle and express x and y in terms of the height h and the area A of the triangle.

17. Gerbert expressed the area of an equilateral triangle of side a as $\left(\frac{a}{2}\right)\left(a - \frac{a}{7}\right)$. Show that this is an approximation that is equivalent to the approximation $\sqrt{3} \approx \frac{12}{7}$. Use a calculator to check its accuracy.

Note that the problems studied by Gerbert are much less sophisticated than those considered by the Greeks.

3F. Mathematics in the 16th Century

The 16th century saw great strides in the development of algebraic symbolism and notation. For example, the French mathematician François Viète used A to denote the unknown, A *quadratum* to denote A^2, and A *cubum* for A^3. He also used "+" and "−" the way we do, but he had no symbol for "=". What is now written $5bx^2 - 2cx + x^3 = d$, Viète would have written

$b5$ in A quad $-c$ plano 2 in $A+A$ cubum aequatum d solido.

So *aequatum* is =. It had been a common practice earlier to work with the "vernacular." In other words, the problem would have been written out: "Take a quantity times itself and multiply this by five times b, subtract from this two times c times the quantity …" and then it would have been solved without the algebraic manipulations that we use today!!

The 16th century also saw work on the solutions of cubic and quartic equations, e.g., equations of the form $x^3 + 3x - 6 = 0$ and $x^4 + 3x^2 + 5x - 8 = 0$. In fact, there were public competitions and challenges in Italy between some very interesting characters (Tartaglia, Cardano, Ferrari—not of sports car fame) to determine who was best at the game of solving such equations. (This required lots of ingenuity.) Instead of going into the details, we point out a related fact:

Let $p(x)$ be a polynomial. Recall that a number d such that $p(d) = 0$ is called a *root* of $p(x)$. Note that if $p(x)$ has a factorization of the form $p(x) = (x - d)q(x)$, then d is a root of $p(x)$. This fact works also in reverse: namely, if d is a root of $p(x)$, then $x - d$ divides $p(x)$.

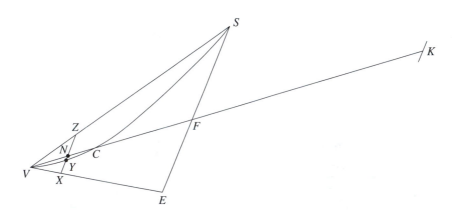

Figure 3.37

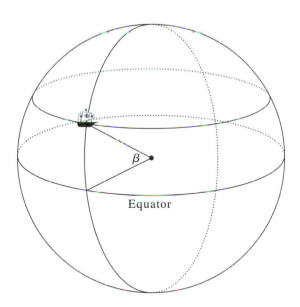 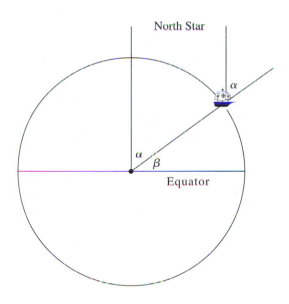

Figure 3.38

18. Use this idea to factor the polynomials $x^3 + 3x - 4$ and $x^3 - 7x - 6$ as completely as possible.

3G.   Celestial Navigation

Elementary celestial navigation was common in the 16th century. The considerations of Eratosthenes and a star catalogue were used to determine the position—at least, the latitude—of a ship at sea.

19. Suppose a ship is positioned on the globe, as shown on the left in Figure 3.38. Note that the angle β determines its north-south position. This is the ship's *latitude*. As the diagram on the right shows, β can be determined if one knows α. But α can be measured: it is the angle between a plumb line and the line of sight to the North Star. Express the distance from the ship to the equator (along the arc) in terms of α and the radius of the Earth.

20. Suppose that the angle α is measured to be 53°. How many miles is the ship north of the equator? (Use the modern value of 3950 miles for the radius of the Earth.)

Why doesn't this strategy work for determining the *longitude*, i.e., the east-west position, of a ship? Without the solution of this "problem of longitude," a ship's position cannot be fully determined. This problem turned out to be very elusive and was not solved until the 18th century. Large cash prizes were offered for a solution by the governments of seafaring states like Spain, Portugal, Venice, Holland, and England. The British "Board of Longitude" offered the largest amount, a sum of 20,000 pounds. Note that as the Earth turns, the stars travel across the sky at a rate of 360° per day, or 15° per hour. Thus, if time can be measured accurately aboard a ship, the sky will provide its position. A ship's rolling motion rendered pendulum clocks useless. The issue, therefore, centered on the construction of a clock accurate under the severe conditions on board. The matter was not settled until 1761–1762. A marine chronometer devised by the Englishman John Harrison (it had taken him nineteen years to complete) was tested aboard one of His Majesty's ships and found to have lost only about 5 seconds in eighty days at sea.[3]

Notes

[1]Ptolemy's *Geographia*, translated into Latin at the beginning of the 15th century, contains geographical coordinates of 8000 localities. It laid the foundation of the science of cartography and was thus of direct importance to navigation in the 16th century. Its map of the known world, in particular of Europe and North-

ern Africa, is astounding. Columbus used Ptolemy's map (in part because its dimensions were a little too small) as evidence for his contention that it would be feasible to reach Asia by sailing west.

[2]Apparently, Leonardo da Vinci was fascinated by lunes. See J.L. Coolidge, *The Mathematics of Great Amateurs*, Clarendon Press, a Division of Oxford University Press, 1991.

[3]The interested reader may wish to engage Dava Sobel's *Longitude: The True Story of a Lone Genius Who Solved the Greatest Scientific Problem of His Time*. Walker, New York, 1995.

4

A New Astronomy and a New Geometry

The achievements of the Greeks—you are now in a position to judge for yourself—are remarkable. They developed trigonometry and geometry and applied both in spectacular ways to estimate the sizes and distances of objects in the universe and to create mathematical models that simulate the motion of these objects. They also developed subtle mathematical methods for computing areas and volumes, and they contributed to number theory and algebra. Given what we know now, however, it is clear that important elements were missing. The Greeks did not have the coordinatized line, and the correspondence between points and real numbers that it implies. Since they had no comprehensive coordinatized geometry, they could not make use of the powerful interaction between algebra and geometry that it implies. In summary, one can say that Greek mathematics is fundamental, substantial, and also brilliant,[1] but that it is incomplete. As for Greek physics and astronomy, however, the matter was different: the basic points of view were flawed and the underlying principles were in error. In time, the scholars of Europe absorbed the learning of Greece and reflected about its science. When they started to question and challenge it, a scientific revolution had begun.

4.1 **A New Astronomy**

Copernicus (1473–1543) was born in the Polish town of Torun. He studied astronomy in Bologna and medicine in Padua. After returning to Poland, he settled into his position as canon of the Church. He found ample time to pursue astronomy in his little study, which is said to have been in the tower of the cathedral. He was not satisfied with the very complicated nature of Ptolemy's astronomy and so published his own comprehensive theory of the solar system, *De Revolutionibus Orbium Celestium*. It placed the Sun at the center of the universe and the Earth and the other planets in orbit around it. The introduction of an early edition of *De Revolutionibus* makes mention of Aristarchus: "...some say that Aristarchus of Samos

was of the same opinion." Later editions apparently dropped this reference. Realizing that it might be controversial (after all, in Genesis it is the Earth that is at the center of God's creation), Copernicus delayed the appearance of this work until 1543, the year of his death.

With the Sun at the center of the system, the retrograde motion of the planets no longer required an elaborate explanation. Therefore, Ptolemy's convoluted orbital clockwork of circles upon circles became unnecessary. However, Copernicus did retain basic elements of Ptolemy's mathematics. For instance, his model for the Earth's orbit was Ptolemy's off-centered circle. Furthermore, when discrepancies between his theory of simple circular orbits and observations persisted, Copernicus was later led back to the epicycloids of Ptolemy.

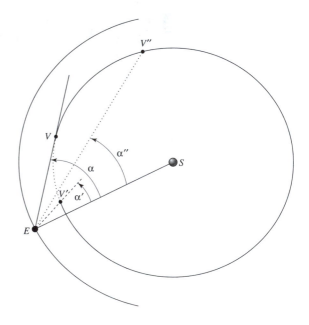

Figure 4.1

But there were also new mathematical elements in Copernicus's work. The planet Venus provides an example. Copernicus determined its distance from the Earth with the following argument. The angle α

between the lines of sight from an observer on Earth to the Sun on the one hand, and to Venus on the other, is greatest when the line of sight to Venus is tangent to its circular orbit around the Sun. (See Figure 4.1.) But then the angle between the line of sight to Venus and the radius *VS* of the orbit is a right angle. Therefore,

$$\sin \alpha = \frac{VS}{ES}.$$

The greatest value for α that Copernicus observed was $\alpha = 46°$. Since $\sin 46° = 0.719$ (recall that $\sin 45° = \frac{\sqrt{2}}{2} = 0.707$), he obtained

$$VS = 0.719\, ES.$$

Let's define the distance between the Earth and the Sun to be one astronomical unit (AU). So $ES = 1\,\text{AU}$. Therefore, Copernicus has shown that the distance between Venus and the Sun is 0.719 AU. In a similar manner (this is more complicated in the case of the outer planets such as Mars), Copernicus was able to compute the distances in AUs between the Sun and all the planets that were known at the time. As you can see from Table 4.1, he did so quite accurately.

Table 4.1

Estimated Average Distances of the Planets from the Sun in Astronomical Units

	Copernicus	Kepler	Modern
Mercury	0.376	0.389	0.387
Venus	0.719	0.724	0.723
Earth	1.000	1.000	1.000
Mars	1.520	1.523	1.524
Jupiter	5.219	5.200	5.202
Saturn	9.174	9.510	9.539

The Danish astronomer Tycho Brahe (1546–1601) designed and constructed new calibrated precision instruments that allowed him to observe celestial phenomena with a much higher degree of accuracy

than was previously possible. In 1572 he observed a very bright star in a location where there had been no star before. When this "nova stella" disappeared about a year and a half later, it was clear to him that the sphere of stars was not as immutable as the Greeks had claimed. (What Tycho had seen was no doubt a supernova, an exploding star. Ironically, it was apparently an explosion that also killed Tycho. It is said that his bladder burst after some heavy drinking.) In 1577 he observed a comet that appeared to have no problem moving through the crystal spheres that supposedly held the planets in their orbits. Tycho realized that this explanation of the Greeks was also in error. Tycho developed his own "Tychonic" system of the universe: the Moon and the Sun are in orbit about the Earth, but all other planets are in orbit about the Sun. For over a century, the Ptolemaic, Copernican, and Tychonic explanations of the universe existed side by side.

Johannes Kepler (1571–1630) was born near Stuttgart, in southern Germany. His father was a mercenary whose only pleasures seemed to have been drink and war. He deserted his family soon after Johannes Kepler's birth. Young Kepler was sickly and poor, but he was also a brilliant student. Being deeply religious, he enrolled in the University of Tübingen to study theology. While there, he learned about the principles of the Copernican system from a professor of mathematics (privately, since the university administration frowned upon such heretical studies). He graduated in 1591 and took the dual post of Mathematician of the Province and teacher at the Evangelical Seminary in Graz, Austria. He continued his speculations about astronomy, with particular interest in the relationship between the distances and velocities of the planets, as well as the driving forces that propel them in their orbits.

In the meantime, Tycho Brahe had accepted an appointment as astronomer to the royal court of the emperor of Austria in Prague. He brought with him some of his instruments as well as the wealth of accurate data that he had collected. When Kepler's work came to his attention, Tycho arranged for Kepler to become his assistant. After Tycho's death, Kepler was free to use Tycho's data in the pursuit of his own investigations and turned his attention to the orbit of Mars. Convinced of the correctness of Copernicus's system, he set out to verify that this orbit is a circle. A circle, not in the simple sense, but in the sense of Ptolemy; see Figure 2.26 of Section 2.6. It was Kepler's goal to determine for the orbit of Mars the precise positions of the Sun S (S replaces E in the figure just mentioned), the center C of the orbit, and the point of reference A for the velocity. He computed orbital distances and angles for many different positions for S, C, and A, and checked and rechecked his computations against Tycho's observations. But after three years of grueling work, he was forced to conclude that the orbit of Mars could *not* be a circle. He wrote:

> The planet's [Mars's] orbit is not a circle; but [starting at aphelion] it curves inward little by little and then [returns] to the amplitude of the circle at [perihelion]. An orbit of this kind is called an oval.[2]

Circles and spheres satisfied Kepler's sense of aesthetics, but the oval did not. "If only the orbit were an ellipse," he wrote, "then the problem would have been solved already by Archimedes and Apollonius." But an oval! After all his efforts he was left, as he put it, with "only a single cartful of dung."

Kepler discarded his work and started afresh. New calculations and comparisons with observation finally brought success. He concluded that the orbit of Mars curves inward at *both* aphelion and perihelion and—at last—that it has the shape of an ellipse. In 1609 he published his first two laws of planetary motion in the book *Astronomia Nova*.

1. Each planet P moves in an elliptical orbit with the Sun S at one of the focal points of the ellipse. (See Figure 4.2.)
2. A given planet sweeps out equal areas in equal times.

Stated more precisely, this second law says the following: Suppose that it takes a planet time t_1 to

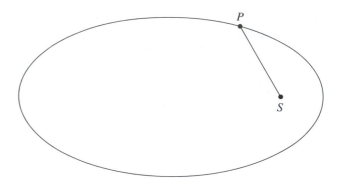

Figure 4.2

move from position P_1 to Q_1 in its orbit, and that it traces out the area A_1 in the process. See Figure 4.3. Suppose that it later moves from P_2 to Q_2 in time t_2 while tracing out the area A_2. According to Kepler's second law,

$$\text{if } t_1 = t_2, \text{ then } A_1 = A_2.$$

A moment's thought shows that the second law implies that a planet moves faster when it is closer to the Sun than when it is farther away.

The second law also implies that if $t_1 = t_2 + t_2$, then $A_1 = A_2 + A_2$. So if $t_1 = 2t_2$, then $A_1 = 2A_2$. Similarly, if $t_1 = 3t_2$, then $A_1 = 3A_2$, and if $t_1 = \frac{1}{2}t_2$, then $A_1 = \frac{1}{2}A_2$. In general, if $t_1 = kt_2$ for some positive constant k, then $A_1 = kA_2$. Now let t_1 and t_2 be any two time intervals and let $\frac{t_1}{t_2} = k$. So $t_1 = kt_2$, and, as just asserted, $A_1 = kA_2$. So $\frac{A_1}{t_1} = \frac{kA_2}{kt_2} = \frac{A_2}{t_2}$. Therefore, Kepler's law second can be reformulated as follows:

Let t be any time interval and let A_t be the area traced out by the planet during time t. Then the ratio $\frac{A_t}{t}$ is the same constant no matter what time t is taken and no matter where in the orbit the motion occurs.

Kepler published his third and final law in 1619.

3. Let M and J be any two planets and consider their elliptical orbits. Let a_M be the *semimajor axis* of the ellipse of M. This is $\frac{1}{2}$ the length of the major axis, i.e., the diameter through the two focal points. See Figure 4.4. Observe that the

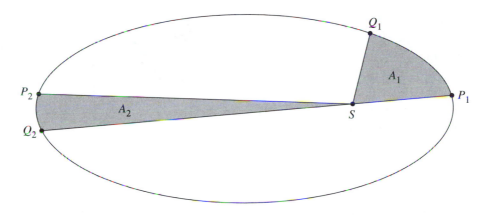

Figure 4.3

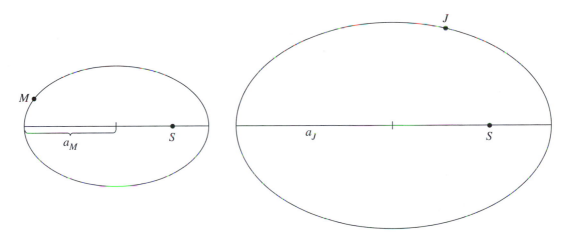

Figure 4.4

semimajor axis is a number and not a segment. Let T_M be the *period* of the orbit of planet M. This is the length of time required for one complete revolution. Let a_J and T_J be the semimajor axis and the period of planet J. Then Kepler's third law states that

$$\frac{a_M^3}{T_M^2} = \frac{a_J^3}{T_J^2}.$$

Notice that this law relates the sizes and periods of the various planetary orbits.

Kepler had hit upon the three laws that are fundamental to the explanation of the workings of the solar system. It should be pointed out that his scientific discourse also contains strange (by today's standards) theological and mystical elements.

Kepler's last years were troubled. His religious beliefs—which saw no contradiction between his astronomy and the Bible—were attacked, and he was labeled a heretic. Europe had become embroiled in the Thirty Years' War. This conflict between Catholic and Protestant forces for the control of Europe devastated the population and left many cities in ruins.

The new astronomy did not find general acceptance. Ptolemy's system of the universe was difficult to dislodge, because it rested on Aristotelian physics. This in itself was part of the comprehensive system of Aristotelian thought, which consists of a tight weave of philosophical discourse, explanations, and observations, deeply rooted and difficult to refute. It was not until Aristotelian learning was drawn into question that the new astronomy could make any serious headway. This task fell in large measure to Galileo Galilei.

4.2 The Studies of Galileo

Galileo Galilei (1564–1642) was born in Pisa. He exhibited an independence of outlook at an early age. He was sent to the University of Pisa, but he left four years later when his father could no longer afford the considerable fees. (Evidently, this was already a problem in those days.) A family friend who was a distinguished mathematician taught Galileo the elements of Euclid and Archimedes. After a time, Galileo mastered the basics of mathematics and physics. In time, however, he began to have questions. He devised experiments both in reality and in his mind to test what he was learning.

Aristotle's theory of motion proclaimed that the heavier a body is, the faster it falls. Galileo had problems with this, and he engaged in the following "thought" experiment. He contemplated a cannonball in free fall. Next, he considered it to be cut into two pieces, and he thought of both pieces falling side by side. See Figure 4.5. From the point of view of the fall, nothing had changed. Therefore, the cannonballs in the two situations must drop with the same speed. According to Aristotle, however, each of the two separate halves of the ball—being lighter—must drop more slowly than the ball as a whole. Aristotle had to be wrong. Galileo had discovered that all bodies fall in the same way, regardless of their weight, assuming that they are heavy enough for air resistance to be ignored. (Incidentally, most historians of science are of the opinion that Galileo did not drop cannonballs from the leaning tower of Pisa.)

Figure 4.5

Galileo next turned his attention to quantitative aspects of falling objects. Instead of considering a ball in free fall, he let it roll down an inclined plane. This slowed things down and made it easier to quantify observations. He fixed a unit of time and a unit of length. In an experiment undertaken in 1604, Galileo let a metal ball start from the top of the inclined plane at time $t = 0$. Careful measurements of times and distances (in repeated trials) revealed the following pattern. See Figure 4.6. The rolling ball covered a distance d between $t = 0$ and $t = 1$; a distance $3d$ between $t = 1$ and $t = 2$; a distance $5d$ between $t = 2$ and $t = 3$; a distance $7d$ between $t = 3$ and $t = 4$; and so on. Since average velocity is distance divided by time, Galileo observed that the successive average velocities of the ball were

$$\frac{d}{1}, \frac{3d}{1}, \frac{5d}{1}, \frac{7d}{1}, \frac{9d}{1}, \dots,$$

and he noticed that the velocity increased at a constant rate. What he had discovered was this: the increase in the velocity of the ball divided by the time over which the increase occurs is the same constant no matter when the measurement is taken. This can be expressed algebraically as follows: Let $v(t)$ denote the velocity of the rolling ball at any time t. Let t_1 and t_2 with $t_1 < t_2$ be any two moments of time. Then

$$\frac{v(t_2) - v(t_1)}{t_2 - t_1} = C$$

for a constant C that is the same no matter what t_1 and t_2 are picked. The constant C does depend on the inclination of the plane, however.

Later, measuring time more accurately using water clocks (Galileo measured time in *tempi* with 1 *tempo* equal to 0.011 seconds) and linking free fall

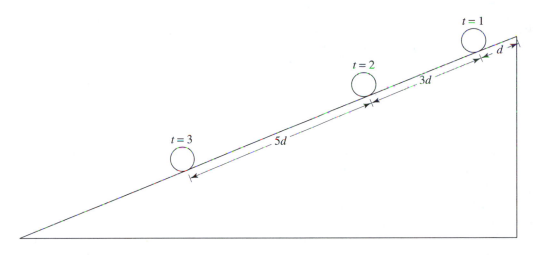

Figure 4.6

to properties of the motion of the pendulum, he discovered that in the case of a ball in free fall, the ratio

$$\frac{v(t_2) - v(t_1)}{t_2 - t_1}$$

is also equal to a constant. Again, this ratio (nowadays written g) is the same regardless of the moments t_1 and t_2 at which the measurements are taken.

Beginning in 1608, Galileo put his inclined plane on a table and measured the horizontal and vertical distances between the point at which the ball left the edge of the table and the point at which it impacted the floor. After many trials with varying velocities, he found that the trajectory of the fall is a parabola. We will analyze one of these experiments[3] in detail in Chapter 9.

Aristotle's physics had also proclaimed that all objects have a natural motion toward the center of the universe, i.e., the center of the Earth, and that the motion of an object in any other direction is possible only as long as a propelling agent is in contact with it and drives it forward. Once this ceases to operate, the motion stops and the object falls straight to the ground or comes suddenly to rest. The propelling agent in the case of a projectile is the disturbance produced in the air at the time of the initial impulse. The air that is being pushed and compressed in front of the object rushes around behind it to prevent a vacuum (which cannot exist). In particular, for Aristotle, motion in a vacuum was inconceivable.

In summary, the fundamental principle of Aristotle's theory of motion is this: An object that is in motion will stop unless an external force continues to propel it.

Galileo, however, was skeptical and returned to his experiments. He released a ball at a certain elevation on an inclined plane. He let it roll up another inclined plane and observed that it reached its original height. See Figure 4.7. He took a second inclined plane with a smaller inclination and noticed that the same was true. Galileo continued this experiment in his mind, imagining frictionless inclined planes of smaller and smaller inclinations. Carrying his thinking to its logical end, he concluded that on a flat (and frictionless) surface, a perfectly spherical ball would roll forever.

Galileo had thus discovered the *law of inertia*: A body that is in a state of either rest or motion will continue in this same state, unless it is acted upon by some external force. This stands in sharp contrast to the hypothesis of Aristotle.

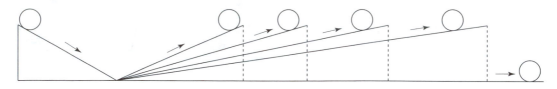

Figure 4.7

Aristotle's thinking does seem to be consistent with ordinary observation and the data supplied by common sense. (After all, a real spherical ball would not roll forever.) The laws of motion are not something likely to be discovered by experimental methods alone. At some point, a transposition in the mind of the scientist is also required. Galileo was aware of the "material impediments" of air resistance, friction, and the fact that the balls that rolled down his inclined planes were not perfectly round. He knew that these factors would lead to discrepancies between the conclusions of an experiment and the assertion of the basic laws of motion. But he knew also that these impediments must be separated from the fundamental principles. In ordinary experience, there are no perfectly spherical balls moving on perfectly smooth, frictionless inclined planes and rolling away to infinity. But by imagining them, Galileo stripped away the inessentials and revealed the laws at the core.

The problem of motion was a problem of fundamental nature, one that could not be solved by better observation within the framework of the older systems of ideas. A new frame of mind was required. In other words, to understand what is going on it is necessary to think not about real bodies as we actually observe them in the real world, but about idealized bodies moving in an idealized world, one without friction and resistance. It calls for the contemplation of geometrical shapes moving in abstract space. Upon arriving at the basic principles in this abstracted context, the variables of drag, friction, etc., can be reinserted and analyzed. Observe that Greek geometry had done a similar thing in a much simpler context. Its focus was not on the wheel and the ball, but on their abstractions, the circle and the sphere.

Galileo also had a considerable impact on astronomy. He was an advocate of the Copernican system. He heard about the invention of the telescope, fashioned his own, and turned it towards the sky with spectacular results. In January of 1610 he observed Jupiter. Diagrams in his book the *Starry Messenger* describe what he observed. See Figure 4.8. He discovered, sometimes on the left, sometimes on the right, and at times hidden from view altogether, four satellites in orbit around Jupiter. The existence of this system, which certainly does not have the Earth at its center, gave support to the viewpoint of Copernicus.

Galileo studied the phases of Venus and concluded that the pattern they exhibit is consistent with the heliocentric hypothesis. In the Ptolemaic system of

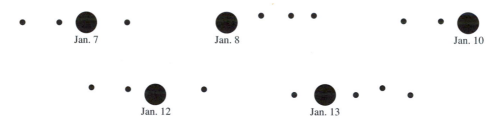

Figure 4.8

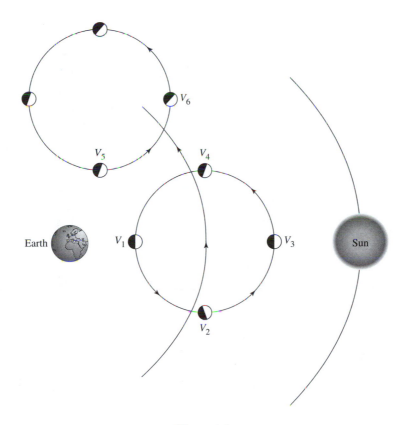

Figure 4.9

Figure 4.9, Venus (as seen from Earth) always appears more or less crescent-shaped. In the Copernican system of Figure 4.10, however, Venus has the full range of phases that Galileo observed. Therefore, the basic configuration of the Ptolemaic model had to be in error and that of Copernicus correct.

Galileo became professor of mathematics at the University of Padua. He was an eloquent speaker, an excellent teacher, and his fame soon spread. When a great hall with a capacity of more than a thousand proved too small, his astronomy lecture would be moved outside. He decided to publish his findings in pamphlets, dialogues, letters, and books. He wrote, not in the Latin of the academy, but in Italian, the language of his country. He became embroiled in a lengthy controversy with the Church. The Copernican system as a mathematical theory was one thing, but

as a physical reality it was contradictory to scripture, undermined the authority of the church, and it had to be fought. After 20 years of hearings, misunderstandings, charges, revisions, and injunctions, Galileo was forced to recant and the Inquisition sentenced him to house arrest for life. After a lengthy illness, he died in 1642.

As we have seen, Galileo's contributions to science were twofold: His experiments established some of the basic laws of motion, and his discoveries with the telescope gave credibility to the Copernican heliocentric point of view. Perhaps even more important was his new approach to the understanding of natural phenomena, which facilitated the transition to modern science.

The definitive explanations of the discoveries of Galileo and Kepler require not only a new approach

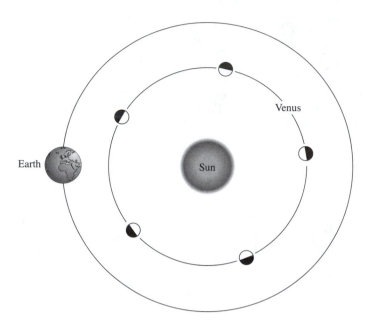

Figure 4.10

to science, but also a new mathematics. It is possible to extract considerable and essential quantitative information from Galileo's discovery that for falling bodies,

$$\frac{v(t_2) - v(t_1)}{t_2 - t_1} = \text{constant}.$$

The same is true with Kepler's laws. The fact that a planet's orbit is an ellipse with the Sun at one of the focal points, together with the equality

$$\frac{A_t}{t} = \text{constant}$$

contains precise quantitative information about the orbits of the planets. In both cases, however, the extraction of this information requires a new mathematics: differential and integral calculus.

The development of both of these branches of calculus required a new geometry. It is impossible to credit a single individual for the discovery of this "analytic geometry." Some of its basic principles have earlier origins. They can be seen, for example, in the work of Apollonius and in the coordinate system

of Ptolemy's maps. Most historians agree, however, that the primary credit for analytic geometry as a modern subject goes to the French mathematician René Descartes (1596–1650).

Descartes received his early education from the Jesuits, who recognized both his physical weakness and his mental alertness. They allowed him to read and meditate in bed past rising hours. This became a lifelong habit that Descartes found very conducive to intellectual output. At the age of 22, amazingly, he became a professional soldier. The lulls in military activity during the winter months allowed him to pursue mathematics and philosophy. One day in 1619, while in service in Bavaria, he escaped the cold by shutting himself into a "stove." He recounted that, during his meditations in this especially warm room, a divine spirit revealed a new philosophy to him. When he emerged from the "stove," he had conceived the principles of analytic geometry and the idea of applying the mathematical method to philosophy.

His famous work *Discours de la méthode*, written in 1629 and published in 1637, recalls that

Archimedes, in order that he might draw the terrestrial globe out of its place and transport it elsewhere [recall that Archimedes supposedly asserted, in reference to the lever, that if he had a place to stand he could move the world], demanded that only one point should be fixed and immovable; in the same way I shall have the right to conceive high hopes if I am happy enough to conceive one thing only which is certain and indisputable.

He then hit upon the most famous assertion in all of philosophy: Je pense, donc je suis. Cogito ergo sum. I think, therefore I am.

Our interest, however, is in a hundred-page appendix to the *Discours de la méthode*, namely *La Géométrie*. In it, Descartes experimented with algebraic notation: he used a, b, c to denote constant quantities; he wrote a^2, a^3, a^4, and $\sqrt{a^2 + b^2}$ as we do today, and he wrote x, y, z for variable quantities. But more importantly in the current context, he fixed a point's position in the plane by assigning to it two "coordinates," namely, its distances to two perpendicular lines. He also discussed the relationship between graphs and equations that this connection provided. We now turn to a detailed description—in modern notation—of what Descartes achieved.

4.3 The Geometry of Descartes

Just as the points on a line can be identified with the real numbers by using a coordinatized line, so the points in a plane can be identified with pairs of real numbers. Start by drawing two perpendicular coordinatized lines that intersect at their origins. One line is placed horizontally with the positive direction to the right. The other is placed vertically with the positive direction upward. These are the *coordinate axes*. The horizontal axis is often called the *x-axis*, and the vertical axis the *y-axis*. However, these designations depend on the variables used. For example, in a context where a variable time t or angle θ is considered, an axis might be labeled by t or θ.

Any point P in the plane determines a unique ordered pair of numbers as follows. Refer to Figure 4.11. Draw the line through P parallel to the y-axis. This line intersects the x-axis at some point and this point has a coordinate, say a, on this number line. Similarly, draw a line through P parallel to the x-axis to get a coordinate, say b, on the y-axis. In this way, Descartes assigns to the point P the pair of numbers (a, b). The first number, a, is called the *x-coordinate* of P, and the second number, b, is called the *y-coordinate* of P. We say that P is the point with coordinates (a, b), and we will often write (a, b) in place of P. For example, the point S in Figure 4.11 has coordinates $(3, -3.2)$, so $S = (3, -3.2)$. By reversing the process we can start with an ordered pair, say (c, d), and arrive at the corresponding point R. For example, $(2, 4\frac{1}{2})$ corresponds to the point Q. The point $(0, 0)$ is called the *origin* and is often denoted by O. The notation 0 (zero) will be used for this point if the attention is on the x-axis or the y-axis.

This setup is called the *rectangular* or *Cartesian* (for Descartes) *coordinate system*. It provides the connection between geometric objects—points, collections of points, and curves—and algebra—pairs of numbers, collections of such pairs, and equations. The plane supplied with a coordinate system is called the *Cartesian plane*. The coordinate axes divide the Cartesian plane into four *quadrants*, labeled **I**, **II**, **III**, and **IV** in Figure 4.11. Notice that the first quadrant consists of the points whose x- and y-coordinates are both positive.

Check that the distance between the points -5 and 3 on the number line is 8. Observe that this is equal to the absolute value of $-5 - 3$ or, equivalently, that of $3 - (-5)$. Similarly, the distance between 6.4 and 4.6 is 1.8. This is the absolute value of either $6.4 - 4.6$ or $4.6 - 6.4$. In general, the distance between the points a and b on a number line is equal to $a - b$ if $b \leq a$ and $b - a$ if $b > a$. See Figure 4.12. In the first case, $a - b \geq 0$ and hence $|a - b| = a - b$. In the second case, $a - b < 0$ and hence $|a - b| = -(a - b) = b - a$. Observe,

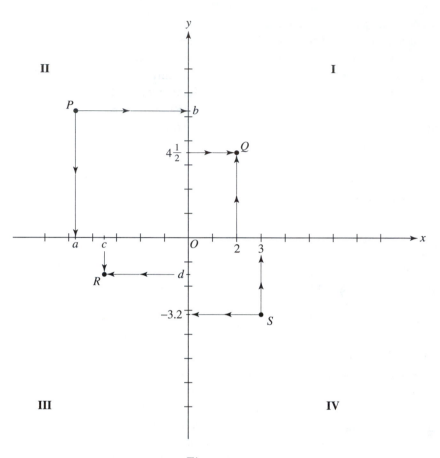

Figure 4.11

therefore, that the distance between a and b is equal to the absolute value $|a - b| = |b - a|$ in either case.

Consider any two points $P_1 = (x_1, y_1)$ and $P_2 = (x_2, y_2)$ in the plane and refer to Figure 4.13. By the remarks just made, the distance between the two x-coordinates is $|x_1 - x_2|$ and the distance between the two y-coordinates is $|y_1 - y_2|$. It follows that the segment $P_1 P_2$ is the hypotenuse of the right triangle of Figure 4.14. Therefore, by the Pythagorean theorem,

$$P_1 P_2 = \sqrt{|x_1 - x_2|^2 + |y_1 - y_2|^2}.$$

Since $|x_1 - x_2|^2 = (x_1 - x_2)^2$ and $|y_1 - y_2|^2 = (y_1 - y_2)^2$, the distance between the points $P_1(x_1, y_1)$ and $P_2(x_2, y_2)$ is equal to

$$\boxed{P_1 P_2 = \sqrt{(x_1 - x_2)^2 + (y_1 - y_2)^2}}$$

Observe that we are continuing an earlier practice: When a segment, here $P_1 P_2$, occurs in a mathe-

Figure 4.12

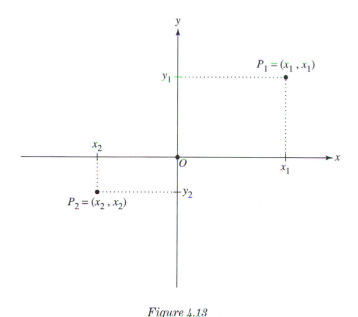

Figure 4.13

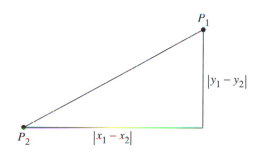

Figure 4.14

for example, we get $y^2 = 3$, and therefore $y = \pm\sqrt{3}$. In the same way, the *x-intercepts* of a graph are the *x*-coordinates of the points of intersection of the graph with the *x*-axis. They are found by setting $y = 0$ and solving for *x*. For $xy^3 + y^2 - x^3 + 3 = 0$, we get $x^3 = 3$; so $x = \sqrt[3]{3}$.

matical expression, the reference will be to the length of the segment.

Example 4.1. The distance between $(1, -2)$ and $(5,3)$ is

$$\sqrt{(1-5)^2 + (-2-3)^2} = \sqrt{4^2 + 5^2} = \sqrt{41}.$$

Since $\sqrt{(5-1)^2 + (3-(-2))^2}$ is also equal to $\sqrt{4^2 + 5^2} = \sqrt{41}$, the order in which the points are considered has no effect on the result.

The *graph* of an equation in *x* and *y* is the set of all points (a, b) in the plane with the property that the values $x = a$ and $y = b$ satisfy the equation. For example, the point $(1, -3)$ is a point on the graph of the equation $3x^2 + y^2 = 12$, since $3(1)^2 + (-3)^2 = 12$. The graph of the equation $x = 5$ is the collection of all points (x, y) with $x = 5$. This is the vertical line through the point $x = 5$ on the *x*-axis. Similarly, the graph of $y = -7$ is the horizontal line through the point $y = -7$ on the *y*-axis.

The *y-intercepts* of a graph are the *y*-coordinates of the points of intersection of the graph with the *y*-axis. They are found by setting $x = 0$ and solving for *y*. When this is done for the graph of $xy^3 + y^2 - x^3 + 3 = 0$,

A graph is a pictorial representation of an equation. It provides a picture of the algebra and clarifies what is going on. In the other direction, when a curve has to be understood, it is often of great advantage to have an equation of the curve, that is to say, an equation whose graph is the curve in question. This translates the geometric concern into algebra. The algebra allows for precise computations that can reveal exact numerical information. In other words, if a curve can be represented by an algebraic equation, then the rules of algebra can be used to analyze it. This complementary duality between geometry on the one hand and numbers and algebra on the other is the essence of analytic geometry. This interplay is of crucial importance in the solution of many mathematical problems.

The parabola provides an example that illustrates the point. Take any parabola and move it so that its directrix *D* is horizontal. Refer to Figure 4.15. Suppose *D* crosses the *y*-axis at the point *c* and observe that $y = c$ is the equation of *D*. Let $F = (a, b)$ be the focus. Since *F* is not on *D*, it follows that $b \neq c$.

Let $P = (x, y)$ be any point in the plane. Then the distance from *P* to *D* is $|y - c|$ and the distance from

P to F is

$$PF = \sqrt{(x-a)^2 + (y-b)^2}.$$

So P is on the parabola precisely when $|y - c| = \sqrt{(x-a)^2 + (y-b)^2}$. Squaring both sides gives $(y-c)^2 = (x-a)^2 + (y-b)^2$. After multiplying things out and simplifying, we have

$$y^2 - 2cy + c^2 = x^2 - 2ax + a^2 + y^2 - 2by + b^2$$

$$2by - 2cy = x^2 - 2ax + a^2 + b^2 - c^2$$

$$2(b-c)y = x^2 - 2ax + (a^2 + b^2 - c^2).$$

Since $b \neq c$, $b - c \neq 0$. So we can divide through by $2(b-c)$ to get

$$(*) \quad y = \left(\frac{1}{2(b-c)}\right)x^2 - \left(\frac{a}{b-c}\right)x + \left(\frac{a^2+b^2-c^2}{2(b-c)}\right).$$

Taking $A = \frac{1}{2(b-c)}$, $B = -\frac{a}{b-c}$, and $C = \frac{a^2+b^2-c^2}{2(b-c)}$, we see that any parabola with a horizontal directrix has an equation of the form

$$y = Ax^2 + Bx + C.$$

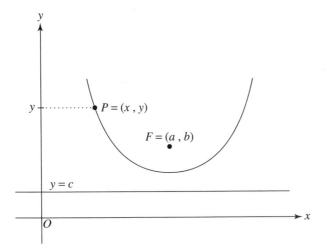

Figure 4.15

Example 4.2. Find an equation of the parabola with directrix the line $y = 5$ and focus $(2, -3)$.

Solution. Here $c = 5$, $a = 2$, and $b = -3$. Since

$$\frac{1}{2(b-c)} = \frac{1}{2(-3-5)} = -\frac{1}{16},$$

$$-\frac{a}{b-c} = -\frac{2}{-3-5} = \frac{1}{4}, \quad \text{and}$$

$$\frac{a^2+b^2-c^2}{2(b-c)} = \frac{4+9-25}{-8} = \frac{3}{4},$$

we see that $y = -\frac{1}{16}x^2 + \frac{1}{4}x + \frac{3}{4}$ is an equation of this parabola.

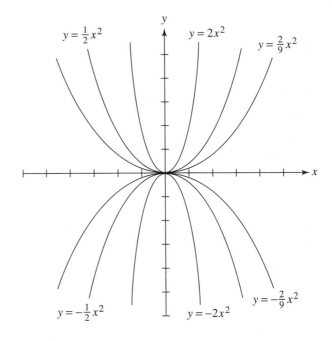

Figure 4.16

We have just established the fact that any parabola with a horizontal directrix has an equation of the form $y = Ax^2 + Bx + C$. This argument can be reversed to show that the graph of any equation of the form $y = Ax^2 + Bx + C$ (with $A \neq 0$) is a parabola with horizontal directrix.

Example 4.3. Show that the graph of the equation $y = 2x^2 + 5x - 4$ is a parabola and find its directrix and focus.

Solution. Work backwards using equation (∗). Since $\frac{1}{2(b-c)} = 2$, it follows that $b - c = \frac{1}{4}$. Because $-\frac{a}{b-c} = 5$, we now get that $a = -\frac{5}{4}$. From the equality $-4 = \frac{a^2 + b^2 - c^2}{2(b-c)}$ we get $a^2 + b^2 - c^2 = -4 \cdot 2(b - c)$. So $b^2 - c^2 = -a^2 - 4 \cdot 2(b - c) = -\frac{25}{16} - \frac{8}{4} = -\frac{57}{16}$. Use of the factorization $b^2 - c^2 = (b + c)(b - c)$ shows that $b + c = -\frac{57}{16} \cdot 4 = -\frac{57}{4}$. Since $b - c = \frac{1}{4}$, it is now easy to solve for b and c to get $b = -7$ and $c = -7\frac{1}{4}$. It follows from equation (∗) that $y = 2x^2 + 5x - 4$ is the equation of the parabola with directrix $y = -7\frac{1}{4}$ and focus $\left(-\frac{5}{4}, 7\right)$.

Consider the equation $y = Ax^2 + Bx + C$, with $A \neq 0$. For very large positive or negative x the term Ax^2 will dominate the others. So if $A > 0$, then y is positive when $|x|$ is large, and if $A < 0$, then y is negative when $|x|$ is large. It follows that the parabola $y = Ax^2 + Bx + C$ opens upward if $A > 0$ and downward if $A < 0$.

Example 4.4. Draw the graph of the parabola $y = 2x^2$.

Solution. Set up a table of values, plot points, and join them by a smooth curve to obtain the appropriate graph in Figure 4.16, which also shows two other parabolas of the form $y = Ax^2$.

Example 4.5. The parabola $y = x^2 + 3x + 4$ opens upward. For which value of x does y attain its smallest value?

Solution. Complete the square to rewrite the equation as

$$y = x^2 + 3x + 4 = x^2 + 3x + \left(\frac{3}{2}\right)^2 - \left(\frac{3}{2}\right)^2 + 4$$

$$= \left(x + \frac{3}{2}\right)^2 + 4 - \frac{9}{4}.$$

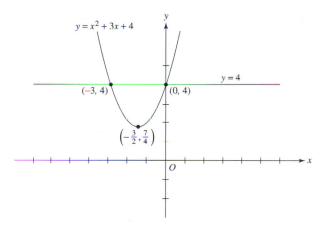

Figure 4.17

So $y = \left(x + \frac{3}{2}\right)^2 + \frac{7}{4}$. Notice that the smallest possible value of y is $y = \frac{7}{4}$ and it occurs for $x = -\frac{3}{2}$.

Example 4.6. Cut the parabola $y = x^2 + 3x + 4$ with the horizontal line $y = 4$ and compute the area of the resulting parabolic section. See Figure 4.17.

Solution. This can be done using Archimedes's theorem. By Example 4.5 $\left(-\frac{3}{2}, \frac{7}{4}\right)$ is the lowest point on the parabola. Since the tangent line hugs the curve near the point of contact, the tangent to the parabola at $\left(-\frac{3}{2}, \frac{7}{4}\right)$ must be horizontal. So this tangent is parallel to the cut $y = 4$. It follows that $\left(-\frac{3}{2}, \frac{7}{4}\right)$ is the vertex of the parabolic section. (See Section 3.1.) The points of intersection of the parabola and the line $y = 4$ are needed next. If (x, y) is on both the parabola and $y = 4$, then $y = x^2 + 3x + 4$ and $y = 4$. So $x^2 + 3x + 4 = 4$, and $x^2 + 3x = 0$. By factoring, $x^2 + 3x = x(x + 3) = 0$, so $x = 0$ or $x = -3$. Therefore the points of intersection are $(-3, 4)$ and $(0, 4)$. By Archimedes's theorem, the area of the parabolic section is $\frac{4}{3}$ times that of the inscribed triangle. Refer to Figure 4.17 again. The inscribed triangle has base 3 and height $4 - \frac{7}{4} = \frac{9}{4}$; so the triangle has area $\frac{1}{2} \cdot 3 \cdot \frac{9}{4} = \frac{27}{8}$. Thus the area of the parabolic section is $\frac{4}{3} \cdot \frac{27}{8} = \frac{9}{2}$.

Consider the general equation $y = Ax^2 + Bx + C$ of the parabola. If the variables x and y are interchanged, then the roles of the x-axis and the y-axis are interchanged. It follows that any parabola with vertical directrix has an equation of the form $x = Ay^2 + By + C$ with $A \neq 0$. It opens to the right if $A > 0$ and to the left if $A < 0$.

In the same way that curves in the Cartesian plane are represented by algebraic equations, regions in the plane are given by algebraic expressions involving inequalities.

Example 4.7. Describe and sketch the regions given by the following sets of points in the Cartesian plane.

(a)   The set $\{(x,y) \mid x \geq 1\}$. The notation means that this is "the set of all points (x,y) with the property that $x \geq 1$."

(b)   The set $\{(x,y) \mid |x| < 1 \text{ and } |y| < 1\}$. This is "the set of all points (x,y) with the property that $|x| < 1$ and $|y| < 1$."

Solution. (a) All points on or to the right of the line $x = 1$. (b) All points inside the box formed by the lines $x = 1$, $x = -1$, $y = 1$, and $y = -1$.

Example 4.8. Let $P = (x,y)$ be a point in the plane. What relationship between x and y places P into the parabolic section of Example 4.6?

Solution. The parabola has equation $y = x^2 + 3x + 4$. Refer back to Figure 4.17 and start with any point (x,y) on the parabola. So $y = x^2 + 3x + 4$. Increasing the y-coordinate moves the point above the parabola. It follows that the points (x,y) on or above the parabola are precisely those with $y \geq x^2 + 3x + 4$. Therefore, the points $P = (x,y)$ in question are those that satisfy $4 \geq y \geq x^2 + 3x + 4$. So the parabolic section is $\{(x,y) \mid x^2 + 3x + 4 \leq y \leq 4\}$.

4.4 Circles and Trigonometry

$\boxed{C}$ onsider a circle with center $C = (3,2)$ and radius 4. By definition, a point $P = (x,y)$ in the plane

is on the circle precisely if the distance from P to the center $(3,2)$ is equal to 4. See Figure 4.18. Put another way, $P = (x,y)$ is on the circle exactly when

$$PC = \sqrt{(x-3)^2 + (y-2)^2} = 4.$$

Squaring both sides, we see that this is so precisely when

$$(x-3)^2 + (y-2)^2 = 16.$$

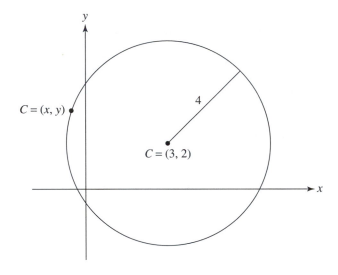

Figure 4.18

Proceeding in the same way, a point $P = (x,y)$ is on the circle with radius r and center (h,k) exactly when its coordinates satisfy

$$(x-h)^2 + (y-k)^2 = r^2.$$

This is an equation of the circle with radius r and center (h,k). If the center is the origin $O = (0,0)$, the equation is $x^2 + y^2 = r^2$.

Example 4.9. Find an equation of the circle with radius 3 and center $(2, -5)$.

Solution. Putting $r = 3$, $h = 2$, and $k = -5$, we obtain $(x-2)^2 + (y+5)^2 = 9$.

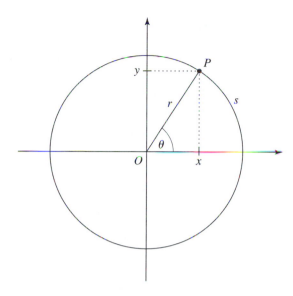

Figure 4.19

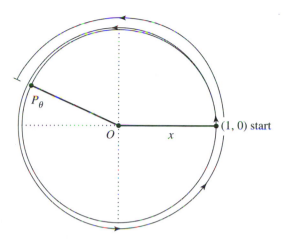

Figure 4.20

Example 4.10. Show that the graph of the equation $x^2 + y^2 + 2x - 6y + 7 = 0$ is a circle and find its center and radius.

Solution. Rewrite the equation as

$$(x^2 + 2x) + (y^2 - 6y) = -7.$$

After completing the square for each variable and adding the appropriate constants to the right side of the equation, this equation becomes

$$(x^2 + 2x + 1) + (y^2 - 6y + 9) = -7 + 1 + 9,$$

or

$$(x + 1)^2 + (y - 3)^2 = 3.$$

Comparing this equation with the standard equation of the circle, we see that $h = -1$, $k = 3$, and $r = \sqrt{3}$. So the given equation is that of a circle with center $(-1, 3)$ and radius $\sqrt{3}$.

Consider a circle of radius r with center the origin. Let $P = (x, y)$ be a point on the circle in the first quadrant. Draw a radius to P and let θ be its angle with the positive x-axis. Denote by s the length of the arc that this angle cuts from the circumference; see

Figure 4.19. Observe that $\sin \theta = \frac{y}{r}$, $\cos \theta = \frac{x}{r}$, and that $\theta = \frac{s}{r}$ (in radians). So

$$x = r \cos \theta \quad \text{and} \quad y = r \sin \theta.$$

When $r = 1$, these relationships have the simpler form

$$x = \cos \theta, \quad y = \sin \theta, \quad \text{and} \quad \theta = s.$$

We will therefore take $r = 1$. This circle with center at the origin and radius 1 is called the *unit circle*. Its equation is $x^2 + y^2 = 1$ and its circumference is 2π.

So far—see Chapter 1—only angles θ with radian measure $0 \le \theta \le 2\pi$ have been considered. In fact, in the definitions of $\sin \theta$ and $\cos \theta$ it was required that $0 \le \theta \le \frac{\pi}{2}$. We will now expand the concepts of angle, sine, and cosine. After our discussion is complete, we will have achieved the following: Any real number θ can be interpreted as an angle, and $\sin \theta$ and $\cos \theta$ will make sense for any real number θ.

Let θ be any real number. Assume first that $\theta \ge 0$. Measure off a distance of θ along the perimeter of the unit circle: start at the point $(1, 0)$ and proceed in a *counterclockwise* direction. Think of winding a string of length θ around a cylinder of radius 1. It may be necessary to go around the circle many times. After the entire distance θ is rolled out, place a point on the circle to mark the measurement and label the point P_θ. Recall that the circumference of the unit circle is

2π; so $\frac{\pi}{2}$ is the length of a quarter of the circle. To measure off $\theta = \frac{\pi}{2}$, start at $(1, 0)$, go around (always counterclockwise) the first quarter of the circle, and stop at $P_\theta = (0, 1)$. To measure off $\theta = \pi = 2 \cdot \frac{\pi}{2}$, go around the first two quarters of the circle and stop at $P_\theta = (-1, 0)$. For $\theta = 1\frac{1}{2}\pi = 3 \cdot \frac{\pi}{2}$, go around the first three quarters to $P_\theta = (0, -1)$. For $\theta = 2\pi = 4 \cdot \frac{\pi}{2}$, go around all four quarters to $P_\theta = (1, 0)$. Note that this was the starting point. For $\theta = 2\frac{1}{2}\pi = 5 \cdot \frac{\pi}{2}$ continue for one more quarter, for a total of five quarters, to $P_\theta = (0, 1)$. For $\theta = 5\pi = 10 \cdot \frac{\pi}{2}$, go around ten quarters. After four quarters, and again after eight quarters, you'll reach the starting point $(1, 0)$, then you'll stop at $P_\theta = (-1, 0)$. Consider $\theta = 8.691$. Since $5 \cdot \frac{\pi}{2} \approx 7.854 \leq 8.691 \leq 9.425 \approx 6 \cdot \frac{\pi}{2}$, go around the perimeter for five quarters to $(0, 1)$, and then for another $8.691 - \frac{5}{2}\pi \approx 0.837$ units past this point to get to P_θ. See Figure 4.20. It should now be clear how any positive number θ can be interpreted as an angle: it is the opening generated by the segment from O to $(1, 0)$ as it rotates from $(1, 0)$ to the point P_θ.

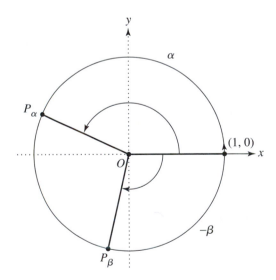

Figure 4.21

If θ is negative, then $-\theta$ is positive. To get the point P_θ in this case, measure off $-\theta$ on the circumfer-

ence in a *clockwise* direction, again starting at $(1, 0)$. Therefore, when the segment from O to $(1, 0)$ rotates counterclockwise, it generates *positive* angles and when it rotates in a clockwise fashion, it generates *negative* angles. In either case, we will refer to the real number θ as the radian measure of the angle obtained in this way. Convince yourself that for $0 \leq \theta \leq 2\pi$ this use of the concept radian measure is the same as that already introduced in Section 1.2. Figure 4.21 gives an example of a positive angle α and a negative angle β. What is the radian measure of the angle obtained by letting the segment from O to $(1, 0)$ rotate clockwise for three and a quarter revolutions? Since three and a quarter revolutions are $3\frac{1}{4} \times 4 = 13$ quarter turns, this angle is $-\frac{13}{2}\pi$. All angles can be given in degrees also. For example, the angle just discussed is equal to $\left(-3\frac{1}{4}\right) \cdot 360° = -1170°$. However, in this book, angles will virtually always be given in radians.

We have defined angles in terms of the length of the arc that they cut from the unit circle. A circle of any radius r can also be used for this purpose, but then division by r is necessary.

We now turn to the definition of $\sin\theta$ and $\cos\theta$ for any real number θ. Continue to consider the unit circle. Take θ and locate the point P_θ. Let the coordinates of P_θ be x and y and define

$$\cos\theta = x \quad \text{and} \quad y = \sin\theta$$

as illustrated in Figure 4.22. To repeat, $\sin\theta$ and $\cos\theta$ are defined to be the y-coordinate and the x-coordinate of the point P_θ, respectively. For $0 \leq \theta \leq \frac{\pi}{2}$, this agrees with the definition in Section 1.4. Refer to Figure 1.16 in that section and take $h = 1$.

Suppose now that we are given a circle of radius r and a point Q on the circle and that θ is the angle between the radius OQ and the positive x-axis. What are the coordinates of Q? Consider the corresponding point $P_\theta = (\cos\theta, \sin\theta)$ on the unit circle. See Figure 4.23. By similar triangles,

$$\frac{x}{\cos\theta} = \frac{y}{\sin\theta} = \frac{r}{1} = r.$$

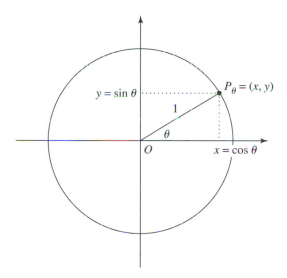

Figure 4.22

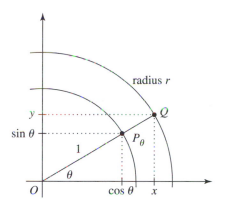

Figure 4.23

Example 4.11. Let θ be any real number. Observe that P_θ and $P_{\theta+2\pi}$ are the same point. So

$$\sin(\theta + 2\pi) = \sin\theta \quad \text{and} \quad \cos(\theta + 2\pi) = \cos\theta.$$

Why are these formulas valid with -2π in place of 2π? Note that the points P_θ and $P_{\theta+\pi}$ are on the opposite ends of a diagonal of the unit circle. After thinking for a moment (and perhaps making a sketch of the situation), you will see that

$$\sin(\theta + \pi) = -\sin\theta \quad \text{and} \quad \cos(\theta + \pi) = -\cos\theta.$$

Why are these formulas valid with $-\pi$ in place of π? Think about how the points P_θ and $P_{-\theta}$ are related and observe that $\sin(-\theta) = -\sin\theta$ and $\cos(-\theta) = \cos\theta$.

Therefore, the coordinates of Q are

$$x = r\cos\theta \quad \text{and} \quad y = r\sin\theta.$$

The sine and cosine satisfy a number of basic formulas. These can often be "read off" from Figure 4.22. Let θ be any real number. Since the point $P_\theta = (\cos\theta, \sin\theta)$ is on the unit circle, it satisfies the equation $x^2 + y^2 = 1$. Therefore,

$$\sin^2\theta + \cos^2\theta = 1.$$

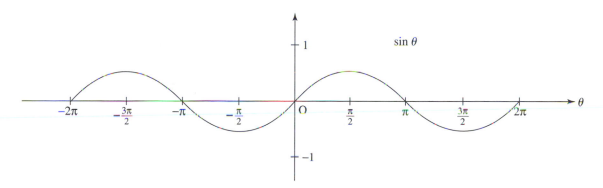

Figure 4.24

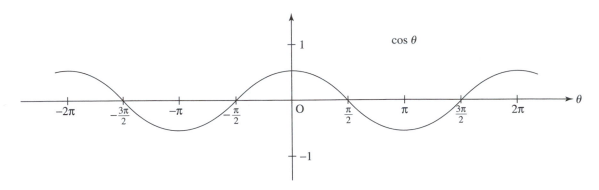

Figure 4.25

Continue to refer to Figure 4.22. Notice that as θ varies from 0 to $\frac{\pi}{2}$, $\sin \theta$ varies from 0 to 1; as θ moves from $\frac{\pi}{2}$ to π, $\sin \theta$ varies from 1 to 0; and so on. Plotting the various points $(\theta, \sin \theta)$ provides the graph in Figure 4.24. Doing the same thing for the cosine, we get the graph in Figure 4.25.

Now turn to $\tan \theta = \frac{\sin \theta}{\cos \theta}$. Suppose first that $0 \leq \theta \leq \frac{\pi}{2}$. If $\theta = 0$, then $\tan = 0$. Again turn to Figure 4.22. As θ increases, the sine increases and the cosine decreases. Since both are positive, $\tan \theta$ increases. When θ is close to $\frac{\pi}{2}$, the sine is close to 1 and the cosine is close to 0, so $\tan \theta$ is very large. If $\theta = \frac{\pi}{2}$, then $\cos \theta = 0$, so that $\tan \theta$ is not defined. When $-\frac{\pi}{2} \leq \theta \leq 0$, the situation is similar. Since

the sine is negative and the cosine is positive, $\tan \theta$ is now negative. The rest of the graph of the tangent is simply a repitition of the pattern for $-\frac{\pi}{2} \leq \theta \leq \frac{\pi}{2}$. It is sketched in Figure 4.26.

Example 4.12. Let θ be any angle. Since $\sin(\theta + \pi) = -\sin \theta$ and $\cos(\theta + \pi) = -\cos \theta$, it follows that $\tan(\theta + \pi) = \tan \theta$. Similarly, $\tan(-\theta) = -\tan \theta$.

4.5 The Ellipse

C onsider an ellipse given by the focal points F_1 and F_2 and the constant $k \geq F_1 F_2$. Recall from

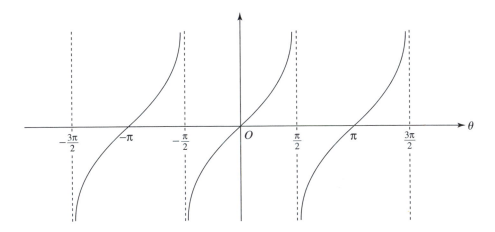

Figure 4.26

Section 3.1 that k is the sum of the distance from a point on the ellipse to F_1 and the distance from that point to F_2. Move the ellipse in the plane in such a way that F_1 and F_2 lie on the x-axis equidistant from the origin. Refer to Figure 4.27. Then the origin is the center of the ellipse. The ellipse is now in what we call *standard position*. Note that in this position, there is a number $e \geq 0$ such that

$$F_1 = (e, 0) \quad \text{and} \quad F_2 = (-e, 0)$$

and $k \geq 2e$. We already know from Section 3.1 that a point $P = (x, y)$ falls on the ellipse exactly when $PF_1 + PF_2 = k$. The concern will now be the precise conditions on the coordinates x and y of P that will guarantee this. Let $a = \frac{k}{2}$. So $2a = k \geq 2e$, and hence $a \geq e$. If $a = e$, then $k = 2e$. In this case, the ellipse consists of the points on the segment F_1F_2. (Why?)

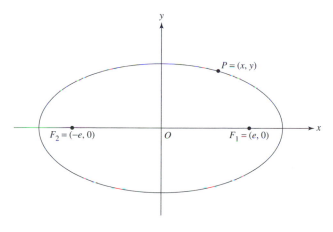

Figure 4.27

Since the ellipse consists of all points $P = (x, y)$ such that $PF_1 + PF_2 = k$, the point $P = (x, y)$ is on the ellipse precisely if

$$\sqrt{(x - e)^2 + (y - 0)^2} + \sqrt{(x - (-e))^2 + (y - 0)^2} = 2a.$$

After simplifying, squaring both sides, canceling, moving things around, and simplifying once more, this

equation is transformed in successive steps to

$$\sqrt{(x - e)^2 + y^2} + \sqrt{(x + e)^2 + y^2} = 2a,$$

$$\sqrt{(x - e)^2 + y^2} = 2a - \sqrt{(x + e)^2 + y^2},$$

$$(x - e)^2 + y^2 = 4a^2 - 4a\sqrt{(x + e)^2 + y^2}$$
$$+ (x + e)^2 + y^2,$$

$$(x - e)^2 = 4a^2 - 4a\sqrt{(x + e)^2 + y^2} + (x + e)^2,$$

$$x^2 - 2ex + e^2 = 4a^2 - 4a\sqrt{(x + e)^2 + y^2}$$
$$+ x^2 + 2ex + e^2,$$

$$a\sqrt{(x + e)^2 + y^2} = a^2 + ex,$$

$$a^2\left((x + e)^2 + y^2\right) = a^4 + 2a^2ex + e^2x^2,$$

$$a^2(x^2 + 2ex + e^2 + y^2) = a^4 + 2a^2ex + e^2x^2,$$

$$a^2x^2 + 2a^2ex + a^2e^2 + a^2y^2 = a^4 + 2a^2ex + e^2x^2,$$

$$a^2x^2 - e^2x^2 + a^2y^2 = a^4 - a^2e^2,$$

$$(a^2 - e^2)x^2 + a^2y^2 = a^2(a^2 - e^2).$$

Now set $b = \sqrt{a^2 - e^2}$. Since $b^2 = a^2 - e^2$, this last equation becomes $b^2x^2 + a^2y^2 = a^2b^2$. In the case $a = e$, it was already pointed out that the ellipse consists of the points on the segment F_1F_2. So we will now assume that $a > e$ and hence that $b > 0$. Dividing the equation $b^2x^2 + a^2y^2 = a^2b^2$ through by a^2b^2 now shows that

$$\frac{x^2}{a^2} + \frac{y^2}{b^2} = 1.$$

This is the *standard equation* of the ellipse. Since $b = \sqrt{a^2 - e^2}$ notice that $a \geq b$. Observe that if $a = b$, then this ellipse is a circle with radius a.

Setting $y = 0$ gives $x^2 = a^2$. So the x-intercepts are $\pm a$. It follows that $k = 2a$ is the length of the major axis. Therefore a is equal to the *semimajor* axis. Similarly, the y-intercepts are $\pm b$, the length of the minor axis is $2b$, and b is called the *semiminor* axis.

The distance e between the center and either of the focal points is called the *linear eccentricity* of the ellipse. Refer to Figure 4.28. Since the point $B = (0, b)$ is on the ellipse, $2BF_1 = BF_1 + BF_2 = k = 2a$, and hence $BF_1 = a$. In the same way, $BF_2 = a$. By Pythagoras's theorem, $a^2 = b^2 + e^2$. Hence, $e^2 = a^2 - b^2$. So the linear eccentricity is equal to $e = \sqrt{a^2 - b^2}$. The ratio

$$\varepsilon = \frac{\text{linear eccentricity}}{\text{semimajor axis}} = \frac{e}{a}$$

is the *astronomical eccentricity*. Since $e \le a$, it is clear that $0 \le \varepsilon \le 1$. The constant ε measures the flatness of the ellipse. If $\varepsilon = 0$, then $e = 0$, in which case $F_1 = F_2$ and the ellipse is a circle. If $\varepsilon = 1$ then $e = a$. This is the case where the ellipse is the line segment between the two focal points. The "general ellipse" has a shape between these two extremes. If it is close to being a circle, then b is close to a, the eccentricity e is small, and ε is close to zero. If the ellipse is flat, then b is small relative to a, the quantities e and a are close to each other, and ε is close to 1.

Example 4.13. Show that the graph of $9x^2 + 16y^2 = 144$ is an ellipse. Determine the semimajor and semiminor axes, the two eccentricities, the two focal points, and the constant k.

Solution. Divide both sides of the equation by 144 to get

$$\frac{x^2}{16} + \frac{y^2}{9} = 1.$$

This is the standard equation of an ellipse. Since $a^2 = 16$ and $b^2 = 9$, we get $a = 4$ and $b = 3$. The linear and astronomical eccentricities are, respectively, $e = \sqrt{16 - 9} = \sqrt{7}$ and $\varepsilon = \frac{\sqrt{7}}{4} = 0.66$. The focal points are $(-\sqrt{7}, 0)$ and $(\sqrt{7}, 0)$, and $k = 2a = 8$.

With the ellipse placed with the focal points on the x-axis and the origin at the center of the segment connecting them, its equation—as we have seen—is of the form

$$\frac{x^2}{a^2} + \frac{y^2}{b^2} = 1$$

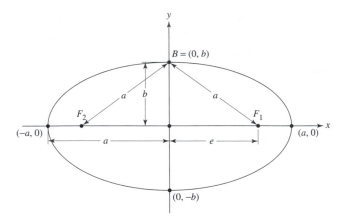

Figure 4.28

with $a \ge b$. If we line up the ellipse so that the focal points are on the y-axis (again with the center of the connecting segment at the origin), then the ellipse has an equation of exactly the same form, but this time with $b \ge a$.

4.6 Cavalieri's Principle

We begin with an important general principle. Kepler made use of it, but it is known today as Cavalieri's principle. Bonaventura Cavalieri (1598–1644) was a student of Galileo.

Consider any two regions with areas C and D. Place both of them above the x-axis as in Figure 4.29. Suppose that for every x, the vertical cross-sectional cut through D has the same length as the vertical cross-sectional cut through C, that is $d_x = c_x$ for all x. In such a situation it is certainly reasonable to conclude that $C = D$. To reach this conclusion regard the two regions as sliced into thin strips. Therefore the underlying principle is the same as that of the *Method of Archimedes*. It is equally reasonable to conclude: If $d_x = 2c_x$ for all x, then $D = 2C$; if $d_x = 3c_x$ for all x, then $D = 3C$; if $d_x = \frac{1}{2}c_x$ for all x, then $D = \frac{1}{2}C$, and so on. We have now arrived at

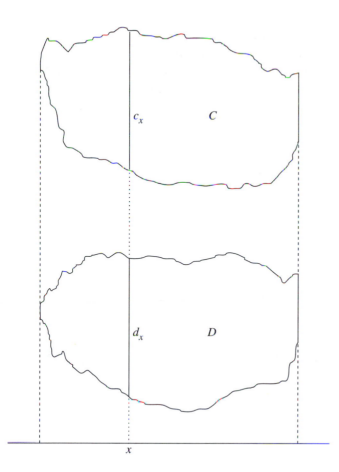

Figure 4.29

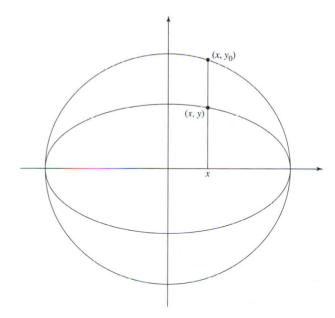

Figure 4.30

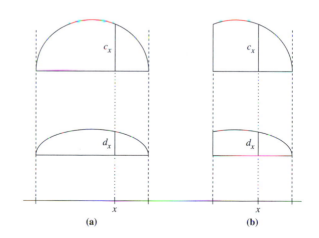

(a)   (b)

Figure 4.31

Cavalieri's principle: If $d_x = kc_x$ for all x and for a fixed positive number k, then $D = kC$.

Observe, in particular, that if the area C and the constant k can both be determined, then the area D can be determined. As an application of Cavalieri's principle, we will compute the area of the ellipse

$$\frac{x^2}{a^2} + \frac{y^2}{b^2} = 1$$

where $a > 0$, $b > 0$, and $a \geq b$. Consider simultaneously the graph of the ellipse and that of the circle $x^2 + y^2 = a^2$. Let x satisfy $-a \leq x \leq a$ and, as shown in Figure 4.30, let (x, y) and (x, y_0) be the indicated points on the ellipse and circle, respectively. Since (x, y_0) satisfies $x^2 + y^2 = a^2$ and $y_0 \geq 0$, it follows that

$y_0 = \sqrt{a^2 - x^2}$. Because (x, y) is on the ellipse,

$$\frac{y^2}{b^2} = 1 - \frac{x^2}{a^2} = \frac{a^2 - x^2}{a^2}$$

$$y^2 = \frac{b^2 a^2 - b^2 x^2}{a^2} = \frac{b^2}{a^2}(a^2 - x^2).$$

and after taking square roots,

$$y = \frac{b}{a}\sqrt{a^2 - x^2} = \frac{b}{a}y_0.$$

Separate the upper semicircles and the upper part of the ellipse, as shown in Figure 4.31(a). Notice that we just demonstrated that $d_x = \frac{b}{a}c_x$ for any x. Since the area of a semicircle of radius a is $\frac{1}{2}\pi a^2$, it follows by Cavalieri's principle that the area of the upper half of the ellipse is equal to $\frac{b}{a}\left(\frac{1}{2}\pi a^2\right) = \frac{1}{2}\pi ab$. Therefore, the full ellipse with semimajor axis a and semiminor axis b has area

$$\pi ab.$$

Note that Cavalieri's principle also applies to Figure 4.31(b). In particular, the area of the elliptical section has area $\frac{b}{a}$ times that of the semicircular section.

Example 4.14. Refer to Figure 4.32. Consider the section QPN of the ellipse $\frac{x^2}{4^2} + \frac{y^2}{3^2} = 1$ and the sector $P'ON$ of the circle $x^2 + y^2 = 4^2$. Determine the area of the elliptical section in terms of the angle β.

Solution. The area of the circular sector $P'ON$ is $\frac{1}{2}\beta \cdot 4^2 = 8\beta$. (See Section 3.2.) Since $OP' = 4$, we have $\sin(\pi - \beta) = \frac{QP'}{4}$ and $\cos(\pi - \beta) = \frac{QO}{4}$. Therefore, the area of the triangle $QP'O$ is $\frac{1}{2}QO \cdot QP' = 8\cos(\pi - \beta)\sin(\pi - \beta)$. Adding the area of the triangle to the area of the sector shows that the area of the circular section $QP'N$ is $8\beta + 8\cos(\pi - \beta)\sin(\pi - \beta)$. Since $a = 4$ and $b = 3$, we find by Cavalieri's principle that the area of the elliptical section QPN is

$$\frac{3}{4} \cdot 8\left[\beta + \cos(\pi - \beta)\sin(\pi - \beta)\right]$$

$$= 6\left[\beta + \cos(\pi - \beta)\sin(\pi - \beta)\right].$$

Making use of Example 4.11, we can conclude that $\cos(\pi - \beta) = \cos(\beta - \pi) = -\cos\beta$ and $\sin(\pi - \beta) = -\sin(\beta - \pi) = \sin\beta$. Therefore the area of the elliptical section can be written as $6(\beta - \sin\beta \cdot \cos\beta)$. What happens if the angle β is less than $\frac{\pi}{2}$? Then the area of the circular section $QP'N$ is equal to the area of the circular sector $P'ON$ minus the area of the

triangle QOP'. Check that the area of this triangle is $8\sin\beta \cdot \cos\beta$. So the area of the circular section is $QP'N$ is $8\beta - 8\sin\beta \cdot \cos\beta$ and by another application of Cavalieri's principle, the area of the elliptical section QPN is $6\beta - 6\sin\beta \cdot \cos\beta$ in this case also.

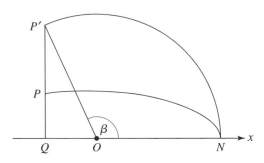

Figure 4.32

4.7 Kepler's Analysis of the Orbits

Let's now turn to Kepler's mathematical analysis of the orbits of the planets. The starting point is Kepler's first law, which states that every planet is in elliptical orbit around the Sun with the Sun at a focal point. The discussion that follows retains the essence of Kepler's work; but it is different in notation as well as in detail, and it adds more recent information.

Recall that one astronomical unit (1 AU) was defined to be equal to the distance between the Earth and the Sun. However, since the Earth's orbit about the Sun is an ellipse, this distance varies, and so more precision is necessary. Let a be the semimajor axis of the Earth's orbit and let e be its linear eccentricity. Since the Sun is at a focus of the ellipse, we see from Figure 4.28 that $a + e$ is the farthest distance from the Earth to the Sun and $a - e$ the closest. So the average distance

$$\frac{(a+e)+(a-e)}{2} = \frac{2a}{2} = a$$

is also equal to the semimajor axis. We will now define 1 AU to be equal to the semimajor axis of

Table 4.2

	Orbital Data of Planets				
Planet	**Semimajor axis in AUs**	**Period of the orbit in years**	**Astronomical eccentricity**	**Angle of orbital plane with that of the Earth**	**Average speed in miles/sec**
Mercury	0.3871	0.2408	0.2056	7.00°	29.6
Venus	0.7233	0.6152	0.0068	3.39°	21.7
Earth	1.0000	1.0000	0.0167	0.00°	18.5
Mars	1.5237	1.8809	0.0934	1.85°	15.0
Jupiter	5.2028	11.8622	0.0484	1.31°	8.1
Saturn	9.5388	29.4577	0.0557	2.49°	6.0

the Earth's orbit. The actual value of 1 AU was not known with any precision until the latter part of the 18th century, when it was shown to be equal to 93,300,000 miles. This was calculated by use of parallax and information about the path that Venus traces out across the Sun. A recent determination by the Jet Propulsion Laboratory in Pasadena, California, is

$$1\ \text{AU} = (499.004784\ \text{seconds}) \times (\text{speed of light})$$

$$= 92,955,807.2\ \text{miles}.$$

It relies on measuring the round-trip travel time of radio waves (radar) from the Earth to the Sun.

Let's return to Kepler's great adversary, the planet Mars. Tycho Brahe's observations, as well as his own, supplied Kepler with essential quantitative information about the orbit. In his *Astronomia Nova* he recorded the farthest distance from Mars to the Sun as 1.6678 AU and the closest as 1.3850 AU. Let a be the semimajor axis of Mars's ellipse and let e be its linear eccentricity. Refer to Figure 4.28 again and observe that $a + e = 1.6678$ AU and $a - e = 1.3850$ AU. So $2a = 3.0528$ AU and $2e = 0.2828$ AU. Hence

$a = 1.5264$ AU and $e = 0.1414$ AU. The astronomical eccentricity of Mars's orbit is $\varepsilon = \frac{e}{a} = 0.0926$. Kepler had precise numerical information about the orbits of the other planets as well.

We will use today's version of these data. See Table 4.2. Only the six planets known at the time of Kepler are listed.

Notice that the astronomical eccentricities are small for all the orbits listed. Since an ellipse with a small astronomical eccentricity is close to being a circle (see Section 4.5), it follows that the orbits of the planets are nearly circular. The orbit of Mercury has the largest astronomical eccentricity and is the most "elliptical."

Example 4.15. Let's continue to look at Mars. Its semimajor axis a is approximately

$$a = (1.5237)(92,960,000)\ \text{miles}$$

$$= 141.64\ \text{million miles}.$$

Since the linear eccentricity is $e = \varepsilon a$, the semiminor axis b equals

$$b = \sqrt{a^2 - e^2} = \sqrt{a^2 - \varepsilon^2 a^2}$$

$$= a\sqrt{1 - (0.0934)^2} = a\sqrt{1 - 0.0087}$$

$$= a\sqrt{0.9913} = 0.9956a.$$

So $b = (0.9956)(141.64) = 141.02$ million miles. Observe that the maximum and minimum distances from Mars to the Sun are, respectively,

$$a + e = a(1 + \varepsilon) = a(1 + 0.0934) = 1.0934a$$

$$= (1.0934)(141.64) = 154.87 \text{ million miles,}$$

$$a - e = a(1 - \varepsilon) = a(1 - 0.0934) = 0.9066a$$

$$= (0.9066)(141.64) = 128.41 \text{ million miles.}$$

Example 4.16. The semimajor axis of the Earth is $a = 1$ AU, and its astronomical eccentricity is $\varepsilon = 0.0167$. It follows that the semiminor axis is

$$b = \sqrt{a^2 - \varepsilon^2 a^2} = \sqrt{1 - (0.0167)^2}$$

$$= \sqrt{1 - 0.0003} = \sqrt{0.9997} = 0.9998 \text{ AU.}$$

The distance from the Earth to the Sun varies from a minimum of

$$a - e = a - \varepsilon a = 1 - 0.0167 = 0.9833 \text{ AU}$$

to a maximum of $a + e = 1.0167$ AU. In miles, the variation is from about 91.41 million miles to about 94.51 million miles.

Examples 4.15 and 4.16 show how close the orbits of both the Earth and Mars are to being circles. Note that the semimajor and semiminor axes are nearly equal in each case. It seems remarkable that Kepler was able to conclude that they were not circular. The variations in the maximum and minimum distances from the Sun are significant, however, especially in the case of Mars. Therefore, the issue is not so much the fact that the orbits are elliptical, but rather that the

Sun is off-center at a focus. This is the reason, in retrospect, why Ptolemy's off-centered circle simulates the Earth's orbit around the Sun reasonably well (see Section 2.5).

Kepler's first law, in combination with the data of Table 4.2 and the analysis of the ellipse in Section 4.5, gives a precise picture of the shape of the orbits of the planets. Now, however, comes a fundamental question: How exactly, does a given planet trace out its orbit? In other words, where in its orbit is a planet at a given time?

This question can be reformulated with more precision as follows. Refer to Figure 4.33, which depicts a planet in orbit around the Sun, S. Suppose that at a certain time, which we will designate as $t = 0$, the planet is at the perihelion position N, the position closest to the Sun. Let $P = (x, y)$ be the position of the planet at some later time t (in Earth days, for example). Let r be the distance from S to P, and let α be the indicated angle. The problem now is this: can both α and r, and therefore the position P of the planet, be determined in terms of the elapsed time t? Reflecting over Kepler's second law suggests that this should be possible. And indeed it is, as we shall now see! The key to the answer to this problem is a construction that was already employed in the computation of the area of the ellipse. Surrounding the elliptical orbit with a circle (which is purely a mathematical tool with no physical significance) will allow us to separate a difficult problem into manageable components.

Let's begin by placing a Cartesian coordinate system in such a way that the ellipse is in standard position with the focus S at $(e, 0)$ on the positive x-axis. So the equation of the planet's orbit is

$$\frac{x^2}{a^2} + \frac{y^2}{b^2} = 1,$$

where, as before, a and b are the semimajor and semiminor axes, respectively. Recall that $b = \sqrt{a^2 - e^2}$. So $a \geq b$, $b^2 = a^2 - e^2$ and $e^2 = a^2 - b^2$.

We will now supplement Figure 4.33 as follows. Refer to Figure 4.34. Let the coordinates of the point P be (x, y) and let $X = (x, 0)$ be the point on the x-

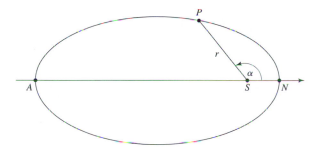

Figure 4.33

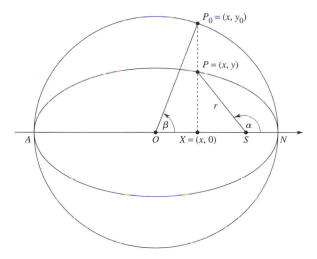

Figure 4.34

axis directly below P. Now superimpose the circle $x^2 + y^2 = a^2$ of radius a, let $P_0 = (x, y_0)$ be the point on the circle directly above P, and let β be the angle determined by ON and OP_0. Finally, let A be the aphelion position of the planet's orbit. Observe that this is the position where its distance from the Sun is greatest.

The solution of the problem requires three steps. Step 1 expresses both r and α in terms of the angle β. Step 2 produces an equation that links β to the time t. Step 3, finally, solves this equation for β in terms of t. The three steps together determine r and α in terms of t. The planet's position P is then given in terms of t, and Kepler's objective is achieved.

Let's turn to Step 1. Note from Figure 4.34 that $\cos \beta = \frac{OX}{OP_0} = \frac{x}{a}$, so that

$$x = a \cos \beta.$$

Similarly, $\sin \beta = \frac{y_0}{a}$, and hence $y_0 = a \sin \beta$. Since $y = \frac{b}{a} y_0$, as already observed in Section 4.6,

$$y = b \sin \beta.$$

Applying the Pythagorean theorem to the triangle SPX and using the equalities $SX = e - x, e^2 = a^2 - b^2$ and $\sin^2 \beta + \cos^2 \beta = 1$, Kepler calculated as follows:

$$r^2 = y^2 + (e - x)^2$$

$$= b^2 \sin^2 \beta + (e - a \cos \beta)^2$$

$$= b^2 \sin^2 \beta + a^2 \cos^2 \beta - 2ae \cos \beta + e^2$$

$$= (a^2 - e^2) \sin^2 \beta + a^2 \cos^2 \beta - 2ae \cos \beta + e^2$$

$$= a^2 - e^2 \sin^2 \beta - 2ae \cos \beta + e^2$$

$$= a^2 + e^2(1 - \sin^2 \beta) - 2ae \cos \beta$$

$$= a^2 + e^2 \cos^2 \beta - 2ae \cos \beta$$

$$= (a - e \cos \beta)^2.$$

Since $a > e \geq e \cos \beta$, the term $a - e \cos \beta$ is positive. Because r is also positive, we can take the square root of both sides of the equality $r^2 = (a - e \cos \beta)^2$ just derived to get $r = a - e \cos \beta$. Factor out a to get $r = a(1 - \frac{e}{a} \cos \beta)$. Recalling that the astronomical eccentricity is $\varepsilon = \frac{e}{a}$, we see that Kepler has derived the equation

$$\boxed{r = a(1 - \varepsilon \cos \beta)}$$

Therefore, r is now determined in terms of β.

By Example 4.12 in Section 4.4,

$$\tan \alpha = -\tan(-\alpha) = -\tan(\pi - \alpha).$$

Another look at the triangle SPX now shows that

$$\tan \alpha = -\tan(\pi - \alpha) = -\frac{y}{e - x} = \frac{b \sin \beta}{a \cos \beta - e}.$$

With the substitution $a\varepsilon = e$, this becomes

$$\tan \alpha = \frac{b \sin \beta}{a(\cos \beta - \varepsilon)}.$$

Thus Kepler has determined $\tan \alpha$, and hence α, in terms of β. It is important to observe that the formulas for both r and $\tan \alpha$ are valid for any position of P along its orbit. This follows from the basic trig formulas of Examples 4.11 and 4.12 of Section 4.4 for each of the three cases (P in the second, third, and fourth quadrants) that must be checked. Refer to Example 4.14, where this was done in a related situation.

Starting with the expression for $\tan \alpha$ that was just determined (and using lots of trigonometry), it is possible to derive the equation

$$\tan \frac{\alpha}{2} = \sqrt{\frac{1 + \varepsilon}{1 - \varepsilon}} \tan \frac{\beta}{2}$$

This formula is easier to use because it involves fewer computations. It is known as Gauss's formula. Its discoverer, the German mathematician-astronomer Carl Friedrich Gauss (1777–1855) is considered to be, along with Archimedes and Newton, the most famous and brilliant of all mathematicians.[4] Exercises 4F will "walk you through" the derivation of this formula.

Step 2 follows next. Recall that its purpose is to connect the angle β with the elapsed time t. We will see that this connection is a consequence of Kepler's second law.

Let A_t be the area swept out by the segment SP during time t the planet moves from N to P. See Figure 4.35. Step 2 begins with the computation of A_t in terms of β.

Refer back to Figure 4.34 and note that

Area section P_0XN

$$= \text{Area sector } P_0ON - \text{Area } \triangle P_0XO$$

Because P_0ON is a circular sector with radius a and angle β, its area is $\frac{1}{2}a^2\beta$. See Section 3.2. Since the area of the right triangle P_0XO is $\frac{1}{2}xy_0$ and $y_0 = a \sin \beta$, it

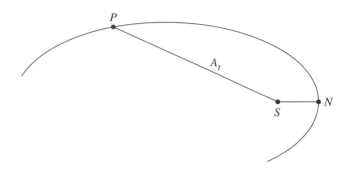

Figure 4.35

follows that

$$\text{Area section } P_0XN = \frac{1}{2}\beta a^2 - \frac{1}{2}xa \sin \beta.$$

Refer next to Section 4.6, in particular to Figure 4.31(b) and deduce that the elliptical section PXN has area $\frac{b}{a}$ times that of the section P_0XN just computed. Therefore,

$$\text{Area section } PXN = \frac{b}{a}\left(\frac{1}{2}\beta a^2 - \frac{1}{2}xa \sin \beta\right)$$

$$= \frac{1}{2}\beta ab - \frac{1}{2}xb \sin \beta.$$

Now return to A_t and observe that

$$A_t = \text{Area section } PXN - \text{Area } \triangle PXS$$

$$= \left(\frac{1}{2}\beta ab - \frac{1}{2}xb \sin \beta\right) - \frac{1}{2}(e - x)y$$

$$= \left(\frac{1}{2}\beta ab - \frac{1}{2}xb \sin \beta\right) - \frac{1}{2}(e - x)b \sin \beta$$

$$= \frac{1}{2}\beta ba - \frac{1}{2}eb \sin \beta = \frac{1}{2}\beta ab - \frac{1}{2}\varepsilon ab \sin \beta$$

$$= \frac{ab}{2}(\beta - \varepsilon \sin \beta).$$

The quantity A_t can be computed in another way. As consequence of Kepler's second law (refer to Section 4.1), the ratio $\frac{A_t}{t}$ is the same for any t. It can therefore be evaluated by taking t to be the period T of the orbit of the planet. Since the area of the full ellipse is $ab\pi$, it follows that $\frac{A_t}{t} = \frac{ab\pi}{T}$. So $A_t = \frac{ab\pi t}{T}$. By setting the two expressions for A_t equal to each other, $\frac{ab}{2}(\beta - \varepsilon \sin \beta) = \frac{ab\pi t}{T}$. After a simplification,

$$\beta - \varepsilon \sin \beta = \frac{2\pi t}{T}$$

This is a famous equation, known appropriately as Kepler's equation. It provides the link between t and β. It should be noted that Kepler's equation and its derivation are valid for any position P of the planet and any elapsed time t (even though our argument relies on Figures 4.34 and 4.35, which place the planet in the first quadrant of its first orbit). Step 2 is now complete.

The problem that remains is the solution of Kepler's equation for β in terms of t. This objective of Step 3 is taken up in the next section.

4.8 **The Method of Successive Approximations**

T o solve the equation $\beta - \varepsilon \sin \beta = \frac{2\pi t}{T}$ for β is not easy. To do so, we will use the method of *successive approximations*. If you have ever determined your weight on a scale that has a sliding pointer, then you already understand the essence of this method. The start is made with an educated guess (an initial estimate of your weight), and then an approximation step (slide the pointer) is used to successively close in on the desired value (your weight). The game "played with squares" in Section 3.3 is also an example of the method of successive approximations. There the educated guess (not so educated, actually) is the choice of the black L as a first approximation of the area of the square. The approximation step is this: reduce the dimension of L by a factor of four and insert this reduced L into the white square that remains. Applied over and over, this procedure homes in on the area of the square.

We will illustrate the method of successive approximations in another context. Suppose that $p(x)$ is

a polynomial. For example, take

$$p(x) = x^4 + 2x^3 - 12x - 1.$$

Notice that $p(0) = -1$ and that

$$p(2) = 2^4 + 2 \cdot 2^3 - 12 \cdot 2 - 1 = 16 + 16 - 24 - 1 = 7.$$

Since $p(0)$ is negative and $p(2)$ is positive, it follows that $p(c) = 0$ for some number c between 0 and 2. Is there a way to determine c? Figure 4.36 illustrates how this can be done.

1. Make an educated guess, say c_1, for c.
2. Use the following approximation step: from a point on the x-axis go up to the graph of $p(x)$ and down on the tangent to a new point on the x-axis.

Begin by applying the approximation step to the guess c_1 to get c_2; apply the approximation step to c_2 to get c_3; apply it to c_3 to get $c_4, \dots$. Continue in this way, and observe that the sequence of numbers $c_1, c_2, c_3, c_4, \dots$ zeroes in on the root c. This method of determining the root of a polynomial is known as *Newton's method*. It can be made numerically precise (Exercise 8K of Chapter 8 describes how this is done), but this is not the point now.

We will now turn to Step 3 in the determination of the position of the planet in its orbit after a given

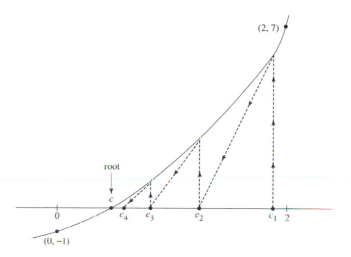

Figure 4.36

elapsed time t. The task in this final step is to consider the time t as given and to solve the equation

$$\beta - \varepsilon \sin \beta = \frac{2\pi t}{T}$$

for β in terms of t. A key ingredient in the solution is the inequality

$$|\sin \varphi - \sin \theta| \leq |\varphi - \theta|$$

for any two angles φ and θ in radian measure. The validity of this inequality ultimately depends on calculus (it will be verified in Section 8.7), but it can be made plausible as follows. Let $(\theta, \sin \theta)$ and $(\varphi, \sin \varphi)$ be any two points on the graph of the sine. Figure 4.37 shows two such pairs of points. In each case, let v and h be the vertical and the horizontal distance between the points. Since the graph of the sine is rather flat, it turns out that the vertical distance v is less than or equal to the horizontal distance h. This is so regardless of where the two points are taken on the graph. Therefore $v \leq h$ in all cases. From Section 4.3 we know that $v = |\sin \varphi - \sin \theta|$ and $h = |\varphi - \theta|$. Therefore,

$$|\sin \varphi - \sin \theta| \leq |\varphi - \theta|$$

for any θ and φ.

We now solve $\beta - \varepsilon \sin \beta = \frac{2\pi t}{T}$ for β by the method of successive approximations.

1. The first stab at β is $\beta_1 = \frac{2\pi t}{T}$. Since $|\beta - \beta_1| = |\beta - \frac{2\pi t}{T}| = |\varepsilon \sin \beta| \leq \varepsilon$ and because the astronomical eccentricity $\varepsilon = \frac{e}{a}$ is small (see Table 4.2), β_1 approximates β. So $\beta_1 = \frac{2\pi t}{T}$ is a good educated guess.

2. The approximation step is this: place an angle (in radians) into the "blank" of the expression $\frac{2\pi t}{T} + \varepsilon \sin(\ \ )$ to get a new angle (in radians).

    Start by putting $\beta_1 = \frac{2\pi t}{T}$ into the blank. This gives the new angle $\beta_2 = \frac{2\pi t}{T} + \varepsilon \sin \beta_1$. Repeating (2) with β_2, we get $\beta_3 = \frac{2\pi t}{T} + \varepsilon \sin \beta_2$. Doing this again and again, we get $\beta_4 = \frac{2\pi t}{T} + \varepsilon \sin \beta_3$, $\beta_5 = \frac{2\pi t}{T} + \varepsilon \sin \beta_4, \dots$, and $\beta_i = \frac{2\pi t}{T} + \varepsilon \sin \beta_{i-1}, \dots$. The big question is this: Does the sequence of numbers

$$\beta_1, \ \beta_2, \ \beta_3, \dots, \ \beta_i, \dots$$

close in on the solution β of $\beta - \varepsilon \sin \beta = \frac{2\pi t}{T}$? This is indeed the case. Notice that $\beta = \frac{2\pi t}{T} + \varepsilon \sin \beta$. It follows that

$$\beta - \beta_2 = \left(\frac{2\pi t}{T} + \varepsilon \sin \beta\right) - \left(\frac{2\pi t}{T} + \varepsilon \sin \beta_1\right)$$

$$= \varepsilon(\sin \beta - \sin \beta_1).$$

Therefore, using (1),

$$|\beta - \beta_2| = \varepsilon|\sin \beta - \sin \beta_1| \leq \varepsilon|\beta - \beta_1| \leq \varepsilon^2.$$

In the same way,

$$\beta - \beta_3 = \left(\frac{2\pi t}{T} + \varepsilon \sin \beta\right) - \left(\frac{2\pi t}{T} + \varepsilon \sin \beta_2\right)$$

$$= \varepsilon(\sin \beta - \sin \beta_2),$$

so that

$$|\beta - \beta_3| = \varepsilon|\sin \beta - \sin \beta_2| \leq \varepsilon|\beta - \beta_2| \leq \varepsilon^3.$$

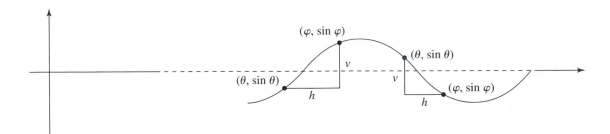

Figure 4.37

Repeating this computation again and again shows that

$$|\beta - \beta_4| \le \varepsilon^4, \ |\beta - \beta_5| \le \varepsilon^5, \ldots, \ |\beta - \beta_i| \le \varepsilon^i, \ldots .$$

Since ε is small, the powers ε^2, ε^3, $\varepsilon^4, \ldots$ become progressively smaller. Therefore, the distances $|\beta - \beta_1|$, $|\beta - \beta_2|$, $|\beta - \beta_3|, \ldots$ between β_1, β_2, $\beta_3, \ldots$ and β get progressively smaller. So the numbers β_1, β_2, $\beta_3, \ldots$ close in on β as required. Since $0 \le \varepsilon < 1$ for any ellipse, this successive approximation process will always converge to a solution β. If ε is small (and this is so for all the planets) it will converge very quickly.

Let's consider the case of the Earth. Here

$$\varepsilon = 0.0167, \ \varepsilon^2 = 0.000279,$$

$$\varepsilon^3 = 0.00000467, \ldots, \ \varepsilon^6 = 2.17 \times 10^{-11}.$$

Since $|\beta - \beta_6| \le \varepsilon^6 < 10^{-10}$, the sixth approximation β_6 of β is already an extremely good estimate of β.

4.9 Computing Orbital Information

We will now compute the position of the Earth at vernal equinox and at summer solstice of 1994 using the method developed in Sections 4.7 and 4.8. All of the data we use was supplied by the U.S. Naval Observatory. We will work with an accuracy of four decimal places. The Earth's orbit is $T = 365.2422$ days and the four seasons have the following lengths:

Spring: 92 days, 18 hours, 20 minutes
          = 92.7639 days.
Summer: 93, 15 hours, 31 minutes
          = 93.6465 days.
Autumn: 89 days, 20 hours, 4 minutes
          = 89.8361 days.
Winter: 88 days, 22 hours, 54 minutes
          = 88.9938 days.

In 1994, perihelion occurred on January 2 at 5 hours 55 minutes Greenwich Mean Time (GMT).[5] This is $t = 0$. Since vernal equinox occurred that year on March 20 at 20 hours 28 minutes, the elapsed time in days from perihelion to vernal equinox for 1994 was

$$t_{ve} = 77 \text{ days, 14 hours, 33 minutes} = 77.6023 \text{ days.}$$

Adding the 92.7639 days of spring to this shows that the elapsed time to summer solstice is

$$t_{ss} = 77.6023 \text{ days} + 92.7639 \text{ days} = 170.3662 \text{ days.}$$

Adding the lengths of summer and autumn gives the elapsed times to autumnal equinox and winter solstice as $t_{ae} = 265.0132$ days and $t_{ws} = 354.8493$ days, respectively.

To compute the position of the Earth at vernal equinox, we will determine r and α for $t = t_{ve} = 77.6023$. The strategy has already been laid out. First, determine β for $t = 77.6023$, and then substitute this value into the equations

$$r = a(1 - \varepsilon \cos \beta) \quad \text{and} \quad \tan \frac{\alpha}{2} = \sqrt{\frac{1 + \varepsilon}{1 - \varepsilon}} \tan \frac{\beta}{2}.$$

So the first step is to solve $\beta - \varepsilon \sin \beta = \frac{2\pi t}{T}$ for β. Since $t = t_{ve} = 77.6023$,

$$\frac{2\pi t}{T} = \frac{2\pi(77.6023)}{365.2422} = \frac{(6.2832)(77.6023)}{365.2422} = 1.3350.$$

The solution of $\beta - \varepsilon \sin \beta = 1.3350$ by successive approximation proceeds as follows: $\beta_1 = \frac{2\pi t}{T} = 1.3350$. Successive substitution into $\frac{2\pi t}{T} + \varepsilon \sin(\ )$ starting with $\beta_1 = \frac{2\pi t}{T}$ gives

$$\beta_2 = \frac{2\pi t}{T} + \varepsilon \sin \beta_1 = 1.3350 + (0.0167)(\sin 1.3350)$$

$$= 1.3350 + (0.0167)(0.9723) = 1.3350 + 0.0162$$

$$= 1.3512,$$

$$\beta_3 = \frac{2\pi t}{T} + \varepsilon \sin \beta_2 = 1.3350 + (0.0167)(\sin 1.3512)$$

$$= 1.3350 + (0.0167)(0.9760) = 1.3350 + 0.0163$$

$$= 1.3513,$$

$$\beta_4 = \frac{2\pi t}{T} + \varepsilon \sin \beta_3 = 1.3350 + (0.0167)(\sin 1.3513)$$

$$= 1.3350 + (0.0167)(0.9760) = 1.3350 + 0.0163$$

$$= 1.3513.$$

Since the sequence of betas has stabilized at 1.3513, this is the value of β within the four decimal accuracy with which we are working. The value $\beta = 1.3513$ must be placed into the equations

$$r = a(1 - \varepsilon \cos \beta) \quad \text{and} \quad \tan \frac{\alpha}{2} = \sqrt{\frac{1+\varepsilon}{1-\varepsilon}} \tan \frac{\beta}{2}$$

to get r and α for the vernal equinox position. Since $a = 1$ AU,

$$r = 1 - (0.0167)(\cos 1.3513)$$

$$= 1 - (0.0167)(0.2177) = 1 - 0.0036$$

$$= 0.9964 \text{ AU}$$

and

$$\tan \frac{\alpha}{2} = \sqrt{\frac{1+\varepsilon}{1-\varepsilon}} \tan \frac{\beta}{2}$$

$$= (\tan 0.6757)\sqrt{\frac{1.0167}{0.9833}} = (1.0168)(0.8016)$$

$$= 0.8151.$$

Doing an inverse tangent operation with the calculator shows that $\frac{\alpha}{2} = 0.6839$. So $\alpha = 1.3678$ radians, or $78.3690°$.

Are our conclusions reasonable? The length of the year is about 365.25 days. So during an elapsed time of $t = 77.6$ days from its perihelion position, the Earth should trace out roughly $\frac{77.6}{365.25} \times 360 = 76.5°$. Therefore, α ought to be close to $76.5°$. Since the Earth moves faster near perihelion, it should be somewhat more than this. And it is! Consider the fact that α is less than $90°$ and refer back to Figure 4.27. Note that $r = 0.9964$ AU ought to be less than the semimajor axis b of the Earth's orbit. And it is! By Example 4.16, $b = 0.9998$ AU.

Now on to summer solstice. In this case $t = t_{ss} = 170.3662$ days. Solving $\beta - \varepsilon \sin \beta = \frac{2\pi t}{T}$ for β this time, start with

$$\beta_1 = \frac{2\pi t}{T} = \frac{(6.2832)(170.3662)}{365.2422} = 2.9308.$$

Computing as before, we get

$$\beta_2 = \frac{2\pi t}{T} + \varepsilon \sin \beta_1$$

$$= 2.9308 + (0.0167)(\sin 2.9308) = 2.9343,$$

$$\beta_3 = \frac{2\pi t}{T} + \varepsilon \sin \beta_2 = 2.9342, \quad \text{and}$$

$$\beta_4 = \frac{2\pi t}{T} + \varepsilon \sin \beta_3 = 2.9342.$$

Therefore, $\beta = 2.9342$.

Substituting $\beta = 2.9342$ into $r = 1 - \varepsilon \cos \beta$ gives us

$$r = 1 - (0.0167)(\cos 2.9342) = 1 + (0.0167)(0.9786)$$

$$= 1.0163 \text{ AU}.$$

This is close to the maximum of 1.0167 AU. Since in 1994 summer solstice occurred on June 21 and aphelion on July 5, this is to be expected. This also suggests that α should be close to $180°$. Substituting $\beta = 2.9342$ into $\tan \frac{\alpha}{2} = \sqrt{\frac{1+\varepsilon}{1-\varepsilon}} \tan \frac{\beta}{2}$, we get

$$\tan \frac{\alpha}{2} = (\tan 1.4671)\sqrt{\frac{1.0167}{0.9833}} = (1.0168)(9.690)$$

$$= 9.8528.$$

Now $\frac{\alpha}{2} = 84.2047°$, so $\alpha = 168.4094°$.

Lean back from this book and think about what Kepler has achieved. Think about Mars in orbit about the Sun. It is over one hundred million miles from Earth. It is visible on some clear nights, but only as a minuscule point of light. And Kepler has determined exactly how it moves!! He did so with an impressive combination of careful observations, the formulations of the fundamental laws, and very delicate mathematics. Incidentally, Kepler's mathematics can be fed into a desktop computer to provide a simulation of the motion of the planets in their elliptical orbits.[6]

4.10 Postscript

Recall that one of the principal goals set out in this chapter was the extraction of precise quantitative information about the orbit of a planet from Kepler's first two laws. This has now been accomplished. In sharp contrast with the systems of both Ptolemy and Copernicus, only a single theory is required. This theory applies not only to every planet, but also to comets and asteroids around the Sun. Indeed, it applies to Jupiter and its satellites. Analytic geometry plays a prominent role, but calculus—in the guise of Cavalieri's principle and the important inequality $|\sin \varphi - \sin \theta| \leq |\varphi - \theta|$—is present also. Put more explicitly, the rigorous justifications of both Cavalieri's principle and this inequality require—as we shall see—the methods of calculus.

Another important question that Kepler grappled with, but ultimately did not solve, concerns the driving force that keeps a planet in its orbit. Kepler realized that the Sun plays a role and suspected that such a force diminishes with distance. Indeed, it is a consequence of his third law that the average velocities of the planets fall off as their distances from the Sun increase (see Table 4.2). However, he was unable to determine the precise quantitative measure of this force. This and related matters remain—as we shall see—for Isaac Newton to accomplish.

A more immediate question concerns a matter as yet untouched, namely, the observation of Galileo that the rate of increase

$$\frac{v(t_2) - v(t_1)}{t_2 - t_1}$$

of the velocity of a falling body is a constant. What quantitative information does this fact contain and how can it be brought to the surface? This too requires the methods of calculus.

The time is ripe. The Italians Cavalieri and Torricelli; the Frenchmen Descartes, Fermat, Pascal, and Robeval; the Dutchman Huygens; the Englishmen Willis and Barrow; Gregory from Scotland; and a number of other mathematicians have set the stage and have put many of the components into place. It remains for Newton and Leibniz to add some decisive touches but, more importantly, to synthesize special cases and examples into a unified theory and into an integrated system that can be applied to a wide variety of mathematical problems: The Differential and Integral Calculus.

Exercises
4A. Basic Analytic Geometry

1. Find the distance between the points $(1,1)$ and $(4,5)$, and then between $(1, -6)$ and $(-1, -3)$.

2. Consider the triangle with vertices $A = (6, -7)$, $B = (11, -3)$, and $C = (2, -2)$.

   i. Show that it is a right triangle by using Pythagorean theorem.
   ii. Find the area of the triangle.

3. Show that the points $A = (-1, 3)$, $B = (3, 11)$, and $C = (5, 15)$ all lie on the same line by showing that $AB + BC = AC$.

4. Sketch the graphs of the equations

   i. $x = 3$
   ii. $y = -2$

5. Sketch the graphs of

   i. $xy = 0$
   ii. $|y| = 1$

In Exercise 6–8, sketch the given region in the x-y plane.

6. $\{(x, y) \mid xy < 0\}$

7. $\{(x, y) \mid 0 \leq y \leq 4 \text{ and } x \leq 2\}$

8. $\{(x, y) \mid |x| < 3 \text{ and } |y| < 2\}$

9. Show that the midpoint of the line segment for $P_1 = (x_1, y_1)$ to $P_2 = (x_2, y_2)$ is

   $$\left(\frac{x_1 + x_2}{2}, \frac{y_1 + y_2}{2} \right).$$

10. Find the midpoint of the line segment joining the given points:

    i. $(1, 3)$ and $(7, 15)$
    ii. $(-1, 6)$ and $(8, -12)$

4B. Circles, Parabolas, and Ellipses

In Exercises 11–15, identify the graph of the equation. Is it a parabola, circle, or ellipse? Sketch the graph in each case.

[Hint: For some problems it will be helpful to complete the square.]

11. $y = -x^2 + 3x + 4$

12. $x^2 + 4y^2 = 16$

13. $x = 2y^2$

14. $(x - 3)^2 + (y + 5)^2 = 7$

15. $9x^2 + 2y^2 = 12$

16. Sketch the graphs of the parabolas $y = -x^2$ and $y = x^2 + 1$. Then shade in the region R in the plane given by $R = \{(x, y) \mid -x^2 \leq y \leq x^2 + 1\}$.

17. Consider the parabolic section determined by the parabola $y = 3x^2 + 6x + 7$ and the line $y = 8$. Let $P = (x, y)$ be any point in the plane. What are the conditions on the coordinates x and y to guarantee that P lies in the parabolic section?

18. Consider the parabola $y = x^2 + 4x + 7$ and find the coordinates of the lowest point. Cut the parabola with the line $y = 7$ and compute the area of the parabolic section obtained.

19. Determine the directrix and focus of the parabola $y = 3x^2 - 2x + 5$.

20. Find an equation of a circle that has center $(3, -1)$ and radius 5.

21. Show that the graph of the equation

$$x^2 + y^2 - 4x + 10y + 13 = 0$$

is a circle. Find its center and radius.

22. Under what condition on the coefficients a, b, and c does the equation $x^2 + y^2 + ax + by + c = 0$ represent a circle? When this condition is satisfied, find the center and radius of the circle.

23. Find the semimajor axis, the semiminor axis, and the linear and astronomical eccentricities of the ellipse $\frac{x^2}{25} + \frac{y^2}{4} = 1$.

24. To draw an ellipse with focal points F_1 and F_2, semimajor axis a and semiminor axis b, proceed as follows. Notice that the distance between the focal points is $2\sqrt{a^2 - b^2}$. Tie a string into a loop of length $2a + 2\sqrt{a^2 - b^2}$. Take a board, place thumbtacks at F_1 and F_2, and put the loop around the tacks. Take a pencil and stretch the string taut. Keep the string stretched and draw a complete revolution. Explain why this results in an ellipse and why it has semimajor axis a and semiminor axis b.

25. Fix a Cartesian coordinate system. Take a line segment of fixed length and let P be a fixed point on it.

Slide it around in the plane over all positions with the property that one endpoint is always on the x-axis and the other on the y-axis. Show that in the process the point P describes an ellipse.

26. Place a hyperbola in the x-y plane in such a way that the focal points F_1 and F_2 are on the x-axis and the origin is the midpoint of the segment F_1F_2. Show that the equation of this hyperbola has the form $\frac{x^2}{a^2} - \frac{y^2}{b^2} = 1$. [Hint: Refer to Section 4.5 and follow the development of the equation $\frac{x^2}{a^2} + \frac{y^2}{b^2} = 1$ for the ellipse.]

4C. Some Geometry and Trigonometry

27. Refer to Figure 4.19 in Section 4.4. Suppose that $r = 1$ and $s = 0.693$. What is the radian measure of θ and what are the coordinates of the point P?

28. Suppose that in Figure 4.19 of Section 4.4, $r = 2$ and $\theta = 5$ radians. What is the length s of the arc?

29. Refer to Figure 4.20 in Section 4.4. Consider the angles $\theta = 17.52$ and $\theta = -21.83$. In each case place the point P_θ carefully on the unit circle. What are the coordinates of the point in each case?

30. The trigonometric function secant is defined by $\sec\theta = \frac{1}{\cos\theta}$. Sketch the graph of this function by analyzing the graph of the cosine. Refer to Figure 4.25 in Section 4.4.

31. Verify the identities $\sec(\theta + 2\pi) = \sec\theta$, $\sec(-\theta) = \sec\theta$, $\sec(\theta + \pi) = -\sec\theta$, and the formula $\sec^2\theta = \tan^2\theta + 1$.

4D. Computing Orbital Information

32. Determine the following quantities for the orbit of Jupiter: the linear eccentricity, semiminor axis, and the greatest and least distances from the Sun. Do so first in AUs and then convert to miles. (Use the data of Table 4.2.)

33. Use the data in Table 4.2 to verify Kepler's third law for Mars, Jupiter, and Saturn. Why are all these ratios equal to 1?

34. Consider a planet (or a comet) in elliptical orbit around the Sun S, as pictured in Figure 4.38. We already know that its speed is greater when it is nearer the Sun than when it is farther away. Can we say more? Observe the planet from P to P' and then again from A to A'. In the figure, arc PP' has the perihelion position as its midpoint, and arc AA' has the aphelion position as its midpoint. Suppose that both arcs are relatively small and regard the regions PSP' and ASA' to be sectors of a circle. Notice that their radii are $a - e$ and $a + e$, respectively where a is the semimajor axis and e the linear eccentricity of the ellipse.

   i. Use the formula for the area of a circular sector (from Section 3.2) to show that the areas of the two sectors are $\frac{1}{2}(a - e)(\text{arc}PP')$ and $\frac{1}{2}(a + e)(\text{arc}AA')$, respectively.

  ii. Suppose that the time it takes the body to travel from P to P' is the same as that from A to A'. Let v_P be the average velocity of the body from P to P' and let v_A be its average velocity from A to A'. Show that $\frac{v_P}{v_A} = \frac{a+e}{a-e}$. What is this ratio for the Earth? For Saturn?

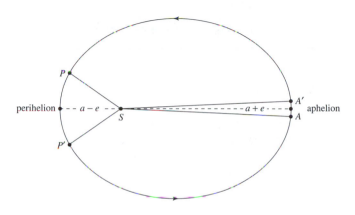

Figure 4.38

35. Do the analysis of Section 4.9 with the data from 1995. In that year perihelion occurred on January 4 at 11 hours 5 minutes GMT, and vernal equinox was on March 21 at 2 hours 14 minutes. Refer to the text for the lengths of the seasons.

   i. Determine t_{ve} and t_{ss}.

  ii. Determine β for t_{ve}.

 iii. Substitute into the formulas $r = a(1 - \varepsilon \cos \beta)$ and $\tan \frac{\alpha}{2} = \sqrt{\frac{1+\varepsilon}{1-\varepsilon}} \tan \frac{\beta}{2}$ to compute r in AUs and α in radians. (Note that $a = 1$, since the Earth is being considered.)

  iv. Repeat steps (ii) and (iii) for t_{ss}.

36. Suppose $\varepsilon = 0$ in the analysis of Sections 4.7 and 4.8. Explain why the following are true: the ellipse and the circle in Figure 4.34 coincide, $S = O$, $\alpha = \beta$, $r = a$, and $\beta = \frac{2\pi t}{T}$. Is successive approximation needed?

4E.   The Orbit of Halley's Comet

In 1705, Edmund Halley, an astronomer and a contemporary of Isaac Newton, published calculations about the orbits of the comets.[7] He noted that properties of the orbits of the bright comets of 1531, 1607, and 1682 were so similar that the three comets could well be the same. This was later verified, and the comet was named in Halley's honor. The elliptical orbit of Halley's comet has an average period of 76 years. (In fact, the period varies from 74 to 78 years due to irregularities caused by the gravitational attraction of Jupiter and Saturn.) The last time Halley (we will refer to the comet simply as Halley, as confusion with the astronomer is unlikely) was near the Earth was in 1986. The minimal distance of Halley from the Sun, i.e., the distance at perihelion, is known to be $d = 0.59$ AU. The unit of distance throughout Exercises 37–39 is the AU.

Let a and b be the semimajor and semiminor axes of Halley's elliptical orbit and draw the ellipse on an x-y plane such that the equation is $\frac{x^2}{a^2} + \frac{y^2}{b^2} = 1$. See Figure 4.39. The Sun S is at a focal point.

37. Use Kepler's third law with Halley and the Earth to compute a in AUs. Use the fact that $d = 0.59$ AU to compute the linear eccentricity e of Halley's orbit and the semiminor axis b. What is Halley's astronomical eccentricity ε? What is the greatest distance from Halley to the Sun? Refer to Exercise 34ii and compute the ratio of the velocities for Halley.

Figure 4.40 is an enlargement of Halley's orbit near the Sun S. (Why is $S = (e, 0)$?) It also shows the orbit of the Earth. For Exercises 38 and 39 the Earth's orbit is taken to be a circle of radius 1 AU and center the Sun S. In the figure it is assumed that the orbits of Halley and the Earth lie in the same plane. The points H_1 and H_2 are the points of intersection of the two orbits, and x is their common x-coordinate.

38. Determine an equation of the Earth's circular orbit. Show that the x-coordinate of the points of intersection H_1 and H_2 is $x = \frac{a^2-a}{e}$. Determine the numerical values of the x- and y-coordinates of the points H_1 and H_2. Does Figure 4.40 give an accurate picture of Halley's path inside the Earth's orbit? [Hints: In reference to the equation of the ellipse, it is better to work with the parameters a, b, e, etc., rather than their numerical values. Use the identity $e^2 = a^2 - b^2$. Since $\frac{a^2+a}{e} = \frac{a^2}{e} + \frac{a}{e} > a$, note that $x = \frac{a^2+a}{e}$ is not possible.]

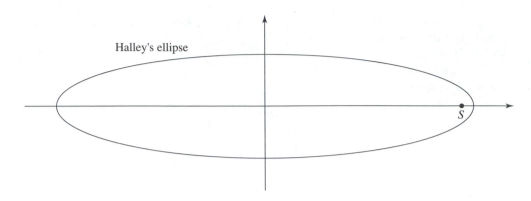

Figure 4.39

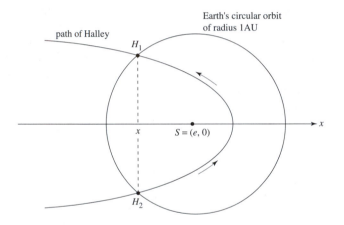

Figure 4.40

39. Consider Figure 4.34 in Section 4.7 in the context of Halley's orbit and take $P = H_1$. What is r equal to? Use the formulas of Sections 4.7 and 4.8 to compute β and t (in years). For how many days will Halley be inside the Earth's orbit?

40. Consider the successive approximation method in reference to computing an angle β for Halley's orbit. Give an estimate for the number of iterations necessary to do so with an accuracy of 0.0002. [Hint: Keep squaring ε with a calculator.]

41. Use the method of Exercise 24 to make a sketch of the elliptical orbit of Halley's comet on a "legal" size sheet of paper. (Use the scale 1 AU = $\frac{3}{8}$ inch.) Use the data from Table 4.2 of Section 4.7 and give indication in your sketch of the location of the orbits of the planets (draw them as circles or parts thereof).

4F.   More Trigonometry and Gauss's Formula

Verify Gauss's equation $\tan \frac{\alpha}{2} = \sqrt{\frac{1+\varepsilon}{1-\varepsilon}} \tan \frac{\beta}{2}$ with the following sequence of steps:

42. Start with the equation $\tan \alpha = \frac{b \sin \beta}{a \cos \beta - e}$ that was already established. Square both sides; add 1 to both sides; take common denominators; and use the identities $a^2 = b^2 + e^2$, $\sin^2 \beta + \cos^2 \beta = 1$, $\tan^2 \beta + 1 = \sec^2 \beta$, and $\sec \alpha = \frac{1}{\cos \alpha}$ to get

$$\cos^2 \alpha = \frac{(a \cos \beta - e)^2}{(a - e \cos \beta)^2}.$$

Because $a - e \cos \beta$ is always positive (why?) and $a \cos \beta - e = x - e$ and $\cos \alpha$ have the same sign (study Figure 4.34 in Section 4.7 and the graph of the cosine carefully), conclude that

$$\cos \alpha = \frac{a \cos \beta - e}{a - e \cos \beta}.$$

Since $\varepsilon = \frac{e}{a}$, you now get

$$\cos \alpha = \frac{\cos \beta - \varepsilon}{1 - \varepsilon \cos \beta}.$$

43. Take $\theta = \varphi$ in the formula

$$\cos(\varphi + \theta) = (\cos \varphi)(\cos \theta) - (\sin \varphi)(\sin \theta)$$

to get $\cos 2\varphi = \cos^2 \varphi - \sin^2 \varphi$. Deduce the formulas $2 \cos^2 \varphi = 1 + \cos 2\varphi$ and $2 \sin^2 \varphi = 1 - \cos 2\varphi$.

44. Deduce that

$$\tan^2 \frac{\alpha}{2} = \frac{1 - \cos \alpha}{1 + \cos \alpha} = \frac{1 + \varepsilon}{1 - \varepsilon} \frac{1 - \cos \beta}{1 + \cos \beta}$$

$$= \frac{1 + \varepsilon}{1 - \varepsilon} \tan^2 \frac{\beta}{2}$$

and therefore that $\tan \frac{\alpha}{2} = \sqrt{\frac{1+\varepsilon}{1-\varepsilon}} \tan \frac{\beta}{2}$.

4G.   A Study of Kepler's Formulas

45. Redraw Figure 4.34 of Section 4.7, this time with P in the third quadrant instead of the first. Study the derivations of the formulas $r = a(1 - \varepsilon \cos \beta)$, $\tan \alpha = \frac{h \sin \beta}{a(\cos \beta - e)}$ and $\beta - \varepsilon \sin \beta = \frac{2\pi t}{T}$ of Section 4.7 and adapt them to this situation. [Hint: Use the trig formulas of Examples 11 and 12, and refer to Example 14.]

46. Consider Figure 4.34 with P as shown in the first quadrant, but suppose that the planet is in its second orbit at time t. Explain why the formulas referred to in Exercise 45 are valid in this case also.

Notes

[1]"Without the concepts, methods and results found and developed by previous generations right down to Greek antiquity, one cannot understand either the aims or the achievements of mathematics in the last fifty years." Thus wrote Hermann Weyl (1885–1955), one of the leading mathematicians of the 20th century.

[2]The word oval comes from the Latin word *ovum*, meaning egg. The "lengthwise" cross-section of most eggs is a curve that is "rounder" at one end and "more pointed" at the other. Such a curve is what Kepler had in mind. Aphelion and perihelion are the positions at which a planet or comet (in this case Mars) are, respectively farthest and closest to the Sun.

[3]For the details about this and other experiments of Galileo, see Stillman Drake, *Galileo: Pioneer Scientist*, University of Toronto Press, Toronto, 1990. It appears that Galileo did not measure g explicitly. After Galileo's time, g was generally thought to be about 24 ft/sec^2. Later, in the 17th century, the Dutch scientist Huygens measured g to be the equivalent of 32.18 ft/sec^2, or 9.8 meters/second^2. This is in close agreement with today's value.

[4]A teacher who apparently wanted to punish Gauss (a schoolboy of age 10) and his classmates for some misdeed told the class to add up the first 100

positive integers. Gauss, to the astonishment of the teacher, had the answer in seconds. Here is how he did it:

$$
\begin{array}{ccccccc}
1+ & 2+ & 3+ & \cdots + & 99+ & 100 \\
100+ & 99+ & 98+ & \cdots + & 2+ & 1 \\
\hline
101+ & 101+ & 101+ & \cdots + & 101+ & 101
\end{array}
$$

He had added the first 100 integers twice, in pairs, and gotten $101 \cdot 100 = 10,100$. So half this number, or 5,050, is the answer.

[5]In 1884 a conference of nations in Washington D.C. declared the meridian through Greenwich, England, to be the prime meridian, i.e., the line of $0°$ longitude. Greenwich had been the seat of an astronomical observatory since Newton's time. The time of day in Greenwich is known as Greenwich Mean Time (GMT).

[6]See the link A Computer Model of Elliptical Orbits (Generated with Kepler's Equations) on the web site http://www.nd.edu:80/ hahn/part1.html

[7]Comets contain matter left over from the formation of the solar system. They are studied by astronomers, among other reasons, for the information that they reveal about this formation process.

The Calculus of Leibniz

The factual understanding of our physical reality necessarily involves the clarification of numerical relationships. For example, Galileo's efforts to understand the path of a projectile require answers to questions such as: How high will it go? How far away will it strike the ground? What will its velocity be at the point of impact? Can its parabolic trajectory be calculated? An initial response to Galileo's questions is this: Information about the tangent lines to the path gives information about the path itself. Figure 5.1 illustrates this. For instance, the tangent at the initial point A provides the initial direction of the projectile, and the tangent at the terminal point C gives the angle of impact. At the point B, where the tangent is horizontal, the projectile reaches its highest elevation. The first concern of calculus is the study of tangent lines and the information that they provide about the curve.

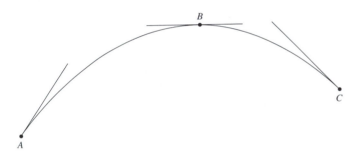

Figure 5.1

The second concern of calculus is the analysis of large sums of small numbers. An example is Archimedes's procedure for computing the area of the parabolic section by adding the areas of more and more triangles that are progressively smaller and smaller. (Refer to Section 3.4.) The principle of Cavalieri (see Section 4.6) is based, at least implicitly, on similar considerations. Consider a curve from A to B and a line segment L (as shown in Figure 5.2). The area between the curve and the segment can be computed to any degree of accuracy as follows: Place rectangles as shown and notice that their areas, when added together, ap-

proximate the area in question. To achieve a very high degree of accuracy, just take a huge number of such rectangles. Since each rectangle is very thin, its area is some very small number. Therefore, the area between the curve and the segment can be computed by an addition of lots of numbers, all very small.

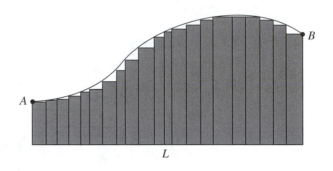

Figure 5.2

Now comes a surprise. The two basic concerns of calculus, as they have just been described, are closely related to each other. Namely, the relationship between a curve and its tangent line provides an effective strategy for the addition of a large number of very small terms.

Recall at this point the important lesson of analytic geometry, that numerical relationships can be interpreted and studied as graphs and hence in the context of curves. To understand the importance of this connection to the current discussion, consider the following: The calculation of the greatest tension in the cable of a suspension bridge or the maximum pressure generated by the explosion that propels a shell in a gun barrel both reduce to the determination of the highest point on a curve. And the ability to gain certain quantitative insights into the nature of gravity, or the energy generated by a force, is the ability to add the areas of a large array of very small numbers.

The points just made can be combined and summarized as follows. The domain of calculus is the quantitative study of certain properties of curves. It is therefore a discipline that concerns itself with the analysis of basic numerical relationships and consequently with quantitative aspects of much of the physical world around us.

Leibniz and Newton succeeded in creating calculus by building their insights, as well as those of their contemporaries and predecessors, into a general theory of curves and applications of that theory. This chapter will begin with a focus on the work of Leibniz.[1]

Gottfried Wilhelm Leibniz (1646–1716) was born in Leipzig, in the German state of Saxony. After obtaining a doctor of laws degree from the University of Nuremberg in 1667, he accepted a diplomatic post. In 1672 he was sent on a mission to Paris. Its purpose was to deflect the growing threat of France to the German states by attempting to persuade the government of Louis XIV to occupy Egypt. The rationale was that this would allow a strengthened, united Christian Europe to control the Eastern trade routes and repel the attacks of the Turkish Empire against Eastern Europe. This scheme failed, and a French attack on the German states soon followed. Leibniz's stay in Paris proved to be most fortuitous for mathematics, however.

In Paris, Leibniz came into contact with a number of brilliant scholars. A brief diplomatic journey early in 1673 brought him to London, where he met leading English scientists. The death of his patron ended his career as a diplomat, but he remained in Paris and resumed his earlier interests in mathematics with great intensity. Under the guidance of Christiaan Huygens (1629–1695), the Dutch mathematician and physicist, he studied Huygens's analysis of the pendulum, the *Géométrie* of Descartes, the work of Isaac Barrow (1630–1677), who was Newton's professor at Cambridge, and the *Geometria Indivisibilibus* and *Exercitationes Sex*[2] of Cavalieri. Leibniz mastered all of this, and more, very quickly. In the period from 1673 to 1676, he laid the foundations of calculus.

Unable to secure permanent employment in Paris, Leibniz reluctantly accepted the post of librarian and councilor to a German duke. In the service of the duke and his successors, he undertook a remarkable variety of tasks. He worked on a hydraulic press, on gears and steering devices for carriages, on water pumps driven by windmills, and he proposed a process for the desalinization of water. He speculated about the origins and the age of the Earth. His *Meditationes de Cognitione, Veritate, et Ideis* (Reflections on Knowledge, Truth, and Ideas) and *Discours de Métaphysique* (Discourse on Metaphysics) developed his philosophy. For Leibniz the world was infinite from both the qualitative and quantitative perspective; the number of substantial units, or *monads*, in the smallest particle was as boundless as the entire universe. Summarizing his philosophical thinking, he wrote, "It is as if God had wanted to create the universe in as many different ways as there are souls, or as if he created as many microcosms which are identical in their deepest foundations, but which are manifold in their appearance. This in a few words is my entire philosophy."

5.1 Straight Lines

Before considering Leibniz's mathematics, we first turn to the quantitative analysis of lines. Lines are automatically assumed to be straight. A line that is not straight is called a curve. We will be working with a plane that is equipped with an x-y coordinate system.

Consider the equation $2x - 3y - 4 = 0$. It can be rewritten in many equivalent ways, for example, $2x - 3y = 4$, or $3y = 2x - 4$, or $y = \frac{2}{3}x - \frac{4}{3}$. If the values $x = a$ and $y = b$ satisfy one of these equations, they satisfy all of them. Therefore, all of these equations have the same graph. We will work with

$$y = \frac{2}{3}x - \frac{4}{3}.$$

If $x = 5$, then $y = \frac{10}{3} - \frac{4}{3} = \frac{6}{3} = 2$. So $x = 5$ and $y = 2$ satisfy the equation, and the point $(5, 2)$ is on the graph. In the same way, $(4, \frac{4}{3})$, $(3, \frac{2}{3})$, $(2, 0)$, $(1, -\frac{2}{3})$, $(0, -\frac{4}{3})$, $(-2, -\frac{8}{3})$, and so on, are all on the graph. Plot these points, and observe that all fall on the single line pictured in Figure 5.3. Call this line L. Move along L from left to right and notice—compare the points

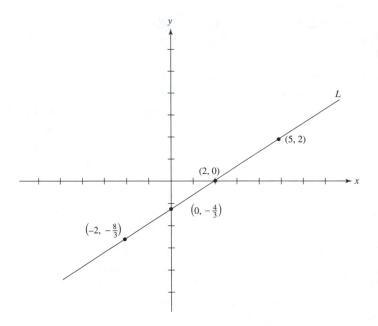

Figure 5.3

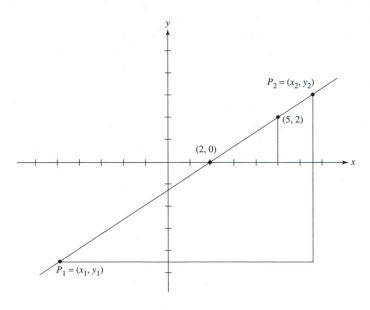

Figure 5.4

$(2,0)$ and $(5,2)$ for example—that for every 3 units of increase in the direction of the positive x axis, there is a corresponding increase of 2 units in the direction of the positive y-axis. So there is an increase of $\frac{2}{3}$ units in the y-coordinate for every increase of 1 unit in the x-coordinate. So the number $\frac{2}{3}$ measures the pitch, or steepness, of the line. It is called the *slope* of the line and is often denoted by m. In the case under consideration now, $m = \frac{2}{3}$.

The slope m can be determined as follows: Consider any two points on L, say, $(5, 2)$ and $(2, 0)$; form the difference $2 - 0 = 2$ between the y-coordinate of the first point and the y-coordinate of the second; form the difference $5 - 2 = 3$ between the x-coordinate of the first point and the x-coordinate of the second; and let $m = \frac{2}{3}$ be the ratio of the two differences. Any two distinct points $P_1 = (x_1, y_1)$ and $P_2 = (x_2, y_2)$ on the line can be used to compute the slope. This is a consequence of the basic property of similar triangles. Refer to Figure 5.4 and notice that the right triangle with hypotenuse given by $(5, 2)$ and $(2, 0)$ is similar to the right triangle with hypotenuse $P_1 P_2$. It follows that the ratios of the lengths of the vertical over the horizontal legs of the two triangles are equal. So 2 is to 3 as $y_2 - y_1$ is to $x_2 - x_1$, or

$$m = \frac{2}{3} = \frac{y_2 - y_1}{x_2 - x_1}.$$

Again, the slope of a line is the ratio of the change in y over the corresponding change in x.

It should be clear that two nonvertical lines are parallel precisely if they have the same slope. Figure 5.5 shows several lines and their slopes. Notice that a line with a positive slope slants upward from left to right and that a line with a negative slope slants downward, again from left to right. Notice also that the steepest lines are the ones for which the absolute value of the slope is largest, and that a horizontal line has slope zero. Consider a line with slope $m \neq 0$. It is not very hard to verify that the line perpendicular to it has slope $-\frac{1}{m}$. Which pairs of lines in Figure 5.5 are perpendicular?

Example 5.1. Sketch the graph of the equation $5x + 3y = 15$ and determine its slope.

Solution. It suffices to find two points on the line and to draw the line through both. It is easiest to find the intercepts. Substituting $y = 0$ into the equation, we get $5x = 15$. So $x = 3$, and $(3, 0)$ is on the graph.

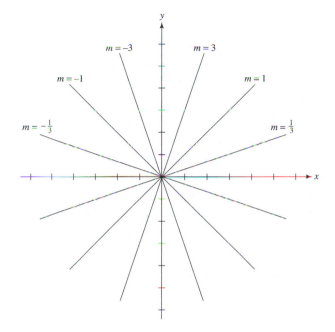

Figure 5.5

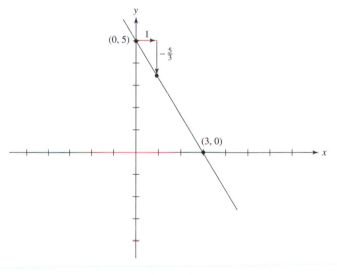

Figure 5.6

Substituting $x = 0$, we see that $(0, 5)$ is on the graph. The graph is shown in Figure 5.6. Using these two points to compute the slope, we get

$$m = \frac{(5-0)}{(0-3)} = -\frac{5}{3}$$

So a change of $+3$ in x results in a change of -5 in y. Equivalently, a change of 1 unit in x produces a change of $-\frac{5}{3}$ units in y. Again, see Figure 5.6.

Consider the equation $2x + 10 = 0$ or, equivalently, $x = -5$. The points (x, y) that satisfy it are precisely the points $(-5, y)$, where y can be anything. Notice, therefore, that the graph of the equation $x = -5$ is the vertical line through $(-5, 0)$. For any two points $P_1 = (x_1, y_1)$ and $P_2 = (x_2, y_2)$ on this line, $x_1 = x_2 = 5$, so the ratio $\frac{y_2 - y_1}{x_2 - x_1}$ is meaningless. This line has no slope. For the same reason, no vertical line has a slope.

Suppose that L is a nonvertical line. Select any two distinct points $P_1 = (x_1, y_1)$ and $P_2 = (x_2, y_2)$ on L. As before, its slope is $m = \frac{y_2 - y_1}{x_2 - x_1}$. A look at Figure 5.7 shows that a point $P = (x, y)$ is on L precisely if the segment $P_1 P$ is parallel to L. So $P = (x, y)$ is on L precisely if the slope of the segment $P_1 P$ is equal to m. Since the slope of $P_1 P$ is $\frac{y - y_1}{x - x_1}$, the point $P = (x, y)$ is on L precisely if $\frac{y - y_1}{x - x_1} = m$. This equation can be

rewritten in the form

$$y - y_1 = m(x - x_1).$$

Observe that it is also satisfied when $x = x_1$ and $y = y_1$. Therefore, it is an equation of the given line L. An equation of a line arranged in this way is said to be in *point-slope* form.

Example 5.2. Find an equation of the line through $(1, -7)$ with slope $-\frac{1}{2}$.

Solution. Using the point-slope form of the equation with $m = -\frac{1}{2}$, $x_1 = 1$, and $y_1 = -7$, we get

$$y + 7 = -\frac{1}{2}(x - 1).$$

This, as well as $2y + 14 = -x + 1$ or $x + 2y + 13 = 0$, are equations of the given line.

Example 5.3. Find an equation of the line through the points $(-1, 2)$, and $(3, -4)$.

Solution. The slope of the line is

$$m = \frac{-4 - 2}{3 - (-1)} = -\frac{3}{2}.$$

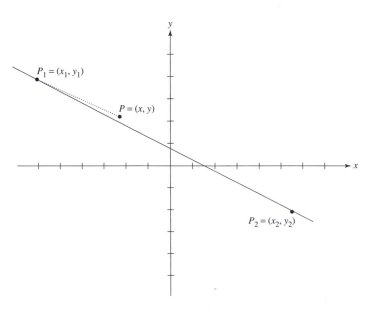

Figure 5.7

Using the point-slope form with $x_1 = -1$ and $y_1 = 2$, we obtain

$$y - 2 = -\frac{3}{2}(x + 1).$$

This equation can be rewritten as $3x + 2y = 1$.

Suppose a nonvertical line has slope m and y-intercept b. This means that it intersects the y-axis at the point $(0, b)$. Taking $x_1 = 0$ and $y_1 = b$, the point-slope form of the equation of the line is $y - b = m(x-0)$. Therefore, an equation of the line with slope m and y-intercept b is

$$y = mx + b.$$

An equation of a line arranged in this way is said to be in *slope-intercept* form. If a line is horizontal, its slope is $m = 0$. So the equation of such a line has the form $y = b$, where b is the y-intercept.

Take any equation of the form $Ax + By + C = 0$, where A, B, and C are constants. If $A = B = 0$, then $C = 0$. Disregard this case, and suppose that either $A \neq 0$ or $B \neq 0$. Suppose that $B \neq 0$. Then $Ax + By + C = 0$ is equivalent to $y = -\frac{A}{B}x - \frac{C}{B}$. A look at the point-slope form shows that this is an equation of the line with slope $-\frac{A}{B}$ and y-intercept $-\frac{C}{B}$. If $B = 0$, then $A \neq 0$. Now the equation is $Ax + C = 0$ or, equivalently, $x = -\frac{C}{A}$. As in the case $x = -5$, this is an equation of the vertical line through $(-\frac{C}{A}, 0)$. Therefore, the graph of any equation of the form $Ax+By+C$ is a line (unless $A = B = C = 0$). Any such equation is therefore called a *linear* equation.

Lines are important in the analysis of the hyperbola, the one conic section we have not as yet had a closer look at. (Refer to Sections 4.3 and 4.5.) Take any hyperbola and move it in such a way that its focal points (see Section 3.1) lie on the x-axis with the origin at the midpoint of the segment that joins the focal points. Doing what was done in Section 4.5 for the ellipse, it can be verified that such a hyperbola has an equation of the form

$$\frac{x^2}{a^2} - \frac{y^2}{b^2} = 1$$

with $a > 0$ and $b > 0$. To find the x-intercepts, set $y = 0$ and obtain $x^2 = a^2$ and $x = \pm a$. If $x = 0$, then $y^2 = -b^2$. But this is impossible, so there is no y-intercept. In fact, since

$$\frac{x^2}{a^2} = 1 + \frac{y^2}{b^2} \geq 1,$$

$x^2 \geq a^2$ and so $|x| = \sqrt{x^2} \geq a$. Therefore, $x \geq a$ or $x \leq -a$. This means that the hyperbola consists of two separate parts, one to the left of the line $x = -a$ and the other to the right of the line $x = a$. What else is going on?

Since $\frac{x^2}{a^2} - \frac{y^2}{b^2} = 1$, it follows that $\frac{y^2}{b^2} = \frac{x^2}{a^2} - 1 = \frac{x^2 - a^2}{a^2}$. So $y^2 = \frac{b^2}{a^2}(x^2 - a^2)$ and, taking square roots, $y = \pm\frac{b}{a}\sqrt{x^2 - a^2}$. Observe that if x^2 is large relative to a^2, then the effect of a^2 can be discounted and $y = \pm\frac{b}{a}\sqrt{x^2 - a^2}$ is approximately equal to $\pm\frac{b}{a}\sqrt{x^2} = \pm\frac{b}{a}|x|$. Consider a point (x, y) on the hyperbola with x large and positive. If $y > 0$, then, by the observation just made, $y \approx \frac{b}{a}x$, so that the point is close to the line $y = \frac{b}{a}x$. If $y < 0$, then in the same way, the point is close to the line $y = -\frac{b}{a}x$. This analysis can be continued and refined to show that the graph of the hyperbola $\frac{x^2}{a^2} - \frac{y^2}{b^2} = 1$ has the general form shown in Figure 5.8. Notice that if a point (x, y) is on the portion of the curve above the x-axis, then $y \geq 0$, so that $y = \frac{b}{a}\sqrt{x^2 - a^2}$. If (x, y) lies below that x-axis, then $y = -\frac{b}{a}\sqrt{x^2 - a^2}$.

By interchanging the roles of x and y we get the equation

$$\frac{y^2}{a^2} - \frac{x^2}{b^2} = 1.$$

This also represents a hyperbola. Its graph is obtained by rotating that of Figure 5.8 by $90°$.

Example 5.4. Sketch the curve $4x^2 - 4y^2 = 8$.

Solution. Dividing both sides by 8, we obtain

$$\frac{x^2}{(\sqrt{2})^2} - \frac{y^2}{(\sqrt{2})^2} = 1.$$

Letting $a = \sqrt{2}$ and $b = \sqrt{2}$, we see that this equation has the form discussed earlier. So its graph

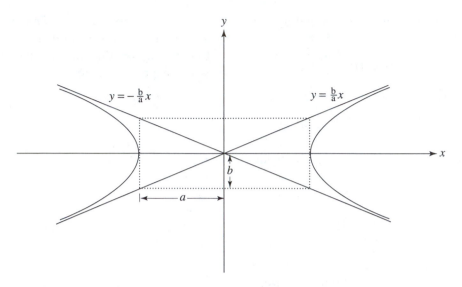

Figure 5.8

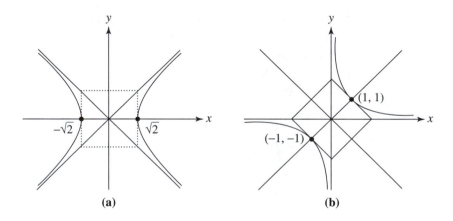

Figure 5.9

is the hyperbola obtained by taking $a = b = \sqrt{2}$ in Figure 5.8. It is sketched in Figure 5.9(a). Rotating this hyperbola by $45°$ gives the hyperbola in Figure 5.9(b). It is the graph of the equation $xy = 1$.

An equation of the form

$$Ax^2 + Bxy + Cy^2 + Dx + Ey + F = 0$$

where A, B, C, D, E, and F are constants with at least one of A, B, or C not zero, is called a *quadratic*

equation. Observe that the hyperbolas of this section, the parabolas of Section 4.3, and the ellipses of Section 4.5 are all given by quadratic equations. Proceeding in the other direction, it can be shown that the graph of a quadratic equation is either a parabola, an ellipse, a hyperbola (all rotated and shifted in any possible way), or a pair of lines (for example, $xy = 0$), or a point (for example, $x^2 + y^2 = 0$), or it has no points at all (for example, $x^2 + 1 = 0$).

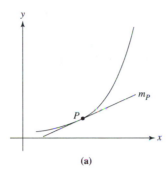

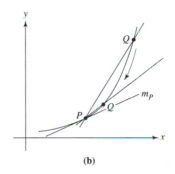

Figure 5.10

Now that we know a few things about lines, we can turn to tangent lines and the work of Leibniz.

5.2 Tangent Lines to Curves

Leibniz's first publication of his work on calculus is the article *Nova methodus pro maximis et minimis, itemque tangentibus … calculi genus* ("A new method for maxima and minima as well as tangents, which is impeded neither by fractional nor by irrational quantities, and a remarkable type of calculus for this"), which appeared in 1684. It presents a new notation, introduces the term "calculus differentialis," contains the general rules of differentiation … and is nearly impossible to read. Even Leibniz's friends, the brothers Bernoulli (famous mathematicians themselves), commented that this article is "an enigma rather than an explication." We advisedly turn instead to a later manuscript. In it, Leibniz provides details and corrections of his earlier exposition. The discussion of the work of Leibniz that follows in this and subsequent sections is a much-expanded version of his original manuscripts. It also adds concepts and notational elements that were not introduced until later. These serve to clarify some of the remaining "enigmas." However, the essence of Leibniz's contributions are retained.

Let a curve in the x-y plane be given. Regard it as the graph of some equation in x and y. Fix a point P on the curve and consider the tangent line to the curve at P. Assume that it is not vertical and let m_P be its slope. See Figure 5.10(a). Now take some other point $Q \neq P$ on the curve. While keeping P fixed, push Q towards P. Refer to Figure 5.10(b) and notice that as Q closes in on P, the line through P and Q closes in on the tangent line at P. So as Q is pushed to P, the slope of the line through P and Q homes in on the slope m_P of the tangent.

Let's have a more algebraic look at what is going on. Let x and y be the coordinates of P and let x' and y' be the coordinates of Q. Let $\Delta x = x' - x$ and $\Delta y = y' - y$ be the differences between the x-coordinates and y-coordinates of P and Q. So the coordinates of Q are $x' = x + \Delta x$ and $y' = y + \Delta y$. Again draw the line through P and Q and observe that the slope of this line is equal to the ratio

$$\frac{(y + \Delta y) - y}{(x + \Delta x) - x} = \frac{\Delta y}{\Delta x}.$$

Again, keeping P fixed push Q towards P. Observe from Figure 5.11 that pushing Q to P is the same thing as pushing Δx to zero. Therefore, when Δx is pushed to zero, the slope $\frac{\Delta y}{\Delta x}$ of the line through P and Q closes in on the slope of the tangent. Rewritten in limit notation, what we have observed is

$$\lim_{\Delta x \to 0} \frac{\Delta y}{\Delta x} = m_P.$$

Notice that Leibniz cannot simply set $\Delta x = 0$, for then (see Figure 5.11) Δy is also equal to zero. Thus

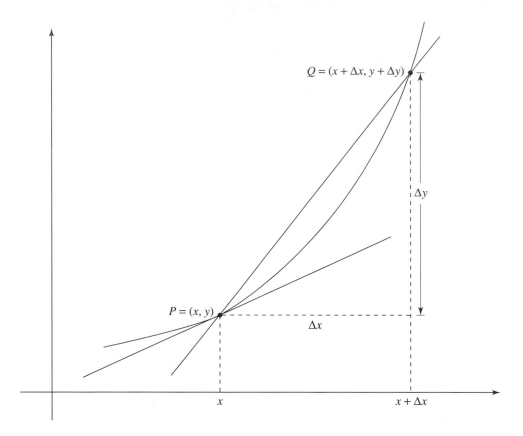

Figure 5.11

$\frac{\Delta y}{\Delta x}$ would be $\frac{0}{0}$, which is not defined, i.e., it does not make sense.

Leibniz turns to the parabola to illustrate his method. He takes $y = Ax^2$, where A is any constant. Let's be more specific and consider $y = x^2$. Instead of a general point $P = (x, y)$ on the parabola, let's take $P = (3, 9)$. Now $Q = (3 + \Delta x, 9 + \Delta y)$. Since Q is on the graph,

$$9 + \Delta y = (3 + \Delta x)^2 = 3^2 + 2 \cdot 3\Delta x + (\Delta x)^2.$$

Subtracting 9 from both sides gives us $\Delta y = 6\Delta x + (\Delta x)^2$, and after dividing both sides by Δx, we get

$$\frac{\Delta y}{\Delta x} = 6 + \Delta x.$$

By pushing Δx to zero,

$$\lim_{\Delta x \to 0} \frac{\Delta y}{\Delta x} = 6.$$

Therefore, the slope of the tangent line to the graph of $y = x^2$ at the point $P = (3, 9)$ is 6. So in this case $m_P = 6$. See Figure 5.12.

Now let $P = (x, y)$ be any point on the parabola $y = x^2$. Since

$$Q = (x + \Delta x, y + \Delta y)$$

is on the graph, $y + \Delta y = (x + \Delta x)^2$. We continue in Leibniz's own words (in English translation):

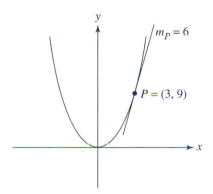

Figure 5.12

Then, since $y = x^2$, by the same law, we have

$$y + \Delta y = (x + \Delta x)^2 = x^2 + 2x\Delta x + (\Delta x)^2;$$

and taking away the y from the one side and the x^2 from the other, we have left

$$\frac{\Delta y}{\Delta x} = 2x + \Delta x.$$

Make use of modern limit notation and observe that Leibniz can now conclude that

$$\lim_{\Delta x \to 0} \frac{\Delta y}{\Delta x} = \lim_{\Delta x \to 0}(2x + \Delta x) = 2x.$$

So the slope of the tangent line at any point $P = (x, y)$ on the parabola $y = x^2$ is $m_P = 2x$. Taking $x = 3$, we get the point $P = (3, 9)$ and $m_P = 6$ (as we already saw).

Example 5.5. Find the slope m_P of the tangent line of the parabola $y = x^2$ at the point $P = (-4, 16)$.

Solution. The slope of the tangent at any point (x, y) is $2x$ for any x. Since $x = -4$, it follows that $m_P = 2 \cdot (-4) = -8$ at $P = (-4, 16)$.

Leibniz next turns to the graph of the equation $y = Ax^3$ where A is a constant. As before, $P = (x, y)$ is a fixed point and $Q = (x + \Delta x, y + \Delta y)$ any other point on the graph. Another look at Figure 5.11 will let you visualize what is going on. Since both P and Q

are on the graph, $y = Ax^3$ and

$$y + \Delta y = A(x + \Delta x)^3$$

$$= A\left(x^3 + 3x^2\Delta x + 3x(\Delta x)^2 + (\Delta x)^3\right).$$

After subtracting $y = Ax^3$ from both sides, he gets $\Delta y = 3Ax^2\Delta x + 3Ax(\Delta x)^2 + A(\Delta x)^3$. Dividing both sides by Δx, he has

$$\frac{\Delta y}{\Delta x} = 3Ax^2 + 3xA\Delta x + A(\Delta x)^2.$$

Pushing Δx to zero, Leibniz can conclude that

$$\lim_{\Delta x \to 0} \frac{\Delta y}{\Delta x} = \lim_{\Delta x \to 0}\left(3Ax^2 + 3xA\Delta x + A(\Delta x)^2\right) = 3Ax^2.$$

He has shown that the slope of the tangent line to the curve $y = Ax^3$ at any point $P = (x, y)$ is equal to $m_P = 3Ax^2$.

Example 5.6. Consider the curve $y = x^3$ and the point $P = (2, 8)$ on it. Taking $A = 1$ and $x = 2$ in the discussion just completed, we see that the slope of the tangent at P is $m_P = 3 \cdot 2^2 = 12$. Notice that the equation of the tangent is $y - 8 = 12(x - 2)$, or $y = 12x - 16$.

What happens when Leibniz's method is applied to a line? Let's take $y = \frac{1}{2}x + 1$. See Figure 5.13. Since P and Q both satisfy this equation, we get

$$y + \Delta y = \frac{1}{2}(x + \Delta x) + 1 = \frac{1}{2}\Delta x + \frac{1}{2}x + 1,$$

and hence $\Delta y = \frac{1}{2}\Delta x$. So $\frac{\Delta y}{\Delta x} = \frac{1}{2}$. Pushing Δx to zero has no effect on $\frac{1}{2}$, and therefore $\lim_{\Delta x \to 0} \frac{\Delta y}{\Delta x} = \frac{1}{2}$. This confirms the obvious: The tangent at any point P on the line is the line itself, and it has slope $\frac{1}{2}$.

Consider the parabola $x = y^2$ next. As before, fix a point $P = (x, y)$ on the parabola and let $Q = (x + \Delta x, y + \Delta y)$ be any other point on it. Since the coordinates of Q satisfy the equation of the parabola,

$$x + \Delta x = (y + \Delta y)^2 = y^2 + 2y\Delta y + (\Delta y)^2.$$

Since P does also,

$$\Delta x = 2y\Delta y + (\Delta y)^2 = \Delta y(2y + \Delta y)$$

by subtracting and factoring. Recall that the focus is on $\frac{\Delta y}{\Delta x} = \frac{1}{2y + \Delta y}$. Push Δx to zero and refer to Figure 5.11. As Q approaches P, the quantity Δy goes to zero. It follows, therefore, that the slope of the tangent to the parabola at $P = (x, y)$ is

$$m_P = \lim_{\Delta x \to 0} \frac{\Delta y}{\Delta x} = \frac{1}{2y}.$$

Example 5.7. What is the slope of the tangent of the parabola $x = y^2$ at $P = (3, \sqrt{3})$? What is the slope of the tangent at the origin?

Solution. Since $y = \sqrt{3}$ it follows from the formula just established that $m_P = \frac{1}{2\sqrt{3}}$. The origin is the point $P = (0, 0)$. When $y = 0$, the formula $m_P = \frac{1}{2y}$ does not make sense. This is explained by the fact that the tangent to the parabola $x = y^2$ at the origin is vertical. So this tangent has no slope. Confirm this by sketching a graph of $x = y^2$.

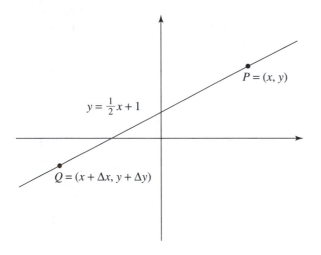

Figure 5.13

Finally, consider the circle $x^2 + y^2 = 5^2$. Once more, let $P = (x, y)$ be a fixed point on the circle and let $Q = (x + \Delta x, y + \Delta y)$ be another point on the circle. Since Q is on the circle, $(x + \Delta x)^2 + (y + \Delta y)^2 = 5^2$, so

$$x^2 + 2x\Delta x + (\Delta x)^2 + y^2 + 2y\Delta y + (\Delta y)^2 = 25.$$

Since $x^2 + y^2 = 5^2$, we get that

$$2x\Delta x + (\Delta x)^2 + 2y\Delta y + (\Delta y)^2 = 0.$$

After collecting the Δy terms on the left and the Δx terms on the right and factoring, we see that

$$\Delta y(2y + \Delta y) = -\Delta x(2x + \Delta x).$$

We need to concentrate on the ratio

$$\frac{\Delta y}{\Delta x} = -\frac{2x + \Delta x}{2y + \Delta y}.$$

Now push Δx to zero. Refer back to Figure 5.11 once more. Since Q goes to P, note that Δy goes to zero. So it follows that the slope of the tangent to the circle at the point $P = (x, y)$ is

$$m_P = \lim_{\Delta x \to 0} \frac{\Delta y}{\Delta x} = -\frac{2x}{2y} = -\frac{x}{y}.$$

Example 5.8. Observe that the slope of the tangent to the circle $x^2 + y^2 = 5^2$ at the point $P = (3, 4)$ is $-\frac{3}{4}$. What is the problem at the points $(5, 0)$ and $(-5, 0)$?

5.3 Areas and Differentials

In a manuscript written in October 1675, Leibniz considers a coordinate system and a curve **C**, as shown in Figure 5.14(a). He places a few points on the x-axis between a and b, starting with a and ending with b. He denotes a typical one of these points by x and lets dx be the distance between it and the next point. Consider any two successive points that Leibniz has placed on the axis. Let x be the first one; notice that the second one is $x + dx$. Let y be the y-coordinate of the point on **C** that lies above x. Draw the rectangle determined by x and $x + dx$, along with the point (x, y). This rectangle is shown in Figure 5.14(b). Observe that its area is $y \cdot dx$. Figure 5.14(c) shows all of the rectangles that are obtained in this way from the points that Leibniz has selected. Observe that these rectangles, when taken together, fill out—in an approximate way—the region under the curve **C** over the interval from a to b.

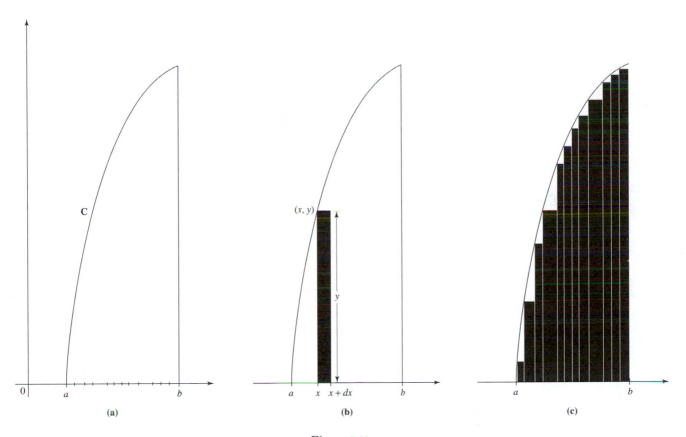

Figure 5.14

Let's have a look at some very specific examples of what Leibniz had in mind.

Example 5.9. Consider the area under the line $y = \frac{2}{3}x + 1$ and over the interval from 1 to 5. Insert the points

$$1 \le 1.7 \le 2.6 \le 3.3 \le 3.8 \le 4.4 \le 5$$

between 1 and 5. The first point is $x = 1$, so the corresponding y is $y = \frac{2}{3} \cdot 1 + 1 = \frac{5}{3}$; the first $dx = 1.7 - 1 = 0.7 = \frac{7}{10}$, and the first $y \cdot dx = \frac{5}{3} \cdot \frac{7}{10} = \frac{35}{30} = \frac{175}{150}$. The second point is $x = 1.7 = \frac{17}{10}$, so the corresponding y is $y = \frac{2}{3} \cdot \frac{17}{10} + 1 = \frac{32}{15}$; the second $dx = 2.6 - 1.7 = 0.9 = \frac{9}{10}$, and the second $y \cdot dx = \frac{32}{15} \cdot \frac{9}{10} = \frac{288}{150}$. The third point is $x = 2.6 = \frac{26}{10}$, so the corresponding y is $y = \frac{2}{3} \cdot \frac{26}{10} + 1 = \frac{41}{15}$; the third $dx = 3.3 - 2.6 = 0.7 = \frac{7}{10}$, and the third $y \cdot dx = \frac{41}{15} \cdot \frac{7}{10} = \frac{287}{150}$. The fourth point is

$x = 3.3 = \frac{33}{10}$, so the corresponding y is $y = \frac{2}{3} \cdot \frac{33}{10} + 1 = \frac{48}{15}$; the fourth $dx = 3.8 - 3.3 = 0.5 = \frac{5}{10}$, and the fourth $y \cdot dx = \frac{48}{15} \cdot \frac{5}{10} = \frac{240}{150}$. The fifth point is $x = 3.8 = \frac{38}{10}$, so the corresponding y is $y = \frac{2}{3} \cdot \frac{38}{10} + 1 = \frac{53}{15}$; the fifth $dx = 4.4 - 3.8 = 0.6 = \frac{6}{10}$, and the fifth $y \cdot dx = \frac{53}{15} \cdot \frac{6}{10} = \frac{318}{150}$. The sixth point is $x = 4.4 = \frac{44}{10}$, so the corresponding y is $y = \frac{2}{3} \cdot \frac{44}{10} + 1 = \frac{59}{15}$; the sixth $dx = 5 - 4.4 = 0.6 = \frac{6}{10}$, and the sixth $y \cdot dx = \frac{59}{15} \cdot \frac{6}{10} = \frac{354}{150}$. All the points have been dealt with and the process is complete. The sum of all the rectangular areas $y \cdot dx$ that the points $1 \le 1.7 \le 2.6 \le 3.3 \le 3.8 \le 4.4 \le 5$ determine is

$$\frac{175}{150} + \frac{288}{150} + \frac{287}{150} + \frac{240}{150} + \frac{318}{150} + \frac{354}{150} = \frac{1662}{150} = 11.08.$$

What is the actual area under the line between 1 and 5? If $x = -\frac{3}{2}$, then $y = \frac{2}{3}(-\frac{3}{2}) + 1 = -1 + 1 = 0$, and so the line $y = \frac{2}{3}x + 1$ crosses the x-axis at

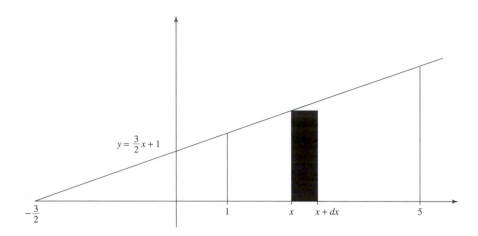

Figure 5.15

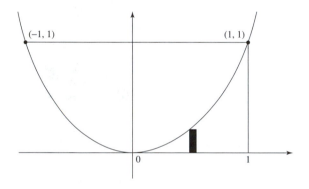

Figure 5.16

$x = -\frac{3}{2}$, as shown in Figure 5.15. So the actual area is the difference between two triangles: a larger triangle with base $5 + \frac{3}{2} = \frac{13}{2}$, height $\frac{2}{3} \cdot 5 + 1 = \frac{13}{3}$, and area $\frac{1}{2} \cdot \frac{13}{2} \cdot \frac{13}{3} = \frac{169}{12}$; and a smaller triangle with base $\frac{3}{2} + 1 = \frac{5}{2}$, height $\frac{2}{3} \cdot 1 + 1 = \frac{5}{3}$, and area $\frac{1}{2} \cdot \frac{5}{2} \cdot \frac{5}{3} = \frac{25}{12}$. The actual area is the difference, $\frac{169}{12} - \frac{25}{12} = \frac{144}{12} = 12$. So 11.08 is an approximation, but only a rough one.

Example 5.10. Next let's look at the parabola $y = x^2$ over the interval $0 \le x \le 1$, as pictured in Figure 5.16. Consider the points

$$0 \le 0.1 \le 0.3 \le 0.5 \le 0.8 \le 1$$

between 0 and 1. The first point is $x = 0$, so the corresponding y is 0; the first dx is $0.1 - 0 = 0.1$, and the first $y \cdot dx = (0)(0.1) = 0$. The second point is $x = 0.1$, so the corresponding y is $(0.1)^2 = 0.01$; the second $dx = 0.2$, and $y \cdot dx = (0.01)(0.2) = 0.002$. The third point is $x = 0.3$, so the corresponding y is $(0.3)^2 = 0.09$; the third dx is 0.2, and $y \cdot dx = (0.09)(0.2) = 0.018$. The fourth point is $x = 0.5$, so the corresponding y is $(0.5)^2 = 0.25$; the fourth dx is 0.3, and $y \cdot dx = (0.25)(0.3) = 0.075$. The fifth point is $x = 0.8$, so the corresponding y is $(0.8)^2 = 0.64$; the fifth dx is $1 - 0.8 = 0.2$, and $y \cdot dx = (0.64)(0.2) = 0.128$. The sum of all the $y \cdot dx$ is equal to

$$0 + 0.002 + 0.018 + 0.075 + 0.128 = 0.223.$$

How good is this approximation of the area under the parabola from 0 to 1? The actual area can be computed as follows. The area of the triangle determined by the points $(-1, 1)$, $(1, 1)$, and the origin, has base 2 and height 1. So its area is 1 and hence by Archimedes's theorem, the area of the parabolic sector cut by the segment from $(-1, 1)$ to $(1, 1)$ is $\frac{4}{3} \cdot 1 = \frac{4}{3}$. It follows that the area *under* the parabola from -1 to 1 is $\frac{2}{3}$. So the area under the parabola from 0 to 1 is $\frac{1}{3} \approx 0.333$.

In Example 5.9, Leibniz's method provided the approximation 11.08 for an area that is actually 12. Since $\frac{12 - 11.08}{12} = \frac{0.92}{12} \approx 0.08$, this approximation is off by about 8%. In the case of Example 5.10, note that $\frac{0.333-0.223}{0.333} \approx \frac{0.11}{0.33} \approx 0.33$. So this error is a substantial 33%. Is there a flaw in Leibniz's method? No. The problem was simply that not enough rectangles were taken! For example, if many more points than just $0.1 \leq 0.3 \leq 0.5 \leq 0.8$ are inserted between 0 and 1, then much more satisfactory approximations can be achieved for the area of the parabolic region.

Return to Figure 5.14 and to Leibniz's discussion. Suppose now that there are lots of points between a and b on the x-axis and that they are all packed very tightly together. Think big here, as Archimedes did when he considered the universe packed with grains of sand. These points will give rise to a huge number of very thin rectangles under the curve. These rectangles, one next to the other, will together fill out the area under the curve **C**. Figure 5.17 illustrates this for the portion of the parabola that is considered in Example 5.10. Now, the sum of the areas of the rectangles will, for all practical purposes, be numerically equal to the area under the parabola, in the same way that, say, 0.33333333 is "equal" to $\frac{1}{3}$.

After various experiments with notation, Leibniz finally uses an elongated S (the Latin word for *Sum* is *Summa*) and writes $\int y \, dx$ to denote[3] the sum of all the $y \cdot dx$ determined by the very tightly packed set of points between a and b. In 1822, the French mathematician Fourier inserted a and b into the notation—to indicate that the terms run from $x = a$ to $x = b$. He thus wrote this sum as

$$\int_b^a y \, dx.$$

This is the modern notation, and we will make use of it here.

Example 5.11. Return to Figure 5.15 and the area under the line $y = \frac{2}{3}x + 1$ over the interval from 1 to 5. This time, insert the points

$$1 \leq 1.0001 \leq 1.0002 \leq 1.0003 \leq \ldots$$

$$\leq 4.9997 \leq 4.9998 \leq 4.9999 \leq 5$$

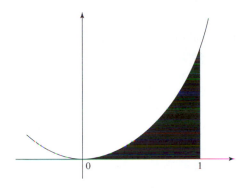

Figure 5.17

between 1 and 5. The first x is 1, the second is $1.0001 = 1 + \frac{1}{10,000}$, the third is $1.0002 = 1 + \frac{2}{10,000}$, the fourth is $x = 1.0003 = 1 + \frac{3}{10,000}, \ldots$, and the last is $x = 4.9999 = 1 + 3.9999 = 1 + \frac{39999}{10,000}$. Therefore, the typical x has the form $1 + \frac{i}{10,000}$ for i between 0 and 39,999. (Notice that tens of thousands of points have been packed between 1 and 5.) Since $y = \frac{2}{3}x + 1$, it follows that the typical $y \cdot dx$ is

$$y \cdot dx = \left[\frac{2}{3}\left(1 + \frac{i}{10,000}\right) + 1 \right] \cdot \frac{1}{10,000}.$$

This can be rewritten as follows:

$$y \cdot dx = \frac{1}{10,000} \cdot \left[\frac{2}{3}\left(1 + \frac{i}{10,000} + \frac{3}{2}\right) \right]$$

$$= \frac{2}{30,000} \cdot \left(1 + \frac{i}{10,000} + \frac{3}{2}\right)$$

$$= \frac{1}{15,000} \cdot \left(\frac{5}{2} + \frac{i}{10,000}\right)$$

$$= \frac{1}{15,000} \cdot \left(\frac{25,000}{10,000} + \frac{i}{10,000}\right)$$

$$= \frac{1}{1.5 \times 10^8} \cdot (25,000 + i).$$

By substituting $i = 0, 1, 2, \ldots, 39,999$ into this expression and adding everything, we see that the sum of all the $y \cdot dx$ is equal to

$$\frac{1}{1.5 \times 10^8} \cdot (25,000 + 25,001 + 25,002 + \cdots + 64,999).$$

The terms inside the parentheses can be added by the technique of Gauss the schoolboy (see Note 4 at the end of Chapter 4), as follows:

$$25{,}000 \; + \; 25{,}001 \; + \; 25{,}002 \; + \; \cdots \; + \; 64{,}998 \; + \; 64{,}999$$
$$64{,}999 \; + \; 64{,}998 \; + \; 64{,}997 \; + \; \cdots \; + \; 25{,}001 \; + \; 25{,}000$$

$$89{,}999 \; + \; 89{,}999 \; + \; 89{,}999 \; + \; \cdots \; + \; 89{,}999 \; + \; 89{,}999$$

Recall that i ranges from 0 to 39,999; therefore, each line in the preceding addition has 40,000 terms. It follows that twice

$$25{,}000 + 25{,}001 + 25{,}002 + \cdots + 64{,}999$$

is equal to $(40{,}000)(89{,}999) = 35.9996 \times 10^8$. So

$$25{,}000 + 25{,}001 + 25{,}002 + \cdots + 64{,}999 = 17.9998 \times 10^8.$$

Inserting this into the earlier expression for the sum of all the $y \cdot dx$, we get

$$\frac{1}{1.5 \times 10^8} \cdot \left(17.9998 \times 10^8\right) = \frac{17.9998}{1.5} = 11.9999$$

with four-decimal accuracy. Refer back to Example 5.9 for the fact that the area under the line $y = \frac{2}{3}x+1$ over the interval from 1 to 5 is 12. So we now have what we wanted: A good approximation of the area under the curve. The equation

$$\int_1^5 y\,dx = \int_1^5 \left(\frac{2}{3}x + 1\right)\,dx = 11.9999 = 12$$

summarizes what was done in Leibniz's notation.

It is important to think of the symbol $\int_a^b y\,dx$ as more than just a number equal to the area under the curve. Include in your thinking the summation process—of a very large number of very small terms—that produces this number. This way of thinking about area will soon have important consequences.

The formula

$(*)$
$$\boxed{\int_a^b cy\,dx = c\int_a^b y\,dx}$$

where c is a positive constant, is obtained by factoring out a c from each term $y\,dx$ of the sum on the left. As the following example shows, this equation is Cavalieri's principle expressed in Leibniz's notation.

Example 5.12. Consider the circle $x^2+y^2 = 5^2$. Solve for y to get $y = \pm\sqrt{25 - x^2}$. Taking the positive value for y gives the equation $y = \sqrt{25 - x^2}$ for the upper half of this circle. See Figure 5.18(a). It follows that $\int_{-5}^5 y\,dx = \int_{-5}^5 \sqrt{25 - x^2}\,dx$ is equal to the area under the upper half of the circle. Since the radius is 5, we get $\int_{-5}^5 \sqrt{25 - x^2}\,dx = \frac{1}{2}\pi 5^2 = \frac{25\pi}{2}$.

Now consider the ellipse $\frac{x^2}{5^2} + \frac{y^2}{2^2} = 1$. Solving for y gives us $\frac{y^2}{2^2} = 1 - \frac{x^2}{5^2} = \frac{5^2-x^2}{5^2}$, or $y^2 = \frac{2^2}{5^2}(5^2 - x^2)$. Take $y = \frac{2}{5}\sqrt{5^2 - x^2}$ and note that the graph of this equation is the upper half of the ellipse. (See Figure 5.18(b).) So $\int_{-5}^5 \frac{2}{5}\sqrt{25 - x^2}\,dx$ is the area under the upper half

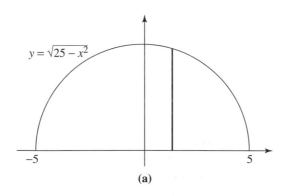

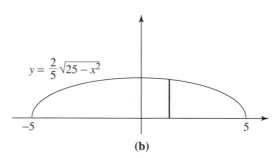

(a)

(b)

Figure 5.18

of the ellipse. By formula (∗) above,

$$\int_{-5}^{5} \frac{2}{5}\sqrt{25-x^2}\,dx = \frac{2}{5}\int_{-5}^{5}\sqrt{25-x^2}\,dx$$

$$= \frac{2}{5} \cdot \frac{25\pi}{2} = 5\pi.$$

Therefore, the area of the full ellipse is $2(5\pi) = 10\pi$. This is a result that was achieved in Section 4.6 by use of Cavalieri's principle.

5.4 **The Fundamental Theorem of Calculus**

W e turn next to a publication of Leibniz[4] from 1693. In it, Leibniz discusses a problem posed to him by a Parisian scholar:

Claude Perrault ... who has ... distinguished himself with his studies in mechanics and architecture has presented to me the following problem. He had presented it to many others before me and openly admitted that he had not as yet been successful in solving it himself. A weight is placed on a horizontal plane. One end, say B, of a piece of string or small chain AB is attached to the weight. When the other end A of the string is moved along the fixed straight line AA' in the plane, so that the string AB lies in the plane for the duration of the motion, the weight is pulled along, and describes a curve BB'. The problem concerns the determination of this curve. [See Figure 5.19.] Perrault instead made use (as an illustration) of a pocketwatch with silver housing B, which he pulled across the table with the attached chain AB by leading the end A along a straight line AA'. In the process the lowest point of the housing (the midpoint of the bottom) described the curve BB'. I observed this curve with some care (I had been occupying myself primarily with the consideration of tangents) and immediately made the relevant remark that the string always touches the curve tangentially, e.g., the straight line $A''B''$ is the tangent of the curve BB' at B'' ... . I saw, therefore, that the question reduced to the following: To find a curve satisfying the condition that the section of the tangent AB between the axis AA' and the curve BB' is equal to a fixed constant. And it was not difficult for me to figure out that the description of this curve could be reduced to the quadrature of the hyperbola.

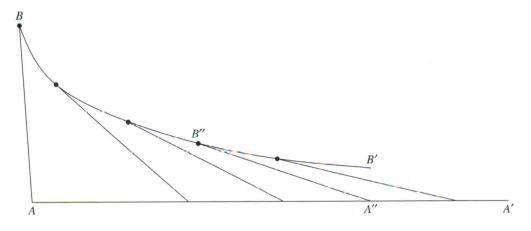

Figure 5.19

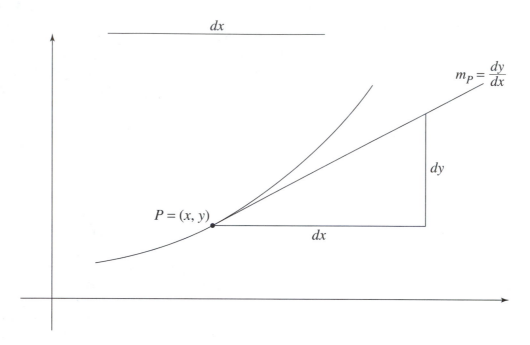

Figure 5.20

Since quadrature, or "squaring," refers to the determination of area. Leibniz has made reference to a fundamental relationship between and slopes of tangent lines and areas under curves. It is the purpose of the remainder of his article to bring this connection to light. We will describe the essence of what Leibniz does.

Let a curve be given and let $P = (x, y)$ be a point on the curve. Consider the tangent to the curve at P and let m_P be its slope; refer to Figure 5.20. Let dx be any length and form the triangle shown. As in the earlier discussion about areas, dx is often considered to be small. Observe that $\frac{dy}{dx} = m_P$, so that

$$dy = m_P \cdot dx.$$

Leibniz calls dx the *differential of* x and the related quantity $dy = m_P \cdot dx$ the *differential of* y. He calls the triangle formed by the segments of lengths dx and dy and the tangent the *characteristic triangle*. For example, in the case of the parabola $y = x^2$, the slope of the tangent is $m_P = 2x$, so that $dy = 2x \cdot dx$. In the case

of the curve $y = x^3$ the slope of the tangent is $m_P = 3x^2$ and $dy = 3x^2 \cdot dx$. The construction pictured in Figure 5.20 will provide Leibniz with the link between areas under curves and slopes of tangent lines.

Leibniz considers a system of Cartesian coordinates and fixes a curve **C** above an interval from a to b on the x-axis. He now assumes that he has another curve **A** with the following property for all points x between a and b:

The slope m_P of the tangent at the point P on **A** lying above x is equal to the y-coordinate of the point on **C** lying above x.

Figure 5.21 illustrates the connection between **C** and **A**: For any number x between a and b, let P and Q be the points on **A** and **C** respectively, both with first coordinate x. Then the slope m_P of the tangent to **A** at P is equal to the second coordinate of the point Q on **C**. So

$$m_P = y.$$

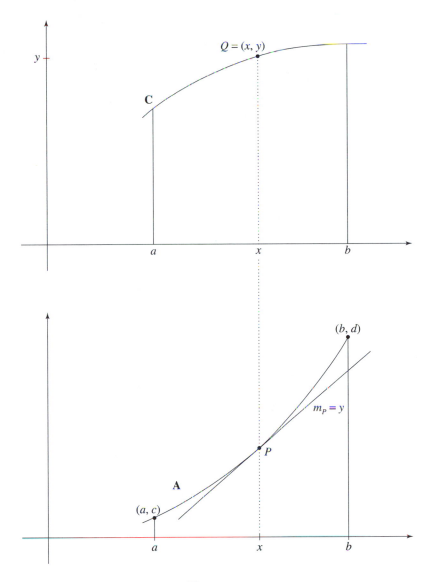

Figure 5.21

Next Leibniz takes any dx and places a character-istic triangle with sides dx and dy at the point P on **A**; see Figure 5.22. Observe that $\frac{dy}{dx} = m_P$ and therefore, that

$$\frac{dy}{dx} = y.$$

He then places a rectangle under the curve **C**, as shown (in black) in Figure 5.23. Since $y = \frac{dy}{dx}$, observe

that the area $y \cdot dx$ of the rectangle is equal to the length dy of the vertical leg of the triangle at P. (How, you ask, is this equality possible, given that the rectangle is large and the segment tiny?)

Now pack lots of points between a and b. These determine, exactly as described in Section 5.3, many very thin rectangles under the curve **C**. As we just saw, the area $y \cdot dx$ of each rectangle is numerically

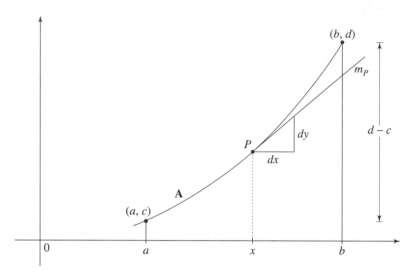

Figure 5.22

equal to the length of the corresponding segment dy. It is now plausible for Leibniz to conclude that the sum of the areas $y \cdot dx$ of all of the rectangles under **C** is equal to the sum of the lengths of all of the segments dy from c to d; see Figure 5.24. So Leibniz has shown that the area under **C** above the interval from a to b is equal to $d - c$. In his notation,

$$\int_a^b y \, dx = d - c.$$

This formula is known as the *Fundamental Theorem of Calculus*.

Let's consider a specific example to illustrate what Leibniz has accomplished.

Example 5.13. Consider the parabola $y = x^2$ over the interval $0 \le x \le 1$. This is the curve **C**. Recall from Section 5.2 that the slope of the tangent of $y = Ax^3$ at any point $P = (x, y)$ is equal to $m_P = 3Ax^2$. Taking $A = \frac{1}{3}$, we see that $m_P = x^2$. Therefore, the graph of $y = \frac{1}{3}x^3$ has the property that the slope of the tangent m_P at x is equal to the y-coordinate of the point on **C** at x. In other words, the graph of $y = \frac{1}{3}x^3$ satisfies what is required of the curve **A**. Since the point on **A** corresponding to $x = 0$ is $(0, 0)$ and that corresponding

to $x = 1$ is $(1, \frac{1}{3})$, the roles of the points (a, c) and (b, d) are now played by $(0, 0)$ and $(1, \frac{1}{3})$. It follows that the area under $y = x^2$ over the interval $0 \le x \le 1$ is $d - c = \frac{1}{3} - 0 = \frac{1}{3}$. Notice that the discussion that follows Example 5.10 has already confirmed this.

5.5 Functions

Leibniz's analysis of the area under a curve implicitly required that for a given x there is at most one point on the curve above that x. Refer to Figure 5.14(b) and notice that if there were more than one such point, the entire matter would be ambiguous. In terms of the equation of the curve, the requirement is that for a given x, there is at most one y with the property that the pair (x, y) satisfies the equation. This condition is met by the equation $y = x^2$. For example, the only point (x, y) on this graph with $x = 2$ is $(2, 4)$, since $y = 2^2 = 4$. Similarly, when $x = 3$, y must be equal to $3^2 = 9$, and so on. A similar thing is true for the equations $y = x^3$, or $y = \sqrt{x}$, or $y = \frac{1}{x}$, or $y = \frac{1}{x^2}$. In each of these cases, we see that for a given x, there is at most one y that satisfies the equation. So a given x determines a single y. This property is the essen-

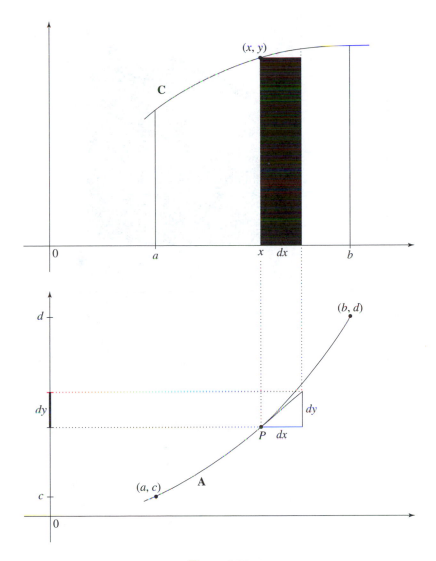

Figure 5.23

tial element in the concept of a function. Leibniz used the notion of a function only in the context of quantities related to curves. The more definitive concept presented next was not established until later.

A *function* is a rule that assigns exactly one real number to each number from a set of real numbers. Such a rule is often given by an algebraic expression. The rule $f(x) = 3x - 7$ is an example of a function. Note that $f(2) = 3 \cdot 2 - 7 = -1$ and $f(4) = 3 \cdot 4 - 7 = 5$. So

f assigns -1 to the number 2, and 5 to the number 4, and $3x - 7$ to a typical real number x. Functions of the form $f(x) = 3x - 7$, $g(x) = -\frac{1}{2}x + 6$, or more generally, $h(x) = mx + b$, where m and b are constants, are called *linear functions*. Functions of the form $f(x) = 6$, or $g(x) = -7$, or $h(x) = b$, with b any constant, are called *constant functions*.

The *graph* of a function f is the set of all points (x, y) in the Cartesian plane that satisfy the equation

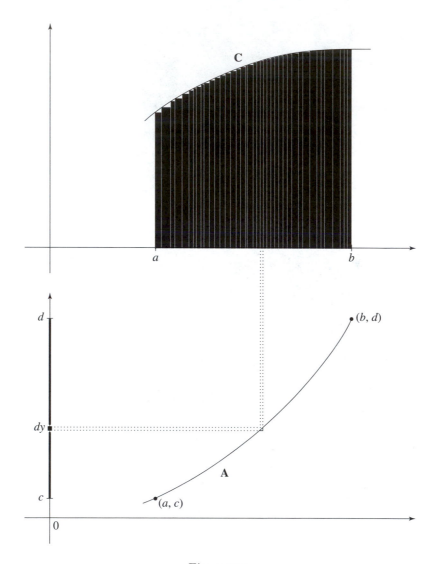

Figure 5.24

$y = f(x)$. So the graphs of the linear functions $f(x) = 3x - 7$, $g(x) = -\frac{1}{2}x + 6$, and $h(x) = mx + b$ are the graphs of the equations $y = 3x - 7$, $y = -\frac{1}{2}x + 6$, and $y = mx + b$, respectively. Observe that they are all lines.

Example 5.14. (a) The rule f that assigns the number x^2 to any x is the function $f(x) = x^2$. This rule is *defined*, i.e., makes sense, for all x. For ex-

ample, $f(2) = 4$, $f(3) = 9$, and $f(-2) = 4$. So the points $(2, 4)$, $(3, 9)$, and $(-2, 4)$ are all on the graph.

(b) The rule $f(x) = \sqrt{x} = x^{\frac{1}{2}}$ is a function. Observe that $f(0) = 0$, $f(1) = 1$, $f(3) = \sqrt{3}$, and $f(4) = 2$. So the points $O = (0, 0)$, $(1, 1)$, $(3, \sqrt{3})$, and $(4, 2)$ are on the graph. This function is defined only for $x \geq 0$. For example, $f(-2) = \sqrt{-2}$ is not defined, i.e., it does not make sense.

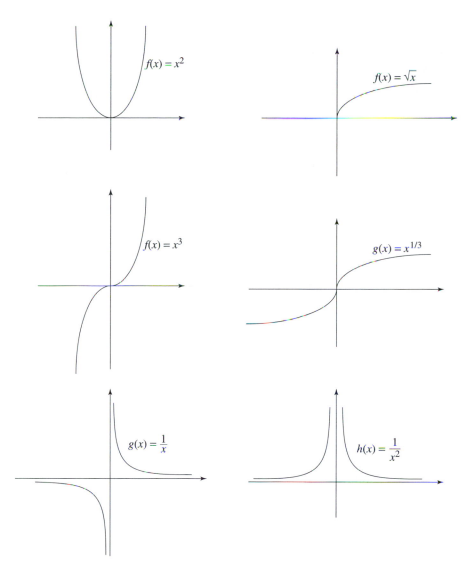

Figure 5.25

(c) The rule $f(x) = x^3$ is a function. It is defined for all x. Note that $(1, 1)$, $(-1, -1)$, and $(-2, -8)$ are on the graph.

(d) The rule $g(x) = \sqrt[3]{x} = x^{\frac{1}{3}}$ is a function defined for all x. Check that $(-1, -1)$, $(8, 2)$, and $(-27, -3)$ are on its graph.

(e) The rule g given by $g(x) = \frac{1}{x} = x^{-1}$ is a function. It is defined for all x except $x =$ 0. The points $(1, 1)$, $(2, \frac{1}{2})$, and $(\frac{1}{2}, 2)$ are on the graph.

(f) The rule $h(x) = \frac{1}{x^2} = x^{-2}$ is a function defined for all $x \neq 0$. Since $h(2) = \frac{1}{4}$, $h(-1) = 1$, and $h(3) = \frac{1}{3}$, the points $(2, \frac{1}{4})$, $(-1, 1)$, and $(3, \frac{1}{3})$ are on the graph.

The graphs of the functions considered in (a)–(f) are sketched in Figure 5.25.

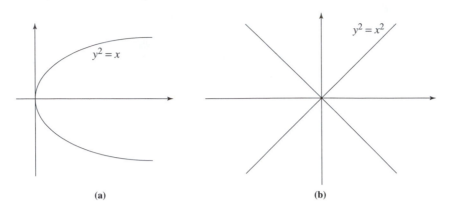

$y^2 = x$

$y^2 = x^2$

(a)

(b)

Figure 5.26

The functions $f(x) = x^2$, $g(x) = -x^2 + 3x - 4$, and, in general $h(x) = Ax^2 + Bx + C$, where $A \neq 0$, are called *quadratic functions*. Their graphs are the graphs of the equations $y = x^2$, $y = -x^2 + 3x - 4$, and $y = Ax^2 + Bx + C$, respectively. Note that the graph of any quadratic function is a parabola. (Refer to Section 4.3.)

Consider the equation $y^2 = x$. The pairs of numbers $x = 4, y = 2$ and $x = 4, y = -2$ are both solutions, so this equation does not determine y uniquely in terms of x. In general, for a given $x \geq 0$, both $y = \sqrt{x}$ and $y = -\sqrt{x}$ are solutions. Similarly, in reference to the equation $y^2 = x^2$, for a given x, both $y = x$ or $y = -x$ are solutions. Again, for a given x, there are two possibilities for y. Therefore neither equation defines a function of x.

A look at the graph of an equation in x and y makes this more explicit. If any vertical line crosses the graph at most once, then the equation defines a function of x. But if at least one vertical line crosses the graph twice or more, then it does not define a function of x. A comparison of the graphs of Figure 5.25 with those of Figure 5.26 illustrates the point.

The function concept clarifies the topics of this chapter. We will now incorporate it into our discussion and use it to reformulate some of the matters that were already developed. We begin with another look at the computation of the slope of a tangent line.

A. The Derivative. Let f be a function and fix a point $P = (x, y)$ on its graph. Since $y = f(x)$, it follows that $P = (x, f(x))$. Suppose the graph has a nonvertical tangent at P. Since P is the only point on the graph with this x-coordinate, we will designate the slope of the tangent by m_x instead of m_P as was done in Section 5.2. Now let $Q = (\bar{x}, \bar{y})$ be some other point on the graph; then $\bar{y} = f(\bar{x})$. Set $\bar{x} - x = \Delta x$ and $\bar{y} - y = \Delta y$. Note that $\bar{x} = x + \Delta x$ and $\bar{y} = f(\bar{x}) = f(x + \Delta x)$. So $Q = (x + \Delta x, f(x + \Delta x))$. See Figure 5.27. Observe that the slope of the line through P and Q is

$$\frac{f(x + \Delta x) - f(x)}{\Delta x}.$$

Since $\Delta y = \bar{y} - y = f(x + \Delta x) - f(x)$, this is the ratio $\frac{\Delta y}{\Delta x}$ that was studied in Section 5.2. When Δx is pushed to zero, the slope of this line closes in on the slope m_x of the tangent to the graph at $P = (x, f(x))$. In the notation of limits, this can be expressed as

$$m_x = \lim_{\Delta x \to 0} \frac{f(x + \Delta x) - f(x)}{\Delta x}.$$

Now consider the rule that assigns the number m_x to a given number x. This rule defines a function

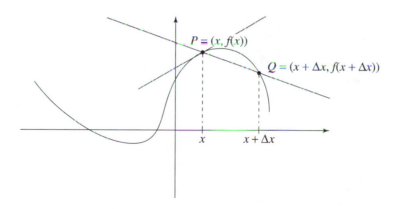

Figure 5.27

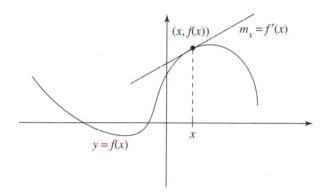

Figure 5.28

called the *derivative of f*, denoted by f'. To repeat, f' is the function given by the rule

$$f'(x) = m_x,$$

where m_x is the slope of the tangent to the graph of f at the point $(x, f(x))$. See Figure 5.28. Once more, the derivative of the function f is also a function. Its rule f' is derived from the rule of f in the manner just described.

Suppose that f is a linear function. Then it has the form $f(x) = mx + b$, with m and b constants. Its graph is a line with slope m. Take any point (x, y) on the graph. The tangent to the line is the line itself, so it follows that $m_x = m$. Therefore, $f'(x) = m$ for all x. Suppose that $m = 0$. Then $f(x) = b$, and $f'(x) = 0$ for all x. So the derivative of any constant function is the zero function.

Example 5.15. The derivative of $f(x) = \frac{1}{2}x + 1$ is $f'(x) = \frac{1}{2}$. The derivative of $f(x) = -3x + 6$ is $f'(x) = -3$. If $g(x) = 5$, then $g'(x) = 0$.

What are the derivatives of the functions of Examples 5.14(a)–14(f)?

Example 5.16. (a) Consider $f(x) = x^2$. Since

$$\frac{f(x + \Delta x) - f(x)}{\Delta x} = \frac{(x + \Delta x)^2 - x^2}{\Delta x}$$

$$= \frac{x^2 + 2x\Delta x + (\Delta x)^2 - x^2}{\Delta x}$$

$$= \frac{2x\Delta x + (\Delta x)^2}{\Delta x} = \frac{\Delta x(2x + \Delta x)}{\Delta x}$$

$$= 2x + \Delta x,$$

it follows that

$$f'(x) = \lim_{\Delta x \to 0} \frac{f(x + \Delta x) - f(x)}{\Delta x} = 2x.$$

(b) Let $y = f(x) = \sqrt{x} = x^{\frac{1}{2}}$. We will not compute $f'(x) = \lim_{\Delta x \to 0} \frac{f(x + \Delta x) - f(x)}{\Delta x}$ directly (try it and you will probably encounter difficulties) but instead refer to Section 5.2. By squaring both sides, we get $y^2 = x$. The graph of $f(x) = \sqrt{x} = x^{\frac{1}{2}}$ is the upper half of the graph of the parabola $x = y^2$; see Figure 5.26(a). It was shown in Section 5.2, using Leibniz's tangent method, that the slope of the tangent to this parabola at any point $P = (x, y)$ is $m_P = \frac{1}{2y}$. If the point P is on the graph of f, then $y = x^{\frac{1}{2}}$ and hence $m_x = m_P = \frac{1}{2}x^{-\frac{1}{2}}$. Therefore, $f'(x) = \frac{1}{2}x^{-\frac{1}{2}}$.

(c) Let $f(x) = x^3$. It was shown in Section 5.2, using Leibniz's tangent method, that the slope of the tangent to the curve $y = x^3$ at any point $P = (x, y)$ is $m_P = 3x^2$. It follows that $f'(x) = 3x^2$. This derivative can also be established by proceeding as in (a). See Exercise 10.

(d) The derivative of $g(x) = x^{\frac{1}{3}}$ is $g'(x) = \frac{1}{3}x^{-\frac{2}{3}}$. This is best shown by applying Leibniz's tangent method to the equation $y^3 = x$. See Exercise 8.

(e) The derivative of $g(x) = x^{-1}$ is $g'(x) = -\frac{1}{x^2} = -(x^{-2})$. This can be done as in (a) or with Leibniz's tangent method. See Exercise 7.

(f) The derivative of $h(x) = \frac{1}{x^2} = x^{-2}$ is $h'(x) = -2\frac{1}{x^3} = -2x^{-3}$. Do this by computing the limit $h'(x) = \lim_{\Delta x \to 0} \frac{h(x + \Delta x) - h(x)}{\Delta x}$ directly. See Exercise 11.

Example 5.17. What is the slope of the tangent to the graph of $f(x) = \sqrt{x} = x^{\frac{1}{2}}$ at the point $(2, \sqrt{2})$? This is computed by substituting $x = 2$ into $f'(x) = \frac{1}{2}x^{-\frac{1}{2}}$. So it is $f'(2) = \frac{1}{2}2^{-\frac{1}{2}} = \frac{1}{2\sqrt{2}}$. What about the slope of the tangent to the graph of $h(x) = \frac{1}{x^2}$ at $(-1, 1)$? This slope is computed by evaluating $h'(x) = -2x^{-3}$ at $x = -1$. Hence it is $h'(-1) = (-2)(-1) = 2$.

The notation $\frac{dy}{dx}$ of Section 5.4 is also used to denote the derivative of a function. So if $y = x^3$, then $\frac{dy}{dx} = 3x^2$. And if $y = \frac{1}{x^2}$, then $\frac{dy}{dx} = -2x^{-3}$.

Leibniz develops many of the basic properties of the derivative. We will turn to these in detail in Chapter 8. A few examples will do for now. Consider the function $f(x) = 5\sqrt{x} = 5x^{\frac{1}{2}}$. The derivative of f is equal to 5 times the derivative of the function $y = \sqrt{x}$. So by Example 5.16(b), $f'(x) = 5\left(\frac{1}{2}x^{-\frac{1}{2}}\right) = \frac{5}{2\sqrt{x}}$. The point is that when computing the derivative of a function, constants can be factored out first. Another basic property asserts that the derivative of the sum (or difference) of two functions is equal to the sum (or difference) of their derivatives.

Example 5.18. Using Example 5.16, we see that the derivative of $y = \frac{1}{x^2} + x^3 + 4x$ is equal to $\frac{dy}{dx} = (-2)\frac{1}{x^3} + 3x^2 + 4$. In a similar way, if $f(x) = 2x^2 - \frac{5}{x} - 7$, then $f'(x) = 4x + 5\frac{1}{x^2}$.

B. Antiderivatives.   Let f be a function. A function F that has the property that the derivative F' is equal to f is called an *antiderivative* of f.

Example 5.19. Let $F(x) = x^2$ and recall that $F'(x) = 2x$. Therefore $F(x)$ is an antiderivative of $f(x) = 2x$. Note that $G(x) = \frac{1}{3}x^3$ is an antiderivative of $g(x) = x^2$. Similarly, if $F(x) = x + 225\frac{1}{x}$, then $F'(x) = 1 - 225\frac{1}{x^2} = \frac{x^2 - 225}{x^2}$. Therefore, $F(x) = x + 225\frac{1}{x}$ is an antiderivative of $f(x) = \frac{x^2 - 225}{x^2}$.

Let's consider the situation of a function h that has the property that $h'(x) = 0$ for all x. This means that the slope of the tangent at any point P on the graph is zero. So all tangents are horizontal. If the graph were to curve upward or downward at any point P, the tangent line would either rise or fall; see Figure 5.29. Since this does not happen, the only possibility is that the graph is a horizontal line. Because every horizontal line has the form $y = C$ for some constant C, it follows that $h(x) = C$ for all x. In words, a function whose derivative is zero must be a constant function.

There is an important consequence of this. Suppose that $y = F(x)$ and $y = G(x)$ are both antiderivatives of the same function $f(x)$. Since

$$F'(x) = f(x) \quad \text{and} \quad G'(x) = f(x),$$

it follows that the derivative of $G(x) - F(x)$ is equal to

$$G'(x) - F'(x) = 0.$$

Therefore, by the fact just pointed out, $G(x) - F(x)$ is equal to a constant C; so

$$G(x) = F(x) + C.$$

Suppose that one antiderivative of a function f is known. We have shown that any other antiderivative

Figure 5.29

of f can obtained by adding a constant to the known antiderivative.

Let's return to the Fundamental Theorem of Calculus of Section 5.4. Let f be a function. Restrict x to lie in an interval $a \leq x \leq b$ and suppose that $y = f(x) \geq 0$ for all such x. As in Section 5.3, pack lots of points into the interval from a to b and form the products $y \cdot dx = f(x) \cdot dx$. Recall that

$$\int_a^b y \, dx = \int_a^b f(x) \, dx$$

is the sum of all these products.

Suppose that the function F is an antiderivative of f. So $F'(x) = f(x)$ for all x. Let **A** be the graph of F and let $P = (x, y)$ be any point on **A**. Since $F'(x) = m_P$ is the slope of the tangent to **A** at the point P, notice that

$$F'(x) = m_P = f(x).$$

Therefore, the graph **A** of F fulfills the requirements of Leibniz's development of the Fundamental Theorem of Calculus. Refer back to Figure 5.21. Because (a, c) and (b, d) are on the graph **A** of F, it follows that $F(a) = c$ and $F(b) = d$. The equality $\int_a^b y \, dx = d - c$ of Section 5.4 can therefore be rewritten

$$\boxed{\int_a^b f(x) \, dx = F(b) - F(a)}$$

In summary, this equality means this: To evaluate a sum of the form $\int_a^b f(x) \, dx$, find an antiderivative F of f and compute the difference $F(b) - F(a)$. The expression $\int_a^b f(x) \, dx$ is called the *definite integral* of

the function f from a to b. It is now customary to use the notation $F(x)\big|_a^b$ for the difference $F(b) - F(a)$.

5.6 **Some Applications**

W e continue with a glimpse at some applications of the derivative. Consider a function f and its graph. Have a look at Figure 5.30 and consider a high point on the graph (the "top of a peak"). If c_1 is the x-coordinate of such a point, then the y-coordinate $f(c_1)$ is larger than $f(x)$ for any x near c_1. Observe that the slope $f'(c_1)$ of the tangent is zero. In the same way, if c_2 is the x-coordinate of a low point on the graph (the "bottom of a valley"), then the y-coordinate $f(c_2)$ is smaller than $f(x)$ for any x near c_2. Here too, $f'(c_2)$ is zero. This suggests that the solution of a problem that calls for the determination of a maximum or minimum value of a function should focus on the values of x at which the derivative is equal to zero. We will illustrate this by considering two problems of the type considered early in the 17th century.

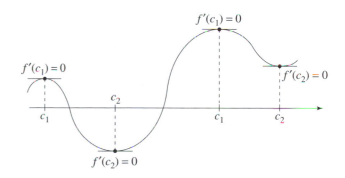

Figure 5.30

A. **Finding Maximum and Minimum Values.**

Consider a segment that is cut into two pieces as shown in Figure 5.31.

Figure 5.31

Example 5.20. Suppose that the total length of the segment is 1200 units. How is the segment to be cut so that the product of the lengths of the two pieces is as large as possible?

Solution. If x is the length of one of the pieces, then the other has length $1200 - x$. The product of the lengths is given by the function $f(x) = x(1200 - x) = 1200x - x^2$. The question is: What is the largest value of this function? Since $f'(x) = 1200 - 2x$, we see that $f'(x) = 0$ when $x = 600$. Because the graph of f is a parabola that opens downward, it follows that the product $f(x) = x(1200 - x)$ is largest when $x = 600$. Since $1200 - 600 = 600$, the length of the other piece is also 600 units.

Example 5.21. Suppose that the product of the lengths of the two pieces in Example 5.20 is specified to be 225. What is the shortest segment for which this is possible?

Solution. Let y be the length of the entire segment. Given that one piece has length x, the second piece has length $y - x$. The product of the lengths is $(y - x)x = 225$. Therefore, $yx - x^2 = 225$. So $yx = x^2 + 225$, and hence $y = x + \frac{225}{x}$. We need to find the minimum value of $f(x) = x + 225\frac{1}{x}$. By Example 5.16(e) and the properties of derivatives, $f'(x) = 1 - 225\frac{1}{x^2} = \frac{x^2 - 225}{x^2}$. In order for this to be equal to zero, the numerator $x^2 - 225$ must be zero. This happens when $x = \pm 15$. Since x is length, $x = 15$, and $f(15) = 15 + \frac{225}{15} = 15 + 15 = 30$. Experiment with other lengths and observe that 30 is the smallest possible length of the segment.

We conclude our discussion of the calculus of Leibniz by showing that the summation process that gives rise to the definite integral can be applied to problems other than the computation of area. We will illustrate this by computing volumes and lengths of curves.

B. Volumes. Begin with the fact that the volume of a cylinder is equal to the area of the base times the height. So if the height is h and the circular base has radius r, then the volume is $\pi r^2 h$; see Figure 5.32.

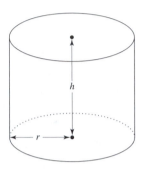

Figure 5.32

Now consider a semicircle of radius r and place a coordinate system as shown in Figure 5.33. As we have already done a number of times, pump the interval from 0 to $2r$ full of points, and consider the corresponding thin rectangles under the circle. A typical one is shown. It is placed at x, it is y units high, and it has thickness dx. Rotate the semicircular region one complete revolution about the x-axis and note that it sweeps out a sphere S of radius r. The thin rectangle sweeps out a thin cylindrical disc of radius y and thickness (height) dx. The circular base of this thin cylinder has area πy^2 and its volume is $\pi y^2 dx$. Since the point (x, y) satisfies $(x - r)^2 + y^2 = r^2$, observe that $y^2 = r^2 - (x - r)^2 = 2rx - x^2$. So the volume of the thin cylinder is

$$\pi y^2 \, dx = (2\pi r x - \pi x^2) \, dx.$$

Do this for all the rectangles from 0 to $2r$. Since all of the thin rectangles together fill out the semicircular region, it follows that the sum of the volumes of all the corresponding thin cylinders is the volume of the entire *solid of revolution*, i.e., the volume of the sphere S of radius r. It is now important to observe that this

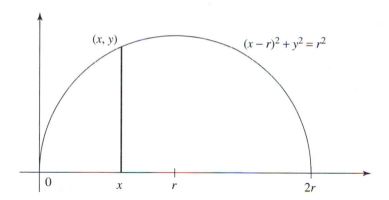

Figure 5.33

sum, namely, the sum of all the terms $(2\pi rx - \pi x^2)\,dx$, is precisely a sum of the type considered in Section 5.3. Therefore,

$$\text{Volume of } S = \int_0^{2r} (2\pi rx - \pi x^2)\,dx$$

$$= \pi \int_0^{2r} (2rx - x^2)\,dx.$$

Refer to the end of the previous section for the fact that this sum can be evaluated as follows: If $y = G(x)$ is an antiderivative of the function $g(x) = 2rx - x^2$, then

$$\pi \int_0^{2r} (2rx - x^2)\,dx = \pi \left[G(x) \big|_0^{2r} \right] = \pi[G(2r) - G(0)].$$

Because the function rx^2 is an antiderivative of $2rx$, and $\frac{1}{3}x^3$ is an antiderivative of x^2, it follows that $G(x) = rx^2 - \frac{1}{3}x^3$ is an antiderivative of $g(x) = 2rx - x^2$. Since

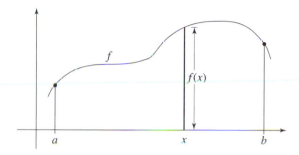

Figure 5.34

$$G(2r) - G(0) = r(2r)^2 - \tfrac{1}{3}(2r)^3,$$

$$\text{Volume of } S = \pi \int_0^{2r} (2rx - x^2)\,dx$$

$$= \pi \left(r(2r)^2 - \frac{1}{3}(2r)^3 \right) = 4\pi r^3 - \frac{8}{3}\pi r^3$$

$$= \frac{4}{3}\pi r^3.$$

Leibniz's summation method combined with the Fundamental Theorem of Calculus has shown the volume of a sphere S of radius r to be $\frac{4}{3}\pi r^3$.

Using this basic argument, many volumes of revolution can be computed. Let f be a function that satisfies $f(x) \geq 0$ for all x between a and b on the x-axis. A typical situation is sketched in Figure 5.34. Proceed exactly as in the situation of the semicircle. The interval from a to b is packed with points, and these determine thin rectangles under the curve. A typical one is shown. It is placed at x, it has height $f(x)$ and thickness dx. Now rotate the region bounded by the graph, the x-axis, and the lines $x = a$ and $x = b$ for one complete revolution about the x-axis. The volume of the resulting solid is obtained by adding the volumes of all of the thin cylindrical discs determined by the rotation of the rectangles. The typical disc shown has circular area $\pi \cdot (f(x))^2$ and height dx, so it has volume $\pi \cdot (f(x))^2 \cdot dx$. Since the volume V of the solid is the sum of all of the volumes of these discs, it is given by

$$V = \int_a^b \pi (f(x))^2 \, dx$$

Notice that this definite integral is also equal to the area under the graph of the function $\pi(f(x))^2$ from a to b.

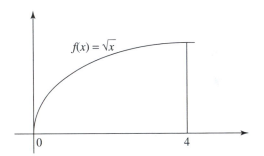

$f(x) = \sqrt{x}$

0   4

Figure 5.35

Example 5.22. Revolve the region shown in Figure 5.35 one complete revolution about the x-axis and compute the volume V of the solid thus obtained. Substituting $f(x) = \sqrt{x}$ into the formula just developed,

we find that

$$V = \int_a^b \pi (f(x))^2 \, dx = \int_0^4 \pi (\sqrt{x})^2 \, dx = \pi \int_0^4 x \, dx.$$

Since $G(x) = \frac{1}{2}x^2$ is an antiderivative of $g(x) = x$, it follows that $V = G(x)\big|_0^4 = G(4) - G(0) = \frac{1}{2} \cdot 4^2 = 8$.

C. Lengths of Curves. The definite integral can also be applied to the computation of the lengths of curves. This is done as follows. Let f be a function, and let $P = (a, c)$ and $Q = (b, d)$ be two points on its graph; see Figure 5.36. Here is the strategy for computing the length L of the curve between the points P and Q.

Again pump the interval from a to b full of points, packing them tightly together. Take a typical point x and let $x + dx$ be the very next one. Let (x, y) be the point on the graph above x, and use a segment of length dx to construct a characteristic triangle at (x, y). Figure 5.37 shows what is happening under a microscope. Notice that the slope of the tangent at (x, y) is $\frac{dy}{dx}$. In particular, $f'(x) = \frac{dy}{dx}$. By the Pythagorean theorem, the length of the hypotenuse of the characteristic triangle is $\sqrt{(dx)^2 + (dy)^2}$. Factoring out $(dx)^2$, we get

$$\sqrt{(dx)^2 + (dy)^2} = \sqrt{\left[1 + \left(\frac{dy}{dx}\right)^2\right](dx)^2}$$

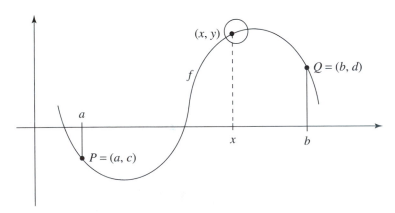

Figure 5.36

$$= \sqrt{1 + \left(\frac{dy}{dx}\right)^2}\, dx = \sqrt{1 + (f'(x))^2}\, dx.$$

Since dx is extremely small, the hypotenuse of the triangle is for all computational purposes equal to the length of the portion of the graph of $f(x)$ inside the triangle. The length L of the curve from P to Q is the sum of the lengths of all of these arcs as x ranges from a to b. Thus, the length L is given as the sum of all of the terms $\sqrt{1 + (f'(x))^2}\, dx$ as x ranges from a to b. Since this sum is of the type considered in Section 5.3, we conclude that

$$L = \int_a^b \sqrt{1 + (f'(x))^2}\, dx$$

By the Fundamental Theorem of Calculus, L can now be determined by finding an antiderivative G of the function $g(x) = \sqrt{1 + (f'(x))^2}$ and evaluating $L = G(b) - G(a)$.

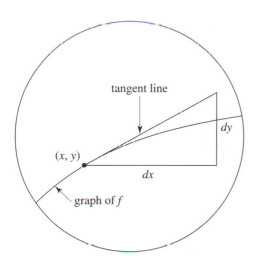

Figure 5.37

Let's apply our formula to the simplest case: the length of a line segment.

Example 5.23. Consider the function $f(x) = 2x - 3$ with the points $P = (-1, -5)$ and $Q = (5, 7)$ on its

graph, as pictured in Figure 5.38. We already know that the distance between P and Q is

$$\sqrt{(-1 - 5)^2 + (-5 - 7)^2} = \sqrt{36 + 144} = \sqrt{180} = 6\sqrt{5}.$$

Let's see whether definite integrals provide the same result. Since $f'(x) = 2$, $a = -1$, and $b = 5$,

$$\int_a^b \sqrt{1 + (f'(x))^2}\, dx = \int_{-1}^5 \sqrt{1 + 2^2}\, dx = \int_{-1}^5 \sqrt{5}\, dx.$$

Because $G(x) = \sqrt{5}x$ is an antiderivative of $\sqrt{5}$,

$$\int_{-1}^5 \sqrt{5}\, dx = G(5) - G(-1) = 5\sqrt{5} - (-1)\sqrt{5} = 6\sqrt{5},$$

as expected.

Example 5.24. Consider the circle $x^2 + y^2 = 5^2$ of radius 5 and with center the origin. Since $y = \pm\sqrt{5^2 - x^2}$, any point (x, y) on the upper half of the circle satisfies $y = \sqrt{5^2 - x^2}$. It follows that the graph of the function $f(x) = \sqrt{5^2 - x^2}$ is the upper half of the circle. See Figure 5.39. Consider the points P and Q on the circle with x-coordinates $-\frac{5}{2}\sqrt{2}$ and $\frac{5}{2}\sqrt{2}$, respectively. What is the length L of the circular arc from P to Q? By the formula already developed,

$$L = \int_a^b \sqrt{1 + (f'(x))^2}\, dx = \int_{-\frac{5}{2}\sqrt{2}}^{\frac{5}{2}\sqrt{2}} \sqrt{1 + (f'(x))^2}\, dx.$$

Recall from the discussion preceding Example 5.8 of Section 5.2 that the slope of the tangent line to this circle at any point (x, y) is $-\frac{x}{y}$. It follows that $f'(x) = -\frac{x}{y} = -\frac{x}{\sqrt{5^2 - x^2}}$. So $(f'(x))^2 = \frac{x^2}{5^2 - x^2}$, and hence $1 + (f'(x))^2 = \frac{5^2 - x^2 + x^2}{5^2 - x^2} = \frac{5^2}{5^2 - x^2}$. Therefore,

$$L = \int_{-\frac{5}{2}\sqrt{2}}^{\frac{5}{2}\sqrt{2}} \sqrt{\frac{5^2}{5^2 - x^2}}\, dx = \int_{-\frac{5}{2}\sqrt{2}}^{\frac{5}{2}\sqrt{2}} \frac{5}{\sqrt{5^2 - x^2}}\, dx.$$

This definite integral is not easy to evaluate. We can, however, compute the length of arc PQ directly. Look again at Figure 5.39 and consider the angle determined by the segment OQ and the positive x-axis. The cosine of this angle is $\frac{\frac{5}{2}\sqrt{2}}{5} = \frac{\sqrt{2}}{2}$, so the angle is $\frac{\pi}{4}$. By symmetry, the angle determined by the segment OP and the negative x-axis is also $\frac{\pi}{4}$. It follows that $\angle POQ = \frac{\pi}{2}$. It is now clear that arc PQ is precisely one quarter of the circle of radius 5; so its length is $\frac{1}{4} \cdot 2\pi \cdot 5 = \frac{5}{2}\pi$. We have verified that

$$L = \int_{-\frac{5}{2}\sqrt{2}}^{\frac{5}{2}\sqrt{2}} \frac{5}{\sqrt{5^2 - x^2}}\, dx = \frac{5}{2}\pi.$$

Notice that this definite integral is also equal to the area under the graph of the function $\frac{5}{\sqrt{5^2 - x^2}}$ from $-\frac{5}{2}\sqrt{2}$ to $\frac{5}{2}\sqrt{2}$.

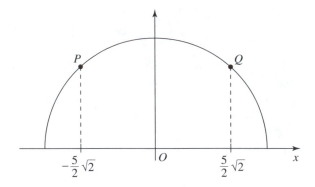

Figure 5.39

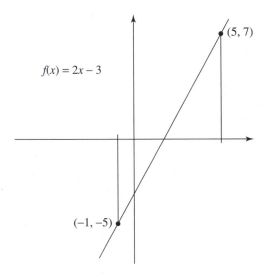

Figure 5.38

We have discussed three applications of definite integrals, namely, the computation of areas, volumes of revolution, and lengths of arcs. At least in principle, these can all be solved by using the Fundamental Theorem of Calculus. A variety of problems of mathematics and physics can be solved in this way. A word of caution is in order, however; the required antiderivatives are not always easily determined.

5.7 Postscript

The difference between the approaches of Leibniz (and also Newton as we will see in the next chapter) and those of Archimedes should now be clear: Archimedes developed ingenious methods (many of which contain the elements of what is now called calculus) to solve specific problems. His computation of the area of a parabolic section is a case in point. Leibniz and Newton, on the other hand, put together a general theory that is applicable to a wide range of problems. It must be pointed out that their achievements incorporated the work of many of their predecessors, among them Descartes, Cavalieri, and Kepler.

The description of Leibniz's work given in this chapter has retained the essence of his arguments, but it incorporates clarifications and improvements that came later. Functions, as already indicated, are implicit rather than explicit in Leibniz's work. They began to move to a central position in mathematics in the 1740s with Leonhard Euler. For Euler, a function was an "analytical expression composed in any way" from x and from "numbers or constant quantities." Later, in the work of Lagrange, the slope, in the role of the derivative, was treated as a function, and the notation $f'(x)$ was first used. The notion of the limit, which D'Alembert emphasized in the 1750s, and the notation $\lim_{\Delta x \to 0} \frac{\Delta y}{\Delta x}$, first used by Cauchy in the 1820s, both added clarity.

Leibniz's fame as a philosopher and scientist spread throughout Europe. He was in correspondence with most of the important European scholars of the day. He was named to the French Academy of Sciences, and the Habsburg emperor bestowed on him the title of baron. But Leibniz's last years were unhappy ones. His long dispute over the priority of the creation of the calculus with Newton and his English colleagues became increasingly bitter. When the son of his patron became King George I of England in 1714 and moved his court to London, Leibniz remained behind in Hanover in isolation. His work was neither understood nor appreciated. He suffered increasingly from illness and died in 1716.

Exercises
5A. Lines and Their Equations

1. What is the slope of the line determined by the points $(2, -3)$ and $(-6, 2)$. Write an equation for this line.

2. Write an equation for the line with slope -3 and y-intercept 4.

3. Write an equation for the line that has slope $\frac{1}{2}$ and has the point $(3, -2)$ on its graph.

4. What are the slope and y-intercept of the line that has equation $2x + 7y + 2 = 0$?

5B. Computing Slopes of Tangents

Go through Leibniz's tangent method ("push Δx to zero") to compute the slope m_P of the tangent to the given curve at the given point P.

5. The parabola $y = x^2$ at the point $P = (2, 4)$.

6. The curve $y = x^3$ at the point $P = (2, 8)$.

7. The hyperbola $y = \frac{1}{x}$ at any point $P = (x, y)$.

8. The curve $x = y^3$ at any point $P = (x, y)$. [Hint: What happens to $\frac{\Delta y}{\Delta x} \Delta y$ when Δx is pushed to zero? Refer to Figure 5.11 in Section 5.2 and the surrounding discussion.]

9. The ellipse $\frac{x^2}{5^2} + \frac{y^2}{4^2} = 1$ at $P = (x, y)$. [If you get $m_P = -\frac{4^2}{5^2} \cdot \frac{x}{y}$, you are right.]

5C. Derivatives

10. Let $f(x) = x^3$. Determine $f'(x)$ by evaluating

$$\lim_{\Delta x \to 0} \frac{f(x + \Delta x) - f(x)}{\Delta x}.$$

11. Consider $h(x) = \frac{1}{x^2}$. Show that $h'(x) = \frac{-2}{x^3}$ by evaluating $\lim_{\Delta x \to 0} \frac{h(x + \Delta x) - h(x)}{\Delta x}$. [Hint: Take common denominators and cancel.]

12. Use the derivative formulas already developed to compute the slopes of the tangents to the graphs of the following functions at the given points.

    i. $f(x) = x^3$ at $(-2, -8)$.
    ii. $g(x) = \sqrt[3]{x}$ at $(-3, \sqrt[3]{-3})$.
    iii. $f(x) = \frac{1}{x}$ at $(-\frac{1}{3}, -3)$.
    iv. $f(x) = \frac{1}{x^2}$ at $(-2, \frac{1}{4})$.

13. By using facts from the text, compute the derivatives of each of the following functions:

    i. $f(x) = -10$.
    ii. $y = 4x + 7$.
    iii. $f(x) = 7x^2 - 5x + 2$.
    iv. $y = 2\sqrt[3]{x} + \pi x^3$.
    v. $g(x) = \frac{3}{x} + 3x - 6$.
    vi. $f(x) = 2x^3 + 3x + 4 - \frac{1}{x^2}$.

14. Consider the parabolic section determined by the parabola $y = -x^2 + 8x$ and the x-axis.

    i.   Determine the point on the parabola with the property that the slope of the tangent line is zero.
    ii. Use Archimedes's theorem (not calculus) to compute the area of the parabolic section.

5D. Definite Integrals

15. Consider the graph of the equation $y = \frac{1}{x}$ over the interval from 2 to 4. Insert the points

$$2 \le 2.3 \le 2.5 \le 2.9 \le 3.4 \le 3.6 \le 4$$

and compute the sum of all of the rectangles $y \cdot dx$.

16. Consider the graph of the equation $y = \sqrt{x}$ over the interval from 0 to 2. Insert the points

$$0 \le \frac{1}{9} \le \frac{2}{9} \le \frac{4}{9} \le \frac{7}{9} \le \frac{11}{9} \le \frac{16}{9} \le 2$$

and compute the sum of all of the rectangles $y \cdot dx$. What is the actual value of the area under the graph of $y = \sqrt{x}$ from $x = 0$ to $x = 2$?

17. Explain with a diagram the full meaning of $\int_0^3 x^2 \, dx$ and $\int_3^{12} \sqrt{x} \, dx$.

18. Draw a graph of the hyperbola $\frac{x^2}{5^2} - \frac{y^2}{4^2} = 1$. Then graph the equation $y = \frac{4}{5}\sqrt{(x^2 - 5^2)}$ and explain the full meaning of $\int_7^{10} \frac{4}{5}\sqrt{x^2 - 5^2} \, dx$.

5E. The Tractrix

19. Recall Perrault's problem from the beginning of Section 5.4. Tie a weight B to the end of a string and place it and the weight on a horizontal Cartesian plane. Stretch the string along the positive x-axis so that one end is at the origin O and the other end, where B is tied, is at $(a, 0)$. So a is the length of the string. Now move the end of the string (the one at the origin) up the positive y-axis. The curve that B traces out in the process is called a *tractrix*. Let (x, y) represent a typical position of the weight B. See Figure 5.40. Leibniz observed that the string is tangent to the curve at (x, y). Study the figure and express the slope m_x of the tangent to the curve at the point (x, y) as a function of x. Suppose that the tractrix is the graph of a function $y = f(x)$. What equation must the derivative of this function satisfy?

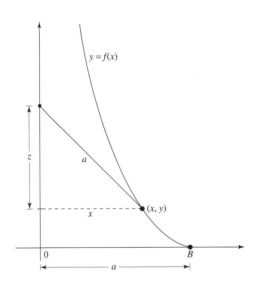

Figure 5.40

20. Use the results of Exercise 19 to express as a definite integral the length of the tractrix from the point $(a, 0)$ to some other point (c, d).

5F. Maximum and Minimum Values

In Exercises 21 and 22 make use of Figure 5.31.

21. If the segment is 1200 units long and the product of the squares of the lengths of the two pieces is to be as large as possible, how long should the two pieces be?

22. The product of the lengths of the two pieces is specified to be 300. What is the shortest segment with which this can be done?

23. Find the dimensions of the rectangle of largest area whose perimeter is 1000 cm.

24. Find that point on the parabola $y = x^2 + 1$ that is closest to the point $(3, 1)$. [Hints: Find that point on the parabola such that the square of its distance to $(3, 1)$ is minimal. Also, recall that if a is a root of a polynomial, then $x - a$ divides it.]

25. Consider the triangular region determined by the graph of the line $y = 3 - x$. Inscribe a rectangle as shown in Figure 5.41. When both the triangle and the rectangle are revolved one complete revolution about the x-axis, a cone and an inscribed cylinder result. Determine the largest volume this cylinder can have.

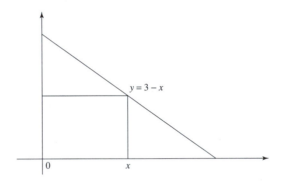

Figure 5.41

5G. Areas and Definite Integrals

26. Use the Fundamental Theorem of Calculus to evaluate the definite integrals

$$\int_0^3 x^2 \, dx, \quad \int_{-8}^{-2} \frac{1}{x^2} \, dx, \quad \text{and} \quad \int_3^{12} \sqrt{x} \, dx.$$

Sketch the areas that you have determined.

27. The equation of the circle with radius 2 and centered at the origin is $x^2 + y^2 = 4$. Why is $y = \sqrt{4 - x^2}$ an equation whose graph is the upper semicircle only? Evaluate the definite integral

$$\int_0^2 \sqrt{4 - x^2} \, dx$$

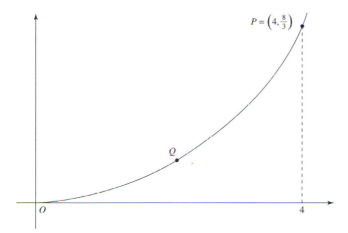

Figure 5.42

after drawing a picture of the area that it represents. Let a be a positive number and evaluate $\int_0^a \sqrt{a^2 - x^2}\, dx$ in the same way.

28. Evaluate the definite integral $\int_0^5 \frac{5}{2}\sqrt{5^2 - x^2}\, dx$ by making use of the area of an appropriate circle.

29. Let a and b be positive numbers. Use Exercise 27 and a basic property of the definite integral to evaluate $\int_0^a \frac{b}{a}\sqrt{a^2 - x^2}\, dx$. What area does this definite integral represent?

30. The graph in Figure 5.42 is that of the parabola $y = \frac{1}{6}x^2$ over the interval from 0 to 4.

   i. Determine the slope of the segment from O to $P = (4, \frac{8}{3})$. Determine the coordinates of the

point Q with the property that the tangent to the curve at Q is parallel to OP.

   ii. Determine first the slope and then the equation of the line perpendicular to OP that goes through the point Q.

   iii. Consider the triangle OQP. Taking OP as the base, determine its height. What is the area of the triangle OQP?

   iv. Use Archimedes's theorem for the area of a parabolic segment to determine the area under the parabola from $x = 0$ to $x = 4$.

   v. Verify this answer by using the Fundamental Theorem of Calculus.

5H. Definite Integrals as Areas, Volumes, and Lengths of Curves

31. Use the Fundamental Theorem of Calculus to compute the volume of a cone of height h and base of radius r. [Hint: Rotate the triangular region in Figure 5.43.]

32. Determine the volume of the solid obtained by rotating the region above the x-axis and under the graph of the function $y = \sqrt{x}$, $0 \le x \le 3$, one complete revolution about the x-axis.

33. Show that the length of the parabolic arc on the parabola $y = x^2$ from the point $(2, 4)$ to the point $(5, 25)$ is equal to the area under the upper half of the hyperbola $y^2 - 4x^2 = 1$ from $x = 2$ to $x = 5$.

34. Consider the upper half of the ellipse $\frac{x^2}{5^2} + \frac{y^2}{4^2} = 1$ in Figure 5.44. Express as definite integrals:

   i. The area of the upper half of the ellipse.
   ii. The volume of the solid obtained by rotating it one revolution about the x-axis.
   iii. The length of the arc from $(1, \frac{8}{5}\sqrt{6})$ to $(3, \frac{16}{5})$.

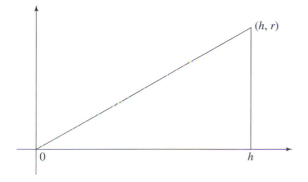

Figure 5.43

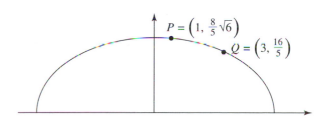

Figure 5.44

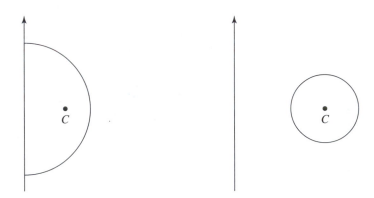

Figure 5.45

Don't forget to put in the limits of integration, but do not evaluate the integrals. In case (iii) use the fact from Exercise 9 that the slope of the tangent to the ellipse at any point (x, y) is $-\frac{4^2}{5^2}\frac{x}{y}$. Integral (iii) is a so-called "elliptic integral." It is difficult to solve.

35. Consider the hyperbola $\frac{x^2}{3^2} - \frac{y^2}{2^2} = 1$.

    i. Use Leibniz's tangent method to evaluate the slope m_P at any point $P = (x, y)$, $y \neq 0$, on the hyperbola.

    ii. Find a function whose graph is precisely the upper half of the hyperbola. Determine the derivative of this function.

    iii. Express as a definite integral (do not evaluate) the area under the upper half of this hyperbola and over the interval from 3 to 7.

    iv. Express as a definite integral (do not evaluate) the volume of the solid obtained by rotating the region of (iii) one complete revolution about the x-axis.

    v. Express as a definite integral (do not evaluate) the length of the hyperbolic arc from the point $(4, \frac{3}{2}\sqrt{7})$ to the point $(7, \frac{3}{2}\sqrt{40})$.

5I. Theorems of Pappus of Alexandria

Pappus of Alexandria (ca 300 A.D) was the last of the great Greek mathematicians. Theorems A and B that follow are named after him. Let S be either a curved segment in the plane or a region in the plane. In either case, think of S as made of a very thin homogeneous material. The *centroid* of S is its center of mass. For example, if the arc is a complete circle, the centroid is the center of the circle.

The centroid of the region inside the circle is also the center of the circle.

A. If an arc that lies in a plane is revolved one complete revolution about an axis that lies in the plane but does not cross the arc, then the area of the surface that is formed is equal to the product of the length of the arc and the length of the path traced by the centroid of the arc. (To see what is going on, refer to the semicircular arc in Figure 5.45. The axis lies on the diameter of the circle. If the semicircular arc is allowed to rotate one complete revolution about this axis, the surface that is formed is a sphere. The centroid C of the arc will trace out a circle.)

B. If a region lying in a plane is revolved one complete revolution about an axis that lies in the plane but does not intersect the region, then the volume of the solid that is formed is equal to the product of the area of the region and the length of the path traced out by its centroid. (Again refer to Figure 5.45. The centroid of the circular disc on the right lies at the center C of the circle. If the disc is rotated one revolution about the axis, the volume formed is that of a donut. This is a "perfect" donut whose cross-sections are circles. The centroid C will trace out a circle.)

36. Use Pappus's theorem A to determine the area of the surface of a donut, given that the radius of its inner circle is r and that of its outer circle is R. (See Figure 5.46.)

37. Use Pappus's theorem A to determine the centroid of a semicircular arc. [Hint: Use the fact that the surface area of a sphere of radius r is $4\pi r^2$.]

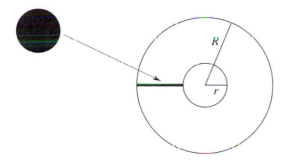

Figure 5.46

38. Use Pappus's theorem B to determine the centroid of a semicircular region. [Hint: Use the fact that the volume of a sphere of radius r is $\frac{4}{3}\pi r^3$.]

Notes

[1]Historians of science recognized long ago that Leibniz and Newton invented the calculus independently. Newton did so in the years 1665 and 1666 but did not publish his findings until much later. Leibniz made his initial discoveries later, in the 1670s, but was first to publish them. Why did we start with Leibniz's version of the calculus instead of Newton's? Because this tactic provided an early opportunity to introduce Leibniz's superior notation. This is the notation that prevailed and is in essence the notation used today.

[2]Just so there is no misunderstanding, this title translates to *Six Exercises*.

[3]Like perhaps no other mathematician before him, Leibniz realized the fundamental importance of notation: "With the notation one has to consider the fact that it should be convenient as regards the process of invention. This is especially the case if it expresses the innermost nature of the thing with economy. In this way, the effort involved in the thought process is minimized in a wonderful way." To this end, he engaged in notational "experiments." For instance, he had originally written $\frac{d}{x}$ for dx and used omn. (from the Latin *omnes*, meaning "all") to denote $\int$. You will appreciate the importance of notation, if you try

to multiply two large numbers written in Greek or Roman numerals.

[4]The development of the Fundamental Theorem of Calculus is based on Leibniz's article in the journal *Acta Eruditorum* in 1693, as translated into German (from the Latin) by Gerhard Kowalewski. See pages 24–34 of Band 162 of *Oswalds Klassiker der Exakten Wissenschaften*, Verlag von Wilhelm Engelmann, Leipzig, 1906. (The excerpts in English were provided by the author.) Other descriptions of Leibniz's work have made use of discussions in J. M. Child, *The Early Mathematical Manuscripts of Leibniz*, The Open Court Publishing Company, Chicago, London, 1920; and J. E. Hofmann, *Leibniz in Paris 1672–1676*, Cambridge University Press, Cambridge, 1974.

The Calculus of Newton

Isaac Newton (1642–1727) was born on Christmas day in Lincolnshire, England, the son of a farmer. He was educated at local schools and was a determined and successful student, but there was little to suggest that this quiet boy had exceptional promise. Since he had little inclination to be a farmer, his mother at last agreed with the schoolmaster to let him go off to the university. He passed his entrance examinations for Trinity College in Cambridge and began his studies there in 1661. He studied under Barrow and read Kepler's *Optics*, Euclid's *Elements* (which he at first found disappointingly simple), and works of Descartes and Galileo. In 1665, Newton took his degree of B.A. without any apparent distinction. Later in the year when the plague began to rage in London, Cambridge was closed as a precaution and he was forced to return to his birthplace in the country. Pondering alone with intense concentration, the twenty-three-year-old Newton hit upon the initial formulations of the ideas that would change many of the fundamental concepts of "natural philosophy," as science was referred to at that time. "I keep the subject of my inquiry constantly before me, and wait till the first dawning opens gradually, by little and little, into a full clear light," he wrote later. He had penetrating ideas about the nature of light, about the mathematical methods of a calculus of "fluents and fluxions," and about the motion of the heavenly bodies. "All this," he would write in his old age, "was in the two plague years of 1665 and 1666, for in those days I was in the prime of my age of invention, and minded mathematics and philosophy more than at any other time since." He became convinced that the motion of the planets around the Sun, as well as that of the Moon around the Earth, obeys the same physical laws as moving objects on Earth. A famous legend finds him meditating under an apple tree when, at the very moment that he heard the thud of a fallen apple, he was struck by the realization that the force that pulled the apple to Earth is one and the same as that which keeps the Moon in orbit around the Earth and from flying off into space.

Newton returned to Cambridge in 1667 and became absorbed with investigations into the nature of light. In experiments with prisms, he separated ordinary "white" light into its spectrum of colors from red to violet. The telescopes of the time relied on the principle of refraction. Lenses of appropriate shapes and configurations can be used to bend and redirect light rays in a way as to produce an enlarged image. (We will study the basic principles involved in Chapter 9.) However, the edges of a lense also act like prisms. As components of a refracting telescope, they create spectra of colors on the fringes of the field of vision that interfere with the image. Newton solved this problem by constructing a telescope in which the principal lens is replaced by a concave mirror, which redirects light by reflection rather than refraction. Newton's reflecting telescope had another advantage: it produced an image nine times larger than those of earlier refracting telescopes that were four times longer.

The description of the new telescope that was published by the *Royal Society of London for the Promotion of Natural Knowledge* generated such interest that Newton became famous overnight. The Royal Society was founded in 1660, and it was given a royal charter two years later. It counted the scientist Robert Hooke (1635–1703), the famous architect and mathematician Christopher Wren (1632–1723), and the astronomer Edmund Halley (1656–1742) among its members. It would play an important role throughout Newton's scientific life. In 1672, the Royal Society published Newton's first scientific work, which described his experiments on the nature of light. This article was very favorably received, but there was some criticism by Hooke and Huygens. Newton regarded these objections as deliberate efforts to annoy him: he published only very reluctantly thereafter.

6.1 **Areas Under Simple Curves**

I n 1669 Newton first informed his colleagues about his mathematical discoveries by circulating the manuscript *De Analysi per Aequationes Numero Terminorum Infinitas*. It would not be published until 1711. Newton's teacher, Barrow, was so impressed by the vast abilities displayed in the "The Analysis by Means of Equations with an Infinite Number of Terms" that he resigned his professorship at Cambridge and arranged for Newton to succeed him. We shall now take a look at the study of areas under curves that the *De Analysi* undertakes. An early English translation[1] of this work (recall that the scientific language of the 17th century was Latin) begins as follows:

> The General Method which I had devised some considerable time ago for measuring the quantity of curves by an infinite series of terms you have, in the following, rather briefly explained than narrowly demonstrated.

> To the base *AB* of some curve *AD* let the coordinate *BD* be perpendicular [see Figure 6.1] and let *AB* be called *x* and *BD y*. Let again . . .

Newton next lets *c* be any constant, *n* and *m* integers, and he formulates his RULE I:

> If $cx^{\frac{m}{n}} = y$ then will $\frac{cn}{m+n} x^{\frac{m+n}{n}}$ equal the area *ABD*.

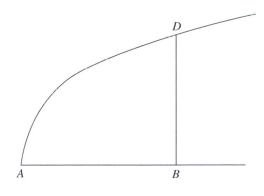

Figure 6.1

Without actually saying so, he assumes that $c > 0$, as well as $m > 0$ and $n > 0$. We will continue to follow Newton's argument closely, but we will use the language of functions, which clarifies what is going on. (Newton did not use functions explicitly.)

Denote the rational number $\frac{m}{n}$ by r. So $\frac{m+n}{n} = \frac{m}{n} + 1 = r + 1$. Now consider the function

$$f(x) = cx^r$$

Recall that $x^r = x^{\frac{m}{n}}$ is equal to $\sqrt[n]{x^m} = (x^m)^{\frac{1}{n}}$, i.e., it is the *n*th root of the *m*th power of *x*. Equivalently, $x^r = x^{\frac{m}{n}}$ is $(\sqrt[n]{x})^m = (x^{\frac{1}{n}})^m$, the *m*th power of the *n*th root of *x*. The functions

$$f(x) = 3x^{\frac{1}{2}} = 3\sqrt{x}, \; f(x) = 6x^{\frac{3}{4}},$$

$$f(x) = 9x^{\frac{7}{6}}, \; f(x) = 4x^2, \text{ and } f(x) = 7x^3$$

are specific examples of the type of function that Newton is considering.

The general shape of the graph of $f(x) = cx^r$ depends on *r*. If $r < 1$, then the graph is the shaped like the one sketched by Newton; see Figure 6.2. The function $f(x) = 3x^{\frac{1}{2}} = 3\sqrt{x}$ is an example. If $r = 1$, then the graph of $f(x) = cx$ is the line through the origin with slope *c*. See Figure 6.3(a). Finally, if $r > 1$, then the general shape of the graph of $f(x) = cx^r$ is shown in Figure 6.3(b). The function $f(x) = 4x^2$ is a specific example. Recall from Section 4.3 that its graph is a parabola.

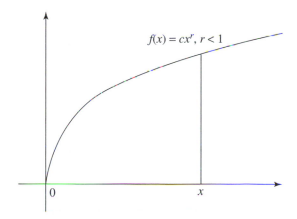

Figure 6.2

Focus on any function of the form $f(x) = cx^r$. We will assume with Newton that $x \geq 0$. Let $A(x)$ be the

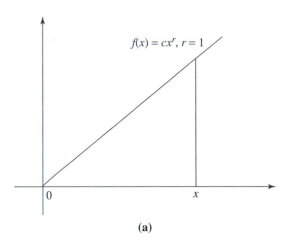

$f(x) = cx^r, r = 1$

0

x

(a)

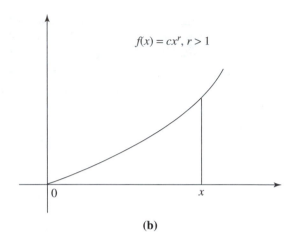

$f(x) = cx^r, r > 1$

0

x

(b)

Figure 6.3

area under the graph from 0 to x. This is the *area function* for the function f. Newton's first rule can now be restated as follows:

RULE I. If $f(x) = cx^r$, then its area function is $A(x) = \frac{c}{r+1}x^{r+1}$.

Example 6.1. Consider the function $f(x) = 3x^{\frac{1}{2}} = 3\sqrt{x}$. Since $r = \frac{1}{2}$ and $c = 3$, the area function is

$$A(x) = \frac{3}{\frac{3}{2}}x^{\frac{3}{2}} = 3 \cdot \frac{2}{3}x^{\frac{3}{2}} = 2x^{\frac{3}{2}}.$$

Taking $x = 5$, for example, we find that the area under the graph of $f(x) = 3x^{\frac{1}{2}}$ from 0 to 5 is $A(5) = 2 \cdot 5^{\frac{3}{2}} = 2(11.18) = 22.36$.

Example 6.2. Let $f(x) = 4x^2$. Now $r = 2$ and $c = 4$. So the area function is $A(x) = \frac{4}{3}x^3$. For instance, the area under the graph of $f(x) = 4x^2$ from 0 to 6 is $A(6) = \frac{4}{3} \cdot 6^3 = \frac{4}{3}(216) = 288$.

Let's now turn to Newton's proof, or verification, of his first rule. Figure 6.4(a) accompanies his argument. Newton "prepares" the proof by first considering a special case:

Preparation for demonstrating the first rule. Let then $AD\delta$ be any curve whose base $AB = x$, the perpendicular ordinate $BD = y$, and

the area $ABD = z$, as at the beginning. Likewise put $B\beta = o$, $\beta K = v$; and the rectangle $B\beta HK$ (ov) equal to the space $B\beta\delta D$. Therefore it is $A\beta = x + o$, and $A\delta\beta = z + ov$; which things being premised, assume any relation betwixt x and z that you please ...

This is the beginning of Newton's "preparation." We will follow Newton's argument closely. Consider a curve **C** as shown in Figure 6.4(b). Let x be any number on the horizontal axis with $x \geq 0$ and let $D = (x, y)$ be the corresponding point on C. In the discussion that follows, x and D are held fixed. As before, let $A(x)$ be the area under **C** from 0 to x. Now let Δx be a positive number and consider the point $x + \Delta x$. Note that Newton's z has been replaced by $A(x)$ and his o by Δx. These are the only changes in his original argument (other than the verbal aspects). Draw vertical lines through both x and $x + \Delta x$. Place the segment HK parallel to the horizontal axis; now raise it until the precise point is reached at which the area of the indicated rectangle is equal to the area under **C** from x to $x + \Delta x$. Let v be the y-coordinate of the point of intersection of HK and **C**. So v is the height of this rectangle. This is illustrated in Figure 6.4(b).

Continuing the preparation for his proof, Newton now assumes that the area under **C** from 0 to x is given

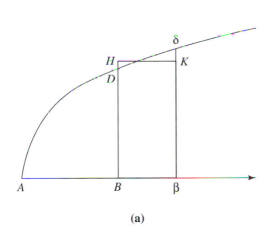

(a)

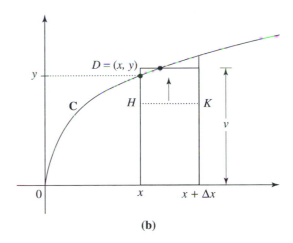

(b)

Figure 6.4

by $A(x) = \frac{2}{3}x^{\frac{3}{2}}$ (for any $x \geq 0$) and asks the following questions: Does this assumption determine the curve **C**? If so, exactly what curve is it? His answers follow.

He observes that the area under the curve from 0 to $x + \Delta x$ is equal to $A(x) + \Delta x \cdot v$. Given his assumption about the area function, on the other hand, this area is also equal to $A(x + \Delta x) = \frac{2}{3}(x + \Delta x)^{\frac{3}{2}}$. So

$$A(x) + \Delta x \cdot v = A(x + \Delta x) = \frac{2}{3}(x + \Delta x)^{\frac{3}{2}}$$

Squaring both sides, he gets

$$(A(x))^2 + 2A(x) \cdot \Delta x \cdot v + (\Delta x)^2 \cdot v^2$$

$$= \frac{4}{9}(x + \Delta x)^3$$

$$= \frac{4}{9}\left(x^3 + 3x^2\Delta x + 3x(\Delta x)^2 + (\Delta x)^3\right)$$

$$= \frac{4}{9}x^3 + \frac{4}{3}x^2\Delta x + \frac{4}{3}x(\Delta x)^2 + \frac{4}{9}(\Delta x)^3.$$

Since $(A(x))^2 = \frac{4}{9}x^3$, he obtains

$$2A(x) \cdot \Delta x \cdot v + (\Delta x)^2 \cdot v^2 = \frac{4}{3}x^2\Delta x + \frac{4}{3}x(\Delta x)^2 + \frac{4}{9}(\Delta x)^3$$

by canceling. Dividing both sides by Δx gives him

$$2A(x) \cdot v + \Delta x \cdot v^2 = \frac{4}{3}x^2 + \frac{4}{3}x\Delta x + \frac{4}{9}(\Delta x)^2.$$

Newton can now push Δx to zero on both sides. In the process, observe in Figure 6.4(b) that v closes in on y, and therefore that

$$2A(x)y = \frac{4}{3}x^2.$$

Since $A(x) = \frac{2}{3}x^{\frac{3}{2}}$, it follows that $\frac{4}{3}x^{\frac{3}{2}}y = \frac{4}{3}x^2$. So $y = \dfrac{x^2}{x^{\frac{3}{2}}}$. Therefore, the x- and y-coordinates of the typical point $D = (x, y)$ on the curve **C** satisfy the equation $y = x^{\frac{1}{2}}$. It follows that **C** is the graph of the function $f(x) = x^{\frac{1}{2}}$.

Newton's "preparation" has accomplished the following: He started with a curve **C** with area function given by $A(x) = \frac{2}{3}x^{\frac{3}{2}}$ and showed that **C** is the graph of the function $f(x) = x^{\frac{1}{2}}$. He could thus conclude that the area function $A(x)$ for the graph of $f(x) = x^{\frac{1}{2}}$ is $A(x) = \frac{2}{3}x^{\frac{3}{2}}$. This is precisely the assertion of his first rule for the case $c = 1$ and $r = \frac{1}{2}$.

Newton goes on to prove RULE I in its general form using the identical strategy. Instead of following Newton's argument directly, we will dissect it into its two main components. The first of these concerns derivatives.

Let f be any function given by a rule of the form $f(x) = kx^s$, where k is any constant and s is any

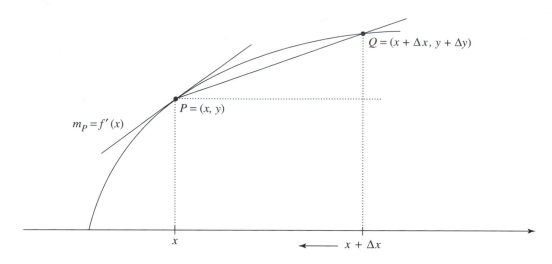

$$m_P = f'(x)$$

$$P = (x, y)$$

$$Q = (x + \Delta x, y + \Delta y)$$

$$x \qquad x + \Delta x$$

Figure 6.5

positive rational number. Newton needs to determine the derivative of f. He proceeds in the same way as Leibniz does in Section 5.2. (To be fair, he discovered this procedure before Leibniz.) He fixes any point $P = (x, y)$ on the graph of f; takes another point $Q = (x + \Delta x, y + \Delta y)$ on the graph; pushes Q to P and evaluates $\lim\limits_{\Delta x \to 0} \dfrac{\Delta y}{\Delta x} = m_P$. This is the slope of the tangent to the graph at $P = (x, y)$. Therefore $f'(x) = m_P$. See Figure 6.5.

Let's carry out the details. Put $s = \frac{m}{n}$ with m and n positive integers. Then the graph of f is the graph of the equation $y = kx^{\frac{m}{n}}$. By raising both sides to the nth power, $y^n = k^n x^m$. Because $Q = (x + \Delta x, y + \Delta y)$ is on the graph,

$$(y + \Delta y)^n = k^n (x + \Delta x)^m.$$

To multiply the n factors

$$(y + \Delta y)^n = (y + \Delta y)(y + \Delta y) \cdots (y + \Delta y)$$

out, both the y and Δy from any one group must be multiplied by the y and Δy from each of the other groups. The product $y \cdot y \cdots y = y^n$ is one term that arises in this way. Fixing any Δy and multiplying it by the y from each of the other $n - 1$ groups gives the product $y^{n-1} \Delta y$. Since n different Δy's can be picked

to do this, $y^{n+1} \Delta y$ will occur n times. The result of the multiplication will therefore be

$$(y + \Delta y)^n = y^n + n y^{n-1} \Delta y + \dots \text{ more terms.}$$

Notice that each of these additional terms must contain at least two factors of Δy (because the terms that contain no or one Δy are already accounted for). Doing the same thing with $(x + \Delta x)^m$, Newton arrives at an equation of the form

$$y^n + n y^{n-1} \Delta y + \text{terms with } (\Delta y)^2 \text{ as factor} =$$

$$k^n \left(x^m + m x^{m-1} \Delta x + \text{terms with } (\Delta x)^2 \text{ as factor} \right).$$

Because $P = (x, y)$ is on the graph, $y^n = k^n x^m$. Therefore, after a subtraction,

$$n y^{n-1} \Delta y + \text{ terms with } (\Delta y)^2 \text{ as factor} =$$

$$k^n \left(m x^{m-1} \Delta x + \text{terms with } (\Delta x)^2 \text{ as factor} \right).$$

Now divide both sides of this equation by Δx to get

$$n y^{n-1} \frac{\Delta y}{\Delta x} + \text{ terms with } \Delta y \cdot \frac{\Delta y}{\Delta x} \text{ as factor} =$$

$$k^n \left(m x^{m-1} + \text{terms with } \Delta x \text{ as factor} \right).$$

Finally, push Δx to zero. Since Δy also goes to 0 in the process (see Figure 6.5), all terms with $\Delta y \cdot \frac{\Delta y}{\Delta x}$ as factor go to zero, and because $\lim\limits_{\Delta x \to 0} \dfrac{\Delta y}{\Delta x} = m_P = f'(x)$, it follows that

$$(*) \qquad ny^{n-1}f'(x) = k^n m x^{m-1}.$$

Recalling that $y = kx^{\frac{m}{n}}$, we get by simple algebra that

$$y^{n-1} = k^{n-1}x^{\frac{m}{n}(n-1)} = k^{n-1}x^{m-\frac{m}{n}} = k^{n-1}x^m x^{-\frac{m}{n}}.$$

By substituting this for y^{n-1} in equation $(*)$ above, we obtain

$$nk^{n-1}x^m x^{-\frac{m}{n}}f'(x) = k^n m x^{m-1}.$$

After canceling, we have $nx \cdot x^{-\frac{m}{n}}f'(x) = km$. Since $\frac{m}{n} = s$, we get $nx^{1-s}f'(x) = km$. So

$$f'(x) = k\frac{m}{n} \cdot \frac{1}{x^{1-s}} = ksx^{s-1}.$$

Therefore, Newton has shown that

The derivative of $f(x) = kx^s$ is $f'(x) = ksx^{s-1}$.

We will see later that this formula is valid for any rational number s, not only positive s. Taking $k = 1$, the formula asserts that the derivative of the function $f(x) = x^s$ is $f'(x) = sx^{s-1}$. Turn to Section 5.5A and observe that all of the cases of Example 5.16 follow from this fact.

Example 6.3. (a) Let $f(x) = 3x^{100}$. Then $f'(x) = 300x^{99}$. (b) If $g(x) = 2x^{-5} - 4x^{\frac{3}{5}}$, then $g'(x) = -10x^{-4} - \frac{12}{5}x^{-\frac{2}{5}}$.

The second component in Newton's verification of his RULE I is the Fundamental Theorem of Calculus. We already studied Leibniz's proof of this theorem in Section 5.4. We will describe Newton's proof next.

6.2 The Fundamental Theorem of Calculus (Again)

$\boxed{C}$onsider any function f that satisfies $f(x) \geq 0$ for all x. For any $x \geq 0$, let $A(x)$ be the area under the graph from 0 to x. As was done earlier in Figure 6.4(b), let $x \geq 0$ be any number and Δx any distance, and construct the rectangle of height v in such a way that its area is equal to the area under the graph from x to $x+\Delta x$. Refer to Figure 6.6 and observe that

$$A(x + \Delta x) = A(x) + v\,\Delta x.$$

Therefore, $A(x + \Delta x) - A(x) = v\,\Delta x$, and hence

$$\frac{A(x + \Delta x) - A(x)}{\Delta x} = v.$$

Keeping x fixed, push Δx to zero, and observe that v gets pushed to $f(x)$ in the process. Thus

$$\lim_{\Delta x \to 0} \frac{A(x + \Delta x) - A(x)}{\Delta x} = f(x).$$

The left side is the derivative $A'(x)$ of $A(x)$. So Newton's argument has shown that the derivative $A'(x)$ of the area function $A(x)$ is the original function $f(x)$. Put another way:

The area function A of a function f is an antiderivative of f.

What are the consequences of this fact? Refer to Figure 6.7 and observe that the area under the graph of f from a to b is equal to the area under the graph from 0 to b minus the area under the graph from 0 to a. So the area under the graph of f from a to b is equal to $A(b) - A(a)$. Recalling that Leibniz had written this area as $\int_a^b f(x)\,dx$, we see that

$$\int_a^b f(x)\,dx = A(b) - A(a).$$

Now let F be any antiderivative of f. Recall from Section 5.5B that any two antiderivatives of the same function differ only by a constant. Thus, $F(x) = A(x) + C$. So $F(b) - F(a) = (A(b) + C) - (A(a) + C) = A(b) - A(a)$. It follows that

$$\int_a^b f(x)\,dx = F(x)\big|_a^b = F(b) - F(a).$$

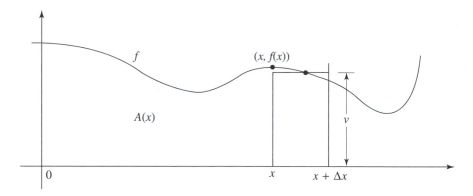

Figure 6.6

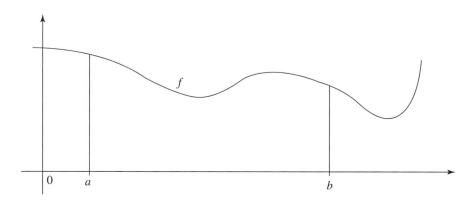

Figure 6.7

This is the Fundamental Theorem of Calculus already discussed in Chapter 5.

Newton's RULE I (slightly reformulated) is now easy. Let $F(x) = \frac{c}{r+1}x^{r+1}$. By the differentiation rule of Section 6.1, $F'(x) = \frac{c}{r+1}(r+1)x^r = cx^r$. So $F(x) = \frac{c}{r+1}x^{r+1}$ is an antiderivative of the function cx^r. Therefore, by the Fundamental Theorem of Calculus:

RULE I.

$$\int_a^b cx^r \, dx = \frac{c}{r+1}x^{r+1}\Big|_a^b = \frac{c}{r+1}b^{r+1} - \frac{c}{r+1}a^{r+1}.$$

Example 6.4. The area under the graph of $f(x) = 5x^3$ from $x = 1$ to $x = 4$ is equal to

$$\int_1^4 5x^3 \, dx = \frac{5}{4}x^4\Big|_1^4 = \frac{5}{4}4^4 - \frac{5}{4}1^4 = 5 \cdot 4^3 - \frac{5}{4} = 318\frac{3}{4}.$$

Newton calls a function *simple* if it is of the form $f(x) = cx^r$, where $c > 0$ and $r > 0$. The theory he has developed allows him to find the area under the graph of a simple function. He next turns his attention to the area problem for functions obtained by adding terms of the form cx^r and he formulates his RULE II:

If the value of y be made up of several such terms, the Area likewise shall be made up of

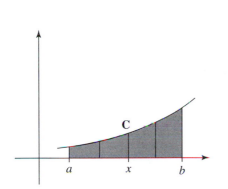 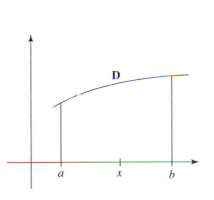 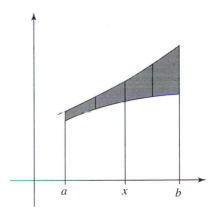

Figure 6.8

the Areas which result from every one of the terms.

What he means is this. Let $f(x) = cx^r$ and $g(x) = dx^s$ be simple functions, and let **C** and **D** be their respective graphs. Then the area under the graph of the function $y = cx^r + dx^s$ from a to b is equal to the sum of the area under **C** from a to b plus the area under **D** from a to b. This is illustrated in Figure 6.8, where the graph of $y = cx^r + dx^s$ is the upper curve in the diagram on the right. The area underneath it is obtained by adding the shaded area (area under **C**) to the area under **D**.

In a more precise way, this addition rule is a consequence of the Fundamental Theorem of Calculus. Let f and g be functions with graphs above the x-axis. Let F and G be antiderivatives of f and g respectively. Then the areas under the respective graphs from a to b are given by

$$\int_a^b f(x)\,dx = F(b) - F(a)$$

and

$$\int_a^b g(x)\,dx = G(b) - G(a).$$

By the sum rule for derivatives, $(F(x) + G(x))' = F'(x) + G'(x) = f(x) + g(x)$. So $F(x) + G(x)$ is an antiderivative of $f(x) + g(x)$. It follows that

$$\int_a^b (f(x) + g(x))\,dx = (F(b) + G(b)) - (F(a) - G(a))$$

$$= (F(b) - F(a)) + (G(b) - G(a))$$

$$= \int_a^b f(x)\,dx + \int_a^b g(x)\,dx.$$

Therefore, Newton's second rule can be reformulated as follows:

RULE II.

$$\int_a^b (f(x) + g(x))\,dx = \int_a^b f(x)\,dx + \int_a^b g(x)\,dx.$$

Example 6.5. Determine the area under the graph of $y = 4x^3 + 2x^{\frac{3}{2}}$ from 0 to 3.

Solution. Using RULE II in combination with RULE I, we get

$$\int_0^3 \left(4x^2 + 2x^{\frac{3}{2}}\right) dx = \int_0^3 4x^2\,dx + \int_0^3 2x^{\frac{3}{2}}\,dx$$

$$= \frac{4}{3}3^3 + \frac{4}{5}3^{\frac{5}{2}} = 36 + \frac{4}{5}(15.59)$$

$$= 36 + 12.47 = 48.47.$$

6.3 **Computing Definite Integrals**

H aving dealt with simple functions and sums of
simple functions, the *De Analysi* turns to the
problem of determining areas under more complicated
curves. Newton begins this discussion as follows:

> But if the value of y, or any of its terms
> be more compounded than the foregoing, it
> must be reduced into more simple terms; by
> performing the operation in letters, after the
> same manner as arithmeticians divide in dec-
> imal numbers, extract the square root, or
> resolve affected equations; and afterwards by
> the preceeding rules you will discover the
> Superficies of the curve sought.

What Newton has in mind is the reduction of the
area problem for complicated functions to that for
simple functions. But how? Consider, for example, the
function $f(x) = \frac{1}{1+x}$. Here the "operation in letters" is
polynomial division. Dividing $1+x$ into 1, Newton gets

$$
\begin{array}{r}
1 \\
\hline
1+x \enclose{longdiv}{1} \\
\text{subtract} \longrightarrow \ 1+x \\
\hline
-x
\end{array}
$$

after the first step, and

$$
\begin{array}{r}
1-x \\
\hline
1+x \enclose{longdiv}{1} \\
1+x \\
\hline
-x \\
\text{subtract} \longrightarrow \ -x-x^2 \\
\hline
-x^2
\end{array}
$$

after the second. Continuing in this way,

$$
\begin{array}{r}
1-x+x^2-x^3+x^4 \\
\hline
1+x \enclose{longdiv}{1} \\
1+x \\
\hline
-x \\
-x-x^2 \\
\hline
x^2 \\
x^2+x^3 \\
\hline
-x^3 \\
-x^3-x^4 \\
\hline
x^4 \\
x^4+x^5 \\
\hline
-x^5
\end{array}
$$

This division process can be continued indefinitely.
If $|x| < 1$, the successive remainders

$$-x, x^2, -x^3, x^4, -x^5, \ldots, \pm x^i, \ldots$$

become smaller and smaller, so that $\frac{1}{1+x}$ is approximated by $1 - x + x^2 - x^3 + x^4 - x^5 + \ldots$. The more
terms that are included, the smaller the $\pm x^i$, and the
better the approximation. This is similar to the way
that the sum $1 + \frac{1}{4} + \left(\frac{1}{4}\right)^2 + \left(\frac{1}{4}\right)^3 + \left(\frac{1}{4}\right)^4 + \ldots$ approximates $\frac{4}{3}$ more and more accurately as more and more
terms are added in. Refer to Section 3.3.

For example, let $x = 0.50$. Then $\frac{1}{1+x} = \frac{1}{1.5} = \frac{2}{3}$.
Working with four-decimal accuracy (for instance) we
get the following pyramid:

$$1-x = 1.000-0.5000 = 0.5000$$

$$1-x+x^2 = 0.5000+0.2500 = 0.7500$$

$$1-x+x^2-x^3 = 0.7500-0.1250 = 0.6250$$

$$1-x+x^2-x^3+x^4 = 0.6250+0.0625 = 0.6875$$

$$1-x+x^2-x^3+x^4-x^5 = 0.6875-0.0313 = 0.6562$$

$$1-x+x^2-x^3+x^4-x^5+x^6 = 0.6562+0.0156 = 0.6718$$

$$1-x+x^2-x^3+x^4-x^5+x^6-x^7 = 0.6718-0.0078 = 0.6640$$

$$1-x+x^2-x^3+x^4-x^5+x^6-x^7+x^8 = 0.6640+0.0039 = 0.6679$$

$$1-x+x^2-x^3+x^4-x^5+x^6-x^7+x^8-x^9 = 0.6679-0.0020 = 0.6659$$

$$1-x+x^2-x^3+x^4-x^5+x^6-x^7+x^8-x^9+x^{10} = 0.6659+0.0010 = 0.6669$$

Observe that the numbers running down the pyramid on the right close in on $\frac{1}{1+x} = \frac{2}{3} = 0.6666\ldots$ as asserted. For a positive number x smaller than $0.5 = \frac{1}{2}$, the terms $\pm x^i$ will get smaller faster, and the convergence of

$$1 - x + x^2 - x^3 + x^4 - x^5 + \ldots \quad \text{to} \quad \frac{1}{1+x}$$

will be more rapid. Take $i = 5$, for example. It follows from the pyramid that $\frac{1}{1+x}$ is approximated by the polynomial $1 - x + x^2 - x^3 + x^4 - x^5$ to within $0.6667 - 0.6562 = 0.0105$ over the entire interval from 0 to $\frac{1}{2}$.

Newton has therefore arrived at an approximation

$$\int_0^{\frac{1}{2}} \frac{1}{1+x}\, dx \approx \int_0^{\frac{1}{2}} (1 - x + x^2 - x^3 + x^4 - x^5)\, dx.$$

The definite integral on the right is easily evaluated by Newton's earlier analysis. In particular, taking antiderivatives term by term shows that

$$F(x) = x - \frac{1}{2}x^2 + \frac{1}{3}x^3 - \frac{1}{4}x^4 + \frac{1}{5}x^5 - \frac{1}{6}x^6$$

is an antiderivative of $1 - x + x^2 - x^3 + x^4 - x^5$. Therefore, by the Fundamental Theorem of Calculus,

$$\int_0^{\frac{1}{2}} \frac{1}{1+x}\, dx \approx \int_0^{\frac{1}{2}} (1 - x + x^2 - x^3 + x^4 - x^5)\, dx$$

$$= F\left(\tfrac{1}{2}\right) - F(0) = F\left(\tfrac{1}{2}\right)$$

$$= \tfrac{1}{2} - \tfrac{1}{2}\left(\tfrac{1}{2}\right)^2 + \tfrac{1}{3}\left(\tfrac{1}{2}\right)^3 - \tfrac{1}{4}\left(\tfrac{1}{2}\right)^4 + \tfrac{1}{5}\left(\tfrac{1}{2}\right)^5 - \tfrac{1}{6}\left(\tfrac{1}{2}\right)^6$$

$$\approx 0.5 - 0.125 + 0.042 - 0.016 + 0.007 - 0.003$$

$$= 0.405.$$

If greater accuracy is required, we can use more terms in the approximation for $\frac{1}{1+x}$ and carry more decimals. Take, for instance,

$$1 - x + x^2 - x^3 + x^4 - x^5 + x^6 - x^7 + x^8 - x^9 + x^{10}.$$

Continuing the pattern of the preceeding computation for five more steps, we get

$$\int_0^{\frac{1}{2}} \frac{1}{1+x}\, dx$$

$$\approx \frac{1}{2} - \frac{1}{2}\left(\frac{1}{2}\right)^2 + \frac{1}{3}\left(\frac{1}{2}\right)^3 - \ldots + \frac{1}{11}\left(\frac{1}{2}\right)^{11}.$$

Round off at the sixth decimal place and show that this sum is equal to

$$0.500000 - 0.125000 + 0.041667 - 0.015625$$

$$+ 0.006250 - 0.002604 + 0.001116 - 0.000488$$

$$+ 0.000217 - 0.000098 + 0.000044$$

$$= 0.405479.$$

To evaluate $\int_0^{\frac{1}{2}} \frac{1}{1+x}\, dx$ with an accuracy of up to, say, four decimal places, use the following rule of thumb. Keep computing (and adding and subtracting terms) until the term is reached that is zero when rounded to four-decimal places. The approximation obtained at this point should be accurate up to four decimal places. In this example, the term that rounds to zero is 0.000044. So the approximation 0.405479 should be accurate up to four decimals. This is correct since the value of the integral $\int_0^{\frac{1}{2}} \frac{1}{1+x}\, dx$ up to six-decimal-place accuracy is 0.405465. (The determination of the precise value of this integral makes use of logarithms and will be discussed in Chapter 10.) A word of caution is in order. This rule of thumb is not always accurate; this is due to the phenomenon known as *roundoff error*.

Observe that $\int_0^{\frac{1}{2}} \frac{1}{1+x}\, dx$ represents the area under the graph of $f(x) = \frac{1}{1+x}$ between $x = 0$ and $x = \frac{1}{2}$. See Figure 6.9. So up to four-decimal accuracy, this area is equal to 0.4055.

Newton also derives the approximation

$$\sqrt{1+x} = (1+x)^{\frac{1}{2}}$$

$$\approx 1 + \binom{\frac{1}{2}}{1} x + \binom{\frac{1}{2}}{2} x^2 + \binom{\frac{1}{2}}{3} x^3$$

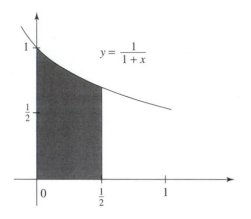

$$y = \frac{1}{1+x}$$

Figure 6.9

$$+ \binom{\frac{1}{2}}{4} x^4 + \cdots + \binom{\frac{1}{2}}{i} x^i + \cdots,$$

where

$$\binom{\frac{1}{2}}{1} = \frac{1}{2}, \quad \binom{\frac{1}{2}}{2} = \frac{\left(\frac{1}{2}\right)\left(\frac{1}{2}-1\right)}{2!}, \quad \binom{\frac{1}{2}}{3} = \frac{\left(\frac{1}{2}\right)\left(\frac{1}{2}-1\right)\left(\frac{1}{2}-2\right)}{3!}$$

and in general,

$$\binom{\frac{1}{2}}{i} = \frac{\left(\frac{1}{2}\right)\left(\frac{1}{2}-1\right)\left(\frac{1}{2}-2\right)\dots\left(\frac{1}{2}-(i-1)\right)}{i!},$$

for any positive integer i. Recall that

$$i! = i \cdot (i-1) \cdot (i-2) \cdots 3 \cdot 2 \cdot 1$$

for any integer $i > 0$. So $3! = 3 \cdot 2 \cdot 1 = 6$; $5! = 5 \cdot 4 \cdot 3 \cdot 2 \cdot 1 = 120$; and so on. Computing the first few terms shows that

$$\sqrt{1+x} \approx 1 + \frac{1}{2}x - \frac{1}{8}x^2 + \frac{1}{16}x^3 - \frac{5}{128}x^4 + \cdots.$$

This approximation works for any x with $|x| \leq 1$. Again, the inclusion of more terms increases the accuracy.

To get a feeling for the validity of this approximation for $\sqrt{1+x}$ let's consider the first five terms, $1 + \frac{1}{2}x - \frac{1}{8}x^2 + \frac{1}{16}x^3 - \frac{5}{128}x^4$. If Newton is right, then the square

$$\left(1 + \frac{1}{2}x - \frac{1}{8}x^2 + \frac{1}{16}x^3 - \frac{5}{128}x^4\right)$$

$$\cdot \left(1 + \frac{1}{2}x - \frac{1}{8}x^2 + \frac{1}{16}x^3 - \frac{5}{128}x^4\right)$$

of this polynomial should be a quantity "close to" $1+x$. Carry out this product as follows: Multiply the 1 on the top by all the terms on the bottom, then the $\frac{1}{2}x$ on the top by all the terms on the bottom, and so on. This results in

$$1+\tfrac{1}{2}x-\tfrac{1}{8}x^2+\tfrac{1}{16}x^3-\tfrac{5}{128}x^4$$
$$+\tfrac{1}{2}x+\tfrac{1}{2}\cdot\tfrac{1}{2}x^2-\tfrac{1}{2}\cdot\tfrac{1}{8}x^3+\tfrac{1}{2}\cdot\tfrac{1}{16}x^4-\tfrac{1}{2}\cdot\tfrac{5}{128}x^5$$
$$-\tfrac{1}{8}x^2-\tfrac{1}{8}\cdot\tfrac{1}{2}x^3+\tfrac{1}{8}\cdot\tfrac{1}{8}x^4-\tfrac{1}{8}\cdot\tfrac{1}{16}x^5+\tfrac{1}{8}\cdot\tfrac{5}{128}x^6$$
$$+\tfrac{1}{16}x^3+\tfrac{1}{16}\cdot\tfrac{1}{2}x^4-\tfrac{1}{16}\cdot\tfrac{1}{8}x^5+\tfrac{1}{16}\cdot\tfrac{1}{16}x^6-\tfrac{1}{16}\cdot\tfrac{5}{128}x^7$$
$$-\tfrac{5}{128}x^4-\tfrac{5}{128}\cdot\tfrac{1}{2}x^5+\tfrac{5}{128}\cdot\tfrac{1}{8}x^6-\tfrac{5}{128}\cdot\tfrac{1}{16}x^7+\tfrac{5}{128}\cdot\tfrac{5}{128}x^8$$

After adding terms column by column and computing the coefficients, this becomes

$$1 + x + \left(-\frac{14}{256}x^5 + \frac{14}{1024}x^6 - \frac{10}{2048}x^7 + \frac{25}{16,384}x^8\right).$$

A moment's thought reveals that for $-1 \leq x \leq 1$, the largest value of

$$-\frac{14}{256}x^5 + \frac{14}{1024}x^6 - \frac{10}{2048}x^7 + \frac{25}{16,384}x^8$$

occurs when $x = -1$. This value is $\frac{896+224+80+25}{16,384} = \frac{1225}{16,384} = 0.075$. So over the interval $-1 \leq x \leq 1$,

$$1 + x \approx \left(1 + \frac{1}{2}x - \frac{1}{8}x^2 + \frac{1}{16}x^3 - \frac{5}{128}x^4\right)^2$$

to within 0.075. Taking square roots, gives

$$\sqrt{1+x} \approx 1 + \frac{1}{2}x - \frac{1}{8}x^2 + \frac{1}{16}x^3 - \frac{5}{128}x^4.$$

As already remarked, approximations such as this can be made as accurate as required by adding on more terms.

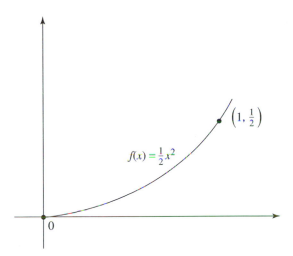

$$f(x) = \tfrac{1}{2}x^2$$

$$\left(1, \tfrac{1}{2}\right)$$

0

Figure 6.10

Example 6.6. Compute the length of the arc of the parabola $y = \frac{1}{2}x^2$ from $(0,0)$ to $(1, \frac{1}{2})$. Let $f(x) = \frac{1}{2}x^2$; then $f'(x) = x$ and $(f'(x))^2 = x^2$. See Figure 6.10. The length of the arc is provided by the definite integral in Section 5.6C:

$$\int_0^1 \sqrt{1 + (f'(x))^2}\, dx = \int_0^1 \sqrt{1 + x^2}\, dx$$

How can we compute it? Replacing x by x^2 in the approximation

$$\sqrt{1 + x} \approx 1 + \frac{1}{2}x - \frac{1}{8}x^2 + \frac{1}{16}x^3 - \frac{5}{128}x^4$$

provides the approximation

$$\sqrt{1 + x^2} \approx 1 + \frac{1}{2}x^2 - \frac{1}{8}x^4 + \frac{1}{16}x^6 - \frac{5}{128}x^8.$$

Taking antiderivatives term by term shows that

$$F(x) = x + \frac{1}{6}x^3 - \frac{1}{40}x^5 + \frac{1}{112}x^7 - \frac{5}{1152}x^9$$

is an antiderivative of $1 + \frac{1}{2}x^2 - \frac{1}{8}x^4 + \frac{1}{16}x^6 - \frac{5}{128}x^8$. Therefore,

$$\int_0^1 \sqrt{1 + x^2}\, dx$$

$$\approx \int_0^1 \left(1 + \frac{1}{2}x^2 - \frac{1}{8}x^4 + \frac{1}{16}x^6 - \frac{5}{128}x^8\right) dx$$

$$= \left(x + \frac{1}{6}x^3 - \frac{1}{40}x^5 + \frac{1}{112}x^7 - \frac{5}{1152}x^9\right)\Big|_0^1$$

$$= 1 + \frac{1}{6} - \frac{1}{40} + \frac{1}{112} - \frac{5}{1152}$$

$$= 1 + 0.1667 - 0.0250 + 0.0089 - 0.0043$$

$$= 1.1466.$$

Again, this approximation can be made accurate to any desired decimal place by continuing the pattern and taking enough terms.

In addition to the approximations

$$\frac{1}{1 + x} \approx 1 - x + x^2 - x^3 + x^4 - x^5 + \dots$$

and

$$\sqrt{1 + x} \approx 1 + \frac{1}{2}x - \frac{1}{8}x^2 + \frac{1}{16}x^3 - \frac{5}{128}x^4 + \dots,$$

Newton also developed the trigonometric approximations

$$\sin x \approx x - \frac{x^3}{3!} + \frac{x^5}{5!} - \frac{x^7}{7!} + \dots$$

and

$$\cos x \approx 1 - \frac{x^2}{2!} + \frac{x^4}{4!} - \frac{x^6}{6!} + \dots .$$

Such "infinite" polynomials are called *power series* in modern mathematics.[2] Power series have remained in a central position in mathematics ever since Newton first made use of them.

We have given an illustration of Newton's power series method for computing definite integrals. Definite integrals can represent areas, volumes, lengths of arcs, indeed—as we will see in subsequent chapters—they can represent many other physical quantities as well. Newton, therefore, produced a powerful method of calculation that is applicable to a large variety of mathematical problems.

6.4 Moving Points

In 1671 Newton gave a second, more extensive, exposition of his ideas in the treatise *Methodus Fluxionum et Series Infinitarum* (Method of Fluxions and Infinite Series). It would not be published until 1736. After discussing the role of power series as a tool for computing areas under curves, Newton goes on to say

> So much for computational methods of which in the sequel I shall make frequent use. It now remains, in illustration of this analytic art, to deliver some typical problems and such especially as the nature of curves represent. But first of all I would observe that difficulties of this sort may all be reduced to these two problems alone, which I may be permitted to propose with regard to the space traversed by any local motion however accelerated or retarded:
>
> **(1)** Given the length of the space continuously (that is at every time), to find the speed of motion at any time proposed.
>
> **(2)** Given the speed of motion continuously, to find the length of the space described at any time proposed.

What is Newton saying here? Consider a point or a particle moving in the Cartesian plane. Take a stopwatch and let it start at time $t = 0$. After any time $t \geq 0$ it will have some position in the plane. Let the x-coordinate of this position be $x(t)$ and the y-coordinate $y(t)$. Both $x(t)$ and $y(t)$ are functions of t. Newton calls such functions *fluents*. The position of the point and hence the coordinates $x(t)$ and $y(t)$ will vary with t. Figure 6.11 shows a typical situation. At $(x(t), y(t))$ the point will have a certain velocity in the x-direction and also in the y-direction. Newton calls these velocities the *fluxions* of $x(t)$ and $y(t)$. These velocities are also functions of t. The two problems that Newton has singled out are these:

(1) Given that the position of the point is known at any time, determine its velocity at any time. For example, given that the function $x(t)$ is known explicitly, determine the function that gives the velocity in the x-direction.

(2) Given that the velocity of the point is known at any time, determine its position at any time. For example, given the function that describes the velocity in the y-direction, determine the function $y(t)$.

The primary objective of Newton's abstract discussion is the development of the mathematics that underlies the phenomenon of motion. This is precisely what is necessary for a fundamental understanding of the basic observations of Galileo that we discussed in Section 4.2. Newton's mathematics of motion will be the focus of the rest of the chapter. In reference to Figure 6.11, observe that if the functions $x(t)$ and $y(t)$ are both understood, then the motion of the point is understood. The mathematics of motion in the plane therefore reduces to the situation of a point moving along a coordinatized line.

Suppose a point is moving along a coordinatized line. Begin to observe it at time $t = 0$. Denote its coordinate at any time $t \geq 0$ by $p(t)$. Note that p is a function of time t. It is the *position function* of the point. (See Figure 6.12.) The position $p(0)$ is the *initial position* of the point. The units of distance and time can be miles and hours, meters and seconds, etc., as required in a specific problem.

Example 6.7. Suppose that p is given explicitly by $p(t) = t$. So the position of the point at time $t = 0$ is $p(0) = 0$; at time $t = 1$, it is $p(1) = 1$; at time $t = 2$, it is $p(2) = 2$, and so on. Next suppose that the position function is $p(t) = t^2$. Then $p(0) = 0$, $p(1) = 1$, $p(2) = 4$, and so forth. The motion of the point is shown in Figure 6.13(a) in the first case and in Figure 6.13(b) in the second. In each case, the location of the point is listed above the axis and the time at which the point is at a given location is indicated below the axis.

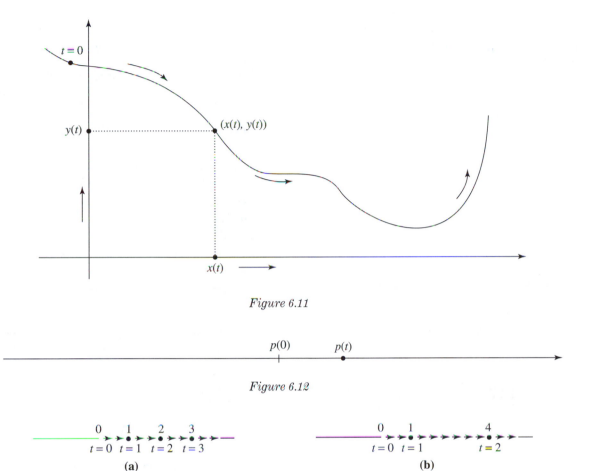

Figure 6.11

Figure 6.12

(a)   (b)

Figure 6.13

Example 6.8. Suppose that $p(t) = t^2 - 4t + 3$. Before considering the motion of a point with position function p, we will investigate the graph of p (with horizontal axis the t-axis). Note that the graph is a parabola that opens upward. The derivative $p'(t) = 2t - 4$ is 0 when $t = 2$. Since $p(2) = 4 - 8 + 3 = -1$, it follows that the parabola has a horizontal tangent at $(2, -1)$. This point is at the bottom of the parabola, so the smallest value that $p(t)$ can have is -1 and it occurs when $t = 2$. Since $t^2 - 4t + 3 = (t - 1)(t - 3)$, it follows that $p(t) = 0$ precisely when $t = 1$ or $t = 3$. The graph of p is sketched in Figure 6.14. Be aware of the fact that it is the graph of the po-

sition function; it is not the trajectory of the moving point.

We now turn to a point that moves on a horizontal axis with position given by

$$p(t) = t^2 - 4t + 3 = (t - 1)(t - 3)$$

at any time $t \geq 0$. Since $p(0) = 3$, the point is at 3 on the axis at $t = 0$. Since $p(t)$ decreases initially (the graph in Figure 6.14 falls), the point moves to the left from this position and reaches the point 0 at $t = 1$ and the point -1 at $t = 2$. There it stops, changes direction (the graph begins to rise), and at $t = 3$ it is back at 0. At $t = 4$ it is at 3 and continues moving to the right.

The motion of the point is described schematically in Figure 6.15. The points singled out on the axis are -1, 0, and 3.

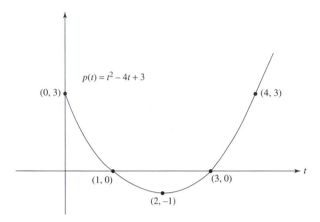

$$p(t) = t^2 - 4t + 3$$

$(0, 3)$ $(4, 3)$

$(1, 0)$ $(3, 0)$

$(2, -1)$

Figure 6.14

$t = 3$ $t = 4$

$t = 2$ $t = 1$ $t = 0$

Figure 6.15

Let's return to the general case of a point moving on a coordinatized axis. Fix some instant $t_1 \geq 0$. Let an additional amount of time Δt elapse and consider $t_1 + \Delta t$. We will denote the time interval from t_1 to $t_1 + \Delta t$ by $[t_1, t_1 + \Delta t]$. We have

$p(t_1)$ = the position at time t_1

$p(t_1 + \Delta t)$ = the position at time $t_1 + \Delta t$

$p(t_1 + \Delta t) - p(t_1)$ = the change in position over the time interval $[t_1, t_1 + \Delta t]$

= the net distance (possibly negative) covered by the point in the time interval $[t_1, t_1 + \Delta t]$

$\dfrac{p(t_1 + \Delta t) - p(t_1)}{\Delta t}$ = the net distance covered divided

by the time it took to cover that distance

= the average velocity during the time interval $[t_1, t_1 + \Delta t]$

Example 6.9. Return to the point with position function $p(t) = t^2 - 4t + 3 = (t - 1)(t - 3)$ for $t \geq 0$. To make things more concrete, suppose that distance is measured in inches and time in seconds. Take $t_1 = 1$ second. Now the point is exactly one second into its trip. Let another $\Delta t = 3$ seconds elapse. So $t_1 + \Delta t = 4$ seconds. Since $p(t_1 + \Delta t) - p(t_1) = p(4) - p(1) = 3 - 0 = 3$, the point has covered a net distance of 3 inches in these 3 seconds. Because

$$\frac{p(t_1 + \Delta t) - p(t_1)}{\Delta t} = \frac{p(4) - p(1)}{3} = \frac{3 - 0}{3} = 1,$$

the point has an average velocity of 1 inch per second over the time interval $[t_1, t_1 + \Delta t] = [1, 4]$. Next take $\Delta t = 2$. Now $t_1 + \Delta t = 3$, and

$$\frac{p(t_1 + \Delta t) - p(t_1)}{\Delta t} = \frac{p(3) - p(1)}{2} = \frac{0 - 0}{2} = 0.$$

The net distance covered by the point is zero and it has an average velocity of zero. With $\Delta t = 1$, we have $t_1 + \Delta t = 2$, and the average velocity of

$$\frac{p(2) - p(1)}{1} = \frac{-1 - 0}{1} = -1 \text{ inches per second.}$$

The minus sign indicates that the point has moved in the negative direction (from $p(1) = 0$ to $p(2) = -1$) over the time interval $[t_1, t_1 + \Delta t] = [1, 2]$.

For any moving point, consider the ratio

$$\frac{p(t_1 + \Delta t) - p(t_1)}{\Delta t}$$

for progressively smaller elapsed times Δt. This is the average velocity of the point computed over smaller and smaller time frames $[t_1, t_1 + \Delta t]$ near t_1. (See Figure 6.16.) In the process, $\frac{p(t_1 + \Delta t) - p(t_1)}{\Delta t}$ closes in on a number called the *velocity at the instant* t_1; it is denoted by $v(t_1)$. This is the rate at which the position changes at the instant t_1. It is obtained by pushing

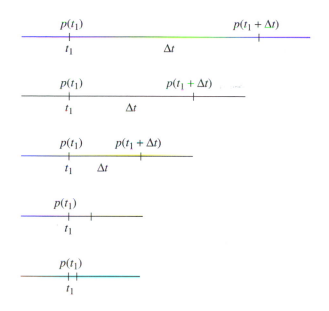

Figure 6.16

Δt to zero and observing what happens to the ratio $\frac{p(t_1+\Delta t)-p(t_1)}{\Delta t}$. Therefore,

$$v(t_1) = \lim_{\Delta t \to 0} \frac{p(t_1 + \Delta t) - p(t_1)}{\Delta t}.$$

The velocity $v(0)$ at $t = 0$ is called the *initial velocity* of the point. It is possible for Δt to be negative. This means that $t_1 + \Delta t$ occurs before t_1 and that the time interval in question is $[t_1 + \Delta t, t_1]$.

Example 6.10. Let's see what happens when this is done with $t_1 = 1$ second for the point with position $p(t) = t^2 - 4t + 3 = (t-1)(t-3)$ for $t \geq 0$. Taking $\Delta t = \frac{1}{2}, \frac{1}{4}, \frac{1}{8}, \ldots, \frac{1}{64}$, we get

$$t_1 + \Delta t = \frac{3}{2}, \frac{5}{4}, \frac{9}{8}, \ldots, \frac{65}{64}.$$

The successive average velocities are

$$\frac{p\left(\frac{3}{2}\right) - p(1)}{\frac{1}{2}} = \frac{\left(\frac{3}{2}-1\right) \cdot \left(\frac{3}{2}-3\right) - 0}{\frac{1}{2}} = \frac{1}{2}\left(-\frac{3}{2}\right)2$$

$$= -1.5 \text{ inches per second,}$$

$$\frac{p\left(\frac{5}{4}\right) - p(1)}{\frac{1}{4}} = \frac{\left(\frac{5}{4}-1\right) \cdot \left(\frac{5}{4}-3\right) - 0}{\frac{1}{4}} = \frac{1}{4}\left(-\frac{7}{4}\right)4$$

$$= -1.75 \text{ inches per second,}$$

$$\frac{p\left(\frac{9}{8}\right) - p(1)}{\frac{1}{8}} = \frac{\left(\frac{9}{8}-1\right) \cdot \left(\frac{9}{8}-3\right) - 0}{\frac{1}{8}} = \frac{1}{8}\left(-\frac{15}{8}\right)8$$

$$= -1.88 \text{ inches per second,} \ldots,$$

$$\frac{p\left(\frac{65}{64}\right) - p(1)}{\frac{1}{8}} = \frac{\left(\frac{65}{64}-1\right) \cdot \left(\frac{65}{64}-3\right) - 0}{\frac{1}{64}} = \frac{1}{64}\left(-\frac{127}{64}\right)64$$

$$= -1.98 \text{ inches per second.}$$

Continuing to push Δt to zero, we see that

$$v(1) = \lim_{\Delta t \to 0} \frac{p(1 + \Delta t) - p(1)}{\Delta t} = \lim_{\Delta t \to 0} \frac{\Delta t(\Delta t - 2) - 0}{\Delta t}$$

$$= \lim_{\Delta t \to 0} (\Delta t - 2) = -2 \text{ inches per second.}$$

Again, the minus sign means that the point is moving in the negative direction at time $t_1 = 1$. Look again at Figure 6.15.

Refer back to Section 5.5A and observe that the limit

$$\lim_{\Delta t \to 0} \frac{p(t_1 + \Delta t) - p(t_1)}{\Delta t}$$

has three different meanings simultaneously:

(1) It is the velocity $v(t_1)$ at time t_1 of the point whose position function is p;

(2) It is the value $p'(t_1)$ of the derivative of the position function p;

(3) It is the slope of the tangent line of the graph of the function p at the point $(t_1, p(t_1))$.

To repeat, if p is the position function of a moving point, then its derivative

$$p'(t_1) = \lim_{\Delta t \to 0} \frac{p(t_1 + \Delta t) - p(t_1)}{\Delta t}$$

is the velocity $v(t_1)$ of the point at time t_1. Changing notation from t_1 to t, we see that $p'(t)$ is the velocity $v(t)$ of the point at any time t. If $v(t)$ is positive, then the point is moving to the right at time t, and if $v(t)$ is negative, the point is moving to the left at time t. The *speed* of the point at time t is the absolute value $|v(t)|$

of the velocity. Note that velocity is the combination of speed and direction. The concepts velocity and speed are often used interchangeably in this book when the direction of the motion is clear from the context.

Example 6.11. The point in Example 6.10 has a velocity of $v(t) = p'(t) = 2t - 4 = 2(t - 2)$ and a speed of $|2t - 4|$ at any time t. Since $v(t)$ is negative for $0 \leq t < 2$, the point moves to the left during this time; since $v(2) = 0$, the point stops at $t = 2$; and since $v(t)$ is positive for $t > 2$, it moves to the right for any such t. Compare this analysis with the information in Figure 6.15. When $t = 1$, the point's velocity is -2 inches per second (as already observed) and its speed is $|-2| = 2$ inches per second.

Let's see whether the definition of the velocity as the derivative of the position function agrees with what we know about velocity intuitively. Consider a point moving on the line and let its position be $p(t)$ at any time $t \geq 0$. Suppose that after some time t_1, when its position is $p(t_1)$, it continues with the constant velocity $p'(t_1)$. This means that $p'(t) = p'(t_1)$ for any $t \geq t_1$. So for $t \geq t_1$, the graph of $p(t)$ has constant slope $p'(t_1)$. Since the graph of $p(t)$ goes through the point $(t_1, p(t_1))$, we find from the point-slope form of the equation of a line that

$$p(t) - p(t_1) = p'(t_1)(t - t_1),$$

or

$$p(t) = p(t_1) + p'(t_1)(t - t_1),$$

for $t \geq t_1$. For an elapsed time Δt after t_1, the point will have moved a distance of

$$p(t_1 + \Delta t) - p(t_1)$$

$$= p(t_1) + p'(t_1)\big((t_1 + \Delta t) - t_1\big) - p(t_1)$$

$$= p'(t_1)\,\Delta t.$$

But since distance = velocity × time when velocity is constant, this is exactly what we would expect. Figure 6.17 shows what is going on from the point of view of the graph of the position function p.

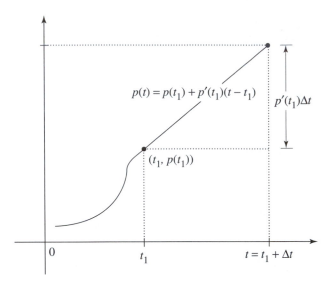

Figure 6.17

Now let's focus on the velocity function $v(t)$ of a point moving on a coordinatized axis. See Figure 6.12. Fix some instant $t_1 \geq 0$. Let an additional amount of time Δt elapse and consider $t_1 + \Delta t$. Then

$$v(t_1) = \text{the velocity at time } t_1$$

$$v(t_1 + \Delta t) = \text{the velocity at time } t_1 + \Delta t$$

$$v(t_1 + \Delta t) - v(t_1) = \text{the change in velocity over the time interval } [t_1, t_1 + \Delta t]$$

$$\frac{v(t_1 + \Delta t) - v(t_1)}{\Delta t} = \text{the average change in velocity per unit time over } [t_1, t_1 + \Delta t]$$

$$= \text{the average acceleration over } [t_1, t_1 + \Delta t]$$

Consider the average acceleration $\frac{v(t_1+\Delta t)-v(t_1)}{\Delta t}$ over the time interval $[t_1, t_1 + \Delta t]$ for progressively smaller elapsed times Δt. In the process, the ratio $\frac{v(t_1+\Delta t)-v(t_1)}{\Delta t}$ closes in on the number called the *acceleration at the instant* t_1. It is denoted by $a(t_1)$ and is the rate at which the velocity is changing at the instant t_1. In mathematical shorthand, the acceleration at time

t_1 is

$$a(t_1) = \lim_{\Delta t \to 0} \frac{v(t_1 + \Delta t) - v(t_1)}{\Delta t},$$

Changing notation from t_1 to t, we find that $a(t)$ is the derivative $v'(t)$ of the velocity function $v(t)$. Since $v(t) = p'(t)$, $a(t)$ is the derivative of $p'(t)$. This *second derivative* of p is denoted $p''(t)$. So $a(t) = p''(t)$. What units is acceleration measured in? For example, if distance is given in inches and time in seconds, then velocity is given in inches per second, and acceleration in (inches per second) per second, or inches per second^2.

Example 6.12. Let's have one more look at the point with position function $p(t) = t^2 - 4t + 3$. Since $v(t) = 2t - 4$, the point's acceleration is $a(t) = v'(t) = 2$. Suppose that the point is a particle of mass m. By one of Newton's laws of motion, an acceleration a on a body of mass m is the consequence of the action of a force F. The connection between the force and the acceleration is given by the famous equation $F = ma$. So the force on this particle is $F = 2m$. It is positive and hence "pushes" the particle in the positive direction. The motion of the particle has the following interpretation. It starts at time $t = 0$ at $p(0) = 3$ with an initial velocity of $v(0) = -4$. So it begins by moving to the left. Since the force acts to the right, the particle must slow down. It eventually comes to a halt. Since $v(2) = 0$, this happens at time $t = 2$. From this time onward, the force accelerates the particle to the right. Check this against Figure 6.15.

We have analyzed the motion of a point on a line and observed how its position function determines both its velocity and acceleration. We will now see that this reasoning can be reversed. Given the acceleration of a point, it is possible to determine first its velocity and then its position. This will make important use of a fact from Section 5.5B: If two functions have the same derivative, then one of them is equal to the other plus a constant. In particular, if the derivative of a function is always zero, the function is a constant.

Example 6.13. Let a point move on an axis with acceleration $a(t) = v'(t) = p''(t) = 0$ for all $t \geq 0$. By

the fact just referred to, this means that $v(t) = C$, for some constant C. So $v(t) = C = v(0)$. Notice that the derivatives of both $v(0) \cdot t$ and the position function are equal to $v(0)$. Therefore, $p(t) = v(0)t + D$ for some constant D. Taking $t = 0$ shows that $D = p(0)$ is the initial position of the point. The position function is now explicitly given by

$$p(t) = v(0)t + p(0).$$

Example 6.14. Suppose a point moves along an axis with constant acceleration $a(t) = v'(t) = p''(t) = k$ for all $t \geq 0$. Observe that $v(t)$ and kt both have the derivative k. It follows that $v(t) = kt + C$. Setting $t = 0$, we see that $C = v(0)$ is the initial velocity; hence

$$v(t) = kt + v(0).$$

Since the derivative of $\frac{k}{2}t^2 + v(0)t$ is equal to

$$kt + v(0) = v(t),$$

it follows that $\frac{k}{2}t^2 + v(0)t$ and the position function $p(t)$ have the same derivative. Hence $p(t) = \frac{k}{2}t^2 + v(0)t + D$. Plugging in $t = 0$ shows that $D = p(0)$ is the initial position of the point. So the position function is equal to

$$p(t) = \frac{k}{2}t^2 + v(0)t + p(0).$$

If the initial position is $p(0) = 0$ and the initial velocity $v(0) = 0$, then $p(t) = \frac{k}{2}t^2$. In this case, the constant acceleration k produces a displacement of $\frac{k}{2}t^2$ in the time interval $[0, t]$.

When a point moves on the x-axis, we will denote its position function by $x(t)$. The initial position is $x(0)$; the velocity function is $v(t) = x'(t)$ and the acceleration is $a(t) = v'(t) = x''(t)$. If it moves along the y-axis, its position function will be written $y(t)$; its initial position is $y(0)$ and its velocity and acceleration are $v(t) = y'(t)$ and $a(t) = v'(t) = y''(t)$, respectively.

Suppose an object is in vertical free fall near the Earth's surface. Place a y-axis along the path of the fall with the origin at ground level. Let $y(t)$ be the position of the object at any time $t \geq 0$. The initial position of the object is $y(0)$ and its initial velocity is $v(0)$. See Figure 6.18. Recall from Chapter 4.2 that Galileo discovered that if the object has velocity $v(t)$ at time t and velocity $v(t+\Delta t)$ at a later time $t+\Delta$, then the ratio

$$\frac{v(t + \Delta t) - v(t)}{(t + \Delta t) - t} = \frac{v(t + \Delta t) - v(t)}{\Delta t}$$

is equal to the same constant C regardless of the time t of the first observation and the elapsed time Δt. Therefore, this ratio remains fixed at C when Δt is pushed to zero. It follows that for an object in free fall, the acceleration $a(t)$ at any time $t \geq 0$ is given by

$$a(t) = v'(t) = \lim_{\Delta t \to 0} \frac{v(t + \Delta t) - v(t)}{\Delta t} = C.$$

Since the acceleration is in the negative y-direction, the constant C is negative. Nowadays, the constant C is written as $-g$, and g is known to be equal to 32.17 feet/second^2, or 9.80 meters/second^2. The force F that produces the acceleration of $-g$ on the falling object is known as the *force of gravity*. By Newton's law, $F = -mg$, so that F depends on the mass m of the object. On the other hand, as Galileo observed, the constant g is the same regardless of the mass of the object.

Since the acceleration $a(t) = -g$ is a constant, the velocity and position functions of the objects are

$$v(t) = -gt + v(0) \quad \text{and} \quad y(t) = -\frac{g}{2}t^2 + v(0)t + y(0).$$

This follows from Example 6.14 by replacing k by $-g$ and $p(t)$ by $y(t)$.

Example 6.15. Suppose Galileo had actually dropped a cannonball from the leaning tower of Pisa. Measure distance in feet and time in seconds. Assume that the initial velocity of the cannonball was $v(0) = 0$ and, since the tower is 177 feet high, that the initial position was $y(0) = 177$. Therefore,

$$y(t) = -\frac{g}{2}t^2 + 177.$$

Suppose the cannonball hit the ground at the instant t_1. So $y(t_1) = 0$ and $\frac{g}{2}t_1^2 = 177$. Since $g = 32$ feet/sec^2, it follows that $t_1^2 = \frac{354}{32}$, and $t_1 = \sqrt{\frac{354}{32}} = 3.3$ seconds.

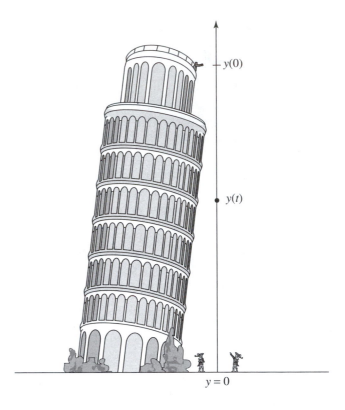

Figure 6.18

The velocity of the cannonball at impact was $v(t_1) = -gt_1 + v(0) = -32(3.3) = -106$ feet per second.

6.5 The Trajectory of a Projectile

We will now turn to the analysis of the trajectory of a projectile, which applies in principle to any object that is thrown and hence to projectiles fired by guns and cannons. The basic understanding is that the projectile is "left to its own devices" after being given an initial velocity at launch. In other words, it is not subject to any propelling forces thereafter.

Set up a coordinate system as shown in Figure 6.19. The projectile starts its motion at time $t = 0$, when its initial position is the point $(0, y_0)$. It has an initial velocity that is represented by the arrow in the

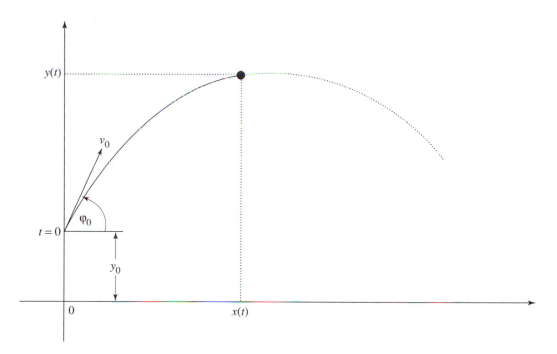

Figure 6.19

diagram. Such arrows are called *vectors*. The length of the vector is equal to the initial speed v_0 of the projectile. The direction of the vector with respect to the horizontal is known as the *angle of elevation* or *angle of departure*. It is given by the angle φ_0, where $0 \leq \varphi_0 \leq \frac{\pi}{2}$.

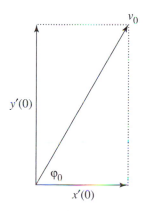

Figure 6.20

For any $t \geq 0$, the coordinates $x(t)$ and $y(t)$ are the perpendicular projections of the position of the projectile onto the x- and y-axes respectively. Note that $x(0) = 0$ and $y(0) = y_0$. By the analysis of Section 6.4, $x'(t)$ is the velocity of the projectile in the x-direction and $y'(t)$ is the velocity in the y-direction. So $x'(0)$ and $y'(0)$ are the initial velocities in the x- and y-directions. Together they determine the initial velocity vector by the *parallelogram law*. Place a horizontal vector of length $x'(0)$ and a vertical vector of length $y'(0)$ "tail to tail" as shown in Figure 6.20. The initial velocity vector is the diagonal of the parallelogram (in this case it is a rectangle) that results. The length of this diagonal is the initial speed v_0 of the projectile.

Refer to Figure 6.20 again and observe that $\cos \varphi_0 = \frac{x'(0)}{v_0}$ and $\sin \varphi_0 = \frac{y'(0)}{v_0}$. Therefore,

$$x'(0) = v_0 \cos \varphi_0 \quad \text{and} \quad y'(0) = v_0 \sin \varphi_0.$$

The only force on the projectile during its flight is gravity. It acts in the negative y-direction. In particular, the force in the x-direction is zero. Therefore, by Newton's law $F = ma$, the acceleration in the x-direction is also zero. Thus $x''(t) = 0$. Observe that Example 6.13 applies to the projection of the projectile onto the x-axis with $C = v_0 \cos \phi_0$. Thus

$$x'(t) = v_0 \cos \varphi_0 \quad \text{and} \quad x(t) = (v_0 \cos \varphi_0)t. \qquad (6a)$$

Because gravity produces an acceleration of $-g$ in the y-direction, $y''(t) = -g$. Since $y'(0) = v_0 \sin \varphi_0$ and $y(0) = y_0$, we find by the discussion at the end of Section 6.4 that

$$\boxed{\begin{aligned} y'(t) &= -gt + v_0 \sin \varphi_0 \quad \text{and} \\ y(t) &= -\frac{g}{2}t^2 + (v_0 \sin \varphi_0)t + y_0 \end{aligned}} \qquad (6b)$$

By the second equation of (6a), $t = \frac{x(t)}{v_0 \cos \varphi_0}$, so $t^2 = \frac{(x(t))^2}{v_0^2 \cos^2 \varphi_0}$. By a substitution into the expression for $y(t)$ in (6b), we get

$$\begin{aligned} y(t) &= \frac{-g}{2v_0^2 \cos^2 \varphi_0}(x(t))^2 + \frac{v_0 \sin \varphi_0}{v_0 \cos \varphi_0}x(t) + y_0 \\ &= \frac{-g}{2v_0^2 \cos^2 \varphi_0}(x(t))^2 + (\tan \varphi_0)x(t) + y_0. \end{aligned}$$

Observe, therefore, that the position $(x(t), y(t))$ of the projectile at any time t satisfies the following equation:

$$\boxed{y = \left(\frac{-g}{2v_0^2 \cos^2 \varphi_0}\right)x^2 + (\tan \varphi_0)x + y_0} \qquad (6c)$$

By Section 4.3, this is the equation of a parabola. Therefore, the trajectory of the projectile is a parabola, indeed, a very specific parabola determined by the constants v_0, y_0, and φ_0.

What is the maximal height reached by the projectile? This maximum occurs at the instant at which the vertical component of the velocity is zero, i.e., at the instant t_1 at which $y'(t_1) = 0$. By the first equation of (6b), $0 = y'(t_1) = -gt_1 + v_0 \sin \varphi_0$, or $t_1 = \frac{v_0 \sin \varphi_0}{g}$. Inserting

this t_1 into the second equation of (6b) shows that

$$\begin{aligned} y(t_1) &= -\frac{g}{2}t_1^2 + (v_0 \sin \varphi_0)t_1 + y_0 \\ &= -\frac{1}{2g}v_0^2 \sin^2 \varphi_0 + \frac{1}{g}v_0^2 \sin^2 \varphi_0 + y_0. \end{aligned}$$

Therefore the maximal height attained by the projectile is

$$\frac{1}{2g}v_0^2 \sin^2 \varphi_0 + y_0. \qquad (6d)$$

For the rest of our discussion we will assume that the terrain is flat and the x-axis lies along the ground. At what time and how far downrange will the projectile hit the ground? Impact occurs precisely at the time t_{imp} for which $y(t_{\text{imp}}) = 0$. Refer to the second equation of (6b) and observe that the time of impact can be found by solving

$$-\frac{g}{2}(t_{\text{imp}})^2 + (v_0 \sin \varphi_0)t_{\text{imp}} + y_0 = 0$$

for t_{imp}. By the quadratic formula and the fact that $t_{\text{imp}} \geq 0$,

$$t_{\text{imp}} = \frac{v_0 \sin \varphi_0 + \sqrt{v_0^2 \sin^2 \varphi_0 + 2gy_0}}{g}. \qquad (6e)$$

The *range* of the projectile is the horizontal distance from the initial position of the projectile to the point of impact. So the range is equal to the x-coordinate R of the projectile at the time of impact t_{imp}. It follows from (6a) that

$$\begin{aligned} R = x(t_{\text{imp}}) &= (v_0 \cos \varphi_0)t_{\text{imp}} \\ &= \frac{v_0^2}{g}(\sin \varphi_0)(\cos \varphi_0) \\ &\quad + \frac{v_0}{g}\sqrt{(\cos^2 \varphi_0)v_0^2(\sin^2 \varphi_0) + 2gy_0(\cos^2 \varphi_0)}. \end{aligned}$$

The trigonometric formula $\sin 2\varphi_0 = 2 \sin \varphi_0 \cos \varphi_0$ (see Exercises 2C) now provides the following expression for the range:

$$R = \frac{v_0^2}{2g} \sin(2\varphi_0) + \frac{v_0}{g} \sqrt{\frac{v_0^2}{4} \sin^2(2\varphi_0) + 2gy_0 \cos^2 \varphi_0}.$$

(6f)

What angle φ_0 gives the maximal range, i.e., the largest possible R? Suppose $y_0 = 0$. A look at the expression in (6f) in this case shows that R is maximal when $\sin(2\varphi_0)$ is largest. Since $0 \le \varphi_0 \le \frac{\pi}{2}$, it follows that $0 \le 2\varphi_0 \le \pi$. The largest value of $\sin(2\varphi_0)$ in this interval is 1, and it occurs for $2\varphi_0 = \frac{\pi}{2}$, i.e., for $\varphi_0 = \frac{\pi}{4}$. So if $y_0 = 0$ the maximal range is

$$R_{\max} = \frac{v_0^2}{2g} + \frac{v_0}{g} \sqrt{\frac{v_0^2}{4}} = \frac{v_0^2}{2g} + \frac{v_0^2}{2g} = \frac{v_0^2}{g}.$$

(6g)

If $y_0 \ne 0$, then $\varphi_0 = \frac{\pi}{4}$ does not provide the greatest range R. However, the determination of the angle φ_0 that does so is beyond our current mathematical firepower.

The velocity of the projectile at any time t is determined by the velocities $x'(t)$ and $y'(t)$, by the parallelogram law (Figure 6.20), and the equations (6a) and (6b). Therefore, its speed at any time t is

$$\sqrt{(x'(t))^2 + (y'(t))^2}$$

$$= \sqrt{v_0^2 \cos^2 \varphi_0 + g^2 t^2 - 2(g v_0 \sin \varphi_0)t + v_0^2 \sin^2 \varphi_0}.$$

Since $v_0^2 \cos^2 \varphi_0 + v_0^2 \sin^2 \varphi_0 = v_0^2 (\cos^2 \varphi_0 + \sin^2 \varphi_0) = v_0^2$, this is equal to

$$\sqrt{v_0^2 + g^2 t^2 - 2g(v_0 \sin \varphi_0)t}.$$

(6h)

The speed at impact is obtained by plugging the moment of impact t_{imp} into this formula. How can the angle of impact be determined? See Exercise 20.

A look at equations (6a)–(6h) shows that the initial conditions y_0, v_0, and φ_0 and the constant g determine everything about the motion. We conclude our discussion by illustrating the theory with a look at a numerical example.

Newton tosses an apple in the direction of his younger colleague, the astronomer Edmund Halley. When the apple leaves his hand, it is 3 feet above the ground, has an initial velocity of 25 feet per second, and has an angle of departure of 45°. The ground is level and Halley stands 18 feet away. Is he in a position to catch it? (See Figure 6.21.)

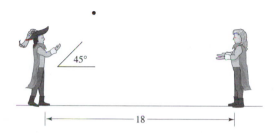

Figure 6.21

We assume that the toss occurs at time $t = 0$. The question that has to be asked is this: how high will the apple be when it reaches the spot where Halley is standing? This is a question about $y(t)$ or, more precisely, about the value of $y(t)$ at the time t when the apple reaches that spot. If we can determine the time when this occurs, then the use of the second formula in (6b) will provide the answer. But how do we get t? We know that $v_0 = 25$ feet per second and $\varphi_0 = 45°$. We also know that we are looking for t such that $x(t) = 18$. So by (6a),

$$18 = x(t) = (v_0 \cos \varphi_0)t = (25 \cos 45°)t = 25\frac{\sqrt{2}}{2}t,$$

and we get[3]

$$t = \frac{18}{25}\frac{2}{\sqrt{2}} = \frac{18}{25}\frac{2}{\sqrt{2}} \cdot \frac{\sqrt{2}}{\sqrt{2}} = \frac{18}{25}\sqrt{2} \approx (0.72)(1.41)$$

$$= 1.02 \text{ seconds}.$$

Plugging $t = 1.02$, $g = 32$, $y_0 = 3$, and $\varphi_0 = 45°$ into $y(t) = -\frac{g}{2}t^2 + (v_0 \sin \varphi_0)t + y_0$, gives us

$$y(1.02) \approx -16(1.02)^2 + (25 \sin 45°)(1.02) + 3$$

$$\approx -16(1.04) + 25\frac{\sqrt{2}}{2}(1.02) + 3$$

$$\approx -16(1.04) + 25(0.71)(1.02) + 3$$

$$\approx -16.65 + 18.03 + 3 = 4.39 \text{ feet.}$$

So the apple will be about $4\frac{1}{2}$ feet above the ground when it gets to Halley. Alas, Halley ducks! How far behind him will the apple land? With what velocity will it hit the ground? Since the angle of departure is $\phi_0 = 45°$, formula (6f) can be used to determine where the apple will hit the ground:

$$R = \frac{v_0^2}{2g} + \frac{v_0}{g}\sqrt{\frac{v_0^2}{4} + gy_0} = \frac{25^2}{64} + \frac{25}{32}\sqrt{\frac{25^2}{4} + (32)(3)}$$

$$\approx 9.77 + 0.78(\sqrt{156.25 + 96}) \approx 9.77 + 0.78(15.88)$$

$$\approx 9.77 + 12.41 \approx 22.17 \text{ feet.}$$

Since Halley stands 18 feet from Newton, the apple lands a little over 4 feet behind Halley. The time of impact is given by the formula

$$t_{\text{imp}} = \frac{v_0 \sin\varphi_0 + \sqrt{v_0^2 \sin^2\varphi_0 + 2gy_0}}{g}.$$

Inserting the constants, $v_0 = 25$, $y_0 = 3$, $\varphi_0 = 45°$, and $g = 32$, we get

$$t_{\text{imp}} = \frac{1}{32}\left(25 \cdot \frac{\sqrt{2}}{2} + \sqrt{25^2 \cdot \frac{2}{4} + 2 \cdot 32 \cdot 3}\right)$$

$$\approx \frac{1}{32}\left(17.68 + \sqrt{312.5 + 192}\right)$$

$$\approx \frac{1}{32}(17.68 + 22.46) \approx \frac{40.14}{32} \approx 1.25 \text{ seconds.}$$

Inserting t_{imp} and the other constants into (6h) tells us that the speed of the apple at impact is

$$\sqrt{25^2 + 32^2(1.25)^2 - 2 \cdot 32 \cdot 25 \cdot \frac{\sqrt{2}}{2}(1.25)}$$

$$\approx \sqrt{625 + 1600 - 1414.21}$$

$$\approx 28.47 \text{ feet/second.}$$

What is the maximal height reached by the apple? This is given by (6d):

$$\frac{1}{2g}v_0^2 \sin^2\varphi_0 + y_0 = \frac{1}{64}25^2 \cdot \frac{1}{2} + 3 = \frac{625}{128} + 3$$

$$\approx 4.88 + 3 \approx 7.88 \text{ feet.}$$

Let's reflect over what Newton's calculus has achieved. It has isolated the gravitational aspect of the motion of a projectile and provided full understanding of it. In particular, equations (6a)–(6h) provide a complete description of the motion of a projectile in any situation where gravity is the only influencing factor. This is so, for instance, on the Moon (for a different constant g), where there is no atmosphere. Equations (6a)–(6h) also provide an accurate picture of what happens on Earth for any relatively heavy object moving at a low speed. In such cases gravity is often the dominant influence, and other factors are negligible. This is so for the apple just thrown by Newton and, as we will see in Section 9.3, for some of the important experiments of Galileo.

For a light object, or for an object moving at a high velocity, air resistance becomes a significant factor. Anyone who has ever stuck his hand out of the window of a car (moving at a speed of, say, 60 miles per hour = 88 feet per second) and felt the stiff resistance of the air has already experienced this. The equations (6a)–(6h) are based on the fundamental observation of Galileo that the increase in the velocity of a falling body over time is a constant. At high speeds, however, this observation is valid only in an idealized, air-resistance-free setting. The question that arises is this: Will these equations retain at least some of their value in a resistant medium, or will the impact of air resistance render them completely useless?

6.6 Applications to Ballistics?

Ballistics is the investigation of the motion of projectiles, primarily in the context of both light and heavy firearms. This aspect of the mod-

KIND OF ORDNANCE	Powder	Ball	Elevation	Range	REMARKS
	LBS		° '	YARDS	
6-PDR. FIELD GUN	1.25	Shot	0 00	318	
		"	1 00	674	
		"	2 00	867	
		"	3 00	1138	
		"	4 00	1256	
		"	5 00	1523	
	1.00	Sph. case shot	2 00	650	Time of flight 2 secs
		"	2 30	840	" " 3 "
		"	3 00	1050	" " 4 "
12-PDR. FIELD GUN	2.50	Shot	0 00	347	
		"	1 00	662	
		"	1 30	785	
		"	2 00	909	
		"	3 00	1269	
		"	4 00	1455	
		"	5 00	1663	
	1.50	Sph. case	1 00	670	Time 2 secs
		"	1 45	950	" 3 "
		"	2 30	1250	" 4 "

Figure 6.22

ern science of artillery has its origins in the middle of the 14th century, when gunpowder came into use in western Europe. At first, artillery was deployed mainly against fortifications. It was used extensively in the Thirty Years War of 1618–1648, and it has played an increasingly important role in warfare ever since.

In the 18th and 19th centuries, cannons were classified by the weight of the cannonball that they fired. For example, there were 6-pdr. (pounder), 9-pdr., 12-pdr., 24-pdr., and 32-pdr. cannons. Figures 6.22 and 6.23 contain data for 19th century American 6-pdr. and 12-pdr. field guns. These tables come from the appendix of *The Artillerist's Manual*,[4] originally published in 1860. In Figure 6.22, the column "Powder." refers to the quantity of powder used, the column "Ball." to the types of ammunition (made mostly of iron), and the column "Elevation" to the angle of departure.

Figure 6.23 (refer to the column Initial Velocity) lists the initial speed of the projectiles in feet per second for different types of ammunition and different quantities of powder used. These initial speeds, also called *muzzle velocities*, were determined by experiments with the ballistic pendulum at the Washington Arsenal. The operative principle

KIND OF ORDNANCE	PROJECTILE		Charge of Powder	Initial Velocity
	Kind	Weight		
6-PDR. FIELD GUN......	Shot............	Lbs. 6.15	Lbs. 1.25 1.50 2.00	Feet 1439 1563 1741
	Sph. case.........	5.50	1.00	1357
	Canister.........	6.80	1.00	1230
12-PDR. FIELD GUN.....	Shot............	12.30	2.50 3.00 4.00	1486 1597 1826
	Sph. case.........	11.00	2.00	1262
	Canister.........	13.50	2.00	1392

Figure 6.23

behind a ballistic pendulum is as follows: A shell is fired at close range at a heavy pendulum, the maximal deviation of the pendulum is measured, and the muzzle velocity of the projectile is determined from the amount of the deviation and the weight of the projectile. (We will study the ballistic pendulum in Section 14.3.)

The data in Figures 6.22 and 6.23 are based on actual measurements and therefore take air resistance into account. The problem that now concerns us is this: How do these data compare to the predictions of the theory of Section 6.5, which ignores air resistance?

Let's consider, for instance, the 6-pdr. field gun standing on horizontal ground. Let's put in 1.25 pounds of powder, load a cannonball, set the angle of departure at $1°$, and fire. From Figure 6.23 we see that the muzzle velocity is $v_0 = 1439$ feet per second. The muzzle of this gun in firing position is about $y_0 = 3.6$ feet above the ground. (This information comes from the appendix of *The Artillerist's Manual*.)

Let's first compute the range of the projectile using the theory of Section 6.5. We need to apply formula (6f),

$$R = \frac{v_0^2}{2g} \sin(2\varphi_0) + \frac{v_0}{g}\sqrt{\frac{v_0^2}{4}\sin^2(2\varphi_0) + 2gy_0 \cos^2 \varphi_0},$$

with $v_0 = 1439$, $y_0 = 3.6$, and $\varphi_0 = 1°$. Plugging in these values, we obtain

$$R = \frac{(1439)^2}{2 \cdot 32} \sin 2°$$

$$+ \frac{1439}{32}\sqrt{\frac{(1439)^2}{4}\sin^2 2° + 2(32)(3.6)(\cos^2 1°)}$$

$$\approx 1129.17 + 44.97\sqrt{630.52 + 230.33}$$

$$\approx 1129.17 + (44.97)(29.68)$$

$$\approx 2448.57 \text{ feet.}$$

The corresponding observed value from Figure 6.22 is 674 yards = 2022 feet. The prediction of the theory deviates from the observed value by about 20%.

Let's continue firing the 6-pdr. field gun. Put in 1.00 pound of powder, load a cannonball, set the angle of departure at 3°, and fire. From Figure 6.23, we see that the muzzle velocity is $v_0 = 1357$ feet per second.

Let's compute the time t_{imp} at which the cannonball will impact. By formula (6e),

$$t_{imp} = \frac{v_0 \sin \varphi_0 + \sqrt{v_0^2 \sin^2 \varphi_0 + 2gy_0}}{g},$$

where $v_0 = 1357$, $y_0 = 3.6$, and $\varphi_0 = 3°$. Therefore,

$$t_{imp} \approx \frac{(1357)(0.05) + \sqrt{((1357)(0.05))^2 + 2(32)(3.6)}}{32}$$

$$\approx \frac{67.9 + \sqrt{5043.83 + 230.40}}{32}$$

$$\approx \frac{71.02 + 72.62}{32} = \frac{143.64}{32}$$

$$\approx 4.49 \text{ seconds.}$$

Notice that this is in reasonable agreement with the observed 4 seconds that Figure 6.22 provides.

Consider formula (6f) next. Taking

$$R = \frac{v_0^2}{2g} \sin(2\varphi_0) + \frac{v_0}{g} \sqrt{\frac{v_0^2}{4} \sin^2(2\varphi_0) + 2gy_0 \cos^2 \varphi_0}$$

with $v_0 = 1357$, $y_0 = 3.6$, and $\varphi_0 = 3°$, we get

$$R = \frac{(1357)^2}{2 \cdot 32} \sin 6°$$

$$+ \frac{1357}{32} \sqrt{\frac{(1357)^2}{4} \sin^2 6° + 2(32)(3.6)(\cos^2 3°)}$$

$$\approx 3007.56 + 42.41\sqrt{5030.01 + 229.77}$$

$$\approx 6083.04 \text{ feet.}$$

The observed range, on the other hand, is 1050 yards = 3150 feet. This time the gap between theory and observation is big.

While air resistance has little effect when velocities are small, the lesson is that its effect at larger velocities can be substantial. In the context of artillery, the formulas of Section 6.5 are of little or no value. Indeed, at high velocities and short ranges, the effect of air resistance is much greater than that of gravity. The shape or, more precisely, the *aerodynamics* of the projectile play a crucial role. By relating this factor to the velocity, it is possible to take it into account by modifying the basic air-resistance-free theory of the trajectory of a projectile. In simple situations, such as free fall, this is relatively simple (see Section 13.6). In general, however, the mathematical relationship between the aerodynamics and the velocity depends on the velocity. It is hardly surprising that mathematical and computer models of trajectories that take air resistance into account are extremely complicated. See Exercises 6F for yet another complicating factor.

6.7 Postscript

Hooke and Newton, in spite of disagreements and quarrels, continued to correspond about matters of science and mathematics. In one letter, Hooke suggested that the planets are kept in orbit by a central force that decreases with increasing distance; indeed he suggested that the central force decreases with the square of the distance. It seems clear, however, that Hooke had no mathematical proof. This correspondence revived Newton's interest in these fundamental questions. Edmund Halley became interested as well. Halley visited Cambridge in 1684 and asked Newton specifically what the shape of the orbit of a planet would be under the assumption that it is attracted by a force centered at the Sun and inversely proportional to the square of its distance from the Sun. Newton immediately answered that the orbit would be an ellipse, indicating that he had proved this long ago. Later he supplied Halley with two proofs. Halley realized the importance of this work at once and, with the support of the Royal Society, decided to persuade

Newton to publish a treatise setting out his discoveries in detail. In spite of his reluctance to publish, Newton finally agreed, and started this task in 1685. In spite of a number of ongoing problems—including disputes about the priority of the discovery of the inverse square law (principally with Hooke) and the financing of the printing (all expenses were met by Halley)—the book appeared in the summer of 1687. Entitled *Philosophiae Naturalis Principia Mathematica*, and known simply as the *Principia*, it is regarded by most scientists as the greatest book of science ever written. We'll look at it in the next chapter.

After the publication of the *Principia*, Newton's scientific output declined. In 1689 he was elected as Cambridge University's Member of Parliament.[5] In 1696 he moved to London and assumed the post of Warden of the Mint. He took his duties there very seriously and oversaw the replacement of the coinage of England. In 1699 he became the head, or Master, of the Mint, a position that he retained until his death in 1727. In 1703 he was elected president of the Royal Society, and in 1704 his work *Optiks* appeared. By that time he was famous throughout the scientific world. In 1705, Queen Anne conferred knighthood on Newton.

Disputes continued, especially with Leibniz. The point of contention was as to whether Leibniz had invented his calculus of differentials independently or whether he had "borrowed" from Newton. The battle, fought out in the scientific literature, was contentious and bitter. It was waged not only by the two scientists themselves, but also by their supporters. Today's scientific community is in unanimous agreement that the two men had invented their versions of the calculus independently. A comparison of the two theories brings the conclusion that while Leibniz had the better theory (from the point of view of notation and basic concepts), Newton's was much more powerful computationally.

The reviews on Newton the man are mixed. A look at his personal library reveals substantial interests in religion as well as alchemy. A look at his character finds him to be noble and sensitive, but also petty,

suspicious, and withdrawn. He was a man who was admired and revered, but he aroused little affection.

Exercises
6A. Derivatives

1. Compute the derivatives of the following functions:

   i. $f(x) = 3\sqrt{x}$

   ii. $g(x) = \frac{4}{x^5}$

   iii. $h(x) = -\frac{2}{\sqrt[3]{x}}$

   iv. $f(x) = \frac{5}{x^{100}} - 4x^{-\frac{1}{3}}$

   v. $g(x) = -2x^{\frac{1}{3}} + 3x^5 - 6$

   vi. $y = x^{-\frac{2}{7}} + 30x^4 - \frac{1}{4}x^{\frac{5}{3}}$

6B. Antiderivatives and Definite Integrals

2. Determine antiderivatives of the following functions:

   i. $f(x) = 2x^3$

   ii. $f(x) = 5x^{\frac{1}{3}}$

   iii. $f(x) = 3x^5 + \frac{1}{4}x^{\frac{2}{7}}$

   iv. $f(x) = 6x^4 - \frac{3}{8}x^{\frac{5}{3}}$

3. Compute the areas under the graphs of the following functions:

   i. $f(x) = 5x^2$ between 0 and 4; between 4 and 6.

   ii. $f(x) = 5x^2$ between 0 and 6.

   iii. $f(x) = 3\sqrt{x}$ between 0 and 5.

   iv. $f(x) = 4x^3 + 2x^{\frac{1}{3}}$ between 2 and 4.

4. Make use of the approximation

$$\frac{1}{1+x} \approx 1 - x + x^2 - x^3 + x^4 - x^5 + x^6 - x^7 + \dots$$

for $|x| < 1$ to compute the definite integrals

$$\int_0^{\frac{3}{4}} \frac{1}{1+x}\,dx \quad \text{and} \quad \int_0^{\frac{3}{4}} \frac{1}{1+x^2}\,dx$$

with an accuracy of four decimal places.

5. The approximation

$$\frac{1}{1+x} \approx 1 - x + x^2 - x^3 + x^4 - x^5 + x^6 - x^7 + \dots$$

for $|x| < 1$, can be used to compute related definite integrals. For example, multiplication by $x^{\frac{1}{2}}$ shows that

$$\frac{x^{\frac{1}{2}}}{1+x} \approx x^{\frac{1}{2}} - x^{\frac{3}{2}} + x^{\frac{5}{2}} - x^{\frac{7}{2}} + x^{\frac{9}{2}} - \dots.$$

Use this approximation to show that

$$\int_0^{\frac{1}{2}} \frac{x^{\frac{1}{2}}}{1+x}\, dx \approx 0.184.$$

6. Make use of the approximation

$$\sin x \approx x - \frac{x^3}{3!} + \frac{x^5}{5!} - \frac{x^7}{7!} + \cdots$$

to evaluate the definite integrals

$$\int_0^{\pi} \sin x\, dx \quad \text{and} \quad \int_0^1 \sin \sqrt{x}\, dx$$

with four decimal accuracy.

7. Try to make use of the approximation

$$\sqrt{1+x} = (1+x)^{\frac{1}{2}}$$

$$\approx 1 + \binom{\frac{1}{2}}{1} x + \binom{\frac{1}{2}}{2} x^2 + \binom{\frac{1}{2}}{3} x^3$$

$$+ \binom{\frac{1}{2}}{4} x^4 + \cdots$$

to compute the area under the graph of $f(x) = \sqrt{1+x}$ between 0 and 5. Why is there a problem?

8. Use the formula $L = \int_a^b \sqrt{1 + (f'(x))^2}\, dx$ together with a power series approximation to compute the length of the graph of $f(x) = \frac{1}{3}x^3$ between $(-1, -\frac{1}{3})$ and $(1, \frac{1}{3})$ with three decimal accuracy.

6C. Moving Points

Exercises 9–12 all involve the following situation: A particle is moving on a coordinatized line. Its position function $p(t)$ and the starting time are given. In each case, determine the particle's velocity $v(t)$ and acceleration $a(t)$ as functions of t, and give a description of the motion of the particle (including comments about the initial position, velocity, acceleration, and force; when and where the particle stops, etc.). Sketch the motion in each case (as in Figure 6.15 of the text).

9. $p(t) = 2t - 5$; start at $t = 0$.

10. $p(t) = 2t^2 + 2t + 12$; start at $t = -10$.

11. $p(t) = t^3 - 4t^2 - 21t$; start at $t = -6$.

12. $p(t) = \frac{3}{t} = 3t^{-1}$; start at $t = 1$.

13. A point is moving on a coordinatized line. It starts at the origin at time $t = 0$ with an initial velocity of zero. Its acceleration is given by $a(t) = 6t - 12$. (a) Determine its velocity function $v(t)$ and its position function $p(t)$. (b) Draw a careful diagram of the motion over the time interval $[0, 7]$. Indicate the direction of

the motion and point out where and at what time(s) the point stops.

14. The acceleration $a(t)$ of a moving point is given by $a(t) = 2t - 6$ for $t \geq 0$. The initial velocity is $v(0) = 5$, and its initial position is $p(0) = 6$. Determine the velocity and position functions $v(t)$ and $p(t)$ for $t \geq 0$, and describe the motion of the point.

6D. Projectiles

15. Isaac Newton throws an apple in the direction of his nemesis, the scientist Robert Hooke, who stands on level ground 35 feet away. Newton throws the apple with an initial velocity of 40 feet per second and an angle of elevation of 20°. The apple is 5 feet above the ground at the moment of release. What is the maximal height above the ground reached by the apple? Will the apple hit Hooke? If so, with what speed?

16. Newton's pet parakeet escapes and flies over his garden straight towards his house. It has a speed of 20 feet per second and flies at an angle of 30° with the horizontal when it suddenly releases, at a height of 50 feet … a dropping. How long after the release will the dropping reach its maximal height? What is the maximal height? Will it splatter against the white stucco wall of Newton's house, which is 20 feet high and 30 feet away (at the time of the release)? If so, with what velocity?

17. At the opening ceremonies of the 1992 Olympic games in Barcelona, an archer lit the Olympic flame by shooting a burning arrow into a circular pool of flammable liquid. Suppose the circular pool was affixed to a structure that put the center of the surface of the liquid 55 meters downrange from the archer and 25 meters above the ground. Now suppose the archer shot the arrow from 1.5 meters above ground level at an angle of 70° with the horizontal and that the arrow struck the center of the pool on its descent. With what initial speed did the arrow leave the bow? [Hint: Ignore air resistance and use Formula (6c) of Section 6.5.]

18. Suppose that the Olympic archer in Exercise 17 shot the arrow from 6 feet above ground level and that the center of the circular pool was 240 feet downrange from the archer and 62 feet above the ground. If the initial speed of the arrow was 120 feet per second, determine the angle of elevation with which the marksman would have had to shoot the arrow in order to hit the center of the surface of the liquid. [Hint: Use the identity $\sec^2 \varphi = 1 + \tan^2 \varphi$.]

19. Suppose that a basketball player's jump shot is most accurate when he releases the ball from 8 feet above

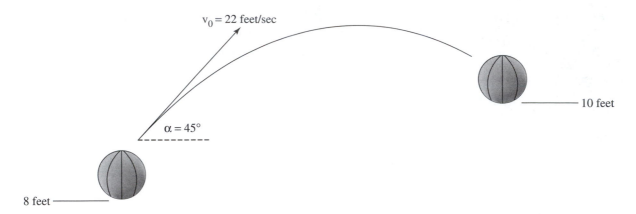

Figure 6.24

the floor at an angle of 45° and a velocity of 22 feet per second. See Figure 6.24. How far from the basket should he be taking his shots?

20. Using the formulas and notation developed for the motion of a projectile, derive a formula for the angle of impact of a projectile. Assume that the ground is horizontal.

6E. Ballistics

21. Load a 6-pdr. field gun with 1.25 pounds of powder. Put in a cannonball and fire the gun with an angle of departure of 0°, i.e., point blank. By Figure 6.23, the muzzle velocity of the cannonball is 1439 feet per second. Use the formulas developed in Section 6.5 to compute the range. Compare your result with the observation recorded in Figure 6.22. Why is your answer inconsistent with the data of Figures 6.22 and 6.23? What is the problem?

22. Now load a 6-pdr. field gun with sperical case shot and 1.00 pounds of powder. Fire with an angle of departure of 2°30′. By Figure 6.23, the muzzle velocity of the projectile is 1357 feet per second. Using the theory of Section 6.5 determine how many seconds after the firing the projectile will impact, and compute the range. Compare your figures with the observed values in Figure 6.22.

23. Fire a 12-pdr. field gun with a cannonball using 2.50 pounds of powder. By Figure 6.23, the muzzle velocity is 1486 feet per second. Fire three successive times with angles of departure of 0°, 1°, and 5°. Compare the ranges predicted by theory with those given in Figuree 6.22.

6F. Connections with Probability Theory

Another difficulty encountered in ballistics is illustrated by data in the appendix of *The Artillerist's Manual*. A certain type of rifle-musket was testfired a total of fifteen times on July 11 and again on July 23 of 1856. The initial velocities, measured with a ballistic pendulum, were found to be 809, 901, 878, 884, 850, 873, 927, 943, 822, 827, 870, 900, 897, 920, and 914 feet per second. Therefore, each firing of the musket produced a different initial velocity, even though the conditions (the weight of the powder, the type of ball, etc.) were kept the same! The number 884 has the property that there are seven velocities below it (809, 878, 850, 873, 822, 827, and 870) and seven above it (914, 901, 927, 943, 900, 897, 920, and 914). Thus 884 is the so-called *median* of the numbers on the list.

The probability that a certain event, often called a *favorable outcome*, occurs is a number between 0 and 1: It is the ratio of the number of all possible favorable outcomes to the number of all possible outcomes. To say that the probability is 0 means that the event will definitely not occur; to say that it is 1 means that it is certain to occur. In reference to the data just given, observe that 9 out of the 15 firings produced a velocity between 860 and 920 feet per second, so the probability that the initial velocity of the ball lies in this range is $\frac{9}{15} = 0.6$.

The data can be organized as in Figure 6.25. Observe that the probability just computed is equal to the sum of the areas of the rectangles between 860 and 920 over the total area of all the rectangles.

If the probability of 0.6 is to be reliable, many more tests than these 15, say several thousand, would be neces-

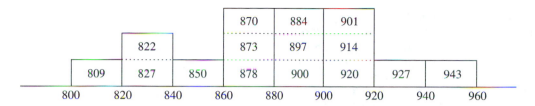

Figure 6.25

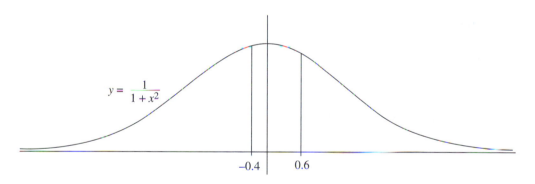

Figure 6.26

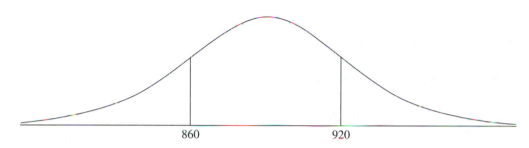

Figure 6.27

sary. If this were done, the array of Figure 6.25 would look something like the "bell-shaped" curve of Figure 6.26. The probability that a given firing of the rifle generates an initial velocity between 860 and 920 feet per second is the ratio of the area under the curve between 860 and 920 to the total area under the curve.

This discussion illustrates the fact that any accurate and useful mathematical theory of ballistics is necessarily complicated and necessarily involves probability theory. Such a theory will be able to assert, for example, that a projectile fired from a certain gun under certain conditions will land within 5 feet from a certain point with a probability of 0.9, or 0.95, or 0.99. Probability theory is beyond the purpose and aim of a basic text such as this.

24. In reference to the test firings just discussed, assume that after one thousand firings, the results can be modeled as follows. The curve of Figure 6.27 is the graph of the function $y = \frac{1}{1+x^2}$, where the value $x = 0$ corresponds to the median velocity of 884 feet per second, and every 0.1 unit on the x-axis represents 1 foot per second. So the region under the graph for $-0.4 \leq x \leq 0$ corresponds to all the firings with velocities v satisfying $880 \leq v \leq 884$, and the region under the graph for $0 \leq x \leq 0.6$ corresponds to all the firings with veloc-

ities v satisfying $884 \leq v \leq 900$. Assuming that the area under the entire graph is π, compute the probability that a particular test yields a velocity v such that $880 \leq v \leq 884$. Then compute the probabilities that $884 \leq v \leq 900$ and $880 \leq v \leq 900$. Is there a reasonable correlation between your findings and the data in Figure 6.25? [Hint: Compute the required definite integrals by using Newton's approximation method.]

Notes

[1] All references to Newton's work can be found in D.T. Whiteside, *The Mathematical Papers of Isaac Newton, Volumes I–VII* (Cambridge University Press, Cambridge, 1969).

[2] Incidentally, your calculator relies on power series in computations of logarithms and trigonometric quantities.

[3] In all calculator computations in this book, the answers will be rounded off only after the final step. This will minimize roundoff errors.

[4] John Gibbon, *The Artillerist's Manual*, Greenwood Press, Westport CT, 1971. Originally published by D. Van Nostrand Co., New York, and Trübner & Co., London, 1860.

[5] Isaac Newton spoke in the House of Commons only once. As the great man arose a hush descended on the House in anticipation of what he would have to say. Newton observed that there was a window causing a draft and asked if it could be closed.

7

The *Principia*

It is one of Newton's major achievements (although not his achievement alone) that he discovered the fundamental concepts that underlie motion. The *Principia* begins with these concepts. The *mass* of an object, or body (which is the word Newton uses), is defined to be density times volume. *Inertia* is the "power of resisting, by which every body as much as in it lies, perseveres in its state, either of rest, or of uniform motion, in a right [straight] line."[1] A *force* is an "action exerted upon a body, in order to change its state, either at rest, or of uniform motion, in a right line." The basic laws of motion follow next:

> **LAW I**. That every body perseveres in its state of resting, or of moving uniformly in a right line, as far as it is not compelled to change that state by an external force impressed upon it.

> **LAW II.** That the change of motion is proportional to the moving force impressed; and is produced in the direction of the right line, in which that force was impressed.

> **LAW III.** That reaction is always contrary and equal to action; or, that the mutual actions of two bodies upon each other are always equal, and directed to contrary parts.

Let's elaborate on what Newton is saying. We will take the meaning of force from everyday experience as any action of pulling or pushing. A force has a magnitude that can be measured. For example, an amount of push or pull can be measured by the amount of the displacement it produces on some standardized steel spring. A force also has a direction. Its magnitude and direction together determine the force.

Let a force act on a body. Newton's second law says that the magnitude F of the force is directly proportional to the acceleration a that it produces in the motion of the body, and the mass m is the constant of proportionality. In other words, force = mass × acceleration, or $F = ma$. Measure F and measure a, and the mass m of the body is determined by $m = \frac{F}{a}$. Since $a = \frac{F}{m}$, an increase (or decrease) in F produces a corresponding increase (or decrease) in a. The second law also says that the direction of the acceleration is the same as that of the force. Galileo discovered (see Sections 4.2 and 6.4) that the acceleration produced on a body by the force of gravity F is equal to the constant $-g$, regardless of the mass m of the body. In this case, Newton's equation is $F = -mg$. The direction of the force is downward (towards the center of the Earth) and its magnitude mg is called the *weight* of the body. Since g is known (32 feet per second^2, or 9.8 meters per second^2), the mass m of the body can be determined from its weight mg.

Notice that Newton's first law is a direct consequence of his second law: If the magnitude of a force is zero, then the acceleration it produces in the motion of the body must be zero, and hence its velocity is constant. Observe that this is Galileo's law of inertia.

The third law is the assertion that for every force there is always an equal and opposite force. If you lean against a wall, the wall will push back on you with an equal and opposite force. If this force did not exist you would fall. So forces always occur in pairs. It is important to note that these two forces never act on the same body. Your leaning against a wall is a force on the wall. The wall pushing back is a force on you.

After a discussion of the "method of prime and ultimate ratios," in other words some elements of differential calculus, the *Principia* turns to the study of centripetal force. A *centripetal force* is one "by which bodies are drawn, impelled, or any way tend towards a point, as to a centre." So a centripetal force is one that always acts in the direction of a single fixed point. This fixed point is called the *center of force*. What is going on is most easily illustrated as follows. Take a string and tie an object P to one end. Hold the other end of the string and twirl the object as shown in Figure 7.1. Keep this end fixed at S and note that P will move in a circle. The force with which the string pulls on P in the process always acts in the direction of S. It is therefore a *centripetal force*. Newton's third law is illustrated also. As the object is twirled, there is a force pulling on the hand at S in the direction of P. It

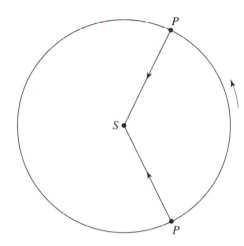

Figure 7.1

acts in the opposite direction but is equal in magnitude to the force pulling on P.

Everything described thus far is in Book I of the *Principia*. Book I goes on to develop a theory of bodies moving under the action of a centripetal force. This theory is completely abstract. It applies to any centripetal force acting on an object, be it the force exerted by a string on a twirling object, or the gravitational force of the Sun on a planet in its orbit. Book II of the *Principia* deals with motion in a resisting medium. But the high point of the *Principia* comes in Book III, the *System of the World*. It applies the theory of Book I to the analysis of the orbits of the planets.

It is the goal of this chapter to provide a detailed look at some of the central elements of Book I and then to illustrate the incredible insights that they provide into the workings of the solar system. The *Principia's* arguments are subtle, difficult, indeed often terse and opaque. It is, in short, quite a challenge to understand it. One reason for this is that Newton often used delicate Greek geometry known to the trained mathematicians of his time but not today's reader. In order to give a glimpse at Newton's explanations and also to facilitate the understanding of what he does, this chapter takes a two-fold approach. Newton's original arguments (from Thorp's 1777 translation) will appear in italic (*slanted*) print, and explanations are added in regular print. Figures labeled "Newton" are Newton's original diagrams (or parts thereof). Others have been added to explain and clarify.

7.1 **Equal Areas in Equal Times**

Consider an object or body in motion. Assume that it is acted upon by only one force, a continuously acting centripetal force. The magnitude of the force can vary. With later application to the motion of the planets around the Sun in mind, Newton denotes the body by P and the center of force by S. For a similar reason, we will refer to the path traced out by P as the *orbit* of P.

Newton begins his analysis by showing that the motion of the body obeys Kepler's second law. In other words, the single assumption that P moves under the action of a centripetal force with center S is enough to guarantee that the segment PS sweeps out equal areas in equal times. This single assumption also implies that the orbit of P lies in a plane that also contains S. See Figure 7.2.

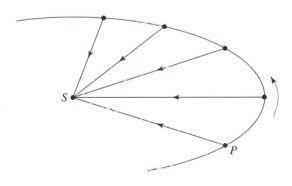

Figure 7.2

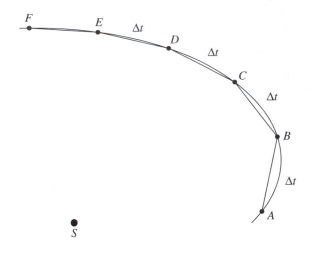

Figure 7.3

Newton takes a fixed interval of time Δt. He considers the position of the body at some point A, then Δt later at B, then another Δt later at C, another Δt later at D, etc. Newton regards the force to be acting only at the points A, B, C, D, etc., as a sequence of "impulses." Each impulse deflects the body in the same way that a billiard ball is deflected by the cushion of the table. In particular, he thinks of the actual curved path of the body as being approximated by a sequence of line segments. See Figure 7.3. If Δt is small, this sequential action of the force and the resulting approximation of the path of the body is a simulation of what is going on. The *Principia* now turns to the verification of Kepler's second law.

PROPOSITION I.   THEOREM I.   *That the areas, which revolving bodies describe by radii drawn to an immovable centre of force, do both lie in the same immovable planes, and are proportional to the times in which they are described.*

For suppose the time to be divided into equal parts (this is the Δt in the discussion above), *and in the first part of that time, let the body by its innate force describe the right line AB.* (Figure 7.4) *In the second part of that time, the same would, (by Law I.) if not hinder'd, proceed directly to c, along the line Bc equal*

to AB. Since the force is (for the moment) considered not to be acting at B, there is no force acting on the body during the entire motion between A and c. So the velocity remains constant between A and c. The elapsed time from A to B is Δt and that from B to c is Δt also. It follows that $AB = Bc$. So that by the radii AS, BS, cS drawn to the center, the equal areas ASB, BSc, would be described. The triangles $\triangle SAB$ and $\triangle SBc$ have equal bases $AB = Bc$. They also, see Figure 7.5, have the same height h. So Area $\triangle SAB =$ Area $\triangle SBc$.

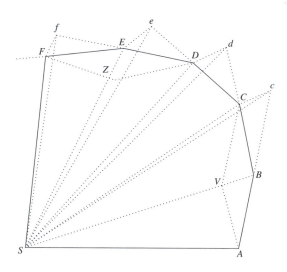

Figure 7.4 (Newton)

But when the body is arrived at B, suppose that a centripetal force acts at once with a great impulse, and turning aside the body from the right line Bc, compells it afterwards to continue its motion along the right line BC. In other words, the motion of the body from B to C is regarded as occurring in two components: first, the motion from B to c determined by the velocity of the body at B, and then the motion from c to C, which is determined by the action of the impulse at B along BS. *Draw cC parallel to BS meeting BC in C; and at the end of the second part of the time, the body (by Cor. I. of the laws) will be found in C, in the same plane with the triangle ASB.* Since the impulse at B lies in the plane of

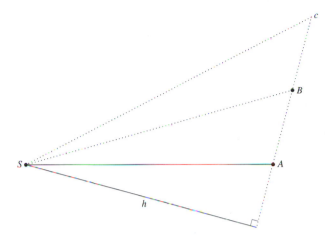

Figure 7.5

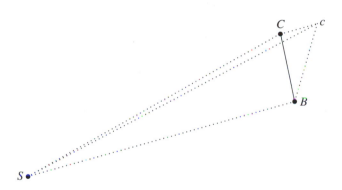

Figure 7.6 (Newton)

the triangle $\triangle SAB$, there is no component of this force that could move the body outside this plane. *Joyn SC, and, because SB and Cc are parallel, the triangle SBC will be equal to the triangle SBc and therefore also to the triangle SAB.* The two triangles $\triangle SBC$ and $\triangle SBc$ have a common base SB. See Figure 7.6. Since Cc is parallel to SB, these triangles have the same height. Therefore, Area $\triangle SBc =$ Area $\triangle SBC$. Since it was already shown that Area $\triangle SBc =$ Area $\triangle SAB$, it follows that Area $\triangle SAB =$ Area SBC, as asserted by Newton.

By the like argument, if the centripetal force acts successively in C, D, E, &c. and makes the body in each single particle of time, to describe right lines CD, DE, EF, &c. they will all lye in the same plane; and the triangle SCD will be equal to the triangle SBC, and SDE to SCD, and SEF to SDE. And therefore in equal times, equal areas are describ'd in one immoveable plane: and, by composition, any sums SADS, SAFS, of those areas, are one to the other, as the times in which they are describ'd. Now let the number of those triangles be augmented, and their breadth diminished in infinitum; and (by Cor. 4. Lemma 3) their ultimate perimeter ADF will be a curve line: and therefore the centripetal force, by which the body is perpetually drawn back from the tangent of this curve, will act continually; and any describ'd areas SADS, SAFS, which are always proportional to the times of description, will, in this case also, be proportional to those times.

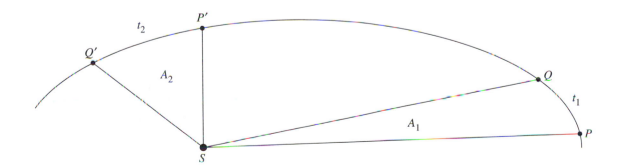

Figure 7.7

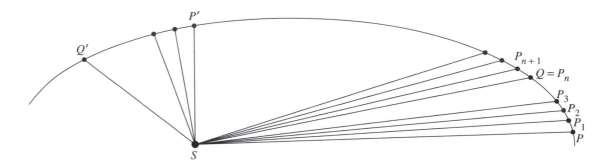

Figure 7.8

This argument, which completes Newton's proof of Kepler's law, requires explanation. Suppose that t_1 and t_2 are any two time intervals. Suppose that the body moves from P to Q during t_1 and from P' to Q' during t_2. Let the areas of the regions SPQ and $SP'Q'$ be A_1 and A_2 respectively. See Figure 7.7. Newton must demonstrate that $t_1 = t_2$ implies $A_1 = A_2$. What he does is this. Choose a small time interval Δt and partition t_1 into, say, n (possibly a huge number) of equal intervals, each of length Δt. So, $t_1 = n(\Delta t)$. Starting from P insert into the diagram the positions P_1, P_2, $P_3, \ldots$ of the body in its orbit at times Δt, $2\Delta t$, $3\Delta t, \ldots$ from when it was at P, and continue this process until Q has been reached (Figure 7.8). Since n steps are needed to get from P to Q, Newton's construction inscribes the n triangles SPP_1, SP_1P_2, $SP_2P_3, \ldots$ into the region SPQ. By the step already verified, the triangles SPP_1 and SP_1P_2 have the same area. By the same step, the areas of the triangles SP_1P_2 and SP_2P_3 are the same. Continuing in this way, all the n triangles inscribed in the region SPQ have the same area, say ΔA. Since Δt is small, P and P_1, P_1 and P_2, P_2 and P_3, etc., are close together, and the n inscribed triangles together essentially fill out the area of the region SPQ. The area of this region is, therefore, $A_1 = n(\Delta A)$. Continue to insert points into the orbit at time intervals Δt, from Q to P', and finally from P' to Q'. Since $t_2 = t_1 = n(\Delta t)$, n steps are also required to get from P' to Q'. The triangles inscribed into the sector $SP'Q'$ in the process continue to have

area ΔA. So the region $SP'Q'$ is also filled out by n triangles each with area ΔA. It follows that the area A_2 of this region is also equal to $n(\Delta A)$. So Newton has established Kepler's second law.

Strictly speaking, $A_1 = n(\Delta A)$ is not an equality, but only an approximation. This is because the triangles will not fill out the region SPQ "on the nose." The same is true for $SP'Q'$ and $A_2 = n(\Delta A)$. However, if the above thought experiment is repeated, each time for smaller and smaller Δt (and therefore larger and larger n) the approximations $A_1 = n(\Delta A)$ and $A_2 = n(\Delta A)$ get better and better: a rapid-fire sequence of impulses will approximate the continuously acting force, a myriad of connected line segments will approximate the curved trajectory, and the approximation $n(\Delta A)$ will (for all computational purposes) produce the equality $A_1 = nA = A_2$.

Let's formulate Newton's conclusion in another way as follows: Let the motion of a body P be produced by the action of a centripetal force acting towards the fixed point S (as in Figure 7.2). Let t be any time interval, and let A_t be the area of the sector that the segment SP sweeps out during the time t. Then the ratio $\frac{A_t}{t}$ is the same, regardless of what t is taken. To see this (we have in fact already seen this in Section 4.1 in the context of Kepler's second law), let t_1 and t_2 be any two time intervals and let A_1 and A_2 be the respective areas that are swept out. Put $\frac{t_1}{t_2} = c$. So $t_1 = ct_2$. Since A_2 is swept out in t_2, cA_2 is swept out in

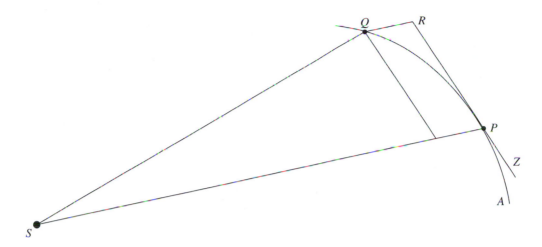

Figure 7.9 (Newton)

ct_2. Since $t_1 = ct_2$, we know that $A_1 = cA_2$. Therefore,

$$\frac{A_2}{t_2} = \frac{cA_2}{ct_2} = \frac{A_1}{t_1},$$

as asserted. So $\frac{A_t}{t}$ is the same constant for any t. We will call this constant the *Kepler constant* of the orbit and denote it by κ (the Greek letter kappa).

7.2 Analyzing Centripetal Force

Continuing the assumption that a body is being propelled by a single centripetal force, Newton begins to turn his attention to the connection between the magnitude of the force and the shape of the orbit.

Let the body be in a typical position P and let F_P be the magnitude of the centripetal force at P. In order to study the centripetal force at P, it is necessary to study its dynamics *near* P. To understand why this is so, just observe that basic properties of a curve at a particular point, for example the slope of the tangent line, can only be assessed by studying the "flow" of the curve near the point.

Again let $\triangle t$ be a very small interval of time, and suppose that the body has moved from P to Q during $\triangle t$. (See Figure 7.9.) During the body's motion from P to Q, the centripetal force will not vary much,

and Newton takes it to be constant. So its magnitude is equal to F_P throughout the motion from P to Q. Consider the tangent RPZ to the orbit at P, choosing R such that QR is parallel to the line connecting P to the center of force S. Complete QR and RP to the parallelogram shown in the figure. Newton now imagines the motion of the body from P to Q to be separated into two parts:

(1) The motion from P to R. This part takes place along the tangent line at P. The velocity is constant and equal to the velocity of the body at P. This part of the motion has zero acceleration and is free of the action of the centripetal force.

(2) The motion from R to Q. This part takes place in the direction from P to S and has zero initial velocity. This is the accelerated part of the motion. It is entirely determined by the centripetal force.

As the body proceeds from P to Q these two motions take place simultaneously. But by regarding the motion from P to Q as occurring in the two parts separately, Newton is able to isolate the effect of the centripetal force and can proceed with its analysis.

Let the mass of the body be m units. Since force = mass × acceleration, the force F_P will provide the body with a constant acceleration of $\frac{1}{m}F_P$ for the

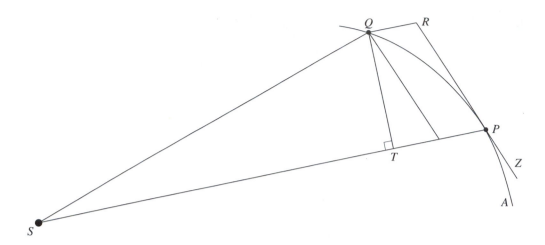

Figure 7.10 (Newton)

duration Δt of its action from R to Q. Now refer to Example 6.14 of Section 6.4 for the following fact. A body that starts from rest and moves with constant acceleration $k = \frac{1}{m}F_P$ will, after a time t, have a velocity of

$$v(t) = kt + v(0) = \frac{1}{m}F_P t + 0 = \frac{1}{m}F_P t$$

and will cover a distance of

$$p(t) - p(0) = \frac{k}{2}t^2 = \frac{1}{2m}F_P t^2.$$

It follows that F_P produces a displacement of $\frac{1}{2m}F_P(\Delta t)^2$ in the body during the time Δt. Therefore, $\frac{1}{2m}F_P(\Delta t)^2 = QR$. So Newton has determined that

$$F_P = \frac{2m\,QR}{(\Delta t)^2}.$$

As in earlier situations, observe that the above equality is only an approximation, but that it improves when Δt is taken smaller.

Newton now expands the diagram in Figure 7.9 to that in Figure 7.10 and makes the following assertion:

COROLLARY I. *If a body P, revolving about the centre S, describes APQ and a right line ZPR touches that curve at any point P; and, from any other point Q of the curve, QR is drawn parallel to the distance*

SP, meeting the tangent in R; and QT is drawn perpendicular to the distance SP; the centripetal force will be reciprocally as the solid $\frac{SP^2 \times QT^2}{QR}$; if the solid is taken of that magnitude which it ultimately acquires, supposing the points P and Q continually to approach to each other.

What Newton does here is this. He drops the perpendicular QT from Q to SP. Since Δt is small, and hence Q is close to P, he observes that the area of the sector SPQ is in essence equal to that of the triangle SPQ. So, Area sector SPQ = Area $\triangle SPQ = \frac{1}{2}(SP \times QT)$. Since SP traces out the sector SPQ in the time interval Δt, it follows from the definition of Kepler's constant κ that

$$\frac{\frac{1}{2}(SP \times QT)}{\Delta t} = \frac{\text{Area sector } SPQ}{\Delta t} = \kappa.$$

So, $\frac{1}{\Delta t} = \frac{2\kappa}{SP \times QT}$. Inserting $\frac{1}{(\Delta t)^2} = \frac{4\kappa^2}{(SP \times QT)^2}$ into the formula $F_P = \frac{2m\,QR}{(\Delta t)^2}$, Newton concludes that

$$F_P = 8\kappa^2 m \frac{QR}{QT^2 \times SP^2}.$$

In particular, F_P is inversely proportional to $\frac{QT^2 \times SP^2}{QR}$, as he asserts in his Corollary I.

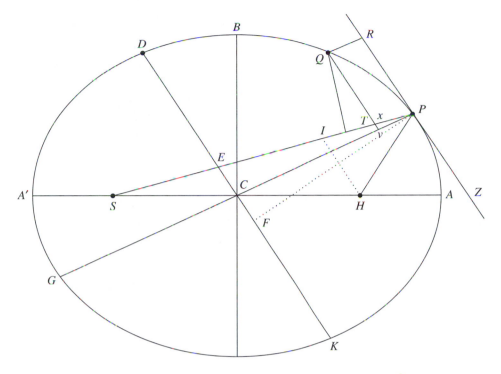

Figure 7.11 (Newton)

Consider the rewritten version

$$F_P = \frac{8\kappa^2 m}{SP^2} \frac{QR}{QT^2}$$

of this equality. Since the area of $\triangle SPQ$ is only an approximation of that of the sector SPQ, the above expression for F_P is only an approximation. However, it improves if Δt is taken to be smaller, in other words, if Q is taken closer to P. For then the area of $\triangle SPQ$ approximates that of the sector SPQ with greater precision. Indeed, to obtain an exact formula for F_P it remains to push Q to P and determine what happens to $\frac{8\kappa^2 m}{SP^2} \frac{QR}{QT^2}$ in the process. Since κ and m are constants and SP is the distance from S to P, it is clear that the term $\frac{8\kappa^2 m}{SP^2}$ does not depend on Q. It follows that

$$F_P = \frac{8\kappa^2 m}{SP^2} \cdot \lim_{Q \to P} \frac{QR}{QT^2}.$$

The limit $\lim_{Q \to P} \dfrac{QR}{QT^2}$ is therefore the remaining mystery. Return to Newton's diagram, and notice that the answer to the question of what happens to the ratio $\frac{QR}{QT^2}$ when Q is pushed to P depends entirely on the shape of the orbit between P and Q. At this point therefore, Newton must assume something about the orbit.

7.3 The Inverse Square Law

With later applications in mind, Newton assumes that the orbit is an ellipse and that the center of force S is one of its focal points. Now he can come to grips with the limit $\lim_{Q \to P} \frac{QR}{QT^2}$. What follows is a very subtle argument that is surely one of the most famous in all of science. The diagram on which it is based—see Newton's Figure 7.11—was depicted on the British one-pound note of the 1960s.

Focus on this diagram. The points S, P, Q, R, Z, and T have the same meaning as in Figure 7.10. In addition, C is the center of the ellipse, $A'A$ is the major axis, CB the upper part of the minor axis, DK the diameter of the ellipse parallel to the tangent RPZ, PCG is the indicated diameter, x is the intersection of PS and the parallel to RP through Q, and v is the intersection of this parallel with PCG. Note that $A'A = 2AC$ with $AC = a$ the semimajor axis, and that $CB = b$ is the semiminor axis. Newton explains the rest of his diagram in the course of his argument.

PROPOSITION XI.   PROBLEM VI. *Let a body revolve in an ellipsis: it is required to find the law of the centripetal force tending to the focus of the ellipsis.*

Let S be the focus of the ellipsis. Draw SP, cutting the diameter DK of the ellipsis in E, and the ordinate Qv in x; and let the parallelogram $QxPR$ be completed. It is evident that EP is equal to the greater semi-axis AC: for, drawing HI from the other focus H of the ellipsis, parallel to EC, because CS, CH are equal, ES, EI will be also equal; so that EP is the half sum of PS, PI, that is, (because of the parallels HI, PR, and the equal angles IPR, HPZ) of PS, PH; which taken together are equal to the whole axis $2AC$.

Newton arrives at these conclusion as follows. Since CS and CH are equal and the triangles $\triangle SIH$ and $\triangle SEC$ are similar, $ES = EI$. Since $EP = PI + EI$, $PS = PI + 2EI$. So $PS + PI = 2(PI + EI) = 2EP$. Therefore, $EP = \frac{1}{2}(PS + PI)$. Since IH and RZ are parallel, $\angle PHI = \angle HPZ$ and $\angle RPS = \angle PIH$. By a basic property of the ellipse known to the Greeks, $\angle HPZ = \angle RPS$. See Figure 7.12 and Proposition 3.4 of Section 3.1. Therefore, $\angle PHI = \angle PIH$, and hence $\triangle HPI$ is isosceles. By properties of isosceles triangles, $PI = PH$. Therefore, $EP = \frac{1}{2}(PS + PH)$. That $PS + PH = AC + A'C = 2AC$ follows from the "string" construction of the ellipse discussed in Exercises 4B of Chapter 4. So as Newton had asserted,

$$EP = AC. \tag{7a}$$

Newton next observes that the triangles $\triangle PCE$ and $\triangle Pvx$ are similar. So $\frac{Px}{Pv} = \frac{EP}{PC}$. Since $RQxP$ is a parallelogram, $QR = Px$. Therefore, $\frac{QR}{Pv} = \frac{EP}{PC}$. Using (7a), Newton has shown that

$$\frac{QR}{Pv} = \frac{AC}{PC}. \tag{7b}$$

Now Newton makes use of another classical property of the ellipse. Let V be any point on a diameter GP of an ellipse. Choose X such that XV is parallel to the tangent at P. According to a classical and basic property of the ellipse, see Figure 7.13 and Proposition 5 of Section 3.1, the ratio $\frac{GV \times PV}{XV^2}$ is the same, no matter which point V is chosen on the diameter GP. Applying this fact first with $V = v$ and again with $V = C$, Newton has shown that $\frac{Gv \times Pv}{Qv^2} = \frac{PC^2}{CD^2}$. Therefore,

$$\frac{1}{Qv^2} = \frac{PC^2}{CD^2} \frac{1}{Gv \times Pv}. \tag{7c}$$

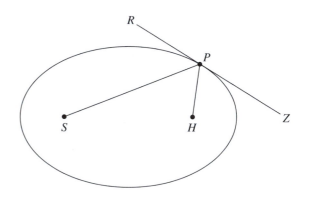

Figure 7.12

Newton next constructs PF perpendicular to DK. Since the segments Qv and EF are parallel, $\angle QxE = \angle PEF$. So the right triangles $\triangle QTx$ and $\triangle PFE$ are similar, and $\frac{Qx}{QT} = \frac{EP}{PF}$. Therefore, by (7a), $\frac{Qx^2}{QT^2} = \frac{EP^2}{PF^2} = \frac{AC^2}{PF^2}$. For Q very close to P, he can take $Qv = Qx$. So $Qv^2 = Qx^2$, and $\frac{Qv^2}{QT^2} = \frac{AC^2}{PF^2}$. Therefore,

$$\frac{1}{QT^2} = \frac{AC^2}{PF^2} \frac{1}{Qv^2}. \tag{7d}$$

By yet another classical property of the ellipse, the areas of any two parallelograms constructed by

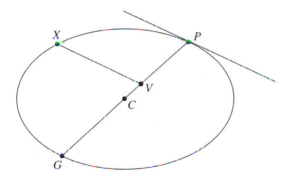

Figure 7.13

starting with any diagonal and taking the indicated tangents are equal. Refer to Figure 7.14. Applying this first to the diagonal DK and then to the diagonal AA', it follows (recall that the area of a parallelogram is equal to base times height) that $2CD \times PF = 2AC \times CB$. Therefore, $\frac{AC}{PF} = \frac{CD}{CB}$, and

$$\frac{AC^2}{PF^2} = \frac{CD^2}{CB^2}. \tag{7e}$$

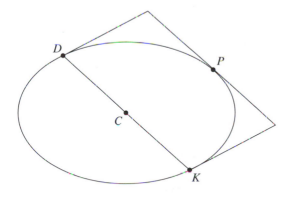

Figure 7.14

Newton has arrived at the last part of his computation. He lets $L = \frac{2CB^2}{AC} = \frac{b^2}{a}$. This quantity is known as the principal latus rectum of an ellipse. He then continues as follows: *and compounding all those ratios together, $L \times QR$ is to QT^2 as $AC \times L \times PC^2 \times CD$, or $2CB^2 \times PC^2 \times CD^2$, to $PC \times Gv \times CD^2 \times CB^2$, or as $2PC$*

to Gv. *But, the points Q and P continually approaching without end, $2PC$ and Gv are equal. Therefore the quantities $L \times QR$ and QT^2, proportional to these, are also equal. Let these equals be multiplied into $\frac{SP^2}{QR}$, and $L \times SP^2$ will become equal to $\frac{SP^2 \times QT^2}{QR}$. Therefore (by Cor. I. and 5. Prop. VI) the centripetal force is reciprocally as $L \times SP^2$, that is, reciprocally in the duplicate ratio of the distance SP. Which was to be found.*

Let's elaborate on this argument. Newton inserts both $L = \frac{2CB^2}{AC}$ and (7d) into $\frac{L \times QR}{QT^2}$, and then uses (7e), (7c), and (7b), to get

$$\frac{L \times QR}{QT^2} = \frac{2CB^2}{AC} \cdot QR \cdot \frac{AC^2}{PF^2} \cdot \frac{1}{Qv^2}$$

$$= \frac{2CB^2}{AC} \cdot QR \cdot \frac{CD^2}{CB^2} \cdot \frac{PC^2}{CD^2} \frac{1}{Gv \times Pv}$$

$$= \frac{2}{AC} \cdot QR \cdot PC^2 \cdot \frac{1}{Gv \times Pv}$$

$$= \frac{2}{AC} \cdot AC \cdot PC^2 \frac{1}{Gv \times PC}$$

$$= \frac{2PC}{Gv}.$$

Therefore, $\frac{QR}{QT^2} = \frac{2PC}{Gv} \frac{1}{L}$. When Q is pushed to P, $\frac{1}{L}$ does not change, and Gv closes in on $2PC$. Therefore,

$$\lim_{Q \to P} \frac{QR}{QT^2} = \frac{1}{L}.$$

Newton has arrived at his goal. Inserting this result into the formula at the end of Section 7.2, he gets

$$F_P = \frac{8\kappa^2 m}{L} \cdot \frac{1}{(SP)^2}.$$

So Newton has proved[2] the following: Suppose a body is propelled by a single centripetal force and that its orbit is an ellipse with the center of force S at one of the focal points. Then the magnitude F_P of the centripetal force at any point P in its orbit is given by the *inverse square law*

$$F_P = \frac{8\kappa^2 m}{L} \frac{1}{r_P^2},$$

where κ is Kepler's constant of the orbit, m is the mass of the body, $L = \frac{2CB^2}{AC} = \frac{2b^2}{a}$ is the latus rectum, a and b are the semimajor and semiminor axes of the ellipse, and $r_P = SP$ is the distance from P to S.

Let T be the time it takes the body to complete one revolution about S, i.e., let T be the *period* of the orbit. Since the area of the ellipse with semimajor axis a and semiminor axis b is $ab\pi$ (this was verified in Section 4.6), it follows that Kepler's constant is equal to $\kappa = \frac{ab\pi}{T}$. Therefore, Newton's formula can be written as

$$F_P = \frac{4\pi^2 a^3 m}{T^2} \frac{1}{r_P^2}$$

where, to repeat, m is the mass of the body and r_P is the distance from the body to the center of force S. Note that both F_P and r_P vary with time, but that everything else is constant.

Is the orbit of every body moving in the solar system an ellipse? No! When a comet passes appropriately close to a larger planet—Jupiter is a prime example—the gravitational pull of the planet will deflect the comet along a hyperbolic arc or (under certain ideal circumstances) a parabolic arc until this pull becomes negligible with increased distance. Newton addresses such a possibility, again in a completely abstract setting.

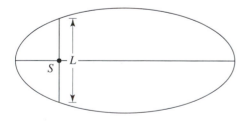

Figure 7.15

It turns out that the latus rectum L of an ellipse is equal to the length of the segment through a focus and perpendicular to the major axis (Figure 7.15).

See Exercise 9. The latus rectum L of a parabola or hyperbola can be defined in the same way, as the length of the segment through the focus and perpendicular to the axis (Figure 7.16). Newton shows that the centripetal force on a body of mass m is given by the inverse square law

$$F_P = \frac{8\kappa^2 m}{L} \frac{1}{r_P^2}$$

also if its orbit is a parabola or a hyperbola. As in the case of the ellipse, the verification depends on the fact that $\lim_{Q \to P} \frac{QR}{QT^2} = \frac{1}{L}$.

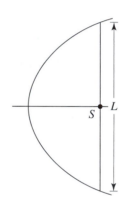

Figure 7.16

Newton next considers the problem in the other direction. If a body moves under the action of a centripetal force with the property that its magnitude is inversely proportional to the square of the distance of the body from the center of force, is the orbit necessarily a conic section, i.e., an ellipse, a parabola, or a hyperbola, and is the center of force at a focus? In particular, if a body subject to such a centripetal force is a closed curve, does this curve have to be an ellipse? The *Principia's* PROPOSITION XVII answers these questions in the affirmative. Newton reasons in the following way. The body in question will, at some fixed point in time t_0, be in a certain position and move in a certain direction with a certain speed. Newton hypothetically considers a second body with the same mass and subject to the same centripetal force as the first that is in elliptical, parabolic, or hyperbolic orbit about the center of force. By setting the various parameters, he obtains the situation where this second body has, at time t_0, the same position and velocity as the first. Since the essential data determining the motion of the two bodies (force, mass, position, and velocity) are the same at time t_0, their orbits must be the same from time t_0 onward. Therefore, the path of the first body is either an ellipse, a parabola, or a hyperbola, at least from time t_0 on. In this way, Newton demonstrates, in particular, that Kepler's first law of elliptical orbits is a mathematical consequence of a more basic law: his inverse square law.[3]

Our look at some of the most important elements of Book I of the *Principia* is now complete. Let's summarize what Newton has accomplished:

Suppose that a body P is propelled by a centripetal force that has center of force S. See Figure 7.17.

Conclusion A. The segment SP sweeps out equal areas in equal times. In particular, if A_t is the area swept out by SP during some time t, then $\frac{A_t}{t}$ is the same constant κ, no matter what t is equal to and no matter where in the orbit this occurs.

Conclusion B. If it is assumed in addition that the orbit of the body is either an ellipse, a parabola, or a hyperbola, then the magnitude F_P of the centripetal force is given by the equation

$$F_P = \frac{8\kappa^2 m}{L} \frac{1}{r_P^2},$$

where m is the mass of the body, L is the latus rectum of the orbit, and r_P is the distance between P and S. Finally,

Conclusion C. If throughout its orbit, the magnitude F_P of the centripetal force is given by an inverse square law, in other words, by an equation of

the form $F_P = c\frac{1}{r_P^2}$, where c is a constant, then this orbit is either an ellipse, a parabola, or a hyperbola, and the center of force S is at a focal point.

In the remaining sections of this chapter, we turn to Newton's primary purpose: The application of his abstract mathematical theory of centripetal forces to the explanation of the workings of the solar system. This is the aim of Book III of the *Principia*, the *System of the World*.

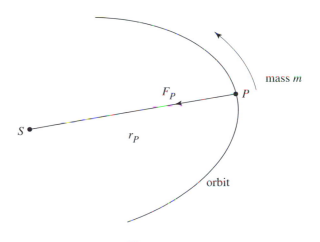

Figure 7.17

7.4 **Test Case: The Orbit of the Moon**

To demonstrate that his conclusions apply to the real world, Newton must test them against the existing evidence. He begins this task by verifying that very basic observations about the Moon's orbit around the Earth are consistent with his theory.

Let's begin with some numerical data about the Moon's orbit. By Newton's time, these were more accurate than the earlier estimates achieved by the Greeks. The radius of the Earth was understood to be about $R = 4000$ miles, and the distance from the Earth to the Moon about 240,000 miles, or $60R$. It was known that the Moon completes an orbit in 27 days, 7 hours,

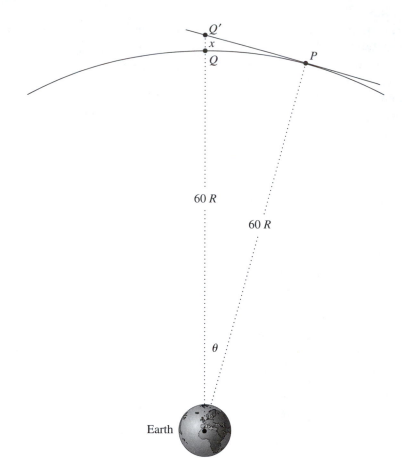

Figure 7.18

and 43 minutes, or 39,343 minutes. We will make the simplifying assumption that the Moon's orbit is a circle of radius $60R$ and that it traces out its orbit at constant speed. Consider the Moon in a typical position P and let it be at Q exactly 1 minute later. In Figure 7.18, the motion of the Moon from P to Q is decomposed into the tangential component PQ' and the component $Q'Q$ in the direction of the Earth. Observe that the angle θ is equal to $\frac{360}{39343}$ degrees. By trigonometry, $\frac{60R}{x+60R} = \cos\theta = 0.999999987$. So $\frac{x+60R}{60R} = 1.000000013$. Solving for x, we get $x+60R = (1.000000013)60R$, and therefore

$$x = (1.000000013 - 1)60R = 0.000000078\,R.$$

This estimate for the distance of the "fall" of the Moon towards the Earth in 1 minute is a consequence of observational data alone. Is the value provided by Newton's theory at least approximately the same?

Consider Newton's equation $F_P = \frac{4\pi^2 a^3 m}{T^2}\frac{1}{r_P^2}$ as applied to the gravitational pull of the Earth on the Moon. Since the orbit is assumed to be a circle of radius $60R$, the semimajor axis a and the distance r_P are both equal to $60R$. Since $T = 39,343$ minutes, we get

$$F_P = \frac{4\pi^2(60R)^3 m}{(39,343)^2}\frac{1}{(60R)^2} = \frac{4\pi^2 60Rm}{(39,343)^2}.$$

This force provides the component of the Moon's motion from Q' to Q with an acceleration of

$$\frac{F_P}{m} = \frac{4\pi^2 60 R}{(39{,}343)^2} = \frac{4(9.869604401)(60)}{(39{,}343)^2}R$$

$$= 0.00000153R.$$

Notice that the initial velocity of this motion is zero. An application of Example 6.14 of Section 6.4 shows that after t minutes, the velocity of the Moon and the distance of its "fall" towards the Earth are respectively

$$v(t) = (0.00000153R)t \quad \text{and}$$

$$p(t) = \frac{1}{2}(0.00000153R)t^2.$$

In the current situation, $t = 1$. So the predicted distance of the fall is

$$x = p(1) = \frac{1}{2}(0.00000153R) = 0.000000765R.$$

Taking $R = 4000$ miles and 1 mile $= 5280$ feet, we can calculate that the observed value of the distance per minute the Moon falls towards the Earth is $x = 16\frac{1}{2}$ feet per minute and that the theoretical value is $x = 16\frac{1}{6}$ feet per minute.

Newton's theory has passed the test. The agreement between observation and theory is good. The discrepancy that does exist can be explained by the fact that simplifying assumptions were made. After all, the Moon's orbit was taken to be circular and not elliptical, and the gravitational effects of the Sun were ignored.

7.5 The Law of Universal Gravitation

Newton now has evidence that his inverse square law $F_P = \frac{8\kappa^2 m}{L}\frac{1}{r^2}$ provides a valid quantitative description of the gravitational force with which a body S attracts a body P of mass m that is a distance r from S. He is convinced that his theory applies not only to the orbit of the Moon about the Earth, but also to the orbits of the planets around the the Sun and to the motion of the moons of Jupiter. Indeed, he believes that it is valid anywhere in the universe!

Consider any two bodies S and P in the universe with P in orbit around S. Rewrite the expression for the magnitude of the force on P as

$$F_P = C_P m \frac{1}{r^2}$$

where C_P is the constant $\frac{8\kappa^2}{L}$. If S exerts a pull on P, why should not also P exert a pull on S? Indeed, by Newton's third law, the force F_P with which S pulls on P has an equal and opposite reaction. In other words, P pulls on S with a force of magnitude $F_S = F_P$ in the opposite direction of that with which S pulls on P. See Figure 7.19. For instance, a planet attracts the Sun with a force equal to that which the Sun exerts on the planet. The effects of the two forces are different, of course, because the Sun's mass is much greater than that of the planet.

Figure 7.19

The symmetry of the situation requires that F_S can be expressed as

$$F_S = C_S M \frac{1}{r^2},$$

where M is the mass of S and C_S is a constant. In the *System of the World*, Newton puts the matter this way:

Since the action of the centripetal force upon bodies attracted is, at equal distances, proportional to the quantities of matter in those bodies, reason requires that it should be also proportional to the quantity of matter in the body attracting.

Since $F_S = F_P$, we see that $C_S M = C_P m$ and hence that $\frac{C_P}{M} = \frac{C_S}{m}$. Now let $G = \frac{C_P}{M} = \frac{C_S}{m}$ and notice that $GM = C_P$ and $Gm = C_S$. Let $F = F_P = F_S$, and substitute to get

$$F = G\frac{mM}{r^2}$$

This is Newton's *Law of Universal Gravitation*. It asserts that any two bodies in the universe—whether one is in orbit about the other or not—attract each other with a force that is given by the above formula. Newton recognizes that the masses m and M of the two bodies and the distance r between them are the essential parameters determining the force. He recognizes, in particular, that G should be a *universal constant*, in other words, a constant that is the same for any two bodies, separated by any distance, anywhere in the universe.

This reciprocal nature of gravitational attraction has important consequences for the motion of the bodies in the solar system. As Newton explains:

> I have hitherto explained the motions of bodies attracted towards an immoveable centre, though perhaps no such motions exist in nature. For attractions are made towards bodies; and the actions of bodies attracting and attracted are always mutual and equal, by the third law of motion: so that, if there are two bodies, neither the attracting nor the attracted body can really be at rest; but both as it were by a mutual attraction, revolve about the common center of gravity.

To understand what Newton is saying, consider the Earth–Moon system for example. The center of mass (or gravity) of this system, this is the so-called *barycenter* of the system, is about 2900 miles from the center of the Earth, or, about 1050 miles below its surface. The centers of mass of both the Moon and the Earth travel along ellipses with the barycenter at a focus of each ellipse. To understand what is going

on, think of the centers of the Earth and Moon as being connected with a horizontally placed lever with fulcrum at the barycenter **B**. See Figure 7.20 and note that the lever is balanced. Now let the lever revolve in the horizontal plane. This simulates the dynamics of the Earth–Moon system. The Moon is in a month-long orbit around **B**. This is a circular orbit in the simulation, but elliptical in fact. The center C of the Earth is also in "orbit" around **B**. In other words, the Moon's gravitational pull on the Earth causes it to wobble about its center C in a monthly cycle, as indicated in Figure 7.21.

Figure 7.20

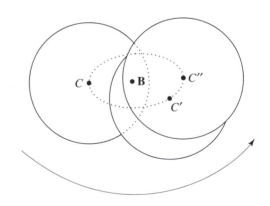

Figure 7.21

Let's apply these considerations to the solar system as a whole. It is the barycenter **B** of the Earth–Moon system (rather than the center of the Earth) that describes an elliptical orbit about the Sun. The center of force is not the Sun, but rather the center of mass of the entire system of the Sun and planets.

This point is at a focus of the ellipses of all the planets. Newton's explanations thus constitute a subtle refinement of Kepler's laws. Since the Sun comprises 99.85% of the mass of the solar system, the difference is only minor, however. In reference to the orbit of the revolving Earth–Moon system about the Sun, observe that the center of the Earth is at times ahead of the Moon and at times behind it.[4] Note also that when the Earth is closer to the Sun, the gravitational tug of the Sun on the Earth–Moon system will be greater than when the Moon is closer to the Sun (because the Earth's mass is greater than that of the Moon). The Sun's gravitational force on the Earth–Moon system thus depends on the dynamics of this system. (Section 14.5 considers subtleties about gravity that are related to the point just made.) It should now be apparent that the mathematics of the "three-body problem" of the gravitational interaction of the Sun, Earth, and Moon is very complicated[5] (and still awaits a complete solution).

These considerations explain other phenomenona as well. The shape of the Earth is essentially spherical, but it is a sphere that is flattened at the poles. This flattening was brought about by the rotation of the Earth about its axis. Consider the plane through the Earth's equator. Since the Moon is near this plane at every point in its orbit (the Moon's orbital plane and the Earth's equatorial plane only differ by up to 5°), it follows that the Moon's gravitational tug on the Earth is a little greater at the equator than at the poles. The difference is substantial enough to produce a very slow gyration of the Earth's axis of rotation. (Why is the analogous effect of the Sun's gravitational force on the Earth negligible?) Translated to the Earth-centered point of view of Figure 2.1 of Section 2.1, this gyration is equivalent to a slight rotation of the axis that joins the two equinox positions. This "precession of the equinoxes" was already observed by the Greeks. A similar effect explains the tides of the oceans: they are produced by the tug of the Moon on the Earth's surface.

We turn next to some of the incredible implications of Newton's Laws of Universal Gravitation.

7.6 Incredible Consequences

Consider the situation of a body P in elliptical orbit about another body S. Suppose that S is much more massive than P. Therefore, in the context of the explanations just given, S is in essence the center of the centripetal force, and the mass M of S is the same as that of the system. Recall that Kepler's constant for the elliptical orbit is $\kappa = \frac{\pi ab}{T}$ and the latus rectum is $L = \frac{2b^2}{a}$, where as before, a is the semimajor axis, b the semiminor axis, and T the period of the orbit. Combining this with the equations

$$C_P = \frac{8\kappa^2}{L} \quad \text{and} \quad C_P = GM$$

from Section 7.5 shows that $GM = \frac{8\kappa^2}{L} = \frac{8a\kappa^2}{2b^2} = \frac{4a\pi^2 a^2 b^2}{b^2 T^2} = \frac{4\pi^2 a^3}{T^2}$. Therefore,

$$\boxed{\frac{a^3}{T^2} = \frac{GM}{4\pi^2}}$$

Notice that the term $\frac{GM}{4\pi^2}$ on the right has nothing to do with the particulars of the body P and its orbit. In other words, it is the same for any P in orbit around S. It follows that the ratio $\frac{a^3}{T^2}$ of the cube of the semimajor axis a to the square of the period T of the orbit is the same for any body P in orbit around S. In the case of the planets of the solar system, this is precisely what Kepler's third law asserts. In other words, Kepler's third law is a consequence of Newton's theory of gravitation! Refer to the conclusions A and C of Section 7.3, and observe therefore that all three of Kepler's laws are consequences of Newton's theory.

Consider the rewritten version

$$\boxed{M = \frac{4\pi^2 a^3}{GT^2}}$$

of the previous equation. If, somehow, the universal constant G could be determined, then the mass of the Sun could be computed by inserting a and T for the orbit of the Earth!! In the same way, the mass of the Earth could be computed by inserting data about the orbit of the Moon. Newton, however, seems to think that the determination of G is out of reach:

> Perhaps it may be objected, according to this philosophy all bodies should mutually attract one another, contrary to the evidence of experiments in terrestrial bodies. But I answer, that the experiments in terrestrial bodies come to no account. For the attraction of homogeneous spheres near their surface are as their diameters. Whence a sphere of one foot in diameter, and of like nature to the Earth, would attract a small body placed near its surface with a force of 20,000,000 less than the Earth would do if placed near its surface. But so small a force could produce no sensible effect.

In other words, Newton regards the experimental determination of the constant G to be an impossible enterprise. This time Newton was wrong! In the latter part of the 18th century, in 1798 to be exact, Henry Cavendish[6] succeeded in measuring G in the laboratory with an extremely delicate experiment. What he did in essence was this. Suspend a fine wire from a fixed point A and attach a rigid "crossbar" BC. Refer to Figure 7.22. From BC in turn, suspend two heavy iron balls. They are shown in black. Now move two more heavy iron balls (shown lightly shaded) in place near the two others. If this apparatus is very delicately balanced and controlled, then the gravitational force F between the two pairs of balls will bring about a rotation of the axis BC. This allows F to be measured. Since Cavendish knew the masses of the balls and the distances between them, the equation $F = G\frac{mM}{r^2}$ allowed him to compute

$$G = 6.67 \times 10^{-11} \frac{\text{meters}^3}{\text{kilograms} \cdot \text{seconds}^2}.$$

Using the fact that the semimajor axis of the Earth's orbit about the Sun is 149.6×10^6 kilometers, or 149.6×10^9 meters and that its period T is 365.25

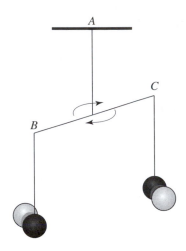

Figure 7.22

days, Cavendish, after expressing T in seconds, could now compute the mass M of the Sun:

$$M = \frac{4\pi^2 a^3}{T^2 G} = \frac{4(3.14)^2(149.6 \times 10^9)^3}{[(365.25)(24)(60)(60)]^2(6.67 \times 10^{-11})}$$

$$\approx \frac{1320 \times 10^{32}}{66 \times 10^3} \approx 2 \times 10^{30} \text{ kg}.$$

The same calculation (making use of information of the Moon's orbit about the Earth) shows that the mass of the Earth is about 6×10^{24} kilograms. Incredibly, Newton's theory of universal gravitation together with Cavendish's value of G has "served up" estimates for the masses of the Sun and the Earth!

Consider, finally, a projectile in motion near the Earth's surface. In view of their universal nature, the above considerations should apply to this situation as well. Do they? Figure 7.23 shows a typical trajectory and typical position P of the projectile and the surface of the Earth. Observe that the projectile is subject to a centripetal force whose center is the center of mass C of the Earth. Neglecting air resistance, we get as a consequence of Newton's theory that the trajectory must be an arc of an ellipse. The parabola of the discussion of Section 6.5 is only an approximation. (What is the difference between the underlying assumptions about the action of gravity?) Now suppose that the projectile is in free-fall and let F_P be the magnitude of the force of gravity. By the universal law of gravitation, $F_P = G\frac{mM}{r^2}$, where M is the mass of the Earth, m the mass of the projectile, and r the radius of the Earth. By Newton's second law, $F_P = mg$, where g is the acceleration produced by the Earth's gravity. By putting the two equations for F_P together, we get

$$g = G\frac{M}{r^2}.$$

After inserting the values $G = 6.67 \times 10^{-11} \frac{\text{meters}^3}{\text{kg}\cdot\text{sec}^2}$, $M = 6 \times 10^{24}$ kg, and $r = 6360$ kilometers (this is equivalent to 3950 miles), the result is

$$g = 6.67 \times 10^{-11} \cdot \frac{6 \times 10^{24}}{(6360 \times 10^3)^2}$$

$$\approx 6.67 \times 10^{-11} \cdot \frac{6 \times 10^{24}}{4 \times 10^7 \times 10^6} \approx 10.$$

The unit is meters/second². Note the close agreement with the observed value of $g = 9.80$ meters/second². Newton's theory of gravitation has provided the explanation for Galileo's observation that all bodies falling near the Earth's surface are subject to the same acceleration. The fact that his theory is valid not only in "outer space" but also "here at home" is further indication of its universality.

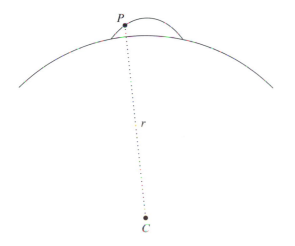

Figure 7.23

7.7 Postscript

Perhaps the greatest triumph of Newton's theory is described by the following sequence of events. In 1781, the British astronomer William Herschel discovered what he originally thought to be a comet, but what later proved to be a planet. The planet Uranus thus joined Mercury, Venus, Earth, Mars, Jupiter, and Saturn—already known since antiquity—and became the seventh planet of the solar system. Efforts were made to compute its orbit, but discrepancies appeared between observed positions and those predicted by theory, and these grew year by year. Several astronomers hypothesized that the gravitational pull of yet another planet on Uranus would explain the discrepancies. By 1845, the British astronomer Adams, and later in 1846, the Frenchman Leverrier, had both computed the hypothetical orbit of the hypothetical planet. Later in 1846, the Berlin Observatory discovered a new planet in a position close to that predicted by these calculations! This eighth planet was later named Neptune. A similar sequence of events led to the discovery of the ninth (and last known) planet Pluto in 1930.

The fact that the planets are attracted not only by the Sun, but also by each other, especially by the more massive Jupiter and Saturn, raises an important concern, of which Newton was certainly aware. Would the effects of the gravitational forces of the planets on each other accumulate over time to threaten the stability of the solar system? Might there be a catastrophic collapse? Consider any of the planets P and let m_P be its mass. It was indeed discovered that neither the semimajor axis a_P of its orbit nor the linear eccentricity e_P nor the angle i_P of inclination of its orbit with its axis of rotation are constants! They are functions of time driven by the gravitational effects of the other planets. The French scientists Lagrange and Laplace carried out a series of investigations in the last part of the 18th century and came to the following remarkable conclusions: The sums

$$(e_{P_1})^2 m_{P_1}\sqrt{a_{P_1}} + (e_{P_2})^2 m_{P_2}\sqrt{a_{P_2}} + \ldots$$

and

$$(i_{P_1})^2 m_{P_1} \sqrt{a_{P_1}} + (i_{P_2})^2 m_{P_2} \sqrt{a_{P_2}} + \dots$$

taken over all the planets are both constants.[7] In other words, while the individual parameters a_P, e_P, and i_P vary, this variation is controlled and limited in the context of the solar system as a whole. So the wobbles and irregularities of the planetary clockwork of the solar system are minor and self-correcting, and do not pose a threat to the stability of the system. End of the story? Hardly! See Section 14.7.

We have reached the end of the first part of this book. It has emerged from our discussion that by thinking about this question and the other, by solving this problem and that, scientists developed both a new outlook and new methods of inquiry. We have traced some of the main themes—especially the mathematical themes—of this *Scientific Revolution*. We saw that this revolution had Greek origins, both from the point of view of the nature of the questions that were asked and the mathematical strategies that were developed in response to them. We saw that it was tied closely to the study of motion, both on the Earth and in the heavens. This scientific revolution reached its culmination in the latter part of the 17th century with the synthesis of astronomy and mechanics and the development of the necessary mathematics, and in particular with Newton's *Principia*. The role played by mathematics in this development was both pervasive and crucial. The explanation of basic physical phenomena by a combination of fundamental laws and mathematics, and the concurrent confirmation of the explanation by observation and experiment, became the basis of modern science.

The success of this outlook and the methods that it spawned energized efforts to extend them to newly emerging fields of science (chemistry, biology, and geology, for example). More generally, however, it contributed to the philosophical underpinnings of an intellectual movement that encouraged the overhaul of every kind of traditional teaching, questioned our entire intellectual heritage, and ultimately brought changes that had significance for the history of humanity in its broadest sense. The transmission of the results of the scientific revolution to the outside world linked the scientific revolution with a new age, the *Enlightenment*. The discoveries of 17th century science were translated into a new outlook and a new world-view, not by scientists themselves, but by a literary movement energized by the *philosophes*. These were radical intellectuals (Voltaire, Rousseau, Diderot, Kant, Locke, Hume, ...) who held the secular view that human beings are mature enough to find their own way. Human beings are able to understand the natural world, indeed their own nature, by reason and the methods of mathematics and science. Therefore, they should think for themselves, shake off the hand of authority in the realms of both religion and politics, and insist on the right to freedom of thought and expression. On the commercial side, the philosophes expressed the view that merchants should be free to buy and sell as they chose, that free competition was essential to trade, and that open markets would free commercial energies and benefit everyone. The philosophes made these appeals—as Galileo had tried to do earlier—to a new arbiter of human thought: a wider general reading public.[8]

The *Industrial Revolution*—as epitomized by James Watt's steam engine (invented in 1769) and the power that it provided to machines of production—is also directly linked with the earlier scientific revolution. The spreading criticism of authority, the accelerating pace of the industrialization of the Western world, and the steady increase in population—all pointed to the prospect of instability. Indeed, in the last half of the 18th century, country after country was afflicted by social unrest. The American Revolution and the French Revolution soon followed ... . But it is time to return to mathematics.

Exercises
7A.   Equal Areas in Equal Times

1. Suppose that an object P moves along a straight line L with constant velocity. Let S be a point not on L. Regard P to be subject to a certripetal force of zero

magnitude centered at S. Show that the segment SP sweeps out equal areas during equal times. Note the connection between the magnitude of the force and the shape of the orbit.

Exercises 2–5 refer to Newton's proof of Kepler's second law in Section 7.1. Recall that the semimajor axis of the Earth's orbit is $a = 1$ AU. By Example 4.16 of Section 4.7, the semiminor axis is $b = 0.9998$ AU. For purposes of Exercises 2–5 we will therefore assume that the orbit is a circle of radius 1 AU. In Figure 7.24, C is the center of the Earth's orbit and S is the position of the Sun. In addition, P is the position of the Earth when it is nearest the Sun and Q is its position $t_1 =$ two months $= 61$ days later. Let $\Delta t = 1$ day. So P_1 is the position of the Earth one day after it is at P (note that the figure is not drawn to scale). The linear eccentricity is $e = 0.0167$ AU, but take $e = 0.0200$ AU. So $SP = a - e = 0.9800$ AU. Recall also that the Earth completes one orbit in 365.2422 days. Compute with 4 decimal accuracy.

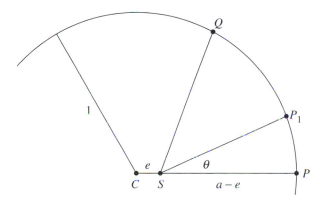

Figure 7.24

2. How many triangles are inscribed into the region SPQ? In other words, what is n equal to in this situation?

3. The area of the Earth's orbit is $\pi \cdot 1^2 = \pi$ in AU^2. Use this to compute Kepler's constant κ for Earth. Making use of κ, compute the area of the region SPQ in AU^2. This is A_1. In the same way, compute the area of the region SPP_1 in AU^2.

4. The region SPP_1 is approximately equal to the circular sector with radius $a - e$ and angle θ. See Figure 7.25. Assume equality and use this and the results of Exercise 3 to compute θ. Show next that the isosceles

triangle $\triangle SPP_1$ has area $(a - e)^2 \sin \frac{\theta}{2} \cos \frac{\theta}{2}$ in AU^2. This corresponds to $\triangle A$.

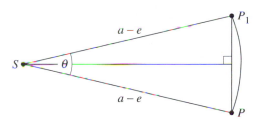

Figure 7.25

5. Compute $n(\Delta A)$ in AU^2 and compare this to A_1. Is $n(\Delta A)$ an acceptable approximation of A_1? Before you answer, convert these areas into square miles using 1 AU $= 93 \times 10^6$ miles.

6. Using the data in Table 4.2 of Section 4.7, compute Kepler's constant for Mercury and Jupiter in AU^2/year.

7. Consider a planet in orbit around the Sun S. Let κ be its Kepler constant. Begin to time the planet when it is at the perihelion position P. This is time $t = 0$. Let $P(t)$ be its position at any time t thereafter, and let $A(t)$ be the area of the region that P, S, and $P(t)$ determine. See Figure 7.26, but allow for the possibility that the planet may already have completed several orbits during time t. Compute the function $A(t)$ in terms of κ.

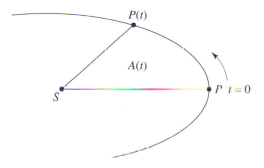

Figure 7.26

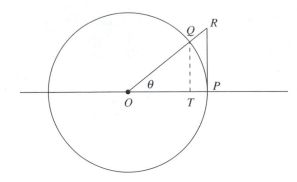

Figure 7.27

7B.   Computations Related to the Inverse Square Law

8. Consider Newton's result $\lim_{Q \to P} \dfrac{QR}{QT^2} = \dfrac{1}{L} = \dfrac{a}{2b^2}$ for the ellipse. See Figure 7.27. Give an interpretation of this limit in the case of the circle of radius 1. Write QT in terms of $\sin \theta$ and QR in terms of $\sec \theta$. Then express $\dfrac{QR}{QT^2}$ in terms of $\cos \theta$. Use the equalities $\sec^2 \theta - 1 = \tan^2 \theta = \dfrac{\sin^2 \theta}{\cos^2 \theta}$. Now push Q to P, or what is equivalent, θ to 0. Make use of the fact that $\cos \theta = OT$, and hence that $\lim_{\theta \to 0} \cos \theta = \lim_{\theta \to 0} OT = 1$. Does your result agree with Newton's?

9. Consider an ellipse with equation $\dfrac{x^2}{a^2} + \dfrac{y^2}{b^2} = 1$. So a is the semimajor axis and b the semiminor axis. Determine the coordinates of the two focal points in terms of a and b. (A review of some basics from Section 4.5 may be necessary.) Show that the latus rectum $\dfrac{2b^2}{a}$ is equal to the length of the segment that goes through the focus, is perpendicular to the major axis, and has both endpoints on the ellipse.

7C.   The Satellites of Jupiter

10. In the *System of the World*, Newton considers the following data:

$$5.578, \quad 8.876, \quad 14.159, \quad 24.903.$$

They are the maximal distances from Jupiter (as measured by Newton's contemporary the astronomer Flamsteed) of the four satellites of Jupiter discovered by Galileo. Here 1 unit is the radius of Jupiter. The corresponding periods of the orbits of these satellites in hours are respectively

$$42.48, \quad 89.30, \quad 172, \quad 402.09.$$

Both sets of values are from the *System of the World*. Newton checked his theory of gravitation against the data. Which part of his theory did he test, and how good is the fit?

7D.   Overview of the Formulas

Inverse Square Law. Suppose that a body of mass m is subject to a centripetal force F_P and that the orbit is an ellipse, a parabola, or a hyperbola. Then

$$F_P = \frac{8\kappa^2 m}{L} \frac{1}{r_P^2},$$

where κ is Kepler's constant, r_P is the distance from the orbiting body to the center of force, and L is the latus rectum. If the orbit is an ellipse, then this formula becomes

$$F_P = \frac{4\pi^2 a^3 m}{T^2} \frac{1}{r_P^2},$$

where a is the semimajor axis and T is the period of the orbit.

Law of Universal Gravitation. The gravitational force F between two bodies is given by

$$F = G \frac{mM}{r^2},$$

where $G = 6.67 \times 10^{-11} \dfrac{\text{meters}^3}{\text{kilograms·seconds}^2}$, m and M are the masses of the two bodies in kilograms, and r is the distance between them in meters.

If a body of mass m is in elliptical orbit around another body of mass M, then

$$\frac{a^3}{T^2} = \frac{GM}{4\pi^2} \quad \text{or} \quad M = \frac{4\pi^2 a^3}{GT^2}$$

where a, T, and G are as before.

7E.   Systems of Units

Study the following system of units:

Length: 100 centimeters = 1 meter, 1000 meters = 1 kilometer.

Mass: 1000 grams = 1 kilogram. 1 kg is defined to be the mass of 1000 cubic centimeters, i.e., one liter, of water (at 4 degrees centigrade). 1000 kilograms = 1 metric ton.

Force: 1 newton = the (constant) force required to accelerate in 1 second a 1 kilogram mass from rest to a velocity of 1 meter per second.

What is the weight of a mass of 1 kilogram on the surface of the Earth? On the surface of the Earth, weight = mass ×

gravitational acceleration. Therefore,

$$\text{weight} = (1 \text{ kilogram})\left(9.80\frac{\text{meters}}{\text{second}^2}\right) = 9.80 \times 1\frac{\text{kg} \cdot \text{m}}{\text{sec}^2}$$

$$= 9.80 \text{ newtons.}$$

So on the surface of the Earth, 1 kg weighs 9.8 newtons. The above is the M.K.S. (meter-kilogram-second) system.

The commonly used American system is as follows:

Length: 1 foot, 5280 feet = 1 mile.
Mass: 1 pound.
Force: 1 slug = the mass that a 1 pound force accelerates in 1 second from rest to a velocity of 1 foot per second.

How many pounds does a mass of 1 slug weigh on the surface of the Earth? Again, weight = mass × gravitational acceleration, or

$$\text{weight} = (1 \text{ slug})\left(32.17\frac{\text{feet}}{\text{second}^2}\right) = 32.17 \times 1\frac{\text{slug} \cdot \text{feet}}{\text{sec}^2}$$

$$= 32.17 \text{ pounds.}$$

The connection between the American system and the M.K.S. system is given by:

Length: 1 foot = 0.30 meters, 1 meter = 3.28 feet, 1000 meters = 1 kilometer = 0.62 miles, 1 mile = 1.61 kilometers.
Mass: 1 pound = 4.45 newtons, 1 newton = 0.22 pounds.
Force: 1 slug = 14.59 kilograms, 1 kilogram = 0.07 slugs.

11. Recall from Table 1.4 of Section 1.5 that the radius of the Earth is 3950 miles and that of the Moon is 1080 miles. Show that this corresponds to 6360 kilometers for the Earth and 1740 kilometers for the Moon.

12. For a basketball to satisfy regulations, its weight must fall between $1\frac{1}{4} = 1.25$ and $1\frac{3}{8} = 1.375$ pounds. (Information supplied by Rawlings Sporting Goods.) What is the mass in slugs of a basketball that weighs 1.3 pounds?

13. Convert the universal gravitational constant $G = 6.67 \times 10^{-11}\frac{(\text{meters})^3}{\text{kilograms(seconds)}^2}$ to $\frac{(\text{feet})^3}{\text{slugs(seconds)}^2}$.

7F. Applying Newton's Formulas to the Moon

14. Show that the mass of the Earth is 6.0×10^{24} kilograms by using information about the Moon's orbit. Use Ex-

ercise 11 to compute the average density of the Earth in kilograms per cubic meter.

15. Show that the mass of the Moon is 12.0×10^{22} kilograms under the assumption that its density is the same as that of the Earth. Use the fact that its radius is 1740 kilometers. [Note: A better estimate for the mass of the Moon is 7.4×10^{22} kg. The Moon is therefore less dense than the Earth.]

16. Refer to Figure 7.20. Use the fact that the masses of the Earth and the Moon are 6.0×10^{24} kilograms and 7.4×10^{22} kilograms respectively and that the distance from the Earth to the Moon is 240,000 miles in order to estimate the distance in miles from the barycenter **B** of the Earth–Moon system to the center C of the Earth.

17. Gravity provides all bodies falling near the surface of the Moon with the same acceleration regardless of the mass of the body. Use the fact that the mass of the Moon is 7.4×10^{22} kilograms and that its radius is 1740 kilometers to show that this acceleration is about 1.63 meters/second^2. Since 1 meter is 3.28 feet, this is equal to about 5.35 feet/second^2.

18. How much does the basketball of Exercise 12 weigh on the Moon?

19. Return to Galileo, his cannonball, and the 177-foot-high tower of Pisa. Assume that such a tower has been constructed on the Moon. How long would it take for a cannonball to hit the ground if it is dropped with zero initial velocity from this tower? Compare your result with that of Example 6.15 of Section 6.4. [Note: Recall that on the Earth it had to be assumed that there is no air resistance. No assumption of this kind is needed for the Moon because it has no atmosphere.]

20. Make a calculation to show that the gravitational force of the Sun on the Moon is greater than the force of the Earth on the Moon. This being so, how is it that the Moon goes around the Earth and not around the Sun? Or does it? [Note: Use the following data: the mass of the Sun is 2.0×10^{30} kg; the mass of the Earth is 6.0×10^{24} kg; the mass of the Moon is 7.4×10^{22} kg; the average distance from the Earth to the Sun is 1.5×10^{11} meters; the average distance from the Moon to the Earth is 3.8×10^8 meters.]

7G. Computing Masses and Forces

21. Suppose that the Earth is shrunk to the size of a basketball. Keeping the density constant, compute the mass of this sphere. How many times more massive is it than that of the basketball of Exercise 12? [Hints:

Recall from Exercise 25 of Chapter 1 that the radius of the Earth must be shrunk by a factor of $\frac{1}{50,000,000} = 2 \times 10^{-8}$ to obtain a sphere of the size of a basketball. This sphere has volume $\frac{1}{50,000,000^3} = 8 \times 10^{-24}$ of that of the Earth. Use the fact that the Earth has a mass of 6.0×10^{24} kilograms.]

22. On October 5, 1957, Sputnik, the first artificial satellite of the Earth, was launched by the (then) Soviet Union. It was reported to encircle the Earth at an altitude of 560 miles above the Earth's surface, to travel 18,000 miles per hour, and to make one revolution in 95 minutes. Are these data consistent? Assume that the satellite's orbit is a circle. Use these data to estimate the mass of the Earth in slugs. How does this compare to 6.0×10^{24} kilograms?

23. Space probes are used to determine the masses of the planets by studying the orbital data from their hyperbolic "flybys." Draw a sketch and explain what makes such conclusions theoretically possible. [Hint: Use the formulas $F_P = \frac{8\kappa^2 m}{L} \frac{1}{r^2}$ and $F_P = G \frac{Mm}{r^2}$.]

24. Refer to the Exercises of Chapter 4 that deal with Halley's comet. Compute the latus rectum for Halley's comet. Use data from Halley's orbit to compute the mass of the Sun.

25. Consider an object of mass 250 grams. Determine its weight in newtons. This object is twirled on a string of length 80 centimeters. It describes a circle of this radius and revolves at constant angular velocity of 3 revolutions per second. Determine the centripetal force in newtons necessary to keep it in motion.

7H.   A Speculation of Newton

Newton makes the following statement in the *System of the World*:

> For the attraction of homogeneous spheres near their surfaces are as their diameters. Whence a sphere of one foot in diameter, and of like nature to the Earth, would attract a small body placed near its surface with a force of 20,000,000 less than the Earth would do if placed near its surface. But so small a force could produce no sensible effect. If two such spheres were distant but by $\frac{1}{4}$ inch, they would not even in spaces void of resistance, come together by the force of their mutual attraction in less than a month's time.

Is what Newton is saying here correct? The circles in Figure 7.28 represent two spheres in a weightless situation. For example, think of them in the cargo bay of an orbiting

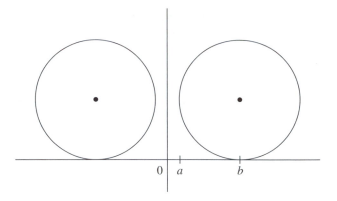

Figure 7.28

space shuttle. The spheres are 1 foot in diameter. The radii are thus $\frac{1}{2}$ foot. The spheres are $\frac{1}{8}$ of an inch, or 0.01 feet, from the vertical axis. So $a = 0.01$ and $b = 0.51$. Both spheres are assumed to have the same density as the average density of the Earth. (Use Exercise 14.) What is the mass of each sphere in slugs? What is the gravitational force with which each sphere attracts the other when they are in the indicated position; when they touch each other? Why is the force in the second situation greater? Suppose that this greater force acts on the sphere (say, the one on the right) when it is in the position indicated in the figure. How long will it take for this force to move the sphere so that it touches the vertical axis? Do the same computation with the weaker force. How do your findings mesh with Newton's assertions?

Notes

[1]All quotations are taken from Isaac Newton, *Mathematical Principles of Natural Philosophy*, Translated into English and Illustrated with a Commentary by Robert Thorp, London, Printed for W. Strahan and T. Cadell, in the Strand, 1777. Reprinted by Dawsons of Pall Mall, London, 1969.

[2]Observe that the considerations of calculus come into play both in the proof of Kepler's second law and also in the computation of $\lim_{Q \to P} \frac{QR}{QT^2}$. While Newton's approach is geometric, it does not make use of analytic geometry. Hence it does not "look" like modern calculus. This is the reason why some historians of science

make a point of saying that Newton does not make use of calculus in the *Principia*.

[3]It has already been pointed out that Newton's arguments are often terse and difficult to penetrate. In reference to the demonstration just described he outdoes even himself, devoting all of two sentences to this matter. So Newton provides, at best, an outline of a demonstration. Indeed, the reconstruction of a proof based on this outline is a very delicate task. In the article: Reading the Master: Newton and the Birth of Celestial Mechanics, *Amer. Math. Monthly 104 (1997)*, 1–19, B. Pourcia takes up this task and provides the relevant (and interesting) historical perspectives as he does. In the 1960s Richard Feynman, the Nobel prize winning physicist, well aware of the very demanding nature of Newton's arguments, developed his own "elementary" approach to Newton's theory. In lecture notes that were "lost" for many years in an unoccupied office at the California Institute of Technology (where Feynman had served on the faculty), Feynman combined Newton's ideas with delicate properties of the ellipse, to deduce from the inverse square law that the planetary orbits are ellipses. For the full story and its background refer to D.L. Goodstein and J.R. Goodstein, *Feynman's Lost Lecture*, W.W. Norton and Co., New York, 1996.

[4]This explains the slight variations (of up to ten minutes, or 0.007 days) in the lengths of the seasons. Refer to the data given in Section 4.9.

[5]To get an insight into just how complicated this problem is and what an interesting history it has, refer to the book F. Diacu and P. Holmes, *Celestial Encounters, The Origins of Chaos and Stability*, Princeton University Press, Princeton, NJ, 1996.

[6]The famous Cavendish laboratories in Cambridge, England, would later be named in his honor.

[7]Laplace's celebrated five-volume treatise *Mécanique Céleste*, which appeared between 1799 and 1825, included these results and much more. Its purpose was to provide a comprehensive treatment of all that had been done in gravitational astronomy since the time of Newton. This work earned Laplace the title of "Newton of France," but it owes a considerable debt to the profound mathematical discoveries of Lagrange, who was the superior mathematician.

[8]The classic account, Herbert Butterfield, *The Origins of Modern Science*, Free Press, New York, 1965, provides the details behind these developments.

Part II

Calculus and the Sciences

Analysis of
Functions

The work of Newton and Leibniz was carried on with increasing energy and success, especially in continental Europe. Newton was withdrawn, but Leibniz attracted students. With the Swiss brothers Jakob (1654–1705) and Johann (1667–1748) Bernoulli, from Basel, leading the way, they learned the new differential and integral calculus eagerly, developed the theory further, and transmitted their knowledge to others. The younger Bernoulli, in turn, passed the torch to another Swiss from Basel, who would become the most prolific and important figure among 18th century mathematicians: Leonhard Euler (1707–1783). In a career spent primarily in St. Petersburg and Berlin, Euler made decisive contributions to many areas of mathematics. Moving the function concept to a central position (where it has remained ever since), he took the calculus of Newton and Leibniz and recast it into a more general and extensive theory that became known as "analysis." His 1748 treatise *Introductio in analysin infinitorum* became the cornerstone of the new theory and a catalyst for many advances.

The Frenchmen Joseph Louis Lagrange (1736–1813), Adrien-Marie Legendre (1752–1833), Jean Baptiste Joseph Fourier (1768–1830), Augustin Louis Cauchy (1789–1857), the priest Bernhard Bolzano (1781–1848) from Prague, the German Lejeune Dirichlet (1805–1859), and others began to clarify the basic concepts (e.g., limit, continuity and differentiability of functions, and power series) and generated new branches of analysis. Especially noteworthy are Cauchy's basic texts *Cours d'analyse de l'Ecole Royale Polytechnique* (1821), *Résumé des leçons sur le calcul infinitésimal* (1823), and *Leçons sur le calcul différential* (1829). They gave calculus the character—both in concept and notation—that it has today. The new analysis was applied with great success to physics, especially to mechanics and the astronomy of the solar system. This included the important work of Pierre Simon Laplace (1749–1827) mentioned in the postscript of the previous chapter. Karl Friedrich Gauss (1777–1855), the greatest mathematician of his time, made definitive contributions to existing areas of mathematics and its applications (e.g., the theory

of numbers, geometry, and astronomy) and initiated entirely new ones (e.g., differential geometry: the analysis of curves and surfaces in space). Georg Bernhard Riemann (1826–1866) developed the new theory of differential geometry into a remarkable study of "manifolds" of any dimension. At the same time that the web of mathematics expanded, the mesh was made tighter and more reliable. The formulations of Cauchy in the 1820s were still not definitive and remained flawed in a number of subtle logical points. Riemann and Karl Weierstrass (1815–1897) among others were instrumental in the movement to solidify the logical foundations of mathematics. The essence of what was achieved is this: The real number system was characterized by a careful set of postulates, and with it as basis the entire analysis was derived by a careful selection of basic concepts and tight logical argumentation.

The developments just discussed are not only vast in scope, but also extremely abstract and sophisticated. They are, therefore, beyond the limits of any elementary text. However, today's basic calculus course concerns itself with the most elementary of these issues and therefore provides a glimpse of what is involved. What is involved? A careful look at the work of Leibniz and Newton reveals that they have not dealt with some very basic and essential matters. The following concerns illustrate this point. For example, the assertion that a certain quantity closes in on another quantity may be clear intuitively and pictorially, but does it have sufficient mathematical precision? Or, in reference to several of the graphs considered in Chapters 5 and 6, wouldn't the accompanying discussion run into difficulties if these curves were to have gaps or breaks at relevant points? Finally, should the question as to whether a curve is smooth at a point be simply a matter of visual inspection? In each case, the matter was either not considered at all and left to lie below the surface, or taken up only intuitively in terms of pictorial features of graphs. Because pictures can often mislead and misinform, the state of mathematical affairs was not entirely satisfactory. These questions and concerns point to one of the important achievements of the mathematics of the 19th century: Single out the important fundamental concepts (for example, limit, continuity, differentiability, and integrability), clarify and formulate them with exacting precision, and finally, recast and rebuild the entire theory in a rigorous way by justifying every step with irrefutable logic. In an appropriately limited scope, this will also be one of the goals of this chapter.

8.1 **Putting a Limit to the Test**

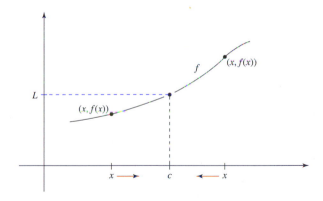

Figure 8.1

R ecall from Section 5.5 that a *function* is a rule that assigns exactly one real number to each number of a set of real numbers. The *domain* of a function is the set of all real numbers for which the rule is defined, i.e., for which it makes sense. Consider, for example, the functions f, g, h, k, and p given respectively by the rules

$$f(x) = x^2, \quad g(x) = \sqrt{x}, \quad h(x) = \sqrt[3]{x},$$

$$k(x) = \frac{1}{x}, \quad \text{and} \quad p(x) = \frac{1}{\sqrt{x+2}}.$$

The domain of f is the set of all real numbers x; that of g is the set of all $x \geq 0$; the domain of h is all x; that of k all x with $x \neq 0$; and that of p consists of all x with $x + 2 > 0$, i.e., all $x > -2$. A function f will also be denoted by $f(x)$ or $y = f(x)$. For example, in reference to the last case, the notations p, $p(x)$, and $y = \frac{1}{\sqrt{x+2}}$ will all refer to the function $p(x) = \frac{1}{\sqrt{x+2}}$.

We begin with a look at limits. Figure 8.1 shows a point (c, L) and the graph of a function f. Observe from the graph that as x is pushed into the number c, the values $f(x)$ close in on the number L. We express this in limit notation by writing $\lim_{x \to c} f(x) = L$. In particular, if x is taken close to c, then the value $f(x)$ is an approximation of L. We now take the point of view that L is not known precisely and that $\lim_{x \to c} f(x) = L$ should be an "approximation machine" for L. In other words, we will "feed" numbers x into f that are closer

and closer to c, but not equal to c, to get better and better approximations for L in terms of its decimal expansion. We would like to be able to compute L in this way to within any required degree of accuracy. Suppose, for instance, that it is required to compute L to within an accuracy of, say, $\frac{1}{1,000,000}$. Can this be done by computing $f(x)$ with x sufficiently close to c (but not equal to c)? Using the fact that $|a - b| = |b - a|$ is the distance between any two numbers a and b on the number line, our goal can be reformulated as follows: Can we achieve $|f(x) - L| < \frac{1}{1,000,000}$ by taking any x such that $|x - c|$ is small enough (but not zero). We would like to be able to do this not only for $\frac{1}{1,000,000}$, but for any accuracy requirement $\varepsilon > 0$ (no matter how small) for L. In conclusion, we want our approximation machine to satisfy the following criterion: Given an accuracy requirement of $\varepsilon > 0$ for L, then there must be some constraint δ on x such that

$$|f(x) - L| < \varepsilon \text{ whenever } |x - c| < \delta \text{ with } x \neq c.$$

This says that any given accuracy requirement for L can be achieved by computing $f(x)$, provided that x lies within the corresponding constraint. We repeat this criterion once more by stating the

Limit Test: For the assertion $\lim_{x \to c} f(x) = L$ to be correct, it must meet the following test. For any accuracy requirement $\varepsilon > 0$, it must be possible to determine a $\delta > 0$ such that the accuracy $|f(x) - L| < \varepsilon$ is achieved with any $x \neq c$ that satisfies $|x - c| < \delta$.

Meditate over this subtle point (and the discussion that precedes it) long enough to convince yourself that if $f(x)$ is to close in on L when x is pushed to c, then this criterion ought to be satisfied.

Example 8.1. The assertion $\lim\limits_{x \to 4} 3x = 12$ is obvious. But let's put it to the test. Notice that $|3x - 12| = 3|x - 4| < \varepsilon$ is equivalent to $|x - 4| < \frac{\varepsilon}{3}$. The accuracy requirement $|3x - 12| < \varepsilon$ can therefore be responded to by taking $|x - 4| < \delta$, where $\delta = \frac{\varepsilon}{3}$. For instance, to get $3x$ to within $\varepsilon = \frac{1}{1,000,000}$ of 12, we need to take x to within $\delta = \frac{1}{3,000,000}$ of 4.

This was easy. The next example is a little more difficult.

Example 8.2. That $\lim\limits_{x \to 2} x^2 = 4$ is obvious. But can it pass the Limit Test? Let δ be any positive number. Draw the numbers 2 and -2 on a number line. Let x be any number on the line and convince yourself that if the distance $|x - 2|$ between x and 2 is less than δ, then the distance $|x - (-2)| = |x + 2|$ between x and -2 is less than $\delta + 4$. Since $x^2 - 4 = (x - 2)(x + 2)$, it follows that if $|x - 2| < \delta$, then

$$|x^2 - 4| = |x - 2||x + 2| < \delta(\delta + 4).$$

Now respond to a given accuracy requirement $\varepsilon > 0$ by taking δ to be a positive number that is less than both $\frac{\varepsilon}{5}$ and 1. Observe that if x satisfies $|x - 2| < \delta$, then

$$|x^2 - 4| = |x - 2||x + 2| < \delta(\delta + 4) < \frac{\varepsilon}{5}(1 + 4) = \varepsilon.$$

Therefore the Limit Test has been passed!

You are right if you suspect at this point that the application of the limit test can be a very delicate matter, even in situations where—from the intuitive viewpoint—the limit is completely self-evident.

As already pointed out, the standards of mathematical precision became more stringent in the 19th century. The Limit Test played a fundamental role in this development. It solidified the basis of calculus by recasting the pictorial notion of the limit into the precise algebraic concept formulated in the Limit Test. Basic limit theorems were proved, that is to say verified with irrefutable logic, and with this new theory of limits, concepts that were implicit or intuitive with Leibniz and Newton were made explicit, exact, and definitive. The pursuit of all this would take us much too far afield, indeed, it would be inappropriate for a basic text. The aim of this chapter will therefore be the following. We will work exclusively with the earlier intuitive notion of limits. In other words,

$$\lim\limits_{x \to c} f(x) \text{ exists and is equal to } L, \text{ or equivalently,}$$

$$\lim\limits_{x \to c} f(x) = L$$

will continue to mean that if x is pushed to c from either the left or the right, then $f(x)$ closes in on the finite number L. In particular, our discussion of the fundamental notions of continuity and differentiability in Sections 8.2 and 8.3 will use only the intuitive idea of limit. If you keep in mind, however, that limits can be required to pass the Limit Test, then you will understand how the mathematicians of the 19th century turned these fundamental notions into precise and rigorous concepts. We conclude this section with some additional examples and comments about limits.

The most difficult limit problems are those of "$\frac{0}{0}$" type. These are the questions of the form

$$\lim\limits_{x \to c} \frac{f(x)}{g(x)} = ?$$

where both $\lim\limits_{x \to c} f(x) = 0$ and $\lim\limits_{x \to c} g(x) = 0$. The answer to such limit questions invariably requires that $\frac{f(x)}{g(x)}$ be rewritten in such a way that a common term can be cancelled from the numerator and denominator. Refer back to Section 7.2 and convince yourself that Newton's limit question

$$\lim\limits_{Q \to P} \frac{QR}{QT^2} = ?$$

is a geometric version of a limit problem of "$\frac{0}{0}$" type. Its solution was difficult and did involve cancellations (see Section 7.3).

Example 8.3. Consider the problem $\lim_{x \to -3} \frac{x^2-9}{x+3} = ?$ Observe that it is of "$\frac{0}{0}$" type. Since

$$\frac{x^2-9}{x+3} = \frac{(x-3)(x+3)}{(x+3)} = x - 3,$$

$\lim_{x \to -3} \frac{x^2-9}{x+3} = \lim_{x \to -3} (x-3) = -6.$ Notice that the second of the two equalities above requires that $x \neq -3$ (because cancellation by 0 is not allowed). But this is of no relevance here, because the issue is: What happens to $\frac{x^2-9}{x+3}$ as x is pushed to -3 (it closes in on -6), and not what is the value of $\frac{x^2-9}{x+3}$ at -3 (it is undefined)!

Example 8.4. The limit $\lim_{x \to 6} \frac{\sqrt{x-2}-2}{x-6}$ is also of "$\frac{0}{0}$" type. Before it can be solved, it must be rewritten. This involves $\sqrt{x-2}+2$, the so-called *conjugate* of $\sqrt{x-2}-2$. Notice that

$$(\sqrt{x-2}-2)(\sqrt{x-2}+2) = (x-2) - 4 = x - 6.$$

Therefore,

$$\frac{\sqrt{x-2}-2}{x-6} = \frac{\sqrt{x-2}-2}{x-6} \cdot \frac{\sqrt{x+2}+2}{\sqrt{x-2}+2}$$

$$= \frac{(x-2)-4}{(x-6)\left(\sqrt{x-2}+2\right)} = \frac{1}{\sqrt{x-2}+2}.$$

It now follows that $\lim_{x \to 6} \frac{\sqrt{x-2}-2}{x-6} = \lim_{x \to 6} \frac{1}{\sqrt{x-2}+2} = \frac{1}{4}$.

The method of multiplying an expression on the top and bottom by the conjugate and thereby achieving a simplification is known as *rationalizing*.

Example 8.5. Consider the limit $\lim_{x \to 0} \frac{\sin x}{x}$ and notice that it is also of "$\frac{0}{0}$" type. Recall from Section 1.4, that $\sin x$ is approximated by x for small x. Therefore, the expectation is that $\lim_{x \to 0} \frac{\sin x}{x} = 1$. That this is indeed the case will be verified in Section 8.6.

Not all limit questions have answers. For example, when x is pushed to 0, $\frac{1}{x}$ becomes larger and larger,

and does not close in on a finite number. In such a case, one says that the limit *does not exist*. In particular, the limits $\lim_{x \to 0} \frac{1}{x}$ and $\lim_{x \to 0} \frac{1}{\sin x}$ do not exist.

We conclude the discussion about limits with a notational matter. Observe that $\lim_{x \to -4} \sqrt{x+4} = 0$. But observe also that $\sqrt{x+4}$ is defined only for $x + 4 \geq 0$, and hence only for $x \geq -4$. Therefore, x can be pushed to -4 from the right only, in other words, through values that are greater than -4. The notation $\lim_{x \to -4^+} \sqrt{x+4} = 0$ emphasizes that x approaches -4 from the right.

In a similar way, $g(x) = \sqrt{5-x}$ is defined only for $5 - x \geq 0$. So we must have $5 \geq x$, or $x \leq 5$. Therefore, $\lim_{x \to 5} \sqrt{5-x} = 0$ can only make sense if x approaches 5 from the left. The notation $\lim_{x \to 5^-} \sqrt{5-x} = 0$ makes this explicit.

This notation can be used to emphasize the following point. The assertion that $\lim_{x \to c} f(x) = L$ means that x can be pushed to c from either the right or the left, unless the function f is not defined to the right or left of c. If the function is defined on both sides of c, then $\lim_{x \to c} f(x) = L$ means the same thing as: $\lim_{x \to c^-} f(x)$ exists, $\lim_{x \to c^+} f(x)$ exists, and both are equal to L. If the function f is not defined to the left of c, then only an approach from the right is possible, and $\lim_{x \to c} f(x) = L$ means that $\lim_{x \to c^+} f(x) = L$. In the same way, if the function f is not defined to the right of c, then only an approach from the left is possible, and $\lim_{x \to c} f(x) = L$ means that $\lim_{x \to c^-} f(x) = L$.

8.2 **Continuous Functions**

L et $y = f(x)$ be a function. From the geometric point of view, f is *continuous at a number c* on the x-axis if the graph of f has no gaps or breaks at the point over, or as the case may be, under, c. This can be translated into an algebraic context as follows. Refer to Figure 8.2. Note first, that if a gap over c is to be avoided, then c must be in the domain of f. For otherwise, $f(c)$ does not make sense, and there is no point on the graph of f over c. With c in the domain of f, there is the point $(c, f(c))$ on the graph of f. Next, let x be a number to the left of c. Push x into c, and consider the point $(x, f(x))$ on the graph in the process. Observe that if a gap in the graph is to be avoided, then $(x, f(x))$ must close in on $(c, f(c))$. But this is possible only if $f(x)$ closes in on $f(c)$. Put another way, it must be the case that $\lim_{x \to c^-} f(x) = f(c)$. In the same way, if x is pushed into c from the right, then what is needed to avoid a gap in the graph is the condition $\lim_{x \to c^+} f(x) = f(c)$. Combining these observations with the remarks that conclude Section 8.1 provides the

Continuity Criterion: The algebraic conditions for a function $y = f(x)$ to be *continuous at a number (or point) c* on the x-axis are

(i) $f(c)$ makes sense. In other words, c is in the domain of f.

(ii) $\lim_{x \to c} f(x) = f(c)$.

Example 8.6. Consider the function $f(x) = 3x - 4$. Let's test for continuity at the number 2. Note first that $f(5) = 3 \cdot 5 - 4 = 15 - 4 = 11$. So criterion (i) is satisfied. Since

$$\lim_{x \to 5} f(x) = \lim_{x \to 5}(3x - 4) = 3 \cdot 5 - 4 = 11$$

and $11 = f(5)$, criterion (ii) holds. Therefore, $f(x) = 3x - 4$ is continuous at 5. In the same way, for any number c, $f(c) = 3c - 4$ and

$$\lim_{x \to c} f(c) = \lim_{x \to c}(3c - 4) = 3c - 4 = f(c).$$

Therefore, $f(x) = 3x - 4$ is continuous at any number c. Its graph, as we already know, is the straight line with y-intercept -4 and slope 3. See Figure 8.3.

Remark. The same argument shows that any function of the form $f(x) = mx + b$ is continuous at all real numbers c.

Example 8.7. Consider the function $f(x) = x^2$. Let's check for continuity at -3. Since $f(-3) = 9$, criterion (i) is satisfied. Since

$$\lim_{x \to -3} f(x) = \lim_{x \to -3} x^2 = (-3)(-3) = 9 = f(-3),$$

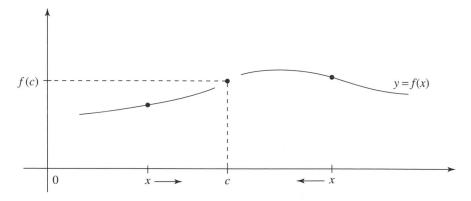

Figure 8.2

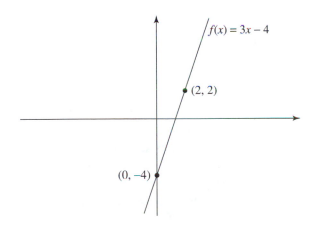

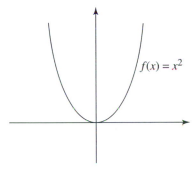

Figure 8.3

Figure 8.4

criterion (ii) holds as well. So $f(x) = x^2$ is continuous at -3. In exactly the same way, $f(c) = c^2$ and

$$\lim_{x \to c} f(x) = \lim_{x \to c} x^2 = c^2 = f(c).$$

Again, both criteria for the continuity of $f(x) = x^2$ at c are satisfied. So this function is continuous at all real numbers c. Its graph is the parabola shown in Figure 8.4.

A similar argument shows that $g(x) = \frac{1}{x}$ is continuous at any $c \neq 0$. Observe, however, that since 0 is not in the domain, g is not continuous at 0. In the same way, the function $h(x) = \frac{1}{x^2}$ is continuous at all $c \neq 0$, but not at $c = 0$. The graphs of these two functions are sketched in Figure 8.5.

Example 8.8. Consider the function $f(x) = \sqrt{x - 2}$. Is f continuous at 5? Since $\sqrt{3}$ makes sense, 5 is in the domain of f. So criterion (i) is satisfied. If x is pushed to 5, it is clear that $\sqrt{x - 2}$ gets pushed to $\sqrt{3}$. This is so whether x is larger than 5 and is pushed to 5 from the right; or whether x is smaller than 5 and is pushed to 5 from the left. So, $\lim_{x \to 5} \sqrt{x - 2}$ exists and is equal

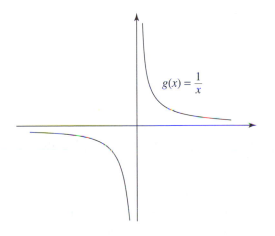

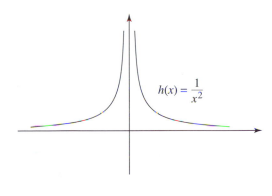

Figure 8.5

to $\sqrt{3}$. Therefore, criterion (ii) is satisfied as well. We can conclude that f is continuous at 5. Now let $c > 2$ be any positive real number. Since $f(c) = \sqrt{c-2}$ makes sense, f is defined at c. As in the case $c = 5$, we see that $\lim_{x \to c} \sqrt{x-2} = \sqrt{c-2}$. Again, all criteria are satisfied. Finally, is f continuous at 2? Since $\sqrt{0} = 0$, 2 is in the domain of f. Now take $x \ne 2$ and push it to 2. Since x must remain in the domain of f in order for this process to make sense, $x > 2$, and x can be pushed to 2 only from the right. Since, $\lim_{x \to 2} \sqrt{x-2} = 0$, the criteria are satisfied, and f is continuous at 2. We have established that $f(x) = \sqrt{x-2}$ is continuous at all points c in its domain. The graph of f is shown in Figure 8.6.

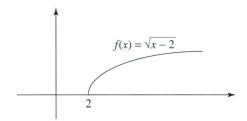

$$f(x) = \sqrt{x-2}$$

2

Figure 8.6

Remark. In the same way, any function of the form $f(x) = x^r$, where r is a rational number, is continuous at any point c in its domain.

Example 8.9. Consider the function $f(x) = \frac{x^2-1}{x+1}$. Let $c \ne -1$. So $c + 1 \ne 0$, and hence c is in the domain of f. Notice that

$$\lim_{x \to c} f(x) = \lim_{x \to c} \frac{x^2 - 1}{x + 1} = \frac{c^2 - 1}{c + 1} = f(c).$$

So both parts of the continuity criterion are satisfied. It follows that f is continuous at c. Since f is not defined at -1, f cannot be continuous at -1. Notice that

$$\frac{x^2 - 1}{x + 1} = \frac{(x - 1)(x + 1)}{x + 1}$$

and hence that $f(x) = \frac{x^2-1}{x+1} = x - 1$ whenever $x \ne -1$. So observe that

$$\lim_{x \to -1} f(x) = \lim_{x \to -1} \frac{(x-1)(x+1)}{x+1} = \lim_{x \to -1} (x-1) = -2.$$

So $c = -1$ is a situation where the limit exists, and yet the function is not continuous. Consider the function $g(x) = x - 1$. By the Remark that follows Example 8.6, g is continuous at all numbers c. Its graph is the line with slope 1 and y-intercept -1. Notice that $f(x) = \frac{x^2-1}{x+1} = x - 1 = g(x)$ for all $x \ne -1$. It follows that the graph of f agrees with that of g, except when $x = -1$. At -1 they differ, because the graph of f has a gap and that of g does not. It follows that except for the gap at -1, the graph of f is the line with slope 1 and y-intercept -1. See Figure 8.7.

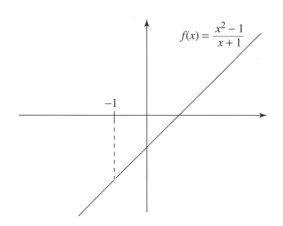

$$f(x) = \frac{x^2-1}{x+1}$$

-1

Figure 8.7

Remark. It can be shown in the same way that any function of the form $k(x) = \frac{f(x)}{g(x)}$, where $f(x)$ and $g(x)$ are polynomials, is continuous at all numbers c for which $g(c) \ne 0$. In particular, such a function is continuous at all numbers c in its domain. Taking $g(x) = 1$, this implies that any polynomial function is continuous at all real numbers.

There are functions that are not continuous at numbers that are in the domain. A concrete but simple example of such a situation follows.

Example 8.10. The prices for some tickets (for example, those for buses, trains, or movie theaters) depend on the age of the user of the ticket. Let $p(x)$ be the price of a certain ticket for someone x years old. Suppose that for $0 \leq x < a_0$, the price is at the level $p(x) = p_0$; for $a_0 \leq x < a_1$, it is the higher amount $p(x) = p_1$; and finally, for $a_1 \leq x$, it is lower again at $p(x) = p_2$. For example, with $a_0 = 16$, the price p_0 might be a youth fare, and with $a_1 = 65$, the price p_2 might be a price discounted for senior citizens. The graph of $p(x)$ is shown in Figure 8.8. Observe that it is not continuous at a_0, a_1, and a_2.

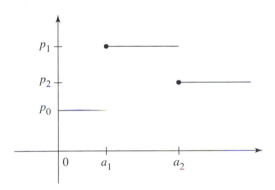

Figure 8.8

Take two numbers $a \leq b$ on a number line and consider the following sets. The set $[a, b]$ consists of

all x on the line with $a \leq x \leq b$. The set (a, b) consists of all x that satisfy $a < x < b$. The set $(a, b]$ consists of all x with $a < x \leq b$, and the set $[a, b)$ is defined in a similar way. Each of these sets is an *interval*. The interval

$[a, b]$ is said to be *closed*, and (a, b) is *open*. For any number a, we will also consider the intervals $(-\infty, a]$ and $[a, \infty)$. The first consists of all x with $x \leq a$, and the second of all x with $x \geq a$. The intervals $(-\infty, a)$ and (a, ∞) can be defined in a similar way. How do they differ from $(-\infty, a]$ and $[a, \infty)$?

The definition of the price function $p(x)$ can now be given as follows:

$$p(x) = \begin{cases} p_0, & \text{for } x \text{ in } [0, a_1), \\ p_1, & \text{for } x \text{ in } [a_1, a_2), \\ p_2, & \text{for } x \text{ in } [a_2, \infty). \end{cases}$$

What part of the continuity criterion is not satisfied by $p(x)$? For what numbers c?

The definition of continuity at a point can be extended to sets of points. A function is *continuous on an interval I* if it is continuous at every point of I. A function is *continuous on its domain* if it is continuous at all points in its domain. It has already been pointed out that the functions of the form $f(x) = x^r$ for r a rational number, as well as those of the form $k(x) = \frac{f(x)}{g(x)}$, where $f(x)$ and $g(x)$ are polynomials, are all continuous on their domains.

Continuity Theorem: Let $y = f(x)$ and $y = g(x)$ be two functions that are continuous on their domains. Then the sum $y = f(x) + g(x)$, difference $y = f(x) - g(x)$, product $y = f(x) \cdot g(x)$, and quotient $y = \frac{f(x)}{g(x)}$ are all continuous on their domains.

For example, we already know that the functions $y = f(x) = 3x^5 - 7x^2$ and $y = g(x) = \sqrt{x} = x^{\frac{1}{2}}$ are both continuous on their domains. So by the continuity theorem, the functions

$$y = 3x^5 - 7x^2 + \sqrt{x},$$

$$y = 3x^5 - 7x^2 - \sqrt{x},$$

$$y = (3x^5 - 7x^2)\sqrt{x}, \quad \text{and}$$

$$y = \frac{3x^5 - 7x^2}{\sqrt{x}}$$

are all continuous on their domains.

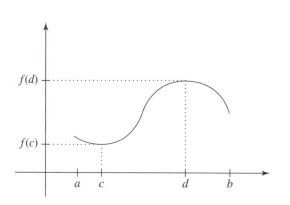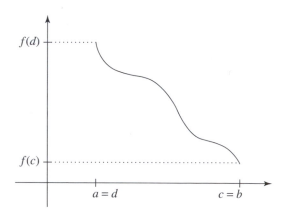

Figure 8.9

The verification, or proof, of the continuity theorem is carried out in the realm of algebra and limits. For instance, let c be any number in the domain of $y = \frac{f(x)}{g(x)}$. So $f(c)$ and $g(c)$ are defined, and $g(c) \neq 0$. If $f(x)$ and $g(x)$ are continuous at c, then $\lim_{x \to c} f(x) = f(c)$ and $\lim_{x \to c} g(x) = g(c)$. It follows that

$$\lim_{x \to c} \frac{f(x)}{g(x)} = \frac{\lim_{x \to c} f(x)}{\lim_{x \to c} g(x)} = \frac{f(c)}{g(c)}.$$

Hence the continuity criterion is satisfied, and $y = \frac{f(x)}{g(x)}$ is continuous at c. The facts used in the proof of the continuity theorem are known as limit theorems: the limit of a sum is equal to the sum of the limits, the limit of a product is the product of the limits, and the limit of a quotient is the quotient of the limits (this is the theorem that was just used). They are proved—as already remarked—by use of the Limit Test.

We close this discussion with some observations about continuous functions. Let $y = f(x)$ be continuous on a closed interval $[a, b]$. Then there are numbers c and d, both between a and b, such that

$$f(d) \leq f(x) \leq f(c), \text{ for all } x \text{ in } [a, b].$$

The values $f(c)$ and $f(d)$ are appropriately called the *maximum* and *minimum* values of the function $y = f(x)$ on $[a, b]$. The point is that a function that is continuous on a closed interval always has a maximum

as well as a minimum value. Figure 8.9 illustrates this in two cases. In the situation on the left, both c and d occur in the interior of $[a, b]$. In the situation on the right they occur at the end points.

Intermediate Value Theorem: Let $y = f(x)$ be continuous on a closed interval $[a, b]$ and let M and m be the maximum and minimum values of f on $[a, b]$. Then for any number v with $m \leq v \leq M$, there is a number u with $a \leq u \leq b$ such that $f(u) = v$.

This theorem can be understood as follows. Let $M = f(c)$ and $m = f(d)$ be the maximum and minimum values of $y = f(x)$ on $[a, b]$. For any number v between m and M, consider the horizontal segment through v over $[a, b]$. Because the graph of $y = f(x)$ is in one continuous, unbroken piece and the points (c, M) and (d, m) are both on the graph, this horizontal segment must intersect the graph at some point. Refer to Figure 8.10. Let (u, v) be this point. Since (u, v) is on the graph of f, it follows that $f(u) = v$. This is what had to be verified. This theorem, while intuitively clear, is proved by use of limits and the Limit Test.

The considerations that led to the Intermediate Value Theorem also give information about the zeros of a continuous function f. For example, let f be a function, and let $a < b$ be numbers such that f is continuous on $[a, b]$. If $f(a) > 0$ and $f(b) < 0$, then there

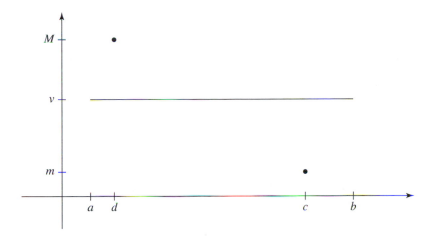

Figure 8.10

must be some number c, with $a < c < b$, such that $f(c) = 0$. Sketch a graph that illustrates that this is so.

Example 8.11. Consider the polynomial

$$f(x) = x^5 + 10x^4 - 12x^3 + 14x^2 + 16x - 20.$$

Check that $f(1) = 9$ and $f(0) = -20$. Because polynomials are continuous everywhere, it follows that $f(c) = 0$ for some c with $0 < c < 1$.

The Intermediate Value Theorem—invisible below the surface—is an important component in Newton's verification of the Fundamental Theorem of Calculus. Refer to the discussion at the beginning of Section 6.2, especially to Figure 6.6. Take any x on the horizontal axis and a positive Δx, and make the necessary assumption that f is continuous on the closed interval $[x, x + \Delta x]$. Now let M and m be the maximum and minimum values of f on $[x, x + \Delta x]$ respectively. Since the area of the region under the graph of f between x and $x + \Delta x$ is $A(x + \Delta x) - A(x)$, it follows that $m\Delta x \leq A(x + \Delta x) - A(x) \leq M\Delta x$. Therefore,

$$m \leq \frac{A(x + \Delta x) - A(x)}{\Delta x} \leq M.$$

Let $v = \frac{A(x+\Delta x)-A(x)}{\Delta x}$. By the Intermediate Value Theorem, there is a number u with $x \leq u \leq x + \Delta x$ such that $f(u) = v = \frac{A(x+\Delta x)-A(x)}{\Delta x}$. So $A(x + \Delta x) -$

$A(x) = f(u)\Delta x$. By using the continuity of f (in the last step), it follows that

$$A'(x) = \lim_{\Delta x \to 0} \frac{A(x + \Delta x) - A(x)}{\Delta x} = \lim_{\Delta x \to 0} \frac{f(u)\Delta x}{\Delta x}$$
$$= \lim_{\Delta x \to 0} f(u) = f(x).$$

This is an illustration of the achievement of the mathematics of the 19th century: The pictures and intuitive elements of Newton's argument have given way to theorems and proofs.

8.3 Differentiability

The definition of the derivative of a function f given in Section 5.5A was in essence this. A nonvertical tangent line was assumed to exist at a point $(x, f(x))$ on the graph of f; the assertion was that the slope of this tangent line could be computed; and the derivative $f'(x)$ was defined to be the slope of this tangent. See Figure 8.11. Intuitively, the tangent line to a curve at a point P is a line that "hugs" the curve closely near P. Intuitively, it seems clear when a tangent exists, and which line it is if it does. But is it possible to formulate the matter in precise mathematical terms? Given the central importance of the derivative, this is an important order of business.

Let f be a function and fix a point $(c, f(c))$ on its graph. It is our concern to formulate a definitive condition that insures that the graph has a nonvertical tangent at a point $(c, f(c))$. To start, take $x < c$. Push x to c (from the left) and assume that

$$\lim_{x \to c^-} \frac{f(x) - f(c)}{x - c} \text{ exists and is equal to, say, } m_-.$$

What does this mean? Fix the line T_- through $(c, f(c))$ with slope m_-. See Figure 8.12. Take any point $(x, f(x))$ on the graph to the left of $(c, f(c))$ and consider the line S through both $(c, f(c))$ and $(x, f(x))$. Note that the slope of S is $\frac{f(x)-f(c)}{x-c}$. Now push x to c and observe the segment S in the process. A look at Figure 8.12 shows that $\lim_{x \to c^-} \frac{f(x)-f(c)}{x-c} = m_-$ means that the slope of S closes in on the slope m_- of T_-, so that S rotates into the line T_-. Now take $x > c$. Push x to c (this time from the right) and assume that

$$\lim_{x \to c^+} \frac{f(x) - f(c)}{x - c} \text{ exists and is equal to, say, } m_+.$$

This situation is analogous to that just considered. Let T_+ be the line through $(c, f(c))$ with slope m_+. See Figure 8.13. The line S connecting the two points $(x, f(x))$ and $(c, f(c))$ has slope $\frac{f(x)-f(c)}{x-c}$. Since $\lim_{x \to c^+} \frac{f(x)-f(c)}{x-c} = m_+$, it follows that S rotates into T_+ as x is pushed to c from the right. Suppose that $m_- = m_+$. This means that the lines T_- and T_+ have the same slope. Since they both go through $(c, f(c))$, they are the same line. Label this line T. Gluing Figures 8.12 and 8.13 together provides Figure 8.14, and, as consequence, the nonvertical tangent line T to the graph at $(c, f(c))$. In summary, therefore, the definitive condition for the existence of a nonvertical tangent line to the graph of $y = f(x)$ at the point $(c, f(c))$ is this:

The limits $\lim_{x \to c^-} \dfrac{f(x) - f(c)}{x - c}$ and $\lim_{x \to c^+} \dfrac{f(x) - f(c)}{x - c}$ both exist, and they are equal.

The common value of these limits is the slope $m_- = m_+$ of the tangent. Observe that the existence of the two limits is equivalent to the existence of the single

limit $\lim_{x \to c} \frac{f(x)-f(c)}{x-c}$. Refer to Figure 8.12 and 8.13 and observe that the point $(x, f(x))$ "glides" into the point $(c, f(c))$ and hence that the graph is smooth at $(c, f(c))$. So the geometric and intuitive condition:

> The graph of f is smooth at $(c, f(c))$ with a nonvertical tangent

has now been recast into the algebraic condition

$$\lim_{x \to c} \frac{f(x) - f(c)}{x - c} \text{ exists.}$$

Set $x = c + \Delta x$. In reference to Figures 8.12 and 8.13, notice that if $\Delta x < 0$, then $x = c + \Delta x$ is to the left of c, and if $\Delta x > 0$, then x is to the right of c. Pushing x to c is equivalent to pushing Δx to 0. Since $f(x) - f(c) = f(c + \Delta x) - f(c)$, and $\Delta x = x - c$, the preceding observations translate into the

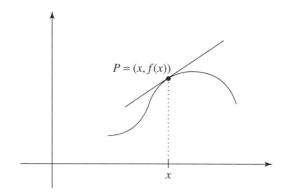

$$P = (x, f(x))$$

Figure 8.11

Differentiability Criterion: The condition for the graph of $y = f(x)$ to be smooth at $(c, f(c))$ with a nonvertical tangent is that

$$\lim_{\Delta x \to 0} \frac{f(c + \Delta x) - f(c)}{\Delta x} \text{ exists.}$$

In this case, f is said to be *differentiable at c*. The limit, denoted by $f'(c)$, is equal to the slope of the tangent.

Replacing c by x in the equation above provides the definition of the *derivative* of $f(x)$. This is the

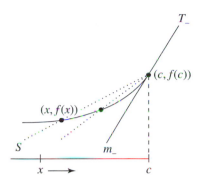

Figure 8.12

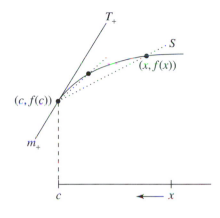

Figure 8.13

function f' defined by the rule

$$f'(x) = \lim_{\Delta x \to 0} \frac{f(x + \Delta x) - f(x)}{\Delta x}.$$

The domain of f' is the set of all x for which this rule makes sense, in other words, for which this limit exists. It is the set of all x such that the graph of f is smooth at $(x, f(x))$ and has a nonvertical tangent at that point. The value $f'(x)$ is the slope of the nonvertical tangent.

As in the case of continuity, the definition of differentiability can be extended from points to sets of real numbers. In particular, f is *differentiable on an* *interval I* if it is differentiable at all numbers in the interval I. And f is *differentiable on its domain* if it is differentiable at all points in its domain. Note that if f is differentiable on its domain, then the domain of $y = f'(x)$ coincides with the domain of $y = f(x)$.

The derivative $f'(x)$ of a function $y = f(x)$ is also commonly designated by

$$\frac{dy}{dx}, \quad \frac{d}{dx}(f(x)), \quad \frac{df(x)}{dx}, \quad \text{or} \quad \frac{df}{dx}.$$

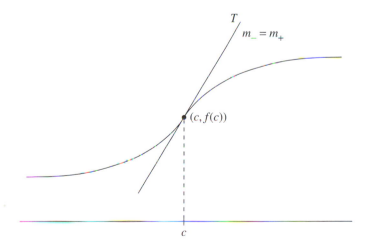

Figure 8.14

This notation originated with Leibniz. All these symbols have the same meaning. They all stand for the derivative of the function $y = f(x)$.

The point of this section can be summarized as follows: The notion of a tangent line and the question as to when it exists had been matters that were dealt with by "drawing a picture." Now the picture has given way to a definitive concept that is formulated with algebraic precision in terms of limits.

8.4 Derivatives as Rates of Change

$\boxed{\text{L}}$et f be a function. It has been established that the derivative f' of f always has the following geometric meaning: f' is the function whose value at a point c is the slope of the nonvertical tangent line to the graph of $y = f(x)$ at the point $(c, f(c))$. If no such line exists, $f'(c)$ is not defined.

Historically, the development of calculus was driven by the requirements of science and engineering, and in particular by efforts to provide a more lucid understanding of these fields. If the function f represents a specific physical qauntity, one would therefore expect for the derivative f' to have (in addition to its geometric meaning) a meaning derived from the context in question. For example, the following was already observed in Section 6.4. For any time $t \geq 0$, let $p(t)$ be the position of a point or particle moving on a number line. Then $p'(t)$ is the velocity $v(t)$ of the point at time t, and $v'(t)$ is the acceleration at time t.

We will now consider other examples. When time $t \geq 0$ is under consideration, think of it as being elapsed time as given by a stopwatch that is started at time $t = 0$.

A. Growth of Organisms. Consider a culture of microorganisms growing in a test-tube and suppose that the number of microorganisms increases as time increases. Let $y(t)$ be this number at any time $t \geq 0$. So $y(0)$ is the number of microorganisms when they are first observed. Fix a time t. Now let an additional time Δt elapse. Then $y(t + \Delta t)$ is the number of organisms at time $t + \Delta t$, and $y(t + \Delta t) - y(t)$ is the change in this number during the time interval $[t, t + \Delta t]$. Observe that the change $y(t + \Delta t) - y(t)$ is positive and that it takes place during a time Δt. So

$$\frac{y(t + \Delta t) - y(t)}{\Delta t}$$

is the rate at which this change unfolds. It is the *average rate of change of the number of organisms* per unit time during the time interval $[t, t + \Delta t]$. Since $y(t + \Delta t) - y(t)$ is positive, the change is an increase. Hence, $\frac{y(t+\Delta t)-y(t)}{\Delta t}$ is the *average rate of increase* during the time interval $[t, t + \Delta t]$. Note that it is a positive number. By taking this average over smaller and smaller Δt, in other words, by pushing Δt to 0, we get what is called the *rate of change of the number of organisms at the instant t*. Observe that this is nothing but the derivative

$$y'(t) = \lim_{\Delta t \to 0} \frac{y(t + \Delta t) - y(t)}{\Delta t}$$

of $y(t)$. Since $y(t)$ increases with time, $y'(t)$ is the *rate of increase at the instant t*. It is positive for any t. The growth of cultures of microorganisms will be studied in Section 11.6.

B. Radioactive Decay. Consider a sample of a radioactive substance. Let $y(t)$ be the amount (either the number of atoms or some unit of mass) of the decaying substance at any time $t \geq 0$. The initial amount is $y(0)$. Fix a time t and then let an additional

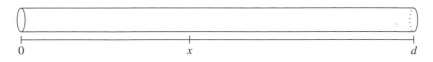

Figure 8.15

Δt elapse. Since $y(t)$ is the amount of the substance at time t and $y(t + \Delta t)$ is the amount at time $t + \Delta t$, it follows that

$$y(t + \Delta t) - y(t)$$

is the change in the amount of the substance during the time interval $[t, t + \Delta t]$. The ratio

$$\frac{y(t + \Delta t) - y(t)}{\Delta t}$$

is the *average rate of change of the amount* per unit time over the time interval $[t, t + \Delta t]$. Since the quantity $y(t)$ is decreasing, the change is a decrease. So $y(t + \Delta t) - y(t)$ is negative, and $\frac{y(t+\Delta t)-y(t)}{\Delta t}$ is the *average rate of decrease* during $[t, t + \Delta t]$. It is a negative quantity. Pushing Δt to 0 as in the preceeding situation, we get the *rate of change of the amount at the instant t*. As before, this is the derivative

$$y'(t) = \lim_{\Delta t \to 0} \frac{y(t + \Delta t) - y(t)}{\Delta t}$$

of $y(t)$. Since $y(t)$ decreases with time, $y'(t)$ is the *rate of decrease at the instant t*. It is negative for any t. Radioactive decay will be studied in detail in Section 11.1.

C. Temperature. Consider a thin metal rod d units long. Suppose that heat is applied to one end, which is kept at a constant temperature. Place an x-axis as shown in Figure 8.15 and assume that the heat is applied at the origin. The rod will be hottest at the left end, and its temperature drops off as we move along the rod to the right. Let $T(x)$ be the temperature of the rod at any point x. Fix x and consider the temperature $T(x)$ at x. Go a distance Δx from x to the right and consider the temperature $T(x + \Delta x)$ at $x + \Delta x$. The difference $T(x + \Delta x) - T(x)$ is the change in temperature from x to $x + \Delta x$, and $\frac{T(x+\Delta x)-T(x)}{\Delta x}$ is the average rate of the change in temperature per unit distance over the interval $[x, x + \Delta x]$. Observe that both the change and the rate of change are negative numbers. Pushing Δx to zero pushes this ratio to the *rate of change of the temperature at x*. This is the

derivative

$$T'(x) = \lim_{\Delta t \to 0} \frac{T(x + \Delta x) - T(x)}{\Delta x}$$

of the temperature function $T(x)$. It is negative for any x between 0 and d.

D. Electric Current. Suppose that an electric current is flowing through a wire. This means that certain particles called electrons are passing through it. Each electron has a charge of -1. Let S be a cross section of the wire and suppose that the current starts to flow through S at time $t = 0$ (Figure 8.16). For any time t later, let

$$Q(t) = \text{total charge that has passed through } S$$
$$\text{from time 0 to } t.$$

Fix some time t and let an additional Δt elapse. Then $Q(t + \Delta t)$ is the total charge that has passed through S from time 0 to $t + \Delta t$, and $Q(t + \Delta t) - Q(t)$ is the charge that has passed through S during the time interval $[t, t + \Delta t]$. The ratio

$$\frac{Q(t + \Delta t) - Q(t)}{\Delta t}$$

represents the average rate of change of charge over time during $[t, t + \Delta t]$. Taking this average over smaller and smaller Δt, in other words, pushing Δt to 0, pushes the ratio $\frac{Q(t+\Delta t)-Q(t)}{\Delta t}$ to the *rate of change of charge at the instant t*. This is called the *current* at time t. Observe that it is the derivative

$$Q'(t) = \lim_{\Delta t \to 0} \frac{Q(t + \Delta t) - Q(t)}{\Delta t}$$

of the charge function $Q(t)$.

E. Cost. Consider a manufacturing company and suppose that the total cost of producing x units per year of a certain commodity is $C(x)$. Suppose that the company raises the production level of this commodity by Δx units from x to $x + \Delta x$ units per year. The cost of producing the additional Δx units per year is $C(x + \Delta x) - C(x)$, and the average cost of these additional Δx units is

$$\frac{C(x + \Delta x) - C(x)}{\Delta x}.$$

Figure 8.16

Consider this ratio when Δx is pushed to 0. This is the derivative

$$C'(x) = \lim_{\Delta x \to 0} \frac{C(x + \Delta x) - C(x)}{\Delta x}$$

of $C(x)$, called the *marginal cost* at x. With Δx a small number of units,

$$C'(x) \approx \frac{C(x + \Delta x) - C(x)}{\Delta x}.$$

Therefore, the marginal cost $C'(x)$ should be thought of as the average cost per unit of a small increase in production from the base level of x units. Cost and marginal cost will be studied in detail in Section 12.4.

Examples A through E illustrate the basic fact that for any function, the derivative measures the rate at which the value of the function changes relative to change in the variable. Observe also that when a function occurs in a specific applied context, then the derivative has a meaning that is relevant in that context. The Exercises at the end of this chapter provide additional examples of this.

8.5 About Derivatives

This section will consider various strategies with which derivatives can be computed. It will also discuss several theoretical concerns involving derivatives.

A. Computing Derivatives.   As we have seen, the evaluation of the derivative of a function necessarily involves the computation of a limit. Such limit computations can give rise to general rules that apply to many specific derivatives. One such rule tells us how to differentiate powers of x. (It was already discussed

in Sections 5.5 and 6.1.) The next several examples illustrate this rule again and point out connections with the discussion of Section 8.3.

Example 8.12. The derivative of $f(x) = x^2$ is $f'(x) = 2x$.

Example 8.13. The derivative of $f(x) = x^3$ is $f'(x) = 3x^2$.

In both of these examples the derivative is defined for all x. It follows that in each case the graph of the function f is smooth, with a nonvertical tangent at every point. (Draw a graph of these two functions and convince yourself of this.)

Example 8.14. The derivative of $g(x) = \sqrt[3]{x} = x^{\frac{1}{3}}$ is $g'(x) = \frac{1}{3}x^{-\frac{2}{3}} = \frac{1}{3}\frac{1}{x^{\frac{2}{3}}}$. Note that $g'(x)$ is not defined for $x = 0$. This means that the graph of $g(x) = x^{\frac{1}{3}}$ at $(0,0)$ cannot be smooth with a nonvertical tangent. A look at Figure 8.17 confirms this. The graph is smooth at the origin and it has a tangent there, but this tangent is vertical.

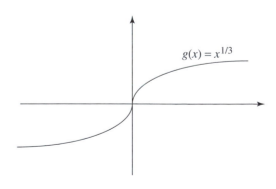

$g(x) = x^{1/3}$

Figure 8.17

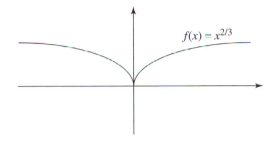

$$f(x) = x^{2/3}$$

Figure 8.18

Example 8.15. The derivative of $f(x) = x^{\frac{2}{3}}$ is $f'(x) = \frac{2}{3}x^{-\frac{1}{3}} = \frac{2}{3}\frac{1}{x^{\frac{1}{3}}}$. Observe that $f'(x)$ is not defined for $x = 0$. Here too, the graph of $f(x) = x^{\frac{2}{3}}$ at $(0,0)$ cannot be smooth with a nonvertical tangent. A look at the graph corroborates this. Since $x^{\frac{2}{3}} = (x^{\frac{1}{3}})^2$, it follows from Figure 8.17 that the graph of $f(x) = x^{\frac{2}{3}}$ has the form shown in Figure 8.18. Note that it comes to a sharp point at the origin and is therefore not smooth there.

There are rules that tell us how to compute the derivatives of the sum, difference, product, and quotient of two functions $y = f(x)$ and $y = g(x)$ in terms of the derivatives of these functions.

Sum Rule: $\big(f(x) + g(x)\big)' = f'(x) + g'(x)$.

Difference Rule: $\big(f(x) - g(x)\big)' = f'(x) - g'(x)$.

Constant Rule: $\big(cf(x)\big)' = cf'(x)$, where c is any constant.

Product Rule: $\big(f(x) \cdot g(x)\big)' = f'(x)g(x) + f(x)g'(x)$.

Quotient Rule: $\left(\dfrac{f(x)}{g(x)}\right)' = \dfrac{f'(x)g(x) - f(x)g'(x)}{(g(x))^2}$.

The proofs of these rules—as was already suggested—involve limit computations. We will carry out the essential details for the product rule in Section 8.5B.

The following fact is a direct consequence of the rules: Let $y = f(x)$ and $y = g(x)$ be two functions that are both differentiable on their domains. Then the sum $y = f(x) + g(x)$, the difference $y = f(x) - g(x)$, the product $y = f(x) \cdot g(x)$, and the quotient $y =$

$\frac{f(x)}{g(x)}$ are all functions that are differentiable on their domains. For instance, let c be in the domain of $\frac{f(x)}{g(x)}$. This means that $f(c)$ makes sense, $g(c)$ makes sense, and $g(c) \neq 0$. Since f and g are both differentiable on their domains, both $f'(c)$ and $g'(c)$ make sense. It now follows that $\frac{f'(c)g(c) - f(c)g'(c)}{(g(c))^2} = \left(\frac{f(c)}{g(c)}\right)'$ makes sense. So $\frac{f(x)}{g(x)}$ is differentiable at c. We have verified that $\frac{f(x)}{g(x)}$ is differentiable at every point in its domain.

Power Rule: For any rational number s, the derivative of $f(x) = x^s$ is $f'(x) = sx^{s-1}$.

In the case $s \geq 0$, the power rule was already established in Section 6.1. Suppose now that s is negative. Observe that $f(x) = x^s = \frac{1}{x^{-s}}$ where $-s$ is positive. By using the quotient rule and the case of the power rule that we already have, we get

$$f'(x) = \frac{\frac{d}{dx}(1) \cdot x^{-s} - 1 \cdot \frac{d}{dx}(x^{-s})}{x^{-2s}}$$

$$= \frac{0 - (-s)x^{-s-1}}{x^{-2s}} = sx^{-s-1+2s} = sx^{s-1}.$$

Therefore, the power rule is also valid for a negative s.

Example 8.16. The derivatives of $g(x) = \frac{1}{x} = x^{-1}$ and $h(x) = \frac{1}{x^2} = x^{-2}$ are, respectively, $g'(x) = -x^{-2} = -\frac{1}{x^2}$ and $h'(x) = -2x^{-3} = -\frac{2}{x^3}$. Neither derivative is defined for $x = 0$. Figure 8.19 shows what is going on: There are no tangent lines at the origin.

Example 8.17. Let $f(x) = 3x^5 - 7x^2 + 2$. Using the constant, power, and difference rules, we get $f'(x) = 15x^4 - 14x$.

Example 8.18. Use of the product, constant, power, difference, and sum rules shows that the derivative of $(3x^{100} - 4x^{21})(50x^{10} + 9x^{\frac{1}{2}})$ is

$$\left(\frac{d}{dx}(3x^{100} - 4x^{21})\right) \cdot (50x^{10} + 9x^{\frac{1}{2}})$$

$$+ (3x^{100} - 4x^{21}) \cdot \left(\frac{d}{dx}(50x^{10} + 9x^{\frac{1}{2}})\right)$$

$$= (300x^{99} - 84x^{20})(50x^{10} + 9x^{\frac{1}{2}})$$

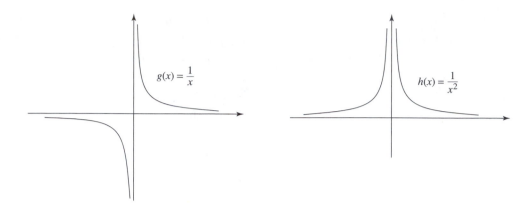

Figure 8.19

$$+ (3x^{100} - 4x^{21})\left(500x^9 + \frac{9}{2}x^{-\frac{1}{2}}\right).$$

Example 8.19. Use of the quotient rule shows that the derivative of $\frac{3x^5 - 7x^2 + 2}{6x^3 + 9x}$ is

$$\frac{(\frac{d}{dx}(3x^5 - 7x^2 + 2)) \cdot (6x^3 + 9x) - (3x^5 - 7x^2 + 2) \cdot \frac{d}{dx}(6x^3 + 9x)}{(6x^3 + 9x)^2}$$

$$= \frac{(15x^4 - 14x)(6x^3 + 9x) - (3x^5 - 7x^2 + 2)(18x^2 + 9)}{(6x^3 + 9x)^2}.$$

The most important differentiation rule is the chain rule. As we will see, its usefulness relies on the fact that many complicated functions are "built up" from two or more simpler functions. The chain rule tells us how to differentiate a complicated function by combining the derivatives of the simple functions from which it is built up.

One way of building up new functions from given functions is the process of composition. The *composite of functions* $f(x)$ and $g(x)$ is obtained by splicing the rule that defines g into the one that defines f, in other words, by forming $y = f(g(x))$. For instance, the composite of $f(x) = 3x^5 - 7x^2 + 2$ and $g(x) = \sqrt{x} = x^{\frac{1}{2}}$ is the function

$$y = f(g(x)) = 3(g(x))^5 - 7(g(x))^2 + 2$$

$$= 3\left(x^{\frac{1}{2}}\right)^5 - 7\left(x^{\frac{1}{2}}\right)^2 + 2 = 3x^{\frac{5}{2}} - 7x + 2.$$

The composite $y = g(f(x))$ of $g(x)$ and $f(x)$ is the entirely different function

$$y = g(f(x)) = g(3x^5 - 7x^2 + 2) = \sqrt{3x^5 - 7x^2 + 2}.$$

It matters, therefore, in which order the composite is taken. It turns out that if both $f(x)$ and $g(x)$ are continuous on their domains, then the composite $y = f(g(x))$ is also continuous on its domain.

The chain rule tells us how the derivative of $y = f(g(x))$ is obtained from the derivatives of the two component functions.

Chain Rule: The derivative of the composite function $y = f(g(x))$ is

$$\left(f(g(x))\right)' = f'(g(x)) \cdot g'(x).$$

So the derivative of $f(g(x))$ is the composite of $f'(x)$ and $g(x)$ times the derivative $g'(x)$.

Example 8.20. Consider the functions

$$f(x) = 3x^5 - 7x^2 + 2$$

and

$$g(x) = \sqrt{x} = x^{\frac{1}{2}}.$$

Let's begin with some evidence for the validity of the chain rule. We already know that

$$f(g(x)) = 3x^{\frac{5}{2}} - 7x + 2.$$

So by the power rule, $f(g(x))' = \frac{15}{2}x^{\frac{3}{2}} - 7$. On the other hand, since $f'(x) = 15x^4 - 14x$ and $g'(x) = \frac{1}{2}x^{-\frac{1}{2}}$, we get by an application of the chain rule that

$$\big(f(g(x))\big)' = f'(g(x)) \cdot g'(x)$$

$$= \big(15(g(x))^4 - 14g(x)\big)\big(\tfrac{1}{2}x^{-\frac{1}{2}}\big)$$

$$= \big(15(x^{\frac{1}{2}})^4 - 14x^{\frac{1}{2}}\big)\big(\tfrac{1}{2}x^{-\frac{1}{2}}\big)$$

$$= \big(15x^2 - 14x^{\frac{1}{2}}\big)\big(\tfrac{1}{2}x^{-\frac{1}{2}}\big) = \tfrac{15}{2}x^{\frac{3}{2}} - 7.$$

Observe that the results of the two computations are the same. Consider next the composite

$$y = g(f(x)) = \sqrt{3x^5 - 7x^2 + 2} = (3x^5 - 7x^2 + 2)^{\frac{1}{2}}.$$

In this case, the chain rule is the only method available. Using it, we get

$$\big(g(f(x))\big)' = g'(f(x)) \cdot f'(x) = \tfrac{1}{2}f(x)^{-\frac{1}{2}} \cdot (15x^4 - 14x)$$

$$= \tfrac{1}{2}(3x^5 - 7x^2 + 2)^{-\frac{1}{2}} \cdot (15x^4 - 14x).$$

A look back at the statement of the chain rule shows that the last function that is applied—that is, the "outside" function f of the formula—is differentiated first, and the function that is applied first—this is the "inside" function g of the formula—is differentiated last. The examples that follow illustrate this "flow" more explicitly.

Example 8.21. Consider $f(x) = (x^2 + 1)^{100}$. Here, the "outside" is $(\quad)^{100}$ and the inside is $x^2 + 1$. Differentiate the outside to get $100(\quad)^{99}$. Now reinsert the inside term and then differentiate it to get

$$f'(x) = 100(\quad)^{99} \cdot \frac{d}{dx}(\quad) = 100(x^2 + 1)^{99} \cdot 2x.$$

Example 8.22. Let's do this for $y = (3x^5 - 7x^2 + 2)^{\frac{1}{2}}$. The derivative of the "outside" is $\frac{1}{2}(\quad)^{-\frac{1}{2}}$. Since the derivative of the "inside" is $15x^4 - 14x$, we get

$$\frac{dy}{dx} = \tfrac{1}{2}(\quad)^{-\frac{1}{2}} \cdot \frac{d}{dx}(\quad)$$

$$= \tfrac{1}{2}(3x^5 - 7x^2 + 2)^{-\frac{1}{2}} \cdot (15x^4 - 14x).$$

As expected, this agrees with the answer already obtained in Example 8.20.

Generalized Power Rule: A combination of the chain rule and the power rule shows directly that if $g(x)$ is a function and s a rational number, then

$$\frac{d}{dx}((g(x)^s)' = s(g(x))^{s-1} \cdot g'(x)$$

The chain rule can be extended to situations that involve more than two functions, simply by applying it more than once. This is done by the "from the outside to the inside" strategy already explained.

Example 8.23. Let $y = \big((x^2 + 6)^{100} + 2x^3\big)^{\frac{3}{2}}$. Differentiating the outside and knowing that the derivative of the inside follows next, we get

$$\frac{dy}{dx} = \tfrac{3}{2}\big((x^2 + 6)^{100} + 2x^3\big)^{\frac{1}{2}} \frac{d}{dx}\big((x^2 + 6)^{100} + 2x^3\big).$$

In the same way,

$$\frac{d}{dx}(x^2 + 6)^{100} = 100(\quad)^{99} \frac{d}{dx}(\quad)$$

$$= 100(x^2 + 6)^{99} \frac{d}{dx}(x^2 + 6)$$

$$= 100(x^2 + 6)^{99}2x,$$

and $\frac{d}{dx}2x^3 = 6x^2$. Putting it all together,

$$\frac{dy}{dx} = \tfrac{3}{2}\big((x^2 + 6)^{100} + 2x^3\big)^{\frac{1}{2}}\big(100(x^2 + 6)^{99}2x + 6x^2\big).$$

The following should now be apparent. With the chain rule, derivatives of complicated functions can be computed efficiently by considering them to be composites of simpler functions that are more easily dealt with. When combined with the other rules of differentiation, the chain rule is a very effective tool for computing derivatives.

Example 8.24. Let $f(x) = (\sqrt{x+1})(x^7 - 1) = (x + 1)^{\frac{1}{2}}(x^7 - 1)$. Then

$$f'(x) = \Big(\frac{d}{dx}(x + 1)^{\frac{1}{2}}\Big)(x^7 - 1) + (x + 1)^{\frac{1}{2}} \cdot \frac{d}{dx}(x^7 - 1)$$

(product rule)

$$= \left(\tfrac{1}{2}(x+1)^{-\frac{1}{2}}(1)\right)(x^7 - 1) + (x+1)^{\frac{1}{2}} \cdot 7x^6$$

(chain rule and power rule)

$$= \tfrac{1}{2}(x+1)^{-\frac{1}{2}}(x^7 - 1) + 7x^6(x+1)^{\frac{1}{2}}.$$

Example 8.25. Let $g(x) = \frac{x^3 + 7x^2 + 2}{(x^4 - 5x^2)^{\frac{2}{3}}}$. Use the quotient rule, and then the sum rule and the chain rule, to get $g'(x) =$

$$\frac{(\frac{d}{dx}(x^3 + 7x^2 + 2)) \cdot (x^4 - 5x^2)^{\frac{2}{3}} - (x^3 + 7x^2 + 2) \cdot \frac{d}{dx}(x^4 - 5x^2)^{\frac{2}{3}}}{((x^4 - 5x^2)^{\frac{2}{3}})^2}.$$

Carrying out the remaining two differentiations, we see that this equals

$$\frac{(3x^2 + 14x) \cdot (x^4 - 5x^2)^{\frac{2}{3}}}{(x^4 - 5x^2)^{\frac{4}{3}}}$$

$$- \frac{(x^3 + 7x^2 + 1) \cdot \frac{2}{3}(x^4 - 5x^2)^{-\frac{1}{3}}(4x^3 - 10x)}{(x^4 - 5x^2)^{\frac{4}{3}}}$$

$$= \frac{(3x^2 + 14x)}{(x^4 - 5x^2)^{\frac{2}{3}}} - \frac{\frac{2}{3}(x^3 + 7x^2 + 2)(4x^3 - 10x)}{(x^4 - 5x^2)^{\frac{5}{3}}}.$$

The meaning of the composite of two functions and the chain rule can be illustrated in intuitive terms. Suppose that a balloon is being inflated. Assume that it retains the shape of a perfect sphere during the process. If r is its radius, then its volume is $V(r) = \frac{4}{3}\pi r^3$. Note that the radius is a function $r = f(t)$ of time t. See Figure 8.20. The composite $V(f(t)) = \frac{4}{3}\pi(f(t))^3$ expresses the volume of the balloon as function of t. The derivative $\frac{dr}{dt} = f'(t)$ is the rate at which the radius is increasing at time t. The derivative $\frac{dV}{dr} = 4\pi r^2$ is the rate at which V is increasing relative to the radius r, and $\frac{dV}{dt} = V(f(t))'$ is the rate at which V is changing relative to t. An application of the chain rule shows that

$$\frac{dV}{dt} = \frac{4}{3}\pi(3(f(t))^2) \cdot f'(t) = (4\pi r^2) \cdot f'(t) = \frac{dV}{dr} \cdot \frac{dr}{dt}.$$

Therefore, $\frac{dV}{dt} = \frac{dV}{dr} \cdot \frac{dr}{dt}$. In this context, the chain rule is nothing but the assertion that the rate at which V changes relative to t is the product of the rate at

which V changes relative to r times the rate at which r changes relative to t. Note the suggestive nature of the notation of Leibniz. Indeed, considering these quantities as differentials (see Section 5.4), the chain rule is the assertion that the differential dr can be canceled.

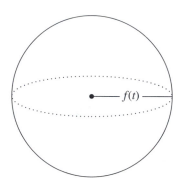

Figure 8.20

B. Some Theoretical Concerns. Suppose that the graph of a function is smooth at a point and that it has a nonvertical tangent line there. Our intuition suggests that the graph should not have a break at such a point. So the expectation is that when a function is differentiable at a number, it should also be continuous at that number. That this is true can be verified as follows. Assume that f is a function that is differentiable at c. The chain of equalities

$$\lim_{x \to c}(f(x) - f(c)) = \lim_{x \to c}\left((x - c) \cdot \frac{f(x) - f(c)}{x - c}\right)$$

$$= 0 \cdot f'(c) = 0$$

forces $\lim_{x \to c} f(x) = f(c)$. Therefore, by the continuity criterion of Section 8.2, f is continuous at c.

To repeat, if a function is differentiable at c, then it must be continuous at c. It follows that if a function f is differentiable on an interval I, then it is continuous on the interval I; and if it is differentiable on its domain, then it is continuous on its domain.

We turn next to the proof of the product rule. Let f and g be two functions. Starting with the definition of the derivative, we get

$$\big(f(x)g(x)\big)' = \lim_{\Delta x \to 0} \frac{f(x+\Delta x)g(x+\Delta x) - f(x)g(x)}{\Delta x}.$$

Adding $-f(x)g(x+\Delta x) + f(x)g(x+\Delta x) = 0$ to the numerator, and then rearranging terms, gives us

$$\lim_{\Delta x \to 0} \left(\frac{f(x+\Delta x)g(x+\Delta x) - f(x)g(x+\Delta x)}{\Delta x} \right.$$

$$\left. + \frac{f(x)g(x+\Delta x) - f(x)g(x)}{\Delta x} \right)$$

$$= \lim_{\Delta x \to 0} \frac{[f(x+\Delta x) - f(x)]g(x+\Delta x)}{\Delta x}$$

$$+ \lim_{\Delta x \to 0} \frac{f(x)[g(x+\Delta x) - g(x)]}{\Delta x}$$

$$= \lim_{\Delta x \to 0} \frac{f(x+\Delta x) - f(x)}{\Delta x} g(x+\Delta x)$$

$$+ \lim_{\Delta x \to 0} f(x) \frac{g(x+\Delta x) - g(x)}{\Delta x}$$

$$= f'(x)g(x) + f(x)g'(x).$$

The proofs of the other rules of differentiation are similar. All rely on facts about limits.

Recall from Section 8.1 that limits of "$\frac{0}{0}$" type are often tricky to compute. With the following rule, many of them can be handled efficiently and effectively. Observe, however, that the rule depends on the derivative, which in turn involves a limit of "$\frac{0}{0}$" type.

L'Hospital's Rule: Suppose that f and g are both differentiable in an interval containing c. If $\lim\limits_{x \to c} \frac{f(x)}{g(x)}$ is of "$\frac{0}{0}$" type and $\lim\limits_{x \to c} \frac{f'(x)}{g'(x)}$ exists, then

$$\lim_{x \to c} \frac{f(x)}{g(x)} = \lim_{x \to c} \frac{f'(x)}{g'(x)}.$$

Example 8.26. Let $f(x) = x^5 + x^4 - x - 1$, $g(x) = x^3 + x^2 - x - 1$, and consider $\lim\limits_{x \to 1} \frac{f(x)}{g(x)}$. Notice that it is of "$\frac{0}{0}$" type. Since $f'(x) = 5x^4 + 4x^3 - 1$, $g'(x) = 3x^2 + 2x - 1$, and $\lim\limits_{x \to 1} \frac{f'(x)}{g'(x)} = \frac{5+4-1}{3+2-1} = \frac{8}{4} = 2$, it follows that $\lim\limits_{x \to 1} \frac{f(x)}{g(x)} = 2$. Notice that $\lim\limits_{x \to -1} \frac{f(x)}{g(x)}$ is also of "$\frac{0}{0}$" type. Since $f'(-1) = 0$ and $g'(-1) = 0$, $\lim\limits_{x \to -1} \frac{f'(x)}{g'(x)}$ is again of "$\frac{0}{0}$" type. So L'Hospital's rule cannot be applied. But why not repeat the process once more? Since $f''(x) = 20x^3 + 12x^2$ and $g''(x) = 6x + 2$, $\lim\limits_{x \to -1} \frac{f''(x)}{g''(x)} = \frac{-8}{-4} = 2$. So by L'Hospital's rule, $\lim\limits_{x \to -1} \frac{f'(x)}{g'(x)} = 2$. By applying it again, $\lim\limits_{x \to -1} \frac{f(x)}{g(x)} = 2$.

We conclude this section with one more fundamental theoretical concern.

Mean Value Theorem: Let $y = f(x)$ be continuous on a closed interval $[a, b]$. If f is differentiable on the open interval (a, b), then there is a number c in (a, b) such that

$$f(b) - f(a) = f'(c)(a - b).$$

Let's consider this assertion from the geometric point of view. A look at Figure 8.21 shows that $\frac{f(b)-f(a)}{b-a}$ is the slope of the segment joining the points $(a, f(a))$ and $(b, f(b))$. The continuity and smoothness properties of the graph imply that at some point $(c, f(c))$ the slope of the tangent is equal to the slope of the segment. Since the slope of this tangent is $f'(c)$, $f'(c) = \frac{f(b)-f(a)}{b-a}$, and the conclusion of the Mean Value theorem follows. The word *mean* means average. In this case, the reference is to the average change in $f(x)$ per change in x over the interval $a \le x \le b$.

Consider a function f with the property that $f'(x) = 0$ for all x in an interval I. Let a and b with $a \le b$ be any two numbers in I. Since f is differentiable on (a, b), it is also continuous there. So the mean value theorem provides a number c with $a < c < b$ such that $f(b) - f(a) = f'(c)(b - a)$. But $f'(c) = 0$, and therefore, $f(b) = f(a)$. It follows that f is constant throughout I. Once again, a theorem has replaced an argument "by

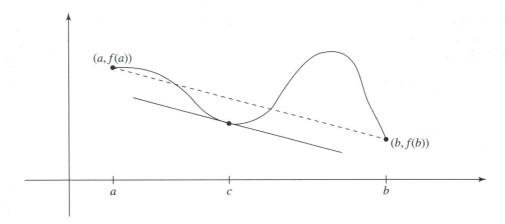

Figure 8.21

picture." Refer to Section 5.5B, primarily to Figure 5.29.

8.6 **Derivatives of Trigonometric Functions**

A ll the trigonometric functions are continuous on their domains. Indeed, we will see shortly that they are all differentiable on their domains. The limit

$$\lim_{\theta \to 0} \frac{\sin \theta}{\theta} = 1$$

is the cornerstone for the differential calculus of the trigonometric functions. Recall from Section 1.4 that $\sin \theta \approx \theta$ if θ is small, so that this limit exists and is equal to 1 is reasonable. In view of the fundamental importance of this fact, we will now verify it.

Let θ be any angle that satisfies $0 < \theta < \frac{\pi}{2}$ and build it into Figure 8.22. The circular arc BC is that on the circle of radius 1 and center O. The segment AC is perpendicular to the radius OB; the radius OC is extended to the point D with the property that BD is perpendicular to OB. Note that $\sin \theta = \frac{AC}{OC} = AC$ and that $\tan \theta = \frac{BD}{OB} = BD$. So the areas of the

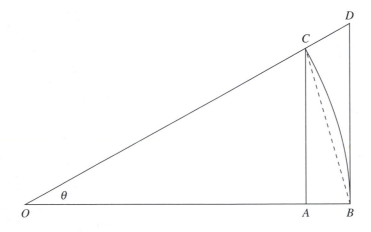

Figure 8.22

triangles $\triangle OBC$ and $\triangle OBD$ are $\frac{1}{2}OB \cdot AC = \frac{1}{2}\sin\theta$ and $\frac{1}{2}OB \cdot BD = \frac{1}{2}\tan\theta$, respectively. Since the radius is 1, the area of the circular sector OBC is $\frac{1}{2}\theta$. This was verified in Section 3.2. From the figure,

$$\text{Area}\triangle OBD \geq \text{Area sector } OBC \geq \text{Area}\triangle OBC,$$

and it follows that $\frac{1}{2}\tan\theta \geq \frac{1}{2}\theta \geq \frac{1}{2}\sin\theta$. So, $\frac{\sin\theta}{\cos\theta} = \tan\theta \geq \theta \geq \sin\theta$. Dividing through by $\sin\theta$ gives $\frac{1}{\cos\theta} \geq \frac{\theta}{\sin\theta} \geq 1$, and after inverting all terms,

$$\cos\theta \leq \frac{\sin\theta}{\theta} \leq 1.$$

This inequality holds for any θ between 0 and $\frac{\pi}{2}$. Now push θ to 0 and recall (from Section 1.4 for instance) that $\lim_{\theta\to 0}\cos\theta = 1$. It follow that $\lim_{\theta\to 0}\frac{\sin\theta}{\theta} = 1$ is the only possible outcome.

The remaining ingredient needed to determine the derivatives of the trigonometric functions is the limit $\lim_{\theta\to 0}\frac{\cos\theta-1}{\theta} = 0$. The verification of this limit is based on the following algebraic step:

$$\frac{\cos\theta - 1}{\theta} = \frac{\cos\theta - 1}{\theta} \cdot \frac{\cos\theta + 1}{\cos\theta + 1}$$

$$= \frac{\cos^2\theta - 1}{\theta} \cdot \frac{1}{\cos\theta + 1}$$

$$= -\frac{\sin^2\theta}{\theta} \cdot \frac{1}{\cos\theta + 1}$$

$$= -\frac{\sin\theta}{\theta} \cdot \frac{\sin\theta}{\cos\theta + 1}.$$

Since $\lim_{\theta\to 0}\frac{\sin\theta}{\theta} = 1$ and $\lim_{\theta\to 0}\sin\theta = 0$ (refer to Section 1.4), it now follows that

$$\lim_{\theta\to 0}\frac{\cos\theta - 1}{\theta} = \lim_{\theta\to 0}\left(-\frac{\sin\theta}{\theta} \cdot \frac{\sin\theta}{\cos\theta + 1}\right)$$

$$= -1 \cdot \frac{0}{1+1} = 0.$$

The determination of the derivative of $f(x) = \sin x$ is now a routine exercise. Use of the addition formula for the sine (from Exercise 6 of Chapter 2) shows that

$$f(x + \Delta x) - f(x) = \sin(x + \Delta x) - \sin x$$

$$= (\sin x)(\cos \Delta x) + (\cos x)(\sin \Delta x) - \sin x$$

$$= \sin x(\cos \Delta x - 1) + (\cos x)(\sin \Delta x).$$

Therefore,

$$\frac{f(x + \Delta x) - f(x)}{\Delta x} = \sin x\frac{(\cos \Delta x - 1)}{\Delta x} + \cos x\frac{\sin \Delta x}{\Delta x}.$$

Inserting the limits $\lim_{\Delta x\to 0}\frac{(\cos \Delta x - 1)}{\Delta x} = 0$ and $\lim_{\Delta x\to 0}\frac{\sin \Delta x}{\Delta x} = 1$ already established yields

$$\frac{d}{dx}\sin x = f'(x) = \lim_{\Delta x\to 0}\frac{f(x + \Delta x) - f(x)}{\Delta x}$$

$$= \sin x \cdot \lim_{\Delta x\to 0}\frac{(\cos \Delta x - 1)}{\Delta x}$$

$$+ \cos x \cdot \lim_{\Delta x\to 0}\frac{\sin \Delta x}{\Delta x}$$

$$= \sin x \cdot 0 + \cos x \cdot 1 = \cos x.$$

We have demonstrated that

$$\boxed{\frac{d}{dx}\sin x = \cos x}$$

A completely analogous argument (which uses the addition formula for the cosine) shows that

$$\boxed{\frac{d}{dx}\cos x = -\sin x}$$

Example 8.27. Let $f(x) = \sin^2 x = (\sin x)^2$. By the chain rule, $f'(x) = 2(\sin x)\frac{d}{dx}(\sin x) = 2(\sin x)(\cos x)$.

Example 8.28. Let $g(x) = \cos^2 x$. Again, by the chain rule, $g'(x) = 2(\cos x)(-\sin x)$.

A combination of Examples 8.27 and 8.28 shows that $\frac{d}{dx}(\sin^2 x + \cos^2 x) = 0$. This confirms the fact that $\sin^2 x + \cos^2 x$ is a constant. (We already know that it is equal to 1.)

Example 8.29. Let $h(t) = \sin(5t)$. By the chain rule, $h'(t) = \cos(5t) \cdot 5 = 5\cos(5t)$.

Example 8.30. Show that for any constant b,

$$\frac{d}{dt}\cos(bt) = -b\sin(bt).$$

Example 8.31. Since

$$\frac{d}{dx}\frac{\sin x}{\cos x} = \frac{(\cos x)(\cos x) - (\sin x)(-\sin x)}{\cos^2 x}$$

$$= \frac{1}{\cos^2 x} = \sec^2 x,$$

and $\frac{\sin x}{\cos x} = \tan x$, we can conclude that

$$\boxed{\frac{d}{dx}\tan x = \sec^2 x}$$

Example 8.32. By the chain rule,

$$\frac{d}{dx}(\cos x)^{-1} = -(\cos x)^{-2}(-\sin x) = \frac{\sin x}{\cos^2 x}$$

$$= \frac{\sin x}{\cos x} \cdot \frac{1}{\cos x}.$$

The substitutions $\sec x = \frac{1}{\cos x}$ and $\tan x = \frac{\sin x}{\cos x}$ now give us the formula

$$\boxed{\frac{d}{dx}\sec x = (\sec x)(\tan x)}$$

Example 8.33. Consider $h(t) = \sec^2 t = (\sec t)^2$. By the chain rule, $h'(t) = 2(\sec t)(\sec t)(\tan t) = 2(\sec t)^2(\tan t)$.

8.7 Increase and Decrease of Functions

Consider a function $y = f(x)$. We will say that f is *increasing on an interval I* if for any numbers x_1 and x_2 in I with $x_1 < x_2$, $f(x_1) < f(x_2)$. Similarly, f is *decreasing on an interval I* if for any numbers x_1 and

x_2 in I with $x_1 < x_2, f(x_1) > f(x_2)$. Observe—see Figure 8.23—that if f is increasing, then the graph rises from left to right over I, and that if f is decreasing, then the graph falls from left to right over I.

Suppose that f is differentiable on I. Let x_1 and x_2 be any numbers in I with $x_1 < x_2$. Applying the Mean Value theorem to f and the interval $[x_1, x_2]$ shows that there is a number x satisfying $x_1 < x < x_2$ such that

$$f(x_2) - f(x_1) = f'(x)(x_2 - x_1).$$

Suppose that $f'(x) > 0$ for all x in I. Since $x_2 > x_1$, $f'(x)(x_2 - x_1) > 0$, and it follows that $f(x_2) > f(x_1)$. Therefore, f is increasing on I. If, on the other hand, $f'(x) < 0$ for all x in I, then $f'(x)(x_2 - x_1) < 0$, and $f(x_2) < f(x_1)$. So f is decreasing on I. Have a look at Figure 8.24 and note that we have confirmed what is intuitively clear from this figure:

First Derivative Test: Consider a function $y = f(x)$ over an open interval (a, b). If $f'(x) > 0$ for all x in (a, b), then f is increasing on (a, b); and if $f'(x) < 0$ for all x in (a, b), then f is decreasing on (a, b).

These considerations can be used to establish inequalities. We will illustrate this by verifying the inequality $|\sin\varphi - \sin\theta| \le |\varphi - \theta|$ for any two angles φ and θ in radian measure. Let a be the smaller of the two constants φ and θ and b the larger. Since

$$|\sin\varphi - \sin\theta| = |\sin\theta - \sin\varphi|$$

$$= |\sin b - \sin a| \text{ and } |\varphi - \theta| = |b - a|,$$

we need to verify the inequality $|\sin b - \sin a| \le b - a$ only for constants a and b with $b \ge a$. Define the functions f and g by

$$f(x) = x - a \quad \text{for } x \ge a$$

and

$$g(x) = \sin x - \sin a \quad \text{for} \quad x \ge a.$$

Observe that $f'(x) = 1$ and $g'(x) = \cos x$. Recalling that $-1 \le \cos x \le 1$, it follows that

$$\frac{d}{dx}(f(x) - g(x)) = f'(x) - g'(x) = 1 - \cos x \ge 0,$$

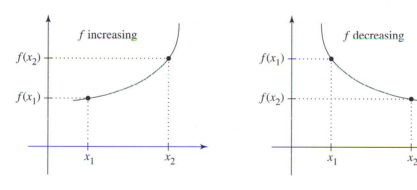

Figure 8.23

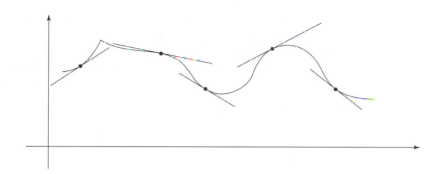

Figure 8.24

and

$$\frac{d}{dx}(f(x) + g(x)) = f'(x) + g'(x) = 1 + \cos x \geq 0.$$

So all the tangent lines of the graphs of both $f(x) - g(x)$ and $f(x) + g(x)$ have slope greater than or equal to 0. So the functions $f(x) - g(x)$ and $f(x) + g(x)$ never decrease. Since $f(a) = 0$ and $g(a) = 0$, it follows that $f(x) - g(x) \geq 0$ and $f(x) + g(x) \geq 0$ for all $x \geq a$. Hence $f(b) - g(b) \geq 0$ and $f(b) + g(b) \geq 0$. So $f(b) \geq g(b)$ and $f(b) \geq -g(b)$. This means that

$$b - a \geq \sin b - \sin a \quad \text{and} \quad b - a \geq \sin a - \sin b.$$

Since $b \geq a$, $|b - a| = b - a$. Since $|\sin b - \sin a|$ equals either $\sin b - \sin a$ or $\sin a - \sin b$, it now follows that $|b - a| \geq |\sin b - \sin a|$ as required.

The inequality just derived was crucial in Kepler's analysis of the planetary orbits. See Section 4.8. The verification of the inequality shows us how calculus underlies the investigations of Kepler. It also makes another point once more: If a mathematical argument is to be conclusive, it cannot rely on "reasoning by picture." Rather, it must be based on a rigorous analytical approach.

Suppose that Figure 8.25 is the graph of the function $y = f(x)$. Note that f is increasing on the intervals (a, c_1) and (c_2, c_3) and decreasing on the intervals (c_1, c_2) and (c_3, b).

Consider the number c_1. Observe that f is increasing immediately to the left of c_1 and decreasing immediately to the right of c_1. So $f(c_1) \geq f(x)$ for all x near c_1. We will say that *f has a local maximum at* c_1 to describe this situation. Notice that f has a local maximum at c_3 also. Now look at the number c_2. Note that f is decreasing immediately to the left of c_2 and

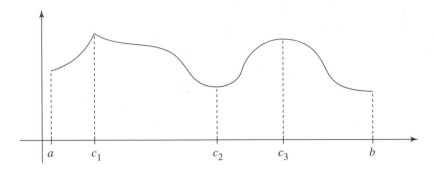

Figure 8.25

increasing immediately to the right of c_2. In this case, we say that f *has a local minimum at* c_2.

The numbers at which a given function f has a local maximum or a local minimum value are important for two related reasons. They are the transition points between the intervals of increase and decrease of f. They are also of immediate interest if f has a maximum or minimum value that is to be determined. We will turn to both concerns shortly.

Max/Min Theorem: If $f'(c)$ exists and if f has either a local maximum or a local minimum at c, then $f'(c) = 0$.

Proof. Suppose that $f'(c)$ exists and consider the case where f has a local maximum at c. Observe that $f(c) \geq f(x)$ for any x close enough to c. This means that $f(x) - f(c) \leq 0$ for any x close enough to c. Suppose $x < c$. So $x - c < 0$. Therefore, $\frac{f(x)-f(c)}{x-c} \geq 0$ for any x to the left and close enough to c, and it follows that

$$\lim_{x \to c^-} \frac{f(x) - f(c)}{x - c} \geq 0.$$

Suppose $x > c$. Now $x - c > 0$, and hence $\frac{f(x)-f(c)}{x-c} \leq 0$ for any x to the right and close enough to c. Therefore,

$$\lim_{x \to c^+} \frac{f(x) - f(c)}{x - c} \leq 0.$$

Since f is differentiable at c, both of these limits exist and are equal. It follows that both must be equal to 0. Since $f'(c) = \lim_{x \to c} \frac{f(x)-f(c)}{x-c}$ is the common value, it is indeed the case that $f'(c) = 0$. The same conclusion,

for entirely similar reasons, also holds if f has a local minimum at c.

A number c in the domain of f where either $f'(c) = 0$ or $f'(c)$ does not exist is called a *critical number* for f. So

$$c \text{ is a } \textit{critical number} \text{ for } f \text{ if } \begin{cases} \text{either } f'(c) = 0, \text{ or} \\ f'(c) \text{ does not exist.} \end{cases}$$

Suppose that f has a local maximum or a local minimum at a number c. If f is not differentiable at c, then c is a critical number. If f is differentiable at c, then by the Max/Min theorem $f'(c) = 0$, and c is a critical number in this case also. It follows that if f has a local maximum or a local minimum at a number c, then c is a critical number for f.

Therefore, the strategy for finding all the numbers c at which f has either a local maximum or a local minimum is the following:

(1) Find all critical numbers for f.

(2) Check each critical number: Does it give a local maximum, a local minimum, or neither?

Example 8.34. What are the critical numbers for the functions $f(x) = x^2$ and $g(x) = x^3$? Since $f'(x) = 2x$ and $g'(x) = 3x^2$, both derivatives exist for all x and are equal to 0 only for $x = 0$. It follows that in each case 0 is the only critical number. Refer to the graphs of the two functions in Figure 8.26 and observe that

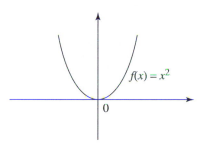

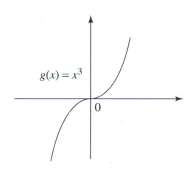

Figure 8.26

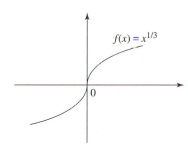

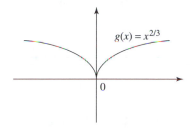

Figure 8.27

$f(x) = x^2$ has a local minimum at 0, but that $g(x) = x^3$ has neither a local minimum nor a local maximum at 0.

Example 8.35. What are the critical numbers for the functions $f(x) = x^{\frac{1}{3}}$ and $g(x) = x^{\frac{2}{3}}$? Since $f'(x) = \frac{1}{3}x^{-\frac{2}{3}} = \frac{1}{3x^{\frac{2}{3}}}$ and $g'(x) = \frac{2}{3}x^{-\frac{1}{3}} = \frac{2}{3x^{\frac{1}{3}}}$, neither is ever equal to 0 and both are undefined at $x = 0$. Therefore, in each case, 0 is the only critical number. Refer to the graphs of the two functions in Figure 8.27 and observe that $f(x) = x^{\frac{1}{3}}$ has neither a local maximum nor a local minimum at 0 and that $g(x) = x^{\frac{2}{3}}$ has a local minimum at 0.

Let c_1 and c_2 be successive critical numbers for a function f and consider the interval $I = (c_1, c_2)$. Observe that f is differentiable on I. If not, there would be a critical number between c_1 and c_2. Observe also that f is either increasing throughout I or decreasing throughout I. (If not, there would be a local maximum or minimum between c_1 and c_2. But this cannot hap-

pen. Why?) Under the assumption that the derivative f' is continuous on I, there is an easy way to determine which of these two possibilities arises. Suppose that $f'(t) > 0$ for a single number t in I. Since c_1 and c_2 are successive critical numbers, it follows that $f'(x) > 0$ for all x in (c_1, c_2). (Why?) So by the First Derivative Test, f is increasing throughout I. In the same way, if $f'(t) < 0$ for a single t in I, then f must be decreasing throughout I. Specific such points t will be referred to as *test points*. Similar strategies can be used for the interval $I = (-\infty, c)$ if c is the smallest critical number and for $I = (c, \infty)$ if c is the largest.

The examples that follow illustrate these principles and show how they can be put to use.

Example 8.36. Consider the function $f(x) = x^3 - 3x$. Since

$$f'(x) = 3x^2 - 3 = 3(x^2 - 1) = 3(x + 1)(x - 1),$$

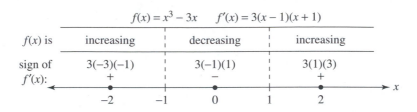

Figure 8.28

$f(x)$ is differentiable for all x. It follows that all critical numbers arise as zeros of $f'(x)$. So the critical numbers are -1 and 1. Placed on the x-axis, they divide it into

the three intervals: $(-\infty, -1), (-1, 1)$, and $(1, \infty)$. Take as test points -2 for $(-\infty, -1)$; 0 for $(-1, 1)$; and 2 for $(1, \infty)$. Since $f'(-2) = 3(-1)(-3) = 9$ is positive, f increases throughout the interval $(-\infty, -1)$. Since $f'(0) = -3$ is negative, f decreases throughout $(-1, 1)$. Finally, the fact that $f'(2) = 9 > 0$ tells us that f is increasing throughout $(1, \infty)$. Figure 8.28 summarizes these findings.

After plotting a few relevant points, we can sketch the graph of f. This is done in Figure 8.29.

Example 8.37. Analyze the function $f(x) = 3x^4 - 4x^3 - 36x^2 + 3$. By easy computations,

$$f'(x) = 12x^3 - 12x^2 - 72x = 12x(x^2 - x - 6)$$

$$= 12x(x - 3)(x + 2).$$

Therefore, f is differentiable for all x. It follows that the critical numbers are those for which $f'(x) = 0$. So they are -2, 0, and 3. They divide the x-axis into the intervals $(-\infty, -2), (-2, 0), (0, 3)$, and $(3, \infty)$.

Take -3, -1, 1, and 4 as the test points. Since $f'(-3) = (-36)(-6)(-1) < 0$, f is decreasing throughout the interval $(-\infty, -2)$. The fact that $f'(-1) = (-12)(-4)(1) > 0$ tells us that f is increasing throughout $(-2, 0)$, and $f'(2) = 24(-1)(4) < 0$ means that f is decreasing throughout $(0, 3)$. Finally, since $f'(4) = 48(1)(6) > 0$, we see that f increases throughout $(3, \infty)$. Figure 8.30 summarizes the matter.

To sketch a meaningful graph, a much larger scale is necessary than that used in Example 8.36. Refer to Figure 8.31.

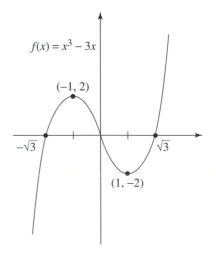

Figure 8.29

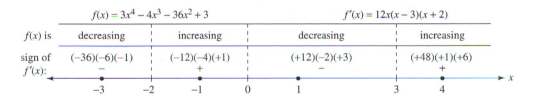

Figure 8.30

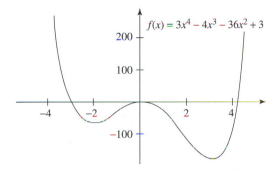

Figure 8.31

$(-\infty, -\sqrt{2})$, $(-\sqrt{2}, 0)$, $(0, \sqrt{2})$, and $(\sqrt{2}, \infty)$. Taking the test points -2, -1, 1, and 2 and proceeding as in the earlier examples provides the information shown in Figure 8.32. Plot a few points and use the information just obtained to see that the graph of g is as shown in Figure 8.33.

Example 8.38. Analyze $g(x) = x^{\frac{1}{3}}(x^2 - 14) = x^{\frac{7}{3}} - 14x^{\frac{1}{3}}$. By routine computations,

$$g'(x) = \frac{7}{3}x^{\frac{4}{3}} - \frac{14}{3}x^{-\frac{2}{3}} = \frac{7x^2}{3x^{\frac{2}{3}}} - \frac{14}{3x^{\frac{2}{3}}} = \frac{7x^2 - 14}{3x^{\frac{2}{3}}}$$

$$= \frac{7(x^2 - 2)}{3x^{\frac{2}{3}}}.$$

Notice that both types of critical numbers occur. The derivative is zero at $-\sqrt{2}$ and $\sqrt{2}$, and the derivative is not defined at 0. Therefore, the critical numbers are $-\sqrt{2}, 0,$ and $\sqrt{2}$. They split the x-axis into the intervals

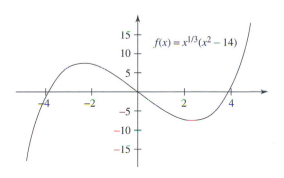

Figure 8.33

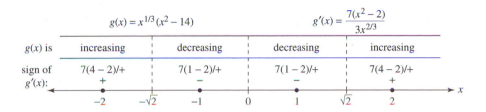

Figure 8.32

8.8 **Maximum and Minimum Values**

Let $y = f(x)$ be any function. If there is a number c such that $f(x) \leq f(c)$ for *all* x in the domain of f, then $f(c)$ is called *the (absolute) maximum value of f*, and we say that f *attains its maximum value at c*. Similarly, there may be a number d such that $f(d) \leq f(x)$ for *all* x in the domain of f. In this case, $f(d)$ is called *the (absolute) minimum value of f*, and we say that f *attains its minimum value at d*. The maximum value of f is the largest of all the y-coordinates that a point on the graph of f can have, and the minimum value is the smallest. So the maximum value is the y-coordinate of the highest point on the graph, and the minimum value is the y-coordinate of the lowest. A look at the graphs in Figures 8.19, 8.25, 8.26, 8.27, 8.29, 8.31, and 8.33 shows that such values may or may not exist. It is possible for the maximum vale of f to be attained at more than one number c (the graph could have more than one highest point), and similarly for the minimum value (the graph could have more than one lowest point).

Refer to Figure 8.9 in Section 8.2 and the discussion that pertains to it for the following fact. If $y = f(x)$ is a continuous function with domain a closed interval $[a, b]$, then there do exist numbers c and d such that

$$f(d) \leq f(x) \leq f(c), \text{ for all } x \text{ in } [a, b].$$

In other words, such a function has both a maximum value and a minimum value. The maximum value must be attained either at one of the endpoints a or b or at a number c in the interior of $[a, b]$. In the latter case, f has a local maximum at c. In the same way, the minimum value is attained at either a or b or at a local minimum. It follows from the "critical point" strategy for finding the local maxima and minima of a function that the maximum value $f(c)$ and the minimum value $f(d)$ can be found as follows:

Determine all the critical numbers $c_1, c_2, c_3, \ldots$ between a and b, and compute the values $f(c_1)$, $f(c_2)$, $f(c_3), \ldots$, as well as $f(a)$ and $f(b)$. The largest of these is the maximum value of f on

$[a, b]$ and the smallest is the minimum value of f on $[a, b]$.

Example 8.39. Consider the function $f(x) = x^2 - 6x^{\frac{1}{3}}$ restricted to the interval $[-1, 3]$. Its derivative is $f'(x) = 2x - 2x^{-\frac{2}{3}} = \frac{2x^{\frac{5}{3}} - 2}{x^{\frac{2}{3}}}$. What are the critical numbers? Note that f' is not defined for $x = 0$ and that f' is 0 for any x that satisfies $2x^{\frac{5}{3}} = 2$, or $x^{\frac{5}{3}} = 1$. In the latter case $x = 1$. Notice that both 0 and 1 fall into the interval $[-1, 3]$. Therefore, both are critical numbers for f. Since

$$f(0) = 0, \quad f(1) = -5, \quad f(-1) = 7, \quad \text{and}$$

$$f(3) = 9 - 6^3\sqrt{3} \approx 9 - 6(1.44) = 9 - 8.64 = 0.36,$$

we can conclude that $f(-1) = 7$ and $f(1) = -5$ are the respective maximum and minimum values of f on $[-1, 3]$.

We close this chapter with some applications of the theory of maxima and minima to some classical (and also very challenging) geometric problems. Examples 8.40–8.42 and 8.44 are taken from the world's very first basic calculus text published by a French marquis in 1696. The Marquis De L'Hospital (L'Hospital's rule was named after him) wrote this elegant text for the scholars of the day. Example 8.43 is taken from a book published in 1748 by the remarkable Italian woman Maria Gaetana Agnesi. It was intended for young students and received much acclaim. Refer to the Postscript of this chapter for more information about these early texts and their authors. Note that the mathematical exactness and rigor discussed in the introduction of this chapter and in Section 8.1 were not yet evident in these two books. The more definitive formulation of calculus would come about one hundred years later (in the 19th century).

Example 8.40. Fix any two positive constants a and h. Of all the rectangular boxes with height h and volume a^3, find the dimensions of the one that has the smallest surface area.

Let x and y be the dimensions of the base and note that $xyh = a^3$. So $y = \frac{a^3}{hx} = \frac{a^3}{h}x^{-1}$. Refer to

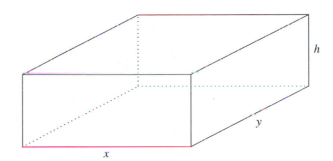

Figure 8.34 and note that the surface area of the box is $2hx + 2hy + 2yx$. Substituting $y = \frac{a^3}{h}x^{-1}$, gives the surface area as the function

$$S(x) = 2hx + 2h\frac{a^3}{h}x^{-1} + 2\frac{a^3}{h}x^{-1}x$$

$$= 2hx + 2a^3x^{-1} + 2\frac{a^3}{h}$$

of x. The question now is this: For which x does $S(x)$ attain its minimum value?

The answer requires the analysis of the derivative $S'(x) = 2h - 2a^3x^{-2} = 2\frac{hx^2-a^3}{x^2}$. Since $hx^2 = a^3$ implies that $x = \sqrt{\frac{a^3}{h}}$, it follows that 0 and $\sqrt{\frac{a^3}{h}}$ are the critical numbers of $S(x)$. Because $x = 0$ gives a box of zero volume, this is ruled out. So the focus is on $x = \sqrt{\frac{a^3}{h}}$. Does this provide the minimum we are looking for? That it does can be seen from the following observation. If the substitution $x = \sqrt{\frac{a^3}{h}}$ makes $hx^2 - a^3$ equal to 0, then substituting any x smaller than $\sqrt{\frac{a^3}{h}}$ must make $hx^2 - a^3$ negative, and substituting any x larger than $\sqrt{\frac{a^3}{h}}$ must make $hx^2 - a^3$ positive. It follows that $S'(x) = 2\frac{hx^2-a^3}{x^2}$ is negative for $x < \sqrt{\frac{a^3}{h}}$ and positive for $x > \sqrt{\frac{a^3}{h}}$. So $S(x)$ is increasing on the left of $x = \sqrt{\frac{a^3}{h}}$ and decreasing on its right. Therefore, as asserted, $S(x)$ has its minimum value when $x = \sqrt{\frac{a^3}{h}}$. Substituting this value of x into $y = \frac{a^3}{h}x^{-1}$ gives $y = \frac{a^3}{h}\left(\frac{a^3}{h}\right)^{-\frac{1}{2}} = \frac{a^3}{h}\frac{a^{-\frac{3}{2}}}{h^{-\frac{1}{2}}} = \frac{a^{\frac{3}{2}}}{h^{\frac{1}{2}}} = \sqrt{\frac{a^3}{h}}$. So the base of the box with minimal surface area is a square. What is the height of this box?

Example 8.41. Inscribe a cone into a sphere of fixed radius R as shown in Figure 8.35. Of all the cones that can be inscribed in this way, find the dimensions of the one whose surface area is the largest. The surface of the cone is understood to be the slanted surface without the circular base.

Let y be the radius of the base of the cone and z the length of the slanted segment. See Figure 8.35. As first order of business, we will compute the surface

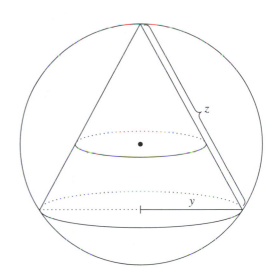

Figure 8.35

area S of the cone. (Remember that this does not include the area of the circular base.) Refer to Figure 8.36 and observe that this cone is obtained by cutting out the circular sector (along the two "perforations") and (after "bending" it) joining the two cuts. It follows from the formula for the area of a circular sector (refer to Section 3.2) that $S = \frac{1}{2}\theta z^2$. Since $\theta = \frac{2\pi y}{z}$, we get that $S = \pi y z$.

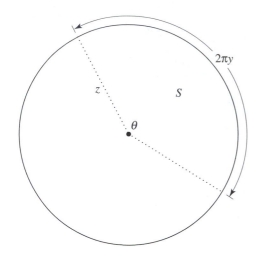

Figure 8.36

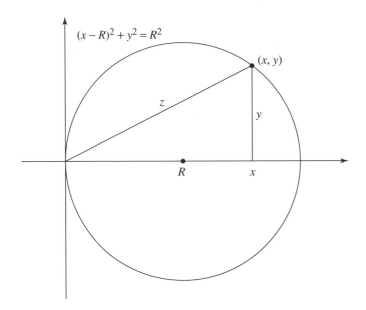

Figure 8.37

Refer to Figure 8.37 next. If the semicircle of radius R and the triangle inscribed in it are revolved one complete revolution around the x-axis, the cone of Figure 8.35 is obtained. (Only its position is different.) The coordinate system is chosen such that the circle has center $(R, 0)$ and equation $(x - R)^2 + y^2 = R^2$. The

problem of finding the dimensions of the required cone can now be reformulated as follows. For which choice of y and z of the triangle inscribed in the circle of radius R is the quantity $S = \pi yz$ a maximum? The strategy that will solve this remaining question is this:

Express both y and z in terms of x, then substitute into S to get S as a function of x, and finally, find the number x at which this function attains its maximum.

Since $y^2 = R^2 - (x - R)^2$, it follows that $y = (R^2 - (x - R)^2)^{\frac{1}{2}} = (2Rx - x^2)^{\frac{1}{2}}$. Note next that $x^2 + y^2 = z^2$, and hence that $z = (x^2 + y^2)^{\frac{1}{2}} = (x^2 + R^2 - (x - R)^2)^{\frac{1}{2}} = (2Rx)^{\frac{1}{2}}$. Therefore,

$$S = S(x) = \pi yz = \pi(2xR - x^2)^{\frac{1}{2}}(2Rx)^{\frac{1}{2}}$$

$$= \pi\big((2Rx - x^2)2Rx\big)^{\frac{1}{2}} = \pi(4R^2x^2 - 2Rx^3)^{\frac{1}{2}}.$$

By the chain rule,

$$S'(x) = \frac{\pi}{2}(4R^2x^2 - 2Rx^3)^{-\frac{1}{2}}(8R^2x - 6Rx^2)$$

$$= \frac{\pi(4R^2x - 3Rx^2)}{(4R^2x^2 - 2Rx^3)^{\frac{1}{2}}}$$

$$= \frac{\pi Rx(4R - 3x)}{(2Rx)^{\frac{1}{2}}(2Rx - x^2)^{\frac{1}{2}}}.$$

In view of the geometry of the situation,

$$0 \leq x \leq 2R.$$

(See Figure 8.37.) Simple calculations show that the numbers for which $S'(x)$ is not defined are $x = 0$ and $x = 2R$, and that $S'(x) = 0$ only for $x = \frac{4}{3}R$. Since $S(x)$ is continuous on the closed interval $[0, 2R]$, it follows that $S(x)$ attains its maximum at one of the endpoints 0 or $2R$, or at the critical number $\frac{4}{3}R$. The value of

$$S(x) = \pi(4R^2x^2 - 2Rx^3)^{\frac{1}{2}} = \pi(2Rx)^{\frac{1}{2}}(4Rx - 2x^2)^{\frac{1}{2}}$$

at both $x = 0$ and $x = 2R$ is zero. So the maximum occurs for $x = \frac{4}{3}R$.

Refer back to Figure 8.35. The dimensions y and z of the cone of maximal surface area are obtained by substituting the value $x = \frac{4}{3}R$ into the formulas $y = (2Rx - x^2)^{\frac{1}{2}}$ and $z = (2Rx)^{\frac{1}{2}}$. The surface area of

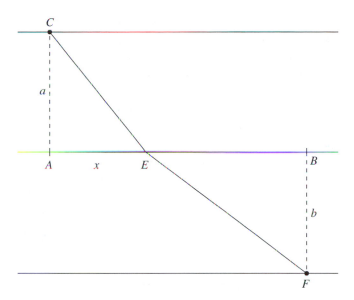

Figure 8.38

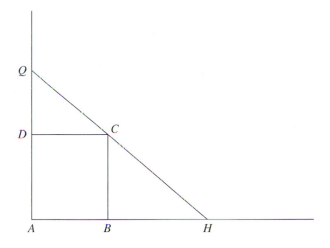

Figure 8.39

this cone is

$$S\left(\tfrac{4}{3}R\right) = \pi\sqrt{2R}\sqrt{\tfrac{4}{3}R}\sqrt{4R\cdot\tfrac{4}{3}R - 2\left(\tfrac{4}{3}R\right)^2}$$

$$= 2\pi\tfrac{\sqrt{2}}{\sqrt{3}}R\sqrt{2\left(\tfrac{4}{3}R\right)^2} = \tfrac{16\pi}{3\sqrt{3}}R^2.$$

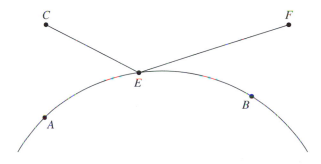

Figure 8.40

Example 8.42. A traveler in going from a place C to a place F has to traverse two fields that are separated by the line AB. See Figure 8.38. We will suppose that he walks through the first field with a speed of v and through the second field with a speed of w. At precisely which point E must he cross the line AB so as to arrive at F in the shortest possible time?

The solution of a reformulated version of this problem is fundamental in the field of geometric optics. The traveler is a light ray and the fields are two media of different densities through which the ray passes. We will study the basics of geometric optics in Section 9.4 of the next chapter and solve this reformulated problem there. In the meantime, study Figure 8.38 and determine a function of x whose minimum provides the answer to the problem.

Example 8.43. Let a rectangle $ADCB$ be given. Draw a line segment QH as shown in Figure 8.39. What is the minimal length that QH can have? The solution will be left to the reader. Start by placing coordinate axes appropriately. The fact that the length of a segment is a minimum exactly when the square of the length is a minimum will simplify the computations.

Example 8.44. An arc AB of a circle with center O is given, as are the points C and F not on the circle. Determine the point E on the arc with the property that the sum of the distances EC and EF is as small as possible. Refer to Figure 8.40.

This is a complicated problem! Make a start towards the solution by placing the center O of the circle at the origin of an x-y coordinate system. Let the radius of the circle be r, label the coordinates of the points C and F, and find an algebraic expression for the quantity that has to be minimized.

8.9 Postscript

The first basic calculus text, the *Analyse des infiniment petits* was published in 1696 by the Marquis De L'Hospital (1661–1704).[1] This volume developed the new methods of Leibniz's differential calculus with elegance and clarity and contained the solution to a number of interesting problems. It soon became well known in intellectual circles. The material, however, was not originally due to the marquis. It appears that the ambitious marquis desired to be one of the first among his French countrymen to understand Leibniz's powerful new theory and also to become a part of these sensational developments of mathematics. To reach this goal efficiently he took the young Johann Bernoulli (then in his mid twenties) into his employ, first as his teacher and then as consultant and collaborator. It became Bernoulli's job to clarify subtle points, to solve (or assist in solving) problems, and to proofread and translate the marquis's manuscripts. For these services Bernoulli received payment, which he thought to be very generous judging by the gratitude that he expressed in his letters to the marquis. It is not impossible to believe, therefore, that the marquis thought in good conscience that he had bought the right to deal with what he had learned from Bernoulli as he saw fit. In particular, he saw no difficulty in using Bernoulli's considerable contributions as the basis for his new book, the *Analyse*. It seems clear that the marquis was not at all interested in publicizing the role played by his secret collaborator. Indeed, there is the suspicion that he suppressed mention of Bernoulli's contribution in order to enhance his own reputation.

In effect, therefore, the *Analyse* and the solutions to the problems that it contains had become his very own brilliant contribution. The correspondence[2] between Bernoulli and De L'Hospital makes it quite clear that the marquis was not the brilliant mathematician that he thus claimed to be. There are no ideas in the *Analyse* that were originally due to him. Most were due to Leibniz, Newton, and the brothers Bernoulli. For example, both the formulation and the proof of L'Hospital's rule were the work of Johann Bernoulli. The final verdict is that while the marquis was certainly a very knowledgeable mathematician and an articulate expositor, he did not have the inventive imagination to enrich the subject with new ideas.

Another important early text was written by a remarkable woman. Maria Gaetana Agnesi (1718–1799) was the first woman in the Western world who can legitimately be called a mathematician. The daughter of a wealthy Milanese and professor of mathematics at the University of Bologna, she began the study of mathematics at an early age. In 1748 she published her *Instituzioni analitiche ad uso della gioventù italiana*. This two-volume text contains in over a thousand pages the newer mathematical concepts of the middle of the 18th century. It presents a complete, comprehensive, and very lucid treatment of algebra, coordinate geometry, and differential and integral calculus. It was written in Italian, rather than the scholarly Latin, and was aimed at students (*gioventù* is the Italian word for youth). The work won the immediate acclaim of academic circles all over Europe.

Exercises
8A. Domains of Functions

1. Determine the domains of $f(x) = \sqrt{7-x}$, $g(x) = \sqrt{x+5}$, and $h(x) = f(x)g(x)$.

2. What are the domains of $f(x) = \sqrt{3x-4}$, $g(x) = \sqrt{2x-3}$ and $k(x) = \frac{f(x)}{g(x)}$?

3. What are the domains of $f(x) = \sqrt{3 - \sqrt{x+6}}$ and $g(x) = \sqrt{\frac{x-5}{x+3}}$?

8B. Evaluation of Limits

Evaluate the following limits. If a limit does not exist, say so.

4. $\lim\limits_{x \to 2}(x^2 + 1)(x^2 + 4x)$

5. $\lim\limits_{x \to 1} \dfrac{x - 2}{x^2 + 4x - 3}$

6. $\lim\limits_{x \to 4} \sqrt{x + \sqrt{x}}$

7. $\lim\limits_{x \to 3} \dfrac{x^2 - x + 12}{x + 3}$

8. $\lim\limits_{x \to -3} \dfrac{x^2 - x + 12}{x + 3}$

9. $\lim\limits_{x \to -3} \dfrac{x^2 - x - 12}{x + 3}$

10. $\lim\limits_{t \to -1} \dfrac{t^3 - t}{t^2 - 1}$

11. $\lim\limits_{x \to 1} \dfrac{x^3 - 1}{x^2 - 1}$

12. $\lim\limits_{h \to 0} \dfrac{(h - 5)^2 - 25}{h}$

13. $\lim\limits_{x \to 9} \dfrac{x^2 - 81}{\sqrt{x} - 3}$

14. $\lim\limits_{x \to 0} \dfrac{x}{\sqrt{1 + 3x} - 1}$

15. $\lim\limits_{s \to 16} \dfrac{4 - \sqrt{s}}{s - 16}$

16. $\lim\limits_{x \to 2} \dfrac{|x - 2|}{x - 2}$. [Hint: Let x approach 2 from the left and right separately.]

17. Given that $\lim\limits_{x \to a} f(x) = 5$ and $\lim\limits_{x \to a} g(x) = 3$, determine

   i. $\lim\limits_{x \to a} \dfrac{f(x)}{g(x)}$

   ii. $\lim\limits_{x \to a} \dfrac{2f(x)}{g(x) - f(x)}$

8C. Continuity

18. Determine the domains of the two functions and explain why they are continuous on their domains.

   i. $G(x) = \dfrac{x^4 + 17}{6x^2 + x - 1}$

   ii. $H(x) = \dfrac{1}{\sqrt{x+1}}$

19. For which value of the constant c is a function f defined below continuous for all real numbers x?

$$f(x) = \begin{cases} cx + 1, & \text{if } x \le 3, \\ cx^2 - 1, & \text{if } x > 3. \end{cases}$$

20. Use the definition of continuity and the properties of limits to show that the given functions are continuous at the given number.

   i. $f(x) = 1 + \sqrt{x^2 - 9}, c = 5$.

   ii. $g(x) = \dfrac{x+1}{2x^2 - 1}, c = 4$.

21. Check whether the following functions are continuous at all numbers in their domains.

   i. $f(x) = \begin{cases} \dfrac{x^2 - 1}{x + 1}, & \text{if } x \ne -1, \\ 6, & \text{if } x = -1. \end{cases}$

   ii. $f(x) = \begin{cases} \dfrac{x^2 - 2x - 8}{x - 4}, & \text{if } x \ne 4, \\ 6, & \text{if } x = 4. \end{cases}$

22. Use the Intermediate Value Theorem to show that $2x^3 + x^2 + 2 = 0$ for some x in the interval $(-2, -1)$.

23. Suppose that a function f is continuous on the interval $[-1, 1]$ and that $f(-1) = 4$ and $f(1) = 3$. Why is there a number r such that $|r| < 1$ and $f(r) = \pi$?

8D. Tangent Lines

24. Let $g(x) = 1 - x^3$. Find $g'(0)$ and use it to determine the equation of the tangent line to the curve $y = 1 - x^3$ at the point $(0, 1)$.

25. Let $h(x) = \dfrac{1}{2x - 1}$. Find $h'(-1)$ and use it to find the equation of the tangent line to the curve $y = \dfrac{1}{2x - 1}$ at the point $(-1, -\frac{1}{3})$.

26. Determine an equation of the tangent to the graph of $y = \dfrac{x}{x - 3}$ at the point $(6, 2)$.

27. Find the equation of the tangent line to the curve $y = x^2 - 1$ that is parallel to the line $x - 2y = 1$.

28. For what values of x does the graph of $f(x) = 2x^3 - 3x^2 - 6x + 87$ have a horizontal tangent?

29. Show that the curve $y = 6x^3 + 5x - 3$ has no tangent lines with slope 4.

30. Draw a diagram to show that there are two lines that are tangent to both of the parabolas $y = -1 - x^2$ and $y = 1 + x^2$. Find the coordinates of the points at which these tangents touch the parabolas.

31. A car is traveling at night along a curved stretch of highway shaped like the parabola $y = \frac{1}{10}x^2$ from left to right. At what point on the highway will the headlights illuminate the point $(10, 5)$?

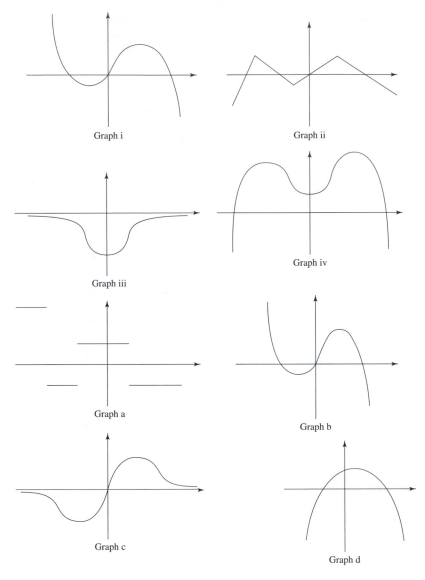

Figure 8.41

8E.   About Derivatives

32.   Each of the three limits below can be interpreted as the derivative of some function f at some number c. In each case, say which function and which number c.

i. $\lim\limits_{h \to 0} \dfrac{(h-5)^2 - 25}{h}$

ii. $\lim\limits_{x \to 1} \dfrac{x^9 - 1}{x - 1}$

iii. $\lim\limits_{x \to 0} \dfrac{\sqrt{1+x} - 1}{x}$

33.   Find the derivative of each of the two functions by using the limit definition of the derivative. In each case, determine the domain of the function and the domain of its derivative.

i. $f(x) = x - \frac{2}{x}$

ii. $f(x) = \sqrt{6 - x}$

34. Match the graph of each function in (i)–(iv) with the graph of its derivative in (a)–(d). See Figure 8.41.
(i) _____. (ii) _____. (iii) _____. (iv) _____.

35. Suppose that $f(3) = 4$, $g(3) = 2$, $f'(3) = -6$, $f'(2) = -3$ and $g'(3) = 5$. Find $\left(\frac{f}{g}\right)'(3)$. What is the value of the derivative of $f(g(x))$ at 3?

36. Sketch the graphs of $f(x) = x^2 - 9$ and $g(x) = |x^2 - 9|$. For what values of x does the function $g(x)$ fail to be differentiable? Sketch the graphs of both f' and g'.

37. For which values of the constants c and d is the function

$$f(x) = \begin{cases} cx^2 + 1, & \text{if } x \le 4, \\ \sqrt{x} + d, & \text{if } x > 4. \end{cases}$$

differentiable (and hence continuous) for all real numbers x?

38. Suppose that f is differentiable at a, where $a \ge 0$. Evaluate the limit

$$\lim_{x \to a} \frac{f(x) - f(a)}{\sqrt{x} - \sqrt{a}}$$

in terms of $f'(a)$.

8F. Rates of Change

39. An experiment in a medical laboratory studies a culture of bacteria. There are 4,000 cells after 4 minutes into the experiment, 62,000 cells after 7 minutes, and 154,000 cells after 9 minutes. Compute the average rates of change in the number of bacteria during the time intervals (given in minutes) $[4, 7]$ and $[7, 9]$.

40. An experiment in a laboratory studies a radioactive substance. There are 70 milligrams of the substance at the beginning of the experiment, 35 milligrams after 5 days, and 17.5 milligrams after 10 days. (Note that 1 milligram $= 10^{-3}$ grams.) Compute the average rates of change in the mass of the substance during the time intervals (given in days) $[0, 5]$ and $[0, 10]$.

41. Consider the rod in Figure 8.15 and take $d = 80$ inches. Suppose that the temperature is $500°$ Fahrenheit at the point 0, and $499°$ Fahrenheit at $x = 20$. Suppose that the temperature is given by a formula $T(x) = 500 - ax$ for any x between 0 and 80. What must a be equal to? Determine the difference in temperature between the point $x = 30$ and $x = 35$. What is the average decrease in the temperature per inch over the interval $[30, 35]$? Compute the rate of change of the temperature at $x = 30$, and then at any x.

42. A tank holds 3000 gallons of water. At time $t = 0$ water begins to drain from the bottom, and after 25 minutes the tank is empty. According to Torricelli's Law, the volume V of water remaining in the tank after t minutes is

$$V(t) = 3000 \left(1 - \frac{t}{25}\right)^2, \text{ where } 0 \le t \le 25.$$

At what average rate per minute did the 3000 gallons drain from the tank? How much water is there in the tank after 10 minutes and after 20 minutes? What is the average rate per minute at which the water drains from the tank during the interval $[10, 20]$? Find the rate at which water is draining from the tank after 10 minutes and after 20 minutes.

43. Express the volume V of a cube as a function of the length x of its side. Find the average rate of change of V with respect to x as x changes from

   i. 3 to 4.
  ii. 3 to 3.1.
 iii. 3 to 3.01.

Find the rate of change of the volume (with respect to x) when $x = 3$. Finally, show that the rate of change of the volume of a cube (with respect to x) is equal to half the surface area of the cube for any x.

44. Find the average rate of change of the area of a circle with respect to its radius r as r changes from

   i. 2 to 3.
  ii. 2 to 2.5.
 iii. 2 to 2.1.

Find the rate of change of the area (with respect to its radius r) when $r = 2$. Finally, show that the rate of change of the area (with respect to r) is equal to the circumference of the circle for any r.

45. Boyle's Law asserts that if the temperature of a confined gas is held fixed, then the product of the pressure P and the volume V is a constant. Suppose that for a certain gas, $PV = 800$, where P is measured in pounds per square inch and V is measured in cubic inches.

   i. Find the average rate of change of P as V increases from 200 in^3 to 250 in^3.
  ii. Express V as a function of P and show that the rate of change of V with respect to P is inversely proportional to the square of P.

46. Recall Newton's Law of Gravitation from Section 7.5. It asserts that the magnitude F of the force exerted by a body of mass m on a body of mass M is $F = \frac{GmM}{r^2}$, where G is the gravitational constant and r is the distance between the bodies. If the bodies are moving, find the rate of change of F with respect to r.

8G. Differentiating Functions

In Exercises 47–55 compute the derivative of the indicated function.

47. $F(x) = (16x)^3$, $\frac{dF}{dx} =$

48. $G(x) = (x^2 + 1)(2x - 7)$, $G'(x) =$

49. $f(u) = \frac{a - u^2}{1 + u^2}$, $f'(u) =$

50. $s = \sqrt[3]{t}(t + 2)$, $\frac{ds}{dt} =$

51. $y = \frac{1}{x^4 + x^2 + 1}$, $\frac{dy}{dx} =$

52. $y = (2x^3 + 4x^5)^6(7x^8 + 9x^{10})^{11}$, $\frac{dy}{dx} =$

53. $y = \frac{x}{\sqrt{9 - 4x}}$, $\frac{dy}{dx} =$

54. $F(x) = \frac{(x^2 + 4x + 6)^5}{\sqrt{x^3 + 4x^5}}$, $F'(x) =$

55. $s(t) = \sqrt[4]{\frac{t^3 + 1}{t^3 - 1}}$, $s'(t) =$

56. Let $y = u^2$ and $u = x^2 + 2x + 3$. Find $\frac{dy}{dx}$ in two ways: Substitute and differentiate; then use the chain rule $\frac{dy}{dx} = \frac{dy}{du} \cdot \frac{du}{dx}$. Let $\frac{dy}{dx}|_{x=1}$ be the value of the derivative $\frac{dy}{dx}$ at $x = 1$. Find this value.

57. Consider the circle with radius r centered at the origin. Rotate the semicircle and the rectangle shown in Figure 8.42 one complete revolution about the y-axis. This rotation generates a sphere of radius r and an inscribed cylinder. Notice that the base of the cylinder has area πx^2.

    i. Show that the volume of this cylinder is $V(x) = 2\pi x^2 \sqrt{r^2 - x^2}$.

    ii. Find the value of x that makes $V(x)$ a maximum.

    iii. What is the ratio of the volume of the sphere to that of the largest inscribed cylinder.

58. Use L'Hospital's rule to evaluate:

    i. $\lim\limits_{x \to -3} \dfrac{x^2 - x - 12}{x + 3}$

    ii. $\lim\limits_{x \to 1} \dfrac{x^3 - 1}{x^2 - 1}$

    iii. $\lim\limits_{x \to 9} \dfrac{x^2 - 81}{\sqrt{x} - 3}$

59. Let $f(x) = x + \sqrt{x}$. Use the Mean Value Theorem to show that there is a number c between 0 and 9 such that $f'(c) = \frac{4}{3}$. Determine the c that works.

60. Suppose that $y = f(x)$ is some differentiable function of x. Use the chain rule to compute the derivatives of the

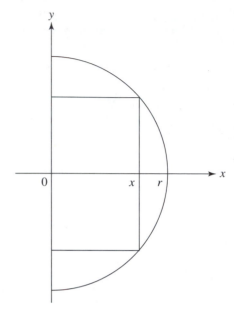

Figure 8.42

functions of x in terms of $f(x)$ and $f'(x)$. This procedure is called *implicit differentiation*.

    i. $F(x) = xy^3 = x(f(x))^3$.

    ii. $G(x) = 3\sqrt{y} + xy = 3(f(x))^{\frac{1}{2}} + xf(x)$.

    iii. $H(x) = \frac{4x}{y^2} = 4x(f(x))^{-2}$

8H. Calculus of Trigonometric Functions

In Exercises 61–69 compute $\frac{dy}{dx}$.

61. $y = \sin \frac{1}{x}$

62. $y = \sin^2(\cos 4x)$

63. $y = \frac{\sin^2 x}{\cos x}$

64. $y = x \sin \frac{1}{x}$

65. $y = \tan 3x$

66. $y = (\cos \sqrt{x^2 + 1})^{-5}$

67. $y = (1 + \sec^3 x)^6$

68. $y = \tan(x^2) + \tan^2 x$

69. $y = \sqrt{1 + 2\tan x}$

70. Express the limit $\lim\limits_{\theta \to \pi/3} \dfrac{\cos \theta - 0.5}{\theta - \pi/3}$ as a derivative and evaluate it.

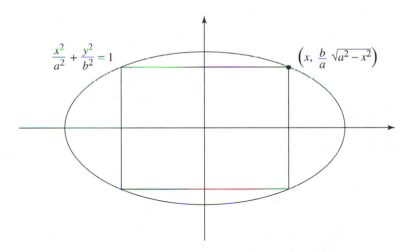

$$\frac{x^2}{a^2} + \frac{y^2}{b^2} = 1 \qquad\qquad \left(x, \frac{b}{a}\sqrt{a^2 - x^2}\right)$$

Figure 8.43

71. Find the equation of the tangent to the curve $y = \tan x$ at the point $(\frac{\pi}{3}, \sqrt{3})$.

8I. Increase and Decrease of Functions

In Exercises 72–74 find the critical numbers of the given function.

72. $f(x) = x^3 - 3x + 1$

73. $F(x) = x^{\frac{4}{5}}(x - 4)^2$

74. $T(x) = x^2(2x - 1)^{\frac{2}{3}}$

In Exercises 75–78 find (a) the intervals on which f is increasing or decreasing and (b) the local maximum and minimum values of f.

75. $f(x) = x^3 - 2x^2 + x$

76. $f(x) = x^4 - 4x^3 - 8x^2 + 3$

77. $f(x) = x\sqrt{1 - x^2}$

78. $f(x) = x\sqrt{x - x^2}$

79. Verify that $a + \frac{1}{a} < b + \frac{1}{b}$ whenever $1 \le a < b$. [Hint: Show that the function $f(x) = x + \frac{1}{x}$ is increasing on $[1, \infty)$.]

80. Aristarchus and Archimedes (refer to Section 1.6) used the inequality

$$\frac{\sin \beta}{\sin \alpha} \le \frac{\beta}{\alpha} \le \frac{\tan \beta}{\tan \alpha}, \text{ whenever } 0 < \alpha < \beta < \frac{\pi}{2}.$$

Verify both inequalities separately using calculus. [Hint: Show that $\frac{\sin x}{x}$ is a decreasing function for $0 < x < \frac{\pi}{2}$.]

In Exercises 81–84 find the intervals on which f is increasing or decreasing and find the local maximum and minimum values of f.

81. $f(x) = x - 2\sin x$, $0 \le x \le 2\pi$.

82. $f(x) = x\sin x + \cos x$, $-\pi \le x \le \pi$.

83. $f(x) = 2\tan x - \tan^2 x$, $-\pi \le x \le \pi$.

84. $g(x) = \sin x - \cos x$, $-\frac{\pi}{2} \le x \le \frac{\pi}{2}$.

In Exercises 85–89 find the maximum and minimum values of f on the given interval.

85. $f(x) = 1 + (x + 1)^2$, $-2 \le x \le 5$.

86. $f(x) = x^3 - 12x + 1$, $[-3, 5]$.

87. $f(x) = 4x^3 - 15x^2 + 12x + 7$, $[0, 3]$.

88. $f(x) = 3x^5 - 5x^3 - 1$, $[-2, 2]$.

89. $f(x) = \sqrt{9 - x^2}$, $[-1, 2]$.

90. Inscribe a rectangle in the ellipse $\frac{x^2}{a^2} + \frac{y^2}{b^2} = 1$ as shown in Figure 8.43. What are the dimensions in terms of a and b of the largest (in terms of area) such rectangle? What is its area?

8J. More Problems from the Books of L'Hospital and Agnesi

91. Place a point E on a fixed line segment AB. Where does E have to be placed so that the product of the square of the length AE times the square of the length EB is a maximum.

Figure 8.44

92. The line segment AB is divided into three parts AC, CF, and FB. Place a point E on the middle segment CF as shown in Figure 8.44 and consider the rectangles $AE \times EB$ and $CE \times EF$. Of all the points E on the middle segment CF, find the location of the one that has the property that the ratio $\frac{AE \times EB}{CE \times EF}$ is smallest. [Hint: Let $AC = a$, $CF = b$, $FB = c$, and $CE = x$. Express the ratio as a function of x and denote this function by $R(x)$. What is the domain of $R(x)$? Using the product and quotient rules, show that the numerator of $R'(x)$ is $(c - a)x^2 + 2a(b + c)x - ab(b + c)$ and that the denominator is positive. Use this to analyze $R'(x)$. Treat the cases, $c > a$, $c = a$, and $c < a$ separately.]

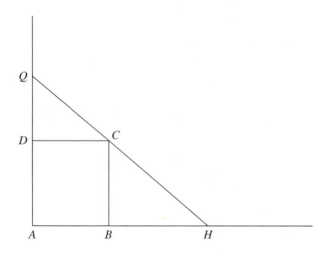

Figure 8.45

93. Consider the rectangle $ADCB$ and the line segment QH of Figure 8.45. What is the minimal length that QH can have? [Hint: Let $AB = b$, $AD = d$, and $x = BH$. Compute the length of the segment as a function of x and let $f(x)$ be the square of this length. Show that $f(x) = d^2(\frac{b}{x} + 1)^2 + (b + x)^2$. Use the fact that the length of a segment is a minimum exactly when the square of the length is a minimum.]

8K.   Newton's Method for Solving Equations

Consider the polynomial $f(x) = x^4 + 2x^3 - 12x - 1$. Notice that $f(0) = -1$ and $f(2) = 2^4 + 2 \cdot 2^3 - 12 \cdot 2 - 1 = 16 + 16 - 24 - 1 = 7$. Since $f(0)$ is negative and $f(2)$ is positive, it follows that $f(c) = 0$ for some c between 0 and 2. (This is an application of the Intermediate Value Theorem.) Newton developed a method of successive approximations that makes it possible to calculate the root c to any desired degree of accuracy. The essence of his method was already described in Section 4.8 and is illustrated in Figure 8.46.

(1) Make an educated guess, say c_1, for c.
(2) Use the following approximation step: From a point on the x-axis go up to the graph and down on the tangent to a new point on the x-axis.

Begin by applying the approximation step to the guess c_1 to get c_2; apply the approximation step to c_2 to get c_3; apply it to c_3 to get $c_4, \dots$. Continue in this way, and observe that the numbers $c_1, c_2, c_3, c_4, \dots$ zero in on the root c.

We will now see how Newton makes this mathematically precise. Just to be on the safe side (and not miss the root), let's start with $c_1 = 2$. To determine c_2, the intersection of the tangent at $(c_1, f(c_1))$ with the x-axis needs to be found. Because $f'(x) = 4x^3 + 6x^2 - 12$, the slope of the tangent is $f'(c_1) = f'(2) = 4 \cdot 2^3 + 6 \cdot 2^2 - 12 = 44$. Since $f(c_1) = f(2) = 7$, the equation of the tangent is

$$y - f(c_1) = f'(c_1)(x - c_1), \quad \text{or} \quad y - 7 = 44(x - 2).$$

The x-coordinate c_2 of the point of intersection is determined by setting $y = 0$ and solving for x. So $-f(c_1) = f'(c_1)(x - c_1)$, $x - c_1 = -\frac{f(c_1)}{f'(c_1)}$, and $x = c_1 - \frac{f(c_1)}{f'(c_1)}$. Therefore, working with an accuracy of four decimal places,

$$c_2 = c_1 - \frac{f(c_1)}{f'(c_1)} = 2 - \frac{7}{44} = \frac{81}{44} = 1.8410.$$

In exactly the same way, $c_3 = c_2 - \frac{f(c_2)}{f'(c_2)}$ and $c_4 = c_3 - \frac{f(c_3)}{f'(c_3)}$. In general, once the approximation c_i of the root has been found, then the next and better approximation is given by

$$c_{i+1} = c_i - \frac{f(c_i)}{f'(c_i)}.$$

In the specific case we are considering, we get

$$c_3 = c_2 - \frac{f(c_2)}{f'(c_2)} = 1.8410 - \frac{f(1.8410)}{f'(1.8410)}$$

$$= 1.8410 - \frac{(1.8410)^4 + 2(1.8410)^3 - 12(1.8410) - 1}{4(1.8410)^3 + 6(1.8410)^2 - 12}$$

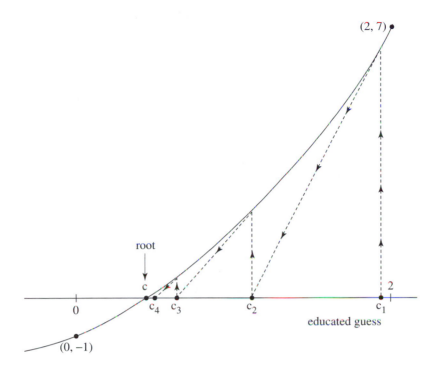

Figure 8.46

$$= 1.8410 - \frac{0.8746}{33.2944} = 1.8147,$$

$$c_4 = 1.8147 - \frac{(1.8147)^4 + 2(1.8147)^3 - 12(1.8147) - 1}{4(1.8147)^3 + 6(1.8147)^2 - 12}$$

$$= 1.8147 - \frac{0.0205}{31.6630} = 1.8141,$$

$$c_5 = 1.8141 - \frac{(1.8141)^4 + 2(1.8141)^3 - 12(1.8141) - 1}{4(1.8141)^3 + 6(1.8141)^2 - 12}$$

$$= 1.8141 - \frac{0.0015}{31.6263} = 1.8141.$$

So 1.8141 ought to be a root of the polynomial $f(x) = x^4 + 2x^3 - 12x - 1$. Because of round-off errors it isn't quite. Our computation has shown that up to four decimal places, $f(c_5) = 0.0015$. Carrying more decimal places would have provided greater accuracy. Since $f(1.8141) = 0.0015$ and $f(1.8140) = -0.0017$, note that the root is somewhere between 1.8141 and 1.8140.

94. Apply Newton's Method to $f(x) = \frac{1}{2}x^2 - 1$. Start with $c_1 = 2$ and use four decimal accuracy to approximate a root.

95. Apply Newton's Method to $f(x) = x^3 + x^2 - 7x - 7$. Start with $c_1 = 3$ and use four decimal accuracy to approximate a root.

Notes

[1] The literature features many different spellings of this name. This is the version that appears in the second edition of the *Analyse*.

[2] This comes from *Der Briefwechsel von Johann Bernoulli*, Band I, Birkhäuser Verlag, Basel 1955, which is a comprehensive study of Johann Bernoulli's correspondence. All the remarks made here are taken from the introduction to Part B, pages 123–157.

9

Connections with Statics, Dynamics, and Optics

All of mathematics, and in particular differential and integral calculus, has been energized throughout its development by the demands of basic science and engineering. In return, mathematics has enlightened these fields with substantial information and has provided a clarifying point of view. Indeed, it is one of the primary purposes of this text to make the point that modern science cannot exist without the input of mathematics. The disciplines of statics, dynamics, optics, electrodynamics, thermodynamics, the theories of propagation of sound and heat, etc. have all played important roles in this ongoing process of cross-fertilization. This chapter will illustrate this interaction with a look at basic elements of statics, dynamics, and optics.

We begin the subject of statics with one of the most interesting problems—the pulley problem—from the Marquis De L'Hospital's *Analyse des infiniment petits*. We will discuss two solutions. One will illustrate the methods of calculus; the other will develop and use very basic properties of forces. Both solutions will be "learning laboratories" for the other concerns of this chapter.

A mathematical analysis of the suspension bridge follows next. Early versions of such bridges using vines and ropes were built by natives of the Himalayas, Equatorial Africa, and South America. Suspension bridges supported by metal chains have spanned rivers in China since the 17th century. The initial mathematical analysis of the suspension bridge was published in 1794 by Nicolaus Fuss (1755–1826), yet another Swiss mathematician from Basel. Fuss's mathematical talents attracted the attention of Leonhard Euler, and in 1772 Fuss accepted Euler's invitation to join him in St. Petersburg. Fuss remained there for the rest of his life, focusing his mathematical writings primarily on questions arising from Euler's work. His study of the suspension bridge was a response to a proposal to erect such a bridge in St. Petersburg. He proved that a suspended cable subject to a uniform horizontal load—this is the condition that the cable of a suspension bridge is subject to—describes a parabola. We will describe this analy-

sis and the information that it provides in the context of the George Washington Bridge. This suspension bridge was constructed over the Hudson River in the period 1927–1931.

To illustrate the connection between mathematics and dynamics we will focus on an experiment of Galileo. As late as the 1970s historians of science could only speculate about the experiments that Galileo actually undertook. Indeed, some very influential historians of science held that Galileo's experiments were primarily thought experiments, in other words, carefully conceived speculations. It is now apparent that this assessment was incorrect. The evidence for this comes from Galileo's working papers, which were recently discovered to contain the results of a number of precise experiments with inclined planes.[1] For instance, a page from the year 1608—certainly one of the most important pages in the history of science—records the numerical data of an experiment into the nature of trajectories. The goal of our discussion will be the analysis of this experiment with particular attention to the mathematics that underlies it.

The final section of this chapter is devoted to optics. Optics is one of the oldest sciences to which mathematical analysis was applied. Experiments undertaken by Euclid (about 300 B.C.) revealed that when a ray of light strikes a mirror, the angle that the approaching ray makes with the perpendicular to the mirror is the same as the angle that the reflected ray makes with the perpendicular. In other words, the angle of incidence is equal to the angle of reflection. Archimedes proved this principle by showing that the inequality of these angles leads to a contradiction. Claudius Ptolemy realized that the path of a light ray coming from a star or planet near the horizon is curved. He had thus discovered the phenomenon of refraction: Light is bent as it travels through media of different densities. Ptolemy attempted to reach a quantitative understanding of this phenomenon but failed. Kepler failed also. Galileo wrestled with the question as to whether light has a finite velocity or whether it is propagated instantaneously and tried to find the answer experimentally.

The French mathematician Pierre de Fermat[2] (1601–1665), assuming that the velocity of light is finite, proposed his Principle of Least Time: In traveling from a point A to another point B, a ray of light will always choose the path of shortest time. He held this to be valid for a reflected ray as well as a refracted ray. From his principle, Fermat deduced Snell's Law of Refraction, which the Dutch mathematician-scientist Willebrord Snell (1580–1626) had discovered earlier experimentally. Our discussion will start with Fermat's principle and pursue its mathematical consequences until we understand the workings of the basic telescope.

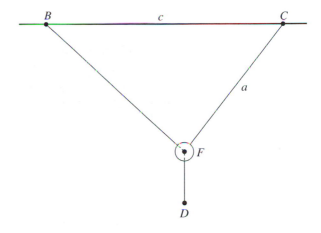

Figure 9.1

9.1 The Pulley Problem of De L'Hospital

Consider two distinct points B and C on the horizontal ceiling of a room. Suppose that the points are a distance c apart. At the point C attach a cord of length a, which has a pulley affixed at the other end, the point F. At the point B attach a cord of length l, pass it through the pulley at F and connect a weight W at the other end, at D. Release the weight and allow the system to achieve its equilibrium position. This is shown in Figure 9.1. The question that L'Hospital asks is this: What is the geometry of this equilibrium configuration? More precisely, what are the dimensions of the triangle BCF in terms of the constants c, a, l, and W? We will see that the solution of this seemingly elementary problem is surprisingly challenging. L'Hospital makes the simplifying assumption that the weight W is heavy enough so that the weights of the cords and the pulley can be ignored. In effect, he assumes that the cords and the pulley are weightless.

The first remark that should be made is this. If $a \geq c$, then the cord CF will not deflect the cord BD because both the pulley and the cord CF that holds it are weightless. So the weight at D will hang directly under B. Figure 9.2 shows a typical situation. We will therefore assume from now on that $a < c$.

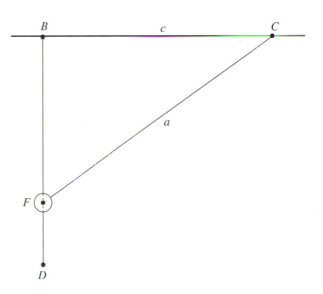

Figure 9.2

A. The Solution Using Calculus. Let E be the intersection of the extension of the segment FD with BC and set x equal to EC (Figure 9.3). Applying the Pythagorean theorem to the right triangle CEF, we see that $EF = \sqrt{a^2 - x^2}$. Note therefore that $x \leq a$. Another application of the Pythagorean theorem shows that

$$BF = \sqrt{BE^2 + EF^2} = \sqrt{(c-x)^2 + (a^2 - x^2)}.$$

Because the length of the string from B to D is l, we see that $FD = l - BF = l - \sqrt{(c-x)^2 + (a^2 - x^2)}$. The fact that the system is in equilibrium means that the weight at D must be at the very lowest point that the geometry of the situation will allow. Put another way, the distance ED must be a maximum. Observe that

$$ED = EF + FD$$

$$= \sqrt{a^2 - x^2} + l - \sqrt{(c-x)^2 + (a^2 - x^2)}$$

$$= (a^2 - x^2)^{\frac{1}{2}} + l - \left((c-x)^2 + (a^2 - x^2)\right)^{\frac{1}{2}}.$$

If L'Hospital can find the value of x for which ED is a maximum in terms of the constants c, a, and l, he will then have determined $EC = x$, $BE = c - x$, as well as EF, BF, EF, and FD in terms of these constants, and he will have solved his problem. It therefore remains to find the value (or, possibly, values) of x for which the function

$$f(x) = (a^2 - x^2)^{\frac{1}{2}} + l - \left((c-x)^2 + (a^2 - x^2)\right)^{\frac{1}{2}},$$

with $0 \leq x \leq a$, attains its maximum value. We will do this by using the strategy of Section 8.8: Compute $f'(x)$; find all the numbers x for which f' is either zero or does not exist (these are the critical numbers of f); and evaluate f at the critical numbers as well as at the endpoints 0 and a. The largest of these values will point to the required x.

By the chain rule,

$$f'(x) = \frac{1}{2}(a^2 - x^2)^{-\frac{1}{2}}(-2x)$$

$$- \frac{1}{2}\left((c-x)^2 + (a^2 - x^2)\right)^{-\frac{1}{2}}(-2(c-x) - 2x)$$

$$= -x(a^2 - x^2)^{-\frac{1}{2}}$$

$$- \frac{1}{2}\left((c-x)^2 + (a^2 - x^2)\right)^{-\frac{1}{2}}(-2c)$$

$$= -x(a^2 - x^2)^{-\frac{1}{2}} + c\left((c-x)^2 + (a^2 - x^2)\right)^{-\frac{1}{2}}$$

$$= \frac{c}{\left((c-x)^2 + (a^2 - x^2)\right)^{\frac{1}{2}}} - \frac{x}{(a^2 - x^2)^{\frac{1}{2}}}.$$

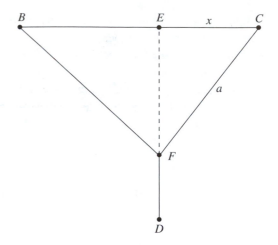

Figure 9.3

To determine the values of x for which $f'(x)$ is equal to zero or for which it does not exist, take common denominators and express $f'(x)$ as the single fraction

$$f'(x) = \frac{c(a^2 - x^2)^{\frac{1}{2}} - x\left((c-x)^2 + (a^2 - x^2)\right)^{\frac{1}{2}}}{(a^2 - x^2)^{\frac{1}{2}}\left((c-x)^2 + (a^2 - x^2)\right)^{\frac{1}{2}}}.$$

Observe that $f'(x) = 0$ precisely when the numerator is zero and that $f'(x)$ is undefined precisely when the denominator is zero. For which values of x with $0 \leq x \leq a$ does this occur?

(i) The denominator of $f'(x)$ is 0 precisely when $(a^2 - x^2)\left((c-x)^2 + (a^2 - x^2)\right) = 0$. This occurs when either

$$(c-x)^2 + (a^2 - x^2) = 0 \quad \text{or} \quad a^2 - x^2 = 0.$$

Because $x \leq a < c$, the first equality is not possible. Therefore, $a^2 - x^2 = 0$. Hence $x^2 = a^2$, and since both $x \geq 0$ and $a \geq 0$, $x = a$.

(ii) The numerator of $f'(x)$ is 0 precisely when

$$c(a^2 - x^2)^{\frac{1}{2}} = x\left((c-x)^2 + (a^2 - x^2)\right)^{\frac{1}{2}}$$

After squaring both sides, we see that

$$c^2(a^2 - x^2) = x^2(c-x)^2 + x^2(a^2 - x^2),$$

and by simplifying,

$$c^2a^2 - c^2x^2 = x^2(c^2 - 2cx + x^2) + x^2a^2 - x^4$$
$$= x^2c^2 - 2cx^3 + x^4 + x^2a^2 - x^4$$
$$= c^2x^2 - 2cx^3 + a^2x^2.$$

It follows that

$$(*) \qquad 2cx^3 - 2c^2x^2 - a^2x^2 + a^2c^2 = 0$$

To solve this equation for x, we will use a strategy from Exercise 3F in Chapter 3. Because

$$2cc^3 - 2c^2c^2 - a^2c^2 + a^2c^2 = 0,$$

it follows that $x = c$ is a root of the polynomial $(*)$ and hence that the term $x - c$ divides it. By a polynomial division,

$$2cx^3 - 2c^2x^2 - a^2x^2 + a^2c^2 = (x - c)(2cx^2 - a^2x - a^2c).$$

Check this by multiplying the factors. The quadratic formula tells us that the roots of $2cx^2 - a^2x - a^2c$ are

$$\frac{a^2 \pm \sqrt{a^4 - 4(2c)(-a^2c)}}{4c} = \frac{a^2 \pm \sqrt{a^4 + 8a^2c^2}}{4c}$$
$$= \frac{a^2 \pm a\sqrt{a^2 + 8c^2}}{4c}$$
$$= \frac{a}{4c}\left(a \pm \sqrt{a^2 + 8c^2}\right).$$

We have now shown that the zeros of the numerator $c(a^2 - x^2)^{\frac{1}{2}} - x\left((c - x)^2 + (a^2 - x^2)\right)^{\frac{1}{2}}$ of $f'(x)$ are

$$x = c, \ x = \frac{a}{4c}\left(a + \sqrt{a^2 + 8c^2}\right), \text{ and}$$
$$x = \frac{a}{4c}\left(a - \sqrt{a^2 + 8c^2}\right).$$

Recall that $x \leq a$ and $a < c$. So $x = c$ cannot occur. Notice also that $\sqrt{a^2 + 8c^2} > \sqrt{a^2} = a$. It follows that $a - \sqrt{a^2 + 8c^2} < 0$. This rules out the possibility that $x = \frac{a}{4c}(a - \sqrt{a^2 + 8c^2})$.

We summarize the results of our computations: The critical numbers of the function

$$f(x) = \sqrt{a^2 - x^2} + l - \sqrt{(c - x)^2 + (a^2 - x^2)}, \ 0 \leq x \leq a,$$

are $x = a$ from **(i)** and $x = \frac{a}{4c}(a + \sqrt{a^2 + 8c^2})$ from **(ii)**.

By Section 8.8, the maximum value of the function f must occur at $x = 0$, $x = a$ or $x = \frac{a}{4c}(a + \sqrt{a^2 + 8c^2})$. Refer back to Figure 9.3. If $x = 0$, then $E = C$ and the weight at D hangs directly under C. But this is impossible, since the weight W pulls on the cord CF that holds the pulley. If $x = a$, then $E = F$. This too is impossible, because the weight W pulls down on the pulley at F. It follows that f attains its maximum value when

$$x = \frac{a}{4c}\left(a + \sqrt{a^2 + 8c^2}\right).$$

So this is the value of x that the configuration in Figure 9.3 has when it is in equilibrium. Note that W does not appear in this expression for x. Therefore, the equilibrium configuration is the same regardless of the magnitude of the weight W. This was not obvious before our analysis began.

Example 9.1. Suppose specifically that $c = 4$ and $a = 3$. Then

$$x = \frac{a}{4c}(a + \sqrt{a^2 + 8c^2}) = \frac{3}{16}(3 + \sqrt{9 + 128})$$
$$\approx \frac{3}{16}(3 + 11.70) = \frac{3}{16}(14.70) = 2.76.$$

B. The Solution by Balancing Forces. Let's start with a few basic facts about forces. As was already pointed out in the introduction to Chapter 7, a force has both a magnitude and a direction. The magnitude of a force can be measured, for example, by the amount of displacement that it produces on some standardized steel spring. The magnitude and direction together determine the force. Forces can, therefore, be represented by arrows, more commonly called *vectors*. This is done as follows. The direction in which the vector is pointing is given by the direction in which the force is acting, and the length of the vector is taken to be quantitatively equal to the magnitude of the force. Forces will typically be denoted by F, F_1, or F_2. When such a symbol appears in a mathematical expression, then it will be understood that it represents the magnitude of the force.

Consider two forces, and let F_1 and F_2 be their magnitudes. If the two forces act in the same direction, then their combined, or net, effect is a force in that direction with magnitude $F_1 + F_2$. If they act in directly opposed directions, then their net effect is a force acting in the direction of the larger force. If, say, F_1 is larger, then the magnitude of the combined force will be $F_1 - F_2$. In general, the combined effect is determined by the *parallelogram law*. The combination of the two forces is represented by the vector determined by the diagonal of the parallelogram in Figure 9.4. If several forces act, then their net effect can be determined by the parallelogram law by combining them two at a time. The combined effect of any number of forces is called their *resultant*. Because of the parallelogram law, the representation by vectors is much more than a convenient way to think about forces. Indeed, the representation of forces by vectors provides a fundamental insight into the way forces act.

combined effect of its two components. Let F_1 and F_2 be the respective magnitudes of the components. Since $\sin\theta = \frac{F_1}{F}$ and $\cos\theta = \frac{F_2}{F}$, it follows that

$$F_1 = F\sin\theta \quad \text{and} \quad F_2 = F\cos\theta.$$

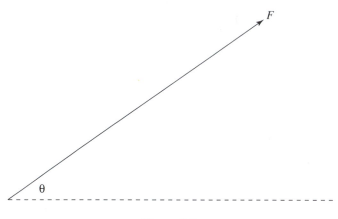

Figure 9.5

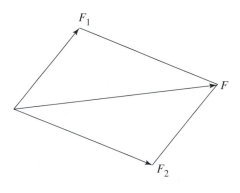

Figure 9.4

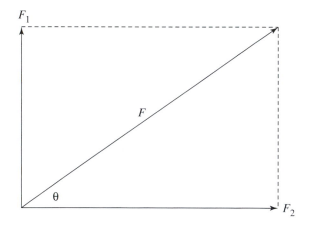

Figure 9.6

The parallelogram law can also be used to separate a force into components. Suppose, for example, that a force of magnitude F acts at an angle θ with the dotted line, as shown in Figure 9.5. Embed the vector corresponding to F as the diagonal of the rectangle indicated in Figure 9.6. The two forces determined by the sides of the rectangles are the *components* of the given force in the indicated directions. By the parallelogram law, the effect of the force is the same as the

Example 9.2. Consider the two vectors from the origin to the points $(1, 4)$ and $(-4, -3)$. What is their resultant? Decompose both vectors into their horizontal and vertical components. These decompositions are shown in Figure 9.7(a). The addition of the two horizontal components gives the vector from the origin

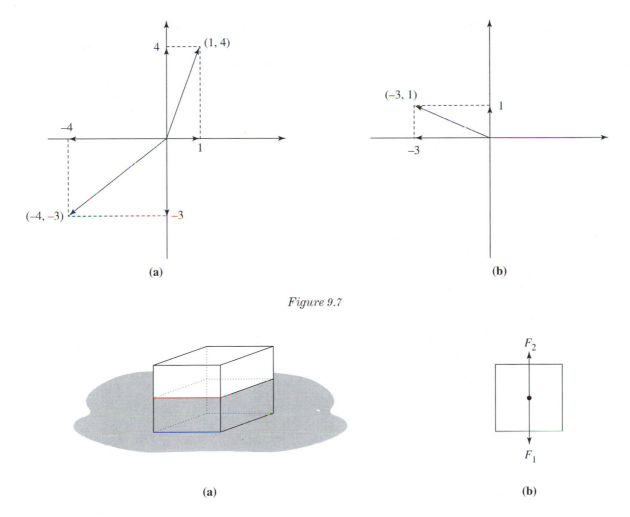

Figure 9.7

Figure 9.8

to the point -3 on the x-axis. Doing the same for the two vertical components gives the vector from the origin to the point 1 on the y-axis. Both of these vectors and the resultant are shown in Figure 9.7(b). Observe that this must be the resultant of the two vectors with which we began.

Suppose that a block of wood is floating motionlessly in a pool of still water. (Figure 9.8.) Consider the vertical forces that act on it. The force of gravity, denote it by F_1, acts downward and has magnitude equal to the weight of the block. The upward force on the block is given by Archimedes's Law of Hydrostatics: A body immersed in a fluid is buoyed up by a force equal to the weight of the fluid that the body displaces. This means that the upward force F_2 on the block is equal to the weight of the water that it displaces. Since the block is not moving, the net vertical force on it is zero. Hence $F_1 = F_2$: The weight of the block is equal to the weight of water that it displaces. Suppose that the block weighs 10 pounds. So the block displaces 10 pounds of water. Since water weighs 62.5 pounds per cubic foot, the block displaces $\frac{10}{62.5} = 0.16$ cubic feet of water. This is the volume of the submersed part of the block. Suppose that the block is a cube with side length 0.6 feet and that it floats with its top surface parallel to the surface of the water. If h is the amount (in feet) of the block that is submersed, then the volume of the submersed part of the block is $(0.6)(0.6)h = 0.36h$. It follows that $0.36h = 0.16$ and hence that $h = \frac{0.16}{0.36} = 0.44$ feet. Because $\frac{0.44}{0.6} \approx 0.73$, about 73% of the block is submersed.

Next consider a 50 pound object suspended at the midpoint of a cable, as in Figure 9.9. The angles at A and B between the cable and the horizontal are both α. What is the tension in the two parts of the cable?

The *tension* in a cable or string is the force with which the cable or string pulls. Let T_1 be the tension in the cable AC and T_2 that in CB. Decompose both T_1 and T_2 into their horizontal and vertical components as shown in Figure 9.10. The system is assumed to be in equilibrium; so nothing moves. So $T_1\cos\alpha = T_2\cos\alpha$, and hence, $T_1 = T_2$. Set $T = T_1 = T_2$. Since the vertical components of the tension are both $T\sin\alpha$, it follows that $2T\sin\alpha = 50$. So $T = \frac{25}{\sin\alpha}$. This answers the question. Note that if $\sin\alpha$ is small, then T will be large. For example if $\alpha = 1°$, then $\sin\alpha = 0.01745$ and T is a hefty 1432.7 pounds!

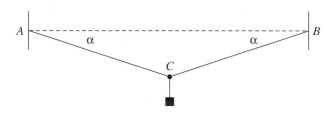

Figure 9.9

We can now return to L'Hospital's pulley problem. Let T_1 be the tension in the cord BF and let T_2 be the tension in the cord CF to which the pulley is affixed. See Figure 9.11. Let θ_1 and θ_2 be the indicated angles. Since the system is in equilibrium, the cord FD pulls with a force equal to the weight W of the suspended

object. And because the pulley at F is free to rotate, the tension in the cord BFD must be a constant. This implies that $T_1 = W$.

Figure 9.12(a) provides a diagram of the forces acting at F. The forces T_1 and T_2 are decomposed into their horizontal and vertical components in Figure 9.12(b). By facts already mentioned, the two horizontal components have magnitudes $T_1\cos\theta_1$ and $T_2\cos\theta_2$ respectively, and the magnitudes of the two vertical components are $T_1\sin\theta_1$ and $T_2\sin\theta_2$. Since the forces are in balance, the two horizontal components must be equal, and the two vertical components must add to the weight W. Therefore,

$$T_1\cos\theta_1 = T_2\cos\theta_2 \quad \text{and} \quad T_1\sin\theta_1 + T_2\sin\theta_2 = W.$$

Using the second equation and the fact that $W = T_1$, we get

$$\frac{T_1\sin\theta_1}{T_1\cos\theta_1} + \frac{T_2\sin\theta_2}{T_1\cos\theta_1} = \frac{T_1\sin\theta_1 + T_2\sin\theta_2}{T_1\cos\theta_1}$$

$$= \frac{W}{T_1\cos\theta_1} = \frac{T_1}{T_1\cos\theta_1}$$

$$= \frac{1}{\cos\theta_1}.$$

Because $T_1\cos\theta_1 = T_2\cos\theta_2$,

$$\frac{T_1\sin\theta_1}{T_1\cos\theta_1} + \frac{T_2\sin\theta_2}{T_2\cos\theta_2} = \frac{1}{\cos\theta_1}.$$

It follows that $\tan\theta_1 + \tan\theta_2 = \frac{1}{\cos\theta_1}$. Refer to Figure 9.11 and recall from the begining of Solution A that $EC = x$, $BE = c - x$, $EF = \sqrt{a^2 - x^2}$, and $BF = \sqrt{(c - x)^2 + (a^2 - x^2)}$. Hence

$$\tan\theta_1 = \frac{EF}{BE} = \frac{\sqrt{a^2 - x^2}}{c - x}, \quad \tan\theta_2 = \frac{EF}{EC} = \frac{\sqrt{a^2 - x^2}}{x},$$

$$\text{and} \quad \cos\theta_1 = \frac{BE}{BF} = \frac{c - x}{\sqrt{(c - x)^2 + (a^2 - x^2)}}.$$

After substituting, we get

$$\frac{\sqrt{a^2 - x^2}}{c - x} + \frac{\sqrt{a^2 - x^2}}{x} = \frac{\sqrt{(c - x)^2 + (a^2 - x^2)}}{c - x}.$$

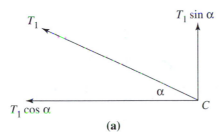 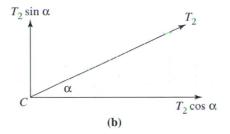

Figure 9.10

Taking common denominators shows that

$$\frac{x\sqrt{a^2 - x^2} + (c - x)\sqrt{a^2 - x^2}}{x(c - x)}$$

$$= \frac{\sqrt{(c - x)^2 + (a^2 - x^2)}}{c - x}.$$

Simplifying the numerator on the left, canceling $c - x$ from both sides, and multiplying the equation through by x gives us

$$c\sqrt{a^2 - x^2} = x\sqrt{(c - x)^2 + (a^2 - x^2)}.$$

By squaring both sides,

$$c^2(a^2 - x^2) = x^2\left((c - x)^2 + (a^2 - x^2)\right).$$

By routine algebra we now see that

$$c^2 a^2 - c^2 x^2 = x^2(c^2 - 2cx + x^2 + a^2 - x^2)$$

$$= x^2(c^2 - 2cx + a^2) = x^2 c^2 - 2cx^3 + a^2 x^2.$$

It follows, finally, that $2cx^3 - 2c^2 x^2 - a^2 x^2 + c^2 a^2 = 0$. But this is equation $(*)$ that was derived in Solution A with the methods of calculus, and we have already seen that the solution $x = \frac{a}{4c}(a + \sqrt{a^2 + 8c^2})$ of this equation solves L'Hospital's pulley problem.

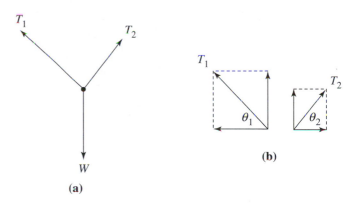

Figure 9.12

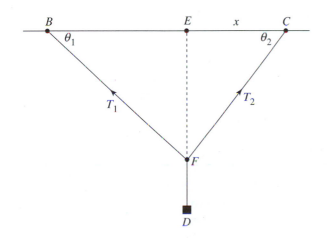

Figure 9.11

9.2 The Suspension Bridge

The basic principle behind the suspension bridge is simple: A road bed is attached by numerous vertical cables to either two or four *main cables*. These are "draped" over two towers, which support

Figure 9.13

the entire load. (See Figure 9.13.) The great pioneer of the modern suspension bridge was John Roebling. He designed and supervised the construction of a number of impressive suspension bridges towards the latter part of the 19th century. The most famous of these is the Brooklyn Bridge over the East River in New York. Opened to traffic in 1883, it has a *center span*—this is the section of the bridge between the two towers—of 1600 feet. The Brooklyn Bridge is generally recognized to be the ancestor of the modern suspension bridge.

A major advance in suspension bridge design came in 1931 with the George Washington Bridge over the Hudson River in New York. Its 3500 foot center span was almost twice as long as that of any previous suspension bridge. This bridge set a new standard for a series of great suspension bridges in America. These included the Golden Gate Bridge in San Francisco (completed in 1937) with its 4200 foot center span, and the Verrazano Narrows Bridge in New York City (completed in 1964) with a 4260 foot center span. Our discussion will focus on the George Washington Bridge from its inception to its completion. In the process, we will focus on essential elements of planning, engineering, and mathematics.

A need for additional bridges and tunnels connecting New Jersey and New York became apparent in the early 1920s. In 1923 the New York Port Authority was authorized to proceed with the planning, and by the mid 1920s various preliminary plans had developed into a concrete project for a bridge spanning the Hudson River between New York City and Fort Lee, New Jersey. Given the requirements of shipping and the fact that a bridge of total length of about 5000 feet was required, a major design question was already answered: Suspension bridges provide sufficient clearance for shipping and span such distances economically. The project began in September of 1927 with chief engineer Othmar Ammann in charge.

The geological properties of the soil and rock formations under the riverbed of the proposed sight were determined by borings, excavations, soil and compression test, etc. At this point, the location of the two main supporting towers was decided upon. This determined the record-setting length of 3500 feet for the *center span*.[3] (Refer to Figure 9.14 for the explanation of the terminology used in the discussion that follows.) In addition to the center span of 3500 feet, the bridge was to consist of two *side spans* of 610 feet and 650 feet each. So the total length of the bridge was to be 4760 feet. The design called for the deck to be about 200 feet over the water surface. This would provide enough clearance for the ships of the Hudson.

The *deck* was designed to accommodate a roadway of eight lanes for a vehicular traffic projected at 25,000,000 vehicles per year. (The plans included provisions for the possible later addition of a second deck. This was added in 1962.) These requirements called for the deck to have a *dead load* (this is the structural weight of the deck without traffic) of about 39,000 pounds per foot for the center span and 40,000 pounds for the two side spans. These figures included

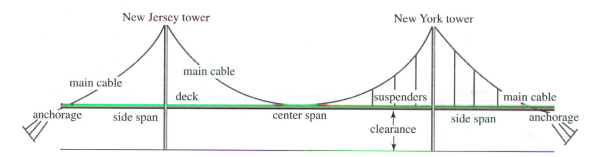

Figure 9.14

not only the weight of the deck, but also the estimated weights of the *main cables* and the *suspender cables* (both of which the main cables have to bear). The design also called for a maximum *live load* capacity (the projected weight of the vehicular traffic at its maximum: cars, buses, and trucks, bumper to bumper, on all eight lanes) of 8,000 pounds per foot for all spans.

It goes without saying that the execution of such a structure cannot proceed simply by sending out a crew of workers with shovels or suitable machinery, by taking some cables off a shelf, or by hoisting some beams into place. Quite obviously, an exacting and detailed mathematical analysis is required before any aspect of the construction can proceed. Automobiles can be crash tested, but an enormous structure such as the suspension bridge under discussion cannot! And catastrophic failure is not an option.

The mathematical analysis must answer a number of fundamental questions: What loads will the towers have to support? How high should the towers be? What tensions will the main cables be subjected to? How much sag should be allowed in the main cables over the center span? How should the side spans be configured? How much should the tension be in the main cables that support the side spans? What are the requirements on the structures—called *anchorages*—on either end of the bridge that affix the main cables to the ground? What about the suspender cables that attach the deck to the main cables? What effect do all these choices have on the stability of the structure as a whole? What materials should be used? What safety factors have to be built in?

Return to the essential elements of the George Washington Bridge. Estimates for the dead and live loads of the decks are already in place. The vertical suspenders connect the deck to the main cables and generate tensions in these cables. These tensions in turn produce a pull on the towers. It follows that the structural demands on the towers can be understood only if it is known what tensions the loads of the deck produce in the main cables. It is one of the purposes of the main cables over the side spans to counterbalance the forces on the towers that the main cables over the center span generate. Therefore, only if the tensions in the main cables over the center span are known can the tension requirements of the main cables over the two side spans be determined. Once these are understood, then the specifications on the anchorages can be considered. It should now be apparent why the main cables over the center span will be the primary focus of the analysis to follow.

Note that the height h of the towers above the water level is largely determined by the *sag s* in the cable of the center span (Figure 9.15). Concentrate on the New York tower of the George Washington Bridge. The deck near this tower is to clear the water level by about 195 feet. Letting r be the vertical distance from the bottom of the deck to the lowest point of the main cable over the center span, we get

$$h = 195 + r + s$$

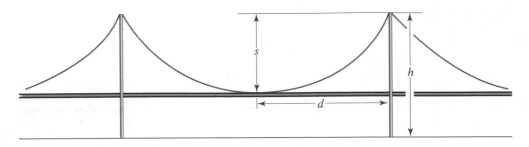

Figure 9.15

Since r is by design relatively small, the sag s in the main cable over the center span is the remaining factor in the height of the towers.

The preceding discussion shows that the requirements that the fundamental components of the bridge must meet depend on an understanding of the tension in the main cables over the center span. This will be the focus of the mathematical analysis that follows.

For the George Washington Bridge it was decided to use a configuration of four parallel main cables, two on each side of the bridge. This was the strategy already deployed in the three large bridges in New York over the East River (the Brooklyn, Williamsburg, and Manhattan Bridges). Given the projected size of the George Washington Bridge, it was realized that the use of only two main cables would have required cables of a size far beyond that of any cables previously constructed.

To facilitate the mathematical analysis we will make certain assumptions that are contrary to actual fact but do not have an appreciable effect on the quantitative considerations in question. These assumptions are the following. The weight of the main cables and the suspenders are considered to be included in the weight of the deck. The main cables are assumed to be completely flexible (they have no resistance to bending) and inextensible (they do not stretch). Finally, we will assume that there is a large number of vertical suspender cables, each very close to the next.

We will see that the important constants—we will also call them *parameters*—are:

$w =$ the dead load of the deck (including the main cables and suspenders) plus the maximum live load capacity *per cable* in pounds per foot. In the case of the George Washington Bridge this is obtained by adding $39{,}000 + 8{,}000 = 47{,}000$ and distributing this over the four cables. So $w = \frac{47{,}000}{4} = 11{,}750$ pounds per foot.

$d =$ the horizontal distance from the bottom of the center span to the center of a tower. In the case of the George Washington Bridge this has already been determined as $d = 1750$ feet.

$s =$ the sag, i.e., the vertical distance from the top of the tower to the bottom of the cable at mid span. (See Figure 9.15.) This is understood to be the sag under dead and live load. In the case of the George Washington Bridge, the sag still has to be determined.

Consider a single main cable over the center span of the bridge. Place an x-y coordinate system in such a way that the origin is at the lowest point of the cable. (See Figure 9.16.) Let C be an arbitrary point on the cable and let x be its coordinate. Let T_0 be the tension in the cable at 0 and let T_x be the tension at x. Notice that the vector T_x is tangent to the cable at x and let θ be the angle that T_x makes with the horizontal. Resolve T_x into its horizontal and vertical components as shown in Figure 9.17. Consider only the part of the cable from 0 to C together with the part of the deck from 0 to x that it carries. Because the system is in balance, the net horizontal force is equal to zero. So $T_x \cos \theta$ is equal to T_0. Since the part of the deck being

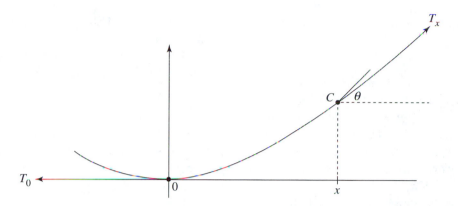

Figure 9.16

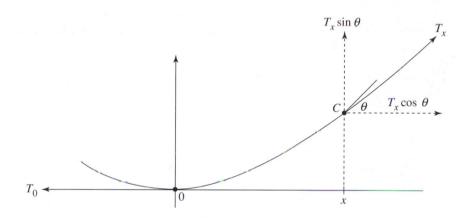

Figure 9.17

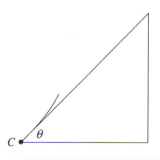

Figure 9.18

considered is x feet long, its weight is wx. Since the net vertical force is also zero, it follows that $T_x \sin\theta$ is

equal[4] to wx. In summary,

$$(*) \qquad T_0 = T_x \cos\theta \quad \text{and} \quad wx = T_x \sin\theta.$$

Therefore, $\frac{wx}{T_0} = \frac{T_x \sin\theta}{T_x \cos\theta} = \frac{\sin\theta}{\cos\theta} = \tan\theta$. A look at the right triangle of Figure 9.18 shows that $\tan\theta$ is equal to the slope of the tangent to the cable at the point C.

Now let $y = f(x)$ be a function whose graph is the curve of the cable. Since $f'(x)$ is the also equal to the slope of the tangent to the curve at C, it follows that

$$f'(x) = \tan\theta = \frac{wx}{T_0} = \frac{w}{T_0}x.$$

Because $\frac{w}{T_0}$ is a constant, f must have the form $f(x) = \frac{w}{2T_0}x^2 + k$ for some constant k. (Why?) Since $f(0) = 0$, we see that $k = 0$. Therefore, the curve of the cable over the center span is the graph of the function

$$f(x) = \frac{w}{2T_0}x^2.$$

We know from Section 4.3 that the graph of such a function is a parabola. Recall that we have made simplifying assumptions (for example about the cable and the suspenders) to achieve this result. In practice, the curve is not quite a parabola, but the greatest deviation in the y-coordinate from the parabolic shape is seldom more than 0.5%.

Refer to Figure 9.15 and observe that the point (d, s) is on the graph of $f(x) = \frac{w}{2T_0}x^2$. Therefore, $s = \frac{w}{2T_0}d^2$. Hence $\frac{s}{d^2} = \frac{w}{2T_0}$ and $T_0 = \frac{1}{2}\frac{wd^2}{s}$. So by a substitution,

$$f(x) = \frac{s}{d^2}x^2 \quad \text{and} \quad f'(x) = \frac{2s}{d^2}x.$$

By squaring both sides of the equations (∗), we get $T_0^2 = T_x^2 \cos^2\theta$ and $(wx)^2 = T_x^2 \sin^2\theta$. An addition of these equations shows that

$$T_0^2 + (wx)^2 = T_x^2 \cos^2\theta + T_x^2 \sin^2\theta$$
$$= T_x^2(\sin^2\theta + \cos^2\theta) = T_x^2.$$

It follows that $T_x = \sqrt{T_0^2 + (wx)^2}$, and, after substituting $T_0 = \frac{1}{2}\frac{wd^2}{s}$, that

$$T_x = \sqrt{\frac{1}{4}\frac{w^2 d^4}{s^2} + w^2 x^2} = w\sqrt{\frac{1}{4}\frac{d^4}{s^2} + x^2}.$$

Because we are considering the cable only from its lowest point to the tower, x is restricted to the interval $0 \le x \le d$. A look at the expression for T_x just derived shows that T_x has its minimum value when $x = 0$ and its maximum value when $x = d$. So the minimal tension in the cable is $T_0 = \frac{1}{2}\frac{wd^2}{s}$. It occurs at its lowest point. The maximal tension that the main cable over the center span is subjected to is

$$T_d = w\sqrt{\frac{1}{4}\frac{d^4}{s^2} + d^2} = wd\sqrt{\frac{1}{4}\frac{d^2}{s^2} + 1}$$
$$= wd\sqrt{\left(\frac{d}{2s}\right)^2 + 1}$$

and this occurs at the point at which the cable meets the tower.

Let α be the angle that the cable makes at the tower with the horizontal. Observe that

$$\tan\alpha = f'(d) = \frac{2s}{d^2}d = \frac{2s}{d}.$$

We have now determined (in terms of the parameters w, d, and s) both the magnitude and the direction of the force with which the main cable pulls on the tower. See Figure 9.19. The downward vertical component

$$T_d \sin\alpha$$

of T_d is the *compression* produced in the tower by a single main cable over the center span. The towers can resist considerable forces of compression, but they have little resistance to horizontal forces. To insure the stability of the towers, the horizontal component

$$T_d \cos\alpha$$

of T_d must be counterbalanced by the horizontal component of the corresponding main cable over the side span. The towers must be able to withstand the changes in the tensions in the cable brought about by the variations in live loads and temperatures that the bridge is subjected to. They must be designed to be flexible enough to allow them to bend slightly as the bridge adjusts to such variations.

It should now be apparent that our analysis has provided important information about essential com-

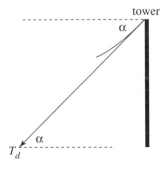

Figure 9.19

ponents of the bridge: the main cables over the center span, the towers, the cables over the side span, and hence also about the demands on the anchorages.

Let's return to the specifics of the George Washington Bridge. We already know that $d = 1750$ feet and $w = 11,750$ pounds. So by a substitution,

$$T_d = wd\sqrt{\left(\frac{d}{2s}\right)^2 + 1} = (11,750)(1750)\sqrt{\left(\frac{1750}{2s}\right)^2 + 1}.$$

The important remaining question is: What allowance should be made for the sag s in the main cable (under dead plus live load)? Taking s larger will make the term $\left(\frac{1750}{s}\right)^2$ smaller. This would decrease the tension T_d in the main cable and the compression in the towers that the design of the bridge has to deal with. On the other hand, a larger sag requires higher towers and longer cables (see Figure 9.15). After weighing all the critical factors (such as the stability of the structure and its cost) carefully, the engineers stipulated a sag of $s = 327$ feet. This would give the George Washington Bridge a *sag-to-center-span ratio* of $\frac{s}{2d} = \frac{327}{3500} = 0.093$ (less than the more conventional $\frac{1}{8} = 0.125$).

This choice of s would result in a tension of

$$T_0 = \frac{1}{2}\frac{wd^2}{s} = \frac{1}{2}\frac{(11,750)(1750)^2}{327} \approx 55,000,000 \text{ pounds}$$

at the bottom of the center span, and a maximal tension in the main cable of

$$T_d = (11,750)(1,750)\sqrt{\left(\frac{1750}{2s}\right)^2 + 1}$$

$$= (11,750)(1750)\sqrt{\left(\frac{1750}{654}\right)^2 + 1}$$

$$= (20,562,500)(2.85659) \approx 58,700,000 \text{ pounds}.$$

This analysis[5] has the following implications for the towers. The four main cables over the center span will subject each tower to a force of

$$4T_d \approx 235,000,000 \text{ pounds}.$$

Because $\tan\alpha = \frac{2s}{d} = \frac{654}{1750} = 0.374$, we get $\alpha = 20.5°$. Since $\sin\alpha = \sin 20.5° = 0.350$, these four cables will produce a compression in each tower of

$$4T_d \sin\alpha \approx (235,000,000)(0.350) \approx 82,000,000 \text{ pounds}.$$

The four main cables over the side span can be expected to roughly double this compression.

Recall that the height h of the towers above the water level is determined by the equation $h = 195 + r + s$, where r is the vertical distance from the lowest point on the main cable to the bottom of the deck. The choice of $s = 327$ feet means that the height of the towers would have to be between 500 and 600 feet. See Figure 9.15.

What has been achieved? An outline of a possible design for the bridge has emerged! In addition, estimates have been derived for the tensions and stresses that a bridge built in accordance with such a design would be subjected to. The engineers determined that such tensions and stresses, as well as the implications that they would have for the suspender cables, the main cables over the side span, the anchorages, and the foundations of the towers, could be dealt with. This analysis and the study of small-scale models formed the basis of a detailed design for the George Washington Bridge. Thereafter, the plans for its construction were drawn up.

Finally, on September 21, 1927, construction of the bridge began. It was a huge undertaking. The large concrete foundational structures in which the towers were to be lodged were put in place under the riverbed. The two towers, 591 feet high and consisting of a combination of carbon and silicone steels, were erected one section at a time.

The main cable was manufactured on-site from galvanized steel wires with a diameter of 0.196 inches. The wires were laid in big loops from one anchorage, over the two towers, to the other anchorage, and back again. This process is called *spinning*. A *spinning wheel* pulled the wires from one end of the bridge to the other. Special tongs then compressed 217 loops at a time into *strands* of about 4.5 inches in diameter.

So each strand consisted of 434 wires. A total of 61 strands was spun. (Why 61? See Exercise 9D.) Squeezers, consisting of twelve hydraulic jacks mounted on a circular frame, pressed the 61 strands together into the main cable. Thus the cable has 61 strands of 434 wires each, for a total of $61 \times 434 = 26,474$ individual wires! After a coat of lead paste was applied, the cable was wrapped with additional wires. The cable was now complete. It had a smooth, round surface, a diameter of about 3 feet, and a weight of 11,120 pounds per foot.

A single wire has an *ultimate strength* (the tension supported just before the point of failure) of 220,000 pounds per square inch. Because a single wire has cross-sectional area $\pi \left(\frac{0.196}{2}\right)^2 = 0.03$ square inches, it can bear an ultimate load of $0.03(220,000) = 6600$ pounds. Therefore, the main cable, with its 26,474 wires, can sustain an ultimate tension of

$$(6600)(26,474) \approx 175,000,000 \text{ pounds.}$$

A comparison of this number with the estimated maximum tension of $T_d = 58,700,000$ pounds computed earlier shows that a *safety factor* of about 3 is built into the main cable.

The strands at the end of each cable were locked into the anchorages with the appropriate tensions. Then the vertical suspender cables were put into place. Each consists of a single wire rope of 2.875 inches in diameter. They occur every 60 feet in sets of 16, four for each main cable. Then the floor steel for the deck of the center span was hoisted into position by cranes. This was done in sections, starting from the towers and proceeding towards the center of the bridge. The floor sections for the side spans followed. Stiffening girders were built into the deck. Running parallel to the deck, they serve to distribute the live load more evenly over the main cables. This controls the amount of deformation in the cables at heavy live loads.

The construction of the bridge was completed on October 24, 1931, four years after it began. The total cost was about 60 million dollars. (Because of inflation, these dollars were worth much more than today's

dollars.) The famous architect Le Corbusier referred to the George Washington Bridge as

> "the most beautiful bridge in the world .... It is blessed. It is the only seat of grace in the disorderly city."

The record for center span length was held by the Humber Bridge in Hull, England. It was completed in 1981 and has a center span of 4625 feet. This record was eclipsed by two "super" suspension bridges both completed in 1998. One is the "missing link" of a highway system that connects Copenhagen with mainland Denmark. It has a center span of 5300 feet. The second bridge with an enormous center span of 6500 feet spans the Akashi Straits in Japan and is an integral part of a highway connecting the main island of Honshu to the islands Awaji and Shikoku. See Exercise 9C.

Our mathematical study of the suspension bridge represents only a start. The complete theory is very complicated. It includes the investigation of the stresses, strains, and bending moments in the girders and towers. (The concept of *moment of force*, i.e., the capacity of a force to rotate an object or a structure, will be discussed in the next section.) The complete theory also includes the analysis of the deformations of the structure under various load and temperature conditions. These can be considerable (especially under large live loads) and must be clarified to insure a safe and stable structure. A so-called *deformation theory* has been developed in response to these concerns.[6] Such mathematical theories are supplemented with computer analyses. Because wind conditions can destabilize a bridge, wind tunnel tests are undertaken on scale models. A number of failures testify to the complexity of the design and construction of suspension bridges. The most dramatic of these occured in 1940 when the Tacoma Narrows Bridge collapsed just four months after its completion. Winds of 40 miles per hour caused the deck of the 2800 foot center span to sway and roll. Later, the deck began to rotate back and forth so violently that a section of the deck dropped into the water some 208 feet below. The wind had subjected the structure to motions that its steady force progressively amplified, until failure occurred.[7]

9.3 An Experiment of Galileo

In 1608, Galileo took an inclined plane with a groove running along it and placed it on top of a table. He let a bronze ball (of about 2 centimeters $= 0.02$ meters in diameter) roll repeatedly down the groove. The ball always started from rest, but Galileo let it descend from various heights. Each time it would roll down the plane, then briefly along the horizontal table, before falling to the ground along a parabolic arc. Galileo measured his observations very carefully and recorded his findings in his notebook. His unit of length was the *punto* (about 0.94 millimeters) and

his unit of time was the *tempo* (about 0.011 seconds). Figure 9.20 depicts one of Galileo's diagram.[8] His data have been converted to meters and the point B designating the foot of the table has been added. We see that the edge of the table was 0.778 meters from the ground. The starting heights (above the surface of the table) of the ball appear in the column on the right; the corresponding distances of the point of impact from the foot of the table are listed at the bottom of the diagram. It is the goal of this section to develop the mathematical theory that underlies this experiment and to compare Galileo's observations with the predictions of the theory.

A. Sliding Ice Cubes and Spinning Wheels. Consider Galileo's inclined plane, and let its angle of inclination be β. Let's start by letting an ice cube of mass m slide down the plane, and let's assume that it does so

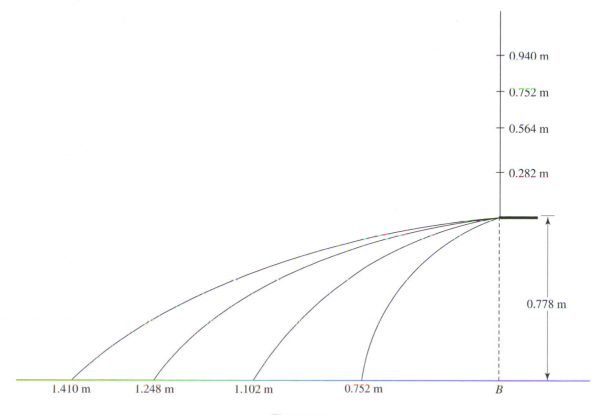

Figure 9.20

in a completely frictionless way. See Figure 9.21. Note that gravity acts on the ice cube. If the ice cube were in free fall, it would experience a downward acceleration of g and a corresponding downward force of mg. (This has already been discussed in Section 6.4.) This force also acts on the sliding ice cube. Consider this force decomposed into the following two components: one in the direction of the inclination of the plane and the other perpendicular to it. The first propels the ice cube down the plane; the second is evenly balanced by the plane pushing up on the ice cube. Refer to Figure 9.22(a). That the indicated angle is equal to β is seen by comparing the two right triangles in Figure 9.22(b). Let $F(t)$ be the magnitude of the force in the direction of the inclination of the plane. From Figure 9.22(a), $\sin \beta = \frac{F(t)}{mg}$ and therefore

$$F(t) = mg \sin \beta.$$

So this is the magnitude of the force that accelerates the ice cube down the plane.

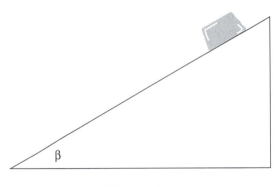

Figure 9.21

We know that Galileo did not experiment with ice cubes and that he did not have frictionless inclined planes. He used metal balls and real planes. But one key element remains the same. If m is the mass of the ball, then the component of the gravitational force that pushes the ball down the plane is the same $F(t) = mg \sin \beta$ already observed for the ice cube. But there is an important difference. The ice cube

on the frictionless surface slides down the plane. On a frictionless plane Galileo's bronze ball would also slide. But his plane is not frictionless, and it is the friction that causes the ball to roll. We will now begin to analyze this rotation of the ball.

Suppose that a point is moving counterclockwise along a circle with center O and radius r. Alternatively, think of a fixed point on the perimeter of a wheel that is spinning counterclockwise. Let A be the indicated point on the circle. Start observing the point at time $t = 0$. Consider the rotating segment determined by O and the position P of the point. See Figure 9.23. For any time $t \geq 0$, let $\theta(t)$ be the angle in radians that this segment has traced out. It is measured counterclockwise from A. Let $s(t)$ be the total distance along the circle from A to the point at time t. This distance is also measured counterclockwise from A. Each time the point goes past A, an amount of 2π is added to $\theta(t)$, and the distance it has covered increases by $2\pi r$. The definition of the radian measure of an angle (review Section 4.4 if necessary) shows that $s(t)$ and $\theta(t)$ are related by the equation $\theta(t) = \frac{1}{r}s(t)$, or $s(t) = r\theta(t)$.

Fix a time t. Then let an additional Δt elapse. Let the point be in position P at time t and in position P' at time $t + \Delta t$. Refer to Figure 9.23. Observe that the angle that the point traces out during the time interval $[t, t + \Delta t]$ is $\theta(t + \Delta t) - \theta(t)$. The ratio

$$\frac{\theta(t + \Delta t) - \theta(t)}{\Delta t}$$

is the *average rate of change of the angular position per unit time* over the time interval $[t, t + \Delta t]$. Taking this average over smaller and smaller Δt pushes $\frac{\theta(t+\Delta t)-\theta(t)}{\Delta t}$ to what is called the *rate of change of the angular position* at the instant t. This rate is the derivative

$$\theta'(t) = \lim_{\Delta t \to 0} \frac{\theta(t + \Delta t) - \theta(t)}{\Delta t}$$

of the function $\theta(t)$ and is called the *angular velocity* of the point at time t. It will be denoted by $\omega(t)$. So $\omega(t) = \theta'(t)$. The derivative $\omega'(t)$ of the angular velocity is the *angular acceleration*. It is denoted by $\alpha(t)$.

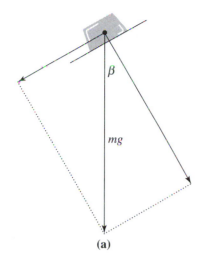

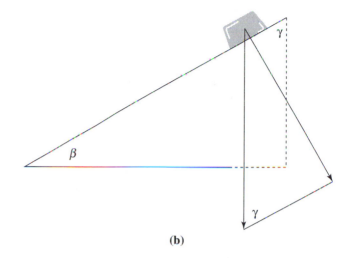

Figure 9.22

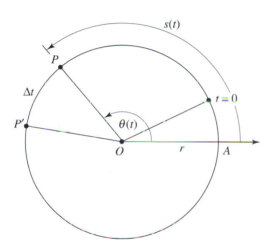

Figure 9.23

Taking derivatives of both sides of the equation, $s(t) = r\theta(t)$, shows that the velocity of the point along the circle at any time $t \geq 0$ is $v(t) = s'(t) = r\theta'(t) = r\omega(t)$. Differentiating $v(t) = r\omega(t)$ shows that the acceleration of the point along the circle at any time $t \geq 0$ is $a(t) = v'(t) = r\omega'(t) = r\alpha(t)$.

Example 9.3. Let time t be given in seconds and suppose that $\theta(t) = (t^2 + 3t)\pi$. Since $\theta(0) = 0$, the point starts at A. Its angular velocity is $\omega(t) = \theta'(t) =$

$(2t + 3)\pi$ radians per second. Its initial angular velocity is $\omega(0) = 3\pi$ radians per second. At time $t = 0.3$ seconds, $\theta(0.3) = ((0.3)^2 + 3(0.3))\pi = (0.09 + 0.9)\pi = 0.99\pi < \pi$, so the point has traced out a little less than 180°. The angular velocity at this time is $\omega(0.3) = (2(0.3) + 3)\pi = 3.6\pi$ radians per second. The angular acceleration of the point is $\alpha(t) = \omega'(t) = 2\pi$ radians per second per second. It is constant. Suppose that the radius of the circle is $r = 1.5$ feet. Then the position of the point along the circle (as measured from the point A) at any time t is $s(t) = 1.5\theta(t) = 1.5(t^2 + 3t)\pi$. Its velocity along the circle is $v(t) = s'(t) = 1.5(2t + 3)\pi$ feet per second, and its linear acceleration is $a(t) = v'(t) = 1.5(3)\pi = 4.5\pi$ feet per second^2.

B. Moments of Force and Inertia. Consider a particle of mass m attached to a rigid rod of length r. Suppose that the mass of the rod is negligible, that its other end is fixed at O, and that it can rotate freely about O. See Figure 9.24. Suppose that at any time $t \geq 0$, a force $f(t)$ is acting on the particle perpendicularly to the rod. This force will cause the particle to rotate in a circle of radius r. If $a(t)$ is the linear acceleration of the particle produced by $f(t)$, then by Newton's second law, $f(t) = ma(t)$. Because $a(t) = r\alpha(t)$, where $\alpha(t)$ is the angular acceleration of the particle, it follows that

$f(t) = (m \cdot r) \cdot \alpha(t)$. After multiplying through by r, we get

$$f(t) \cdot r = (mr^2) \cdot \alpha(t).$$

The product $f(t) \cdot r$ is called the *moment of force*, or *torque*. This is the capacity of a force to produce a rotation. The torque is determined not by the force alone, but by the product of the magnitude of the force times the distance between the point of action of the force and the axis of rotation. What is going on can be illustrated with Archimedes's law of the lever. (See Exercises 3C.) Consider a lever and a fulcrum. Fix a force F a distance R from the fulcrum, as shown in Figure 9.25, and consider the three forces f_1, f_2, and f_3. If $f_1r_1 = f_2r_2 = f_3r_3 = FR$, then each one of these three forces will balance the system. It follows that each of the forces f_1, f_2, and f_3 will have the same capacity to produce a rotation precisely if the torques $f_1r_1 = f_2r_2 = f_3r_3$ are the same. You have probably encountered this phenomenon: In pushing a door shut, you will need to apply much less force if you push near the handle than if you push near the hinges.

The equation $f(t) \cdot r = (mr^2) \cdot \alpha(t)$ that was just established has the form

moment of force $=$ constant $\times$ angular acceleration.

Such an equality holds in any situation where a force acting tangentially to the motion brings about an acceleration in the rotation of an object about an axis. The constant depends on the object being rotated and is called the *moment of inertia*, or *index of inertia*, of the object. It is denoted by I. So the equation above can be written

$$f(t) \cdot r = I \cdot \alpha(t),$$

where r is the distance from the point of action of the force to the axis of rotation. For a fixed moment of force, notice that the larger the moment of inertia I, the less the angular acceleration $\alpha(t)$, and so the greater the resistance of the object to being rotated. Observe that in the situation of the particle of mass m and the rod of length r the moment of inertia is $I = mr^2$.

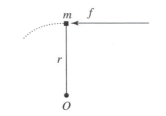

Figure 9.24

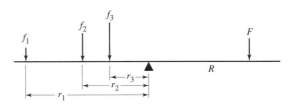

Figure 9.25

Take a circle in a plane with radius r and center a point O. Regard the circle as a physical object with mass m. Suppose that the mass is evenly distributed around the circumference and that the circle can rotate freely about O. Think of a bicycle wheel whose spokes have negligible mass. Let $f(t)$ be a force acting on the circle as shown in Figure 9.26. This force will cause the circle to rotate. Consider the circle to be subdivided into many, say n, equal particles with each particle attached to the center by a "spoke." Instead of having $f(t)$ acting on the entire circle, consider the force $\frac{f(t)}{n}$ as acting tangentially on each of the n particles. Since the mass of each particle is $\frac{m}{n}$, we see by the equality already established for the rotating rod that $r \cdot \frac{f(t)}{n} = \left(\frac{m}{n} \cdot r^2\right) \cdot \alpha(t)$. After multiplying through by n, we get

$$f(t) \cdot r = (m \cdot r^2) \cdot \alpha(t).$$

So the moment of inertia of this circular object is given by the same formula $I = m \cdot r^2$ as the moment of inertia of the point mass and the rod.

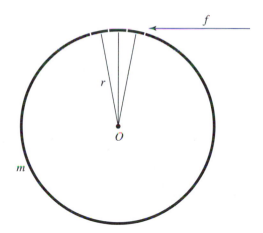

Figure 9.26

Consider next a circular disc that has its mass evenly distributed and can rotate freely about its center. Compared to the case of the circle, more of the mass is now closer to the center of rotation. One would therefore expect less resistance to rotation and hence a smaller moment of inertia. This is indeed so, as the moment of inertia of the disc is $I = \frac{1}{2}m \cdot r^2$. Turn to a solid sphere of radius r and mass m that is rotating about an axis through its center. See Figure 9.27. Think of a marble rolling on a flat surface. The moment of inertia of the sphere is $I = \frac{2}{5}mr^2$. Exercises 9F will lead you through the derivations of the moments of inertia of both the disc and the sphere. They rely on the summation process of integral calculus.

The fact that the moment of inertia of a solid sphere of radius r and mass m is $\frac{2}{5}mr^2$ tells us that the connection between a force $f(t)$ acting tangentially at the surface of the sphere and the angular acceleration $\alpha(t)$ that this force produces is given by the equation $f(t) \cdot r = \left(\frac{2}{5}mr^2\right) \cdot \alpha(t)$, or

$$f(t) = \frac{2}{5}mr \cdot \alpha(t).$$

We are now in a position to undertake the mathematical analysis of Galileo's experiment.

C. The Mathematics for Galileo's Experiment.
So let's return to Galileo and the bronze ball rolling

down the inclined plane. Let m be the mass of the ball and let r be its radius. Place a coordinatized axis as shown in Figure 9.28, with the positive part pointing down and to the left. Suppose that the origin 0 is at a height h above the base of the plane. Position the ball at the origin 0 and release it from a state of rest at time $t = 0$. For any time $t \geq 0$, let $p(t)$ be the position of the ball on the inclined axis. So $p(t)$ is the distance that the ball has moved during the time interval $[0, t]$. As already discussed in Section 9.3A, the ball is propelled down the plane by a force $F(t) = mg \sin \beta$. The vector labeled $f(t)$ represents friction.

The first concern is the explicit determination of the function $p(t)$. The strategy is the same as that already used in Section 6.4. Let $a(t)$ be the acceleration of the moving ball down the inclined plane; determine $a(t)$ by an analysis of forces and Newton's second law $F = ma$; finally, use $a(t)$ to determine the velocity $v(t)$ and then the position $p(t)$.

The frictional force $f(t)$ is produced by the contact of the ball with the plane. Therefore, this is the force that rotates the ball. It follows from the discussion in Section 9.3B that the relationship between $f(t)$ and the angular acceleration $\alpha(t)$ of the ball is $f(t) = \frac{2}{5}mr \cdot \alpha(t)$. Because $a(t) = r\alpha(t)$ by Section 9.3A, we get the connection

$$f(t) = \frac{2}{5}ma(t)$$

between the frictional force $f(t)$ and the acceleration $a(t)$ of the ball. By Figure 9.28 and Newton's second law, $F(t) - f(t) = ma(t)$. By substituting $F(t) = mg \sin \beta$ and $f(t) = \frac{2}{5}ma(t)$, we get

$$ma(t) = F(t) - f(t) = mg \sin \beta - \frac{2}{5}ma(t).$$

So $a(t) = g \sin \beta - \frac{2}{5}a(t)$, and hence $\frac{7}{5}a(t) = g \sin \beta$. Therefore, the acceleration $a(t)$ of the ball down the inclined plane is

$$a(t) = \frac{5g}{7} \sin \beta.$$

The rest is routine! By Example 6.14 of Section 6.4, $v(t) = \left(\frac{5g}{7} \sin \beta\right) t + v(0)$. Because the ball started

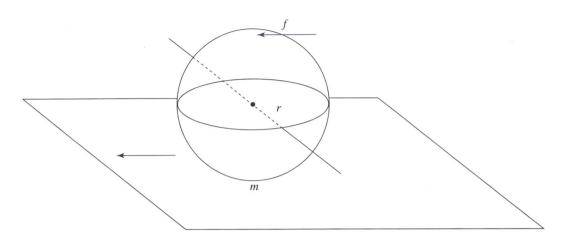

Figure 9.27

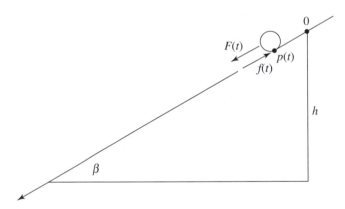

Figure 9.28

from rest, $v(0) = 0$, and hence

$$v(t) = \left(\frac{5g}{7} \sin \beta \right) t.$$

Again by Example 6.14 of Section 6.4,

$$p(t) = \frac{1}{2} \left(\frac{5g}{7} \sin \beta \right) t^2 + p(0).$$

Since the ball started from the origin, $p(0) = 0$, and therefore

$$p(t) = \left(\frac{5g}{14} \sin \beta \right) t^2.$$

We now know the position and the velocity of the ball along the inclined plane at any time $t \geq 0$.

Let's step back from our analysis for a moment. Refer back to Galileo's diagram in Figure 9.20 and recall that his observations relate the starting height of the ball with the distance from the ball's point of impact to the foot of the table. It is the purpose of our discussion to establish a mathematical connection between these two quantities.

To determine analytically the distance between the point of impact of the ball and the foot of the table, we need to know the velocity that the ball has when it reaches the bottom of the inclined plane. Suppose that this occurs at time t_b. Notice that t_b is the moment at which $p(t_b)$ is equal to the length of the hypotenuse of the right triangle of Figure 9.28. It follows that $\sin \beta = \frac{h}{p(t_b)}$. Hence $p(t_b) = \frac{h}{\sin \beta}$. But by the preceding formula, $p(t_b) = \left(\frac{5g}{14} \sin \beta \right) t_b^2$. Therefore,

$$\left(\frac{5g}{14} \sin \beta \right) t_b^2 = \frac{h}{\sin \beta}.$$

So $t_b^2 = \frac{14h}{5g} \frac{1}{\sin^2 \beta}$, and hence

$$t_b = \sqrt{\frac{14h}{5g} \frac{1}{\sin \beta}}.$$

Substituting t_b into the velocity formula $v(t) = \left(\frac{5g}{7}\sin\beta\right)t$ shows us that the velocity of the ball at the instant that it reaches the bottom of the inclined plane is

$$v(t_b) = \left(\frac{5g}{7}\sin\beta\right)t_b = \left(\frac{5g}{7}\sin\beta\right)\left(\sqrt{\frac{14h}{5g}}\frac{1}{\sin\beta}\right)$$

$$= \frac{5g}{7}\cdot\sqrt{\frac{14h}{5g}} = \sqrt{\frac{5^2g^214h}{7^2\cdot5g}} = \sqrt{\frac{10}{7}gh}.$$

Observe that the velocity $v(t_b)$ depends only on the starting height h of the ball. It is independent of the mass of the ball, its radius, even the angle β of the inclined plane!

The situation that we have arrived at is this. When Galileo releases his bronze ball from rest at a starting height h above the table, it arrives at the bottom of the plane with a velocity of $\sqrt{\frac{10}{7}gh}$. See Figure 9.29. At this point, its motion is subject to the theory that was developed in Section 6.5. In particular, the ball will fly to the ground in a parabolic arc. It remains to determine how far from the base B of the table it will land. This distance is the range R of the ball. It is provided by the formula

$$R = \frac{v_0^2}{2g}\sin(2\varphi_0) + \frac{v_0}{g}\sqrt{\frac{v_0^2}{4}\sin^2(2\varphi_0) + 2gy_0\cos^2\varphi_0}$$

from Section 6.5. What constants have to be inserted? The v_0 is the velocity with which the ball flies off the edge of the table. So $v_0 = v(t_b) = \sqrt{\frac{10}{7}gh}$. Because the table top is horizontal, the angle of departure is $\varphi_0 = 0$ and the initial height y_0 is the distance $y_0 = 0.778$ from the top of the table to the ground. Notice that $\sin(2\varphi_0) = \sin 0 = 0$ and $\cos\varphi_0 = \cos 0 = 1$. It follows that the distance from the point of impact of the ball to the point B is

$$R = 0 + \frac{v_0}{g}\sqrt{2gy_0} = \sqrt{\frac{2y_0}{g}}v_0 = \sqrt{\frac{2y_0}{g}}\cdot\sqrt{\frac{10}{7}gh}$$

$$= \sqrt{\frac{20}{7}y_0h}.$$

Taking $y_0 = 0.778$, we get

$$R = 1.491\sqrt{h}.$$

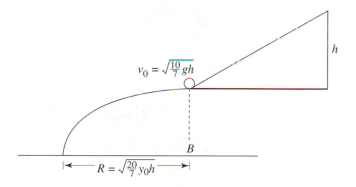

Figure 9.29

We have now come to the key question: Are the data that Galileo displays in the diagram of Figure 9.20 "in tune" with this equation? For starting heights h successively equal to 0.282, 0.564, 0.752, and 0.940 meters, the equation provides the distances

$$R = 0.792, 1.120, 1.293, \text{ and } 1.446 \text{ meters.}$$

Galileo's measurements for these distances are respectively,

$$0.752, 1.102, 1.248, \text{ and } 1.410 \text{ meters.}$$

Recalling that (1 inch = 2.54 cm = 0.0254 meters) we get the following differences between theory and experiment: 0.04 meters (1.6 inches); 0.018 meters (0.7 inches); 0.045 meters (1.8 inches); and 0.036 meters (1.4 inches). All of Galileo's measurements fall a little short of the distances predicted by the theory. But this is to be expected! The ball that Galileo used was certainly not perfectly spherical, the groove in the inclined plane was not perfectly smooth, etc., so that the acutal velocity of the ball would have been less than that given by the equation. Therefore one would expect the ball's actual points of impact to have been closer to the foot of the table than those predicted by the mathematics.

It was pointed out earlier that some prominent historians of science had thought that Galileo's experiments were largely based on speculations. The comparison between the conclusion of the mathematical analysis and Galileo's data suggests quite clearly[9] that these data were the result of a real experiment. So our mathematical analysis has shed light on an important question in the history of science.

9.4 From Fermat's Principle to the Basic Telescope

It seems intuitively clear that if a light ray is reflected from the surface of a flat mirror, then the *angle of incidence* α is equal to the *angle of reflection* β, as illustrated in the diagram of Figure 9.30(a). We will begin our study of the geometric properties of light rays by showing that this conclusion is a mathematical consequence of a basic property of the propagation of light: Fermat's Principle of Least Time.

A. Fermat's Principle and the Path of a Light Ray. Suppose that a light ray proceeds from some source, strikes a mirror, and is reflected. Let A be a point on the ray before it strikes the mirror and let B be a point on the ray after the reflection. See Figure 9.30(b). Suppose that the surrounding medium is air or a vacuum and let v be the speed of light in this medium.[10] The light ray determines a plane that is perpendicular to the mirror. Place a coordinate system in this plane in such a way that the x-axis runs along the mirror's surface and the y-axis goes through A. Let $A = (0, a)$, $B = (b, d)$, and suppose that the ray reflects off the mirror at x. Let α be the angle of incidence and β the angle of reflection. Let D_1 be the distance from A to x and let t_1 be the time it takes for the ray to travel this distance. Similarly, let D_2 be the distance from x to B and let t_2 be the time of travel for this distance. Observe that $D_1 = vt_1$, and hence $t_1 = \frac{D_1}{v}$. In the same way, $t_2 = \frac{D_2}{v}$. By applying the Pythagorean theorem twice, we obtain

$$D_1 = \sqrt{x^2 + a^2} = (x^2 + a^2)^{\frac{1}{2}} \text{ and}$$

$$D_2 = \sqrt{d^2 + (b - x)^2} = \left(d^2 + (b - x)^2\right)^{\frac{1}{2}}.$$

Therefore, the time t that it takes for the ray to travel from A to B is

$$t = t_1 + t_2 = \frac{D_1}{v} + \frac{D_2}{v}$$

$$= \frac{1}{v}(x^2 + a^2)^{\frac{1}{2}} + \frac{1}{v}\left(d^2 + (b - x)^2\right)^{\frac{1}{2}}.$$

So the time of travel is determined by the function

$$t(x) = \frac{1}{v}(x^2 + a^2)^{\frac{1}{2}} + \frac{1}{v}(d^2 + (b - x)^2)^{\frac{1}{2}}$$

where x is the point of incidence. We can now ask: For which x is the travel time $t(x)$ a minimum? To solve this problem, we will analyze the derivative $t'(x)$. By the chain rule,

$$t'(x) = \frac{1}{2v}(x^2 + a^2)^{-\frac{1}{2}}(2x)$$

$$+ \frac{1}{2v}\left(d^2 + (b - x)^2\right)^{-\frac{1}{2}} 2(b - x)(-1)$$

$$= \frac{x}{v(x^2 + a^2)^{\frac{1}{2}}} - \frac{b - x}{v\left(d^2 + (b - x)^2\right)^{\frac{1}{2}}}.$$

What are the critical numbers of $t(x)$? Note that $t'(x)$ is not defined when

$$(x^2 + a^2)^{\frac{1}{2}} = 0 \quad \text{or} \quad (d^2 + (b - x)^2)^{\frac{1}{2}} = 0.$$

In the first case, a must be zero. This means that A is on the x-axis and hence on the mirror. But this is not the case. If $d^2 + (b - x)^2)^{\frac{1}{2}}$ is zero, then d must be zero, and then B is on the mirror. Again, this is not the case. It follows that the only critical numbers are those x for which $t'(x) = 0$. Setting $t'(x) = 0$, we get

$$\frac{x}{v(x^2 + a^2)^{\frac{1}{2}}} = \frac{b - x}{v\left(d^2 + (b - x)^2\right)^{\frac{1}{2}}}.$$

From a look at Figure 9.30(b), we see that

$$\sin \alpha = \frac{x}{D_1} = \frac{x}{(x^2 + a^2)^{\frac{1}{2}}}$$

and

$$\sin \beta = \frac{b - x}{D_2} = \frac{b - x}{\left(d^2 + (b - x)^2\right)^{\frac{1}{2}}}.$$

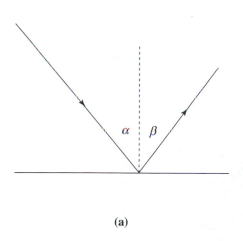

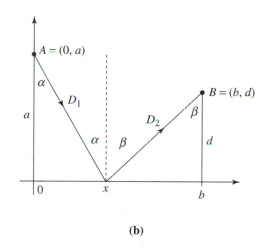

(a)   (b)

Figure 9.30

It follows directly that

$$\frac{\sin \alpha}{v} = \frac{\sin \beta}{v}.$$

Multiplying through by v gives us $\sin \alpha = \sin \beta$. Because both α and β are between 0 and 90°, it follows that

$$\alpha = \beta.$$

A look at the graph of the sine function in Figure 4.24 of Section 4.4 confirms that different angles in the range $0 \leq \theta \leq \frac{\pi}{2}$ have different sines.

The above argument has demonstrated that the property "the angle of incidence is equal to the angle of reflection" is a consequence of the principle that of all neighboring paths, light chooses the one that requires the least time. This fundamental principle is Fermat's *Principle of Least Time*.

Next, consider two homogeneous transparent mediums of different densities. For example, let one of them be air and the other a certain type of glass. Suppose that the boundary between them is a plane. Let A be a point in one medium and let B be a point in the other, say the denser, medium. Suppose that neither A nor B lies on the boundary separating the two mediums. The path of a light ray traveling from A to

B lies in a plane perpendicular to the boundary. At the boundary the ray bends as shown in Figure 9.31. Such a change in direction is known as *refraction*. Let v_A be the speed of light in the medium containing A and let v_B be the speed of light in the medium containing B. The angle α is the *angle of incidence*, and the angle β is the *angle of refraction*.

We saw that in the case of a reflected ray, there is a connection between the relevant angles (the angles of incidence and reflection are equal). Is there also a connection between the angles α and β in the case of a refracted ray? There are many ways to connect A and B with two line segments that meet at the boundary. Of all these possibilites, which one will a light ray pick out? We will see that Fermat's principle provides the answers to both of these questions.

Consider the plane determined by the light ray and place a coordinate system so that the x-axis is on the boundary and the y-axis goes through A. Refer to Figure 9.32. Let $A = (0, a)$ and $B = (b, d)$, and suppose that the light crosses the boundary at x. Let D_1 be the distance from A to x and let t_1 be the time it takes for the ray to travel this distance. Similarly, let D_2 be the distance from x to B and let t_2 be the time of travel through this distance. Note that $D_1 = v_A t_1$, so

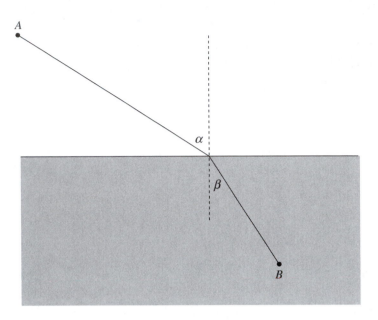

Figure 9.31

$t_1 = \frac{D_1}{v_A}$. In the same way, $t_2 = \frac{D_2}{v_B}$. By the Pythagorean theorem,

$$D_1 = \sqrt{x^2 + a^2} = (x^2 + a^2)^{\frac{1}{2}} \quad \text{and}$$

$$D_2 = \sqrt{d^2 + (b - x)^2} = \left(d^2 + (b - x)^2\right)^{\frac{1}{2}}.$$

So the time it takes for the ray to travel from A to B is

$$t = t_1 + t_2 = \frac{D_1}{v_A} + \frac{D_2}{v_B}$$

$$= \frac{1}{v_A}(x^2 + a^2)^{\frac{1}{2}} + \frac{1}{v_B}\left(d^2 + (b - x)^2\right)^{\frac{1}{2}}.$$

For which x is t a minimum? The calculations are exactly the same as in the earlier situation of reflection:

$$t'(x) = \frac{1}{2v_A}(x^2 + a^2)^{-\frac{1}{2}}(2x)$$

$$+ \frac{1}{2v_B}\left(d^2 + (b - x)^2\right)^{-\frac{1}{2}} 2(b - x)(-1)$$

$$= \frac{x}{v_A(x^2 + a^2)^{\frac{1}{2}}} - \frac{b - x}{v_B\left(d^2 + (b - x)^2\right)^{\frac{1}{2}}}.$$

As before, the only critical points are those where $t'(x) = 0$. Setting $t'(x) = 0$ gives us

$$\frac{x}{v_A(x^2 + a^2)^{\frac{1}{2}}} = \frac{b - x}{v_B\left(d^2 + (b - x)^2\right)^{\frac{1}{2}}}.$$

A look at Figure 9.32 shows that

$$\sin\alpha = \frac{x}{D_1} = \frac{x}{(x^2 + a^2)^{\frac{1}{2}}} \quad \text{and}$$

$$\sin\beta = \frac{b - x}{D_2} = \frac{b - x}{\left(d^2 + (b - x)^2\right)^{\frac{1}{2}}},$$

and it follows that

$$\boxed{\frac{\sin\alpha}{v_A} = \frac{\sin\beta}{v_B}}$$

This is the connection between the angle of incidence α and the angle of refraction β. It is called *Snell's Law of Refraction*. We have just seen that this basic geometric property of light is a consequence of Fermat's principle: Of all neighboring paths, light chooses the one that requires the least time. (Refer

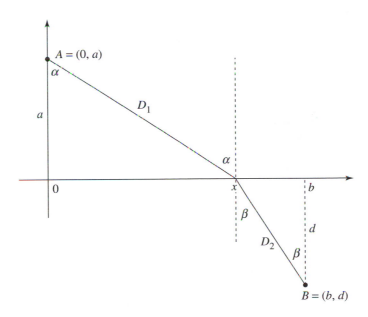

Figure 9.32

for a moment to Example 8.42 and Figure 8.38 of Section 8.8 and notice that this problem from L'Hospital's *Analyse* is a "slower" version of the problem just solved.)

In air, water, glass, in fact in any translucent medium, light travels more slowly than the $c = 186{,}272$ miles per second at which it travels in a vacuum. The *index of refraction* n of any medium is defined to be

$$n = \frac{c}{v},$$

where v is the speed of light in that medium. The index of refraction of a vacuum is $n = \frac{c}{c} = 1$. There is no medium in which light propagates faster than it does in a vacuum. So the index of refraction of any medium is $n \geq 1$. Table 9.1 lists some examples. Think of the index of refraction as a measure of the density of the medium: the denser the medium, the less the speed v at which light will travel through it, and hence the higher its index of refraction n. The index of refraction of a medium can be determined by a careful analysis of the path of light through a prism made of the material in question.

Table 9.1	
substance	**index of refraction**
crown glass	1.52
flint glass	1.66
diamond	2.42
ice	1.31
water	1.33
air at sea level at $0°$ C	1.00029

Example 9.4. What is the speed of light in air? How fast does light travel inside a diamond? The formula $n = \frac{c}{v}$ can be rewritten as $v = \frac{c}{n}$. Now refer to Table 9.1. Because $n = 1.00029$ for air, the speed of light in air is $v = \frac{186{,}272}{1.00029} = 186{,}218$ miles per second. For a diamond, $n = 2.42$, and hence $v = \frac{186{,}272}{2.42} = 76{,}972$ miles per second.

Let the indices of refraction of the two mediums in Figure 9.31 be n_A and n_B, respectively. The speeds of

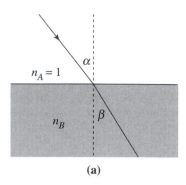

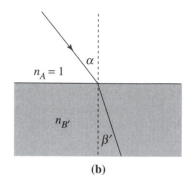

(a)

(b)

Figure 9.33

light in the two mediums are $v_A = \frac{c}{n_A}$ and $v_B = \frac{c}{n_B}$. By substituting into Snell's law, we get $\frac{n_A \sin \alpha}{c} = \frac{n_B \sin \beta}{c}$. So Snell's law can be reformulated as

$$n_A \sin \alpha = n_B \sin \beta.$$

Suppose that a light ray travels from a vacuum into a denser medium with an angle of incidence α. Now $n_A = 1$, and therefore $\sin \alpha = n_B \sin \beta$. Since $n_B > 1$, it follows that $\sin \alpha > \sin \beta$. The sine is an increasing function over the interval from 0 to $\frac{\pi}{2}$ (look at its graph in Section 4.4), and hence $\alpha > \beta$. Suppose next that a light ray enters a medium with index of refraction $n_{B'} > n_B$ with the same angle of incidence α. As before, $\sin \alpha = n_{B'} \sin \beta'$. Because

$$n_{B'} \sin \beta' = \sin \alpha = n_B \sin \beta \quad \text{and} \quad n_{B'} > n_B,$$

it follows that $\sin \beta' < \sin \beta$ and hence that $\beta' < \beta$. Our conclusions are summarized in Figure 9.33. In both cases, the angle of incidence α is greater than the angle of refraction. Therefore both rays are bent to the right. The denser medium, shown in Figure 9.33(b), produces the smaller angle of refraction, that is, it bends the light ray more.

Given the initial angle of incidence, it is now possible to determine the path of a light ray through any sequence of translucent materials, provided that we know the indices of refraction of these materials. For example, suppose that a light ray travels through air, then through a sheet of plate glass, and then through air again, as shown in Figure 9.34. Suppose

that the initial angle of incidence is α and the index of refraction of the glass is n_B. Notice that $\beta = \beta'$. By Table 9.1, the index of refraction of air is $n_A = 1.00029$. Consider the entry of the ray into the glass. By Snell's law, $n_A \sin \alpha = n_B \sin \beta$, and hence

$$\sin \beta = \frac{n_A}{n_B} \sin \alpha = \frac{1.00029}{n_B} \sin \alpha.$$

So $\sin \beta$ is determined by α and n_B, and hence β is determined by α and n_B. By applying Snell's law to the ray as it exits from the glass, we get $n_A \sin \alpha' = n_B \sin \beta'$. Since $\beta = \beta'$, it follows that $n_A \sin \alpha' = n_B \sin \beta' = n_B \sin \beta = n_A \sin \alpha$. Hence $\sin \alpha' = \sin \alpha$, and therefore

$$\alpha' = \alpha$$

So the angle of exit is equal to the initial angle of incidence. As asserted, the path of the light ray is determined by α and the properties of the glass.

Example 9.5. Suppose that the glass is crown glass and that $\alpha = 20°$. From Table 9.1, $n_B = 1.52$. So

$$\sin \beta = \frac{n_A}{n_B} \sin \alpha = \frac{1.00029}{1.52} \sin 20°$$

$$= (0.658)(0.342) = 0.225.$$

A calculator shows that $\beta = 13.00°$. If the glass is the denser flint glass instead, then $n_B = 1.66$, and

$$\sin \beta = \frac{n_A}{n_B} \sin \alpha = \frac{1.00029}{1.66} \sin 20°$$

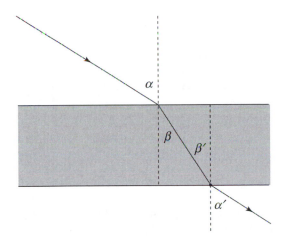

Figure 9.34

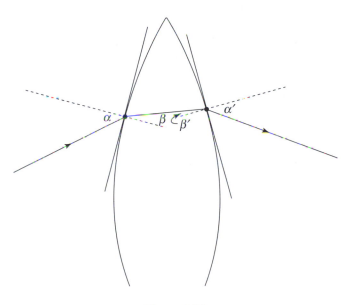

Figure 9.35

as Figure 9.35 shows, the path of light through a lens can be computed from Snell's law, the index of refraction of the glass of the lens, and the geometry of its surfaces (the tangent planes play the role of the flat surfaces of Figure 9.34). It goes without saying that the essence of any optical instrument is its lenses and mirrors and the manner in which they are arranged. It is the specific configuration of these elements that determines the optical properties of the instrument. What makes a lens or a system of lenses work as intended is that it bends all the light rays that pass through it in certain predictable ways.

B. Basic Properties of Lenses. Take a lens of the general shape shown in Figure 9.36. We will suppose that the lens is *symmetric*. This means that its rim is a circle and that the axis perpendicular to the circle and through the circle's center is an axis of symmetry of the lens. Thus, any rotation of the lens about this axis will not shift the position of the lens overall. As a consequence of Snell's law, a collection of incoming light rays that are parallel to the axis of symmetry will be bent inward as indicated in Figure 9.36(a). A lens that is used in a precision optical instrument must be very carefully crafted to meet exacting standards. We will describe these next.

The first requirement is that all rays that come in parallel to the axis of symmetry are bent by the lens in such a way that they meet at a point on this axis. This point is the *focus* of the lens. See Figure 9.36(b). The distance from the center C of the lens to the focus is the *focal length* of the lens. Now turn to Figure 9.37. Consider an object — in the figure the object is the arrow AA' — and position it as shown. Suppose that light emanates from the arrow in all directions (assume, for example, that it is bathed in sunlight). The requirements that the lens must satisfy for any position of the arrow not too close to the lens (the arrow must be further away than the focus on that side) are as follows. All the light rays that emerge from a given point on the arrow and travel through the lens must collect at some point on the other side of the lens. This point is the image of the original point. For

$$= (0.603)(0.342) = 0.206.$$

This time $\beta = 11.89°$. Observe that in the denser glass the angle of refraction is smaller, and hence the ray is bent more.

The point of this section so far is that the path of a single light ray can be calculated from Fermat's fundamental principle and basic mathematics. For example,

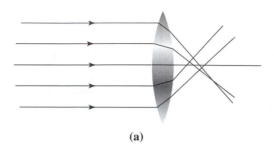 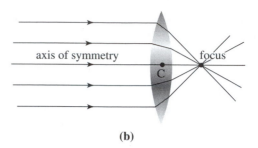

(a)   (b)

Figure 9.36

example, in the figure, the light rays from the point *A* gather at the image point *B*. All the image points taken together must align themselves, so that when the various light rays carry the "pictoral" properties of the arrow *AA′*, for example, the colors, textures, and shadows, to the other side of the lens, an image of the arrow is produced. This image is the arrow *BB′* shown in the figure. Since the image point of *A* is *B* and that of *A′* is *B′*, observe that the image is inverted. If a screen is positioned at the place where the image is formed, then the image will be visible on the screen. Since the image can be realized in this way, we will call it *real*. If the screen is a light sensitive film, then the image will be recorded and what we have described are the essential elements of the simple camera. If the lens is the lens of an eye and the screen is the retina, then we have explained the first step of the act of seeing.

It turns out that lenses that are thin and spherical have the optical properties just described. A lens

is *spherical* if each of its two surfaces is either flat or has the shape of a section of a sphere. A lens is *thin*, if it is thin relative to the size of its surfaces. (Incidentally, the lens as sketched in Figure 9.37 is not thin.) The process of manufacturing such lenses was developed in the 19th century. Molten glass was poured into a lens shaped mold. After various chemicals were added to alter the properties of the glass (for instance the addition of lead oxide or barium oxide raises the index of refraction) the glass was allowed to cool under carefully controlled conditions to keep imperfections to a minimum. The resulting piece of glass was then ground and polished into the desired spherical shape. Such simple lenses perform reasonably well, but they are not distortion free. The lenses used in modern optical instruments (cameras, microscopes, or telescopes) are much more sophisticated than those just described. To increase the sharpness and clarity of the image, today's lenses are crafted by cementing

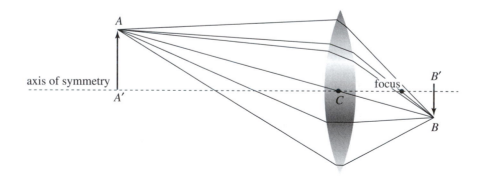

Figure 9.37

together components of different shapes made of different kinds of glasses and plastics. The design of such *compound* lenses is almost entirely computerized.

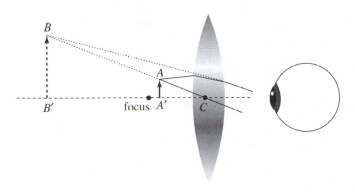

Figure 9.38

We return to the discussion of the basic properties of lenses. A lens of the type described can be used as a magnifying glass. Place the arrow AA' within the focus of the lens. See Figure 9.38. A light ray emanating from the object at A and striking the lens is bent as shown. The eye perceives this ray to be coming from some point B. Since this happens for all the light rays emanating from AA', it follows that the eye sees a noninverted image BB'. A moment's thought shows that a film placed at BB' would not record anything. This noninverted image is therefore not a real image, but a perceived, or *virtual*, image.

The simple telescope combines the two different uses of lenses already discussed. Light from an object is collected by a lens called the *objective lens* and a real inverted image is formed. If we assume that the distance of the object from the lens is very large compared to the size of the object (think of a planet as viewed from Earth), then the light rays coming into the lens from the object will be close to and nearly parallel to the axis. This means — look at a combination of Figures 9.36(b) and 9.37 — that the real image will form very near the focus of the objective lens. When a second lens — the *eyepiece* — is placed in an appropriate position, a magnified virtual image of this real image is produced. See Figure 9.39. Such combination of two lenses placed in a suitable housing is the *refracting telescope* in its most simple form. (The refracting telescope that Galileo constructed was slightly more complicated. It featured an additional lens between the objective and the eyepiece.)

In order to come to an understanding of the preceding discussion in precise terms—this would be necessary if we actually wanted to build a sim-

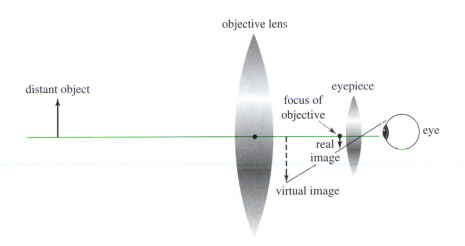

Figure 9.39

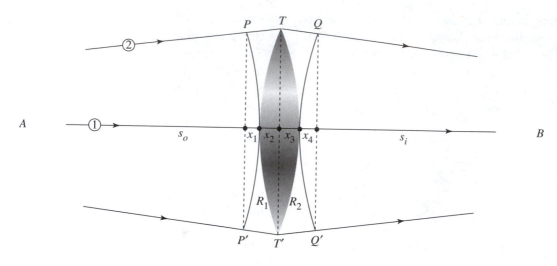

Figure 9.40

ple telescope—it remains to undertake a quantitative analysis of the formation of an image by a lens.

C. Quantitative Analysis of Lenses.

Consider a thin spherical lens. Refer back to Section 9.4A, especially Figure 9.35, for the fact that the index of refraction and the shape of the curved surfaces of the lens determine exactly how light travels through the lens. This means that the index of refraction n of the glass and the radii R_1 and R_2 of the two spherical surfaces together determine all the optical properties of the lens. So there ought to be a precise numerical connection between the parameters n, R_1, R_2 and the focal length f of the lens. But what is this relationship?

Place an object, as represented by the point A, on the axis of symmetry of the lens some distance s_o from the lens. Its image will be a point B on the axis some distance s_i from the lens. Refer to Figure 9.40. All the light rays from A that strike the lens bend and then come together at B. The ray from A to B through the center of the lens (the ray numbered 1) is a straight line. The rays ATB (it is numbered 2) and $AT'B$ both pass through the lens near its rim.[11] The arc PP' is that of the circle with radius s_o and center A, and the arc QQ' lies on the circle with radius s_i and center B. The quantities x_1, x_2, x_3, and x_4 are the distances between the indicated points on the axis of symmetry.

The light rays 1 and 2 both travel from point A to point B through air and glass. By a basic property of thin spherical lenses, these two light rays require the same time to get from A to B. These two rays also travel from A to the circular arc PP' in the same time (because the distances are the same) and they travel from the arc QQ' to B in the same time (for the same reason). It follows that ray 1 will travel the distance $x_2 + x_3$ through the lens in the same time t that it takes ray 2 to travel from P to Q across the top. Because the index of refraction of the glass is n, the velocity of ray 1 through the lens is $v = \frac{c}{n}$ (where c is the speed of light in vacuum). So $x_2 + x_3 = v \cdot t = \frac{c}{n}t$, and hence $n(x_2 + x_3) = ct$. The velocity of ray 2 through the air is very nearly equal to c (see Example 9.4), and therefore $ct = PT + TQ$. Adding the assumption that the diameter TT' of the lens is small (relative to the distance between A and B) and recalling that the lens is thin, allows us to take $PT + TQ = x_1 + x_2 + x_3 + x_4$. Therefore, $ct = x_1 + x_2 + x_3 + x_4$. It follows that

$$n(x_2 + x_3) = ct = x_1 + x_2 + x_3 + x_4.$$

Solving for x_4, we get

$$x_4 = -x_1 + n(x_2 + x_3) - (x_2 + x_3)$$
$$= -x_1 + (n-1)(x_2 + x_3).$$

Now we turn to a fact from the Greeks. Take a diameter of a circle of radius R. Consider a segment perpendicular to the diameter and label the lengths of the segments that arise as in Figure 9.41. By Example 3.1 of Section 3.1, $\frac{x(2R-x)}{h^2} = 1$, and hence $x(2R - x) = h^2$. So $2Rx - x^2 = h^2$. Suppose that x is small. Then x^2 is much smaller than x, and it follows that $2Rx \approx h^2$, and so $x \approx \frac{h^2}{2R}$. We will take $x = \frac{h^2}{2R}$.

Apply this equality to the four circles that give rise to the circular arcs of Figure 9.40. The radii of these circles are s_o, R_1, R_2, and s_i, respectively. Each of these circles has a diameter on the axis of symmetry of the lens. Consider the perpendicular segments from the axis to the points P, T, and Q. Since the lens is thin and its diameter TT' small, we can assume that the lengths of these segments are all equal. Let h be this common length. Applying the approximation just discussed four successive times gives us the equalities

$$x_1 = \frac{h^2}{2s_o}, \; x_2 = \frac{h^2}{2R_1}, \; x_3 = \frac{h^2}{2R_2}, \; \text{and} \; x_4 = \frac{h^2}{2s_i}.$$

Inserting these equalities into the equation $x_4 = -x_1 + (n-1)(x_2 + x_3)$ derived earlier, we get

$$\frac{h^2}{2s_i} = -\frac{h^2}{2s_o} + (n-1)\left(\frac{h^2}{2R_1} + \frac{h^2}{2R_2}\right).$$

Multiplying this through by $\frac{2}{h^2}$ gives us *Descartes's formula* for a thin lens:

$$\boxed{\frac{1}{s_i} = -\frac{1}{s_o} + (n-1)\left(\frac{1}{R_1} + \frac{1}{R_2}\right)}$$

Assume that the object at A is very far from the lens, i.e., assume that s_o is extremely large. Then we can take $\frac{1}{s_o} = 0$. It follows from a basic property of a thin spherical lens (see Figure 9.36(b)) that $s_i = f$, where

f is the focal length of the lens. We can conclude that

$$\boxed{\frac{1}{f} = (n-1)\left(\frac{1}{R_1} + \frac{1}{R_2}\right)}$$

This is the *Lens Maker's* equation. It shows precisely how the focal length of the lens is determined by the curvature of its spherical surfaces and the index of the refraction of the glass.

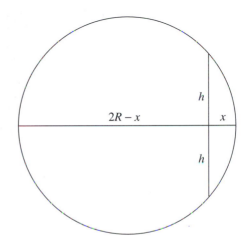

Figure 9.41

Suppose that the lens in Figure 9.36(b) is thin and spherical and suppose that the radii R_1 and R_2 of its two spherical surfaces are different. In the figure, the parallel rays of light come into the lens from the left and the focus is formed to the right of the lens. Suppose now that the rays come in from the right, so that the focus is on the left. The question that arises is this: Is the distance from the center of the lens to this second focus the same as the distance shown in the figure. In other words, does the lens have one focal length or two? The Len's Maker's equation provides the answer. Because

$$\frac{1}{R_2} + \frac{1}{R_1} = \frac{1}{R_1} + \frac{1}{R_2}$$

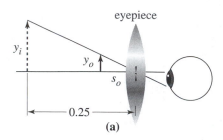

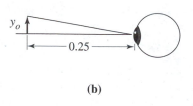

Figure 9.42

it does not matter which of the two spherical surfaces is struck first by a parallel array of light rays; the equation provides the same result for $\frac{1}{f}$ and hence for the focal length f.

Example 9.6. Suppose that a lens is made of crown glass and that its sperical surfaces both have a radius of 0.20 meters. What is the focal length? By Table 9.1, the index of refraction is $n = 1.52$. So by the lens maker's equation,

$$\frac{1}{f} = (1.52 - 1)\left(\frac{1}{0.20} + \frac{1}{0.20}\right) = (0.52)(10.00)$$

$$= 5.20.$$

Hence $f = \frac{1}{5.20} = 0.19$ meters.

Substituting the Len's Maker's equation into Descartes's formula, gives us

(∗) $$\frac{1}{s_i} = -\frac{1}{s_o} + \frac{1}{f}.$$

This equation tells us how the focal length and the position of the object together determine the position of the image.

Example 9.7. Place an object 0.50 meters from the lens of the previous example. Where is the image formed? Move the object closer to a distance of 0.10 meters from the lens. Where is the image formed now? Substituting $s_o = 0.50$ and $\frac{1}{f} = 5.20$ into equation (∗) shows that

$$\frac{1}{s_i} = -\frac{1}{0.50} + 5.20 = -2 + 5.20 = 3.20.$$

So the image is formed $s_i = \frac{1}{3.20} = 0.31$ meters from the lens. With $s_o = 0.10$, we get

$$\frac{1}{s_i} = -\frac{1}{0.10} + 5.20 = -10 + 5.20 = -4.80$$

so that $s_i = -\frac{1}{4.80} = -0.21$ meters.

Equation (∗) has produced a negative distance. What has gone wrong? Nothing, if one interprets this result appropriately. We have already observed that by positioning an object within the focus of a lens, a magnified virtual image is produced on the same side of the lens as the object. See Figure 9.38. With $s_o = 0.10$ meters and $f = 0.19$ meters, the object is withing the focal point. The − in the answer −0.21 that equation (∗) has provided tells us that the image is a virtual image on the same side as the object. The 0.21 tells us that this image is 0.21 meters from the lens. At the original distance of $s_o = 0.50$ meters, the object was outside the focus of the lens. The answer +0.31 that equation (∗) gave us in this case, lets us know that the image is real, that it is on the other side of the lens as the object, and that this image 0.31 meters from the lens.

When it is interpreted in this way, equation (∗) confirms the following fact for a thin, spherical lens: When an object is placed within the focus of a lens, then the image is formed on the same side of the object. In this case, the image is virtual and the lens is being used as a magnifying glass. When the object is placed outside the focus of the lens, then the image is real and it is formed on the other side of the lens.

Refer to Figure 9.36(b) and notice that a lens with a large focal length f will bend a light ray relatively little, but that a lens with small focal length can bend a light ray considerably. Put another way, when $\frac{1}{f}$ is small a light ray will be bent only a little, and when $\frac{1}{f}$ is large it will be bent a lot. So the ratio $\frac{1}{f}$ is a measure of the ability, or *power*, of a lens to bend light. The unit of power is $\frac{1}{\text{meter}}$ and is called *diopter*. So the more diopters a lens has, the more powerful it is. The lens of the previous examples has $\frac{1}{f} = 5.20$ diopters. The *curvature* of a spherical surface is defined to be $\frac{1}{R}$ where R is the radius of the sphere. This definition parallels our intuition: the smaller the radius of the sphere, the more it curves. The formulas just developed show exactly how the curvatures of the lens surfaces together with the index of refraction of the glass determine some of the basic properties of the lens. With the concepts just introduced, we can restate the Lens Maker's equation as follows: The power of a thin, spherical lens is the product of the amount by which the index of refraction of the glass exceeds that of a vacuum (which has index of refraction equal to 1) times the sum of the curvatures of its spherical surfaces.

We will now use a magnifying glass of focal length f to look at an object. The first question that arises concerns the optimal distance s_o of the object from the lens. We already know that s_o must be less than f, but there is another fact that needs be considered. A human eye has to strain to see an object that is too close and it loses the ability to distinguish fine details in objects that are too far away. A normal eye can see an object that is as close as 0.25 meters without having to strain. So we should place the object that we are looking at through the magnifying glass in such a way that the image is perceived to be at least 0.25 meters from the eye. In other words, we need to have $-s_i \geq 0.25$ meters. Recalling the equality $\frac{1}{s_i} = -\frac{1}{s_o} + \frac{1}{f}$, we now get

$$\frac{1}{f} < \frac{1}{s_o} = -\frac{1}{s_i} + \frac{1}{f} \leq \frac{1}{0.25} + \frac{1}{f} = 4 + \frac{1}{f} = \frac{4f+1}{f}.$$

Therefore $\frac{1}{f} < \frac{1}{s_o} \leq \frac{4f+1}{f}$, and hence

$$\frac{f}{1+4f} \leq s_o < f.$$

Focusing is the process of moving the lens (or the object) in such a way that the distance s_o of the object to the lens falls within the bounds given by this inequality. For a normal eye the sharpest and most detailed image is obtained when $-s_i \approx 0.25$ meters. Our discussion shows that this is achieved by placing the object $s_o \approx \frac{f}{1+4f}$ meters from the lens.

What about the magnification? Let y_o be the height of the object and y_i the height of the image. The *magnification* is the ratio $m = \frac{y_i}{y_o}$. For instance, a magnification of $m = 30$ means that $y_i = 30 y_o$ and hence that the image is 30 times as large as the object. To determine the magnification of the lens, place the object at the optimal distance of $s_o \approx \frac{f}{1+4f}$ meters from the lens. Refer to Figure 9.42(a) and notice that it abstracts the essence of Figure 9.38. Using the similar triangles of the figure, we get $\frac{y_i}{0.25} = \frac{y_o}{s_o}$. It follows that

$$m = \frac{y_i}{y_o} = \frac{0.25}{s_o} \approx \frac{1}{4}\frac{4f+1}{f} = \frac{4f+1}{4f} = 1 + \frac{1}{4f}.$$

Therefore, the *magnification* is equal to

$$m \approx 1 + \frac{1}{4f}.$$

What is seen through the lens is an image that is 0.25 meters away and $m \approx 1 + \frac{1}{4f}$ times larger than the object. The object as seen at this distance with the naked eye is shown in Figure 9.42(b).

Example 9.8. Consider a magnifying glass that has $\frac{1}{f} = 30.0$ diopters. The focal length of the lens is $f = 0.033$ meters. After substituting, the inequality $\frac{f}{1+4f} \leq s_o < f$ becomes

$$0.029 = \frac{0.033}{1.132} \leq s_o < 0.033.$$

So an object must be placed between 2.9 and 3.3 centimeters from the lens if it is to be viewed without strain. The magnification obtained is

$$m \approx 1 + \frac{1}{4f} = 1 + \frac{1}{4}\frac{1}{f} = 1 + \frac{30}{4} = 1 + 7.5 = 8.5.$$

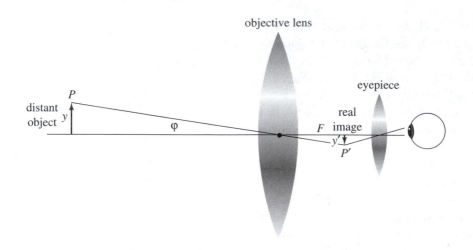

Figure 9.43

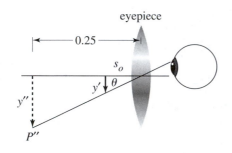

Figure 9.44

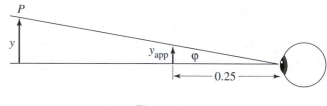

Figure 9.45

We conclude our discussion of optics by considering the question of the magnification of the simple refracting telescope described in Section 9.4B. Figure 9.43 recalls the essence of the matter. The real image of the distant object produced at the focus of the objective lens is enlarged by the eyepiece that acts as a magnifying glass. The object is represented by an arrow of height y and P is the tip of the arrow. The real image is depicted by an arrow of height y' and tip P'. Let the focal length of the objective lens be F meters and observe that this is the distance from the objective lens to the real image that it produces.

We will concentrate first on the enlargement of the real image by the eyepiece. Let f be the focal length of the eyepiece. Focus in such a way that the enlarged virtual image will form at the optimal distance of 0.25 meters from the eyepiece. Figure 9.44 shows what is going on. The magnified image is represented by an arrow of height y'' and P'' is the tip of this arrow. The real image that is being enlarged is the required distance $s_o \approx \frac{f}{1+4f} < f$ from the eyepiece, and it is magnified by a factor of $m \approx 1 + \frac{1}{4f}$. Since we are interested in good magnification, we will now assume that the power $\frac{1}{f}$ of the eyepiece is much larger than 1 and hence that f much smaller that 1. Notice that $s_o + 4s_o f = s_o(1 + 4f) \approx f$. Since both f and s_o are much smaller than 1, the term $4s_o f$ is negligible. This allows us to take $s_o \approx f$. Another look at Figure 9.44 shows that $\frac{y''}{0.25} \approx \tan\theta = \frac{y'}{s_o} \approx \frac{y'}{f}$. Therefore,

$$y'' \approx (0.25)\tan\theta \quad \text{and} \quad \tan\theta \approx \frac{y'}{f}.$$

So the image seen through the eyepiece of the telescope appears to be about $y'' \approx (0.25)\tan\theta$ meters high and 0.25 meters away.

It remains to consider the distant object. It has height y, and its tip is at the point P. Refer to Figure 9.43. If we were to move the object to within 0.25 meters from the eye, then the apparent height y_{app} would be equal to $y_{app} = (0.25)\tan\varphi$. See Figure 9.45. It follows that when we look at the distant object with the naked eye we see something that appears to be $y_{app} = (0.25)\tan\varphi$ high at 0.25 meters away. Another look at Figure 9.43 shows that $\tan\varphi = \frac{y'}{F}$.

A comparison of what is seen through the telescope with what appears to the naked eye shows that the magnification of the telescope is the ratio $\frac{y''}{y_{app}}$. This ratio can be estimated as follows:

$$\frac{y''}{y_{app}} \approx \frac{(0.25)\tan\theta}{(0.25)\tan\varphi} = \frac{\tan\theta}{\tan\varphi} \approx \frac{y'}{f}\cdot\frac{F}{y'} = \frac{F}{f}.$$

In summary, therefore, the magnifying factor of the simple refracting telescope is the ratio $\frac{F}{f}$ of the focal length F of the objective lens to the focal length f of the eyepiece. To achieve large magnification, the idea is to take an objective with a large focal length and an eyepiece with a small focal length.

Example 9.9. Suppose that a simple telescope is made with an objective of focal length $F = 3.00$ meters and an eyepiece with 50 diopters. What is the magnification? Since $\frac{1}{f} = 50, f = 0.02$ meters. Therefore, the magnification is $\frac{F}{f} = \frac{3}{0.02} = 150$. Notice that this telescope is over 3 meters long.

You may recall that Newton developed a telescope based on a different strategy. The objective lens is replaced by a concave mirror. The incoming light rays that it collects are reflected by a second mirror, creating an image that is then magnified by an eyepiece as shown in Figure 9.46. All of the world's most powerful telescopes are such *reflecting* telescopes. The telescope at Mount Palomar has a 200 inch mirror and magnifies by a factor of 3500.

One more comment is in order. Any discussion of telescopes and the accuracy of the image as a

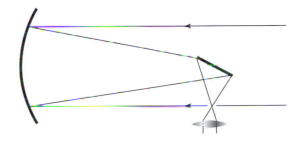

Figure 9.46

representation of the object being observed, has to deal with the fact that light rays are refracted by the Earth's atmosphere. This effect has considerable impact. See Exercise 9.J.

9.5 Postscript

This chapter has discussed the impact of calculus on statics, dynamics, and optics. All of these three concerns have, since their classical beginnings, developed into large and important disciplines. Mathematics, and the precision of the insights that it affords, has continued to play a crucial role. Statics underlies essential aspects of civil engineering and architecture. Dynamics in the context of forces and their effects is fundamental to mechanical engineering and the physical sciences. (This theme will be taken up in Chapter 14.) The case of optics is perhaps the most interesting. The development of the telescope energized astronomy and, in turn, investigations into the nature of light. Newton (1642–1727) developed a *corpuscle theory* of light based on the premise that light consists of minute particles. Huygens (1629–1695) viewed light as propagating itself in *waves* in a universally existing medium of *ether*. Parts of both theories were consistent with observed facts, but they led to contradictory conclusions, and both were flawed. Important discoveries in the years 1800 and 1801 set new agendas for science. The discovery of infrared radiation by William Herschel and ultraviolet radiation by Johann Wilhelm Ritter gave scientists a completely new perspective

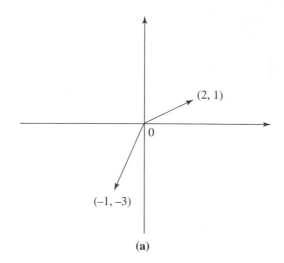

(a)

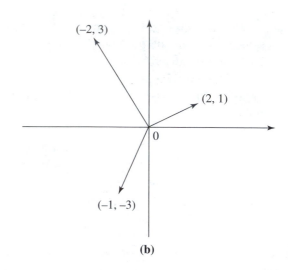

(b)

Figure 9.47

on the nature of light. It was realized that visible light was but one slice of the larger phenomenon of electromagnetic radiation, which is equally real but (except for light) invisible. Euler had grappled with these questions, and he anticipated some of the elements of the modern theory: Light, indeed all electromagnetic radiation, consists of a discontinuous series of individual packages of energy called *photons*, which travel in waves. With powerful telescopes, e.g., those of the Keck observatory in Hawaii and the orbiting Hubble telescope, light from galaxies that are billions of light years away can be gathered and studied. The light that reaches us from these distant objects is therefore billions of years old. By analyzing it, scientists can "look" at the early stages of the development of the universe.

Exercises
9A. Vectors and Forces

1. Refer to Figure 9.6 of the text. Suppose that F has magnitude 5 units and that $\theta = 15°$. Compute F_1 and F_2.

2. The vectors in Figure 9.47 represent forces. In each of the two situations, determine the magnitudes of all forces; resolve all the forces into their horizontal and vertical components, and use this to determine the resultant force.

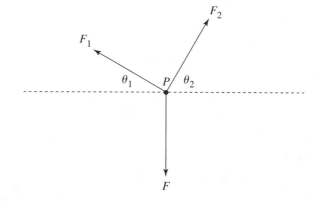

Figure 9.48

3. The three forces in Figure 9.48 are in equilibrium. They all act on the point P. The dotted line is horizontal, and F acts vertically. If F_1 has a magnitude of 10 pounds and if the angles θ_1 and θ_2 are 30° and 60° respectively, determine the magnitudes of F_2 and F.

4. Suppose the hull of a boat has a volume of 200 ft³. If the boat (without load) weighs 8,000 pounds, what is the maximum weight that its cargo can have?

5. How much of the block of wood of Section 9.1B would be submersed if it weighed 5 pounds? What if it were to weigh 15 pounds?

A basketball weighing 1.3 pounds with a radius of $r = 0.39$ feet floats in a pool of calm water. Exercises 6 and 7 contain the following question: How much of the ball is submersed?

6. Let h be the distance in feet from the bottom of the ball to the surface of the water. Use the theory of Section 5.6B to show that volume of the part of the ball that is under water is equal to $\pi r h^2 - \frac{\pi}{3} h^3$. Use this to show that $1.047h^3 - 1.225h^2 + 0.021 = 0$.

7. Consider the function $f(x) = 1.047x^3 - 1.225x^2 + 0.021$. Note that the h of Exercise 6 is a zero of $f(x)$. To find h, turn to Newton's Method in Exercises 8G. A preliminary guess for this zero is $c_1 = 0.39$ (the radius of the ball). Carry out three steps in Newton's method with six decimal accuracy to get $c_4 = 0.139533$. Show that c_4 is extremely close to being a zero of $f(x)$. Is $h \approx 0.14$ feet a reasonable answer to the question? Take a basketball to a tub of water and experiment.

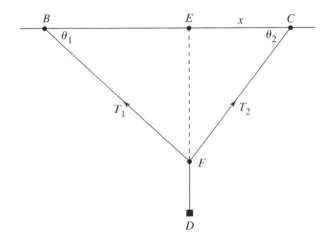

Figure 9.49

8. Consider the pulley system of De L'Hospital and assume that $c = 10$ feet and $a = 8$ feet. So in reference to Figure 9.49, $CB = 10$ and $CF = 8$. Determine the distances EC, EF, and BF and the angles θ_1 and θ_2.

9. In the situation of Exercise 8, suppose that the weight at D is 100 pounds. Determine the tensions T_1 and T_2.

10. A weight of 200 pounds is suspended from a cable as shown in Figure 9.50. The segment AB is horizontal. If the angle α is $5°$, determine the tension in the cable.

Figure 9.50

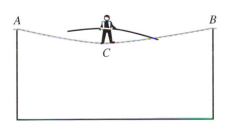

Figure 9.51

11. One of the flying Wallendas walks on a high wire from platform A to platform B. She weighs 120 pounds. When she is at point C, the angle that the wire makes with the horizontal is α at A and β at B. See Figure 9.51. Compute the tensions of the wire segments AC and CB in terms of α and β. If $\alpha > \beta$, which segment is under greater tension?

12. A sphere 8 inches in diameter made of homogeneous material and weighing 50 pounds is held by a cord against a frictionless vertical wall. See Figure 9.52.

   i. Express the tension in the string in terms of α.
  ii. Express the force with which the sphere pushes against the wall in terms of α.
 iii. Compute the tension and the force if $\alpha = 15°$.

9B. Archimedes and the Crown

The ruler of Syracuse commissioned a goldsmith to make a crown and provided an appropriate quantity of gold for the purpose. The finished crown was identical in weight to the amount of gold that he had given, but the ruler suspected that the goldsmith retained some of the gold for himself and that he substituted an amount of silver. He asked Archimedes to analyze the crown and verify whether the goldsmith had stolen some of the gold. When Archimedes hit on the answer (it is based on his law of hydrostatics) he was elated and is said to have run naked through the streets of Syracuse shouting "eureka" (I have found it.)

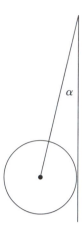

Figure 9.52

13. Let the crown weigh a total of w pounds, and let it be made up of w_1 pounds of gold and w_2 pounds of silver. Suppose that w pounds of pure gold loses f_1 pounds when weighed in water, that w pounds of pure silver loses f_2 pounds when weighed in water, and that the crown loses f pounds when weighed in water. Show that

$$\frac{w_1}{w_2} = \frac{f_2 - f}{f - f_1}.$$

[Hint: Suppose first that $w_1 = \frac{1}{3}w$. How much weight will the gold in the crown lose in terms of f_1 and how much will the silver lose in terms of f_2? Then, more generally: Since w_1 is some fraction of w, suppose that $w_1 = cw$ for some constant c.]

14. Suppose the crown of Exercise 13 displaces a volume of v cubic units when immersed in water, and that lumps of pure gold and pure silver, of the same weight as the crown, displace, respectively, v_1 and v_2 cubic units when immersed in water. Show that

$$\frac{w_1}{w_2} = \frac{v_2 - v}{v - v_1}.$$

Given this formula, all Archimedes has to do is immerse the crown in water and then do the same for quantities of pure gold and silver equal in weight to the crown. He could measure the amount of water that is displaced in each case. He then knew the exact ratio of the weight of gold to the weight of silver in the crown.

9C.   Suspension Bridges

The data below for the Brooklyn, Williamsburg, and Manhattan Bridges, all of which span the East River in New York City, were supplied by the New York City Department of Transportation, Bridges and Roadways. It must be noted that these bridges have undergone a great deal of structural change since they were built.

15. The Brooklyn Bridge was completed in 1883. Some of the important data are as follows. The center span has a length of 1,595.5 feet and the two side spans are 930 feet each. So the total length of the bridge is 3,455.5 feet. The dead loads are 9,200 pounds per linear foot for the center span and 9,100 pounds per linear foot for each of the two side spans. The bridge is designed for a live load capacity of 1,500 pounds per foot. The sag in each of the four main cables over the center span is $s = 128$ feet. Compute the tensions T_0 and T_d. Compute the angle α that one of these cables makes with the horizontal at the point of contact with the tower. What does the horizontal component of the tension in a cable over the side span have to be if the horizontal forces that act on a tower are to be in balance?

16. The Williamsburg Bridge was completed in 1903. Its center span of 1,600 feet exceeded the previous record of the Brooklyn Bridge. The dead load for two decks is 19,210 pounds per foot, and the live load capacity is 7,160 pounds. The sag in each of the four main cables is 177 feet. Compute T_d and α. The Williamsburg Bridge is the only one of the East River suspension bridges for which the cables over the side span do not bear any of the load of the side span. The only purpose of these cables is to counterbalance the forces with which the cables over the center span pull on the towers. Each cable over the side span makes an angle of $22.7°$ with the horizontal at the tower. Compute the tension in a cable over the side span. Compute the compression that all the cables together produce in each of the towers.

17. The Manhattan Bridge was completed in 1909. Its center span has a length of 1,470 feet. The dead load of the center span is 24,000 pounds per foot and the live load capacity is 11,000 pounds per foot. The bridge has four main cables with a sag of 160 feet each. Compute T_d. The main cable consists of 37 strands with 256 steel wires per strand. Each of these wires has a diameter of 0.195 inches and an ultimate strength of 220,000 pounds per square inch. Give an estimate of the ultimate strength of the main cable. What is the safety factor for the main cable?

18. Construction for the Golden Gate Bridge in San Francisco was completed in 1937. It has a center span of 4,200 feet and two sides of 1,125 feet each. The total length of the bridge is 6,450 feet. It has one deck, a dead load of 21,300 pounds per foot, and a live load capacity of 4,000 pounds per foot. The towers have a height of 746 feet and the sag in each of the two main cables is 470 feet. Compute T_0 and T_d. Each of the main cables has a diameter of 36 inches. They consist of 61 strands each with 452 steel wires of diameters 0.196 inches, for a total of 27,572 wires per cable. The ultimate strength of each wire is 220,000 pounds per square inch. Compute the ultimate strength of the cable. What is the safety factor built into the main cable? (These data, as well as the fact that the tension in the cable over the center span at the towers is 54,000,000 pounds under dead load only and 64,100,000 pounds under dead plus live load, were supplied by the Golden Gate Bridge, Highway and Transportation District, San Francisco, California.)

19. The Tacoma Narrows Bridge was completely rebuilt in the years 1948–1950. The data for the new bridge are as follows. It has a total length of 5,000 feet and a center span of 2,800 feet. Its two main cables support a single deck that carries 4 lanes of automobile traffic. The dead load is 8,680 pounds per foot. Assume that it is designed for a live load capacity of 4,000 pounds per foot and that the sag in the main cable over the center span is $s = 280$ feet. Compute T_d and α. Compute the compression that one of the main cables over the center span produces in a tower. (The information is based on data supplied by the Washington State Department of Transportation.)

20. The Verrazano Narrows Bridge, New York City, was completed in 1964. It has two decks, a center span of 4,260 feet, a dead load of 37,000 pounds per foot, and a live load capacity of 4,800 pounds per foot. It has four main cables, and the sag in each of them is 385 feet at mid span. Compute T_d and α. Compute the compression that the four main cables over the center span produce in one of the towers. (Data supplied by the Triborough Bridge and Tunnel Authority.)

21. The Akashi Straits Bridge is part of a highway system that links the islands of Honshu, Awaji, and Shikoku. The gigantic bridge[12] has an overall length of 3910 meters (about 2.5 miles) and a center span of 1990 meters (about 6530 feet). Each of its two towers rises to a height of 297.2 meters (about 975 feet) above sea level. Its two main cables support a deck that carries 6 lanes of traffic. The sag in the main cables is about 201 meters. It is estimated that the total compression in one of the towers is 980 million newtons. Provide estimates for the dead plus live load capacity of the deck in newtons per meter, as well as T_d and α. (Based on published data of the Tarumi Construction Office, First Construction Bureau, Honshu-Shikoku Bridge Authority.)

9D. About the Cables

22. Study the configurations of circles in Figure 9.53. They consist of 1, 7, 19, and 37 circles respectively. The last three have the shape of hexagons. Notice that the horizontal diagonals of the three hexagons contain 3, 5, and 7 circles respectively. Draw a hexagonal configuration of circles that has a horizontal diagonal of 9 circles. Check that the resulting configuration has 61 circles. The main cable of a suspension bridge is constructed as follows: A fixed number of wires (several hundred at a time) are squeezed together into strands with a circular cross-section of a few inches. The strands are then arranged into one of the hexagonal configurations that we have discussed, caulked, and wrapped to form the main cable. The main cables of the Brooklyn Bridge are built with a 19 strand hexagonal arrangement, the main cables of the Williamsburg Bridge and the Manhattan Bridge with a 37 strand arrangement, and the main cables in the George Washington Bridge, the Golden Gate Bridge, and the Verrazano Narrows Bridge with a 61 strand arrangement. The strands are arranged in one of the hexagonal configurations that we have discussed.

23. Consider the "hexagonal" numbers 1, 7, 19, 37, and 61 discussed in Exercise 22. Do you see how one of these numbers leads to the next? Explain why 61 is followed by 91 and why 91 is followed by 127. What is the hexagonal number that follows 127? The design for the main cable of the Akashi Straits Bridge is different from that described in Exercise 22. The strand is a hexagonal configuration of 127 wires with a diameter of 5.23 millimeters each. A cable consists of an arrangement of 290 such strands grouped into a circular arrangement. What advantage does this arrangement have over the one described in Exercise 22?

24. Compare $1 + x$ with the product

$$\left(1 + \frac{1}{2}x - \frac{1}{8}x^2 + \frac{1}{16}x^3 - \frac{5}{128}x^4\right)$$
$$\cdot \left(1 + \frac{1}{2}x - \frac{1}{8}x^2 + \frac{1}{16}x^3 - \frac{5}{128}x^4\right).$$

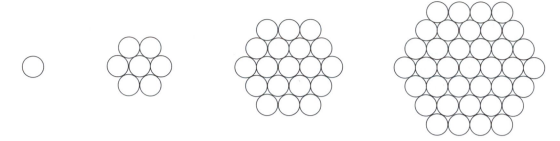

Figure 9.53

Deduce that $\sqrt{1+x}$ is closely approximated by

$$1 + \frac{1}{2}x - \frac{1}{8}x^2 + \frac{1}{16}x^3 - \frac{5}{128}x^4$$

for $|x|$ much smaller than 1. Use this to show that $\sqrt{1+cx^2}$ is approximated by

$$1 + \frac{1}{2}(cx^2) - \frac{1}{8}(c^2x^4) + \frac{1}{16}(c^3x^6) - \frac{5}{128}(c^4x^8)$$

for $|cx^2|$ much smaller than 1.

25. Consider a suspension bridge with parameters d and s. Show that the length of the cable over the center span from its lowest point to the top of a tower is given by the definite integral $\int_0^d \sqrt{1 + \frac{(2s)^2}{d^4}x^2}\, dx$. [Hint: Review Section 5.6C.]

26. Notice that for all the suspension bridges considered, the quantity $\frac{(2s)^2}{d^4}x^2 = (\frac{2s}{d})^2(\frac{x}{d})^2$ is much less than 1 for $0 \le x \le d$. Use the results of Exercise 24 to determine an approximation of the definite integral

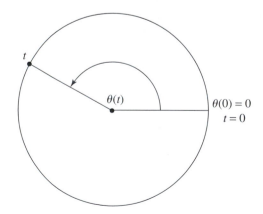

Figure 9.54

of Exercise 25. Use your conclusions to compute the length of one of the cables over the center span of the George Washington Bridge. Do the same for the Brooklyn Bridge.

9E.   Calculus of Rotation

27. A particle rotates counterclockwise starting from rest at time $t = 0$ according to the law $\theta(t) = \frac{t^3}{125} + \frac{t}{5}$, where θ is in radians and t is in seconds. Calculate the angle θ at $t = 10$, the average angular velocity between $t = 0$ to $t = 10$, the angular velocity ω at the instant $t = 10$, the average angular acceleration from $t = 0$ to $t = 10$, and the angular acceleration α at $t = 10$. See Figure 9.54.

28. An automobile tire has an outer diameter of 0.65 meters. At how many revolutions per second does this tire turn when it is on an automobile that is moving at 100 kilometers per hour? [Hint: One solution proceeds as follows. Pick a point on the outer diameter. How fast (in meters per second) is the point moving? Now compute the angular velocity ω of the moving point. How many revolutions per second does this correspond to?]

Figure 9.55 depicts two systems of pulley wheels that are fixed distances apart and loops of cords that run in their grooves. Think of a fan belt system of the engine of a car.

29. Refer to Figure 9.55(a). Pulley A is 8 inches in diameter and rotates at 120 revolutions per minute (rpm). A fan belt connects it with pulley B. What is the angular velocity ω_A of a point moving on the perimeter of pulley A? What is the velocity v in inches per minute of a point moving on the fan belt? At how many rpm is B rotating if it has a diameter of 12 inches. [Hint: What is the connection between v and the angular velocity ω_B of a point moving on the perimeter of pulley B?]

30. The pulleys in Figure 9.55(b) have the following diameters: A, 8 centimeters; B, 20 centimeters; C, 10

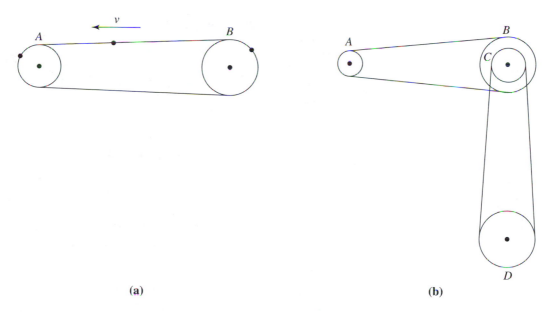

(a) (b)

Figure 9.55

centimeters; and D, 14 centimeters. A rotates at 200 rpm. Pulleys B and C rotate together with the same angular velocity. At how many rpm does D rotate? [Use the strategy of the solution of Exercise 29.]

9F.   The Rotational Effect of Forces

Archimedes's Law of the Lever says in essence that two weights w_1 and w_2 placed on a lever at the respective distances d_1 and d_2 from the fulcrum have the same rotational effect if $w_1d_1 = w_2d_2$. The rotational effect, or *torque*, of a weight or force w that is a distance d from a point is therefore equal to wd.

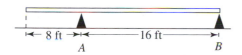

Figure 9.56

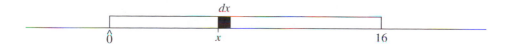

Figure 9.57

31. A uniform and rigid beam weighs 150 pounds and has a length of 24 feet. So the beam weighs $\frac{150}{24} = 6.25$ pounds per foot. It is placed on two supports at points A and B as shown in Figure 9.56. Notice that the part of the beam on the right weighs 100 pounds and that on the left weighs 50 pounds. What is the rotational effect, or torque, around A produced by the right side of the beam? To answer this, think of the beam as being placed on a completely rigid lever with fulcrum at A, and think of the beam as being divided into very small pieces each of length dx. See Figure 9.57. Each piece will weigh $6.25\,dx$ pounds. Let the typical piece be x feet from the fulcrum and note that it contributes a torque of $6.25x\,dx$ around A. Review Sections 5.3–5.6 and write the total torque contributed by all the pieces as a definite integral. Evaluate the integral by using the Fundamental Theorem of Calculus. Conclude that the sum of all the torques that these little pieces generate is equal to 800 foot-pounds. Since $800 = 8 \times 100$, observe that for the purposes of computing the torque of the right side of the beam, it can be assumed that all the weight of the beam is located at its midpoint.

32. Continue to consider the beam of Exercise 31. What is the maximum distance x a person weighing 180 pounds can walk beyond A without tilting the beam? See Figure 9.58.

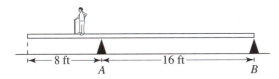

Figure 9.58

33. How much weight has to be placed at A so that the man weighing 160 pounds can walk to the end of the rigid beam in Figure 9.59. The beam weighs 20 pounds per foot. This problem illustrates the principle behind the cantilever bridge.

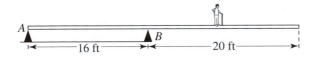

Figure 9.59

Let some units of length and mass be given. Consider a thin circular object of radius r and mass m. Recall from Section 9.3B that its moment of inertia is mr^2, where the axis of rotation is the line perpendicular to the plane of the circle and through its center. Recall also that the verification of this fact was achieved by regarding the circle as subdivided into many particles and by adding the moments of inertia of all the individual particles. That the moment of inertia of an object is equal to the sum of the moments of inertia of all the particles that constitute it is a fact that is generally valid. It will be used in Exercises 34–36. These Exercises will also require the use of Leibniz's summation strategy and the Fundemental Theorem of Calculus. (Refer back to Sections 5.3–5.6.)

34. Consider a thin rigid rod of length r and mass m attached to a fixed point O. See Figure 9.60(a). Suppose that the mass of the rod is distributed uniformly and

that the rod can rotate freely about O. Think of a single spoke of a rotating bicycle wheel. (Since gravity confuses the issue, consider the plane of rotation to be horizontal.) Notice that one unit of length of the rod has a mass of $\frac{m}{r}$. Consider the rod to be subdivided into a large number of pieces each of length dx. Consider a typical piece and let its distance from O be x, where $0 \le x \le r$. Show that the moment of inertia of this piece is $\left(\frac{m}{r}\,dx\right)x^2 = \left(\frac{m}{r}x^2\right)dx$. Express the moment of inertia I of the rod as a definite integral and use the Fundamental Theorem of Calculus to show that $I = \frac{1}{3}mr^2$. (Where is the flaw in the following approach to this problem? Consider the entire mass of the rod concentrated at its midpoint. Because $2f \cdot \frac{r}{2} = f \cdot r$, the rotational acceleration α produced by a force f applied at the end is the same as that produced by $2f$ applied at the midpoint $\frac{r}{2}$. See Figure 9.60(b). Applying the formula for the moment of inertia of a point mass already established, we get $(2f)(\frac{r}{2}) = m(\frac{r}{2})^2\alpha$. So $f \cdot r = \frac{1}{4}mr^2\alpha$, and hence $I = \frac{1}{4}mr^2$.)

35. Consider a thin circular disc of radius r and mass m and suppose that its mass is distributed uniformly. Suppose that the disc is free to rotate about the axis perpendicular to its plane and through its center. Note that one unit of square area of the disc has a mass of $\frac{m}{\pi r^2}$. Place the center of the disc at the origin of a coordinate axis that lies in the plane of the disc. Suppose that the disc is subdivided into lots of very thin circular pieces each with center 0. Let the thickness of every circular piece be dx. Notice that their radii range from 0 to r. Take a typical circular piece and let its radius be x. Show that the mass of this circular piece is closely approximated by $\frac{m}{\pi r^2}(2\pi x\,dx) = \frac{m}{r^2}2x\,dx$ and that the moment of inertia of this circular piece is $\frac{m}{r^2}(2x\,dx)x^2 = \frac{m}{r^2}2x^3\,dx$. After drawing a diagram of what is going on, express the moment of inertia I of the circular disc as a definite integral and use the Fundamental Theorem of Calculus to show that $I = \frac{1}{2}mr^2$.

36. Consider a solid sphere of radius r and mass m rotating about an axis through its center. Refer to Figure 9.27. Suppose that the mass of the sphere is uniformly distributed. Place an x-y coordinate system in such a way that the axis of rotation is the x-axis and that the center of the sphere is at the origin. Observe that the intersection of the sphere and the plane of the coordinate system is the circle $x^2 + y^2 = r^2$. In reference to the x-axis notice that the sphere extends from $-r$ to r. Because the volume of a sphere is $\frac{4}{3}\pi r^3$

(see Chapter 5.6B), one unit of volume has mass $\frac{m}{\frac{4}{3}\pi r^3} =$ $\frac{3}{4}\frac{m}{\pi r^3}$. Consider the sphere to be sliced into very thin parallel discs each of thickness dx, each perpendicular to the axis of rotation, and each with center on the axis. Let x be the coordinate of the center of a typical disc and notice that its radius is $y = \sqrt{r^2 - x^2}$. It follows that the volume of this disc is $\pi y^2 dx = \pi(r^2 - x^2)dx$ and that its mass is $\frac{3}{4}\frac{m}{\pi r^3}\pi(r^2 - x^2)dx = \frac{3}{4}\frac{m}{r^3}(r^2 - x^2)dx$. By applying the conclusion of Exercise 35, we get that the moment of inertia of this typical disc is

$$\frac{1}{2}\left(\frac{3}{4}\frac{m}{r^3}(r^2 - x^2)dx\right)y^2 = \left(\frac{3}{8}\frac{m}{r^3}(r^2 - x^2)dx\right)(r^2 - x^2)$$

$$= \frac{3}{8}\frac{m}{r^3}(r^2 - x^2)^2 dx.$$

Use this expression to write the moment of inertia I of the sphere in terms of a definite integral and use the Fundamental Theorem of Calculus to show that $I = \frac{2}{5}mr^2$.

9G. Galileo's Experiment

37. Recall that Galileo's bronze ball has a radius of 1 centimeter. Determine for each of the starting heights the rotational velocity of the ball at the bottom of the plane in revolutions per second.

38. Suppose that Galileo had released a ball 1.10 meters above the table. How far from the table would the ball have landed as predicted by theory? After considering Galileo's data, where do you think it would have landed?

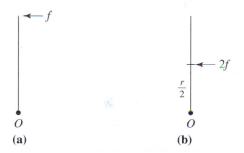

Figure 9.60

39. The historian Stillmann Drake (see Note 8 at the end of this chapter) suggests that Galileo's table was perhaps only 0.75 meters high (instead of the 0.778 meters used

in Section 9.3C). Compute the four distances reached by the bronze ball for this table height.

40. Which equation of the mathematical analysis of Galileo's experiment explains why the results of his experiment did not depend on the angle β of the inclined plane? Would the distances of 0.752, 1.102, 1.248, and 1.410 meters that Galileo measured have been much different had Galileo carried out the same experiment on the Moon? Why might they have been different? But why in fact would they not have been?

9H. Basic Optics

41. Suppose that a light ray passes through the air and strikes a sheet of crown glass at an angle of incidence $\alpha = 30°$. Determine the angle of refraction.

42. A mirror consists of a piece of plate glass supplied with a silver coating as shown in Figure 9.61. Verify that the angle of incidence α is equal to the angle of reflection β.

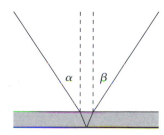

Figure 9.61

43. Suppose that a light ray passes through two successive parallel slabs of transparent mediums that are separated by a slight amount of vacuum. See Figure 9.62. Assume that the initial angle of incidence is α. Suppose the indices of refraction of the mediums are n_1 and n_2. If β_2 is the angle of refraction in the second slab, show that $\sin \alpha = n_2 \sin \beta_2$. Show also that $\alpha_2 = \alpha$.

44. Suppose that a light ray passes through four successive parallel slabs of transparent mediums each separated from the other by a slight amount of vacuum. Assume that the initial angle of incidence is α. Suppose the index of refraction of the last medium is n_4. Show that $\sin \alpha = n_4 \sin \beta_4$ and that $\alpha_4 = \alpha$. See Figure 9.63. Consider a light ray that passes through k successive parallel slabs of transparent mediums each separated from the other by a slight amount of vacuum. Suppose the index of refraction of the last medium is

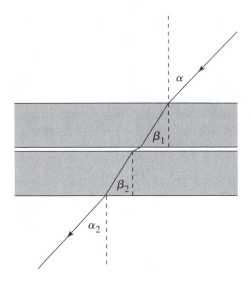

Figure 9.62

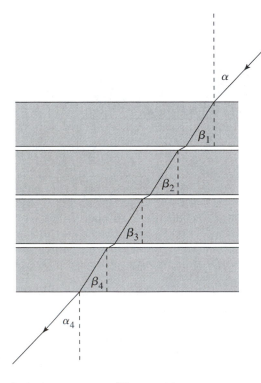

Figure 9.63

n_k. It should now be clear that if the initial angle of incidence is α, and β_k is the last angle of refraction, then $\sin \alpha = n_k \sin \beta_k$.

45. Suppose that the spherical surfaces of a lens made of flint glass both have a radius 0.55 meters. Compute the focal length of this lens.

46. Suppose that the two spherical surfaces of a lens made of crown glass have radii 0.30 and 0.45 meters respectively. Determine the power of the lens in diopters.

47. Suppose that a thin, spherical lens has a focal length of $f = 0.38$ meters and that the radii of its spherical surfaces are both 20 centimeters. Estimate the index of refraction of the glass of the lens.

48. Consider a 12 diopter lens. Suppose that an object 16 centimeters high is placed 55 centimeters from the lens. Is the image real or virtual? How far is the image from the lens and how large is it?

49. A 40 diopter magnifying glass is focused on an ant 7 millimeters long. The lens is focused in such a way that the image appears to be 0.25 meters away. How long is the ant when observed through the magnifying glass?

50. A simple telescope features a large objective with a focal length of 4 meters and an eyepiece of 20 diopters. By what factor is a distant satellite of Jupiter enlarged by the telescope?

51. You wish to construct a simple refracting telescope that has a magnification of 200. Suppose that the most powerful eyepiece available to you has 40 diopters. What is the smallest possible length of your telescope?

9I.   The Speed of Light

In the 1870s, the physicist Michelson measured the speed of light with great accuracy with the following apparatus consisting of a lens and two mirrors. The lens and its two focal points S' and S'' are shown in Figure 9.64. The two mirrors are placed as shown in Figure 9.65. Mirror 1 is free to rotate about an axis perpendicular to its plane. The point S is the reflection on the focal point S'. Now suppose that a point light source (that radiates light in all directions) is placed at S. The light rays that emanate from it strike mirror 1, go through the lens, refocus at S'', are reflected by mirror 2, are reflected again by mirror 1, and finally reconverge at the point S. Let c be the speed of light and let t be the time it takes for the rays to travel from mirror 1 back to mirror 1. Since the rays cover the distance $2d$ during this time, it follows that $2d = ct$. Hence $c = \frac{2d}{t}$. Now suppose that mirror 1 is allowed to rotate. When the axis of rotation is placed

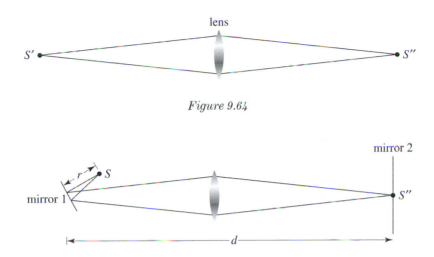

Figure 9.64

Figure 9.65

appropriately and the speed of rotation is high enough, the light rays will hit a different position on mirror 1 on their return. Thus they no longer reconverge at the point S, but at a point that is a small, but measurable, distance δ away. The larger the time t, the larger the deviation δ. For his apparatus Michelson was able to show that $\delta = 4\pi rnt$, where r is the distance indicated in Figure 9.65 and n is the number of revolutions per second of the mirror. Measuring δ allowed him to compute t, and hence c. He was able to conclude that light travels at $299{,}895 \pm 30$ kilometers per second. This corresponds roughly to $186{,}235 \pm 19$ miles per second. For this and other achievements, Michelson was awarded the Nobel prize for physics in 1907.

9J. The Phenomenon of Refraction in Astronomy

The density of Earth's atmosphere diminishes as the height above the Earth increases. Studies show that there is some atmosphere at a height of 800 kilometers above the Earth. The density is still appreciable enough at 150 kilometers for air drag to produce changes in the orbit of a satellite. Refraction of light rays begins to be a factor at about 100 kilometers. We will model these 100 kilometers of the Earth's atmosphere as follows. Above the atmosphere is the vacuum of space. We will neglect the curvature of the Earth. This is not unreasonable since the radius of the Earth is about 6400 kilometers as compared to the relevant layers of atmosphere of 100 kilometers. Then we will regard the atmosphere as being made up of many, say k, layers of different densities, the density being greatest in the

lowest layer and constant within each layer. We will think of consecutive layers as being separated by a tiny strip of vacuum. In particular, we will use Exercise 44 to model what is going on.

A ray of light from a star strikes the top layer at the point T with an angle of incidence of α. Refer to Figure 9.66. It is then refracted by the successive layers of the atmosphere until it reaches the observer at O. The axis OZ points north from the observer (say to the North Star). Draw the line OS parallel to the ray coming in to T. This is the line from the observer in the direction of the actual position of the star. The line OA points in the observed direction of the star. Notice that the star will appear to be higher in the sky than it actually is. The angle $z_{\text{true}} = \angle ZOS$ is called the *true zenith distance* of the star. Observe that $z_{\text{true}} = \alpha$. The angle $z_{\text{app}} = \angle ZOA$ is called the *apparent zenith distance* of the star. The angle

$$\rho = z_{\text{true}} - z_{\text{app}}$$

measures the difference between the true position and the observed position of the star. It is the effect of the refraction of the atmosphere.

Now consider an enlargement of the last layer (Figure 9.67). Refer to the diagram of Exercise 44 and note that α is the angle of incidence of the first layer and that β_k is equal to the angle of refraction of the ray in the last layer. Observe also that $z_{\text{app}} = \beta_k$. By the conclusion of Exercise 44,

$$\sin z_{\text{true}} = \sin \alpha = n_k \sin \beta_k = n_k \sin z_{\text{app}}.$$

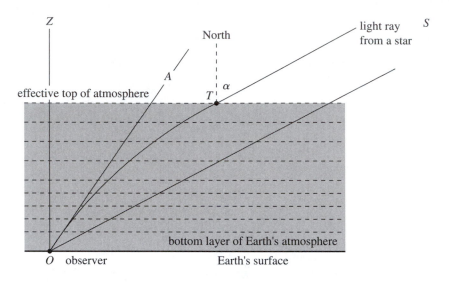

Figure 9.66

z_{app} in°	ρ in″	z_{app} in°	ρ in″	z_{app} in°	ρ in″	z_{app} in°	ρ in″
0	0	50	70	80	319	86	706
10	10	55	84	81	353	87	863
20	21	60	101	82	394	88	1103
30	34	65	125	83	444	89 0′	1481
40	49	70	159	84	509	89 31′	1760
45	59	75	215	85	593	90 0′	2123

Table 9.2

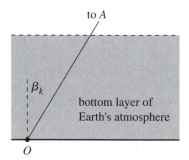

Figure 9.67

Since $z_{\text{true}} = \rho + z_{\text{app}}$, we get by the addition formula for the sine that

$$\sin z_{\text{true}} = (\sin \rho)(\cos z_{\text{app}}) + (\cos \rho)(\sin z_{\text{app}}).$$

So $(\sin \rho)(\cos z_{\text{app}}) + (\cos \rho)(\sin z_{\text{app}}) = n_k \sin z_{\text{app}}$. Assuming that ρ is small and hence that $\sin \rho \approx \rho$ and $\cos \rho \approx 1$, we get the approximation $\rho \cos z_{\text{app}} + \sin z_{\text{app}} \approx n_k \sin z_{\text{app}}$. So $\rho \approx (n_k - 1)\frac{\sin z_{\text{app}}}{\cos z_{\text{app}}} = (n_k - 1)\tan z_{\text{app}}$. The index of refraction of the air at sea level at $10°$ centigrade is 1.00028. Inserting this for n_k gives the approximation $\rho \approx 0.000283 \tan z_{\text{app}}$.

Converting from radians gives $\rho \approx 0.016214 \tan z_{\text{app}}$ in degrees and

$$\rho \approx 58.372941 \tan z_{\text{app}} \approx 58.37 \tan z_{\text{app}}$$

in seconds (″). Let's summarize. To find the true zenith distance z_{true} of a heavenly body in degrees, measure z_{app} in degrees, take the corresponding ρ, and (being careful with units) compute $z_{\text{true}} = z_{\text{app}} + \rho$. The approximation $\rho \approx 58.37 \tan z_{\text{app}}$ for the difference between the true position z_{true} of a star and its observed position z_{app} is very accurate for z_{app} up to $45°$ but becomes progressively less accurate for larger angles z_{app}. Table 9.2 provides, for selected angles z_{app} (in degrees, in the columns on the left), the more accurate values for ρ (in seconds, in the columns on the right).

52. Compare the values for ρ provided by the approximation with those of Table 9.2 for $z_{app} = 20°$, $60°$, and $80°$. Suppose that stars are observed at these apparent zenith distances. What are the true zenith distances?

53. Consider the Sun at sunset, precisely when the apparent zenith position of the bottom of the Sun's disc is $90°$. Consider the angle determined at the eye of an observer by a vertical diameter of the Sun's disc. This apparent "vertical angular diameter" of the Sun is equal to about 29 minutes. Use the data in Table 9.2 to show that the true zenith position of the top of the Sun's disc is greater than $90°$. This means that the Sun is in fact already below the horizon when it is observed at sunset! Draw a diagram of what is happening. Use Table 9.2 again to show that the true vertical angular diameter of the Sun is about 35 minutes. Since the Sun is circular, this is also the true horizontal angular diameter of the Sun. But the horizontal angular diameter will not be affected appreciably by refraction. So the apparent horizontal diameter is also about 35 minutes. Therefore, when the Sun is observed at sunset, its apparent vertical and horizontal diameters will be 29 and 35 minutes respectively. This is why the Sun at sunset is seen as a slightly flattened disc.

Notes

[1] See the book by Stillmann Drake, *Galileo: Pioneer Scientist*, University of Toronto Press, Toronto, 1990.

[2] Fermat was a lawyer by profession (he served as counselor to the court of law of Toulouse), but he achieved lasting fame as a mathematician and is regarded, along with Kepler, Galileo, Descartes, Leibniz, and Newton, as one of the greats of the 17th century mathematics. Along with his contemporary Descartes he is credited with the invention of analytic geometry and he made early contributions to both differential and integral calculus. His great passion was number theory and in particular the questions that grew out of his studies of the *Arithmetica* of Diophantus. (See Exercises 2E of Chapter 2.) The most famous of these questions became known as *Fermat's Last Theorem*. Start with the fact that the equation $a^2 + b^2 = c^2$ can be solved with positive integers. For example, $a = 3$, $b = 4, c = 5$ is one such solution, and $a = 5, b = 12$, and $c = 13$ is another. The problem that Fermat asked was this: Can any of the equations $a^3 + b^3 = c^3, a^4 + b^4 = c^4$, $a^5 + b^5 = c^5$, and so on, be solved with positive integers. This question became a holy grail of mathematics. A significant body of advanced abstract arithmetic and algebra was developed to show that at least some of the equations did not have such a solution. The complete answer was not achieved until 1995 when Andrew Wiles of Cambridge University proved that none (!) of these equations had such a solution. Wiles' answer consists of a formidable web of very sophisticated mathematics that stands in its complexity in sharp contrast to the simplicity of the question itself.

[3] All the data are taken from *The George Washington Bridge across the Hudson River at New York, N.Y.*, American Society of Civil Engineers, Reprinted from Transactions, Am. Soc. C. E., Vol. 97 (1933) by The Port of New York Authority.

[4] If you are unconvinced, consider the portion of the deck below the segment from 0 to x. Think of the deck as having been cut perpendicularly to the flow of the traffic both below 0 and x. This does not affect the forces on the deck, because we have assumed that the deck is attached to the main cable by many suspender cables, each close to the next. Assume that you are a giant and refer to Figure 9.17. Grab the cable at (0,0) with your left hand and at C with your right hand. Instead of having the cable pull to the left at (0,0) with a force of T_0, let your left hand pull in this way, and instead of having the cable pull at C with a force of T_x towards the tower, let your right hand pull in this way. Your two hands are now holding the entire deck under the segment from 0 to x. Your left hand is pulling horizontally, so this pull has zero vertical component; it follows that the vertical force that counterbalances the weight wx of the deck must be supplied by your right hand alone. This vertical force is $T_x \sin \theta$.

[5] Note that the estimates for the maximum and minimum tensions in the main cable that we have derived match well with the values $T_d = 58,500,000$ pounds and $T_0 = 55,000,000$ pounds supplied by the New York Port Authority.

[6]See for instance, D.B. Steinman, *A Practical Treatise on Suspension Bridges*, second edition, John Wiley and Sons, New York, 1953.

[7]A videotape entitled *Resonance* of the series *The Mechanical Universe Series* (Films Incorporated, 5547 N. Ravenswood, Chicago, IL 606640-1199) includes footage of the collapse of this bridge and an explanation of why it did.

[8]See the book Stillmann Drake, *Galileo: Pioneer Scientist*, University of Toronto Press, Toronto, 1990.

[9]It is relevant to note in this context that the mathematics of moments of inertia and rotational motion was first developed by Leonhard Euler and others more than one hundred years after Galileo died.

[10]In the 1670s the Danish astronomer Olaus Römer detected periodic variations in the times at which one of the bright moons of Jupiter went into eclipse and realized that these variations must be caused by the different lengths of time it takes light to travel the changing distance from Earth to Jupiter. Using the distance to Jupiter (as then understood), he calculated the speed of light to within 30% of the accurate value of 186,272 miles per second. The accurate value was obtained in the last part of the 19th century by the American physicist Albert Michelson. See Exercise 9I.

[11]Our argument assumes that we are dealing with an ideal lens. In an actual lens, rays that strike a lens far from its center (such as AT and AT') will deviate from the ideal path and not return to the axis of symmetry precisely at B as we are assuming.

[12]Refer to the very informative account about the bridge and its construction by D. Normile and F. Vizard, A Bridge So Far, *Popular Science*, March 1998, 48–53.

10

Basic Functions and Their Graphs

This chapter will introduce and study a number of important functions. We have seen, primarily in the context of astronomy, that geometric and repetitive phenomena often require the trigonometric functions for their definitive explanation. The quantitative analysis of other phenomena, however, for instance those involving trends, and in particular processes of growth or decline, requires different functions. Of these, the exponential and logarithm functions are the most important. Leonhard Euler in his treatise *Introductio in Analysin Infinitorum* (1748) gives the first comprehensive treatment of these functions (indeed this work contains the modern approach to functions in general). This chapter begins with the exponential functions, then turns to inverse functions, and then to the special cases of the log and inverse trigonometric functions. Finally, now that the arsenal of functions is substantial, it considers the issue of concavity and the various components that constitute the strategy of graphing functions. The chapter builds[1] on the material already developed in Chapter 8.

10.1 **Exponential Functions**

We already know that $3^1 = 3$, $3^2 = 3 \cdot 3 = 9$, and $3^0 = 1$. Recall also that the square root of 3, written $\sqrt{3} = 3^{\frac{1}{2}}$, is that number whose square is 3. Similarly, for any positive integer n, the nth root of three, denoted by $\sqrt[n]{3} = 3^{\frac{1}{n}}$, is that number which when raised to the nth power is 3. We know also that $3^{\frac{2}{3}} = \left(3^{\frac{1}{3}}\right)^2 = \left(3^2\right)^{\frac{1}{3}}$. More generally, for any positive rational number $r = \frac{m}{n}$ (where m and n are both positive integers),

$$3^r = 3^{\frac{m}{n}} = \left(3^{\frac{1}{n}}\right)^m = \left(3^m\right)^{\frac{1}{n}}.$$

If r a negative rational number, then $-r$ is positive, and 3^r is defined to be equal to $\frac{1}{3^{-r}}$. For example,

$$3^{-\frac{5}{4}} = \frac{1}{3^{\frac{5}{4}}} = \frac{1}{\left(3^{\frac{1}{4}}\right)^5} = \frac{1}{\left(3^5\right)^{\frac{1}{4}}}.$$

Suppose now that x is a number that is not rational. For example, take $x = \sqrt{2}$ or $x = \pi$. What does 3^x mean then? In particular, what is the meaning of $3^{\sqrt{2}}$ or 3^π? Here is how we can give meaning to these expressions. Start with the decimal expansion

$$\sqrt{2} = 1.414213562\ldots$$

of $\sqrt{2}$. Observe that this expansion gives rise to the sequence of rational numbers

$$1,\ 1.4 = \frac{14}{10},\ 1.41 = \frac{141}{100},\ 1.414 = \frac{1414}{1000},$$

$$1.4142 = \frac{14{,}142}{10{,}000},\ 1.41421 = \frac{141{,}421}{100{,}000}, \ldots,$$

which closes in on $\sqrt{2}$. Raising 3 to each of these terms makes sense and provides the sequence

$$3^1 = 3,\ 3^{1.4} = 4.655537,\ 3^{1.41} = 4.706965,$$

$$3^{1.414} = 4.727695,\ 3^{1.4142} = 4.728734,$$

$$3^{1.41421} = 4.728786,\ 3^{1.414213} = 4.728801,$$

$$3^{1.4142135} = 4.728804, \ldots.$$

It is apparent (at least intuitively) that this sequence closes in on some number that has a decimal expansion of the form $4.72880\ldots$. We will also say that the sequence *converges* to $4.72880\ldots$ or *has the limit* $4.72880\ldots$ to describe what is going on. This number is what $3^{\sqrt{2}}$ is defined to mean. Its decimal expansion can be found as far as a particular context might require. For example, up to nine-decimal accuracy, $3^{\sqrt{2}} = 4.728804388$.

The same thing can be done with 3^π. Start with $\pi = 3.141592653\ldots$; consider the sequence

$$3^3 = 27,\ 3^{3.1} = 30.135326,\ 3^{3.14} = 31.489136,$$

$$3^{3.141} = 31.523749,\ 3^{3.1415} = 31.541070,$$

$$3^{3.14159} = 31.544189,\ 3^{3.141592} = 31.544258,$$

$$3^{3.1415926} = 31.544278, \ldots$$

and define 3^π to be the limit of this sequence. With an accuracy of seven decimal places, $3^\pi = 31.5442807$. In

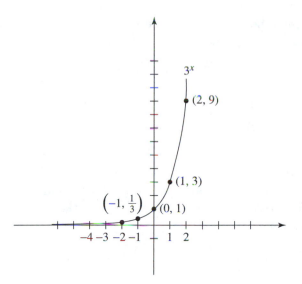

Figure 10.1

this way, the number 3^x can be defined for any real number x.

The rule $f(x) = 3^x$ determines a function. It is an increasing function. Its graph rises slowly through negative x and then rapidly through positive x. The general shape of the graph is shown in Figure 10.1.

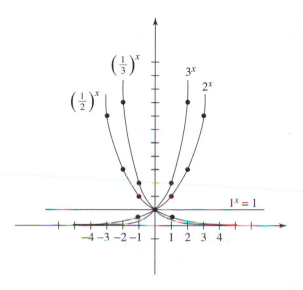

Figure 10.2

The same strategy can be used to define a^x for any positive constant a. Namely, start with the decimal expansion of x to obtain a sequence of rational numbers $r_1, r_2, r_3, r_4, \ldots$ that converges to x. So $x = \lim_{n \to \infty} r_n$. Then define

$$a^x = \lim_{n \to \infty} a^{r_n}.$$

Notice that this definition does not work for $a = 0$ or $a < 0$. For example, $0^{-1} = \frac{1}{0}$ and $(-3)^{\frac{1}{2}} = \sqrt{-3}$ are not defined.

The rule $f(x) = a^x$ determines a function for any constant $a > 0$. These are the *exponential functions*. All have the set of all real numbers as their domain and they are continuous on their domains. The graph depends on whether $a < 1$, $a = 1$, or $a > 1$. The cases $a = \frac{1}{2}$, $a = \frac{1}{3}$, $a = 1$, $a = 2$, and $a = 3$ are sketched in Figure 10.2.

The process of raising positive numbers to powers satisfies a number of basic laws:

Laws of Exponents: Let a and b be any positive constants and x and y any real numbers. Then

  i. $a^x a^y = a^{x+y}$
      (for example, $a^2 a^3 = (aa)(aaa) = a^5$).
  ii. $a^x a^{-y} = a^{x-y} = \frac{a^x}{a^y}$
      (for example, $a^{-5} = a^{2-7} = a^2 \cdot a^{-7} = \frac{a^2}{a^7}$).
  iii. $(a^x)^y = a^{xy}$
      (for example, $(a^6)^3 = (a^6)(a^6)(a^6) = a^{18}$).
  iv. $(ab)^x = a^x b^x$
      (for example, $(ab)^4 = (ab)(ab)(ab)(ab) = a^4 b^4$).

We will now single out a particularly interesting exponential function. Begin with the sequence

$$\left(1 + \frac{1}{1}\right)^1, \left(1 + \frac{1}{2}\right)^2, \left(1 + \frac{1}{3}\right)^3, \left(1 + \frac{1}{4}\right)^4,$$

$$\left(1 + \frac{1}{5}\right)^5, \ldots, \left(1 + \frac{1}{n}\right)^n, \ldots.$$

The first few terms are

$$2^1 = 2, \left(\frac{3}{2}\right)^2 = 2.25000, \left(\frac{4}{3}\right)^3 = 2.37037,$$

$$\left(\frac{5}{4}\right)^4 = 2.44140, \left(\frac{6}{5}\right)^5 = 2.48832\ldots,$$

$$\left(\frac{101}{100}\right)^{100} = 2.70481,\ldots, \left(\frac{1001}{1000}\right)^{1000} = 2.71692,\ldots$$

and shifting into much higher gears,

$$\left(\frac{1,000,001}{1,000,000}\right)^{1,000,000} = 2.718280469,\ldots,$$

$$\left(\frac{100,000,001}{100,000,000}\right)^{100,000,000} = 2.718281815,\ldots.$$

It should be evident that this sequence converges. It does so extremely slowly, however. Notice that even the 1,000,000th term does not produce the correct value for the sixth decimal place. The 100,000,000th term does not give the correct value for the eighth decimal place. The limiting value of the sequence is the number e, so designated in recognition of the work of Leonhard Euler. Its decimal expansion up to nine decimal places is

$$e = 2.718281828\ldots.$$

Incidentally, e is not a rational number. (Refer to Exercises 1D of Chapter 1 and conclude that the pattern "1828" cannot keep repeating.) In limit notation,

$$e = \lim_{n \to \infty} \left(1 + \frac{1}{n}\right)^n.$$

In this limit it is understood that the numbers n are positive integers that are made larger and larger, i.e., are pushed to "infinity." It seems reasonable (and we now assume this to be the case) that $\lim_{z \to \infty} \left(1 + \frac{1}{z}\right)^z = e$ where z is pushed to infinity by taking real numbers as values for z. In particular, for any very large real number z, $e \approx \left(1 + \frac{1}{z}\right)^z$. By raising both sides to the $\frac{1}{z}$ power and using exponential law (iii), we get $e^{\frac{1}{z}} \approx 1 + \frac{1}{z}$. The larger the z is, the better the approximation will be. Now take a very small number Δx and set $z = \frac{1}{\Delta x}$. So z is very large and $\Delta x = \frac{1}{z}$. Therefore, $e^{\Delta x} \approx 1 + \Delta x$. Hence $e^{\Delta x} - 1 \approx \Delta x$, and $\frac{e^{\Delta x} - 1}{\Delta x} \approx 1$. Note that the smaller the Δx, the larger the

z, and the better each of these approximations will be. We can conclude that

$$\lim_{\Delta x \to 0} \left(\frac{e^{\Delta x} - 1}{\Delta x}\right) = 1.$$

We are now in a position to develop a fundamental property of the exponential function

$$f(x) = e^x.$$

Since $2 < e < 3$, the graph of $f(x) = e^x$ fits in between the graphs of 2^x and 3^x of Figure 10.2. What is the derivative of $f(x) = e^x$? Using exponential law (i) and algebra, we get

$$f'(x) = \lim_{\Delta x \to 0} \frac{f(x + \Delta x) - f(x)}{\Delta x} = \lim_{\Delta x \to 0} \frac{e^{x + \Delta x} - e^x}{\Delta x}$$

$$= \lim_{\Delta x \to 0} \frac{e^x e^{\Delta x} - e^x}{\Delta x} = \lim_{\Delta x \to 0} e^x \left(\frac{e^{\Delta x} - 1}{\Delta x}\right)$$

$$= e^x \lim_{\Delta x \to 0} \left(\frac{e^{\Delta x} - 1}{\Delta x}\right) = e^x.$$

Therefore, the function $f(x) = e^x$ is its own derivative! In the notation of Leibniz,

$$\boxed{\frac{d}{dx} e^x = e^x}\ .$$

Combining this fact with the chain rule shows that

$$\boxed{\frac{d}{dx} e^{g(x)} = e^{g(x)} g'(x)}$$

for any differentiable function $g(x)$.

Example 10.1. $\frac{d}{dx} e^{x^3} = e^{x^3} 3x^2$.

Example 10.2. $\frac{d}{dx}(e^{x^2} \cdot \sin x) = e^{x^2} 2x \cdot \sin x + e^{x^2} \cos x$.

10.2 **Inverse Functions**

The rules of some functions $y = f(x)$ can be reversed. Consider, for example, the function

$f(x) = (x-2)^3 + 4$. Notice that $f(-1) = -23, f(0) = -4$, $f(1) = 3, f(2) = 4, f(3) = 5$, and so on. So f takes -1 to -23; 0 to -4; 1 to 3; 2 to 4; etc. The question now is this: Is there a function that reverses not only these assignments, but all the assignments of f? Observe that such an "inverse" function of f would have to take -23 to -1; -4 to 0; 3 to 1; 4 to 2; and 5 to 3. What (if any) number would the inverse assign to, say, 2? If an explicit b can be found such that $f(b) = 2$, and if there is only one such b, then the inverse would have to take 2 to b. So we need to solve $f(b) = (b-2)^3 + 4 = 2$ for b. Because $(b-2)^3 = -2$, $b-2 = (-2)^{\frac{1}{3}}$, and hence $b = 2 + (-2)^{\frac{1}{3}}$. So the inverse must take 2 to $2 + (-2)^{\frac{1}{3}}$. More generally, to find the value (if any) that the inverse assigns to a number c, we need to set $f(b) = (b-2)^3 + 4 = c$ and try to solve for b. Doing so, we get $(b-2)^3 = c - 4$, so $b-2 = (c-4)^{\frac{1}{3}}$ and $b = (c-4)^{\frac{1}{3}} + 2$. Thus we have found *the* number b that f sends to c. Therefore the rule that takes c back to where it comes from under f is

$$c \to (c-4)^{\frac{1}{3}} + 2.$$

This rule reverses the rule f. It is the *inverse function* that we were looking for. We will denote it by f^{-1}. So $f^{-1}(c) = (c-4)^{\frac{1}{3}} + 2$. Replacing c by x gives $f^{-1}(x) = (x-4)^{\frac{1}{3}} + 2$.

Take care to observe that $f^{-1}(x)$ is not the same thing as $\frac{1}{f(x)}$. (It is often the case in mathematics that the same notation has different meanings in different contexts.) The inverse of a function $y = f(x)$ can often be obtained (in those cases where it has one) by solving for x in terms of y. This is precisely what was done above (with b in place of x and c in place of y).

Example 10.3. Let $y = f(x) = \frac{1+3x}{5-2x}$. Does it have an inverse? If yes, what is it? What x does y come from under f? The answers are obtained as before: Try to solve $\frac{1+3x}{5-2x} = y$ for x. Doing so, we see that $1 + 3x = y(5-2x)$, and hence $1 + 3x = 5y - 2yx$ and $3x + 2yx = 5y - 1$. It follows that $x = \frac{5y-1}{3+2y}$. So this is the x that f sends to y. Therefore, the rule for f^{-1} is $f^{-1}(y) = \frac{5y-1}{3+2y}$, or letting x be the variable, $f^{-1}(x) = \frac{5x-1}{3+2x}$.

A *value* of a function f is any number that arises as an output. For example, 2 is a value of $f(x) = \sqrt{x}$ because $f(4) = \sqrt{4} = 2$. The number 3 is also a value because $f(3^2) = \sqrt{3^2} = 3$. A function $y = f(x)$ has an inverse as long as any value c of f comes from *exactly one* number b in its domain. Put another way, if there is a value c such that $f(b_1) = c$ and $f(b_2) = c$ for numbers b_1 and b_2 with $b_1 \neq b_2$, then f does not have an inverse. If f meets the *exactly one* criterion, then f^{-1} is the function that sends the particular c to the b from which it comes. Figure 10.3 shows graphically how the rule of f is reversed in such a situation to give f^{-1}. Observe that a function f has an inverse precisely if any horizontal line crosses the graph at most once. A function whose graph satisfies this horizontal line test is called *one-to-one*: A given value c of f can come from only one number b in the domain. In any such case, the inverse function f^{-1} is defined by the following rule:

$$f^{-1}(c) \text{ is precisely that number } b$$

with the property that $f(b) = c$.

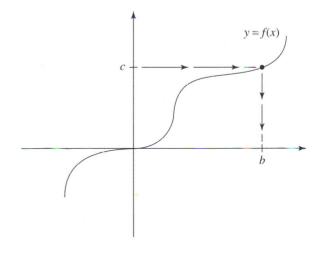

Figure 10.3

Example 10.4. Consider $f(x) = x^2$. Observe from Figure 10.4(a) that the graph does not satisfy the horizontal line test. So it is not possible to define f^{-1}

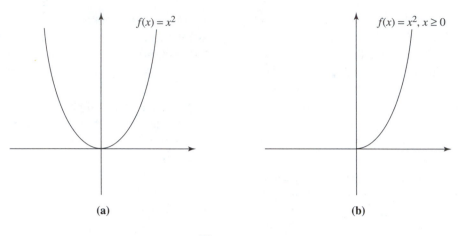

Figure 10.4

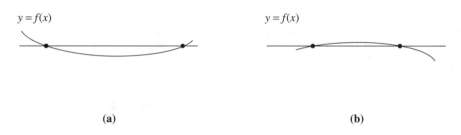

Figure 10.5

for this f. But let's change this function by requiring that $x \geq 0$. The function $f(x) = x^2$, $x \geq 0$ does have an inverse! A look at its graph in Figure 10.4(b) shows that it passes the horizontal line test. Let $c \geq 0$ be any value of this function. What number b does c come from under this f? Let $b^2 = c$ for some $b \geq 0$. So $b = c^{\frac{1}{2}}$, and the inverse rule is $c \to c^{\frac{1}{2}}$. Therefore $f^{-1}(x) = x^{\frac{1}{2}} = \sqrt{x}$. To summarize, the function $f(x) = x^2$ does not have an inverse, but $f(x) = x^2$, $x \geq 0$ does have one. It is the function $f^{-1}(x) = x^{\frac{1}{2}} = \sqrt{x}$.

Figure 10.5 shows that if a function f is continuous on an interval I, then it can have an inverse only if it is either increasing on I or decreasing on I. For if it decreases and then increases as in Figure 10.5(a), or increases and then decreases as in Figure 10.5(b), then the horizontal line criterion is violated: There

are values of f that come from at least two different numbers in the domain.

Suppose that f is a function that has an inverse f^{-1}. Take any number b in the domain of f and let $f(b) = c$. Any such c is in the domain of f^{-1}, and $f^{-1}(c) = b$. Observe that therefore

$$f^{-1}\big(f(b)\big) = f^{-1}(c) = b$$

for any b in the domain of f, and

$$f\big(f^{-1}(c)\big) = f(b) = c$$

for any c in the domain f^{-1}. It follows that

$$f^{-1}(f(x)) = x \quad \text{and} \quad f(f^{-1}(x)) = x.$$

Therefore, both of these composite functions fix all x for which they are defined.

Continue to assume that f has an inverse f^{-1}. Let (b, c) be any point on the graph of f. So $f(b) = c$.

Therefore, $f^{-1}(c) = b$, and (c, b) is on the graph of f^{-1}. Note that whenever (b, c) is a point on the graph of f, then (c, b) is a point on the graph of f^{-1}. Plot both of the points (b, c) and (c, b) and also (b, b) and (c, c). Notice that these four points determine a square and that the segment from (b, c) to (c, b) is a diagonal of the square. See Figure 10.6. It follows that the graph of f^{-1} can be sketched by using the following procedure:

i. Start with any point on the graph of f.
ii. Move perpendicularly towards the line $y = x$. Stop at this line and record the distance moved.
iii. Now continue in the same direction away from the line $y = x$. Stop after you have moved the distance recorded in (ii). The point you have reached is on the graph of f^{-1}.

This "reflection" procedure for sketching the graph of f^{-1} is illustrated in Figure 10.7. Note that the graph of f^{-1} is the mirror image (with the mirror placed on the line $y = x$) of the graph of f. In particular, if the graph of f is continuous, then that of f^{-1} is also continuous.

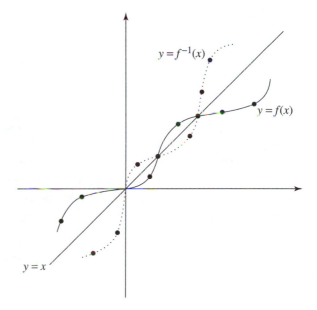

Figure 10.7

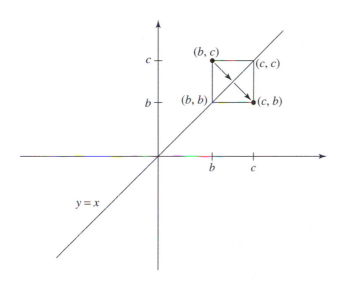

Figure 10.6

If we do this for the graph of the function $f(x) = x^2$, $x \geq 0$ in Figure 10.4(b), we get the graph of the

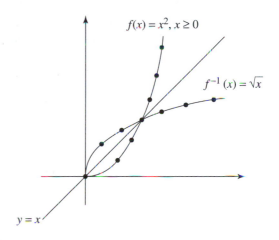

Figure 10.8

inverse function $f^{-1}(x) = x^{\frac{1}{2}} = \sqrt{x}$. Refer to Figure 10.8.

Continue to consider a function f that has an inverse. Suppose that f is differentiable on its domain. Differentiating both sides of the equation $f(f^{-1}(x)) = x$ shows that $\frac{d}{dx} f(f^{-1}(x)) = 1$. By the chain rule, $\frac{d}{dx} f(f^{-1}(x)) = f'(f^{-1}(x)) \frac{d}{dx} f^{-1}(x)$. So

$f'(f^{-1}(x))\frac{d}{dx}f^{-1}(x) = 1$, and it follows that

$$\frac{d}{dx}f^{-1}(x) = \frac{1}{f'(f^{-1}(x))}$$

As one might have expected, f^{-1} and f' together determine the derivative $(f^{-1})'$.

Let's illustrate this formula for the function $f(x) = (x-2)^3 + 4$ and its inverse $f^{-1}(x) = (x-4)^{\frac{1}{3}} + 2$ from the beginning of this section. Because $f'(x) = 3(x-2)^2$, the formula tells us that

$$\frac{d}{dx}f^{-1}(x) = \frac{1}{3(f^{-1}(x)-2)^2} = \frac{1}{3\left((x-4)^{\frac{1}{3}} + 2 - 2\right)^2}$$

$$= \frac{1}{3\left((x-4)^{\frac{1}{3}}\right)^2} = \frac{1}{3(x-4)^{\frac{2}{3}}}.$$

Computing the derivative of $f^{-1}(x) = (x-4)^{\frac{1}{3}} + 2$ again, this time by using the chain rule directly, gives us $\frac{d}{dx}f^{-1}(x) = \frac{1}{3}(x-4)^{-\frac{2}{3}}$. Observe that the two answers agree.

10.3 Logarithms

The logarithm functions are the inverses of the exponential functions. Return to the function $f(x) = a^x$ for $a > 0$ and suppose that $a \neq 1$. The function f is either an increasing function (if $a > 1$) or a decreasing function (if $a < 1$). Refer to Figure 10.2 for instance. Hence its graph satisfies the horizontal line test, and so f has an inverse function. This inverse is written

$$f^{-1}(x) = \log_a x$$

and is called the *logarithm to the base a*. In view of the relationship between any function and its inverse, observe that for a given number $x > 0$

$(*)$ $\log_a x$ is precisely that number y with the property that $a^y = x$.

So the question "$\log_a x = ?$" has the same answer as the question "$a^? = x$." For example,

$$\log_a 1 = 0 \text{ (since } a^0 = 1\text{)}\quad \text{and}$$
$$\log_a a = 1 \text{ (since } a^1 = a\text{)}$$

Because $a > 0$ and $x = a^y > 0$, notice that $\log_a x$ makes sense for positive x only. It follows that the domain of the function $\log_a x$ consists of all $x > 0$. The graph of the function $\log_a x$ is obtained from that of a^x by the "reflection" procedure already described. The case $a > 1$ is sketched in Figure 10.9. Since $f(x) = a^x$ is continuous on its domain, $f^{-1}(x) = \log_a x$ is also continuous on its domain. The equations $f(f^{-1}(x)) = x$ and $f^{-1}(f(x)) = x$ applied in this situation show us that for any $a > 0$ and $a \neq 1$,

$$\log_a(a^x) = x \quad \text{and} \quad a^{\log_a x} = x$$

for all x in the first equation and all $x > 0$ in the second.

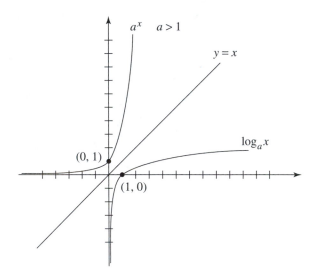

Figure 10.9

It should not come as a surprise that the basic laws that the exponential functions satisfy give rise to corresponding

Laws for Logarithms: Let a and b be positive constants not equal to 1. If $x > 0$ and $y > 0$, then

i. $\log_a(xy) = \log_a x + \log_a y$

ii. $\log_a x^r = r \log_a x$

iii. $\log_a \left(\dfrac{x}{y} \right) = \log_a x - \log_a y$

iv. $\log_a x = \dfrac{\log_b x}{\log_b a}$

The verifications are routine. For (i), let $\log_a x = u$ and $\log_a y = v$. By applying (∗), we get $a^u = x$ and $a^v = y$. By the exponential law (i), $a^{u+v} = a^u a^v = xy$. Applying (∗) again, we see that $\log_a(xy) = u + v$. Therefore, $\log_a(xy) = \log_a x + \log_a y$. For (ii), let $\log_a x = u$. So $a^u = x$. By the exponential law (iii), $a^{ur} = (a^u)^r = x^r$. By (∗) once more, $\log_a x^r = ur = ru = r \log_a x$. Therefore, (ii) is established. By combining (i) and (ii) we get

$$\log_a \left(\frac{x}{y} \right) = \log_a x + \log_a \frac{1}{y} = \log_a x + \log_a y^{-1}$$

$$= \log_a x - \log_a y$$

and (iii) is verified. Finally, let $\log_a x = u$ and $\log_b a = v$. So $a^u = x$ and $b^v = a$. By the exponential law (iii), $b^{vu} = (b^v)^u = a^u = x$. So by (∗), $\log_b x = vu = (\log_b a)(\log_a x)$. Because $a \neq 1$, we know that $\log_b a \neq 0$. Therefore, $\log_a x = \frac{\log_b x}{\log_b a}$ as required.

Formula (iv) asserts that any two log functions $\log_a x$ and $\log_b x$ differ by a constant multiple. The function $\log_e x$ is called the *natural log* function, and $\log_e x$ is denoted by

$$\ln x.$$

Taking $b = e$ in equality (iv) we get

$$\boxed{\log_a x = \frac{\log_e x}{\log_e a} = \frac{1}{\ln a} \cdot \ln x}$$

In this way, any log function $\log_a x$ can be expressed in terms of the natural log function $\ln x$.

Example 10.5. What is $\log_8 2$ equal to? This question has the same answer as $8^? = 2$. Because $\sqrt[3]{8} = 2$ and $\sqrt[3]{8} = 8^{\frac{1}{3}}$, we get $8^{\frac{1}{3}} = 2$. Therefore, $\log_8 2 = \frac{1}{3}$. Law

(ii) allows us to check this answer:

$$\log_8 2 = \log_8 8^{\frac{1}{3}} = \frac{1}{3} \log_8 8 = \frac{1}{3} \cdot 1 = \frac{1}{3}.$$

Example 10.6. Evaluate $\ln 2 = \log_e 2$. This is the number b with the property that $e^b = 2$. Because $e = 2.7182818\ldots$, we see that $e^1 = e > 2 = e^b$. Hence $1 > b$. From $e^{0.5} < 4^{0.5} = \sqrt{4} = 2 = e^b$, it follows that $0.5 < b$. So $0.5 < b < 1$. This process can be continued to show that $\ln 2 = b = 0.693\ldots$.

Example 10.7. Simplify $\ln x + c \ln y - d \ln z$. Using laws (i)–(iii), we get

$$\ln x + c \ln y - d \ln z = \ln x + \ln y^c - \ln z^d$$

$$= \ln xy^c - \ln z^d = \ln \frac{xy^c}{z^d}.$$

Example 10.8. Solve $\ln x^2 = 2 \ln 4 - 4 \ln 2$ for x. By laws (i) and (ii),

$$\ln x^2 = \ln 4^2 + \ln 2^{-4} = \ln(4^2 \cdot 2^{-4}) = \ln 1 = 0.$$

It follows that $x^2 = 1$, and hence $x = \pm 1$.

Example 10.9. Find the inverse function of $f(x) = (\ln x)^2$, $x \geq 1$. We need to set $(\ln x)^2 = y$ and solve for x. Thus $\ln x = \pm \sqrt{y}$. However, since $x \geq 1$, $\ln x \geq 0$, and therefore, $\sqrt{y} = \ln x$. It follows that $e^{\sqrt{y}} = e^{\ln x} = x$. So the rule for the inverse of f is given by $y \rightarrow e^{\sqrt{y}}$. Therefore, $f^{-1}(x) = e^{\sqrt{x}}$.

Any exponential function $f(x) = a^x$ can be expressed in terms of e^x as follows. Since $a > 0$, we let $c = \ln a$. So $a = e^c$. Therefore, $f(x) = a^x = (e^c)^x = e^{(\ln a)x}$. In particular,

$$f(x) = e^{(\ln a)x}.$$

We have not as yet computed the derivative of the function $f(x) = a^x$. This is now a simple matter. Applying the chain rule to the formula just derived, we get

$$f'(x) = \frac{d}{dx} e^{(\ln a)x} = (\ln a)e^{(\ln a)x} = (\ln a)a^x.$$

So we have established that

$$\frac{d}{dx}a^x = \ln a \cdot a^x$$

Compare and contrast the derivative just determined with that of the function $f(x) = x^a$. The exponent is now a constant, and therefore $f'(x) = ax^{a-1}$.

Example 10.10. Determine $\frac{d}{dx}2^x$. By the formula, $\frac{d}{dx}2^x = \ln 2 \cdot 2^x$. Inserting the value $\ln 2 = 0.693$ from Example 10.6, we get $\frac{d}{dx}2^x = (0.693)2^x$.

We turn to consider the derivatives of the logarithm functions. Let's begin with

$$\log_e x = \ln x$$

Let $f(x) = e^x$ and recall that $f^{-1}(x) = \ln x$. Because $f'(x) = e^x$, we get $f'(f^{-1}(x)) = e^{\ln x} = x$. Therefore, by substituting into the formula $\frac{d}{dx}f^{-1}(x) = \frac{1}{f'(f^{-1}(x))}$ from Section 10.2, we obtain

$$\frac{d}{dx}\ln x = \frac{1}{x}$$

Let g be a differentiable function and suppose that $g(x) > 0$ for all x. An application of the chain rule shows that

$$\frac{d}{dx}\ln g(x) = \frac{1}{g(x)} \cdot g'(x) = \frac{g'(x)}{g(x)}$$

Example 10.11. The function $\ln x^4$ can be differentiated in two ways. On the one hand, since $\ln x^4 = 4\ln x$, $\frac{d}{dx}\ln x^4 = \frac{d}{dx}4\ln x = 4\frac{1}{x}$. On the other hand, by a direct application of the formula above, $\frac{d}{dx}\ln x^4 = \frac{1}{x^4} \cdot 4x^3 = 4\frac{1}{x}$.

Example 10.12. By the same formula, $\frac{d}{dx}\ln(\cos x) = \frac{1}{\cos x}(-\sin x) = -\tan x$.

Differentiating the identity $\log_a x = \frac{1}{\ln a} \cdot \ln x$, we get that

$$\frac{d}{dx}\log_a x = \frac{1}{\ln a} \cdot \frac{1}{x}$$

Example 10.13. $\frac{d}{dx}\log_{10} x = \frac{1}{\ln 10} \cdot \frac{1}{x}$.

The computations of derivatives of some nasty functions $g(x)$ can be simplified by making use of the formula $g'(x) = g(x) \cdot \frac{d}{dx}\ln g(x)$. The next example illustrates this method of *logarithmic differentiation*.

Example 10.14. Let

$$g(x) = \frac{(2x^5+6)^{100}(3x^4-7)^{90}}{(3x^3+8)^{80}(5x^2+9x)^{70}}.$$

Taking the natural logarithm of both sides, we get

$$\ln g(x) = \ln(2x^5+6)^{100} + \ln(3x^4-7)^{90}$$
$$- \ln(3x^3+8)^{80} - \ln(5x^2+9x)^{70}$$
$$= 100\ln(2x^5+6) + 90\ln(3x^4-7)$$
$$- 80\ln(3x^3+8) - 70\ln(5x^2+9x).$$

By differentiating each side, we obtain

$$\frac{g'(x)}{g(x)} = 100\frac{10x^4}{2x^5+6} + 90\frac{12x^3}{3x^4-7}$$
$$- 80\frac{9x^2}{3x^3+8} - 70\frac{10x+9}{5x^2+9x}.$$

Therefore, $g'(x) =$

$$\left(100\frac{10x^4}{2x^5+6} + 90\frac{12x^3}{3x^4-7} - 80\frac{9x^2}{3x^3+8} - 70\frac{10x+9}{5x^2+9x}\right)$$
$$\cdot \frac{(2x^5+6)^{100}(3x^4-7)^{90}}{(3x^3+8)^{80}(5x^2+9x)^{70}}.$$

The conventional approach of differentiating $g(x)$ directly using the chain, product, and quotient rules would have been considerably more complicated. (Give it a try!)

Logarithms were invented[2] independently in 1614 by Napier, a Scottsman; in 1620 by Bürgi, a Swiss; and again in 1624 by Briggs, an Englishman. Cavalieri helped spread the concept in Italy, and Kepler did the same in German states. The French mathematician Laplace remarked that logarithms "by shortening the labors, doubled the life of the astronomer." He was referring to the fact that logarithms were of enormous historical importance as a computational tool. What makes them a tool for computation? Observe that law (i) converts the multiplication of two numbers to an addition of two numbers, and law (ii) converts exponentiating into multiplying. In other words, the logarithm converts numerical operations that can be very tedious (especially for large numbers) into more "user-friendly" ones.

Example 10.15. We will give an illustration how logarithms "shortened labors" by computing (or at least estimating) the number $(384{,}937)^{23}$. Observe that

$$\log_{10}(384{,}937)^{23}$$

$$= 23\log_{10} 384{,}937$$

$$= 23 \cdot \log_{10}(3.84937 \times 10^5)$$

$$= 23\big(\log_{10}(3.84937) + \log_{10} 10^5\big)$$

$$= 23\big(\log_{10}(3.84937) + 5 \cdot \log_{10} 10\big)$$

$$= 23\big(\log_{10}(3.84937) + 5\big).$$

Prepared "log" tables (readily available in the old days) contained approximations of $\log_{10}$ for numbers between 0 and 10, and one could determine that $\log_{10}(3.84937) \approx 0.5854$. Therefore,

$$\log_{10}(384{,}937)^{23} \approx 23(5.5854) = 128.4642$$

$$= 128 + 0.4642.$$

$$= \log_{10} 10^{128} + 0.4642.$$

Returning to the log tables, one would find that $0.4642 \approx \log_{10} 2.91$. Therefore,

$$\log_{10}(384{,}937)^{23} \approx \log_{10} 2.91 + \log_{10} 10^{128}$$

$$= \log_{10}(2.91 \times 10^{128}).$$

Since $384{,}937^{23}$ and 2.91×10^{128} have approximately the same $\log_{10}$, they are approximately equal. So $384{,}937^{23} \approx 2.91 \times 10^{128}$. (Try a calculator. If it is powerful enough, it will give you $2.910\ldots \times 10^{128}$ for $384{,}937^{23}$.) In the days before hand calculators (which came into wide use only in the early 1970s) and computers, logarithms provided a method by which complicated and potentially very lengthy arithmetic computations could be reduced to simpler and manageable ones.

10.4 **Returning to a Problem of Leibniz**

W e will begin with a differentiation exercise. Consider the function

$$f(x) = a\ln\left(\frac{a + (a^2 - x^2)^{\frac{1}{2}}}{x}\right) - (a^2 - x^2)^{\frac{1}{2}}$$

where a is a positive constant and $0 < x \le a$. Since $\ln\left(\frac{a+(a^2-x^2)^{\frac{1}{2}}}{x}\right) = \ln(a + (a^2 - x^2)^{\frac{1}{2}}) - \ln x$, the function can be rewritten as

$$f(x) = a\ln\left(a + (a^2 - x^2)^{\frac{1}{2}}\right) - a\ln x - (a^2 - x^2)^{\frac{1}{2}}.$$

By the chain rule and a number of algebraic steps, we get

$$f'(x) = a\frac{1}{a + (a^2 - x^2)^{\frac{1}{2}}}\left(\frac{1}{2}(a^2 - x^2)^{-\frac{1}{2}}\right)(-2x)$$

$$- a\frac{1}{x} - \frac{1}{2}(a^2 - x^2)^{-\frac{1}{2}}(-2x)$$

$$= -ax\frac{(a^2 - x^2)^{-\frac{1}{2}}}{a + (a^2 - x^2)^{\frac{1}{2}}} - a\frac{1}{x} + x(a^2 - x^2)^{-\frac{1}{2}}$$

$$= -ax\frac{(a^2 - x^2)^{-\frac{1}{2}}}{a + (a^2 - x^2)^{\frac{1}{2}}} \cdot \frac{a - (a^2 - x^2)^{\frac{1}{2}}}{a - (a^2 - x^2)^{\frac{1}{2}}}$$

$$-a\frac{1}{x} + x(a^2 - x^2)^{-\frac{1}{2}}$$

$$= -ax\frac{a(a^2 - x^2)^{-\frac{1}{2}} - 1}{a^2 - (a^2 - x^2)} - a\frac{1}{x} + x(a^2 - x^2)^{-\frac{1}{2}}$$

$$= -a\frac{a(a^2 - x^2)^{-\frac{1}{2}} - 1}{x} - a\frac{1}{x} + x(a^2 - x^2)^{-\frac{1}{2}}$$

$$= -\frac{a^2(a^2 - x^2)^{-\frac{1}{2}}}{x} + \frac{x^2(a^2 - x^2)^{-\frac{1}{2}}}{x}$$

$$= -\frac{(a^2 - x^2)(a^2 - x^2)^{-\frac{1}{2}}}{x}$$

$$= -\frac{(a^2 - x^2)^{\frac{1}{2}}}{x}.$$

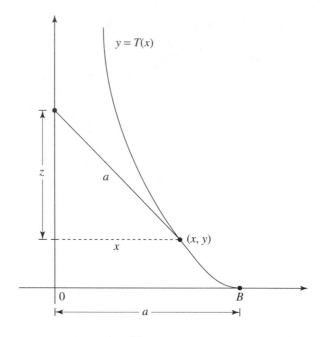

Figure 10.10

We will now return to the "pocketwatch" problem that Claude Perrault had posed to Leibniz. (See Section 5.4.) A weight B is tied to a string, and both weight and string are placed on a horizontal plane that is provided with a Cartesian coordinate system. See Figure 10.10. Stretch the string along the positive x-axis so that the weight is at the point $(a, 0)$ and the other end is at the origin, where a is the length of the string. Now move the end of the string (the end at the origin) up the positive y-axis. The weight B is dragged along and traces out a certain curve. This curve is called a *tractrix*.[3] Figure 10.10 shows it being traced out. The question that Perrault had raised was this: Could this curve be determined? In other words, can an explicit function be found that has the tractrix as its graph? We will now see that the computation at the beginning of this section provides such a function.

Let $y = T(x)$ be a function that has the tractrix as its graph. Take a typical position (x, y) of the weight. So (x, y) is a typical point on the curve and hence on the graph of T. Recall Leibniz's observation that the string is always tangent to the curve. It follows from Figure 10.10 that $T'(x) = -\frac{z}{x}$. By the Pythagorean theorem, $z = \sqrt{a^2 - x^2}$. Therefore,

$$T'(x) = -\frac{(a^2 - x^2)^{\frac{1}{2}}}{x}.$$

Observe next that $T(x)$ and the function $f(x)$ from the beginning of this section have the same derivative. So there is a constant C such that

$$T(x) = f(x) + C$$

$$= a\ln\left(\frac{a + (a^2 - x^2)^{\frac{1}{2}}}{x}\right) - (a^2 - x^2)^{\frac{1}{2}} + C.$$

By Figure 10.10, $T(a) = 0$, and hence

$$0 = T(a) = a\ln\left(\frac{a + (a^2 - a^2)^{\frac{1}{2}}}{a}\right) - (a^2 - a^2)^{\frac{1}{2}} + C$$

$$= a\ln 1 + C = C.$$

Therefore, $C = 0$. It follows that the solution of Perrault's question is provided by the function

$$T(x) = a\ln\left(\frac{a + (a^2 - x^2)^{\frac{1}{2}}}{x}\right) - (a^2 - x^2)^{\frac{1}{2}}.$$

10.5 Inverse Trigonometric Functions

T he function $f(x) = \sin x$ does not satisfy the horizontal line criterion, but the restricted function $f(x) = \sin x$, $-\frac{\pi}{2} \leq x \leq \frac{\pi}{2}$ does. See Figure 10.11. Therefore, this function has an inverse. This inverse is written $f^{-1}(x) = \sin^{-1}(x)$ and is called the *inverse sine*. The reflection procedure of Section 10.2 provides its graph; it is sketched in Figure 10.12. Observe from the graph that the domain of $f^{-1}(x) = \sin^{-1} x$ is the interval $[-1, 1]$.

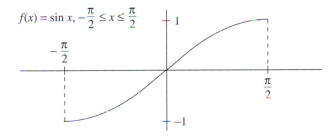

Figure 10.11

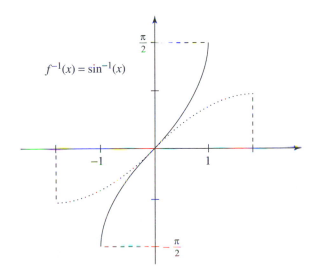

Figure 10.12

For a number c with $-1 \leq c \leq 1$, $\sin^{-1} c$ is precisely that number b between $-\frac{\pi}{2}$ and $\frac{\pi}{2}$ such that $\sin b = c$. The question "$\sin^{-1} c = ?$" translated into words is "what angle between $-\frac{\pi}{2}$ and $\frac{\pi}{2}$ has sine equal to c?" For example, $\sin \frac{\pi}{4} = \frac{\sqrt{2}}{2}$, so the answer to $\sin^{-1} \frac{\sqrt{2}}{2} = ?$ is $\frac{\pi}{4}$. The function $\sin^{-1} x$ is the function behind the $\sin^{-1}$ button on your calculator. Use it to show that $\sin^{-1}(0.750) = 0.848$ and reconfirm this by checking that $\sin(0.848) = 0.750$. (Be sure that the calculator is in "radian" mode.)

Recall from Section 8.6 that for $f(x) = \sin x$, $f'(x) = \cos x$. Combining this with the formula

$$\frac{d}{dx} f^{-1}(x) = \frac{1}{f'(f^{-1}(x))}$$

shows that $\frac{d}{dx} \sin^{-1} x = \frac{1}{\cos(\sin^{-1} x)}$. This expression can be made more transparent. From $\sin^2 x + \cos^2 x = 1$, we get $\cos^2 x = 1 - \sin^2 x$. But $\cos x \geq 0$ for all x with $-\frac{\pi}{2} \leq x \leq \frac{\pi}{2}$, so $\cos x = \sqrt{1 - (\sin x)^2}$. Therefore, $\cos(\sin^{-1} x) = \sqrt{1 - (\sin(\sin^{-1} x))^2} = \sqrt{1 - x^2}$, and hence

$$\frac{d}{dx} \sin^{-1} x = \frac{1}{\sqrt{1 - x^2}}$$

Example 10.16. Find an antiderivative of $f(x) = \frac{1}{\sqrt{1-(3x)^2}}$. By the chain rule,

$$\frac{d}{dx} \sin^{-1}(3x) = \frac{1}{\sqrt{1 - (3x)^2}} \cdot \frac{d}{dx}(3x) = 3\frac{1}{\sqrt{1 - (3x)^2}}.$$

So we get 3 times what we need. It follows that $F(x) = \frac{1}{3} \sin^{-1}(3x)$ is an antiderivative of $f(x) = \frac{1}{\sqrt{1-(3x)^2}}$.

Refer back to the graph of the function $f(x) = \tan x$ in Section 4.4. Because it does not satisfy the horizontal line test, it does not have an inverse. However, the restricted function $f(x) = \tan x$, $-\frac{\pi}{2} < x < \frac{\pi}{2}$, does. See Figure 10.13. This function has an inverse, which is denoted by $f^{-1}(x) = \tan^{-1}(x)$. The graph of the inverse is sketched in Figure 10.14. It is obtained by the reflection procedure. Note that the domain of $f^{-1}(x) = \tan^{-1} x$ consists of all real numbers.

For any number c, $\tan^{-1} c$ is the number b between $-\frac{\pi}{2}$ and $\frac{\pi}{2}$ such that $\tan b = c$. So "$\tan^{-1} c = ?$" is the same question as "what angle between $-\frac{\pi}{2}$ and $\frac{\pi}{2}$ has tangent equal to c?" Because $\tan \frac{\pi}{4} = 1$, the answer to $\tan^{-1} 1 = ?$ is $\frac{\pi}{4}$. Use the $\tan^{-1}$ button on a calculator to verify that $\tan^{-1}(0.5) = 0.46$, $\tan^{-1}(2) = 1.11$, $\tan^{-1}(10) = 1.47$, and $\tan^{-1}(1000) = 1.57$. Locate the corresponding points on the graph in Figure 10.14.

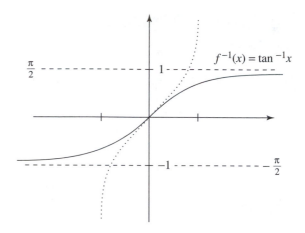

Figure 10.14

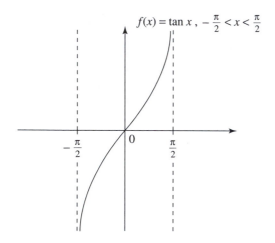

Figure 10.13

By Section 8.6, the derivative of $f(x) = \tan x$ is $f'(x) = \sec^2 x$. An application of the formula $\frac{d}{dx} f^{-1}(x) = \frac{1}{f'(f^{-1}(x))}$ shows that

$$\frac{d}{dx} \tan^{-1} x = \frac{1}{\sec^2(\tan^{-1}(x))}.$$

Recall that $\sec^2 x = \tan^2 x + 1$. So $\sec^2(\tan^{-1} x) = (\tan(\tan^{-1} x))^2 + 1 = x^2 + 1$. So it follows that

$$\boxed{\frac{d}{dx} \tan^{-1} x = \frac{1}{x^2 + 1}}$$

Example 10.17. Differentiate $f(x) = \tan^{-1}(5x^2)$. By the chain rule,

$$f'(x) = \frac{1}{(5x^2)^2 + 1} \cdot \frac{d}{dx}(5x^2) = \frac{10x}{25x^4 + 1}.$$

One of the primary reasons for introducing the new functions of this chapter is that they arise in the solution of a variety of different problems. The tractrix is a case in point. This curve has a simple geometrical description, but its precise analysis requires a complicated antiderivative that is provided by the logarithm. The exponential functions are another important example. They form the mathematical backbone of Chapter 11, which treats topics as diverse as nuclear physics, geology, archeology, and the growth of cultures of bacteria. The functions developed in this chapter often arise as antiderivatives required in the solution of definite integrals. We conclude this section by providing some examples of this.

Consider any function $y = f(x)$. Let $F(x)$ be any antiderivative of $f(x)$. Recall the basic fact that any other antiderivative of $f(x)$ has the form $F(x) + C$ for some constant C. We will refer to the symbol

$$\int f(x)\, dx$$

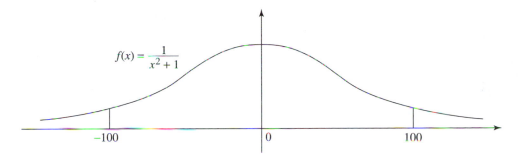

$$f(x) = \frac{1}{x^2 + 1}$$

Figure 10.15

as the *indefinite integral* of $f(x)$ and use it to denote the typical antiderivative of $f(x)$. So

$$\int f(x)\,dx = F(x) + C$$

for an arbitrary constant C. There is a connection between the concept $\int f(x)\,dx$ and the concept

$$\int_a^b f(x)\,dx$$

introduced long ago in Section 5.5B. It is provided by the Fundamental Theorem of Calculus. Recall that this theorem asserts that any sum of the form $\int_a^b f(x)\,dx$ can be evaluated by taking any antiderivative $G(x) = F(x) + C$ of $f(x)$ and computing $G(b) - G(a)$.

The notation just introduced will be in use for other functions, other variables, and other labels for the constants. For example,

$$\int h(t)\,dt = H(t) + C_1$$

also means that the typical antiderivative of $h(t)$ has the form $H(t)$ plus a constant.

With this notation, we can summarize some of the results of this and previous chapters. For example, the fact that the derivative of $\frac{1}{r+1}x^{r+1}$ is x^r can be expressed as

$$\int x^r\,dx = \frac{1}{r+1}x^{r+1} + C$$

Observe that this formula is not valid when $r = -1$. In this remaining case,

$$\int \frac{1}{x}\,dx = \ln x + C$$

This follows from the fact that $\frac{d}{dx}\ln x = \frac{1}{x}$. That

$$\int \frac{\sqrt{a^2 - x^2}}{x}\,dx$$
$$= \sqrt{a^2 - x^2} - a\ln\left(\frac{a + (a^2 - x^2)^{\frac{1}{2}}}{x}\right) + C$$

follows directly from the analysis of the tractrix in Section 10.4. Finally, the formulas

$$\int \frac{1}{\sqrt{1 - x^2}}\,dx = \sin^{-1} x + C$$

and

$$\int \frac{1}{x^2 + 1}\,dx = \tan^{-1} x + C$$

come from the derivative formulas for $\sin^{-1} x$ and $\tan^{-1} x$ that were established earlier in this section.

Example 10.18. Compute the area under the graph of $f(x) = \frac{1}{x}$ from 1 to 8. This area is given by the definite integral $\int_1^8 \frac{1}{x}\,dx$. Since $F(x) = \ln x$ is an antiderivative of $f(x) = \frac{1}{x}$, we find (using the value $\ln 2 = 0.693$ from Example 10.6) that

$$\int_1^8 \frac{1}{x}\,dx = \ln 8 - \ln 1 = \ln 2^3 - 0 = 3\ln 2$$

$$= 3(0.693) = 2.079.$$

Example 10.19. Compute $\int_0^{\frac{1}{2}} \frac{1}{1+x}\,dx$. By the chain rule, $\frac{d}{dx}\ln(1+x) = \frac{1}{1+x}$. Therefore,

$$\int_0^{\frac{1}{2}} \frac{1}{1+x}\,dx = \ln\left(1+\frac{1}{2}\right) - \ln 1 = \ln\frac{3}{2} - 0$$

$$= \ln 3 - \ln 2 = 1.0986 - 0.6931$$

$$= 0.4055.$$

An approximation of this integral was given in Section 6.3 using Newton's power series method.

Example 10.20. Compute the area under the graph of $f(x) = \frac{1}{x^2+1}$ from -100 to 100. The graph of this function is shown in Figure 10.15. Study $f'(x)$ and show that the graph of f is increasing for $x < 0$ and decreasing for $x > 0$. Then plot some points to verify that the graph has the indicated shape. The indicated area is equal to

$$\int_{-100}^{100} \frac{1}{x^2+1}\,dx = \tan^{-1}(100) - \tan^{-1}(-100)$$

$$= 1.56 - (-1.56) = 3.12.$$

We have introduced a number of new functions and their graphs. The time has now come to develop a systematic scheme for analyzing the graph of any function. The next section is an important step towards this goal.

10.6 Concavity

$\boxed{C}$onsider a function f that is differentiable on some interval I. Suppose that the graph of f over I has one of the two shapes sketched in Figure 10.16. While f is increasing in both cases, it is clear that one situation is fundamentally different from the other. The difference is easy to see. In the first case the slopes of the tangents to the graph are increasing, and in the second case these slopes are decreasing. Equivalently, since the derivative measures slope, f' is an increasing function on I in the first case and a decreasing function on I in the second. Both the function f and and its graph are said to be *concave up* on I in the first situation and *concave down* on I in the second.

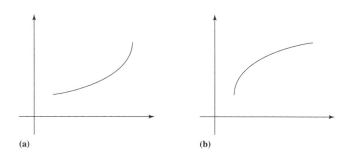

(a)   (b)

Figure 10.16

Example 10.21. Consider the functions $f(x) = x^2$ and $g(x) = \sqrt{x} = x^{\frac{1}{2}}$ on the interval $(0,4)$. A look back to Figure 10.8 shows that both are increasing functions. The graph of $f'(x) = 2x$ is a line with slope 2. So f' is increasing, and $f(x) = x^2$ concave up on $(0,4)$. Since $g'(x) = \frac{1}{2}x^{-\frac{1}{2}} = \frac{1}{2x^{\frac{1}{2}}} = \frac{1}{2\sqrt{x}}$ and $\sqrt{x}$ is increasing on $(0,4)$, it follows that g' is decreasing on $(0,4)$. Therefore, $g(x) = \sqrt{x}$ is concave down on $(0,4)$. Another look at Figure 10.8 confirms our conclusions.

Suppose that the graph of f is as given in Figure 10.17. Observe that f is concave up on the intervals (a,c_1), (c_1,c_2), (c_3,c_4), and (c_5,b) and that f is concave down on (c_2,c_3) and (c_4,c_5). Any point $(c,f(c))$ on the

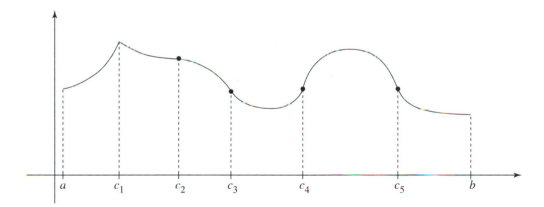

Figure 10.17

graph with the property that f is concave in the one sense to the one side and in the opposite sense to the other side is called a *point of inflection*. The points $(c_2, f(c_2))$, $(c_3, f(c_3))$, $(c_4, f(c_4))$, and $(c_5, f(c_5))$ are the inflection points of f. Note, however, that $(c_1, f(c_1))$ is not.

Suppose now that the derivative f' of f is itself a differentiable function and consider the second derivative $f'' = (f')'$ of f. Recall the connection between the derivative and the matter of increase and decrease of a function, and observe:

Concavity Test: If $f''(x) > 0$ for all x in I, then f' is increasing on I and hence f is concave up on I. Similarly, if $f''(x) < 0$ for all x in I, then f' is decreasing on I and hence f is concave down on I.

The method for determining the intervals over which a function f is concave up or concave down uses the second derivative in the same way that the first derivative is used in determining the intervals of increase or decrease. In other words, the strategy is this: Find the numbers for which f'' is either 0 or does not exist (these are the critical numbers for f'); use these numbers to divide the x-axis into intervals; choose test points in each interval; if f'' is positive at a test point, then f is concave up on the entire interval, and if f'' is negative at the test point, f is concave

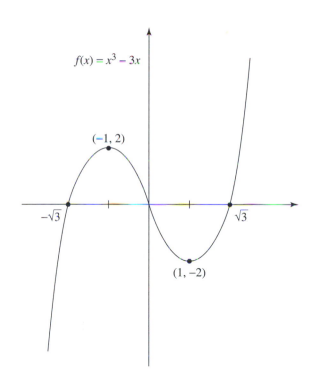

Figure 10.18

down on the interval. The examples that follow will illustrate the method.

Example 10.22. Consider the function $f(x) = x^3 - 3x$. Since $f'(x) = 3x^2 - 3$, we see that $f''(x) = 6x$. It follows that $f''(x) < 0$ for all $x < 0$, and $f''(x) > 0$ for

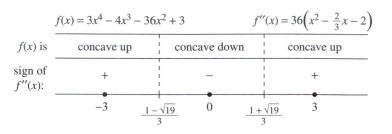

Figure 10.19

all $x > 0$. Therefore, the graph of f is concave down to the left of the origin and concave up to the right of the origin. The point $(0, f(0)) = (0, 0)$ is a point of inflection. Figure 10.18 shows the graph of this function as it was sketched in Example 8.36 of Section 8.7. Observe that it reflects the concavity information just developed.

Example 10.23. Consider

$$f(x) = 3x^4 - 4x^3 - 36x^2 + 3.$$

Because $f'(x) = 12x^3 - 12x^2 - 72x$, we see that

$$f''(x) = 36x^2 - 24x - 72 = 12(3x^2 - 2x - 6)$$

$$= 12 \cdot 3\left(x^2 - \frac{2}{3}x - 2\right).$$

To find the numbers for which $f''(x) = 0$, apply the quadratic formula to $x^2 - \frac{2}{3}x - 2$ to get

$$x = \frac{\frac{2}{3} \pm \sqrt{\frac{4}{9} + 8}}{2} = \frac{\frac{2}{3} \pm \sqrt{\frac{4+72}{9}}}{2} = \frac{1}{2}\left(\frac{2}{3} \pm \frac{\sqrt{76}}{3}\right)$$

$$= \frac{1}{2}\left(\frac{2}{3} \pm \frac{2\sqrt{19}}{3}\right) = \frac{1 \pm \sqrt{19}}{3}.$$

Because $f''(x) = 36(x^2 - \frac{2}{3}x - 2)$ exists for all x, we know that $\frac{1+\sqrt{19}}{3}$ and $-\frac{1-\sqrt{19}}{3}$ are the only critical numbers of f'. The x-axis is split as indicated.

Since $\sqrt{19} \approx 4.36$, these numbers are approximately 1.79 and -1.12, respectively. Take the test points $-3, 0$, and 3, respectively, and check that $f''(-3) = 36 \cdot 9 > 0$, $f''(0) = -72 < 0$, and $f''(3) = 36 \cdot 5 > 0$. It follows that f is concave up on the intervals $\left(-\infty, \frac{1-\sqrt{19}}{3}\right)$ and $\left(\frac{1+\sqrt{19}}{3}, \infty\right)$ and concave down on $\left(\frac{1-\sqrt{19}}{3}, \frac{1+\sqrt{19}}{3}\right)$. The points on the graph with x-coordinates $\frac{1-\sqrt{19}}{3}$ and $\frac{1+\sqrt{19}}{3}$ are both points of inflection. Figure 10.19 summarizes our conclusions.

The graph of this function was drawn in Example 8.37 of Section 8.7. It shows that there must be a change in concavity somewhere to the left of 0, and again somewhere to the right of 0. We now know that these changes occur at $x = \frac{1-\sqrt{19}}{3} \approx -1.12$ and $x = \frac{1+\sqrt{19}}{3} \approx 1.79$. (See Figure 10.20.)

Example 10.24. Let $g(x) = x^{\frac{1}{3}}(x^2 - 14)$. By multiplying, $g(x) = x^{\frac{7}{3}} - 14x^{\frac{1}{3}}$, and hence $g'(x) = \frac{7}{3}x^{\frac{4}{3}} - \frac{14}{3}x^{-\frac{2}{3}}$. Standard computations show that

$$g''(x) = \frac{28}{9}x^{\frac{1}{3}} + \frac{28}{9}x^{-\frac{5}{3}} = \frac{28}{9}\left(x^{\frac{1}{3}} + x^{-\frac{5}{3}}\right)$$

$$= \frac{28}{9}\left(\frac{x^{\frac{6}{3}} + 1}{x^{\frac{5}{3}}}\right) = \frac{28}{9}\left(\frac{x^2 + 1}{x^{\frac{5}{3}}}\right).$$

Because $x^2 + 1$ is never zero and $x = 0$ is the only number for which $g''(x)$ is not defined, it follows that 0 is the only critical number for g'. Since $x^{\frac{5}{3}} = \sqrt[3]{x^5}$, observe that $g''(x) < 0$ when $x < 0$, and $g''(x) > 0$ when $x > 0$. Therefore, the graph of $g(x)$ is concave down to the right of the origin $(0, 0)$ and concave up

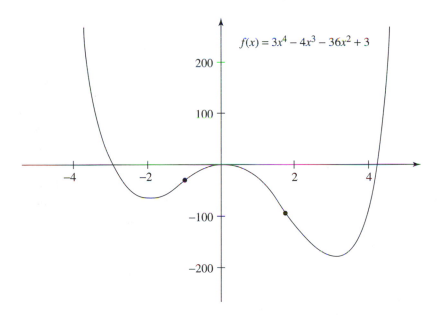

$$f(x) = 3x^4 - 4x^3 - 36x^2 + 3$$

Figure 10.20

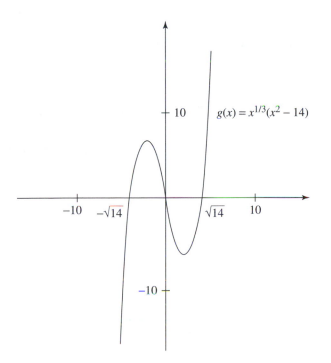

$$g(x) = x^{1/3}(x^2 - 14)$$

Figure 10.21

to the left. The origin is a point of inflection. The graph of $g(x) = x^{\frac{1}{3}}(x^2 - 14) = x^{\frac{7}{3}} - 14x^{\frac{1}{3}}$ was already sketched in Example 8.38 of Section 8.7. It is repeated in Figure 10.21. Observe that it reflects the concavity information just derived.

Suppose that the graph of a function f has a horizontal tangent at the point $(c, f(c))$. A moment's thought shows that if the graph is concave down at $(c, f(c))$, then it must have a local maximum at c, and if it is concave up at $(c, f(c))$, then it must have a local minimum at c. This conclusion is known as the

Second Derivative Test: Let f be a function that is differentiable on an open interval containing the number c and suppose that $f'(c) = 0$. If $f''(c) < 0$, then f has a local maximum at c. If $f''(c) > 0$, then f has a local minimum at c.

Example 10.25. Let $f(x) = x^4 - 4x^2 + 16$. Because $f'(x) = 4x^3 - 8x = 4x(x^2 - 2)$, observe that $f'(0) = 0$, $f'(\sqrt{2}) = 0$, and $f'(-\sqrt{2}) = 0$. Since $f''(x) = 12x^2 - 8$, we get $f''(0) = -8$, $f''(\sqrt{2}) = 24 - 8 = 16$, and $f''(-\sqrt{2}) = 24 - 8 = 16$. Therefore, by the second derivative test, f has a local minimum at both $\sqrt{2}$ and $-\sqrt{2}$, and a local maximum at 0.

Notice that the Second Derivative Test says nothing about the situation $f''(c) = 0$. To see why nothing can be said, analyze each of the three functions $y = x^3$, $y = x^4$, and $y = -x^4$ at $x = 0$.

10.7 **Asymptotes**

An *asymptote* of a curve is a line to which the curve tends for large x or y. (The word comes from the Greek *a* meaning "not" and *symptotos* meaning "falling together.") A look at the graphs of the functions $f(x) = \frac{1}{x}$ and $f(x) = \frac{1}{x^2}$ shows (Figure 10.22) that the x- and y-axes are both asymptotes for each of these graphs.

Consider the function $f(x) = \frac{1}{x+2} + 3$. Let (x, y) be a point on the graph with x large (either positive or negative). Since $\frac{1}{x+2}$ is small (either positive or negative), it follows that $y = \frac{1}{x+2} + 3$ is close to 3. So when x is large, the point (x, y) on the graph is close to the line $y = 3$. It follows that the line $y = 3$ is an asymptote of the graph of f. Since this line is horizontal, it is a *horizontal asymptote*.

Again let (x, y) be a point on the graph of f, but suppose this time that y is large (either positive or negative). Because $y = \frac{1}{x+2} + 3$, it follows that $\frac{1}{x+2} = y - 3$ is large (either positive or negative). So $x + 2$ must be small, and so x must be close to -2. We see, then, that if (x, y) is a point on the graph with y either large and positive or large and negative, then (x, y) is close to the line $x = -2$. Therefore, the line $x = -2$ is an asymptote of the graph of f. Since this is a vertical line, it is a *vertical asymptote*.

How are horizontal and vertical asymptotes detected in general? In the horizontal case, detection is provided by the answer to the following question: What happens to $y = f(x)$ for large (positive or negative) x? In the notation of limits, this translates to the two questions

$$\lim_{x \to +\infty} f(x) = ? \quad \text{and} \quad \lim_{x \to -\infty} f(x) = ?$$

If the answer to either question is a finite number, say L, then the line $y = L$ is a horizontal asymptote. For example,

$$\lim_{x \to +\infty} \frac{1}{x + 2} + 3 = 0 + 3 = 3.$$

Thus $y = 3$ is a horizontal asymptote of $f(x) = \frac{1}{x+2} + 3$.

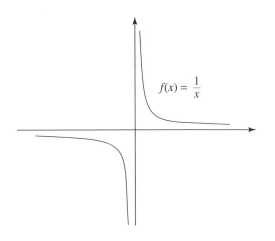

$f(x) = \dfrac{1}{x}$

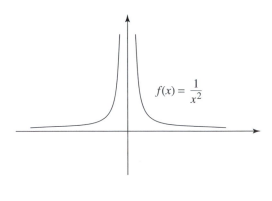

$f(x) = \dfrac{1}{x^2}$

Figure 10.22

Example 10.26. Consider the function

$$f(x) = \frac{x-1}{x^2(x+2)}.$$

To find the horizontal asymptotes we need to answer the questions

$$\lim_{x \to -\infty} \frac{x-1}{x^2(x+2)} = ? \quad \text{and} \quad \lim_{x \to +\infty} \frac{x-1}{x^2(x+2)} = ?$$

One common strategy to compute limits of this type is first to divide both the numerator and the denominator of the fraction by the highest power of the variable. In the current case this power is x^3. Dividing both the numerator and denominator of $\frac{x-1}{x^2(x+2)} = \frac{x-1}{x^3+2x^2}$ by x^3, we get

$$\frac{x-1}{x^2(x+2)} = \frac{x-1}{x^3+2x^2} = \frac{\frac{x}{x^3} - \frac{1}{x^3}}{\frac{x^3}{x^3} + \frac{2x^2}{x^3}} = \frac{\frac{1}{x^2} - \frac{1}{x^3}}{1 + \frac{2}{x}}.$$

Pushing x to either $+\infty$ or $-\infty$ forces this quotient to $\frac{0-0}{1-0} = 0$ in either case. Therefore, the x-axis $y = 0$ is a horizontal asymptote of the graph of $f(x) = \frac{x-1}{x^2(x+2)}$.

What about vertical asymptotes? Suppose, say, that the vertical line $x = d$ is a vertical asymptote of the graph of $y = f(x)$. The first observation that needs to be made is that f cannot be continuous at d. If f were to be continuous at d, then f would be defined at d and $\lim_{x \to d} f(x) = f(d)$. On the other hand, since the line is a vertical asymptote, $y = f(x)$ must be large when x is near d. However, the graph of a function cannot satisfy both of these requirements because it has to satisfy the vertical line test. Figure 10.23 depicts a typical situation. The graph in this figure is in one piece near d and has the line $x = d$ as a vertical asymptote, but it is *not* the graph of a function.

The detection of the vertical asymptotes of a function f proceeds as follows: Collect the numbers, say, d, d_1, d_2, d_3, and so on, for which f is not continuous; then investigate the behavior of the function f near each such number. Consider a discontinuity d for example. Then the behavior of f for x near d—both to the right of d and to the left of d—is determined by

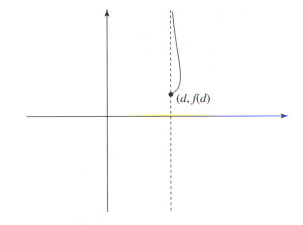

Figure 10.23

answering the questions

$$\lim_{x \to d^+} f(x) = ? \quad \text{and} \quad \lim_{x \to d^-} f(x) = ?$$

If the answer to either question is "plus infinity" or "minus infinity", then $x = d$ is a vertical asymptote of the graph of f.

Return to the function $f(x) = \frac{1}{x+2} + 3$ for example. Notice that it is not continuous at -2 because it is not defined at -2. What about the question

$$\lim_{x \to -2^+} \left(\frac{1}{x+2} + 3 \right) = ?$$

Observe that as x closes in on -2, $\frac{1}{x+2}$ and thus $\frac{1}{x+2} + 3$ becomes large. Does it become large and positive, or large and negative? The -2^+ tells us that x closes in on -2 from the right. So $x > -2$ and hence $x + 2 > 0$. Therefore, $\frac{1}{x+2} + 3$ becomes large and positive. So

$$\lim_{x \to -2^+} \left(\frac{1}{x+2} + 3 \right) = +\infty.$$

In the same way, $\lim_{x \to -2^-} \left(\frac{1}{x+2} + 3 \right) = -\infty$. We have confirmed that the line $x = -2$ is a vertical asymptote of the graph of $f(x) = \frac{1}{x+2} + 3$. Because $x = -2$ is the only number at which f is not continuous, $x = -2$ is the only vertical asymptote.

Example 10.27. Consider $f(x) = \frac{x^2-x-6}{x^2-9}$. Because $x^2 - 9 = (x-3)(x+3)$, it follows that f is not defined and hence not continuous at -3 and 3. Let's examine the behavior of f near 3 first.

$$\lim_{x \to 3^+} \frac{x^2-x-6}{x^2-9} = \lim_{x \to 3^+} \frac{(x-3)(x+2)}{(x-3)(x+3)}$$

$$= \lim_{x \to 3^+} \frac{(x+2)}{(x+3)} = \frac{3+2}{3+3} = \frac{5}{6}.$$

In precisely the same way,

$$\lim_{x \to 3^-} \frac{x^2-x-6}{x^2-9} = \lim_{x \to 3^-} \frac{(x-3)(x+2)}{(x-3)(x+3)}$$

$$= \lim_{x \to 3^-} \frac{(x+2)}{(x+3)} = \frac{3+2}{3+3} = \frac{5}{6}.$$

Because neither limit is "infinity," $x = 3$ is *not* a vertical asymptote of the graph of $f(x) = \frac{x^2-x-6}{x^2-9}$. What about $x = -3$? Note that for $x \neq 3$,

$$\frac{x^2-x-6}{x^2-9} = \frac{(x-3)(x+2)}{(x-3)(x+3)} = \frac{(x+2)}{(x+3)}.$$

For any x near -3, the term $x+2$ is negative, and for any x on the right of -3, $x > -3$ and so $x+3 > 0$. So as x is pushed to -3 from the right, $\frac{x+2}{x+3}$ remains negative. It now follows that

$$\lim_{x \to -3^+} \frac{x^2-x-6}{x^2-9} = \lim_{x \to -3^+} \frac{(x-3)(x+2)}{(x-3)(x+3)}$$

$$= \lim_{x \to -3^+} \frac{x+2}{x+3} = -\infty.$$

In a similar way,

$$\lim_{x \to -3^-} \frac{x^2-x-6}{x^2-9} = \lim_{x \to -3^-} \frac{(x-3)(x+2)}{(x-3)(x+3)}$$

$$= \lim_{x \to -3^-} \frac{x+2}{x+3} = +\infty.$$

We have shown that $x = -3$ is a vertical asymptote of $f(x) = \frac{x^2-x-6}{x^2-9}$ and that it is the only vertical asymptote. Our analysis has also determined the behavior of the function immediately to the left and right of -3.

Consider the function $f(x) = e^x$ and the question $\lim_{x \to -\infty} e^x = ?$ Rewrite this limit as $\lim_{x \to -\infty} \frac{1}{e^{-x}}$ and observe that the denominator is huge for large negative x. Therefore

$$\lim_{x \to -\infty} e^x = \lim_{x \to -\infty} \frac{1}{e^{-x}} = 0.$$

Hence the line $y = 0$ (this is the x-axis) is a horizontal asymptote for $f(x) = e^x$. Because $\lim_{x \to \infty} e^x = \infty$, the line $y = 0$ is an asymptote only for the left side of the graph. Because $f(x) = e^x$ is continuous for all numbers x, its graph has no vertical asymptotes. A look at the graph of a^x in Figure 10.9 confirms the assertions just made.

Recall that the natural logarithm function is the inverse $f^{-1}(x) = \ln x$ of $f(x) = e^x$. So its graph is obtained by applying the reflection procedure in Section 10.2 to the graph of $f(x) = e^x$. Review this procedure in general and convince yourself that it converts any horizontal asymptote of the graph of a function f to a vertical asymptote of the graph of its inverse f^{-1} and that it converts any vertical asymptote of f to a horizontal asymptote of f^{-1}. It follows that the graph of $f^{-1}(x) = \ln x$ has exactly one asymptote. It is the line $x = 0$ (the y-axis). Refer once more to Figure 10.9.

Example 10.28. Consider the hyperbola $\frac{x^2}{a^2} - \frac{y^2}{b^2} = 1$ where $a > 0$ and $b > 0$. It was verified back in Section 5.1 that the lines $y = \frac{b}{a}x$ and $y = -\frac{b}{a}x$ are both asymptotes of this hyperbola. Refer to Figure 5.8 in Section 5.1 and notice that these asymptotes are neither horizontal nor vertical.

10.8 Graphing

$\boxed{\text{W}}$e will now add a few observations to the theory already developed in this chapter and in Chapter 8 and describe a comprehensive strategy for sketching the graph of any function $y = f(x)$.

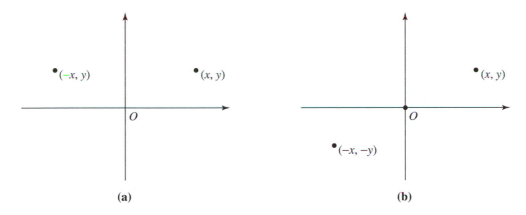

Figure 10.24

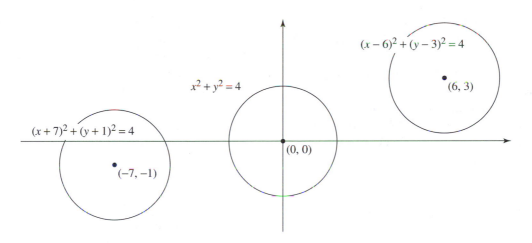

Figure 10.25

A. The Domain. Determine the numbers for which f is defined. This information tells us over which intervals along the x axis the function f has a graph.

B. Symmetry and Shifting. Suppose that for any point (x, y) on the graph of f, the point $(-x, y)$ is on the graph also. This arises if $f(-x) = f(x)$ for all x in the domain of f. Observe that $f(x) = x^2$ is an example of such a function. Refer to Figure 10.24(a) and notice that in such a situation the graph is *symmetric about the y-axis*.

Another type of symmetry occurs if for any point (x, y) on the graph, the point $(-x, -y)$ is also on the graph. This occurs if $f(-x) = -f(x)$ for all x in the domain of f. The function $f(x) = x^3$ satisfies this condition. In this case, we say that the graph of f is *symmetric about the origin*. See Figure 10.24(b). Why did we not mention symmetry about the x-axis?

The matter of shifting graphs is best illustrated by observing what happens with circles. Start with the circle $x^2 + y^2 = 4$ of radius 2 and center the origin $(0, 0)$. Replacing x by $x - 6$ and y by $y - 3$ in this equation gives the equation $(x - 6)^2 + (y - 3)^2 = 4$. This circle also has radius 2, but its center has been shifted to the point $(6, 3)$. In the same way, if we replace x and y by

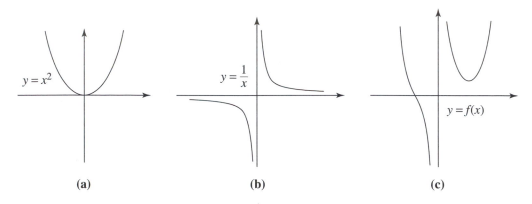

Figure 10.26

$x + 7$ and $y + 1$, respectively, then the center of the circle is shifted from $(0,0)$ to $(-7,-1)$. Figure 10.25 shows what is going on.

The same considerations apply to any graph. For example, the graph of $y - 2 = (x + 3)^2$ has the same shape and orientation as the graph of $y = x^2$. It is obtained from that of $y = x^2$ by shifting the origin $(0,0)$ to the point $(-3,2)$. Let $y = f(x)$ be any function and $p > 0$ any constant. Replacing x by $x - p$ and leaving y as is shifts the graph of $y = f(x)$ by p units to the right. In other words, the graph of $y = f(x - p)$ is obtained from that of $f(x)$ by moving the latter p units to the right. In particular, if $f(x) = f(x - p)$, then these graphs are the same. We say in this case that the graph is *periodic*. The smallest $p > 0$ for which a periodic function satisfies the shift condition $f(x) = f(x - p)$ is called the *period* of f. The trigonometric functions, as we already know (and as we will see again in a moment), exhibit periodic phenomena.

C. Asymptotic Behavior.   The analysis of asymptotic behavior of a function includes much more than the determination of the horizontal and vertical asymptotes. Consider, for example, the function $f(x) = x^2 + \frac{1}{x}$. Observe that when x is large in either a positive or negative sense, then the term x^2 will dominate and $\frac{1}{x}$ will contribute little. So for large x, the function f will behave like $y = x^2$. On the other hand, if x is small, then $\frac{1}{x}$ is large and x^2 is small. Now $\frac{1}{x}$ will

control f, and x^2 will be negligible. It follows that the graph of f is approximated by that of $y = x^2$ when x is large and by $y = \frac{1}{x}$ when x is small. This is illustrated in Figure 10.26.

Study the graph of $f(x) = x^3 - 3x$ in Example 10.22 and convince yourself that it can be analyzed by a comparison with $y = x^3$ and $y = -3x$. Note that for large positive or negative x, the term x^3 will dominate, and the graph of f is approximated by that of $y = x^3$; for very small x on the other hand, $3x$ will be much larger than x^3, and $y = -3x$ will dominate. In reference to Example 10.23, note that $f(x) = 3x^4 - 4x^3 - 36x^2 + 3$ is dominated by $3x^4$ when x is large, so that for large x the graph of f is approximated by that of $y = 3x^4$.

The remaining components of the comprehensive strategy of graphing functions have already been discussed in detail:

D. Intervals of Increase or Decrease and Local Maxima and Minima.

E. Concavity and Points of Inflection.

Finally, some points on the graph of the function $y = f(x)$ should be plotted, especially those that analysis of the concerns **A**–**E** supplies. The last trick is to sketch a graph that takes into account (and is consistent with) *all* the information that has been collected.

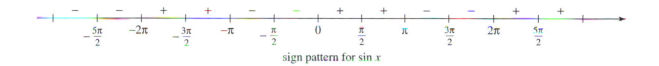

sign pattern for sin x

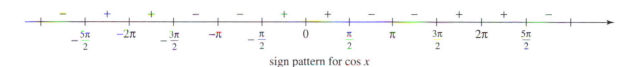

sign pattern for cos x

Figure 10.27

Let's consider some examples. We will begin with the trigonometric functions $\sin x$ and $\cos x$. The basic facts already developed in Section 4.4 will be used.

Example 10.29. We already know that the functions $\sin x$ and $\cos x$ are defined for all x. So we turn to the issue of symmetry and periodicity.

B. Since $\cos(-x) = \cos x$, the graph of $\cos x$ is symmetric about the y-axis. Since $\sin(-x) = -\sin x$, that of $\sin x$ is symmetric about the origin. Recall that $\sin(x - 2\pi) = \sin x$ and $\cos(x - 2\pi) = \cos x$, in each case for any x. This means that if the graph of either $\sin x$ or $\cos x$ is shifted 2π units to the right nothing changes. So both graphs are periodic, and since 2π is the smallest such value, the period is 2π.

C. It follows from the periodicity just observed that if x is pushed to either $+\infty$ or $-\infty$ neither $\sin x$ nor $\cos x$ can close in on a fixed finite number. It follows that there are no horizontal asymptotes. Since both functions are continuous for all x (in fact by Section 8.6 they are differentiable for all x) neither graph can have any vertical asymptotes.

D. We already know that

$$\frac{d}{dx}\sin x = \cos x \quad \text{and} \quad \frac{d}{dx}\cos x = -\sin x.$$

Recall that $\sin x = 0$ precisely for

$$x = 0,\ \pi,\ -\pi,\ 2\pi,\ -2\pi,\ 3\pi,\ -3\pi, \ldots,$$

and that $\cos x = 0$ precisely for

$$x = \frac{\pi}{2},\ \frac{-\pi}{2},\ \frac{3\pi}{2},\ \frac{-3\pi}{2},\ \frac{5\pi}{2},\ \frac{-5\pi}{2}, \ldots.$$

The "sign patterns" for the sine and the cosine, that is to say the intervals over which $\sin x$ and $\cos x$ are positive and negative, are provided by Figure 10.27.

We will now focus on $f(x) = \sin x$. The situation for the cosine is analogous. Since $f'(x) = \cos x$, we see from the sign pattern of the cosine that $\sin x$ is increasing over the intervals $\left(-\frac{\pi}{2}, \frac{\pi}{2}\right)$, $\left(\frac{3\pi}{2}, \frac{5\pi}{2}\right)$, $\left(\frac{-5\pi}{2}, \frac{-3\pi}{2}\right), \ldots$.

E. Since $f''(x) = -\sin x$, it follows that $f(x) = \sin x$ is concave up when its graph is below the x-axis and concave down when its graph is above the x-axis. We can now sketch the graph of $\sin x$ with more authority than was possible in Chapter 4.4. See Figure 10.28. The case of the cosine is analogous. Its graph is shown in Figure 10.29.

Example 10.30. Consider $f(x) = \tan x = \frac{\sin x}{\cos x}$. The only numbers that are not in the domain of $\tan x$ are those for which $\cos x = 0$. This occurs precisely when

$$x = \frac{\pi}{2},\ \frac{-\pi}{2},\ \frac{3\pi}{2},\ \frac{-3\pi}{2},\ \frac{5\pi}{2},\ \frac{-5\pi}{2}, \ldots.$$

Since $\sin(-x) = -\sin x$ and $\cos(-x) = \cos x$, it follows that

$$\tan(-x) = \frac{\sin(-x)}{\cos(-x)} = -\frac{\sin x}{\cos x} = -\tan x.$$

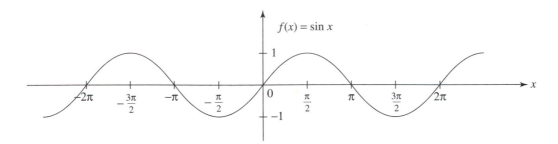

Figure 10.28

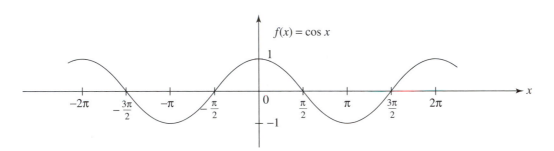

Figure 10.29

Therefore, the graph of $\tan x$ is symmetric about the origin. By Example 4.11 of Section 4.4,

$$\cos(x - \pi) = -\cos x \quad \text{and} \quad \sin(x - \pi) = -\sin x.$$

It follows that $\tan(x - \pi) = \frac{\sin(x-\pi)}{\cos(x-\pi)} = \frac{-\sin x}{-\cos x} = \tan x$. Therefore, the graph is periodic of period π. It follows that we need to consider the graph only for the interval $[\frac{-\pi}{2}, \frac{\pi}{2}]$.

C. Because of the periodicity just observed, $\tan x$ cannot have a horizontal asymptote. Since $\tan x$ is not defined for $x = \frac{\pi}{2}$ and $x = -\frac{\pi}{2}$, there may be vertical asymptotes there. Study the sign pattern of $\sin x$ and $\cos x$ to the left of $\frac{\pi}{2}$ (see Figure 10.27) and notice that

$$\lim_{x \to \frac{\pi}{2}^-} \tan x = \lim_{x \to \frac{\pi}{2}^-} \frac{\sin x}{\cos x} = +\infty.$$

Therefore, the vertical line $x = \frac{\pi}{2}$ is vertical asymptote. In the same way,

$$\lim_{x \to -\frac{\pi}{2}^+} \tan x = \lim_{x \to -\frac{\pi}{2}^+} \frac{\sin x}{\cos x} = -\infty,$$

so $x = -\frac{\pi}{2}$ is a vertical asymptote also.

D. By Example 8.31 in Section 8.6, $f'(x) = \sec^2 x$. Since $\sec x = \frac{1}{\cos x}$, the critical numbers of f are the numbers where $\cos x = 0$. These numbers have already been considered. They are the same numbers for which $\tan x$ is not defined. Since $f'(x) = \sec^2 x > 0$ for all x in $(\frac{-\pi}{2}, \frac{\pi}{2})$, it follows that $f(x) = \tan x$ is increasing over $(\frac{-\pi}{2}, \frac{\pi}{2})$. There are no local maxima or minima.

E. By Example 33 of Section 8.6,

$$f''(x) = 2 \tan x \sec^2 x.$$

It follows that the graph of $\tan x$ is concave up whenever $\tan x$ is positive and concave down whenever $\tan x$ is negative. A look at Figure 10.27 shows that $\tan x$ is negative over $(\frac{-\pi}{2}, 0)$ and positive for $(0, \frac{\pi}{2})$. Combining all the information that has been developed provides the graph of $\tan x$. See Figure 10.30.

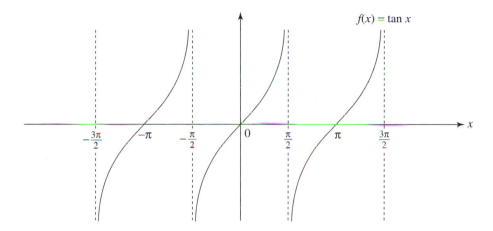

$$f(x) = \tan x$$

Figure 10.30

Example 10.31. Sketch the graph of $f(x) = \frac{2x^2}{x^2-1}$.

A. Domain: All x not equal to ± 1.

B. Symmetry: If x is replaced by $-x$ the function is unchanged. So the graph is symmetric about the y-axis.

C. To check for possible horizontal asymptotes, answer the questions

$$\lim_{x \to +\infty} f(x) = ? \quad \text{and} \quad \lim_{x \to -\infty} f(x) = ?$$

Dividing both the numerator and the denominator of $\frac{2x^2}{x^2-1}$ by x^2 (the largest power of x appearing in the expression), we get $\frac{2x^2}{x^2-1} = \frac{2}{1-\frac{1}{x^2}}$. It follows that

$$\lim_{x \to \infty} f(x) = \lim_{x \to \infty} \frac{2x^2}{x^2-1} = \lim_{x \to \infty} f(x) \frac{2}{1-\frac{1}{x^2}} = \frac{2}{1} = 2.$$

In the same way (or by the symmetry already observed), $\lim_{x \to -\infty} f(x) = 2$. It follows that $y = 2$ is a horizontal asymptote and that it is the only one.

Vertical asymptotes can occur only at those numbers where $f(x)$ is not continuous, in other words only at $x = 1$ and $x = -1$. To verify whether vertical asymptotes actually exist at these points and (if they do) to determine the behavior of $f(x)$ near them, the questions

$$\lim_{x \to -1^-} f(x) = ? \quad \lim_{x \to -1^+} f(x) = ?$$

$$\lim_{x \to 1^-} f(x) = ? \quad \lim_{x \to 1^+} f(x) = ?$$

need answers. Note that $f(x) = \frac{2x^2}{x^2-1} = \frac{2x^2}{(x-1)(x+1)}$. Consider $\lim_{x \to -1} f(x)$. As x goes to -1 from the left, $2x^2$ goes to 2; $x - 1$ goes to -2; and $x + 1$ goes to zero. Therefore in $f(x) = \frac{2x^2}{(x-1)(x+1)}$ the numerator goes to 2 and the denominator to 0. It follows that

$$\lim_{x \to -1^-} f(x) = \text{"infinity."}$$

But is it $+\infty$ or $-\infty$? Since x goes to -1 from the left, $x + 1$ goes to zero from the left, i.e., through negative values. Since $x - 1$ goes to -2, it follows that the denominator of $f(x) = \frac{2x^2}{(x-1)(x+1)}$ goes to zero through positive values. Since the numerator goes to 2, it follows that

$$\lim_{x \to -1^-} f(x) = +\infty.$$

Similar considerations show that

$$\lim_{x \to -1^+} f(x) = -\infty, \quad \lim_{x \to 1^-} f(x) = -\infty, \quad \text{and}$$

$$\lim_{x \to 1^+} f(x) = +\infty.$$

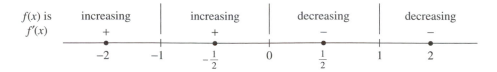

Figure 10.31

D. By the quotient rule,

$$f'(x) = \frac{4x(x^2 - 1) - 2x^2 2x}{(x^2 - 1)^2} = \frac{4x^3 - 4x - 4x^3}{(x^2 - 1)^2}$$

$$= \frac{-4x}{(x^2 - 1)^2}.$$

So the critical numbers are $x = -1, 0$, and 1. Taking $-2, -\frac{1}{2}, \frac{1}{2}$, and 2 as test points, we get the information of Figure 10.31. Notice that there is a local maximum at 0.

E. Using the quotient rule again with $f'(x) = \frac{-4x}{(x^2-1)^2}$ gives us

$$f''(x) = \frac{-4(x^2 - 1)^2 - (-4x)2(x^2 - 1)2x}{(x^2 - 1)^4}$$

$$= \frac{(x^2 - 1)[-4(x^2 - 1) + 16x^2]}{(x^2 - 1)^4}$$

$$= \frac{12x^2 + 4}{(x^2 - 1)^3} = \frac{4(3x^2 + 1)}{(x^2 - 1)^3}.$$

Since the numerator is never 0, the critical points in this case are -1, and 1. Take $-2, 0$, and 2 as test points and check the information of Figure 10.32. After plotting a few points, it is now a routine matter to draw a graph that is consistent with all the data that has been collected. It is sketched in Figure 10.33.

Example 10.32. Sketch the graph of $f(x) = \frac{300x}{(2+x)^3}$.

A. The only x not in the domain is $x = -2$.

B. Replace x by $-x$ and notice that $f(x)$ is equal to neither $f(-x)$ nor $-f(-x)$. So f is not symmetric about the y-axis nor is it symmetric about the origin.

C. Since $(2 + x)^3$ grows much faster than $100x$, it is clear that

$$\lim_{x \to \infty} f(x) = 0 \quad \text{and} \quad \lim_{x \to -\infty} f(x) = 0.$$

A more explicit argument goes as follows: Because $(2 + x)^3 = x^3 + 6x^2 + 12x + 8$,

$$f(x) = \frac{300x}{x^3 + 6x^2 + 12x + 8}.$$

Divide top and bottom by x^3 (the largest power of x that occurs) to get

$$f(x) = \frac{\frac{300}{x^2}}{1 + \frac{6}{x} + \frac{12}{x^2} + \frac{8}{x^3}}.$$

Now push x to ∞ or $-\infty$ and notice that all terms go to zero except for the 1 in the denominator. So either limit is 0 as asserted. Therefore, $y = 0$ (the x-axis) is a horizontal asymptote.

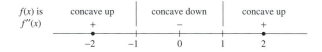

Figure 10.32

The only x for which x is not continuous is $x = -2$. So this is the only possibility for a vertical asymptote. Arguing as in E (the "vertical" part) of Example 10.31, we get

$$\lim_{x \to -2^-} \frac{300x}{(2 + x)^3} = +\infty \quad \text{and} \quad \lim_{x \to -2^+} \frac{300x}{(2 + x)^3} = -\infty$$

In particular, $x = -2$ is a vertical asymptote.

Observe that when x is large, $f(x) = (2 + x)^3$ is approximated by x^3, and hence $f(x) = \frac{300}{(2+x)^3}$ is

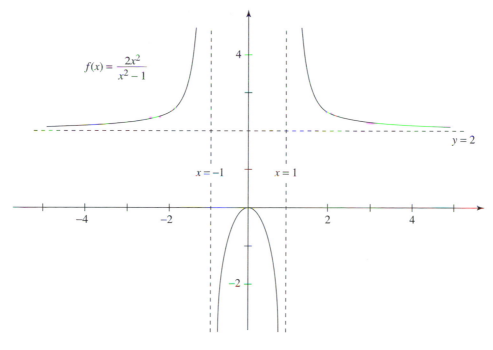

$$f(x) = \frac{2x^2}{x^2 - 1}$$

$y = 2$

$x = -1$ $x = 1$

Figure 10.33

approximated by $\frac{300}{x^2}$. It follows that for large x, the graph of f has the same general shape as the graph of $\frac{1}{x^2}$.

D. Using the quotient rule, we see that

$$f'(x) = \frac{300(2 + x)^3 - 300x3(2 + x)^2}{(2 + x)^6},$$

and after factoring out 300 and canceling $(2 + x)^2$, we obtain

$$f'(x) = \frac{300[(2 + x) - 3x]}{(2 + x)^4} = \frac{300(2 - 2x)}{(2 + x)^4}$$

So $f'(x) = 0$ for $x = 1$. Because $(2 + x)^4$ is always positive, it follows that $f'(x) > 0$ when $x < 1$, and $f'(x) < 0$ when $x > 1$. Therefore, f is increasing over the intervals $(-\infty, -2)$ and $(-2, 1)$, has a local maximum at 1, and is decreasing over $(1, \infty)$. The value of f at 1 is $f(1) = \frac{300}{3^3} = 11.11\dots$.

E. The concavity matter is next. Using the quotient rule again, show that

$$f''(x) = \frac{300(-2)(2 + x)^4 - 300(2 - 2x) \cdot 4(2 + x)^3}{(2 + x)^8}$$

$$= \frac{-600 \cdot (2 + x) - 1200(2 - 2x)}{(2 + x)^5}$$

$$= \frac{1800x - 3600}{(2 + x)^5} = \frac{1800(x - 2)}{(2 + x)^5}.$$

Observe that $f''(2) = 0$, and $f''(-2)$ is undefined. So -2 and 2 are the critical points for the analysis of $f''(x)$. Take the points -3, 0, and 3 as test points and check that f is concave up over $(-\infty, -2)$, concave down over $(-2, 2)$, and concave up over $(2, \infty)$. Note that $(2, f(2)) = (2, \frac{600}{64}) = (2, 9.38)$ is a point of inflection. However there is no point of inflection at $x = -2$ because the function is not defined at -2.

By putting all the information together, we can now sketch the graph of $f(x) = \frac{300x}{(2+x)^3}$. See Figure 10.34.

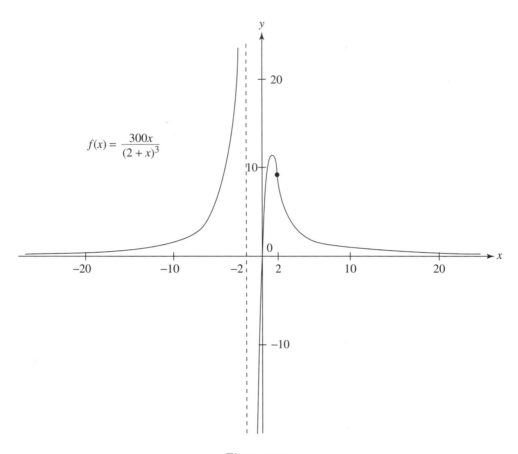

$$f(x) = \frac{300x}{(2+x)^3}$$

Figure 10.34

A function of this type plays an important role in interior ballistics: the portion of the graph from 0 to the right simulates the pressure generated by the explosion in a gun barrel. The details will be developed in Chapter 14.

10.9 Postscript

It has been the purpose of this chapter to introduce and study several new functions and to investigate their graphs. The exponential function and its properties are especially important. The study of many phenomena involving trends, especially processes of growth or decrease, is informed by the exponential function. The next chapter will have a look at a number of different areas of science to illustrate this point.

Exercises
10A.   Exponential Functions

1. Sketch the graphs of the following functions on the same axis system by plotting a few points. Scale the vertical axis so that 1 centimeter corresponds to 1000 units.

    i. $y = 5^x$
    ii. $y = 6^x$
    iii. $y = \left(\frac{1}{5}\right)^x$
    iv. $y = \left(\frac{1}{6}\right)^x$

2. Compare the functions $f(x) = x^2$ and $g(x) = 2^x$ by evaluating each of them for $x = 0, 1, 2, 3, 4, 5, 10, 15$, and 20. Then draw the graphs of f and g on the same set of axes for $-4 \leq x \leq 4$.

3. Differentiate the functions below:

    i. $f(x) = e^{\sqrt{x}}$
    ii. $g(x) = e^{-5x} \cos 3x$
    iii. $y = e^{x + e^x}$
    iv. $f(x) = x^2 e^x$
    v. $y = xe^{x^2}$
    vi. $y = e^{1/(1-x^2)}$
    vii. $y = \tan(e^{3x-2})$
    viii. $y = \frac{e^x + e^{-x}}{e^x - e^{-x}}$

4. Find the equation of the tangent line to the curve $y = x^2 e^{-x}$ at the point $(1, \frac{1}{e})$.

5. Show that the function $y = e^{2x} + e^{-3x}$ satisfies the equation $y'' + y' - 6y = 0$.

6. Find the one hundredth derivative of $f(x) = xe^{-x}$.

7. Use the Intermediate Value Theorem to show that there is a solution of the equation $e^x + x = 0$.

8. Use the graph of $y = e^x$ to draw the graphs of $y = e^{-x}$ and $y = 3 - e^x$.

9. Find the absolute minimum value of the function $g(x) = \frac{e^x}{x}, x > 0$.

10B. Inverse Functions

10. Find the inverses of the functions $f(x) = 4x + 7$ and $f(x) = \frac{x-2}{x+2}$.

In Exercises 11–13 let $g = f^{-1}$ be the inverse function of the given function. For the given number c, evaluate $g(c)$ and $g'(c)$.

11. $f(x) = x^3 + x + 1, c = 1$.

12. $f(x) = x^5 - x^3 + 2x, c = 2$.

13. $f(x) = \sin x + \cos x$ with $-\frac{\pi}{2} < x < \frac{\pi}{2}, c = \sqrt{2}$.

10C. Logarithms

14. Evaluate the following without using a calculator:

    i. $\log_2 64$
    ii. $\log_3 3^{\sqrt{5}}$
    iii. $\log_3 108 - \log_3 4$
    iv. $\log_5 10 + \log_5 20 - 3 \log_5 2$

15. Without using a calculator determine which of the numbers $\log_{10} 99$ or $\log_9 82$ is larger.

16. Sketch, using the same axes, the graphs of the following functions.

    i. $y = \log_2 x$
    ii. $y = \ln x$
    iii. $y = \log_{10} x$

17. Express the given quantity as a single logarithm.

    i. $\log_2 x + 3 \log_2(x + 1) + \frac{1}{4} \log_2(x - 1)$
    ii. $\frac{1}{3} \ln x - 4 \ln(2x + 3)$

18. Solve the given equations for x.

    i. $\log_2 x = 3$
    ii. $2^{x^2 - 5} = 3$
    iii. $5^{x^2 - 1} = 2$
    iv. $4^{x^2 + 1} = 3$
    v. $\log_9(4x^2 - 11) = 7$
    vi. $\log_5(\log_5 x) = 6$
    vii. $\ln(x + 6) + \ln(x - 3) = \ln 5 + \ln 7$
    viii. $\ln(\frac{x-2}{x+1}) = 1 + \ln(\frac{x-3}{x+1})$

19. Solve the inequality $\ln(3x - 2) \leq 0$.

20. Solve the equation $4^x - 2^{x+3} + 12 = 0$.

The geologist C.F. Richter developed a scale for measuring the magnitude of earthquakes. Suppose we a have certain standard *seismograph* (this is an apparatus that measures amplitudes of shock waves). A quake whose largest shock wave registers an amplitude of 1 micron = 10^{-4} cm when the seismograph is 100 km away from the epicenter is considered to be very light and is said to be a zero-level quake. Suppose that the largest shock wave of some earthquake is measured to have an amplitude of x microns by a standard seismograph that is some distance from the epicenter of the quake. Then the magnitude of this quake is defined to be $M = \log_{10}(\frac{x}{a})$, where a is the amplitude that the seismograph would measure for a zero-level quake with epicenter the same distance away (as the quake being measured.) Values of a for various distances have been tabulated, so one only needs to know x and the distance to the epicenter in order to be able to compute the magnitude of a quake. For example, suppose that the largest shock wave of a quake is measured to have an amplitude of 140,000 microns by a standard seismograph that is 100 km from the epicenter of the quake. In this case $a = 1$ micron, and the magnitude of this quake is $M = \log_{10}(\frac{140,000}{1}) = 5.1$ on the *Richter scale*.

21. The 1989 earthquake in San Francisco had a magnitude of 6.9 on the Richter scale. What was the amplitude of the largest shockwave that a seismograph would have measured if placed exactly 100 km from this quake? State your answer first in microns and then in meters.

22. The 1906 San Francisco earthquake generated an amplitude measurement that was 25 times greater than that of the 1989 quake. What was its magnitude on the Richter scale?

23. Find the domains of the functions:

  i. $f(x) = \log_{10}(1-x)$
  ii. $F(t) = \sqrt{t}\ln(t^2 - 1)$

24. Find the inverse functions of the given functions:

  i. $y = \ln(x+3)$
  ii. $y = (\ln x)^2,\ x \geq 1$
  iii. $y = \frac{1+e^x}{1-e^x}$

25. For each of the functions f below find f' and the domains of f and f'

  i. $f(x) = \cos(\ln x)$
  ii. $f(x) = \ln(2 - x - x^2)$
  iii. $f(x) = \ln(\sqrt{x} - \sqrt{x-1})$
  iv. $f(x) = \log_{11}(x^4 + 3x^2)$
  v. $f(x) = \ln\sqrt{x + 3x^2}$

26. Find y' and y'' for each of the functions below.

  i. $y = x\ln x$
  ii. $y = \log_{10} x$
  iii. $y = \ln(\sec x + \tan x)$

27. Differentiate the functions below.

  i. $g(x) = \sqrt{\ln x}$
  ii. $f(t) = \log_7(t^4 - t^2 + 1)$
  iii. $f(x) = e^x \ln x$
  iv. $h(t) = t^3 - 3^t$

28. Find the absolute minimum value of the function $f(x) = x\ln x$.

29. Show that $\int_1^r \frac{1}{x}\,dx = \ln r$. Use this equality and a comparison of areas to show that $\frac{1}{4} < \ln\frac{4}{3} < \frac{1}{3}$.

30. Evaluate $\int_0^{\frac{\pi}{4}} \tan x\,dx$.

31. Consider the tractrix

$$T(x) = 10\ln\left(\frac{10 + (10^2 - x^2)^{\frac{1}{2}}}{x}\right) - (10^2 - x^2)^{\frac{1}{2}}.$$

Determine the length of this curve between the points $(6, 10\ln 3 - 8)$ and $(10, 0)$. [Hint: Use Leibniz's formula in Section 5.6C for the length of an arc.]

10D. Inverse Trigonometric Functions

32. Evaluate without using a calculator

  i. $\sin^{-1}\left(\frac{\sqrt{3}}{2}\right)$
  ii. $\tan^{-1}(-1)$
  iii. $\sin^{-1}\left(-\frac{1}{2}\right)$
  iv. $\tan^{-1}(\sqrt{3})$

33. Show that $\tan(\sin^{-1} x) = \frac{x}{\sqrt{x^2 - 1}}$.

34. Show that $\tan^{-1} x = \frac{1}{\tan x}$ is not a valid identity.

35. Evaluate

  i. $\frac{d}{dx}\sin^{-1} e^{2x}$
  ii. $\frac{d}{dx}\tan^{-1}(5x - 7)$
  iii. $\frac{d}{dx}\left(\frac{1}{x} - \sin^{-1}\frac{1}{x}\right)^3$
  iv. $\frac{d}{dx}\left(\frac{\tan^{-1} x}{x^2 + 1}\right)$
  v. $\frac{d}{dx} x\sin^{-1}\sqrt{4x + 1}$
  vi. $\frac{d}{dx}\frac{1}{\sin^{-1} x}$

36. Evaluate

  i. $\int_{-1000}^{1000} \frac{1}{x^2 + 1}\,dx$
  ii. $\int_{-1}^{1} \frac{1}{\sqrt{1 - x^2}}\,dx$

10E. Concavity and Asymptotes

37. Find the intervals over which the graph of $h(x) = x^4 + 6x^2$ is concave up and those over which it is concave down. Use the second derivative test to determine the local maximum and minimum values of h. What can you say about the absolute maximum and minimum values?

38. On what interval is the curve $y = e^x - 2e^{-x}$ concave up?

39. Find the intervals of concavity and the inflection points of the function $f(x) = \frac{\ln x}{\sqrt{x}}$.

40. Evaluate the following limits. Then deduce what information the values provide about the asymptotes of the functions and the behavior of the graphs near them.

  i. $\lim_{x \to 5^+} \ln(x - 5)$
  ii. $\lim_{x \to 0^+} \log_{10}(4x)$
  iii. $\lim_{x \to \infty} \log_2(x^2 - x)$

iv. $\lim\limits_{x \to \pi/2^-} \log_{10}(\cos x)$

41. Evaluate the given limits. What do they tell us about the graphs of the functions $\tan^{-1}(x)$ and $\tan x$?

i. $\lim\limits_{x \to \infty} \tan^{-1}(x)$

ii. $\lim\limits_{x \to -\infty} \tan^{-1}(x)$

iii. $\lim\limits_{x \to \frac{\pi}{2}^+} \tan x$

iv. $\lim\limits_{x \to -\frac{\pi}{2}^-} \tan x$

10F.   Sketching Graphs

42. Sketch the graphs of $y + 3 = (x - 5)^2 + 3$ and $y = (x + 7)^3 + 4$ by starting with the graphs of $y = x^2$ and $y = x^3$ and using "shifting."

Make careful sketches of the graphs of the following functions. Make use of the **A–E** strategy of the text.

43. $y = x^2 - 3x - 5$

44. $y = x^3 - 3x^2 + 2$

45. $y = 1 - 3x + 5x^2 - x^3$

46. $y = \frac{x}{(2x-3)^2}$

47. $y = \frac{1}{(x-1)(x+2)}$

48. $y = \frac{1+x^2}{1-x^2}$

49. $y = \frac{x^3-1}{x^3+1}$

50. $y = \frac{1}{x^3-x}$

51. $y = e^{\sqrt{x^2+1}}$

52. $y = x^2 e^{-x^2}$

53. $y = \ln(x + 3)$

54. $y = x^2 + \ln x$

55. Graph $g(x) = e^{-x} \sin x$. Before proceeding with the formal analysis, recall that $-1 \le \sin x \le 1$ and observe that therefore

$$-e^{-x} \le e^{-x} \sin x \le e^{-x}$$

Begin by sketching both $-e^{-x}$ and e^{-x} on the same axes. Then think about the effect of $\sin x$.

56. Do a careful analysis of the graph of the function $f(x) = \frac{1}{1+x^2}$. (We have already encountered it in part 6F of the Exercises of Chapter 6.) Maria Agnesi investigated this function in her calculus text. In the English literature it is known as the "witch of Agnesi." Agnesi's name for this curve was the Italian *versiera*, or "turning curve." An early translator of her text

evidently confused *versiera* with *avversiera*, which means "wife of the devil," and thus mistranslated *versiera* as "witch."

57. The function $f(x) = e^{-\frac{x^2}{2\sigma^2}}$, where σ is some positive constant, is used in statistics and probability theory. Graph it and then turn to the discussion in Exercises 6F to get an idea how it might arise in this connection.

10G.   Concavity and Newton's Method

Consider a function $y = f(x)$ and assume that it is differentiable on an interval $[a, b]$. Suppose that c is the unique solution of the equation $f(x) = 0$ in $[a, b]$. Newton's Method is a strategy with which c can be found in concrete situations. Before you turn to the exercises below, review Newton's Method in Exercises 8K concentrating in particular on Figure 8.46.

58. Suppose that $y = f(x)$ is increasing and concave up on $[a, b]$. Start with any guess c_1 where $c < c_1 < b$. Provide a graph similar to Figure 8.46 to show that Newton's Method will converge to c. Now start with a guess c_1 where $a < c_1 < c$. Will Newton's Method necessarily converge to c? Provide a graph of a situation where it will not do so. What condition on c_1 will guarantee that Newton's Method will converge to c?

59. Do Exercise 58 in the situation where $y = f(x)$ is increasing and concave down on $[a, b]$.

Consider the polynomial function $f(x) = x^4 - 3x^2 + 2$. The factorization $x^4 - 3x^2 + 2 = (x^2 - 1)(x^2 - 2)$ shows that $-\sqrt{2}, -1, 1$, and $\sqrt{2}$ are the solutions of the equation $f(x) = 0$.

60. Make a careful to sketch of the graph of the function $f(x) = x^4 - 3x^2 + 2$ paying particular attention to the intervals over which the function is increasing, decreasing, concave up and concave down.

61. Consider the application of Newton's Method to $f(x) = x^4 - 3x^2 + 2 = 0$ with the guess $c_1 = 1.2$. By analyzing the graph (and without doing any computations) predict the solution to which Newton's Method will converge. Do the same for the guess $c_1 = 0.1$. To which solution will Newton's Method converge for any c_1 with $c_1 > \sqrt{2}$? What about $c_1 = \sqrt{2}$? What about $\sqrt{\frac{3}{2}} < c_1 < \sqrt{2}$? What can you say about a guess that satisfies $1 < c_1 < \sqrt{\frac{3}{2}}$ and, finally, $0 < c_1 < 1$?

Notes

[1]It will be our goal to describe the essential aspects of the matters just discussed. The important concern of mathematical rigor (as described in the introduction and first section of Chapter 8) will remain below the surface.

[2]For a history of the development of logarithms, see Howard Eves, *An Introduction to the History of Mathematics*, New York, Holt, Rinehart and Winston, 1976.

[3]The trumpet-shaped surface obtained by rotating any tractrix one complete revolution (about the vertical axis in Figure 10.10) is called a *pseudosphere* or a *hyperbolic surface*. These surfaces have a number of important properties. They have negative curvature: For any triangle drawn on such a surface (the three legs are understood to be curves of minimal lengths between the three vertices), the sum of the angles minus π is negative. This curvature is constant: Geometric figures can be moved on such a surface without distortion. By comparison, the plane has a constant curvature of zero, and a sphere has constant positive curvature.

Hyperbolic surfaces are relevant in the context of Euclidean geometry. Suppose that a point P and a line L (with P not on L) are given on a surface. (A line is understood to extend indefinitely and to connect any two points on it with a path of minimal length.) On a hyperbolic surface, there are infinitely many lines through P that are parallel to (do not intersect) L. By contrast, there is exactly one such line on the plane (in this case lines are ordinary straight lines), and there are no such lines on a sphere (in this case lines are great circles, or "equators"). The statement "For any line L and any point P not on L there exists exactly one line through P parallel to L" is the Fifth Axiom of Euclid's Plane Geometry. Mathematicians thought for centuries that this axiom (an axiom is a fundamental "truth" that underlies a theory) should be a logical consequence of the other axioms of Euclidean Geometry. The discovery of spherical and hyperbolic geometries showed that this is not the case.

Finally, hyperbolic surfaces arise in connection with the space-time geometry of Einstein's Theory of Relativity. (See Section 14.7.)

The Exponential Function and the Measurement of Age and Growth

William Smith, the English canal surveyor and engineer, observed during the course of excavations for canals in southern England in the 1790s that the exposed rock revealed a pattern of distinctive and recognizable layers, or strata. Placed, as he put it, like "slices of bread and butter," he discovered that they were differentiated not only by their general composition, but also by the various sorts of fossils they contained. Working tediously and carefully in this region rich in fossil-bearing rocks, he traced a gently tilting sequence of alternating layers of limestone and shale across the countryside. Smith noted that the same strata always contained the same fossils and were always found in the same order. The identical sequence of strata, each with its unique grouping of fossils, was recognizable at every exposed site in widely separated locations. This indicated that the mechanisms that laid the strata into place had operated on a large scale. Surely, the strata held information about the history of the Earth, but Smith had no idea why this should be so.

The answer was provided by Charles Darwin (1809–1882) some years later. His theory of evolution has at its core the process of *natural selection,* which consists of the following observations. There is some variation in inheritable characteristics within every species of an organism. Some of these characteristics will give individuals an advantage over others in surviving to maturity and in reproducing. These individuals will tend to have more offspring, and these in turn will be more likely than others to survive and reproduce. The expected result is that over many generations, the proportion of individuals that have inherited advantage-providing characteristics will increase. So natural selection is a mechanism that acts over time to produce organisms that are well adapted to survival in their enviroments. Of course, when the enviroment in which a species operates changes, then the acquired characteristics may no longer be advantageous, and the process will start afresh. Thus natural selection does not necessarily result in long-term progress in a set direction. Accord-

ing to Darwin, the operation of natural selection on new characteristics and in new environments, starting with single-celled organisms and unfolding slowly and continuously over time, has created the enormous diversity of species of animals and plants. In *The Origin of Species by Means of Natural Selection* published in 1859, Darwin asserted that through the mechanisms of evolution, plant and animal forms have appeared in a determinable order, persisted for a while, only to be replaced by new forms. Geologic processes, such as erosion, sedimentation, and volcanic activity, have, over time, encrusted the remains of these forms in layers of rock. As species and subspecies evolved and changed, so did the layers that entombed them. In this way, the stratified layers of rock have become the pages of a book that has recorded the history of the development of plant and animal life on Earth. It is a book that has the theory of biological evolution as its central organizing principle.

Darwin understood that the forces that created these pages had taken huge amounts of time to play themselves out. In *The Origin of Species* he puts it this way: "He who ... does not admit how vast have been the past periods of time may at once close this volume." In other words, evolution as the explanation of the development of life can be valid only if the Earth is sufficiently old. The thinking in scientific circles of the 19th century was that the Earth had an age of about twenty to forty million years. But this was not nearly enough.

In the beginning of the 20th century there was a dramatic change in the assessment of the age of the Earth. The nuclear physicist Ernest Rutherford discovered that radioactive materials decay in a measurable and predictable pattern. Measuring the "ticks" of radioactivity in certain samples of uranium ore, he was able to conclude that these samples must already have been ticking for several hundred million years. Rutherford's measurements had suddenly provided enough time for both evolution and geology to do their work. In his own words: "The discovery of the radioactive elements, which in their disintegration liberate enormous amounts of energy, thus increases the possible limit of the duration of life on this planet, and allows the time claimed by the geologists and biologists for the process of evolution."

Within the first two decades of the 20th century the ages of rocks from several geologic periods had been determined by this method. By the 1930s, the validity of the theory of biological evolution and the explanations of the development of life that it provided were beginning to be recognized. The revolution launched by Darwin was (and is) perceived by some as a threat to the fundamental religious conviction that the human species has a unique position at the center of created life. It thus shares a common element with the revolution in astronomy of the 17th century, which had removed the Earth from a unique position at the center of the universe.

It turns out that there are myriads of nuclear clocks all over the surface of the Earth. When these are read off, one at a time or many at once, they provide an incredible amount of information about the development of the geology of the Earth as well as the life that evolved on it. This chapter will study this information as well as the mathematics of the mechanisms of the clocks that provide it. We will also see that the mathematics that underlies the ticking of nuclear clocks also informs us about quantitative aspects of growth in biology.

11.1 Nuclear Activity

How was Rutherford able to compute the age of his samples of uranium ore? The basic idea is this. As a radioactive substance decays, its atoms change from one type to another. The situation is analogous to that of a basket of magic apples in which the apples change spontaneously to oranges at a fixed and known rate. The apples represent unstable radioactive atoms, and the oranges represent the atoms into which these transform. Consider a basket that contains only apples initially. As the basket ages, more and more of its apples become oranges. Since the

change occurs at a computable rate, the ratio of oranges to apples changes in a very specific way over time. In other words, it is possible to determine the age of the basket simply by measuring the ratio of oranges to apples.

We now turn to a description of the process of radioactivity and to its mathematical analysis.

A. The Atom and its Nucleus. Two discoveries, one in 1895 and the other in 1896, marked the beginning of the nuclear age. In 1895 the German physicist Wilhelm Roentgen discovered quite by accident that radiation emanating from an apparatus in his laboratory brought certain crystals to a glow. He tried to shield the crystals from the source of the radiation with various substances, but the mysterious rays kept coming through. Finally, Roentgen found that the metals lead and platinum obstructed their progress. Roentgen had discovered X-rays. In 1896 the French physicist Henri Becquerel placed some photographic plates in a drawer of his laboratory. They were shielded from light by a black paper wrapping. Without giving it a second thought, he put a lump of uranium salts on top of the plates. To his amazement, when he developed the plates later, he found an image of the lump imprinted on them. Becquerel realized that the plates had recorded some kind of radiation coming from the uranium salts. He had discovered nuclear radiation.

The explanation of the phenomena discovered by Roentgen and Becquerel was provided over a period of several decades by the work of a number of teams of scientists, which included several Nobel Prize winners. Foremost among them was the New Zealander Ernest Rutherford (1871–1937). Rutherford was the very first nuclear physicist. In the early years of the 20th century he and his collaborators laid the foundations of the theory of radioactivity.[1] In 1908 he won the Nobel Prize for chemistry for his work. He was subsequently called to head the famous Cavendish Laboratories, at Cambridge, England. By the 1930s the research team at these laboratories had unraveled the basic structure of the atom. (The

Table 11.1

Solar System		Carbon Atom	
$\dfrac{\text{Diameter of Sun}}{\text{Diameter of Pluto's Orbit}} = \dfrac{1}{1700}$		$\dfrac{\text{Diameter of Nucleus}}{\text{Diameter of Electron Orbit}} = \dfrac{1}{100,000}$	
$\dfrac{\text{Mass of Earth}}{\text{Mass of Sun}} = \dfrac{1}{300,000}$		$\dfrac{\text{Mass of Electron}}{\text{Mass of Nucleus}} = \dfrac{1}{22,000}$	

Cavendish Laboratories were named after the man who measured the gravitational constant G. See Section 7.6.)

An atom was shown to consist of a nucleus of protons and neutrons surrounded by electrons. An electron zips around the nucleus about 10^{17} times in one second. It is an elusive blurr in orbit about the nucleus. The particles that make up an atom are extremely tiny. In terms of an *atomic mass unit* (amu), which is a minuscule 1.66×10^{-27} kilograms, the mass of an electron is 0.00055 amu, that of a proton is 1.0076 amu, and that of a neutron is 1.0090 amu. A typical atom has a diameter of about 10^{-10} meters with a nucleus of diameter about 10^{-15} meters. One can think of an atom as a tiny solar system. Table 11.1 makes some comparisons. Both the solar system and an atom consist primarily of vacuous space (the atom being comparatively much more empty) with virtually all the mass concentrated in the Sun, respectively the nucleus.

The number of protons in the nucleus of an atom is the *atomic number*. This is also equal to the number of electrons. The number of protons plus the number of neutrons is the *mass number*. These numbers are incorporated into the chemical symbol of the element. For example, the basic chemical symbol for the element helium is He. When its atomic number and mass number are considered, it is designated as $^4_2$He. This means that it has 2 protons and $4 - 2 = 2$ neutrons. Aluminum, with its 13 protons and 14 neutrons, is written $^{27}_{13}$Al. Chlorine is $^{35}_{17}$Cl, which means that it has 17 protons and $35 - 17 = 18$ neutrons. A number of elements occur in different forms, or *isotopes*. The number of protons is the same in each case, but the number of neutrons varies. For example, there is the isotope $^{37}_{17}$Cl of chlorine. In addition to its 17 protons, it has $37 - 17 = 20$ neutrons. The gas radon occurs in the three isotopes $^{222}_{86}$Rn, $^{220}_{86}$Rn, and $^{219}_{86}$Rn. They are often designated as radon-222, radon-220, and radon-219. There are four isotopes $^{228}_{88}$Ra, $^{226}_{88}$Ra, $^{224}_{88}$Ra, and $^{223}_{88}$Ra of the element radium. The designations radium-228, radium-226, and so on, are also used. Another example is provided by the four isotopes $^{238}_{92}$U, $^{235}_{92}$U, $^{234}_{92}$U, and $^{233}_{92}$U of uranium.

One gram of an element contains

$$\frac{1}{m}(6.02 \times 10^{23}) \text{ atoms,}$$

where m is the mass number of the element and 6.02×10^{23} is *Avogadro's number*. This expression is valid for any element. So one gram of pure aluminum $^{27}_{13}$Al has $\frac{1}{27}(6.02 \times 10^{23}) = 2.23 \times 10^{22}$ atoms, and one gram of pure $^{238}_{92}$U contains $\frac{1}{238}(6.02 \times 10^{23}) = 2.53 \times 10^{21}$ atoms.

B. The Discoveries of Rutherford. Let's turn to Rutherford's explanation of radioactivity. The atoms of some of the heavier elements are unstable, and they decay; that is to say, they are transformed into different, lighter atoms by emitting one or more of the following three packets of matter and energy:

α particle: a combination of two protons and two neutrons. Observe that an α-particle is simply a helium nucleus.

β particle: a high-speed electron.

γ ray: a package of energy of one photon, a certain type of X-ray.

In experiments carried out in early years of the 20th century, Rutherford and his colleagues studied the quantitative aspects of radioactivity. A number of different radioactive substances were investigated. These included the radon isotope $^{222}_{86}$Rn and the radium isotope $^{226}_{88}$Ra, both found in minerals containing uranium. A pure sample of a radioactive substance was chemically isolated and enclosed in a vessel containing a gas. The α, β, or γ radiation packets that the substance propelled into the gas knocked electrons away from the gas atoms. In this way, positively and negatively charged particles, called *ions*, were produced in the vessel. When two opposite locations in the vessel were hooked up to the poles of a battery, these particles began to flow. The greater the rate of disintegration of the atoms in the sample, the greater the rate of ion production, and thus the greater the intensity of the current. Therefore by measuring the current generated by such an *ionization chamber*, Rutherford was able to draw conclusions about the rate of disintegration of the radioactive substance.

Let's introduce some mathematical precision as we observe Rutherford during one of his experiments. Suppose he began his observations at time $t = 0$. As time progresses, more and more atoms in the sample disintegrate. For any time $t \geq 0$, taken in hours (but possibly also in seconds, days, or years), let $y(t)$ be the number of atoms in his sample. So $y(t)$ is the number of atoms that did not disintegrate during the first t hours of the experiment. The initial number of atoms is $y_0 = y(0)$. Recall from Section 8.4 that $y'(t)$ is the rate at which the atoms in the sample are disintegrating at time t. This is the rate at which the packets of α, β, and γ radiation are emitted. This in turn is proportional to the rate at which the ions are created and hence to the current that is produced. In other words, $y'(t) = kc(t)$, where $c(t)$ is the current that is generated and k is a constant. For instance, the number of disintegrations per second at time $t = 0$ is $y'(0)$, and this is equal to $kc(0)$, where $c(0)$ is the current measured at $t = 0$. Rutherford's ionization chamber gave him no direct information about the quantities y_0 or $y(t)$, but it did

allow him to compute the ratio

$$\frac{y'(t)}{y'(0)} = \frac{kc(t)}{kc(0)} = \frac{c(t)}{c(0)}$$

at any time $t > 0$ that he measured the current $c(t)$. He observed that the ratio $\frac{y'(t)}{y'(0)}$ was always less than 1. This meant that the activity $y'(0)$ was greater than the activity $y'(t)$ at any other time. A comparison of the ratios at various times into the experiment suggested to Rutherford that they were related. After he took the natural logs of the ratios, the relationship became clear. Since $\frac{y'(t)}{y'(0)} < 1$, he saw that $\ln \frac{y'(t)}{y'(0)} < 0$. More importantly, when he plotted the points $\left(t, \ln \frac{y'(t)}{y'(0)}\right)$ for different times, say $0 < t_1 < t_2 < t_3$ and so on, he found that they all fell very near a line. Since $\left(0, \ln \frac{y'(0)}{y'(0)}\right) = (0, \ln 1) = (0, 0)$, the line goes through the origin. See Figure 11.1. The vertical axis is labeled z, since the symbol y is already being used. The slope of the line is negative, and Rutherford denoted this slope $-\lambda$; with λ positive. So the equation of the line is $z = -\lambda t$. Since all the points that he plotted were on the line (or very nearly so), he concluded that

$$\ln \frac{y'(t)}{y'(0)} = -\lambda t$$

at any time $t \geq 0$ into the experiment. Observe that λ can be determined by computing the ratio $-\frac{\ln \frac{y'(t)}{y'(0)}}{t}$. Rutherford called the constant λ the *radioactive constant* of the element. Today, it is more commonly referred to as the *decay* or *disintegration constant*. We will see shortly that it provides important information.

Example 11.1. In one experiment, Rutherford isolated a sample of a radioactive substance in radium, which he referred to as "radium emanation." It is now known to be the isotope $^{222}_{86}$Rn of the gas radon. Testing the sample in his ionization chamber, he found:

(1) At $t = 20.8$ hours, $\frac{y'(t)}{y'(0)} = 0.857$.

(2) At $t = 187.6$ hours, $\frac{y'(t)}{y'(0)} = 0.240$.

(3) At $t = 354.9$ hours, $\frac{y'(t)}{y'(0)} = 0.069$.

(4) At $t = 521.9$ hours, $\frac{y'(t)}{y'(0)} = 0.015$.

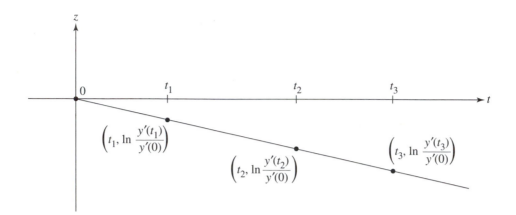

Figure 11.1

(5)   At $t = 786.9$ hours, $\frac{y'(t)}{y'(0)} = 0.0019$.

This information provided him with the following equalities,

(i) $-\dfrac{\ln \frac{y'(20.8)}{y'(0)}}{20.8} = -\dfrac{\ln 0.857}{20.8} = -\dfrac{-0.154}{20.8}$

$\qquad\qquad = 0.0074.$

(ii) $-\dfrac{\ln \frac{y'(187.6)}{y'(0)}}{187.6} = -\dfrac{\ln 0.240}{187.6} = -\dfrac{-1.427}{187.6}$

$\qquad\qquad = 0.0076.$

(iii) $-\dfrac{\ln \frac{y'(354.9)}{y'(0)}}{354.9} = -\dfrac{\ln 0.069}{354.9} = -\dfrac{-2.674}{354.9}$

$\qquad\qquad = 0.0075.$

(iv) $-\dfrac{\ln \frac{y'(521.9)}{y'(0)}}{521.9} = -\dfrac{\ln 0.015}{521.9} = -\dfrac{-4.200}{521.9}$

$\qquad\qquad = 0.0080.$

(v) $-\dfrac{\ln \frac{y'(786.9)}{y'(0)}}{786.9} = -\dfrac{\ln 0.0019}{786.9} = -\dfrac{-6.266}{786.9}$

$\qquad\qquad = 0.0079.$

Each of the numbers given in (i)–(v) is an estimate of the decay constant λ. Rutherford took the average $\lambda = 0.0077$ for the decay constant of radon-222. The unit for λ is hours^{-1}.

Rutherford continued his analysis as follows. The equation $\ln \frac{y'(t)}{y'(0)} = -\lambda t$ leads directly to

$$e^{\ln \frac{y'(t)}{y'(0)}} = e^{-\lambda t}$$

and it follows from the basic relationship between the log and exponential functions (see Section 10.3) that $\frac{y'(t)}{y'(0)} = e^{-\lambda t}$. Therefore,

$$y'(t) = y'(0)e^{-\lambda t}.$$

Because

$$\frac{d}{dt}\left(-\frac{y'(0)}{\lambda}e^{-\lambda t}\right) = -\frac{y'(0)}{\lambda}\frac{d}{dt}e^{-\lambda t} = -\frac{y'(0)}{\lambda}(-\lambda\, e^{-\lambda t})$$

$$= y'(0)e^{-\lambda t} = y'(t)$$

the function $-\frac{y'(0)}{\lambda}e^{-\lambda t}$ is an antiderivative of $y'(t)$. But $y(t)$ is also an antiderivative of $y'(t)$, and it follows that

$$y(t) = -\frac{y'(0)}{\lambda}e^{-\lambda t} + C$$

for some constant C. Since the atoms keep disintegrating, $y(t)$ will be very small if t is very large (say many billions of years). For a very large t the term $e^{\lambda t}$ will be huge since $\lambda > 0$. So $e^{-\lambda t} = \frac{1}{e^{\lambda t}}$ will also be very small. It follows that if t is pushed to infinity on both sides of the equation just established, then

$$0 = \lim_{t \to \infty} y(t) = \lim_{t \to \infty} \left(-\frac{y'(0)}{\lambda} e^{-\lambda t} + C \right)$$

$$= \lim_{t \to \infty} \left(-\frac{y'(0)}{\lambda} \frac{1}{e^{\lambda t}} + C \right) = C.$$

It follows that $C = 0$, and therefore,

$$y(t) = -\frac{y'(0)}{\lambda} e^{-\lambda t}.$$

Since $y_0 = y(0) = -\frac{y'(0)}{\lambda} e^{-\lambda 0} = -\frac{y'(0)}{\lambda}$, Rutherford has obtained the equation

$$\boxed{y(t) = y_0 e^{-\lambda t}} \tag{11a}$$

Since he was able to determine the decay constant λ, Rutherford could now compute the number of atoms $y(t)$ in a sample of a radioactive element at any time t in terms of the initial number $y_0 = y(0)$ of atoms in the sample.

The number of atoms in a sample of a radioactive substance can be converted to a mass equivalent, for instance to grams, by using Avogadro's number. Converting both $y(t)$ and y_0 in the equation (11a) in this way shows that this equation is valid—with the same λ—whether the sample is measured in terms of numbers of atoms or some unit of mass. It is important to note that the rate of the disintegration of a given radioactive element is not affected by changes in temperature, pressure, chemical state, or physical environment. So the decay constant λ and equation (11a) are unaffected by such factors.

Example 11.2. Let's return to Rutherford's experiment with radon-222. Since $\lambda = 0.0077$ hours^{-1} for this isotope, Rutherford knows that the number of atoms in his sample at any time $t \geq 0$ is $y(t) = y_0 e^{-0.0077t}$, where y_0 is the initial amount. What fraction of the initial amount of radon-222 that Rutherford tested was left after 80 hours? After 240 hours? After 400 hours?

By substitution,

$$y(80) = y_0 e^{-0.0077(80)} = y_0 e^{-0.616} = 0.540 y_0$$

$$y(240) = y_0 e^{-0.0077(240)} = y_0 e^{-1.848} = 0.158 y_0$$

$$y(400) = y_0 e^{-0.0077(400)} = y_0 e^{-3.08} = 0.0460 y_0$$

So after 80 hours, 54% of the initial amount remained. After 240 hours about 16%, and after 400 hours about 5%, or $\frac{1}{20}$. Figure 11.2 shows the graph of the function $y(t) = y_0 e^{-0.0077t}$.

Example 11.3. Suppose that the sample of radon-222 that Rutherford was investigating had a mass of 0.01 milligrams at time $t = 0$. What was the initial number y_0 of atoms in the sample? We know that 1 gram of radon-222 contains

$$\frac{1}{222}(6.02 \times 10^{23}) = 2.7 \times 10^{21} \text{ atoms.}$$

Since 1 milligram is equal to 10^{-3} grams, $y_0 = (0.01 \times 10^{-3})(2.7 \times 10^{21}) = 2.7 \times 10^{16}$ atoms.

Differentiating both sides of equation (11a), shows that $y'(t) = -\lambda y_0 e^{-\lambda t} = -\lambda y(t)$. Therefore,

$$\boxed{y'(t) = -\lambda y(t)} \tag{11b}$$

Expressed in words, this says that the rate with which the number of atoms in a sample of a radioactive substance decreases at any time t is in fixed proportion to the number of atoms in the sample at that time.

Example 11.4. Let's return to Rutherford's experiment with radon $^{222}_{86}$Rn one more time. The number of atoms in the sample is $y(t) = y_0 e^{-0.0077t}$ with t given in hours. Let's take $y_0 = 2.70 \times 10^{16}$. So $y(t) = (2.70 \times 10^{16})e^{-0.0077t}$. What was the rate at which the atoms were decaying at the instant $t = 0$? This was

$$y'(0) = -\lambda y(0) = -0.0077(2.70 \times 10^{16})$$

$$= -2.08 \times 10^{14}.$$

So the rate was 2.08×10^{14} atoms per hour. Recall that the minus indicates that the number of atoms is

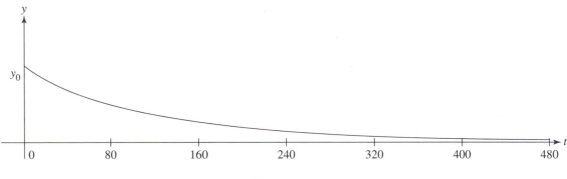

Figure 11.2

decreasing with time. What was the rate of decay at the instant $t = 2$ hours? This was

$$y'(2) = -\lambda y(2)$$

$$= -(0.0077)(2.70 \times 10^{16})e^{-0.0077(2)}$$

$$= -(2.08 \times 10^{14})(0.9847) = -2.05 \times 10^{14}.$$

So this rate was 2.05×10^{14} atoms per hour. How many atoms decayed during the first two hours of the experiment? This was

$$y(0) - y(2)$$

$$= 2.70 \cdot 10^{16}e^0 - 2.70 \cdot 10^{16}e^{-0.0077(2)}$$

$$= 2.70 \cdot 10^{16}(1 - 0.9847)$$

$$= 2.70 \cdot 10^{16}(0.0153) = 4.13 \times 10^{14}.$$

Because $(4.13 \times 10^{14})/(2.70 \times 10^{16}) = 1.53 \times 10^{-2} = 0.0153$, about 1.5% of the original sample disintegrated over the first two hours. Over the first two hours, atoms were decaying at an average rate of $\frac{4.13 \times 10^{14}}{2} = 2.07 \times 10^{14}$ atoms per hour.

Rutherford also studied the radioactive isotope $^{226}_{88}$Ra of radium discovered by Marie and Pierre Curie. He knew that $^{226}_{88}$Ra decays to the radon $^{222}_{86}$Rn that he had been investigating, by shedding an α-particle (recall that this is a packet of two protons and two neutrons) from its nucleus.

Example 11.5. Measurements with a radiation counter gave a value of 3.67×10^{10} disintegrations per second as the activity of one gram of pure radium-226. Use this information to determine the decay constant of this isotope.

Consider this gram of pure radium-226 at time $t = 0$ and let $y(t)$ be the number of atoms at any time $t \geq 0$. By equation (11b), $y'(0) = -\lambda y(0)$. So $\lambda = -\frac{y'(0)}{y(0)}$. This is the equation with which λ will be computed. At time $t = 0$ (when there is precisely one gram), 3.67×10^{10} radium-226 atoms are disintegrating per second. So $y'(0) = -3.67 \times 10^{10}$. It remains to note that $y(0)$ is the number of atoms in one gram of radium-226. Since this number is $\frac{1}{226}(6.02 \times 10^{23}) = 2.67 \times 10^{21}$,

$$\lambda = \frac{-3.67 \times 10^{10} \text{atoms/sec}}{2.67 \times 10^{21} \text{atoms}} = -1.37 \times 10^{-11} \frac{1}{\text{sec}}.$$

In the context of the mass of the sample as given in grams, the year is a more meaningful unit of time. (This will be apparent in a moment.) Since one year has about 3.16×10^7 seconds (check this),

$$\lambda = -1.37 \times 10^{-11}\frac{1}{\text{sec}} \cdot \frac{3.16 \times 10^7 \text{sec}}{\text{year}}$$

$$= -4.33 \times 10^{-4}\frac{1}{\text{year}}.$$

Having determined that $\lambda = 4.33 \times 10^{-4}$ in (years)$^{-1}$ for radium-226, we now know that there will be $y(t) = 1 \cdot e^{-\lambda t} = e^{-4.33 \times 10^{-4}t}$ grams in the sample at any time t (in years) after the initial observation. For instance, after 1000 years, there will be

$$y(1000) = e^{-4.33 \times 10^{-4}(1000)} = e^{-4.33 \times 10^{-1}} = e^{-0.433}$$

$$= 0.649 \text{ grams.}$$

Consider the decay of a radioactive substance. As we have seen, the number of atoms in a sample at any time $t \geq 0$ is given by $y(t) = y_0 e^{-\lambda t}$, where $y(0) = y_0$ is the number of atoms at $t = 0$ and λ is in the decay constant. How long will it take for the initial number of atoms y_0 to decay to one-half this number? In translated form, the question is this: At what time t will $y(t) = y_0 e^{-\lambda t}$ be equal to $\frac{y_0}{2}$? To answer this question, we set

$$y_0 e^{-\lambda t} = \frac{y_0}{2}$$

and solve for t. After canceling y_0, we see that $e^{-\lambda t} = \frac{1}{2}$, and taking natural logs of both sides, we get

$$\ln \frac{1}{2} = \ln e^{-\lambda t} = -\lambda t.$$

Therefore, $t = \frac{1}{-\lambda} \ln \frac{1}{2} = \frac{1}{-\lambda} \ln 2^{-1} = \frac{\ln 2}{\lambda}$. This is the *half-life* of the radioactive substance. We will denote it by h. So

$$h = \frac{\ln 2}{\lambda} = \frac{0.693}{\lambda}$$

is the length of time that it takes for a sample to decay down to one-half its size. Note that $\lambda = \frac{\ln 2}{h}$. So the equation $y(t) = y_0 e^{-\lambda t}$ can be rewritten as

$$\boxed{y(t) = y_0 e^{-\frac{\ln 2}{h} t}} \tag{11c}$$

Example 11.6. What is the half-life of radon-222? Since the decay constant is $\lambda = 0.0077$ hours^{-1}, its half-life is $h = \frac{0.693}{\lambda} = \frac{0.693}{0.0077} = 90$ hours, or 3.75 days.

So if the initial number of atoms in one of Rutherford's samples was, say, $y_0 = 100,000,000$ atoms, then

after 90 hours, there were 50,000,000 atoms, after another 90 hours there were 25,000,000, after another 90 hours 12,500,000, and so on.

In the same way, the half-life of radium-226 is

$$h = \frac{0.693}{\lambda} = \frac{0.693}{4.33 \times 10^{-4}} = 0.1600 \times 10^4 = 1600 \text{ years.}$$

Notice that a given amount of radium-226 will decay to one-half the amount in 1600 years, but a given quantity of radon-222 will do so in 90 hours. So the half-life varies greatly from one radioactive element to another. This reflects the fact that some elements are rather stable and decay extremely slowly, and others are highly radioactive and decay very rapidly. The half-lives of radioactive elements range from billions years—that of uranium-238 is 4.5 billion years for instance—to but fractions of a second. Thus, some elements decay imperceptibly slowly, while others do so with almost instantaneous rapidity.

By the mid 1930s, the decay process for the various radioactive elements was understood. For example, in a step-by-step sequence, uranium $^{238}_{92}$U decays into thorium $^{234}_{90}$Th by ejecting an α particle and emitting γ radiation, thorium $^{234}_{90}$Th in turn decays to protactinium $^{234}_{91}$Pa by the conversion of a neutron to a proton and the emission of a β and more γ radiation, the uranium isotope $^{234}_{92}$U appears in the third step, ..., the radium isotope $^{226}_{88}$Ra discovered by the Curies in the fifth step, the radon isotope $^{222}_{86}$Rn investigated by Rutherford in the sixth step, ..., and finally, after nine more such steps (some proceeding at a snail's pace and others in a flash), the sequence of radioactive transmutations stops with the stable isotope $^{206}_{82}$Pb of lead. The diagram

$$^{238}_{92}\text{U} \rightarrow {}^{206}_{82}\text{Pb} + 8\alpha + 6\beta$$

$$+ \text{ energy (in particular } \gamma \text{ radiation)}$$

summarizes the total process. The basic idea is that unstable nuclei have an excess of neutrons and protons. These are shed in the decay process until a stable configuration is achieved.

This is the decay process that Rutherford had used earlier to determine the age of his samples of uranium ore. He estimated the rate at which the eight α-particles were produced in the samples. He also realized that some of the denser and more compact uranium minerals would be able to retain these α-particles within their mass in the form of helium. By estimating the amount of helium they contained, Rutherford was able to conclude that some of his samples must be several hundred million years old.

The isotope uranium-235 is "at the top" of a second radioactive series. In a sequence of twelve steps, which feature for the most part variants of the radioactive isotopes of the uranium-238 series just described, uranium-235 decays to the stable isotope $^{207}_{82}$Pb of lead. The radioactive isotope $^{232}_{90}$Th of thorium is at the top of another radioactive series. A fourth radioactive series and twelve additional radioactive elements that do not belong to any series complete the list of naturally occurring radioactive elements.

When an element occurs naturally, all the isotopes are present in a "mix." For example, natural uranium consists of 99.28% uranium-238, with uranium-235 at 0.715% comprising most of the rest. The apparatus with which one can make such a determination is called an *accelerator mass spectrometer*. The principle behind it is simple: Since they have different masses, the isotopes of a given element can be separated by passing them through an electromagnetic field. This technique, both delicate and sensitive, has been refined in recent years to give very accurate results. Accelerator mass spectrometers are used to provide very accurate measurements of decay constants. This is done as follows. Count the number of atoms in a tiny sample of a radioactive element at time $t = 0$ and again some time t later. Let $y(0) = y_0$ be the first count and $y(t)$ the second. By equation (11a), $y(t) = y_0 e^{-\lambda t}$. Since $e^{-\lambda t} = \frac{y(t)}{y_0}$, $-\lambda t = \ln \frac{y(t)}{y_0}$. So $\lambda = -\frac{1}{t} \ln \frac{y(t)}{y_0}$. Since all the quantities on the right are known, λ can be computed.

The values of a function $y(t)$ that expresses the number of atoms in the sample of a substance in terms of time are all positive integers. The graph of such a function thus features a sequence of jumps. This means that it is not continuous and therefore not dif-

ferentiable. And yet our analysis has assumed that it is. Indeed, the end result of this analysis is the differentiable function $y(t) = y_0 e^{-\lambda t}$. That Rutherford's approach, which to be sure is an approximation, is valid nonetheless can best be seen by analogy: a set of points in the plane can be studied by fitting a smooth curve through all the points and by analyzing the curve.

The validity of the analysis of radioactive decay undertaken in this section does require the assumption that a sample consists of a large number of atoms. In particular, it says little about an individual atom. Refer to Exercise 6F for an explanation of the fact that the probability that a certain event occurs is some number between 0 and 1. Recall that a probability of 0 means that the event has no chance of occurring and that a probability of 1 means that it will definitely occur. What is the probability that a particular atom in some radioactive substance that has y_0 atoms initially will not decay for the next t years? In other words, what is the probability that at time t the particular atom is among the $y(t)$ atoms still remaining? Since there are y_0 atoms at the beginning, this is

$$\frac{y(t)}{y_0} = e^{-\lambda t} = e^{-(\ln 2)t/h}.$$

The probability that a particular atom of a radioactive substance survives until the half-life $t = h$ is $\frac{y(h)}{y_0} = e^{-(\ln 2)h/h} = e^{-\ln 2} = \frac{1}{e^{\ln 2}} = \frac{1}{2} = 0.5$, as one might have expected. We have taken a glimpse at the fact that sophisticated statistical and probabilistic methods—known as quantum mechanics and quantum electrodynamics—lie at the heart of modern nuclear physics.

We now leave the microscopic world of the atom and turn to a reality of much larger scale.

11.2 The Earth's Geologic History

Scientists have established an outline of the geologic history of the Earth. According to a picture that continues to emerge, its formation began about

4.6 billion years ago when the huge gas cloud of matter that surrounded the Sun early in the formation of the solar systems settled into a flat disc of orbiting particles. The early Earth (and the other planets) developed by a process that took several hundred million years. The orbiting particles collected into clumps ranging in size from pea to small planet. The largest of these grew to planetary size during a period of heavy bombardment by smaller ones, including meteoroids (clumps consisting of combinations of metals and minerals) and comets (clumps consisting largely of ice). This bombardment produced so much heat that the newly formed Earth melted. Subsequent volcanic activity released large quantities of gas and steam from the Earth's interior. This was the beginning of the development of the atmosphere. After the Earth cooled, steam and the continued impact of comets provided the water for the formation of the early oceans.

The Earth's outer layer is composed of a number of huge slabs of dense solid rock, like a cracked shell of a giant egg. Such a slab, called a *tectonic plate* (from the Greek *tektonikos* "pertaining to construction"), is on average about forty miles thick. There are now somewhere between twelve and twenty such plates. The plates drift and shift, gliding on and being propelled by a hot, molten lower layer of rock that churns and simmers like a very thick soup in super slow motion. Heat produced by radioactivity keeps the soup hot and simmering. The plates rub, chafe, and crash against each other, all with the same super slow motion. This activity is called *plate tectonics*. When two plates collide, one usually dives below the other. This process often causes massive earthquakes. As the plunging slab melts to become part of the simmering soup, molten rock, called *lava*, erupts in volcanic outbursts through the overriding plate. Over time this volcanic activity produced small islets of crust. These eventually collected into larger land masses and continents. The oceans and continents ride like puddles and thin clumps on top of the huge shifting plates. Crust is also produced on ocean floors when lava is pushed through a globe-encircling network of cracks. This has led to the formation of a system of mid-oceanic ridges. The

phenomenon of plate tectonics accounts for the birth and death of oceans, the wanderings of continents, and the creation of mountain ranges. This activity goes on imperceptibly slowly. Operating over hundreds of millions of years in an ever evolving way, it has produced the basic features of the Earth's surface.

What evidence is there to confirm the picture just described? Over what time frame did these developments take place? What are the particulars of this general process? How did the current configuration of continents and oceans take shape? Intense and ongoing research has produced some answers to these difficult questions. These include the following information.

While the geologic activity just described has destroyed almost all of the earliest portions of the original crust, a few fragments have survived. Some mineral grains discovered in western Australia were found to be between 4.1 and 4.3 billion years old. An area in North America is 3.96 billion years old. A terrain in western Greenland is between 3.7 and 3.8 billion years old. The drift of a certain region currently part of South Africa has been traced back 3.5 billion years.

The Appalachian mountain range was discovered to have been part of a larger belt that was split and separated by tectonic processes. Parts of it now occupy the East Coast of North America, but other parts of the original belt have been located in Ireland, England, and Norway. This conclusion is based on the identification of the remains of oceanic crust formed on both sides of the Atlantic between 480 and 500 million years ago. There is no doubt that these rocks were all part of a single formation before the North Atlantic Ocean opened about 200 million years ago and North America separated from Europe.

The Pacific Coast was put together by the interaction of the Pacific plate with neighboring plates. It is built of many different blocks stretching from northern Mexico to Alaska. The blocks arrived some 100 to 200 million years ago and extended the continent westward by some 300 miles. The current position of the Pacific plate can be seen as follows. Draw a dot for each one of the hundreds of active volcanoes in the Pacific region, and add more dots for the locations of recorded earthquakes. Then "connect the dots" to obtain a curve that runs up along the coast of California, touches Alaska, sweeps across the Pacific, touches the Asian continent north of Japan, moves down below Australia, and then loops back to Southern California. This "ring of fire" outlines the Pacific plate and shows the seams, called *faults*, between this plate and the others.

The other regions of the world have similar histories. All continents and oceans move. About 80 million years ago, the South Atlantic began to open as South America and Africa separated.[2] India was the last patch added to Asia, arriving about 50 million years ago with a big crash that created the Himalayas. About 34 million years ago the Red Sea began to open up. This rift is one of the youngest regions of continental breakup on Earth. The Red Sea is widening at the rate of about one-half inch per year. Eventually it will become an ocean.

Figure 11.3 provides an overview of the Earth as a moving geologic jigsaw puzzle over the past 700 million years.

How can one reach these conclusions? How is it possible to say how and when the Earth was formed, and to piece together its geologic development? How can the movement of continents be charted? What allows us to look billions of years into the past?

A. Reading Nuclear Clocks. When meteoroides are pulled in by the Earth's gravitational field, they enter the atmosphere. If they are large enough, the burnup caused by atmospheric friction will not consume them entirely. Now called *meteorites* they survive and impact on the Earth's surface. On occasion, they can be recovered and studied. The analysis of the chemical elements that meteorites contain and the ratios in which they contain them has confirmed a similarity with the chemical composition of the Sun and has suggested that they were formed during the initial phases of the development of the solar system. Meteorites contain radioactive elements. Since

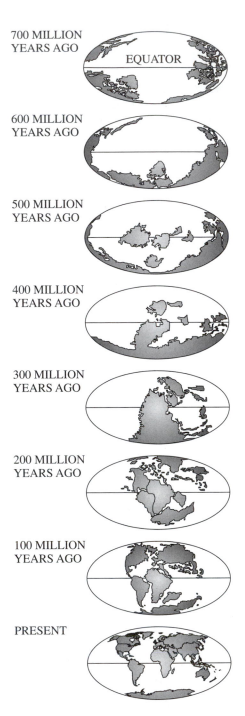

Figure 11.3 From *Life in the Universe*, Scientific American, Special Issue, October 1994, page 69.

these, as we have seen, decay in a mathematically predictable way, they can be used as clocks to determine the time of their origin.

As rocks developed during the initial phase of crust formation, the constituent mineral components solidified. When rocks are later subjected to the high temperatures and pressures generated by volcanic activity or plate tectonics, they return to a liquified state. They can cool later and crystallize again. Where the crust comes under tension, great fissures develop. These serve as channels that allow black lava, called *basalt*, to reach the surface. When the lava cools it solidifies in the fissures. This activity goes on continuously and relentlessly everywhere near the surface of the Earth. Where such regions are exposed in uptilted portions of the crust, the rocks and minerals that they are composed of can be studied. Many of them contain radioactive elements. By reading off these radioactive clocks—note that these clocks are reset when exposed to very high temperatures and pressures—the age of the rocks, as well as insights into their geologic development, can be obtained. In this way, radioactive clocks are an important source of information about the Earth's formation. Before looking at some of the explicit information that these clocks provide, let's first see how they operate.

Think of a mineral mass at the time it cools from its last molten state. This, to repeat, could have occurred during the early phases of one of the processes that formed the Earth's crust, or as a result of geologic activity later. Suppose that the mass contains some radioactive "parent" element that decays to a stable (that is, nonradioactive) "daughter" element. As long as the mineral mass is in a molten state, the daughter atoms produced by the radioactive parent are dispersed. But when it solidifies, the daughter atoms are locked into place in the grains that crystallize. Take this as time $t = 0$. Let $y(t)$ be the number of parent atoms in a fragment of such a grain after any elapsed time $t \geq 0$. So $y(0) = y_0$ is the initial number of parent atoms, and

$$y(t) = y_0 e^{-\lambda t}$$

for any $t \geq 0$. The decay constant λ can be determined. See Section 11.1B. Take natural logs of both sides of $\frac{y(t)}{y_0} = e^{-\lambda t}$ to get $\ln \frac{y(t)}{y_0} = \ln e^{-\lambda t} = -\lambda t$. Therefore

$$t = -\frac{1}{\lambda} \ln \frac{y(t)}{y_0} = \frac{1}{\lambda} \ln \left(\frac{y(t)}{y_0} \right)^{-1} = \frac{1}{\lambda} \ln \frac{y_0}{y(t)}.$$

Notice that the difference $z(t) = y_0 - y(t)$ is the number of "daughter" atoms in the grain fragment at any time t. Since $y_0 = z(t) + y(t)$ and $\frac{y_0}{y(t)} = \frac{z(t)+y(t)}{y(t)} = \frac{z(t)}{y(t)} + 1$, it follows that

$$t = \frac{1}{\lambda} \ln \left(\frac{z(t)}{y(t)} + 1 \right) \qquad (11d)$$

Now consider the present and let t be the time that has elapsed since the cooling occurred. Mass spectrometers can be used to measure the number $y(t)$ of parent atoms and the number $z(t)$ of daughter atoms in the fragment. Inserting this information into the formula $t = \frac{1}{\lambda} \ln \left(\frac{z(t)}{y(t)} + 1 \right)$ provides an estimate of the time t that has elapsed since the formation of the grain.

Such analyses have revealed the information already described. They tell us that meteorites are commonly from 4.5 to 4.6 billion years old. They inform us that the mineral grains discovered in western Australia are between 4.1 and 4.3 billion years old. Episodes of global volcanic activity, separation and collision of continents, and folding of geologic layers can be dated in this way. In addition, geologic formations can be identified by the age patterns of the rocks that they contain. By this method, rock units that were formed at the same time can be determined and reassembled into recognizable regions. For instance, it has been shown that the ages of a sandstone that underlies most of the Canadian province of Nova Scotia cluster around 0.6 billion, 2 to 2.3 billion, and 2.8 to 3 billion years. The age range of 1 to 1.8 billion years common in North America is absent from this sandstone. Similar age patterns of sandstone in areas of North Africa suggest that these regions were once joined to the Canadian formations. This is the method used to trace the split in the Appalachian range.

While the mathematical mechanisms of radioactive clocks are relatively simple, their use involves several subtle points. If one is interested in timing, say, a mile run with a stopwatch, there are three very basic questions that must be addressed: After the watch is stopped, can it be read off with precision? Did the stopwatch run accurately? Was it started at the right time? Dating processes that utilize radioactivity are governed by the same concerns. To read off a nuclear clock with precision, it is necessary to compute the ratio $\frac{z(t)}{y(t)}$ with great precision in samples that often involve only microgram quantities of mineral grains. High-resolution mass spectrometers can count the number of atoms in such minute quantities with great accuracy. So if contamination of the sample during the measurements is avoided or minimized, the ratio $\frac{z(t)}{y(t)}$ can be computed. The second concern is the accuracy of the clock. If a meaningful measurement of t is to be made, the sample being tested must have remained closed to the addition or escape of both parent and daughter atoms since the time of its formation. If this is not the case, the ratio $\frac{z(t)}{y(t)}$ will not lead to the true age. Finally, as to the matter of starting the clock at the right time, it must be possible to correct for atoms identical to daughter atoms that might already have been present when the mineral grain formed. All these are delicate matters and challenges, but there are lots of rocks that contain grains of radioactive minerals for which these challenges can be met. Not surprisingly, this requires the integrated effort of geologists, chemists, and physicists.

The decay process

$$^{87}_{37}\text{Rb} \rightarrow \,^{87}_{38}\text{Sr} + \text{a beta particle}$$

of rubidium-87 to strontium-87 was the first to which this method was widely applied. It has provided much information about the geologic history of the Earth. For example, meteorites can be dated with the rubidium-strontium clock. Rubidium-87 is a relatively abundant element in the Earth's crust and can be found in many common rock-forming minerals. It has a half-life of about 4.7×10^{10} years, and its strontium-87 daughter is stable. The sample prepara-

tion is relatively easy. Typically, rocks believed to have been part of a single homogeneous liquid prior to solidification are collected in samples weighing several kilograms. The samples are crushed, and a fine powder is produced. From this, a fraction of a gram is withdrawn and dissolved, and strontium and rubidium are extracted.

Example 11.7. A sample of the mineral *biotite* was taken from granite in the Grand Canyon. It was found to contain 202 parts per million of rubidium-87 and 3.97 parts per million of the daughter element strontium-87. When was the sample formed?

This is a simple application of the formula $t = \frac{1}{\lambda} \ln \left(\frac{z(t)}{y(t)} + 1 \right)$. Use of the relationship $h = \frac{\ln 2}{\lambda}$ shows that $\frac{1}{\lambda} = \frac{h}{\ln 2} = \frac{4.7 \times 10^{10}}{0.693} = 6.8 \times 10^{10}$. Therefore,

$$t = (6.8 \times 10^{10}) \ln \left(\frac{z(t)}{y(t)} + 1 \right) \qquad (11\text{e})$$

Since the ratio of daughter to parent is known to be $\frac{z(t)}{y(t)} = \frac{3.97}{202} = 0.0197$, we get that the sample was formed

$$t = (6.8 \times 10^{10}) \ln 1.0197 = 1.33 \times 10^9$$

$$= 1.33 \text{ billion years ago.}$$

It is now generally recognized that the uranium-to-lead decay process provides the most precise age information, at least in the high age ranges under consideration now. One reason why uranium-lead dating is superior to other isotopic methods is simple. There are two uranium-lead clocks in a sample: the uranium-238 clock and the uranium-235 clock. Therefore, two uranium-lead ages can be calculated for every analysis, and the readings of the two clocks can be compared. Another reason is the discovery that many rock units contain high-quality mineral grains that have retained closed uranium-lead systems in their interior parts. The mineral *zircon* is particularly relevant here. This mineral resists change to its internal structure even at high temperatures. It can block daughter loss and encode the primary age even if it remains hot, say at 1500° Fahrenheit, for a long time. In other words, it contains uranium-lead clocks that start early and are resistant to change. In advanced laboratories, samples that weigh only a few millionths of a gram can be isolated, and the uranium series can be analyzed by sophisticated spectrometry. Many types of rocks can be dated with relatively high degrees of accuracy in this way. For example, the ages of rocks 3 billion years old can be established within an accuracy of ±2 million years. The mineral grains from western Australia that were found to be between 4.1 and 4.3 billion years old were zircons dated by the uranium-lead clock. Uranium-lead dating was one important strategy used to put the moving jigsaw puzzle of Figure 11.3 together.

The mechanism of the uranium-to-lead clock is more delicate than that of the rubidium-to-strontium clock. The reason is simple. In rubidium-to-strontium decay, the daughter strontium-87 atoms are stable, and the computation of their number at any time t is routine. In uranium-to-lead decay, all the daughter products (with the exception of the lead produced at the last step) decay themselves. Since daughter atoms of each type are simultaneously being created and destroyed, the problem of computing how many there are at any time t is challenging. See the Exercises of Chapter 13.

B. The Potassium-Argon Clock.

One advantage of both the rubidium-strontium and uranium-lead clocks is that once they start, they are often able to resist the heat and pressure that geologic processes generate. However, since they resist change, they are unable to provide information about the later development of the mineral. There is another radioactive clock, based on the potassium-40 to argon-40 decay process, that is more sensitive to change and more easily disturbed. It can therefore provide information about a rock's later stages of development.

Potassium is the seventh most abundant element in the Earth's crust and is a major constituent of many minerals. The chemical symbol of potassium is K (from its Latin name *kalium*). The isotope $^{40}_{19}$K is radioactive and constitutes about 0.012% of the total. There are two competing modes in which $^{40}_{19}$K decays:

$$^{40}_{19}K \rightarrow {}^{40}_{20}Ca + \text{a beta particle}$$

where $^{40}_{20}$Ca is an isotope of calcium, and

$$^{40}_{19}K + \text{an electron} \rightarrow {}^{40}_{18}Ar + \text{gamma rays}$$

where $^{40}_{18}$Ar is an isotope of the gas argon. Both of these decay products of potassium-40 are stable. Measurements have established that about 88.97% of the potassium-40 atoms decay to calcium and the remaining 11.03% to argon.

Suppose a mineral grain containing potassium was formed, in other words, cooled and crystallized, at a certain time that we label $t = 0$. At this time the radioactive $^{40}_{19}$K that it contains will begin to decay. The decay to calcium is generally not useful, since calcium is often already present in the mineral. However, in most cases, argon $^{40}_{18}$Ar is not initially present. But it is created by the decay process and becomes entrapped in the grain. Let $y(t)$ be the number of $^{40}_{19}$K atoms in a fragment of the grain after any time $t \geq 0$. We know that

$$y(t) = y_0 e^{-\lambda t}$$

where $y_0 = y(0)$ is the initial number of $^{40}_{19}$K atoms in the fragment and λ is the decay constant of $^{40}_{19}$K. Notice that $y_0 - y(t)$ of these atoms will have decayed. Of these, 11.03%, or

$$z(t) = 0.1103(y_0 - y(t))$$

will have become $^{40}_{18}$Ar atoms. Solving this equation for y_0, we get $y_0 = 9.07z(t) + y(t)$. By a substitution, $y(t) = (9.07z(t) + y(t))e^{-\lambda t}$, and hence $\frac{y(t)}{9.07z(t)+y(t)} = e^{-\lambda t}$. Therefore,

$$\frac{9.07z(t)}{y(t)} + 1 = \frac{9.07z(t) + y(t)}{y(t)} = e^{\lambda t}.$$

Taking natural logs of both sides shows that $\lambda t = \ln e^{\lambda t} = \ln\left(\frac{9.07z(t)}{y(t)} + 1\right)$. Hence

$$t = \frac{1}{\lambda} \ln\left(\frac{9.07z(t)}{y(t)} + 1\right).$$

The half-life of potassium-40 is known to be $h = 1.31 \times 10^9$ years, and its decay constant is equal to $\lambda = \frac{\ln 2}{h}$. So $\frac{1}{\lambda} = \frac{h}{\ln 2}$, and therefore, $\frac{1}{\lambda} = \frac{1.31 \times 10^9}{0.693} = 1.89 \times 10^9$. It follows that

$$t = (1.89 \times 10^9) \ln \left(\frac{9.07 z(t)}{y(t)} + 1 \right) \qquad (11f)$$

where, to repeat, $y(t)$ is the number of potassium-40 atoms and $z(t)$ the number of argon-40 atoms in the sample after time t. By measuring the quantities $y(t)$ and $z(t)$ and inserting them into the formula, one obtains an estimate of the time t that has elapsed since the formation of the mineral grain.

Observe that the principle behind the potassium-argon clock is in essence the same as that of the rubidium-strontium clock. But there are subtle and important differences. For example, the measurement of the quantities $y(t)$ and $z(t)$ of the parent potassium-40 and the argon-40 daughter is more delicate and involved. The most important difference is the (hardly surprising) fact that gaseous argon-40 is much more easily dislodged from a mineral grain by the high temperatures of geologic events than is the strontium generated by the decay of rubidium. A number of minerals (for example, biotite, feldspar, and hornblende) release such argon when subjected to temperatures beyond $500°$ Fahrenheit. This means that a geologic event that generates temperatures beyond such a level will reset the potassium-argon clocks of the particular mineral grain. This fact and the general abundance of potassium makes the potassium-argon clock a very versatile tool with wide-ranging applicability.

Potassium-argon clocks have been used to establish the ages of some samples of volcanic ash as falling into a range as low as $215,000 \pm 4,000$ years. They have also shown that certain lava formations cooled as many as 3.5 billion years ago. The evidence that suggests that between 80 and 85 percent the Earth's atmosphere was formed about 4.4 billion years ago in the relatively short time frame of 1 million years is based in part on the measurement of potassium-argon decay. Potassium-argon clocks have also supplied direct evidence for the phenomenon of continental drift. When basalt cools and hardens, it becomes magnetized by the Earth's magnetic field. Minerals within the basalt align themselves with the planet's north and south poles, becoming compass needles that are locked into place when the cooling process occurs. It turns out that some basalt has a northern orientation and some a southern orientation. By dating many samples of basalt in different parts of the globe with potassium-argon clocks, scientists have discovered that the Earth's magnetic field has undergone nine sudden reversals from south to north during the last 4 million years. The analysis and dating of lava from the sea floor has allowed these pole flips to be traced there. The study of the pattern confirms the phenomenon of sea floor spreading and hence of continental drift.

11.3 Life Evolves: A Timeline

Let's turn next to the question of life on Earth! Scientists believe that life began about 4 billion years ago soon after the Earth began to cool. We have already observed that fissures and cracks in the seafloor vented volcanic lava into the oceans. In addition to water, the resulting mix contained carbon, hydrogen, methane, and ammonia, among other chemicals. The temperatures ranged from the $1000°$ Fahrenheit of the molten lava, to the $600°$ of the surrounding water, to the $20°$ of the water some distance away. Many kinds of reactions took place in this potent mix, and an inevitable rapid chemical evolution occured. Life began in this enormous laboratory of nature as a consequence of some of the myriad of interactions in this organic chemical brew. There are several plausible speculations (but no definitive accounts) about the biochemical processes by means of which nature may have built the maze of bridges that led from inanimate chemicals to living cells.

The record of early life appears to have been destroyed by geologic activity. The earliest pieces of evidence not in dispute are fossilized remains of blue-green algae (similar to pond scum), the first ancestors of plant life. These were dated by uranium clocks in zircon crystals as being about 3.5 billion years old. Early forms of bacteria also date from this time. Relics of these single-celled microbes have survived to this day, thriving in geysers and volcanic vents in temperatures of up to 235° Fahrenheit (hotter than the 212° of boiling water).

One of the most important factors in the development of more complicated life forms was the composition of the atmosphere. The early atmosphere was most certainly dominated by carbon dioxide. There was little oxygen. Oxygen levels remained low until about 2 billion years ago. Recent studies point to a sudden increase in oxygen between 2.1 and 2 billion years ago (and suggest that the present situation of an oxygen content of 20% was reached about 1.5 billion years ago). Only then did it become possible for more complicated multicelled organisms to develop. Fossil finds confirm that such organisms (leaflike plants that lived on the seafloor) existed about 1.7 billion years ago. After oxygen became more abundant, the oxygen compound ozone began to form. This served the important purpose of blocking the hostile ultraviolet radiation that breaks down many molecules. About 530 million years ago, there was a burst of developments. In an astonishingly brief time, corals, sponges, the first invertebrates, and other more complex life forms made their appearance in what is now known as the *Cambrian Explosion*. Later, life moved onto land; insects, amphibians, and reptiles developed; and so forth. See Figure 11.4 for an overview of the entire picture.

The analysis of the fossil record substantiates this picture of the development of life on our planet. If an organism is covered by soils or sands soon after death, some of its remains can survive the forces of decay and can become entombed in a layer of sediment by the geologic forces of erosion and sedimentation. Such remains are called *fossils*. They are either the remains themselves—for example parts of skeletons or bone fragments—or certain impressions or traces of them.

These can take different forms. Some organic materials, such as bones, shells, and wood, can literally turn into stone. Mineral matter from underground solutions can fill the microscopic voids in such structures. Indeed, these minerals can also replace the original material, replicating the microscopic structures molecule by molecule. Or, in the absence of oxygen, organic tissue can undergo a process that liberates carbon dioxide and water until only free carbon remains. This carbon forms a black film in the rock, a "carbon copy" of the animal or plant. Leaves are copied in this way, and, more rarely, the outline of an animal's fleshy body is preserved, at times surrounding its fossilized skeleton.

By this process parts of the fauna and flora of an age become entombed in a layer of stone. Successive ages form successive strata, each different from the next, and these strata record a history of the development of plant and animal life. In such sequences of layers the same or similar fossils are found to occur in identical order in different geographic locations. By investigating certain "index fossils" (common index fossils are certain microscopic algae, pollen, spores, shells, bones and teeth of vertebrate animals, or remains of primitive humans), it is possible to compare and correlate the layers and to synthesize local and regional details into a broader global picture. In this way, a continuous and composite record of the development of life on Earth can be pieced together. In the lowest layers of this *geologic column* lie the earliest life forms and in the upper ones those that have more in common with current forms. In this way, the history of the development of life on the planet can be pieced together and recorded. The fossil record verifies that life is like a treadmill. Species climb on, revolve on it for a while, then disappear completely and never return. It also gives evidence of progressive change within given animal groups and to the mechanisms of evolutionary development. The fossil record extends over 3.5 billion years. In general, it is more extensive and better understood for younger eras, say the last 600 million years, or about 13% of the Earth's history, because more sedimentary rock has survived the pressures of

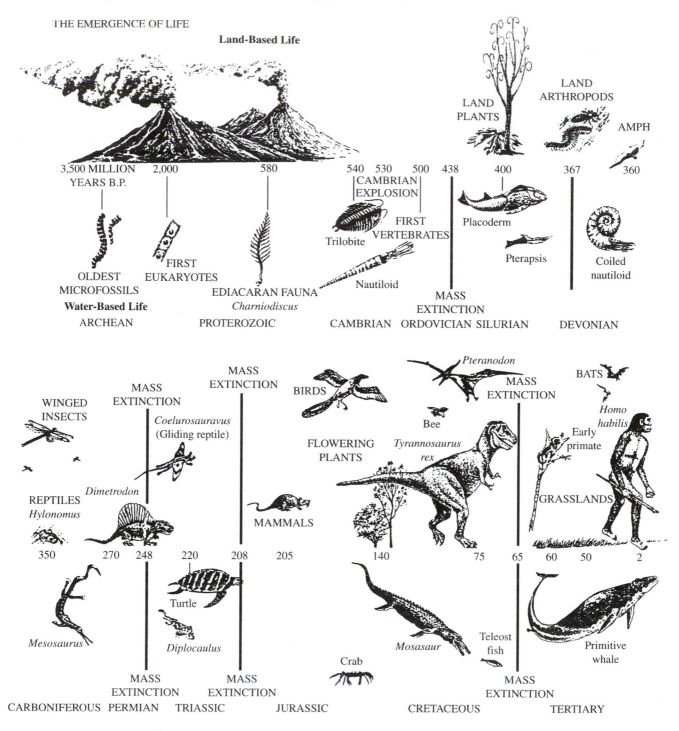

Figure 11.4 From *Life in the Universe,* Scientific American, Special Issue, October 1994, pages 46–47.

geologic processes. The fossil record is more complete for marine environments, which are sites of net deposition of sediment, than for continental environments, which are sites of net erosion.

Nuclear clocks, especially the potassium-argon clock, have provided a timeline for the fossil record. This time line has been incorporated into Figure 11.4. Nuclear clocks have also been instrumental in the process that has made it possible for local fossil records to be pieced together into a fossil record of regional and global scale.

We now turn to the record of human life on our planet. The first humanlike beings—called *Australopithecus* (*australo* is Latin for *southern*, and *pithecus* is from the Greek for *ape*) by anthropologists (*anthropos* is Greek for *man*, and *logos* is Greek for *study*)— appear in the fossil record some 4 million years ago. The species *Homo habilis* (this is Latin for the "handy man" and suggests that he had the ability to shape and use tools) and the species *Homo erectus* (the "upright man" who is thought to have had language ability) occur later in the range from about 2 to 0.5 million years ago. These "links" between ape and human have humanlike bodies typically weighing from 130 to 170 pounds, human skeletal structures, upright postures, teeth of human form, and brains that were relatively large compared to body size.

Many of the most interesting and complete fossils come from East Africa, especially Kenya, Ethiopia, and Tanzania. The fact that the fossil fields of East Africa contain isotope-rich volcanic ash deposits has made it possible to use radioactive clocks, especially the potassium-argon clock (but also the uranium-lead clock) to date the finds. This is based on the reasonable assumption that a fossilized bone dates from about the same time as the volcanic sediment in which it is embedded. The discoveries from East Africa include a nearly complete australopithecus skull shown to be 1.75 million years old. The most remarkable find was the partial skeleton (nearly half of the bones) of an australopithecus female. This skeleton was discovered in 1974. Later dubbed "Lucy" by anthropologists, it was shown to be about 3.75 million years old.

Example 11.8. The fossil Lucy was embedded in sediment that contained volcanic ash. The potassium that this ash contained made it possible to determine the age of the skeleton. Approximate the fraction of the original number of potassium-40 atoms in a fragment of this ash at the time of the discovery.

The age of the fossil was determined by assuming that the sediment was formed at about the same time as Lucy's demise. So the formation of the sediment occurred about 3.75 million years ago. This is time $t = 0$. Let y_0 be the number of potassium-40 atoms in the fragment of the ash at that time. Recall from Section 11.2B that the half-life of potassium-40 is $h = 1.31 \times 10^9$ years. So the decay constant is $\lambda = \frac{\ln 2}{h} = \frac{0.693}{1.31 \times 10^9} = 5.29 \times 10^{-10}$. Therefore, there are

$$y(t) = y_0 e^{-\lambda t} = y_0 e^{-5.29(10^{-10})t}$$

potassium-40 atoms in the fragment t years after the eruption. Taking $t = 3.75 \times 10^6$, we see that the fragment contained

$$y_0 e^{-5.29(10^{-10})3.75(10^6)} = y_0 e^{-1.98(10^{-3})} = 0.998y_0$$

potassium-40 atoms at the time of the discovery. So the fragment had retained about 99.8% of the original potassium-40 atoms. What fraction of the atoms that decayed, decayed to argon-40? Note that $0.002y_0$ potassium-40 atoms disintegrated. It follows from information in Section 11.2B, that about 11% of these, or $0.11(0.002y_0) = 0.0002y_0$, decayed to argon-40.

Example 11.9. Crude stone tools were uncovered in fossil fields in Ethiopia. The tools were dated to be between 2.5 and 2.6 million years old. They were, at the time of their discovery in the early 1990s, the earliest known artifacts created by ancestors of humans. The age of the tools was estimated by applying the potassium-argon method to grains of the volcanic sediment in which they were embedded. Give an estimate of the ratio of the number of argon-40 atoms to the number of potassium-40 atoms in a grain of this ash at the time of the discovery of the tools.

As in the earlier example, we assume that the manufacture of the tools and the formation of the grains occured at the same time, $t = 0$. Recall from Section 11.2B that the connection between the number t of years that elapsed since the formation of the grain and the numbers $z(t)$ of argon-40 atoms and $y(t)$ of potassium-40 atoms that it contained at the time of the discovery is given by

$$t = (1.89 \times 10^9) \ln \left(\frac{9.07z(t)}{y(t)} + 1 \right).$$

We know that at the time of discovery of the tools, t was in the range $2.5 \times 10^6 \le t \le 2.6 \times 10^6$. Therefore,

$$2.5 \times 10^6 \le (1.89 \times 10^9) \ln \left(\frac{9.07z(t)}{y(t)} + 1 \right) \le 2.6 \times 10^6.$$

Hence, $1.32 \times 10^{-3} \le \ln \left(\frac{9.07z(t)}{y(t)} + 1 \right) \le 1.38 \times 10^{-3}$. Since the exponential function e^x is an increasing function, we get

$$e^{1.32(10^{-3})} \le \frac{9.07z(t)}{y(t)} + 1 \le e^{1.38(10^{-3})}.$$

So $1.0013 \le \frac{9.07z(t)}{y(t)} + 1 \le 1.0014$. Hence

$$0.0013 \le \frac{9.07z(t)}{y(t)} \le 0.0014.$$

After dividing by 9.07, we see finally that the ratio $\frac{z(t)}{y(t)}$ lies between 0.00014 and 0.00015.

Examples 11.8 and 11.9 both give indication of the subtle nature of the process of reading potassium-argon clocks. In Example 11.8, we saw that only $\frac{1}{50}$ of one percent of the original potassium decayed to argon. Example 11.9 shows how carefully the quantities $z(t)$ and $y(t)$ must be measured in order to provide an accurate assessment of the age of the grain.

11.4 Dating the More Recent Past

In December 1994 explorers discovered a cave in southern France, now called the Chauvet cave, containing wall paintings and engravings of amazing vitality and power.[3] They depict horses, bison, rhinos, lions, panthers, owls, and mammoths, in masterful strokes and wonderful hues of reds, grays, and ochres. Over three hundred animals in all move and leap across several great panels that stretch along the walls of the cave. Their subject, similar to that of other cave paintings, suggested that this was prehistoric art. But how old was it? At what point in their development were humans able to execute such works with so much confidence and flair?

Between 1947 and 1956 some eight hundred manuscripts were found in caves near the Dead Sea in the Middle East. Later known as the "Dead Sea Scrolls" they include the oldest known manuscripts of books from the Hebrew Bible. Their discovery is considered one of the great archeological finds of the century. The scrolls are thought to be the work of a Jewish sect that lived near the caves between 200 B.C. and 100 A.D. According to another school of thought, however, at least some of the scrolls were written by early Christians and are said to describe a messianic movement that was virtually indistinguishable from the rise of Christianity. This "Christianization" of the scrolls outraged some scholars. Are they Jewish or Christian manuscripts? Could the controversy be resolved?

An old piece of cloth known as the Shroud of Turin (because it has been preserved in the cathedral of this Italian city since 1578) has a negative image of the head and body of a man impressed upon it. The image was thought to be that of Christ. Indeed, the cloth was considered sacred and believed to be Christ's burial cloth. Could this be true? Was the cloth old enough to have originated from the time of Christ?

Can these questions of obvious interest to anthropologists and scholars of religion be answered? Can the time of origin of these articles be determined? The radioactive clocks already considered cannot be used. The half-lives of rubidium-87, uranium-238, uranium-235, and potassium-40 all range between 1 and 50 billion years. Therefore, the clocks that rely on these elements operate much too slowly to age objects that are only a few thousand years old. What is needed is a radioactive substance with a half-life of appropriate length that is abundant enough to be a part of the fabric of the articles whose ages we wish to determine. The American scientist Willard F. Libbey (1908–1980), while at the University of Chicago in the late 1940s, discovered an element that was ideal for the purpose and analyzed the mechanism of the clock that it provided. His accomplishment was rewarded in 1960 with the Nobel Prize for chemistry.

The element carbon in its naturally occuring state on Earth consists of a mix of different isotopes. The most common is carbon-12. It has 6 protons and 6 neutrons in its nucleus and is designated $^{12}_{6}C$. Carbon-12 is stable. Another is $^{13}_{6}C$. It has 7 neutrons and 6 protons and is also stable. Then there is carbon-14. Its designation $^{14}_{6}C$ shows that it has 8 neutrons to go with its 6 protons. Carbon-14 is radioactive. The force binding its nucleus together is not strong enough to hold all the neutrons in place. In a sample of pure carbon 99% consists of carbon-12 and most of the rest is carbon-13. Carbon-14 only occurs in very minute traces: on average only one of 6.463×10^{11} carbon atoms is a radioactive carbon-14. These minute traces of carbon-14 form the mechanism of a very useful radioactive clock. The first indication that this clock might be useful is the fact that the half-life of carbon-14 is 5730 years. This is roughly of the same order of magnitude as the ages of the objects that we wish to determine.

A carbon-14 atom disintegrates according to the scheme

$$^{14}_{6}C \rightarrow {}^{14}_{7}N + \text{a beta particle} + \text{a neutrino},$$

where $^{14}_{7}N$ is a nitrogen atom. The neutrino is a curious particle. It has an extremely small mass if any (the question of the mass of the neutrino is one of the vexing problems of particle physics), travels at the speed of light, and passes very easily through everything that it encounters.

There is a process that replenishes the carbon-14 that is lost to disintegration. Cosmic rays generate high-energy neutrons that bombard nitrogen high in the atmosphere and produce the reaction

$$^{14}_{7}N + \text{a neutron} \rightarrow {}^{14}_{6}C + \text{a proton}.$$

The carbon-14 that is continually being produced in the atmosphere in this way enters the carbon reservoir on the Earth's surface and replaces the carbon-14 that disintegrates. Amazingly, the two processes just described are in equilibrium: in naturally occurring carbon on Earth the ratio of carbon-14 atoms to stable carbon atoms is maintained at about $r_0 = \frac{1}{6.463 \times 10^{11}}$.

Carbon exists throughout the atmosphere, primarily as a component of carbon dioxide. It is taken up by all living organisms. It is absorbed in the growth process by all cells and used up in a continuous cycle (the so-called carbon cycle) and exchange. So any living organism will contain carbon, and this carbon will contain carbon-14 and stable carbon in the equilibrium ratio of r_0. When an organism dies, the metabolic process stops, and the carbon cycle stops. The carbon-14 that is locked into the tissues of the organism at that time will disintegrate. As a consequence, the ratio of carbon-14 atoms to stable carbon atoms will fall progressively below the equilibrium ratio. When this smaller ratio is measured and compared to the equilibrium ratio $r_0 = \frac{1}{6.463 \times 10^{11}}$, the time of death of the organism can be estimated. This is done as follows.

Suppose that an organism died an unknown number of years ago. Designate the time of death by $t = 0$ and let t be the number of years that have passed since that time. Consider a fragment of the remains. Let y_0 be the initial number of carbon-14 atoms in the fragment and let $y(t)$ be the number of carbon-14 atoms remaining when it is tested at time t. By the theory of Section 11.1B, $y(t) = y_0 e^{-\lambda t}$, where λ is the decay constant of carbon-14. Notice that $y(t)e^{\lambda t} = y_0$ and hence that $e^{\lambda t} = \frac{y_0}{y(t)}$. After taking natural logs of both sides, we have $\lambda t = \ln\left(\frac{y_0}{y(t)}\right)$, and therefore

$$t = \frac{1}{\lambda} \ln\left(\frac{y_0}{y(t)}\right).$$

Since the half-life of carbon-14 is $h = 5730$ years and $h = \frac{\ln 2}{\lambda}$, it follows that $\frac{1}{\lambda} = \frac{h}{\ln 2} = \frac{5730}{0.693} = 8.26 \times 10^3$. Can we use this formula together with an analysis of the fragment to determine t? Since $y(t)$ is the number of carbon-14 atoms in the fragment at the time of the analysis, this measurement is no problem in principle. But what about the original number y_0 of carbon-14 atoms? Our earlier discussion provides the key. As already pointed out, at the time of death the fragment contains carbon-14 atoms to stable atoms in the equilibrium ratio of r_0. This means that if k is the number of stable carbon atoms in the fragment (observe that this number does not change with time), then $\frac{y_0}{k} = r_0$, and hence $y_0 = r_0 k$. After inserting $y_0 = r_0 k$ into the earlier expression for t, we get the formula

$$t = (8.26 \times 10^3) \ln \left(r_0 \frac{k}{y(t)} \right) \qquad (11\text{g})$$

Therefore the time t that has elapsed since the death of the organism can be determined by measuring the ratio $\frac{k}{y(t)}$ of the number of stable carbon atoms to carbon-14 atoms in the fragment. The modern accelerator mass spectrometer can count the number $y(t)$ of carbon-14 atoms and the number k of stable carbon atoms in a tiny fragment of a sample with great precision.

Example 11.10. A wood fragment tested with an accelerator mass spectrometer is found to contain carbon-14 atoms to stable carbon atoms in a ratio of 1 to 1.744×10^{12}. How old is the fragment?

Note that $\frac{k}{y(t)} = 1.744 \times 10^{12}$. So

$$r_0 \frac{k}{y(t)} = \frac{1.744 \times 10^{12}}{6.463 \times 10^{11}} = 2.698.$$

By substituting into the formula just derived, we get

$$t = (8.26 \times 10^3)(\ln 2.698) = (8.26 \times 10^3)(0.993)$$

$$= 8.2 \times 10^3 = 8200 \text{ years.}$$

The accuracy of the carbon-14 clock depends on the answer to a crucial question. Was the equilibrium ratio of carbon-14 to stable carbon in the past the same $r_0 = \frac{1}{6.463 \times 10^{11}}$ that it is today? Observe that the accuracy of the estimate of the age of the wood fragment in Example 11.10 depends directly on the assumption

that the equilibrium ratio had this value 8,200 years ago. It is now known that r_0 has varied over time. One reason is the fact that carbon-14 production in the atmosphere fluctuates in accordance with fluctuations in the amount of cosmic radiation. How is it possible to reach an understanding of the variation in the carbon-14 level over time? This comes from an analysis of trees! In one season of growth, any tree will add a recognizable outer layer of wood to its trunk. As a consequence, the age of a tree is equal to the number of rings in the crossection of its trunk. For example, some bristle cone pines have been shown to be as old as 8,200 years by a careful study of their trunks. The next useful fact is that only a few of the outer layers of a living tree are metabolically alive. The others are subject to carbon-14 decay. This has the following consequence. Take a sample from an inner ring of a tree of known age t and measure the ratio $\frac{k}{y(t)}$ of stable to radioactive carbon. Turning to the formula $t = (8.26 \times 10^3) \ln \left(r_0 \frac{k}{y(t)} \right)$ and solving for r_0 provides the value of r_0 at time t. By correlating and comparing the information obtained from many trees, an understanding of the way r_0 has fluctuated over time is achieved. For instance, 8000 years ago, r_0 was about 8% less than it is today. In recent years, r_0 has been affected by the industrial emission of carbon-14-free carbon dioxide into the air, and on the other by the injection of carbon-14 into the air as a result of nuclear tests. This, fortunately, has no effect on the dating process of older samples.

Example 11.11. Let's refine the estimate of the age of the wood sample of Example 11.10. Taking the equilibrium ratio r_0 to be 8% less than the current $\frac{1}{6.463 \times 10^{11}}$, we get

$$r_0 \frac{k}{y(t)} = \frac{0.92}{6.463 \times 10^{11}} (1.744 \times 10^{12})$$

$$= (0.92)(2.698) = 2.482.$$

Therefore,

$$t = (8.26 \times 10^3)(\ln 2.482) = (8.26 \times 10^3)(0.909)$$

$$= 7.5 \times 10^3 = 7500 \text{ years.}$$

Now what about the Chauvet cave? The Center for Radiocarbon Dating of the University of Lyons, France, and the Research Laboratory for Archeology and Art History, at Oxford, England, determined that some of the charcoal pigment used in the paintings is between 30,340 and 32,410 years old. It makes them the oldest known paintings in the world. This prompted the French Culture Ministry to assert that the age of these paintings "overturned the accepted notions about the first appearance of art and its development."

What about the controversy surrounding the Dead Sea Scrolls? In 1995 the Accelerator Mass Spectrometry Laboratory of the University of Arizona tested fingernail-size pieces of parchment and papyrus samples cut from the edges of the margins of some scrolls for their carbon-14 content. The result? The scrolls originate in the period between 150 B.C. and 5 B.C. This finding, therefore, discredits the theory that these texts were the work of early Christians. The same laboratory also tested the Shroud of Turin in the 1980s. The conclusion? The shroud was made in the Middle Ages, somewhere between 1260 and 1390 A.D.

Example 11.12. Estimate the ratio of carbon-14 atoms to stable carbon atoms in the papyrus fragments of the Dead Sea Scrolls at the time that they were tested by the lab at the University of Arizona.

Let t be the age of the sample of the scroll at the time of the analysis in 1995. Since the fragment dates from the period between 150 B.C. to 5 B.C., notice that $2000 \leq t \leq 2150$. Since t is also given by the formula $t = (8.26 \times 10^3) \ln\left(\frac{r_0 k}{y(t)}\right)$, we get

$$2000 \leq (8.26 \times 10^3) \ln\left(\frac{r_0 k}{y(t)}\right) \leq 2150$$

and hence

$$0.242 \leq \frac{2 \times 10^3}{8.26 \times 10^3} \leq \ln\left(\frac{r_0 k}{y(t)}\right) \leq \frac{2.15 \times 10^3}{8.26 \times 10^3} \leq 0.261.$$

Applying the exponential function e^x to both sides shows that

$$1.273 \leq e^{0.242} \leq \frac{r_0 k}{y(t)} \leq e^{0.261} \leq 1.299$$

and hence that

$$\frac{1.273}{r_0} \leq \frac{k}{y(t)} \leq \frac{1.299}{r_0}.$$

Assume that $r_0 = \frac{1}{6.463 \times 10^{11}}$. Since $1.273(6.463 \times 10^{11}) = 8.227 \times 10^{11}$ and $1.299(6.463 \times 10^{11}) = 8.395 \times 10^{11}$ the papyrus fragment tested by the lab contained roughly 8.3×10^{11} stable carbon atoms for every carbon-14 atom.

The delicate nature of carbon-14 dating should be obvious from the example. Notice that the window of error is $32,410 - 30,340 = 2,070$ years for the paintings in the Chauvet cave, $150 - 5 = 145$ years for the Dead Sea Scrolls, and $1390 - 1260 = 130$ years for the shroud of Turin. The fact that the margin for error is greater in older objects parallels the earlier point about the element of uncertainty introduced by the fluctuation in carbon-14 levels over time. The carbon-14 method is used in estimating the ages of substances that date anywhere from 50,000 B.C. to 1,500 A.D. Outside this range this method is ineffective.

Reflect over what has been done in this chapter so far. We started with Rutherford's measurements of the ratios of the decay rates of a radioactive substance and his realization that the logarithms of these ratios are organized in a linear pattern. This linear pattern together with the basic properties of the exponential function made the mathematical simulation of the decay process possible. This mathematical simulation, in turn, provided us with a precise quantitative understanding of the works of the different radioactive clocks. We have seen that such clocks are embedded everywhere on the surface of the Earth. And we have also seen how scientists have read them off and used them to piece together an incredible amount of information about the history of our planet and its inhabitants.

We will now do a similar thing for the growth of populations. In principle, this analysis will apply to the growth dynamics of any population: people in a city or a country, cells in a culture of microorganisms, or individuals of a species of animal in a given habitat. This situation is more complicated than that of radioactive decay. The reason is simple: a radioactive substance—not affected by outside factors—always exhibits the same pattern of decay: a population, on the other hand—influenced by environmental factors—exhibits changing patterns of growth.

11.5 The World's Population

The following strategy is often useful in the study of populations and the dynamics of their growth. A determination is made of a function that represents the statistical data for the population in at least an approximate way. If the function that is constructed is differentiable (so that it has a smooth graph), then its "flow" can clarify the patterns of growth and decline of the population, give indications of general trends, and provide insights into its behavior in the future. We illustrate this strategy for the human population.

A. Statistics and Trends. We begin with a look at the numerical progression of the world's population. The data in the accompanying tables are taken from estimates published in 1992 by the Population Reference Bureau, in Washington D.C.

The world's population increased slowly until about 1650 A.D. when approximately 500 million people inhabited the Earth. During the next 100 years it increased to about 760 million. Population statistics became available from the 1750s onward, and Table 11.2 lists a cross section of these. It provides estimates for the world's population from 1750–1900 in five-year intervals. The years listed are representative of the rest. In each group of data, the first column lists the year, the second column the population (in millions) in that year (for the sake of the discussion, take it to be that at the beginning of the year), the third is the increase (in millions) during that year, and the fourth is the percentage of this increase. Consider the year 1825, for example. The world's population was 1,100 million (or 1.1 billion) at the beginning of the year, and increased by about 5 million during the year. Since

$\frac{5}{1,100} = 0.005$ and 0.5% corresponds to $\frac{0.5}{100} = 0.005$, this is the 0.5% increase listed in Table 11.2 for the year 1825.

Let t be the elapsed time in years from 1750. For example, $t = 0$ corresponds to the year 1750, $t = 35$ corresponds to 1785, and $0 \leq t \leq 130$ corresponds to the period from 1750 to 1880. Let $y(t)$ be a differentiable function defined for any t with $0 \leq t \leq 130$. Let's suppose that $y(t)$ represents the world's population at least in an approximate way. What properties must $y(t)$ satisfy? For example, since the population in 1785 was 921 million, we see that $y(35) \approx 921$. The world's population increased by about 5 million during 1785. So it increased at a rate of about 5 million per year during that year. Since $y(t)$ represents the population at time t, we know that $y'(t)$ represents its rate of increase. (This is discussed in Section 8.4.) It follows that $y'(35) \approx 5$. Therefore, $\frac{y'(35)}{y(35)} \approx \frac{5}{921} \approx 0.005$. Notice that $\frac{y'(35)}{y(35)} \times 100 \approx 0.5\%$ is the percentage increase for the year 1785. See Table 11.2. Our focus on the year 1785 has shown that $y'(35) \approx 0.005y(35)$. Since the population increased by 0.4% in 1815 and 0.6% in 1850, we see in the same way that $y'(65) \approx 0.004y(65)$ and $y'(100) \approx 0.006y(100)$. Note from Table 11.2 that the world's population increased on average by about 0.5% per year from 1750 through 1880. This fact provides the approximation $y'(t) \approx 0.005y(t)$ for $0 \leq t \leq 130$.

Can we determine a specific function $y(t)$ that simulates the world's population for the entire period from 1750 to 1880? The analysis just undertaken shows that if

$$y'(t) = 0.005y(t) \quad \text{for} \quad 0 \leq t \leq 130,$$

then $y(t)$ approximates the trend observed in Table 11.2. So the question is this: Does a function $y(t)$ that satisfies the equation

$$\boxed{y'(t) = \mu y(t)} \tag{11h}$$

where μ is a constant, have a specific form that can be determined? The answer is yes! It is an exponential function. Start with the function $e^{\mu t}$. After recalling

Table 11.2

World Population Estimates 1750–1900 (in millions)															
1750	760	4	0.5	1790	947	5	0.5	1830	1,122	4	0.4	1870	1,363	8	0.6
1755	781	4	0.5	1795	973	5	0.5	1835	1,143	5	0.4	1875	1,404	8	0.6
1760	803	4	0.5	1800	1,000	4	0.4	1840	1,165	5	0.4	1880	1,446	9	0.6
1765	825	5	0.6	1805	1,019	4	0.4	1845	1,188	5	0.4	1885	1,490	9	0.6
1770	848	5	0.6	1810	1,039	4	0.4	1850	1,211	7	0.6	1890	1,534	10	0.7
1775	872	5	0.6	1815	1,059	4	0.4	1855	1,247	8	0.6	1895	1,580	10	0.6
1780	896	5	0.6	1820	1,080	4	0.4	1860	1,285	7	0.5	1900	1,628	11	0.7
1785	921	5	0.5	1825	1,100	5	0.5	1865	1,323	8	0.6				

basic facts from Section 10.1, we see that

$$\frac{d}{dt}e^{\mu t} = e^{\mu t}(\mu) = \mu e^{\mu t}.$$

Notice that $e^{\mu t}$ also satisfies equation (11h). This tells us that $y(t)$ and $e^{\mu t}$ are related. To see what the relationship is, consider $f(t) = \frac{y(t)}{e^{\mu t}} = y(t) \cdot e^{-\mu t}$. By the Product Rule,

$$f'(t) = y'(t) \cdot e^{-\mu t} + y(t) \cdot (-\mu)e^{-\mu t}$$

$$= (\mu y(t))e^{-\mu t} - y(t) \cdot \mu e^{-\mu t} = 0.$$

So $f'(t) = 0$, and it follows that $f(t)$ is a constant. (See Section 5.5B for instance.) Therefore, $y(t) \cdot e^{-\mu t} = C$ for some constant C, and hence $y(t) = Ce^{\mu t}$. By substituting $t = 0$, we get that $y(0) = Ce^0 = C$. We have shown that

$$y(t) = y_0 e^{\mu t} \qquad (11i)$$

where $y_0 = y(0)$.

In the situation of the world's population from 1750 to 1880, we know that $\mu = 0.005$. Taking $y_0 = y(0) = 760$, we obtain

$$y(t) = 760e^{0.005t}.$$

Since the derivation of this function is based on the statistics of the world's population from 1750 to 1880, it should simulate the world's population for that period. Let's check it against the population data of Table

11.2. In the year 1785 (this is $t = 35$) the table lists a population of 921 million, and the function provides $y(35) = 760e^{0.005(35)} = 760e^{0.175} = 905$ million. For the year 1820 (this is $t = 70$), we get a population of 1,080 million from the table and $y(70) = 760e^{0.35} = 1,078$ million from the function. For the year 1860 ($t = 110$), the table has 1,285 million, and $y(110) = 760e^{0.55} = 1,317$ million. Finally, for the year 1880 ($t = 130$), the table lists 1,446 million, and $y(130) = 760e^{0.65} = 1,455$ million.

As expected, the fit between the function $y(t) = 760e^{0.005t}$ for $0 \le t \le 130$ and the statistics for the period 1750–1880 is reasonably good. (The reason that it is not better is the fact that the underlying assumption of a 0.5% increase per year is only an approximation.) Since the population is described by an increasing exponential function, it is said to *grow exponentially*. This function also provides population estimates for (at least) some years beyond 1880. For example, the population in the year 1900 was 1,628 million. Substituting the corresponding $t = 150$ into the function gives the estimate $y(150) = 760e^{0.75} = 1,609$ million. It is common to refer to a mathematical simulation of the type that we have considered as a *mathematical model*.

Table 11.3 provides an overview of the world's population for the years 1900–1995. This table is organized in the same way as Table 11.2. The second column lists the population in millions, the third lists

Table 11.3

World Population Estimates 1905–1995 (in millions)

1905	1,683	12	0.7	1920	1,861	20	1.1	1950	2,517	45	1.8	1975	4,078	71	1.7
1910	1,741	11	0.6	1925	1,963	21	1.1	1955	2,752	51	1.9	1980	4,448	78	1.8
1915	1,800	12	0.7	1930	2,070	22	1.1	1960	3,019	61	2.0	1985	4,852	85	1.8
				1935	2,180	23	1.1	1965	3,337	69	2.1	1990	5,292	92	1.7
				1940	2,296	21	0.9	1970	3,698	73	2.0	1995	5,770	95	1.6
				1945	2,404	22	0.9								

the increase (in millions) during the year, and the fourth is the percentage of this increase. For the first two decades of the 20th century (in particular through the years 1914–1918 of World War I), the world's population increased by about 0.7% per year. In 1920, there was a sudden jump in the increase to 1.1% per year. This is explained by the slower death rates brought about by the fact that powerful drugs (for example, the sulfa drugs were discovered in 1910) were beginning to control diseases such as malaria and cholera. This effect was counterbalanced by the ravages of World War I and hence did not become a noticeable factor until 1920. The rate of 1.1% per year persisted through the 1920s and 1930s and diminished to 0.9% from 1940 to 1949 (and in particular through the years 1940–1945 of World War II). Because the population increases are substantially higher than 0.5%, the population figures for the years beyond 1900 deviate considerably from those given by the formula $y(t) = 760e^{0.005t}$. For example, for the year 1930, that is, for $t = 180$, the formula provides $y(180) = 760e^{0.9} = (760)(2.46) = 1,870$ million. Table 11.3, on the other hand, lists a population of 2,070 million.

In 1950 the increase jumped to 1.8%. This was the beginning of a population explosion. It is explained in large part by the discovery of penicillin, a substance that inhibits the growth of bacteria. It was discovered by accident by the Scottish physician and bacteriologist Alexander Fleming in 1928.[4] By the 1940s penicillin was being mass-produced, and it had a substantial impact on keeping diseases in check. The

increase in world's population continued to climb. It reached a maximum of 2.1% in 1965, where it remained until 1969.

By the year 1970 the rate of increase in the world's population had begun a steady decline. One important reason was the growth of industrialization in many less developed countries, which made it possible to replace "man by machine" and led to a decrease in birth rates. Notice that a trend becomes discernible. The percentage of the yearly increase decreased from the high water mark of 0.21 in 1965 to 0.20 in 1970, to 0.18 in 1980, and to 0.16 in 1995.

Suppose that the trend observed above were to continue. What implication would this have on the future growth of the world's population? Can the steady decline that we have observed be used as the basis for a mathematical model for the world's population that answers this question? This is indeed possible. The mathematical model in question was first developed by the Belgian mathematician Pierre François Verhulst in the 1830s. We will have a look at it next.

B. The Logistics Model. Consider any population. This can be a population of humans, bacteria, insects, and so on. Let $y(t)$ be the number of individuals at any time $t \geq 0$. Suppose that the population is increasing. So $y'(t) \geq 0$. The ratio

$$\frac{y'(t)}{y(t)}$$

is called the *specific growth rate*. It measures the rate of growth relative to the size of the population. Have a look back at the discussion of the previous section and notice that $\frac{y'(t)}{y(t)} \times 100$ is the percentage increase in the population at time t.

We saw in the previous section that the equation $y'(t) = 0.005y(t)$ was the basis for the mathematical model $y(t) = 760e^{0.005t}$ that simulates the world's population over the period 1750–1880. Notice that this is a situation where the specific growth rate $\frac{y'(t)}{y(t)}$ is a constant (the constant 0.005 to be exact). It is possible for a population to maintain such a trend over some stretches of time. This can occur if it has adequate resources (such as space and food) and is not subject to excessive depletion (as brought about by famine, war, and disease). However, if resources begin to stabilize or shrink (or if depletion becomes a factor), then one would expect that the specific growth rate can no longer be maintained at a constant level and that it will begin to decline. In other words, one would expect the specific growth rate to be a decreasing function of the population y. But what function?

Verhulst takes the following approach. He assumes that the specific growth rate decreases linearly with respect to population size. In other words, he assumes that $\frac{y'}{y}$ satisfies an equation of the form

$$\boxed{\frac{y'}{y} = \mu - ky} \qquad (11\text{j})$$

for some positive constants μ and k. The graph of this equation, taking population size on the horizontal axis and specific growth rate on the vertical, is a line of the form shown in Figure 11.5. Note that its slope is $-k$. Observe that μ is the largest value of the specific growth rate. Having made this assumption, Verhulst turns to the crucial questions. What does the equation $\frac{y'}{y} = \mu - ky$ tell us about the function $y = y(t)$? Does it determine the "shape" of the function? (A similar, but simpler, question was answered in Section 11.5A, where we saw that the equation $y' = \mu y$ forced $y(t)$ to have the form $y(t) = y_0 e^{\mu t}$.) After this matter is resolved, we will turn to the more important issue: Can the resulting function $y(t)$ be used to simulate the behavior of populations?

Let $M = \mu k^{-1}$. So $Mk = \mu$, and $k = \frac{\mu}{M}$. The equation $\frac{y'}{y} = \mu - ky$ can now be written as

$$\frac{y'}{y} = \mu - \frac{\mu}{M}y = \mu\left(1 - \frac{y}{M}\right) \qquad (11\text{k})$$

Let's first assess the impact of equation (11k) in general terms. Since the population y is positive and the rate of increase y' is positive, it follows that $\mu - \frac{\mu}{M}y \geq 0$. So $\frac{\mu}{M}y \leq \mu$, and hence $\mu y \leq \mu M$. Therefore,

$$y \leq M.$$

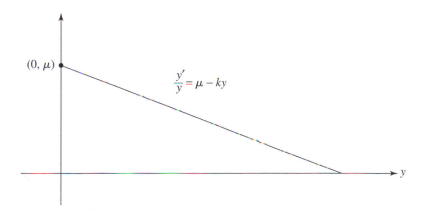

$$\frac{y'}{y} = \mu - ky$$

$(0, \mu)$

Figure 11.5

So Verhulst's linearity condition implies that the population y can never exceed M. Suppose that in its initial phase of growth, the population y is very small relative to this maximal size M. Because $\frac{y'}{y} = \mu\left(1 - \frac{y}{M}\right)$, it follows that in this phase $\frac{y'}{y} \approx \mu$, and hence $y' \approx \mu y$. This implies (see the derivation of equation (11i) from equation (11h) in Section 11.5A) that $y \approx y_0 e^{\mu t}$. Therefore, the population increases exponentially in this phase. Notice, finally, that if the population y is near M, then $\frac{y}{M}$ is close to 1. So $\frac{y'}{y} = \mu\left(1 - \frac{y}{M}\right)$ is close to 0. Therefore, the rate of increase y' is close to zero, and the population stagnates. So Verhulst's approach has features that one would expect at least some populations to exhibit.

A precise analysis of the impact of equation (11k) on the function $y(t)$ follows next. We begin by rewriting equation (11k) as

$$\frac{\frac{dy}{dt}}{y} = \mu - \frac{\mu}{M}y.$$

Multiplying through by yM, we get

$$M\frac{dy}{dt} = yM\left(\mu - \frac{\mu}{M}y\right) = y\mu(M - y),$$

and therefore,

$$\frac{M\,dy}{y(M - y)} = \mu\,dt.$$

This step has separated the variables t and y. Taking antiderivatives of each side, we have

$$\int \frac{M}{y(M - y)}\,dy = \int \mu\,dt. \qquad (11\mathrm{l})$$

(The validity of these steps, including separating the expression $\frac{dy}{dx}$ will be explained in Section 13.5.)

The bulk of the difficulty that remains concerns the antiderivative of $\frac{M}{y(M-y)}$. To find it, start with the expression $\frac{C}{y} + \frac{D}{M-y}$, where C and D are constants. Take common denominators to get

$$\frac{C}{y} + \frac{D}{M - y} = \frac{C(M - y) + Dy}{y(M - y)} = \frac{CM + (D - C)y}{y(M - y)}.$$

In order to get the right side to have the form $\frac{M}{y(M-y)}$

we now set $C = 1$ and $D - C = 0$. Because $D = C = 1$, we see that $\frac{M}{y(M-y)} = \frac{1}{y} + \frac{1}{M-y}$. This algebraic step of common denominators "in reverse" has decomposed the complicated term $\frac{M}{y(M-y)}$ into two simpler ones. It follows that

$$\int \frac{M}{y(M - y)}\,dy = \int \frac{1}{y}\,dy + \int \frac{1}{M - y}\,dy.$$

Taking derivatives confirms that the two integrals (or antiderivatives) on the right are

$$\int \frac{1}{y}\,dy = \ln y + \text{constant}$$

and

$$\int \frac{1}{M - y}\,dy = -\ln(M - y) + \text{constant}.$$

Refer to Section 10.3 if necessary. By combining the constants into the single constant C_1, we get

$$\int \frac{1}{y(M - y)}\,dy = \ln y - \ln(M - y) + C_1.$$

Note that $\int \mu\,dt = \mu t + C_2$. Therefore, by equation (11l),

$$\ln y - \ln(M - y) = \mu t + C_2 - C_1.$$

Recall at this point that we are looking for the "shape" of the function $y(t)$. It thus remains to solve this equation for y. Set $B = C_2 - C_1$. Since $\ln y - \ln(M - y) = \ln\frac{y}{M-y}$, we see that $\ln\frac{y}{M-y} = \mu t + B$. By properties of the exponential function,

$$\frac{y}{M - y} = e^{\ln\left(\frac{y}{M-y}\right)} = e^{\mu t + B} = e^B e^{\mu t} = A e^{\mu t}$$

where $A = e^B$. So $\frac{y}{M-y} = A e^{\mu t}$. Since $y = A e^{\mu t}(M - y) = AMe^{\mu t} - yAe^{\mu t}$, we get $y + yAe^{\mu t} = AMe^{\mu t}$, and therefore $y = \frac{AMe^{\mu t}}{1 + Ae^{\mu t}}$. After dividing both numerator and denominator by A, we obtain

$$y = y(t) = \frac{Me^{\mu t}}{A^{-1} + e^{\mu t}}.$$

Set $t = 0$ to get $y(0) = \frac{M}{A^{-1}+1}$. So $y(0)(A^{-1} + 1) = M$, and hence $A^{-1} = \frac{M}{y(0)} - 1$. Now put $y_0 = y(0)$ and

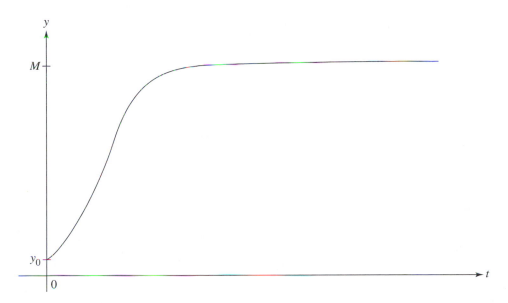

Figure 11.6

substitute into the previous equation to get

$$y(t) = \frac{Me^{\mu t}}{\left(\frac{M}{y_0} - 1\right) + e^{\mu t}}. \tag{11m}$$

Equation (11m) can be written in another form. After multiplying both numerator and denominator of the equation by $e^{-\mu t}$, we get

$$y(t) = \frac{M}{1 + \left(\frac{M}{y_0} - 1\right)e^{-\mu t}} \tag{11n}$$

The function $y(t)$ that the condition $\frac{y'}{y} = \mu - \frac{\mu}{M}y$ determines has now been found explicitly in terms of the parameters μ, M, and y_0.

Example 11.13. When a population is observed for the first time it has 1.2 million individuals. Determine the number of individuals $y(t)$ in the population at any time t if its specific growth rate satisfies the equation $\frac{y'}{y} = 0.06 - 0.015y$.

Refer to equation (11k) and observe that this is the special case of Verhulst's model where $\mu = 0.06$, $\frac{\mu}{M} = 0.015$, and $y_0 = y(0) = 1.2$. Since $0.015M = \mu =$

0.06, $M = \frac{0.06}{0.015} = 4$, we see that $M = 4$ and $\frac{M}{y_0} - 1 = \frac{4}{1.2} - 1 = \frac{2.8}{1.2} = \frac{7}{3}$. Substituting this information into equation (11n) shows that

$$y(t) = \frac{4}{1 + \frac{7}{3}e^{-0.06t}}.$$

Let's have a look at some of the basic properties of the function $y(t)$. Note first that $y(t)$ levels off for large t. This can best be seen from equation (11n). Push $t \to \infty$. Because $e^{-\mu t} = \frac{1}{e^{\mu t}}$ goes to 0, we see that $y(t) \to M$. So $y(t)$ approaches the constant M. Suppose that M is much larger than the initial population y_0. So $\frac{M}{y_0}$ is large and $\frac{M}{y_0} - 1 \approx \frac{M}{y_0}$. If t is small enough, then the term $\frac{M}{y_0}$ will dominate the denominator of equation (11m) and hence $y(t) \approx y_0 e^{\mu t}$. So the population will grow exponentially during its initial phase of growth. The specifics of the graph of the function $y(t)$ depend on the values of the parameters μ, M, and y_0. A typical case is sketched in Figure 11.6. Notice that this graph reflects the observations just made. It certainly seems plausible that real populations under appropriate conditions should follow such a pattern, at least during certain phases of growth.

Verhulst applied his model—now known as the *logistics model*—to the populations of Belgium and France of the 19th century. We will test it instead against the world's population for the period from 1965 to 1995.

C. Applying the Logistics Model. The fundamental concerns of this section will be these: Can the logistics model be applied to the population data for the years 1965 to 1995, and if so, what does it tell us? The population will now be expressed in billions rather than millions. Refer back to Table 11.3 and note, for instance, that the population in 1995 was 5.77 billion. All the data that we will use come from Table 11.3.

Designate the year 1965 as time $t = 0$. Let $y(t)$ for $t \geq 0$ be a function. We start by assuming that $y(t)$ is any differentiable function that represents the world's population in an approximate way. Take a typical year, say 1980, in the period 1965–1995. This is the year $t = 15$. Note that $y(15) \approx 4.448$ billion. During 1980 the population increased by 78 million $= 0.078$ billion and hence at the rate of 0.078 billion per year. Therefore, $y'(15) \approx 0.078$, and it follows that

$$\frac{y'(15)}{y(15)} \approx \frac{0.078}{4.448} \approx 0.018.$$

Repeating this computation for the years 1965, 1970, ..., 1995, in other words, for $t = 5, 10, \ldots, 30$, shows that $\frac{y'(0)}{y(0)} \approx 0.021$, $\frac{y'(5)}{y(5)} \approx 0.020$, $\frac{y'(10)}{y(10)} \approx 0.017$, and so on. Table 11.4 summarizes these findings. We now plot the points

$$(3.34, 0.021), (3.70, 0.020), \ldots, (5.77, 0.016)$$

on a Cartesian coordinate system. The population size y is on the horizontal axis. The specific growth rate is on the vertical axis, labeled with the variable z. Graph the line that the two points $(3.34, 0.021)$ and $(5.77, 0.016)$ determine. See Figure 11.7. The slope of this line is $\frac{0.021-0.016}{3.34-5.77} = -\frac{0.005}{2.43} = -0.002$. Since $(3.34, 0.021)$ is on its graph, we get that the line has

equation

$$z - 0.021 = -0.002(y - 3.34).$$

So $z = 0.021 + 0.007 - 0.002y$, and hence

$$z = 0.028 - 0.002y$$

is the equation of the line in slope-intercept form. Note that all the points that were plotted are close to the line.[5] It follows that $\frac{y'}{y} \approx 0.028 - 0.002y$ for all the data points of Table 11.4.

Table 11.4							
	1965	1970	1975	1980	1985	1990	1995
y	3.34	3.70	4.08	4.45	4.85	5.29	5.77
$\dfrac{y'}{y}$	0.021	0.020	0.017	0.018	0.018	0.017	0.016

Suppose now that the function $y(t)$ satisfies the specific growth rate equation $\frac{y'}{y} = 0.028 - 0.002y$. In view of Table 11.4 and Figure 11.7, this means that $y(t)$ approximates the specific growth rate conditions of the world's population from 1965 to 1995. This suggests in turn that it will simulate the world's population for this period.

The equation $\frac{y'}{y} = 0.028 - 0.002y$ also means that the function $y(t)$ satisfies the basic equation (11j) of the logistics model with $\mu = 0.028$ and $k = 0.002$. It can therefore be explicitly determined from the analysis of Section 11.5B. First note that $M = \mu k^{-1} = \frac{0.028}{0.002} = 14$. Taking $y_0 = y(0) = 3.34$, we get $\frac{M}{y_0} - 1 = \frac{14}{3.34} - 1 = 4.19 - 1 = 3.19$. Feeding this information back into equation (11n), we get

$$y(t) = \frac{14}{1 + 3.19e^{-0.028t}}.$$

Evaluating the function $y(t)$ at $t = 5, 10, 15, 20, 25, 30$ provides the following population figures in billions

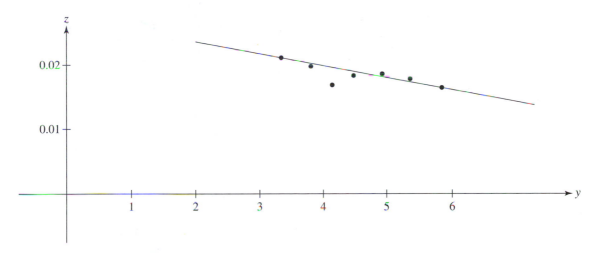

Figure 11.7

for the years 1965, 1970, . . . , 1995:

3.34, 3.71, 4.10, 4.52, 4.96, 5.42, 5.89.

The population estimates from Table 11.3 for these years (converted to billions and rounded off to two-decimal accuracy) are

3.34, 3.79, 4.08, 4.45, 4.85, 5.29, 5.77.

Notice that the population figures predicted by the model are in good correspondence with these estimates.

What has been done so far? We have derived a mathematical model $y(t)$ for the world's population from 1965 onward, based on the statistics and trends for the years 1965 to 1995. Not surprisingly, it fits the statistics upon which it is based rather closely. Now comes the interesting question. Can the model be used to provide meaningful estimates of the world's population into the future? By evaluating $y(t)$ for $t = 35, 45, 55, 85,$ and 130, for example, we get the projections (again in billions)

6.39, 7.37, 8.33, 10.85, 12.96

for the years 2000, 2010, 2020, 2050, 2095, respectively. Since $M = 14$, we can also conclude that the world's population will level off at 14 billion! When considering

these projections, it is important not to forget the assumption on which it is based, namely, that the basic trend displayed by the population figures from 1965 to 1995 will persist. Keep in mind that any population projection must involve an assumption of this sort, simply because the only way we have to predict the future is to extrapolate from the present and the past.

The projections just obtained parallel those by the experts of the Population Reference Bureau! It is their expectation that the pattern of decline of the rates of increase of the world's population will continue for the next century. See Table 11.5. As before, the first column lists the year, the second the population at the beginning of that year (in millions), the third the increase in population during the year (again in millions), and the fourth the percentage of this increase. Note, for instance, that the rate of increase is projected to be a very low 0.1% per year for the years 2080 to 2095. For the years 2000, 2010, 2020, 2050, 2095, this table provides the predictions

6.26, 7.21, 8.09, 9.78, 11.13 billion.

Note that they fall somewhat below those given earlier by the function $y(t)$.

The world's population has been and will continue to be influenced by many variables. Some of these are

Table 11.5

World Population Projections 2000–2095 (in millions)

2000	6,261	93	1.5	2025	8,503	79	0.9	2050	10,019	43	0.4	2075	10,840	18	0.2
2005	6,740	91	1.4	2030	8,874	74	0.8	2055	10,235	39	0.4	2080	10,931	16	0.1
2010	7,205	89	1.2	2035	9,252	52	0.6	2060	10,430	33	0.3	2085	11,009	13	0.1
2015	7,660	84	1.1	2040	9,516	53	0.6	2065	10,598	26	0.2	2090	11,075	12	0.1
2020	8,091	81	1.0	2045	9,784	47	0.5	2070	10,730	22	0.2	2095	11,133	11	0.1

difficult, and others impossible, to control: famines, wars, epidemics, birth control measures, etc. In other words, the growth of human populations, subject as it is to many diverse, competing, and unpredictable forces, is beyond description by simple mathematical formulas. Any projection, therefore, is no more than an educated guess. Recognizing that it depended on the assumption that existing trends persist, Verhulst was certainly aware of the inherent limitations of his model. Most current published population estimates and projections (for instance, those of Table 11.5 from the Population Reference Bureau) do not use the logistics model. Instead, they consider estimates of the birth and death rates and use the fact that the rate of increase of a population is equal to the difference between the birth rate and death rate. This is done on a country-by-country basis (this is necessary because these rates vary greatly from one region of the world to another) and then added up to provide a global projection.

11.6 The Growth of Microorganisms

In spite of the amazing progress of modern medicine, the world is still ravaged by diseases brought on by a variety of different types of microbiological organisms. According to recent statistics, such diseases (notably pneumonia, diarrheal diseases, hepatitis, and tuberculosis) kill over ten million people a year. It must be pointed out, however, that only very few of the hundreds of thousands of species of microbes are pathogenic, i.e., harmful to humans and animals. Most of them are beneficial, indeed essential, to the processes of life. Indeed, life as we know it would be impossible without them. Marine and freshwater microbes form the basis of the food chain in oceans, lakes, and rivers. Soil bacteria and fungi decompose matter from dead organisms and return vital chemicals to the environment. Microbes also play an important role in the oxygen-generating process of photosynthesis. In the context of the evolutionary development of life, there is the remarkable fact that the bacterial mode of life has existed from the very beginning of the fossil record until today. It is hardly surprising, therefore, that the study of microbiological organisms is carried out with intensity in research labs the world over.

Most microbes are one-celled, and the discussion of this section will focus exclusively on these. There are hundreds of thousands of different species, including all *bacteria* and *protozoans*, as well as some *algae* and *fungi*. Most microbes are very small. For instance, the bacterium *E. coli*, which breaks down nutriments in the intestinal tract of humans, has a size of 3 micrometers (1 micrometer = 1μm = 10^{-6} meters).[6] By comparison, a human hair has a thickness of about 100 micrometers. Protozoans typically range in size from 10 to 100 micrometers. They are more complicated and advanced than other microbes. The protozoan *paramecium*, for instance, has separate nuclei for growth regulation and reproduction, oral grooves for the ingestion of food, and a pore for the excretion of wastes. Our discussion does not apply to *viruses*. These agents have no cell structure. In fact, they do not have their own "machinery" for growth and can reproduce only inside the cells of other organisms.

The study of the growth of populations of microorganisms is at the heart of microbial cell physiology. Physiological events in the individual cells are frequently manifested to the investigating scientist only by changes in the growth of populations. By manipulating population growth, microbiologists can come to an understanding of relevant aspects of the functioning of cells. Accordingly, it is important to understand the changes in growth that such populations undergo and to be able to measure such changes accurately. Populations of one-celled microbes grow by cell division. There are different processes for such division. The cell of a bacterium is simple. It does not have its genetic material contained within a nuclear membrane, and it splits by a simple process. Cells that have their genetic information enclosed in a membrane-lined nucleus divide in more complicated ways. Populations of one-celled microbes exhibit the same general growth dynamics that will now be described.

A population of microbes is referred to as a *culture*. When studied in a laboratory, microbes are introduced into a sterile *culture medium*. This contains water and carbon (which are needed for all the organic compounds that make up a living cell), oxygen (necessary for metabolic processes), and other nutrients. The culture medium is held at an appropriate temperature and has an appropriately low acidity level. Under such controlled conditions, mathematical studies of the growth of microbe populations are possible. The growth of a culture is defined to be the increase in its mass. In important phases of growth, the mass per microbe in the culture will be about the same, so that growth can be measured by counting the number of organisms in the culture. The methods for taking count include the repetitive dilution of the culture medium, direct microscopic count, and the use of electronic cell counters.

Right after cells are transferred to a fresh culture medium, there is little or no growth in the number of cells or individual cell mass. This period—the *lag phase*—is the time required for the physiological adaptation to the medium. Cells in the lag phase are not dormant, but in a state of preparation for growth. Indeed, there is intense metabolic activity. The length of the lag phase depends on the status of the cells when introduced into the medium. When cells come from a previous stationary phase of growth, they may need time to recover from products toxic to the metabolism, such as acids and alcohols. The lag phase can be minimized if the cells are taken from a phase of rapid growth and transferred to a fresh culture medium of the same type.

A. The Exponential Phase. Eventually, the lag phase ends and cells will begin to divide. So one cell splits into two, two into four, four into eight, and so on. This doubling occurs at a regular interval called the *doubling time*. The culture is now in its *exponential growth phase*. The doubling time depends on the specifics of the culture medium (types of nutrients, acidity, temperature, etc.). Under ideal conditions, most bacteria have a doubling time of 1 to 3 hours. For some microbes it is over 24 hours and for others only a few minutes. For *E. coli* this maximal doubling time is about 16 minutes.

Suppose that a culture is observed during its exponential growth phase. Designate the beginning of the observation by $t = 0$. Let $y(t)$ be the number of cells in the culture at any time $t \geq 0$. So $y(0) = y_0$ is the number of cells when the culture is first observed. Let d be the doubling time. At times $t = d, 2d, 3d, \ldots, nd, \ldots$, the number of cells will have doubled, doubled twice, doubled three times,..., doubled n times,.... Therefore,

$$y(d) = 2y_0, \ y(2d) = 2(2y_0) = 2^2 y_0, \ y(3d) = 2(2^2 y_0)$$

$$= 2^3 y_0, \ldots, \ y(nd) = 2^n y_0, \ldots.$$

Note that this is the quantitative behavior of the culture as a whole and that it is not the case that all cells divide at the same instant. Take any time t and let $r = \frac{t}{d}$. So $t = rd$. Making use of the pattern just observed, we get that

$$y(t) = y(rd) = 2^r y_0 = 2^{\frac{t}{d}} y_0 = \left(2^{\frac{1}{d}}\right)^t y_0.$$

The last term can be rewritten as follows. Start with the equality $2 = e^{\ln 2}$ and raise both sides to the power $\frac{1}{d}$ to get $2^{\frac{1}{d}} = e^{\frac{\ln 2}{d}}$. Now set $\mu = \frac{\ln 2}{d}$. So $2^{\frac{1}{d}} = e^{\mu}$. By a substitution,

$$y(t) = y_0 e^{\mu t}$$

The constant $\mu = \frac{\ln 2}{d} = \frac{0.693}{d}$ is called the *growth constant* of the culture.

By differentiating $y(t) = y_0 e^{\mu t}$ we get $y'(t) = \mu y_0 e^{\mu t} = \mu y(t)$. The equation

$$y'(t) = \mu y(t)$$

is characteristic of the exponential phase of growth. It asserts that $y'(t)$, the rate of increase in the cell count at time t, is proportional to $y(t)$. The growth constant μ is the constant of proportionality. The same growth dynamics were observed for the human population in Section 11.5A.

If the exponential growth phase of a culture goes on for any length of time, an enormous number of cells can arise.

Example 11.14. A culture of *E. coli* bacteria is growing in its exponential phase under ideal conditions. It has a doubling time of 16 minutes. When it is first observed at time $t = 0$ it has 20,000 cells.

  (i)   Find an expression for the number of cells after t hours.
  (ii) Find the number of cells after 6 hours.
 (iii) When will the population reach 1 billion?

Since $y_0 = 20{,}000$, $y(t) = 20{,}000 e^{\mu t}$ for any time $t \geq 0$. The problem suggests to measure the time t in

hours. The doubling time of 16 minutes equal to 0.267 hours. So $\mu = \frac{0.693}{0.267} = 2.596$. Therefore,

$$y(t) = 20{,}000 e^{2.596t}$$

and part (i) is done. The number of cells in the culture after 6 hours is

$$y(6) = 20{,}000 e^{(2.596)6} = 20{,}000 e^{15.576}$$

$$= (20{,}000)(5{,}815{,}000) = 1.163 \times 10^{11}$$

$$= 116 \text{ billion.}$$

To answer (iii), we need to set $20{,}000 e^{2.596t} = 10^9$ and solve for t. Since, $e^{2.596t} = 0.5 \times 10^5 = 5 \times 10^4$, we get $2.596t = \ln 5 + 4 \ln 10$. So $2.596t = 1.609 + 4(2.303)$. Hence, $2.596t = 10.821$, and $t = 4.17$ hours.

The exponential phase is the phase in which cellular reproduction and metabolic activity are at their peak. It is also the phase during which the cells are most susceptible to adverse conditions. Radiation or antibiotics, for example, exert their effect by interfering with some important step in the growth process and are most effective during this phase.

No culture can sustain its exponential growth phase forever, and growth eventually slows. The accumulation of waste products toxic to cells, the exhaustion of required nutrients, limitation of oxygen, changes in temperature, or increased acidity levels are usually factors. As growth slows, the cells undergo physiological changes and become metabolically less active. They become more resistant to environmental hazards such as heat and chemicals and often decrease in size, sometimes from 5–10 micrometers to 1–2 micrometers. Some bacteria form spores or cysts. These are resting cells that remain viable and will germinate in fresh media. Typically, the population stabilizes and enters into a period of equilibrium called the *stationary phase*. Microbes can generally adapt to nutrient depletion and remain viable for long periods in the stationary phase. Eventually, the number of viable bacteria will begin to decline. In this *death phase* the bacteria will no longer be able to carry out reproduc-

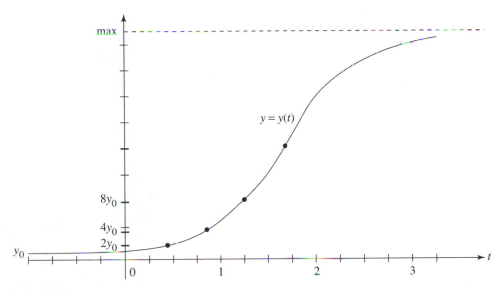

Figure 11.8

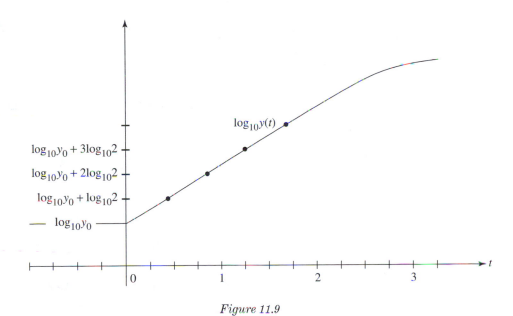

Figure 11.9

tion, and the number of living cells in the culture will tend to zero.

Suppose that a culture is introduced into a medium and that the exponential phase begins at time $t = 0$. Let $y(t)$ be the number of cells at any time t

in hours. The graph of a typical growth function $y(t)$ is shown in Figure 11.8. The lag phase is indicated to the left of the y-axis. The exponential phase begins with $y_0 = y(0)$ cells. In the graph, 4 ticks on the time axis represent one hour. So each tick is $\frac{1}{4}$ of an hour.

A study of the graph shows that the doubling time is about $1\frac{2}{3} = \frac{5}{3}$ ticks and hence about $d = \frac{5}{3} \cdot \frac{1}{4} = \frac{5}{12}$ hours (or 25 minutes). At $t = d = \frac{5}{12}$ there are $2y_0$ cells; at $t = 2d = 2 \cdot \frac{5}{12} = \frac{10}{12}$ there are $4y_0$ cells; and at $t = 3d = 3 \cdot \frac{5}{12} = 1\frac{1}{4}$ there are $8y_0$ cells. The exponential phase ends after about 2 hours, and the transition into the stationary phase begins.

Since the growth during the exponential phase is extremely rapid, it is often visualized in "slowed down" form. In particular, the graph of $\log_{10} y(t)$ is considered instead. See Figure 11.9. Notice that the graph is a line during the exponential phase. For an increase of time d on the horizontal axis, there is vertical increase of $\log_{10} 2$. It follows that the slope of this line is $m = \frac{\log_{10} 2}{d} = \frac{0.301}{d}$. So the doubling time is $d = \frac{0.301}{m}$. The connection between the growth constant μ and the slope m is given by

$$\mu = \frac{0.693}{d} = 0.693 \frac{m}{0.301} = 2.3m.$$

B. Experimenting with *E. coli*.   One of the first mathematical studies of the growth of bacteria was undertaken by A. G. M'Kendrick and M. Kesava Pai, of the Pasteur Institute of southern India. Their research "The rate of multiplication of micro-organisms," published in the Proceedings of the Royal Society of Edinburgh in 1911, considered the growth of *E. coli*.

The two scientists prepared test tubes containing a nutrient broth of meat extract and salt. *E. coli* populations were first incubated and then introduced into the test tubes. Five of the scientists' trials produced the population counts listed in Table 11.6. Initial runs of their experiments (including the first three of the table) gave evidence of the lag phase already discussed. To eliminate it, they placed greater control on the temperatures and incubation times. Both the broth and the test tubes were kept at a constant $37°$ centigrade throughout all phases of the experiment, and the incubation period was reduced. This ensured that the cells were in active growth when introduced into the medium. In Trials 4 and 5 the lag phase had been eliminated.

The general pattern of growth of bacteria cultures (see Figure 11.8) suggested to the investigators that the logistics model (refer to Section 11.5B, especially Figure 11.6) could be used to simulate this growth.

Table 11.6

Time in hours	Populations				
	Trial 1	Trial 2	Trial 3	Trial 4	Trial 5
0	1,760	19,000	176,000	2,850	64,250
$\frac{1}{2}$	4,020	35,900	280,000	7,500	165,000
1	...	88,000	608,500	17,500	357,500
2	72,000	482,000	3,870,000	105,000	2,625,000
3	380,000	3,400,000	28,200,000	625,000	16,250,000
4	2,600,000	26,300,000	74,200,000	2,250,000	45,375,000
5	19,750,000	71,600,000	127,000,000	17,750,000	122,125,000
6	77,000,000	118,000,000	150,000,000	50,000,000	158,500,000
7	120,000,000	...	149,000,000	97,500,000	194,500,000
8	106,000,000	135,000,000	154,000,000	...	...

Accordingly, they began with the hypothesis that the number of cells in a culture of *E. coli* can be represented by a function $y(t)$ of the form

$$y(t) = \frac{M}{1 + \left(\frac{M}{y_0} - 1\right)e^{-\mu t}}$$

for $t \geq 0$. The constant y_0 is the cell count when the culture is first observed at time $t = 0$, and M is the cell count that is reached in the stationary phase. A look at the remarks towards the end of Section 11.5B shows that μ is the growth constant of the culture during the exponential phase.

The goal of the experiment was to determine y_0, M, and μ for each trial and to check the resulting function $y(t)$ against the population counts that they had observed. We will do this for Trial 4. The initial count is $y_0 = y(0) = 2,850$. The data as well as previous trials suggested the value $M = 100,000,000$. Since $\frac{M}{y_0} - 1 = \frac{10^8}{2.85 \times 10^3} - 1 = 35,100$, we see that

$$y(t) = \frac{10^8}{1 + 35,100e^{-\mu t}}.$$

It only remained to compute μ. To do this M'Kendrick and Pai used the formula $\mu = 2.3m$, where m is the slope of the graph $\log_{10} y(t)$ for t restricted to the exponential phase of growth. Using the cell counts at $t = 0$ and $t = 3$, we see that

$$m = \frac{\log_{10} y(3) - \log_{10} y(0)}{3 - 0}$$

$$= \frac{\log_{10} 625,000 - \log_{10} 2,850}{3}$$

$$= \frac{5.80 - 3.45}{3} = 0.8.$$

So $\mu = (2.3)(0.8) = 1.8$. So we will take

$$y(t) = \frac{10^8}{1 + 35,100e^{-1.8t}}.$$

Rounding off the values of $y(t)$ for $t = \frac{1}{2}$, 1, 2, 3, 4, 5, 6, and 7, to the nearest thousand provides the comparison between the mathematical simulation and observation shown in Table 11.7.

Table 11.7

t	$y(t)$	observation
$\frac{1}{2}$	7,000	7,500
1	17,000	17,500
2	104,000	105,000
3	627,000	625,000
4	3,676,000	2,250,000
5	18,756,000	17,750,000
6	58,274,000	50,000,000
7	89,417,000	97,500,000

Because the slope m is equal to $\frac{0.30}{d}$, where d is the doubling time of the culture during its exponential phase, it follows that $d = \frac{0.30}{m} = \frac{0.30}{0.8} = 0.38$ hours, or about 23 minutes. This is greater than the maximal doubling time of *E. coli*, which is 16 minutes.

Observe that the logistics model does reflect the growth of the culture in general terms. But the fit between theory and experiment could be better. Why isn't it? Some relevant questions in this connection are these: Was the experiment refined enough? How precisely were the important parameters of the experiment (temperature, acidity, and nutrients) controlled? How accurate were the cell counts? Was there contamination? Beyond all this, however, is the fact that the assumption

$$\frac{y'(t)}{y(t)} = \mu - \frac{\mu}{M}y(t)$$

that underlies the logistics model is not always flexible enough to capture what is going on with precision. Consider, for instance, the following consequences of this assumption. Since $y'(t) = \mu y(t) - \frac{\mu}{M}y(t)^2$, we get by using the chain rule that

$$y''(t) = \mu y'(t) - 2\frac{\mu}{M}y(t)y'(t) = \mu y'(t)\left(1 - 2\frac{1}{M}y(t)\right).$$

It follows that $y''(t) = 0$ for $y(t) = \frac{M}{2}$. In other words, the inflection point of the graph of $y(t)$ occurs precisely when the population has reached one-half its limiting value of M. A look at Figure 11.8 shows that the inflection point occurs during the transition from the exponential to the stationary phase. One limitation of the logistics model is now apparent. For instance, it cannot simulate a growth pattern of the following sort: An exponential phase during which, say, 90% of the maximum number of cells is reached, followed by a sharp transition to the stationary phase. The logistics model has, nonetheless, been generally successful in describing the population growth of microorganisms (including protozoans and fungi), as well as fruit flies, flour beetles, and other larger organisms.

The mathematical simulation of the growth of a culture of microbes can provide a better understanding of the dynamics of this growth. The next section will attempt to illustrate that this can have important implications in commercial and industrial contexts.

C. Fermentation Processes.

The concept *fermentation process* refers to the production of chemical substances by microbes that are allowed to grow in a culture medium that contains appropriate raw materials. If the microorganism is suitably selected (certain strains of bacteria or fungi, for instance), the raw materials appropriately chosen (these often include plant materials and metals), and if the culture medium is adequately controlled (for example, nutrients, acidity, temperature, and oxygen), then the culture can produce a useful chemical. A large number of commercial and industrial chemicals are produced in this way. These include many antibiotics, vaccines, proteins, vitamins, industrial enzymes (such chemicals are used in the production of foods and textiles, for example, and in the recovery of silver from used photographic film), and organic acids (used in the production of foods, soft drinks, and detergents; in the treatment of leather and fabrics; and to stimulate the growth of plants).

The important first step is the selection of the organisms. Let's consider the production of antibiotics as a case in point. Many soil microorganisms produce antibiotics naturally. For example, the fungi *Penicillium notatum* and *Penicillium chrysogenum* produce penicillin, and the bacteria strains *Strepto-*

myces erythreus, *Streptomyces fradiae*, and *Streptomyces griseus* produce the antibiotics erythromycin, streptomycin, and neomycin. It is thought that antibiotics give the microorganisms an advantage in their competition for nutrients and space. Thousands of different antibiotics have now been isolated and identified. The discovery of new antibiotics used to be mostly a matter of chance. Subsequently, pharmaceutical companies undertook massive screening programs to search for new soil microbes that exhibit antibiotic activity. The strains of microbes must be carefully selected because different strains of microbes may differ greatly in the amounts and quality of the chemical that they produce. The search process for naturally occurring antibiotics has therefore become very targeted and selective. The majority of antibiotics in clinical use today are produced by microorganisms via the fermentation process. (It should be noted that most of them are modified chemically, by genetic engineering, for instance, to improve their medicinal properties.)

The production of chemicals by fermentation processes is analyzed in the laboratories of industrial companies. A sterile culture medium containing nutrients is prepared in a vessel or tank called a *fermentor*, and a strain of a species of microbes is introduced. As in the experiments discussed in Section 11.6B, the lag phase is avoided by taking incubated cultures in phases of exponential growth. The medium is continually supplied with oxygen. It is agitated and constantly mixed. As the culture grows, the chemical is generated. The process is normally stopped when the culture is in the stationary phase and before the decline phase has set in. In the final step, the chemical that is produced is filtered out, extracted, and purified.

The goal of the experiment is to find a combination of microorganisms, types of raw materials and nutrients, temperature, acidity levels, and oxygen concentrations that will maximize the production of the chemical and at the same time minimize costs. In other words, the goal is to combine the right organism, an inexpensive medium, and the proper environment to produce high yields of a desired commercial product as economically as possible. This is usually a very challenging enterprise, which is undertaken in a step-by-step fashion. The start is made with small-scale-laboratory experiments, where the impact of changes in the parameters can be tested by running repeated trials. Small changes in the parameters can have considerable impact on both the product and the yield. Later, the scale is increased from small fermentors to larger fermentors, and finally to large production fermentors that can hold hundreds of thousands of gallons. A fermentation process involves many variables, and a large number of different combinations are possible. The search for the best combinations—those that simultaneously maximize production and minimize cost—is a difficult effort that can be facilitated by mathematical simulations. A large number of mathematical models have been developed for this purpose. We will have a look at a relatively simple model that has given insight into some fermentation processes.

Suppose a fermentation run is started at time $t = 0$ and studied for $t \geq 0$. In view of the fact that the medium is being constantly mixed, we will focus on the concentration of the microbes in the medium rather than their total number in the fermentor. Since the lag phase has been eliminated and the run is stopped when the microbes are in the stationary phase of growth, we can assume that the number of microorganisms per milliliter of medium is simulated by a function $y(t)$ that satisfies the logistics condition. Therefore,

$$\frac{y'}{y} = \mu - \frac{\mu}{M} y \qquad (11\text{o})$$

where M is the limiting number of microbes per milliliter and μ is the growth constant during the exponential phase.

It is in the company's interest to maximize production of the chemical. Let $p(t)$ be the concentration of the chemical being produced in milligrams per milliliter at any time t. What should be maximized? Is it p? To answer this question, think for just a moment about any product that a firm might produce. Is

it the total number of units that it wants to maximize? Perhaps, but what the firm is really interested in maximizing is the *rate*, in, say, units per month, at which the product rolls off the assembly line. The situation is the same with the production of the chemical. So it is the rate of change $p'(t)$ of the concentration that should be maximized.

Lab experiments have shown that in certain situations (for example the production of lactic acid by bacteria called *lactobacilli*), $p'(t)$ is related to both $y(t)$ and $y'(t)$ by an equation of the form

$$p'(t) = ay'(t) + by(t) \qquad (11\text{p})$$

where a and b are constants. The assertion is that the rate of change in the concentration of the product depends linearly on both the microbe count per milliliter and the rate at which the microbe count changes. In a specific fermentation run, a and b can be determined by measurement. The equation $y'(t) = \mu y(t) - \frac{\mu}{M} y(t)^2$ is a direct consequence of equation (11o). By substituting it into (11p), we get

$$p'(t) = a \left(\mu y(t) - \frac{\mu}{M} (y(t))^2 \right) + by(t)$$

$$= (a\mu + b)y(t) - \frac{a\mu}{M} (y(t))^2 \qquad (11\text{q})$$

$$= y(t) \left(a\mu + b - \frac{a\mu}{M} y(t) \right).$$

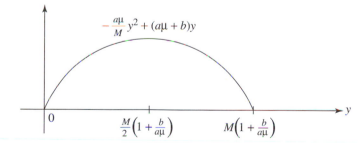

Figure 11.10

When considering $p' = -\frac{a\mu}{M} y^2 + (a\mu + b)y$ as a function of y, the graph of p' is a parabola that opens downward. See Figure 11.10. Since

$$-\frac{a\mu}{M}y^2 + (a\mu + b)y = y\left(a\mu + b - \frac{a\mu}{M}y\right),$$

this parabola crosses the horizontal axis at $y = 0$ and $y = \frac{M}{a\mu}(a\mu + b) = M(1 + \frac{b}{a\mu})$. It follows from elementary calculus that the highest point of the parabola occurs when $y = \frac{M}{2}(1 + \frac{b}{a\mu})$. So p' attains its maximum value when the count stands at $y = \frac{M}{2}(1 + \frac{b}{a\mu})$ microbes per milliliter. At this concentration, the rate at which the chemical is being produced is at a maximum. This maximum rate is obtained by substituting the value $y = \frac{M}{2}(1 + \frac{b}{a\mu})$ into equation (11q):

$$
\begin{aligned}
p' &= \frac{M}{2}\left(1 + \frac{b}{a\mu}\right)\left(a\mu + b - \frac{a\mu}{M}\frac{M}{2}\left(1 + \frac{b}{a\mu}\right)\right) \\
&= \frac{M}{2}\left(1 + \frac{b}{a\mu}\right)\left(a\mu + b - \frac{a\mu}{2} - \frac{b}{2}\right) \\
&= \frac{M}{2}\left(\frac{a\mu + b}{a\mu}\right)\left(\frac{a\mu + b}{2}\right) = \frac{M}{4a\mu}(a\mu + b)^2.
\end{aligned}
$$

Substituting the value $y = \frac{M}{2}(1 + \frac{b}{a\mu})$ into equation (11o) we get

$$\frac{y'}{y} = \mu - \frac{\mu}{M}\frac{M}{2}\left(1 + \frac{b}{a\mu}\right) = \mu - \frac{\mu}{2} - \frac{\mu b}{2\mu a} = \frac{\mu}{2} - \frac{b}{2a}.$$

So this is the specific growth rate of the culture at which the rate of production $p'(t)$ is a maximum.

The study of these equations can provide a better understanding of the process and insight into the problem of deciding which of the many combinations give the best results. This study is often facilitated by analysis with computers, especially in situations that involve complicated models with many equations.

We have focused on what is called the *batch fermentation process*. One "batch" is run until the rate of production of the chemical reaches a level that is too low to warrant continuation of the run. At this point the run is stopped, the product is recovered, and the fermentor is cleaned and prepared for another run. The batch process thus involves considerable downtime, time during which production is stopped. The

use of flow-through processes avoids this inefficiency. This variation of the fermentation process is in principle the same as the batch process, but the raw materials and nutrients are fed into the fermentor continuously, and the product is recovered on a continuous basis. Such fermentation processes can be run at full capacity and with little downtime, but they are more susceptible to contamination. (See Exercise 11M.)

11.7 Postscript

It has been the theme of this chapter to illustrate how the exponential function and its calculus inform—indeed provide crucial information about—elements of nuclear physics, geology, anthropology, and population growth, especially that of cultures of bacteria. As we have already seen, the range of interactions between these subjects and applications in science, technology, and industry is wide. Radioactive isotopes have found extensive use. Most of these isotopes are produced artificially by either particle accelerators (which fire high-velocity charged particles that split atoms) or by research nuclear reactors (see Exercises 11C–11E). In either case, stable atoms are bombarded with high-energy neutrons generating nuclear reactions that result in the desired isotope. (The production of $^{14}_{6}\text{C}$ in the atmosphere is a naturally occurring example of this process!) Radioactive isotopes are effective as tracers in medical diagnostic procedures. Since they are chemically identical to the stable isotopes of the same element, they can substitute for them in physiological processes. Because of their radioactivity, they can be readily traced even in minute quantities. Among the most important tracers are iodine-131, phosphorus-32, and technetium-99m. Physicians employ iodine-131 to determine cardiac output, plasma volume, and fat metabolism, and to measure the activity of the thyroid gland where this isotope accumulates. Phosphorus-32 is useful in the identification of malignant tumors because phosphates tend to accumulate at a higher rate in cancerous cells than in normal cells. Technetium-

99m is also used with radiographic scanning devices and is valuable in the study of the anatomic structure of organs. The radioactive elements cobalt-60 and cesium-137 are widely used to treat cancer. They can be administered selectively to malignant tumors in such a way as to minimize damage to adjacent healthy tissue. Tracing methods also find application in industry. For instance, the effectiveness of motor oils to prevent wear on piston rings and cylinder walls in automobile engines is measured in this way. We have also seen that microbial activity forms the basis of a number of technologies. This includes the production process of synthetic fuels (ethanol and methane are examples), the mining of minerals (such as copper and uranium), and the recovery of oil. It is also instrumental in the removal of pollutants from the environment. For instance, one of the major methods utilized in the cleanup of the vast quantities of oil spilled by the tanker *Exxon Valdez* on the shores of Alaska consisted in adding nutrients that stimulated the growth of bacteria that have the ability to break down oil.

Exercises
11A.   Radioactive Decay

1. In an experiment with the radioactive isotope $^{220}_{86}$Rn of the gas radon, Rutherford made the following measurements of the ratio $\frac{y'(t)}{y'(0)}$ of the decay rates for the sample he was testing: at $t = 40$ seconds, $\frac{y'(t)}{y'(0)} = 0.60$; at $t = 80$ seconds, $\frac{y'(t)}{y'(0)} = 0.36$; and at $t = 120$ seconds, $\frac{y'(t)}{y'(0)} = 0.22$. Plot the points $(0,0)$, $(40, \ln 0.60)$, $(80, \ln 0.36)$, $(120, \ln 0.222)$ carefully on a coordinate plane and check that they fall on a line (at least approximately). How did Rutherford deduce the value $\lambda = 0.0128$ for the decay constant of radium-220? Determine the half-life h of radium-220.

2. Let $y(t)$ for $t \geq 0$ (in seconds) be the number of atoms in the radon-220 sample of Exercise 1. Observe that $y(t) = y_0 e^{-0.0128t}$. What fraction of the initial amount of the sample was left after 5 minutes? Assuming that the initial number of atoms in the sample was $y_0 = 10^9$, how many atoms decayed during the first two seconds? At what average rate did the atoms decay

per second over these two seconds? What was the rate of decay at the $t = 2$ second mark?

3. Rutherford also studied the radium isotope $^{224}_{88}$Ra. In one experiment he made the following measurements of the ratio $\frac{y'(t)}{y'(0)}$ of decay rates for the sample he was testing: at $t = 2$ days, $\frac{y'(t)}{y'(0)} = 0.72$; at $t = 4$ days, $\frac{y'(t)}{y'(0)} = 0.53$; at $t = 7$ days, $\frac{y'(t)}{y'(0)} = 0.295$; at $t = 8$ days, $\frac{y'(t)}{y'(0)} = 0.252$; at $t = 11$ days, $\frac{y'(t)}{y'(0)} = 0.152$; at $t = 13$ days, $\frac{y'(t)}{y'(0)} = 0.111$. Plot the points $(0,0)$, $(2, \ln 0.72)$, $(4, \ln 0.53)$, $(7, \ln 0.295)$, $(8, \ln 0.252)$, $(11, \ln 0.152)$ and $(13, \ln 0.111)$ carefully on a coordinate plane. Note that the line determined by $(0,0)$ and $(13, \ln 0.111) = (13, -2.20)$ has the property that the six points are all very nearly on the line. What value λ for the decay constant of radium-224 could Rutherford deduce from this? Radium-224 appears in the fourth step of the thorium series (Rutherford had called it Thorium X). The isotope radon-220 of Exercises 1 and 2 appears in the fifth step.

4. Assume that the decay equation for radon-222 is $y(t) = y_0 e^{-0.18t}$ with t given in days.

    i.   What value for the half-life is this equation based upon?
    ii. About how long will it take for a sample of pure radon-222 to decay to 90% of the original amount?
    iii. How long will it take for the sample to decay to $\frac{1}{3}$ of the original amount?

5. The last step of the uranium-238 series is the decay of the polonium isotope $^{210}_{84}$Po to lead $^{206}_{82}$Pb. Polonium-210 has a half-life of 138 days.

    i.   Determine the decay constant of polonium-210.
    ii. Compute the number of atoms in a 30 milligram sample of polonium-210.
    iii. Express the number of polonium-210 atoms that remain in this sample after t days in terms of an exponential function.
    iv. How many polonium-210 atoms will remain after 4 weeks?

6. The polonium isotope $^{215}_{84}$Po that occurs in the uranium-235 series has the very short half-life of 1.83×10^{-3} seconds.

    i.   A sample of this isotope has a mass of 75 milligrams at time $t = 0$. Find a formula for the mass that remains after t seconds.

ii. How long will it take for the sample to decay to $\frac{1}{10}$ of the original mass?

7. One series of measurements gave a value of 3.7×10^{10} disintegrations per second as the activity of 1 gram of pure radium-226. What half-life for radium-226 can be deduced from this measurement?

8. An unknown radioactive substance is tested in a lab with a radiation counter. At 8 A.M. the counter gives a reading of 3200 disintegrations per minute and at 5 P.M. 900 counts per minute. Determine the half-life of the radioactive substance being tested.

9. A radiation counter shows that a certain radioactive substance disintegrates at a rate of 8.67×10^{13} atoms per minute at a certain time and at a rate of 7.67×10^{12} atoms per minute 6 minutes later. Determine the half-life of this radioactive substance.

10. A 25 milligram sample of a radioactive substance is found by a radiation counter to disintegrate at a rate of 6.57×10^{15} atoms per hour. The atomic mass of the element is 252. Determine the decay constant of this radioactive substance.

Potassium is present in every living organism. It plays a role in the metabolic process and is constantly being replenished. In naturally occurring potassium, the radioactive isotope $^{40}_{19}$K constitutes 0.012% of the total. The half-life of $^{40}_{19}$K is 1.3×10^9 years.

11. A sample of fresh seawater contains 0.48 grams of potassium per liter. How many grams of potassium-40 are there per liter? At what rate are the potassium atoms in a liter of fresh seawater disintegrating per second?

12. An instrument called a *whole body gamma ray counter* can be used to measure the radioactivity of potassium in a person. Suppose that a person weighs 60 kilograms (about 132 pounds) and that 0.35% of the body mass is potassium.

    i. Calculate the mass of the potassium in the person in grams.
    ii. Calculate the mass of $^{40}_{19}$K in the person in grams.
    iii. Estimate the number of $^{40}_{19}$K atoms in the person.
    iv. Compute the decay constant of $^{40}_{19}$K.
    v. Compute the number of disintegrations of $^{40}_{19}$K atoms per second in the person.

11B. Matter and Energy

At the very heart of matter there is energy. Indeed, matter *is* energy. The protons and neutrons in a nucleus of an atom are bound together by a force known as the *strong nuclear force*. As a consequence, it requires energy to take a nucleus apart. The energy needed to completely separate all the protons and neutrons is called the *binding energy* of the nucleus. Incredibly, the mass of a nucleus is *less* than the sum of the masses of the protons and neutrons that constitute it. This difference in masses, when converted to energy with Einstein's mass/energy equation $E = mc^2$, where c is the speed of light, is equal to the binding energy of the nucleus. (The speed of light is $c = 186,000$ miles/sec $= 3 \times 10^8$ meters/sec.) The instability of a radioactive nucleus is explained by the fact that the excess neutrons and protons that it contains give it more mass than the strong nuclear force can effectively hold together.

Refer next to the systems of units discussed in Exercises 7E. In the M.K.S. system the basic unit of energy is the *joule*. It is defined to be the energy produced by a 1 newton force acting through 1 meter. Recalling that 1 newton is the force that increases the velocity of a mass of 1 kilogram by 1 meter/sec per second, we see that

$$1 \text{ joule} = 1 \text{ newton-meter} = 1 \text{ (kg-meter/sec}^2\text{)-meter}$$

$$= 1 \text{ kg-meter}^2/\text{sec}^2.$$

The corresponding unit in the American system is the foot-pound. Recall that 1 newton $= 0.22$ pounds and 1 meter $= 3.38$ feet. Therefore,

$$1 \text{ joule} = 0.22 \text{ pounds} \times 3.28 \text{ feet} = 0.72 \text{ foot-pounds}.$$

Converting the joule into the American system gives us some "feeling" for this unit of energy: Since 0.72 feet is equal to approximately 9 inches, 1 joule is the energy that is produced by a 1 pound force acting through a distance of 9 inches.

The unit of energy common in atomic physics is the *electron volt*, or eV. One eV is equal to 1.6×10^{-19} joules. One million eV are grouped into a larger unit, the MeV. So

$$1 \text{ MeV} = 10^6 \text{ eV} = 10^6 (1.6 \times 10^{-19}) \text{ joules} = 1.6 \times 10^{-13} \text{ joules}.$$

Recall that the *atomic mass unit* (amu) is given by 1 amu $= 1.66 \times 10^{-27}$ kilograms. How much energy is obtained if 1 amu is converted to energy by Einstein's formula? Because $c = 3 \times 10^8$, we get

$$E = mc^2 = (1.66 \times 10^{-27})(3 \times 10^8) \text{ kg-meter}^2/\text{sec}^2$$

$$= (1.66 \times 10^{-27})(9 \times 10^{16}) \text{ kg-meter}^2/\text{sec}^2$$

$$= 14.94 \times 10^{-11} \text{ joules}$$

$$= (14.94 \times 10^{-11}) \cdot \frac{1}{1.6 \times 10^{-13}} \text{ MeV}$$

$$= 9.33 \times 10^2 \text{ MeV}.$$

The mass of a proton is 1.0076 amu and that of a neutron is 1.009 amu. About 940 MeV are produced if one of these particles is converted to energy. In foot-pounds this is equal to

$$940 \text{ MeV} = 940(1.6 \times 10^{-13}) \text{ joules}$$

$$= 940(1.6 \times 10^{-13})(0.72) \text{ foot-pounds}$$

$$= 1.1 \times 10^{-10} \text{ foot-pounds}.$$

This is a minuscule amount of energy.

13. Determine the sum of the masses (in amu) of the protons and neutrons in the nucleus of a $^{235}_{92}\text{U}$ atom. Work with 4-decimal accuracy.

14. You are given that the mass of a $^{235}_{92}\text{U}$ nucleus is 235.1175 amu. Use the result of Exercise 13 and Einstein's equation $E = mc^2$ to compute the binding energy of a $^{235}_{92}\text{U}$ nucleus first in MeV and then in joules.

11C. Nuclear Fission

As was just discussed, there is energy bound up within the nucleus of an atom. These are small amounts within a single atom, but they can be enormous quantities in larger samples of matter. Scientists learned how to tap this energy in the 1940s, and their discoveries have both benefited and threatened the world ever since. Released steadily in a nuclear power plant, this energy has been the source of considerable amounts of the world's electricity. Released violently in a nuclear bomb, it can destroy cities (and has done so). Indeed, for several decades of the 20th century it brought the world face to face with the prospect of extinction.

We will designate a neutron by $^1_0 n$. (Since a neutron consists of 1 neutron and 0 protons, this is the same notation that is used for isotopes of elements.) When the nucleus of a uranium-235 atom is bombarded with a neutron, it can happen that the nucleus shatters into two smaller nuclei and a number of neutrons. Such an event, now called *nuclear fission*, is accompanied by a release of energy. For instance, the diagram

$$^{235}_{92}\text{U} + ^1_0 n \rightarrow ^{139}_{56}\text{Ba} + ^{94}_{36}\text{Kr} + 3^1_0 n + \text{energy}$$

indicates that the uranium-235 nucleus splits into $^{139}_{56}\text{Ba}$ (an isotope of the metal barium), $^{94}_{36}\text{Kr}$ (an isotope of the gas krypton), and 3 free neutrons. Another possible outcome is

$$^{235}_{92}\text{U} + ^1_0 n \rightarrow ^{141}_{57}\text{Pr} + ^{93}_{42}\text{Nb} + 2^1_0 n + \text{energy}$$

where $^{141}_{57}\text{Pr}$ and $^{93}_{42}\text{Nb}$ are two isotopes of exotic metals. This time only 2 free neutrons occur. A third possibility is

$$^{235}_{92}\text{U} + ^1_0 n \rightarrow ^{139}_{57}\text{La} + ^{95}_{42}\text{Mo} + 2^1_0 n + \text{energy}$$

where $^{139}_{57}\text{La}$ and $^{95}_{42}\text{Mo}$ are two more isotopes of metals. The important common feature in all such reactions is that the bombarded uranium-235 nucleus breaks up into large fragments and some neutrons, and that there is an accompanied release of energy. Some thirty different fragments, called *fission fragments*, have been observed. Many of them are highly radioactive.

In any fission reaction the total number of protons and neutrons is conserved and the energy that is released is that part of the binding energy that had kept the two fragments (and neutrons) together before the splitting. So the amount of energy released can be computed by comparing the total masses before and after the reaction and converting it to energy with Einstein's equation.

15. Consider the reaction

$$^{235}_{92}\text{U} + ^1_0 n \rightarrow ^{139}_{57}\text{La} + ^{95}_{42}\text{Mo} + 2^1_0 n + \text{energy}.$$

A $^{235}_{92}\text{U}$ nucleus has a mass of 235.118 amu, a $^{139}_{57}\text{La}$ nucleus has a mass of 138.950 amu, and a $^{95}_{42}\text{Mo}$ nucleus has a mass of 94.936 amu. Compute the mass of matter in amu before the reaction and again thereafter. Determine the energy released by the reaction in MeV.

11D. Chain Reactions

Consider a sample of uranium. A typical fission reaction releases about 200 MeV. (The solution of Exercise 15 confirms this.) We saw in Exercises 11B that this is a tiny amount of energy. But there is the theoretical possibility that some of the two or three neutrons freed by the fission of a uranium-235 nucleus will bring about fissions in nearby uranium-235 nuclei. These fissions will again produce energy and additional free neutrons. These neutrons could in turn do the same. The continued repetition of this sequence could lead to a *chain reaction* in which, step by step, a huge number of nuclei undergo fission. If this happens, a huge number of tiny bursts of energy would be released, and the total amount of energy that is generated could be enormous.

16. Use Avogadro's number and the fact that 1 pound equals 453.6 grams to show that there are 1.16×10^{24} atoms in 1 pound of pure uranium-235. Take 200 MeV as the average energy release per fission and show that if all the nuclei undergo fission, then a total energy of 3.71×10^{13} joules will be released. How much energy is this? To get a rough idea, use the fact that 1

mile is 5280 feet, and check that

$$3.71 \times 10^{13} \text{ joules} = 2.67 \times 10^{13} \text{ foot-pounds}$$

$$= (1000 \text{ pounds}) \times (5 \times 10^6 \text{ miles}).$$

So the total fission energy contained in 1 pound of uranium is equivalent to the energy produced by a 1000 pound force acting for 5 million miles. This is also equivalent to the heat generated by burning about 2,500,000 pounds of coal. (Refer also to Note 9 at the end of Chapter 12.)

A chain reaction does not happen in naturally occurring uranium. The reason is that in naturally occurring uranium, 99.28% of the nuclei are uranium-238 nuclei and only 0.715% are fissionable uranium-235 nuclei. When a uranium-235 atom fissions, the neutrons released will in all likelihood be absorbed by the surrounding uranium-238 nuclei. The probability that one of them will produce a fission of a uranium-235 nucleus is too low to sustain a chain reaction. But can this environment be modified so that such a chain reaction can occur? This question and its answer were of pivotal concern in human history. The key lies in the fact that uranium-238 nuclei do not absorb slow neutrons. If, therefore, the neutrons released in uranium-235 fission can be slowed to the right speed range, they would pass by the uranium-238 nuclei without being absorbed and would eventually produce fission in the uranium-235 nuclei. It turned out that neutrons can be slowed down by introducing materials called *moderators*. Common moderators are water and graphite.

The first chain reaction was achieved in 1942 at the University of Chicago. This effort was directed by Enrico Fermi (1901–1954), who had won the Nobel prize for physics in 1938. It was part of the "Manhattan project," the U.S. government's (ultimately) successful effort to build a nuclear bomb. About 6.2 tons of uranium was imbedded in a graphite lattice in 19 alternating layers of graphite and uranium. The control of this so-called *pile*, as well as the assurance that it would not "start up" during construction, was accomplished by inserting cadmium control rods. The high neutron-absorbing property of cadmium kept the number of fissions sufficiently low. By slowly withdrawing the control rods, more and more neutrons became available for fission, and the startup of the pile was realized. With all the control rods out, a chain reaction was achieved and sustained. After 28 minutes, the cadmium rods were reinserted and the chain reaction was stopped.

Consider a chain reaction, and let r be the average number of neutrons per fission that go on to produce another fission. If $r < 1$, the chain reaction is not sustained and stops.

Such a reaction is *subcritical*. If $r = 1$, one neutron from each fission goes on to produce another fission. Such a *controlled reaction* is in steady state. This is the operating mode in the reactor of a thermal nuclear power plant. The energy that is produced can be harnessed, for instance by circulating water through the reactor and producing steam. If $r > 1$, more than one neutron from each fission goes on to produce another fission. In such a *supercritical* situation, the number of fissions can increase at a rapid rate and produce an explosion. The chain reaction at the University of Chicago became supercritical with $r = 1.0006$. The insertion of the control rods made it subcritical, and it stopped soon thereafter.

Consider a chain reaction and suppose that N neutrons fission initially. This first step produces rN neutrons that will fission. These in turn will produce $r(rN) = r^2 N$ neutrons that will fission, which in turn will produce $r^3 N$ neutrons that will fission, and so on. The total number of neutrons that will have fissioned after m such steps is

$$N + rN + r^2 N + r^3 N + r^4 N + \ldots + r^m N.$$

If E is the average energy released per fission, then the total energy $E_{m\text{-total}}$ that these m steps produce is

$$E_{m\text{-total}} = EN + rEN + r^2 EN + r^3 EN + \ldots + r^m EN.$$

17. Show that $E_{m\text{-total}} = \frac{(r^{m+1}-1)}{(r-1)} EN$ by computing $rE_{m\text{-total}} - E_{m\text{-total}}$.

18. Suppose that $N = 100$ neutrons start a chain reaction, that $E = 200$ MeV, and that $r = 1.005$. Also assume that the average time in which a newly created free neutron produces a fission is 10^{-12} seconds. Show that the amount of energy produced by the chain reaction in the first 10^{-8} seconds is approximately 1.84×10^{28} MeV. Use Exercise 16 to convert this amount of energy to its "coal equivalent." [Hint: What value of m should be used in the formula of Exercise 17?]

19. One of the waste products of a nuclear explosion is the radioactive isotope strontium-90. It has a half-life of about 25 years. This dangerous isotope behaves chemically like calcium and it takes the place of calcium in the metabolic processes of human beings and animals. If 20 milligrams of the isotope are present in a sample now, find how much will remain in 15 years? In how many years will only 5 milligrams remain?

11E.   Critical Mass

The average number of neutrons produced by a single fission of a uranium-235 nucleus is about 2.5. Any such neutron will do one of three things: Escape from the system; remain in the system, but not produce a fission (e.g., it may be absorbed);

or remain in the system and produce a fission. In order to produce a chain reaction, it is clear that neutron escape must be controlled. An important factor in this regard is the size and shape of the uranium sample in question. Suppose, for instance, that it has the shape of a sphere of radius R. The number of neutrons that escape turns out (not surprisingly) to be proportional to the surface area of the sphere (which is $4\pi R^2$). The number of fissionable uranium-235 nuclei is proportional to the volume of the sphere (which is $\frac{4}{3}\pi R^3$).

20. Suppose that a sample of essentially pure uranium-235 has the shape of a sphere of radius R. Show that the ratio

$$\frac{\text{(number of neutrons that escape)}}{\text{(neutrons that can produce a fission)}}$$

is proportional to $\frac{1}{R}$. So the loss can be minimized by making R large enough. At a radius that is appropriately large, the sample will have what is called *critical size* or *critical mass*.

The critical mass depends on other factors, e.g., the purity of the fissionable material, the type and arrangement of the moderator, and the reflectors that are used to scatter escaping neutrons back into the system.

A uranium bomb is an assembly consisting of two sub-critical masses of uranium-235 that are large enough to become critical when brought together. One of these has the shape of a sphere with a deep depression. When the other mass is propelled by a conventional charge into the hollow of this sphere, critical mass is achieved. The bomb has now been detonated. A supercritical chain reaction begins that must be contained long enough for large numbers of the uranium-235 nuclei to fission. The average number r of neutrons per fission that goes on to produce another fission should now be as large as possible. To achieve a large r, *enriched* uranium is used. Recall that in natural uranium only 0.715% is the fissionable uranium-235, most of the rest being uranium-238. There are processes by which naturally occurring uranium can be converted into a uranium that contains much larger percentages of uranium-235. In "weapon-grade" uranium this can be 95%. It is clear that in such enriched uranium the probability that a neutron will hit and split a uranium-235 nucleus is much greater.

11F. Fusion

There is another nuclear process that produces energy. Consider one atom of the hydrogen isotope $^2_1$H and another of the hydrogen isotope $^3_1$H. Both are "heavy" versions of the gas hydrogen $^1_1$H. Compare this with one helium atom

and one neutron. So we are comparing

$$^2_1\text{H} + {}^3_1\text{H} \quad \text{and} \quad {}^4_2\text{He} + {}^1_0 n.$$

Observe that for each pair, the total atomic number is 2 and the mass number is 5. It turns out that the total binding energy of the pair of hydrogen isotopes is 17.6 MeV less than that of the helium atom. Under extremely high temperatures (in the millions of degrees) the hydrogen isotopes can be fused together in the reaction

$$^2_1\text{H} + {}^3_1\text{H} \rightarrow {}^4_2\text{He} + {}^1_0 n + 17.6 \text{ MeV}.$$

The Sun's energy is generated by such hydrogen-to-helium *fusion reactions*. The conversion of mass to energy in the Sun currently occurs at a rate of 4.3×10^9 kilograms per second. This, while a huge amount, is but a small fraction of the Sun's mass. Recall from Section 7.6 that the Sun has a mass of 2×10^{30} kg. Astrophysicists generally believe that the Sun is a normal star about halfway through a 10 billion year lifetime and that it will produce energy at current levels for about another 4.5 billion years.

Fusion requires the confinement of material sufficiently dense for a long enough time so that the reactions can take place and return more energy than was necessary to initiate the process. The first occurrences of fusion on Earth were the tests of hydrogen bombs in the 1950s. The high temperatures necessary were obtained by the explosion of fission bombs. A fission bomb of the explosive power of several thousand tons of TNT is used as a trigger in a fusion bomb of the explosive power of several million tons of TNT. To date, nuclear physicists have not succeeded in producing a controlled fusion reaction on Earth. This remains one of the great challenges of modern physics. The implications for the world's energy supply are enormous.

11G. About the Moon

The Apollo manned space missions to the Moon during the years 1969–1972 collected 382 kilograms of lunar rocks from six sites.

21. Some lunar rocks formed by meteor activity were collected by the Apollo 16 mission and found to be about 4.53 billion years old by rubidium-strontium analysis. Assume that there was no strontium in the rock when it was formed and determine the current percentage of strontium to rubidium.

22. A rock collected by the Apollo 16 mission from the lunar plains was tested and shown to be approximately 4.19 billion years old by the potassium-argon method. Assume that the rock contained no argon at the time of its formation and estimate the current ratio of argon to potassium in the rock.

The evidence is, therefore, that the Moon formed at the same time as the Earth. The Apollo missions also provided new information about the origins of the Moon. The Moon and the Earth have identical amounts of oxygen isotopes. This indicates that they are closely related. However, data from lunar samples and a network of seismometers left behind on the Moon enable scientists to conclude that the Moon and Earth have different chemical compositions. It appears that certain elements (for example aluminum and calcium) are present in the Moon in concentrations that are 50% higher than on the Earth. Also, the ratio of some common oxygen compounds seems to be about 10% higher on the Moon than on the Earth. There appears now to be a consensus among research scientists on a theory of the Moon's origin. The Moon formed from the Earth by the glancing impact of an object the size of a small planet, a colossal event early in the history of the formation of the solar system. The debris from the collision collected to form the Moon. This projectile must have struck the Earth off center, causing the Earth's rotation to speed up to its current value. The Moon is dry because of the enormous heat that the collision generated. The differences in the chemical composition would be accounted for by the fact that the Moon formed mostly from the debris of the impacting object.

11H.   Geology and Anthropology

23. The Red Sea is between 100 miles and 200 miles wide. Suppose that it were to continue to separate at a rate of $\frac{1}{2}$ inch per year for the next 100 million years. Give an estimate for its width after that time.

24. A fragment of a mineral grain is found to contain 305 parts per million rubidium-87 and 4.67 parts per million strontium-87. Determine the age of the fragment.

25. A fragment of a mineral grain is found to contain 420 parts per million rubidium-87 and 5.3 parts per million strontium-87. What is the age of the fragment?

26. Mass spectrometry readings from a mineral grain in a sample of basalt taken from the Grand Canyon show that it contains 1.1739×10^{12} atoms per gram of argon-40 and 1.7368×10^{16} atoms per gram of potassium-40. How long ago did the grain crystallize?

27. Consider the discovery of the stone tools discussed in Example 11.9 of the text. Assume that the labora-

tory that tested the volcanic sediment was only able to determine the ratio of argon-40 to potassium-40 within the range $0.00011 \le \frac{z(t)}{y(t)} \le 0.00018$. What conclusion about the age of the tools follows from this estimate?

28. A wood fragment tested with an accelerator mass spectrometer is found to contain carbon-14 atoms to stable carbon atoms in a ration of 1 to 1.573^{12}. Compute the age of the fragment. Assume that at the time the metabolic processes in the wood stopped, the equilibrium ratio of radioactive carbon to stable carbon was the same as it is today.

29. Measurements with a radiation counter indicate that 15.30 disintegrations per minute are produced by one gram of carbon. Consider a one-gram sample of carbon at time $t = 0$ and let $y(t)$ be the number of radioactive carbon-14 atoms in this sample at any time t.

    i. How many grams of radioactive carbon-14 does the sample contain at time $t = 0$?

    ii. Compute $y_0 = y(0)$.

    iii. Use the fact that $y'(0) = -\lambda y(0)$ to deduce the value $\lambda = 2.23 \times 10^{-10}$ (in minutes^{-1}) for the decay constant of carbon-14.

    iv. Compute the half-life h of carbon-14 in minutes. Show that this converts to $h = 5727$ years.

30. Tests of fragments of skeletons unearthed near the town of Arella, Pennsylvania, gave evidence of a civilization that existed in this area from around 14,300 to 15,000 years ago. Assume that the equilibrium ratio of carbon-14 atoms to stable carbon atoms in the fragments at that time was the same as today's value. Estimate the ratio of carbon-14 atoms to stable carbon atoms in the fragments when they were tested.

31. The australopithecus skull mentioned in Section 11.3 (it was discovered by the famous anthropologist Richard Leakey in Kenya and later designated "1470") was shown to be about 1,750,000 years old by potassium-argon dating. Suppose that at the time of the potassium-argon test, a fragment of the skull had also been tested for its carbon content. What ratio of carbon-14 atoms to stable carbon atoms would the test have revealed? What does this say about the range of applicability of carbon dating?

11I. Integrals and Equations involving Derivatives

Exercises 32 and 33 require a careful study of the methods of Section 11.5B.

32. Use common denominators "in reverse" to solve the following indefinite integrals

  i. $\int \dfrac{1}{x(x-1)}\,dx$

  ii. $\int \dfrac{1}{(x-2)(x-3)}\,dx$

  iii. $\int \dfrac{x}{(x-2)(x+3)}\,dx$

  iv. $\int \dfrac{x+1}{(x+2)(x-3)}\,dx$

33. Separate variables, antidifferentiate, and use properties of the ln function (and algebra) to find an explicit function $y(t)$ that satisfies the given conditions.

  i. $\dfrac{dy}{dt} = 3(y-2)(y+4)$ and $y(0) = 4$

  ii. $(y+1)\dfrac{dy}{dt} = 2(y-2)(y+4)$ and $y(0) = 3$

  iii. $\dfrac{1}{t}\dfrac{dy}{dt} = (y-2)(y+4)$ and $y(0) = 7$

  iv. $\dfrac{y+1}{t}\dfrac{dy}{dt} = 5(y-2)(y+4)$ and $y(0) = 6$

  v. $\dfrac{dy}{dt} = 2(y-2)^2$ and $y(0) = 1$. [Hint: Start with the question $\int \frac{1}{x^2}\,dx = ?$]

11J. The Logistics Model

34. When some population is observed for the first time, it has $y(0) = y_0 = 2.5$ million individuals. Determine the number of individuals $y(t)$ in the population at any time t, if

  i. Its specific growth rate satisfies the equation $\dfrac{y'}{y} = 0.08 - 0.02y$.

  ii. Its specific growth rate satisfies the equation $\dfrac{y'}{y} = 0.02 - 0.08y$.

35. The analysis of the world's population in Section 11.5C was based on the estimate $\dfrac{y'}{y} = 0.028 - 0.002y$ for the specific growth rate for the years 1965 to 1995. The method of "least squares" that will be discussed later in Section 12.4B provides the estimate

$$\frac{y'}{y} = 0.0261 - 0.0018y,$$

which fits the data points better. Use it to derive the function $y(t)$. What population levels does this $y(t)$ predict for the years 2000, 2010, 2020, 2050, and 2095?

36. The statistics for the population of a hypothetical country show the following:

4.50 million at the beginning of 1950,
and an increase at a rate of 0.24 million per year in 1950;
8.43 million at the beginning of 1960,
and an increase at a rate of 0.44 million per year in 1960;
14.71 million at the beginning of 1970,
and an increase at a rate of 0.73 million per year in 1970;
23.40 million at the beginning of 1980,
and an increase at a rate of 1.13 million per year in 1980;
36.12 million at the beginning of 1990,
and an increase at a rate of 1.56 million per year in 1990.

Let $t = 0$ correspond to the year 1950. Let $y(t)$ be the population of this country at any time $t \geq 0$ in years. Verify that the population data of this country satisfy the basic assumption of the logistics model and determine the parameters y_0, μ, and M. (Work with an accuracy to within 3 decimal places.) Determine the function $y(t)$. What is the limit on the population of this country? [Hint: Use the stategy of Section 11.5C.]

37. The statistics for the population of a certain (hypothetical) country show the following:

8.45 million at the beginning of 1975,
and an increase at a rate of 0.21 million per year in 1975;
9.56 million at the beginning of 1980,
and an increase at a rate of 0.23 million per year in 1980;
10.77 million at the beginning of 1985,
and an increase at a rate of 0.25 million per year in 1985;
12.09 million at the beginning of 1990,
and an increase at a rate of 0.27 million per year in 1990;
13.52 million at the beginning of 1995,
and an increase at a rate of 0.30 million per year in 1995.

  i. Show that the population data of this country satisfy the basic assumption of the logistics

Table 11.8

U.S. Population Estimates 1790–1910 (in millions)														
1790	3.93	0.13	1820	9.64	0.30	1850	23.19	0.90	1880	50.19	1.35	1910	92.23	1.63
1800	5.31	0.18	1830	12.87	0.45	1860	31.44	0.91	1890	62.98	1.38			
1810	7.24	0.22	1840	17.07	0.66	1870	38.56	2.38	1900	76.21	1.37			

model and determine the parameters y_0, μ, and M. [Hint: Use the approach of Section 11.5C.]

  ii.   What limit on the population of this country does the logistics model predict?
 iii.   Determine the function $y(t)$ for $t \geq 0$.
  iv.   What is the predicted population for the year 2020?

38.   The information in Table 11.8 was extracted from official data of the U.S. Bureau of the Census. The table lists the population at the beginning of each year and the increase during that year. For example, in 1860 the U.S. population was 31.44 million and it increased by 0.91 million to 32.35 million in 1861. Designate 1790 as $t = 0$, and let $y(t)$ be the population at any time $t \geq 0$ (in years). So $y(0) = 3.93$ million.

  i.   Reproduce Table 11.4 of Section 11.5C for this data. (Use millions, rather than billions.)
 ii.   Plot the data as in Figure 11.7 of Section 11.5C. Do the data fall on a line?
iii.   Compute the constants μ and k of the logistics model. What is the predicted limiting value of the population?
 iv.   Determine the population function $y(t)$ and evaluate it for $10 \leq t \leq 120$ in ten-year intervals. Compare these values with the corresponding values in Table 11.8.
  v.   The U.S. Bureau of the Census has supplied (See Table 11.9) estimates for the years 1920–1990. Check how well the values $y(140)$, $y(160)$, $y(180)$, and $y(200)$ fit these values.
 vi.   Repeat steps (i) and (ii) with the data from Table 11.9. Do they satisfy the requirements of the logistics model?

Exercise 38 is the work of Pearl and Reed in the 1920s. Unaware of the work of Verhulst, they applied the logistics model to the population statistics of the United States for the years 1780 to 1910. Using slightly different population estimates, they arrived at the result

$$y(t) = \frac{197{,}274{,}000}{1 + 67.3219335e^{-0.0313395t}}.$$

Observe that in their version of the study $M = 197{,}274{,}000$, $\mu = 0.0313395$, and $\frac{M}{y_0} - 1 = 67.3219335$.

11K.   The Growth of Microbes

39.   Suppose that a colony of bacteria is growing exponentially. If the number of bacteria increases from 4000 to 6000 in 12 hours, find the doubling time.

40.   A bacteria culture is in its exponential growth phase. It has 1000 bacteria when it is first observed at time $t = 0$. Two hours later there are 4000 bacteria.

  i.   Express the number $y(t)$ of bacteria at any time $t \geq 0$ in terms of an exponential function.
 ii.   What is the size of the population after 5 hours?
iii.   At what time will the population reach 30,000?

41.   A bacteria culture is in its exponential growth phase. It starts with 10,500 bacteria at time $t = 0$. Two hours later there are 23,000 bacteria.

  i.   Express the number $y(t)$ of bacteria at any time $t \geq 0$ in terms of an exponential function.
 ii.   What is the size of the population after 6 hours?
iii.   At what time will the population reach 130,000?

42.   A bacteria culture is in its phase of exponential growth. The cell count was 5000 exactly 2 hours after it was first observed and 256,000 exactly 7 hours after it was first observed.

  i.   What was the number of bacteria when it was first observed at time $t = 0$?
 ii.   Determine the number $y(t)$ of bacteria in the culture at any time $t \geq 0$ as an exponential function?
iii.   What is the doubling time of the culture?

43.   Repeat the analysis in Section 11.6B of the experiment of M'Kendrick and Pai for Trial 5.

Table 11.9

	U.S. Population Estimates 1920–1990 (in millions)										
1920	106.02	2.52	1940	132.16	1.24	1960	179.32	4.37	1980	226.54	3.43
1930	123.20	0.84	1950	151.33	3.55	1970	203.30	4.36	1990	248.71	3.93

44. In 1934 the Russian biologist G. F. Gause placed 5 protozoans of the species *Paramecium caudatum* in 5 cubic centimeters of a nutritive medium. The number of individuals was counted daily. The average data from 63 separate counts showed that the population $y(t)$ at any time t in days was approximately

$$y(t) = \frac{375}{1 + e^{5.169 - 2.309t}}.$$

Show that this conclusion fits the logistics model and determine the constants y_0, M, and μ.

Gause showed that the population functions for various yeasts also follow the logistics model. He also experimented with two different species of protozoans. He cultured two closely related species of protozoans together on a constant supply of nutrition. His experiments confirmed the hypothesis that two species with similar needs for the same limiting resources cannot coexist, at least in controlled laboratory conditions. Even a slight reproductive advantage would eventually lead to the elimination of the weaker competitor. He simulated this phenomenon mathematically with a pair of differential equations.

45. Recall from Section 11.6C that the rate of change in the concentration of the product in the medium is $p'(t) = y(t)(a\mu + b - \frac{a\mu}{M}y(t))$. Verify that $p'(t)$ attains its maximum value when $y = \frac{M}{2}(1 + \frac{b}{a\mu})$.

11L. Gompertz's Model

Benjamin Gompertz (1779–1865) was a prominent British actuary. Being Jewish, he was barred from a university education. He studied mathematics privately, however, and developed into a mathematician of such distinction that he later became president of the London Mathematical Society.

46. In the 1820s Gompertz developed a population model based on the assumption that the specific growth rate declines exponentially with time. More precisely, he considered a function $y(t)$ that satisfies

$$\frac{y'(t)}{y(t)} = mke^{-kt}$$

for all $t \geq 0$ and some positive constants m and k.

i. Show that $\ln y(t) = -me^{-kt} + C$. [Hint: Differentiate both $\ln y(t)$ and $-me^{-kt}$.]

ii. Deduce that $y(t) = y_0 e^m e^{-me^{-kt}} = y_0 e^{m - me^{-kt}}$, where $y_0 = y(0)$.

iii. Show that $y(t)$ tends to $y_0 e^m$ as $t \to \infty$.

iv. Find $y''(t)$ in terms of $y(t)$ by differentiating the equation $y'(t) = y(t)mke^{-kt}$.

v. Show that the graph of $y(t)$ has an inflection point when $t = \frac{1}{k}\ln m$. Show in particular that it has no inflection point when $m \leq 1$.

47. Suppose that Gompertz's model $y(t) = y_0 e^{m - me^{-kt}}$ simulates both the exponential phase and the stationary phase of Trial 4 of the *E. coli* experiment in Section 11.6B.

i. Use the initial cell count of 2850 and a stable count of 100,000,000 to show that $m = 10.5$. Determine k by using the three hour cell count of 625,000.

ii. Reproduce Table 11.7 for the resulting function $y(t)$. Does this $y(t)$ or the $y(t)$ derived in Section 11.6B fit the observed cell counts better?

48. Return to the study of the world's population from 1965 to 1995 in Section 11.5C and in particular to Table 11.4. Consider a function $y(t)$ that satisfies $\frac{y'}{y} = 0.021$ for $t = 0$, i.e., for the year 1965, and $\frac{y'}{y} = 0.016$ for $t = 30$, i.e., for the year 1995. Suppose also that $y(t)$ satisfies Gompertz's specific growth rate condition $\frac{y'(t)}{y(t)} = mke^{-kt}$.

i. Deduce that $\frac{y'(t)}{y(t)} = 0.021e^{-0.009t}$ and hence that $k = 0.009$ and $m = 2.3$. Conclude that $y(t) = 3.3e^{2.3 - 2.3e^{-0.009t}}$.

ii. Compute $\frac{y'(t)}{y(t)}$ for $t = 5, 10, 15, 20$, and 25, and compare the results with those from Table 11.4. Does the logistics model or that of Gompertz give the better fit for the specific growth rate for the period 1965 to 1995?

iii. What projections for the world's population does Gompertz's model provide for the years 2000, 2020, 2050, and 2095?

iv. At what count will the world's population stabilize according to Gompertz's model?

A comparison of the predictions for the world's population of the logistics model with those of Gompertz's model shows that they yield very different conclusions—even though both fit the population trends of 1965–1995 rather well. The point is that different mathematical interpretations of the same data can lead to very different results.

11M. Monod's Equation

Many antibiotics are produced by *continuous fermentation*. Consider bacteria growing in a culture medium in a tank. Suppose that a reservoir feeds a certain fresh nutrient medium into the tank and that this occurs at a relatively slow but steady rate that can be regulated. The culture medium is constantly mixed, and the excess, including the bacteria that are in it, is allowed to drain from the tank at the same steady rate. On a small laboratory scale, such an apparatus is called a *chemostat*. Changing the concentration of the inflowing nutrient will change the specific growth rate of the culture. For a certain concentration of the nutrient the conditions for growth will be best. The culture is now in an exponential phase of growth, and the number of cells will double in minimal time d. Since $\mu = \frac{\ln 2}{d}$, the specific growth rate $\frac{y'}{y} = \mu$ is now at its maximum value $\mu = \mu_{max}$.

In the 1950s, the French microbiologist Jacques Monod (1910–1976) investigated the genetics of bacteria and the dynamics of their growth. In 1953 he was called to direct the Department of Cellular Biology at the Pasteur Institute in Paris, and in 1965 he was awarded the Nobel Prize for medicine. For a culture growing in a chemostat he observed a connection between the specific growth rate $\frac{y'}{y}$ of the culture and the concentration s of the nutrient in the medium with which it is supplied. More precisely, Monod discovered that

$$\frac{y'}{y} = \frac{\mu_{max} s}{s_{1/2\,max} + s}$$

where $s_{1/2\,max}$ is the nutrient concentration that yields a specific growth rate of $\frac{1}{2}\mu_{max}$, i.e., one half of the maximum. This equation is commonly called *Monod's equation*. Suppose that s is very large relative to $s_{1/2\,max}$. In this case, $s_{1/2\,max} + s \approx s$, and by Monod's equation, $\frac{y'}{y} \approx \mu_{max}$. So, as expected, at such a nutrient concentration the culture will grow at a near maximal rate. If, on the other hand, s is very low, then $\frac{\mu_{max} s}{s_{1/2\,max} + s} \approx 0$.

Monod's equation is important in industrial applications. The maximal production rate of a chemical in a fermentor depends on the specific growth rate $\frac{y'}{y}$ of the particular bacteria in question. It is the purpose of laboratory experiments to determine the specific growth rate at which this occurs. The strategy is to measure the nutrient level s that yields the highest production and to compute the corresponding $\frac{y'}{y}$ by using Monod's equation.

A chemostat is a good model for bacterial growth in open systems such as rivers and lakes. The continuous removal of the culture from the bottom of the chamber simulates the mortality rate of the population. Suppose, for example, that the chamber contains 1 liter of culture medium and that the flow rate of the chemostat is maintained at 0.1 liter per day. Then 10% of the bacteria are removed from the chamber per day, and thus a 10% mortality rate is simulated. As a result, Monod's equation has been used to model the growth of microbe populations in experiments that study the dependence of such growth on a limited resource. See Exercise 49 for an example of such a study.

49. *Asterionella formosa*, a type of algae, needs silicate (SiO_2 in the notation of chemistry) to survive. The dependence of a culture on the silicate was studied in a chemostat. The maximum growth rate of a culture was found to be $\mu_{max} = 0.6$ (days)$^{-1}$, and the concentration of silicate at which half the maximum growth rate occured was $s_{1/2\,max} = 9M$ (micromoles).

   i. Write down Monod's equation for this situation.

   ii. Suppose that after some time $t > 0$, the silicate resource was given by $s(t) = \frac{50}{t^2} + 1$.

   iii. Show that $F(t) = at - a\sqrt{b}\tan^{-1}(\frac{t}{\sqrt{b}})$ is an antiderivative of $f(t) = \frac{at^2}{t^2 + b}$, and use this to determine $y(t)$. [Hint: Review Section 10.5.]

50. Consider a population and let $y(t)$ be the number of individuals at any time $t \geq 0$. Suppose that the specific growth rate is $\frac{y'(t)}{y(t)} = f(t)$ for some function of time $f(t)$.

   i. Show that Monod's equation applies to this general situation. What must the function $s(t)$ be set equal to?

   ii. Consider the logistics model in Section 11.5B. What is $f(t)$ in this case? What is $s(t)$? [Hint: Use equation (11j) to compute $\frac{y'(t)}{y(t)}$.]

Notes

[1] Refer to Rutherford's classic text *Radioactive Substances and their Radiations*, Cambridge: at the University Press, 1913, for the details.

[2] Consult an atlas and check how well the coastline of Brazil fits the coastline of Africa from Ivory Coast to Angola.

[3] See R. Hughes, "Behold the Stone Age," *Time*, February 13, 1995. See also, "Dawn of Art: A New Vista," *New York Times*, June 8, 1995.

[4] There are a number of such accidental discoveries in the history of science. Recall, for example, Roentgen's discovery of X-rays and Becquerel's discovery of radioactivity. It is appropriate here to recall the dictum of Pasteur that "chance favors the prepared mind." The French microbiologist Louis Pasteur (1822–1895) was a pioneer in the study of the connection between microbiological organisms and disease.

[5] There are more precise methods to fit a line to a set of data points than simply to take a line that two of the points determine. We will consider one of these methods in Section 12.4B.

[6] *E. coli* is a primary component of feces and has been used since the early days of bacteriology to measure the sewage contamination of water. Today, it is also a focal point of DNA research and genetic engineering. It received its current name *Escherichia coli* or its shortened version *E. coli* in 1919 to honor its discoverer, the German pediatrician and bacteriologist Theodor Escherich (1857-1911). *E. coli* bacteria are benign; indeed, as already observed, they play a positive role in the digestion process. However, recently a "killer" strain of these bacteria has been tracked down. See the article "Detective Work and Science Reveal a New Lethal Bacteria," The New York Times, Tuesday, January 6, 1998.

12

The Calculus of Economics

In 1985 Ford Motor Company introduced the Taurus, a newly designed, front-wheel-drive automobile. The car proved to be a big success and helped Ford to almost double its profits by 1987. The design and efficient production of this car involved some impressive engineering advances, but the success of the venture also depended on extensive economic analysis.

Ford had to think carefully about the public's reaction to the Taurus's design. Would consumers be impressed by the car? How strong would demand be initially, how fast would it grow, and how would it depend on the car's price? Clearly, Ford needed to be concerned about the costs of manufacturing the car and how these would depend on the number of cars produced per year. How much and how fast would costs decline as managers and workers gained experience with the production process? What sales and pricing strategies would maximize profits? The Taurus program would require a large investment in new capital equipment. Should this be financed by the sale of assets, a stock offering, or by a bond issue? The answer to these broad questions required attention to a lot of details: Should the parts and components for the car be obtained from "upstream" divisions, or should they (or at least some of them) be purchased from outside firms? Would the prices of steel and other raw materials remain stable? What would be the effect on sales of changes in the price of gasoline, in the level of consumers' disposable income, and in the state of the economy as a whole? Ford also had to worry about union wage negotiations, emission standards, health and safety regulations of production line operations, and the effects of governmental regulatory policies in general.

It should be clear by now that a formidable and far-reaching economic analysis was necessary before Ford's executives were able to decide, first on the feasibility, and then on the scope, of the Taurus project. Many of the basic concerns just raised can be quantified and subjected to statistical studies in terms of projections and comparisons with similar projects undertaken in the past. By focusing on one issue at a time, concentrating on the important factors and variables, and omitting those of lesser significance, it is possible to reduce these considerations to certain essential elements. Thus simplified to manageable proportions, mathematical methods can illuminate fundamental relationships and suggest answers to the litany of questions about the project.

Economics, put briefly, is the study of the marketplace. The first systematic exploration of the role of mathematics in economics was the *Researches into the Mathematical Theory of Wealth*, published by Augustin Cournot in 1838. Cournot (1801–1877) was a mathematician and philosopher, who served as a university official in Lyons and Grenoble and held high positions in the French government. This treatise seemed a failure when first published—its reasoning was too sophisticated and advanced for the times—but proved to be very influential later. By the latter part of the 19th century there were a number of economists, most notably the Englishmen Stanley Jevons (1835–1882) and Alfred Marshall (1842–1924), who were scribbling equations with X's and Y's, big deltas and little d's, in an effort to understand and enlighten economics. The quantification and mathematization of economics has expanded ever since and has grown into a sophisticated, powerful, and central tool of economic analysis. The human factor that is obviously involved brings a note of caution. Unlike the situation in engineering, physics, or even biology, the fit between the mathematical theory of the marketplace and its practice is "loose." Rather than a precise predictor, the theory offers only general indications. These, however, can be extremely valuable.

12.1 **Basics of Banking**

Money plays a fundamental role in all of the concerns just discussed. This section will explore how it behaves.

A. Interest. Interest has been a very basic property of money ever since the practice of borrowing and lending began.

Suppose that an account is opened at a bank with an initial investment of A_0 dollars and that it earns 6 per cent per year. Since 6 per cent means 6 per 100, the interest earned after one year is the fraction $\frac{6}{100} = 0.06$ of A_0, or $0.06A_0$. If $A_0 = \$1000$, for instance, then the interest earned in one year is $0.06(\$1000) = \$60$. In view of this computation, interest will be expressed as a fraction r between 0 and 1. So 4% corresponds to $r = 0.04$, 7% to $r = 0.07$, 9% to $r = 0.09$, and so forth.

The process of computing the interest on the amount of money in an account and adding it to the account is referred to as the *compounding* of interest. If the bank compounds interest once a year, a bank's customer faces a potential disadvantage. If the money is needed and has to be withdrawn, say, 10 months after the account is opened, then the money will earn no interest, since the compounding occurs only after 12 months. It is for this reason that banks initiated the practice of compounding interest more than once a year. Consider an annual interest rate of r. What banks do is this: The year is divided into n equal periods. It is common to take quarters, months, or days. This corresponds respectively to $n = 4$, $n = 12$, or $n = 365$ ($n = 366$ in a leap year). Instead of paying interest at a rate of r at the end of each year, the bank will pay at a rate of $\frac{r}{n}$ per period and compound it at the end of each period.

Let's suppose that the bank pays interest at an annual interest rate of $r = 0.06$ compounded monthly. So $n = 12$, and $\frac{r}{n} = \frac{0.06}{12} = 0.005$. The initial investment of

$$A_0 = \$1000$$

will earn interest at a rate of 0.005 over the first month. So after the first compounding period the account will contain

$$A_1 = \$1000 + \mathbf{0.005(\$1000)} = \$1000(1 + 0.005).$$

This amount will earn interest at a rate of 0.005 over the second month. So after the second compounding period, the account will have

$$A_2 = \$1000(1 + 0.005) + \mathbf{0.005(\$1000(1 + 0.005))}.$$

Factoring $1000(1 + 0.005)$ out of both terms gives

$$A_2 = \$1000(1 + 0.005)(1 + 0.005) = \$1000(1 + 0.005)^2.$$

This amount will in turn grow to

$$A_3 = \$1000(1 + 0.005)^2 + \mathbf{0.005(\$1000(1 + 0.005)^2)}.$$

at the end of the third compounding period. After factoring $1000(1 + 0.005)^2$ out of both terms, we get

$$A_3 = \$1000(1 + 0.005)^2(1 + 0.005) = \$1000(1 + 0.005)^3.$$

The pattern should now be apparent: After p compounding periods there will be

$$A_p = \$1000(1 + 0.005)^p$$

in the account. Since 1 year has 12 months and hence 12 compounding periods, the value of the account after one year will be

$$A_{12} = \$1000(1 + 0.005)^{12} = \$1000(1.005)^{12}$$

$$= \$1000(1.06168) = \$1061.68.$$

Recall that when compounded only once a year the same $1000 grew to $1060. So the advantage is $1.68. A more important difference is the fact that if the money were to be withdrawn after 10 months, the bank's client would have

$$A_{10} = \$1000(1 + 0.005)^{10} = \$1000(1.005)^{10}$$

$$= \$1000(1.05114) = \$1051.14$$

instead of just $1000.

The computation just undertaken shows that if A_0 dollars are put into an account that pays at an annual interest rate of r compounded n times per year, then after p compounding periods, there will be

$$\boxed{A_p = A_0 \left(1 + \frac{r}{n}\right)^p}$$

dollars in the account. Since t years consist of nt compounding periods, the account will have

$$A(t) = A_{nt} = A_0 \left(1 + \frac{r}{n}\right)^{nt}$$

dollars after t years.

Example 12.1. Consider an initial investment of $A_0 = \$1000$ in an account that pays at an annual interest rate of 0.06. What will the investment be worth after 3 years, if

(a) Interest is compounded annually? Now $n = 1$ and $\frac{r}{n} = 0.06$. Since $t = 3$,

$$A(t) = A_0 \left(1 + \frac{r}{n}\right)^{nt} = \$1000(1 + 0.06)^{1\cdot3}$$

$$= \$1000(1.06)^3 = \$1191.02.$$

(b) Interest is compounded semiannually? Now $n = 2$ and $\frac{r}{n} = 0.03$. Since $t = 3$,

$$A(t) = A_0 \left(1 + \frac{r}{n}\right)^{nt} = \$1000(1 + 0.03)^{2\cdot3}$$

$$= \$1000(1.03)^6 = \$1194.05.$$

(c) Interest is compounded quarterly? Now $n = 4$ and $\frac{r}{n} = 0.015$. With $t = 3$,

$$A(t) = A_0 \left(1 + \frac{r}{n}\right)^{nt} = \$1000(1 + 0.015)^{4\cdot3}$$

$$= \$1000(1.015)^{12} = \$1195.62.$$

(d) Interest is compounded monthly? Now $n = 12$ and $\frac{r}{n} = 0.005$. This time,

$$A(t) = A_0 \left(1 + \frac{r}{n}\right)^{nt} = \$1000(1 + 0.005)^{12\cdot3}$$

$$= \$1000(1.005)^{36} = \$1196.68.$$

(e) Interest is compounded daily? Since $n = 365$, $\frac{r}{n} \approx 0.0001644$, and $t = 3$,

$$A(t) = A_0 \left(1 + \frac{r}{n}\right)^{nt}$$

$$\approx \$1000(1 + 0.0001644)^{365\cdot3}$$

$$= \$1197.22.$$

A look at the amounts \$1191.02, \$1194.05, \$1195.62, \$1196.65, and finally, \$1197.22, shows that the continued increase in the number of compounding periods makes less and less of a difference.

What happens if the year is divided into more and more periods and interest is compounded more and more frequently (at the same fixed annual interest rate r)? In other words, what happens to the expression

$$A(t) = A_0 \left(1 + \frac{r}{n}\right)^{nt}$$

when n is pushed to ∞? This has the effect of creating infinitely many equal compounding periods. So interest will be compounded at every instant. Will this break the bank? A look at the relatively small increases just observed in Example 12.1 suggests that it will not. But let's analyze the matter carefully.

Refer back to Section 10.1 and recall that when $x = 1, 2, 3, \dots$ is pushed to infinity, the numbers $\left(1 + \frac{1}{x}\right)^x$ close in on $e = 2.718281828 \dots$. In limit notation, $\lim_{x \to \infty} \left(1 + \frac{1}{x}\right)^x = e$. This is also true if r is a fixed number and x is pushed to infinity by taking $x = \frac{1}{r}, \frac{2}{r}, \frac{3}{r}, \frac{4}{r}, \dots$. Returning to our question, observe that

$$A(t) = \lim_{n \to \infty} A_0 \left(1 + \frac{r}{n}\right)^{nt} = \lim_{n \to \infty} \left[A_0 \left(1 + \frac{r}{n}\right)^{\frac{n}{r}}\right]^{rt}$$

$$= A_0 \lim_{n \to \infty} \left[\left(1 + \frac{r}{n}\right)^{\frac{n}{r}}\right]^{rt}.$$

But as just observed, $\lim_{n \to \infty} \left(1 + \frac{r}{n}\right)^{\frac{n}{r}} = e$. It follows that

$$A(t) = A_0 e^{rt}$$

Therefore, when interest is compounded at every instant at an annual rate of r, an initial investment of A_0 dollars will grow to $A(t) = A_0 e^{rt}$ dollars after t years. The strategy of compounding interest at every instant is called *continuous compounding*.

Example 12.2. Consider an initial investment of $A_0 = \$1000$ in an account that pays an annual interest rate of 0.06 compounded continuously. What will the investment be worth after 3 years? Since $A_0 = \$1000$, $r = 0.06$, and $t = 3$,

$$A(t) = A_0 e^{rt} = \$1000 e^{(0.06)3} = \$1000 e^{0.18}$$

$$= \$1000(1.19722) = \$1197.22.$$

Refer back to Example 12.1e and observe that the amount $A_0 = \$1000$ grows to $1197.22 both under continuous compounding (a practice not ordinarily used by banks) and daily compounding (a common practice). The difference becomes visible, but just barely, with an initial investment of $A_0 = \$10,000$. Check that in this case, $10,000 will grow to $11,971.99 when compounded daily, and to $11,972.17 when compounded continuously.[1] The difference is a mere 18 cents! What is the difference when $A_0 = \$1,000,000$? In this case (still at $r = 0.06$ in 3 years) the figure is $1,197,199.57 with daily compounding and $1,197,217.36 with continuous compounding. The difference is $17.79.

B. Investment Plans. An *investment plan* is an account that is paid into on a regular basis. Such accounts can generate lots of money, due to the simple fact that when money remains in an account for a long time, lots of interest can be generated.

Suppose that the amount of A_0 dollars is paid into an account every month and that the account earns interest at an annual rate of r compounded monthly. Assume that the amount A_0 is paid in at the very beginning of each month and that interest is compounded at the end of each month. How much money will be in the account after p months?

Let's first assume that $A_0 = \$1000$ and that $r = 0.06$. Since $n = 12$, $\frac{r}{n} = \frac{0.06}{12} = 0.005$. By the discussion of Section 12.1A, the first payment of $1000 will grow to

$$\$1000(1 + 0.005)^p$$

after p months. The $1000 paid at the beginning of the second month will be in the account for $p - 1$ months and will grow to

$$\$1000(1 + 0.005)^{p-1}.$$

In the same way, the next $1000 will grow to

$$\$1000(1 + 0.005)^{p-2}.$$

Proceeding in this way, we see that the account will contain the sum

$$S_p = \$1000(1 + 0.005)^p + \$1000(1 + 0.005)^{p-1}$$

$$+ \cdots + \$1000(1 + 0.005)$$

after p months. The last term is the $1000 paid into the account at the beginning of the pth and final month together with the interest that it earns. Observe that

$$S_p = \$1000(1 + 0.005) + \$1000(1 + 0.005)^2$$

$$+ \cdots + \$1000(1 + 0.005)^p$$

$$= \$1000(1 + 0.005)[1 + (1 + 0.005) + (1 + 0.005)^2$$

$$+ \cdots + (1 + 0.005)^{p-1}].$$

This quantity can be expressed more compactly as follows. Let x be any number and consider the sum

$$C = 1 + x + x^2 + \cdots + x^{p-1}.$$

Since $xC = x + x^2 + \cdots + x^{p-1} + x^p$,

$$(x - 1)C = xC - C$$

$$= (x + x^2 + \cdots + x^{p-1} + x^p)$$

$$- (1 + x + x^2 + \cdots + x^{p-1})$$

$$= x^p - 1$$

and therefore,

$$C = \frac{x^p - 1}{x - 1}. \qquad (*)$$

Taking $x = 1 + 0.005$, we now get

$$1 + (1 + 0.005) + (1 + 0.005)^2 + \cdots + (1 + 0.005)^{p-1}$$

$$= C = \frac{(1 + 0.005)^p - 1}{0.005}.$$

By substituting this into the earlier expressions for S_p, we see that $S_p = \$1000(1+0.005)C$ and hence that

$$S_p = \$1000(1 + 0.005)\frac{(1 + 0.005)^p - 1}{0.005}.$$

Example 12.3. How much money will be in this account after 3 years? Since three years have 36 months, $p = 36$. So by the formula just developed, the amount in the account after 3 years will be

$$S_{36} = \$1000(1 + 0.005)\frac{(1 + 0.005)^{36} - 1}{0.005}$$

$$= \$1000(1.005)\frac{1.19668 - 1}{0.005}$$

$$= 1000(1.005)(39.33610)$$

$$= \$39{,}532.79.$$

Since a total of $36 \times \$1000 = \$36{,}000$ was paid in, the interest earned is

$$\$39{,}532.79 - \$36{,}000 = \$3{,}532.79.$$

After 10 years, the situation becomes even more interesting for the holder of the account. In this case, the account will contain

$$S_{120} = \$1000(1 + 0.005)\frac{(1 + 0.005)^{120} - 1}{0.005}$$

$$= \$1000(1.005)\frac{1.81940 - 1}{0.005}$$

$$= 1000(1.005)(163.87935)$$

$$= \$164{,}698.74.$$

Of this, $120 \times \$1000 = \$120{,}000$ was paid in and $\$164{,}698.74 - \$120{,}000 = \$44{,}698.74$ is interest earned!

A careful check of the preceeding computations shows the following. If a monthly payment of A_0 dollars is paid into an investment plan that earns interest at an annual rate of r compounded monthly, then after p months, the investment plan will have a total of

$$S_p = A_0\left(1 + \frac{r}{12}\right)\frac{(1 + \frac{r}{12})^p - 1}{\frac{r}{12}}$$

$$= A_0\left(1 + \frac{r}{12}\right)\left(\left(1 + \frac{r}{12}\right)^p - 1\right)\frac{12}{r}$$

dollars. In slightly rewritten form this is equal to

$$\boxed{S_p = \frac{12}{r}A_0\left(1 + \frac{r}{12}\right)\left(\left(1 + \frac{r}{12}\right)^p - 1\right)}$$

C. Annuities and Bonds. We will assume throughout this section that the bank we are dealing with pays interest at an annual rate of r compounding monthly. Under these conditions, we saw in Section 12.1A that $\$A$ grows to $\$A(1 + \frac{r}{12})^p$ in p months. Figure 12.1 expresses this in terms of a "flow" diagram.

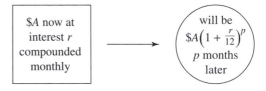

Figure 12.1

Turning this around, suppose that $\$B$ in needed in p months. What amount $\$A$ needs to be invested now? An investment of $\$A$ grows to $\$A(1 + \frac{r}{12})^p$ in p months. Since we need to have $\$B = \$A(1 + \frac{r}{12})^p$, it follows that

$$\$A = \frac{\$B}{(1 + \frac{r}{12})^p} = \$B\left(1 + \frac{r}{12}\right)^{-p}$$

has to be invested now. This is illustrated in Figure 12.2. The amount $\$B(1 + \frac{r}{12})^{-p}$ is called the *present value of B dollars p months from now* (at an annual interest rate of r compounded monthly).

Example 12.4. Suppose $\$10{,}000$ is needed in 3 years. How much should be invested now in an account that pays at an annual interest rate of $r = 0.05$

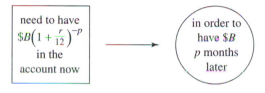

Figure 12.2

(compounded monthly)? Taking $B = 10,000$, $r = 0.05$, and $p = 36$, we get that $A = \$10,000(1 + \frac{0.05}{12})^{-36} = \$10,000(1.00417)^{-36} = \$8609.76$ has to be invested now.

The present value of an account decreases as the number of "months from now" increases. The reason for this is simple: Less is needed now if interest has more time in which to do its work. For instance, at $r = 0.05$ the present value of $B = \$100$

in 6 months is $\$100 \left(1 + \dfrac{0.05}{12}\right)^{-6}$

$= \$100(1.00417)^{-6} = \$100(0.99585)^6$

$= \$97.54;$

in 24 months is $\$100 \left(1 + \dfrac{0.05}{12}\right)^{-24}$

$= \$100(1.00417)^{-24} = \$100(0.99585)^{24}$

$= \$90.50;$

in 60 months is $\$100 \left(1 + \dfrac{0.05}{12}\right)^{-60}$

$= \$100(1.00417)^{-60} = \$100(0.99585)^{60}$

$= \$77.92.$

A higher interest rate reduces the present value. Why? Suppose that some amount of money is needed in the future. At a higher interest rate less money is needed to grow to this amount than would be needed at a lower rate. For instance, if $r = 0.06$, then the

present value of $B = \$100$

in 6 months is $\$100 \left(1 + \dfrac{0.06}{12}\right)^{-6}$

$= \$100(1.005)^{-6} = \$100(0.99502)^6$

$= \$97.05;$

in 24 months is $\$100 \left(1 + \dfrac{0.06}{12}\right)^{-24}$

$= \$100(1.005)^{-24} = \$100(0.99502)^{24}$

$= \$88.72;$

in 60 months is $\$100 \left(1 + \dfrac{0.06}{12}\right)^{-60}$

$= \$100(1.005)^{-60} = \$100(0.99502)^{60}$

$= \$74.14.$

For similar reasons, a lower interest rate increases the present value. See Exercises 12A.

The essence of an investment plan is the flow of money *into* an account at a fixed amount at regular intervals over a specified period of time. An *annuity* is the reverse of this situation. It is a sequence of equal payments *out of* an account at regular intervals over a specified period of time. The central issue with an annuity is this: What amount of money does an account need to have now so that it can pay a certain fixed amount of B dollars for p consecutive months starting exactly one month from now? Thinking of the account as subdivided into p "compartments" (one for each payment) and making use of Figure 12.2, we get the "flow" chart of Figure 12.3. It follows that the amount the account needs to have now is

$$B \left(1 + \frac{r}{12}\right)^{-1} + B \left(1 + \frac{r}{12}\right)^{-2}$$

$$+ B \left(1 + \frac{r}{12}\right)^{-3} + \cdots + B \left(1 + \frac{r}{12}\right)^{-p}.$$

To repeat, this is the amount that the account needs to have *now* so that (starting in one month) it can

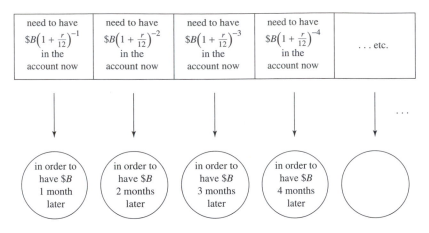

Figure 12.3

generate a payment of B dollars per month for p consecutive months. This sum is called the *present value* of the *income stream* of B dollars per month for p consecutive months at an annual interest rate of r.

Let's designate this present value by PV_p for short. We will see next that PV_p can be written more compactly. Note that

$$PV_p = B\left(1+\frac{r}{12}\right)^{-1} + B\left(1+\frac{r}{12}\right)^{-2}$$

$$+ B\left(1+\frac{r}{12}\right)^{-3} + \cdots + B\left(1+\frac{r}{12}\right)^{-p}$$

$$= \frac{B}{1+\frac{r}{12}} + \frac{B}{\left(1+\frac{r}{12}\right)^2}$$

$$+ \frac{B}{\left(1+\frac{r}{12}\right)^3} + \cdots + \frac{B}{\left(1+\frac{r}{12}\right)^p}$$

$$= \frac{B}{1+\frac{r}{12}}$$

$$\times \left[1 + \frac{1}{\left(1+\frac{r}{12}\right)} + \frac{1}{\left(1+\frac{r}{12}\right)^2} + \cdots + \frac{1}{\left(1+\frac{r}{12}\right)^{p-1}}\right].$$

By applying formula (∗) of Section 12.1B with $x = \frac{1}{1+\frac{r}{12}}$ we get

$$PV_p = \frac{\frac{B}{1+\frac{r}{12}}\left[1 - \left(\frac{1}{1+\frac{r}{12}}\right)^p\right]}{1 - \frac{1}{1+\frac{r}{12}}}.$$

Note that

$$1 - \frac{1}{1+\frac{r}{12}} = \frac{1+\frac{r}{12}-1}{1+\frac{r}{12}} = \frac{\frac{r}{12}}{1+\frac{r}{12}}.$$

Therefore, by substitution,

$$PV_p = \frac{B}{1+\frac{r}{12}}\left[1 - \frac{1}{(1+\frac{r}{12})^p}\right] \cdot \frac{1+\frac{r}{12}}{\frac{r}{12}}$$

$$= \frac{B}{\frac{r}{12}}\left[1 - \frac{1}{(1+\frac{r}{12})^p}\right].$$

In rewritten form,

$$\boxed{PV_p = \frac{12B}{r}\left[1 - \left(1+\frac{r}{12}\right)^{-p}\right]}$$

Example 12.5. What is the present value of the income stream of twelve consecutive monthly payments of \$100 at an interest rate of $r = 0.08$ (compounded monthly). This is

$$PV_{12} = \frac{12(100)}{0.08}\left[1 - \left(1+\frac{0.08}{12}\right)^{-12}\right]$$

$$= 15{,}000\left(1 - (0.99338)^{12}\right)$$

$$= 15{,}000(1 - 0.92336) = 15{,}000(0.07664)$$

$$= \$1{,}149.58.$$

Therefore, this is the amount that an account needs to have now so that it can pay out, beginning in one month, the income stream of $100 per month for 12 months.

Example 12.6. A professor who turned 52 in September of 1995 is planning to retire in September of 2008 at the age of 65. At that time he expects a Social Security payment in the amount of $1,400 per month and a payment of $2,100 per month from a retirement plan set up by his university. But he needs to have an income of $6,000 per month for the first 10 years of his retirement and has decided to set up an annuity with a bank that will supply him with the additional $2,500 per month he needs. The bank compounds interest monthly at an annual rate of $r = 0.06$. How much will he need to have in this account when he retires?

The account has to generate $2,500 monthly payments for 120 months. So $B = 2,500$ and $p = 120$. Since the professor knows the formula that was just developed, he determines that he will need to have

$$PV_{120} = \frac{12(2,500)}{0.06}\left[1 - \left(1 + \frac{0.06}{12}\right)^{-120}\right]$$

$$= 500,000\left(1 - (1.005)^{120}\right)$$

$$= 500,000(1 - 0.54963)$$

$$= 500,000(0.45037)$$

$$= \$225,184.$$

In order to accumulate this hefty sum, the professor began (on his 52nd birthday) to make monthly payments into an investment plan. He will be making $12 \times 13 = 156$ equal monthly payments over 13 years until he retires. The bank has agreed to pay him at the same interest rate of 0.06 for this investment account until he retires. How much are his monthly payments?

Recall from Section 12.1B that if A_0 dollars per month are paid into an account (that earns an interest rate of r), then this account will have

$$S_p = \frac{12}{r}A_0\left(1 + \frac{r}{12}\right)\left(\left(1 + \frac{r}{12}\right)^p - 1\right) \text{ dollars}$$

after p months. So the professor's investment account will have

$$S_{156} = \frac{12}{0.06}A_0\left(1 + \frac{0.06}{12}\right)\left(\left(1 + \frac{0.06}{12}\right)^{156} - 1\right)$$

$$= 200A_0(1.005)\left((1.005)^{156} - 1\right)$$

$$= 200(1.005)(2.17724 - 1)A_0$$

$$= 201(1.17724)A_0$$

$$= 236.62A_0 \text{ dollars.}$$

Because $236.62A_0$ must be equal to $225,184$, the professor is making monthly payments of $A_0 = \$951.67$.

Large firms and corporations have financial needs that are more substantial than those of our professor. These can be met by loans or stock offerings. A stock offering proceeds as follows. A firm's total worth is divided into a certain number of units called *shares*, and all the shares together are the firm's *stock*. To raise needed capital, the firm can sell additional shares of the stock. The purchasers of these shares will become co-owners of the firm. The following example clarifies what is going on. Suppose that a firm with total assets of $1.5 billion needs to raise $120 million. It decides to subdivide its worth into 100 million shares of $15 each. The total value of $1500 million, or $1.5 billion, of these shares is equal to the firm's total assets. To raise the capital it needs, the firm sells 8 million additional shares for the required total of $8 \times 15 = $120 million. Since $120 million is equal to $0.12 billion, the total worth of the firm has increased to $1.62 billion. The purchasers of the 8 million shares are the owners of $0.12 billion of this total. Stocks are bought and sold on *stock markets* all over the world. The New York Stock Exchange is an important example of such a market. The price of a stock will generally reflect the fortunes of the firm. If a firm is perceived to be prosperous, or to have the potential to become prosperous, the price of a share of its stock will rise. If this is not the case, the price of a share will be stable or will fall.

Firms can also raise needed capital by issuing bonds. A *bond* is a loan that is similar to an annuity in its essential aspect. It is a contract in which the borrower of a sum of money agrees to pay the bondholder (the lender) interest in the form of a "stream" of payments. These so-called *coupon payments* are paid at six-month intervals until the time the bond *matures*. At maturity, the borrowed sum—this is the *face value* of the bond—is also repaid.

Let's consider the bonds offered by the Chrysler Corporation on July 15, 1987. Chrysler needed $245,000,000, probably in connection with its purchase of American Motors on August 4, 1987. To raise this amount, Chrysler sold 245,000 bonds at a face value of $1000 each. These bonds were to mature on August 1, 1999. For each $1000 bond Chrysler agreed to pay a semiannual coupon payment of $52 (one on February 1 and another on August 1) until the bond matures, as well as the face value of $1000 at maturity. So the holder of this Chrysler bond would receive $2 \times 12 = 24$ payments of $52 each in addition to the $1000.

At any time between the date of issue and the date of maturity, a bond has a *trading price*, and it is at this price that the bonds are bought and sold (on the New York Stock Exchange and large investment banks). How, precisely, is the trading price determined? What, for instance, was the trading price of one Chrysler bond on November 20, 1995? Since the maturity of the bond was then 4 years away, the holder of a bond at that time was scheduled to receive 8 semiannual payments of $52 until August 1999 as well as the face value of $1000 at maturity on August 1, 1999. So the bond will generate the semiannual income stream

$$\$52, \$52, \$52, \$52, \$52, \$52, \$52, \$52,$$

plus the $1000. The trading price of the Chrysler bond on November 20, 1995, was the present value (for that day) of this income stream plus the present value of the $1000. Refer to the earlier discussion of the mathematics of an annuity, and replace 12 (this represented months) by 2 (for semiannual) to

see that this present value can be computed as follows:

$$52 \left(1 + \frac{r}{2}\right)^{-1} + 52 \left(1 + \frac{r}{2}\right)^{-2} + \cdots$$

$$+ 52 \left(1 + \frac{r}{2}\right)^{-8} + 1000 \left(1 + \frac{r}{2}\right)^{-8}$$

$$= \frac{2(52)}{r} \left[1 - \left(1 + \frac{r}{2}\right)^{-8}\right] + 1000 \left(1 + \frac{r}{2}\right)^{-8}$$

$$= \frac{104}{r} \left[1 - \left(1 + \frac{r}{2}\right)^{-8}\right] + 1000 \left(1 + \frac{r}{2}\right)^{-8}.$$

But what is the interest rate r? This is determined by market forces and depends on the interest rates earned on investments with similar risk factors as the bond (such as certificates of deposit). On November 20, 1995, the interest rate for a Chrysler bond was $r = 0.083$. Therefore, the trading price of this bond was

$$\frac{104}{0.083}(1 - 0.722) + 1000(0.722) = \$347.94 + \$722.31$$

$$= \$1070.25.$$

The bond page of the *New York Times* of November 21, 1995, summarized the status of the bond:

Chrysl 10.40s99   9.7   459   107   ...

This summary contains the following information. The bond pays $52 every six months, for a total of $104 or 10.40% of its face value of $1000. This is the meaning of the 10.40. The 99 refers to the year of maturity 1999. The "s" simply separates the two numbers. The 459 indicates the number of bonds that were traded that day. The 107 means that the trading price of the bond was $1070 when the market closed on November 20, 1995. The ... refers to the fact that the bond's closing price did not change from the day before. Since $0.097(1070) = 104$, the 9.7 tells an investor that the year's return of $2 \times \$52 = \$104$ represents an annual interest of 9.7% of the $1070 investment in the bond.

To summarize, the trading price of a bond on a given day is a sum of the form

$$C\left(1+\frac{r}{2}\right)^{-1} + C\left(1+\frac{r}{2}\right)^{-2} + \cdots$$

$$+ C\left(1+\frac{r}{2}\right)^{-p} + F\left(1+\frac{r}{2}\right)^{-p}$$

$$= \frac{2C}{r}\left[1-\left(1+\frac{r}{2}\right)^{-p}\right] + F\left(1+\frac{r}{2}\right)^{-p}$$

where C is the coupon payment, F is the face value, p is the number of semiannual periods from that day to the bond's date of maturity, and r is the interest that the market factors on that day determine.

If the prevailing interest rate (or return) r for competing investments drops, then the fixed income stream generated by a bond becomes a more attractive investment. So the trading price of the bond will go up. The fact that the computation of the trading price involves division by the term $(1+\frac{r}{2})$ confirms this: If r becomes smaller, then $(1+\frac{r}{2})$ becomes smaller, so that the computed trading price will increase. By similar reasoning, if r goes up, then the trading price of a bond will go down. Suppose that you are an investor who feels that the interest rates paid by interest-paying investments will go down in the future. You are therefore of the opinion that r will tend to go down, and hence that bonds will go up. You might therefore consider an investment in bonds.

A look at the Chrysler bond illustrates this. For example, at the higher interest rate of $r = 0.09$, the price of the Chrysler bond on November 20, 1995, would have been

$$\frac{104}{0.09}\left[1-(1.045)^{-8}\right] + 1000(1.045)^{-8}$$

$$= (1155.56)(0.29681) + 1000(0.70319)$$

$$= 342.99 + 703.19 = \$1046.18$$

instead of the actual closing price of $1070.25. On the other hand, at the lower interest rate of $r = 0.06$, the price would have been

$$\frac{104}{0.06}\left[1-(1.03)^{-8}\right] + 1000(1.03)^{-8}$$

$$= (1733.33333)(0.21059) + 1000(0.78941)$$

$$= 365.02 + 789.41 = \$1154.43.$$

12.2 **Inflation and the Consumer Price Index**

As we are all aware, prices often change, generally in the upward direction. The price increases we have in mind are brought about by the dynamics of the marketplace (see Section 12.3) and not by improvements in a product or a service. When an excess demand for a product or group of products brings about a shortage of the supply, the manufacturer can raise prices. When there is an oversupply of money in a country (because too much is printed), then the value of the country's monetary unit is reduced, leading to an increase in prices. Such price increases are referred to as *inflation*.

The statisticians of the U.S. Department of Labor Bureau of Labor Statistics measure inflation by making use of a number called the *Consumer Price Index*, or CPI. This index is a carefully computed average of the costs of food, clothing, housing, transportation, medical care, personal care, and entertainment. The current index[2] was at 100 in August 1983. It rose to 108.0 in August 1985, to 114.4 in August 1987, to 149.0 in August 1994, and to 152.9 in August 1995. In other words, what you could buy for $1.00 in August 1983 cost $1.08 in August 1985, $1.14 in August 1987, $1.49 in August 1994, and $1.53 in August 1995.

The CPI makes it possible to measure the price increase over a given period in precise numerical terms. It is the official yardstick for inflation. The *inflation* over a given year is the increase in the CPI for the year, and the *rate of inflation* is this increase in the CPI divided by the CPI at the beginning of the year. For instance, since the CPI increased by $152.9 - 149.0 = 3.9$ from August 1994 to August 1995, the inflation for that year was 3.9 points. The CPI stood at 149.0 in August 1994, so the rate of inflation was $\frac{3.9}{149} = 0.026$ or 2.6%. The increase in social security benefits of 2.6% for 1996 (that our government announced in October 1995) was based on such a computation. This increase would prevent inflation from invisibly eroding these benefits for the recipients of social security. The CPI is also used to adjust tax rates.

Let $t \geq 0$ be the time (we will take it to be in years) that has elapsed from some date of reference $t = 0$. Let the CPI be $p(t)$ at time t. For example, a study of the inflation rate starting from August 1985 would take August 1985 as $t = 0$. In this case, $t = 2$ would refer to August 1987, and $t = 9$ to August 1994. So $p(2) = 114.4$ and $p(9) = 149.0$. In order to obtain insight into the connection between inflation and the CPI, we will assume that $p(t)$ is a differentiable function of t. Fix some time t and let a time of Δt elapse from that time on. Take $\Delta t = \frac{1}{2}$ for instance. Observe that over this $\frac{1}{2}$ year, the CPI increased by

$$p(t + \Delta t) - p(t) = p(t + \frac{1}{2}) - p(t)$$

points. Projected over an entire year, this is an increase of $2[p(t + \frac{1}{2}) - p(t)] = \frac{p(t+\frac{1}{2})-p(t)}{\frac{1}{2}}$. It follows that the annual inflation rate for this $\frac{1}{2}$-year period was $\frac{p(t+\frac{1}{2})-p(t)}{\frac{1}{2}}$ divided by $p(t)$. Suppose that $\Delta t = \frac{1}{8}$ is any fraction of a year. Taking $\frac{1}{8}$ in place of $\frac{1}{2}$ shows that inflation occurred at an annual rate of

$$\frac{p(t + \frac{1}{8}) - p(t)}{\frac{1}{8}} \cdot \frac{1}{p(t)} = \frac{p(t + \Delta t) - p(t)}{\Delta t} \cdot \frac{1}{p(t)}$$

over the period from t to $t + \Delta t$. Pushing Δt to zero tells us that inflation[3] is progressing at an annual rate of $\frac{p'(t)}{p(t)}$ at time t. For example, if inflation runs at an annual rate of 4% at time t, then $\frac{p'(t)}{p(t)} = 0.04$. This running rate of inflation is called the *continuous rate of inflation*.

Suppose we know that inflation occured at a constant annual rate of k over some period $0 \leq t \leq t_1$ of years. Therefore, $\frac{p'(t)}{p(t)} = k$ for any t with $0 \leq t \leq t_1$. Since $p'(t) = kp(t)$, it follows that $p(t) = p(0)e^{kt}$. (This was verified for the function $y(t)$ and the constant μ in Section 11.5A.) Taking $t = t_1$, we get $p(t_1) = p(0)e^{kt_1}$.

Consequently, $e^{kt_1} = \frac{p(t_1)}{p(0)}$ and hence $kt_1 = \ln\frac{p(t_1)}{p(0)}$. Therefore,

$$k = \frac{1}{t_1} \ln\left(\frac{p(t_1)}{p(0)}\right).$$

As we might have expected, knowing the CPI at the beginning and also at the end of the period $0 \leq t \leq t_1$ allows us to compute the inflation rate k over this period.

Example 12.7. What was the average annual inflation rate for the decade from August 1985 to August 1995? Let's reformulate this question as follows: If the running inflation rate over this decade would have been a constant k, what would k have been equal to?

The formula just derived provides the answer. Take August 1985 to be $t = 0$ and $t_1 = 10$. Since the CPI was 108.0 in August 1985, $p(0) = 108.0$. Since it was 152.9 in August 1995, $p(t_1) = p(10) = 152.9$. It follows that

$$k = \frac{1}{10} \ln\left(\frac{152.9}{108.0}\right) = \frac{\ln 1.416}{10} = \frac{0.348}{10} = 0.035.$$

So the rate of inflation over this 10-year period was 3.5%.

The rate of inflation was not always this low. For example, in 1974 and again in 1979 and 1980 it was over 12%. The most significant reason for the high inflation rates in those years was the dramatic increase in the price of crude oil. Crude oil is a basic raw material not only in the production of gasoline, diesel fuel, and heating oil, but also asphalt and plastics. Therefore, the increases in the price of crude oil drove up prices on a broad front. What brought the increases in the price of oil about? A number of major oil-producing countries (mostly from the Middle East) banded together and cut their production sharply. This reduced the world's oil supply, and the price of oil jumped. We will analyze this matter in Section 12.3B.

Example 12.8. If inflation were to continue at the rate of 3.5%, what would the CPI be in August 1999 and in August 2008?

Recall that the CPI was 152.9 in August of 1995. Take August 1995 as $t = 0$. So August 1999 is $t_1 = 4$ and August 2008 is $t_1 = 13$. Substituting $p(0) = 152.9$ and $k = 0.035$ into the formula $p(t_1) = p(0)e^{kt_1}$, we get $p(t_1) = 152.9e^{0.035t_1}$. So the CPI for August 1999 would be

$$p(4) = 152.9e^{(0.035)(4)} = 152.9e^{0.14}$$

$$= (152.9)(1.15) = 175.88$$

and that for August 2008,

$$p(13) = 152.9e^{(0.035)(13)} = 152.9e^{0.455}$$

$$= (152.9)(1.576) = 241.00.$$

Let's return for a moment to the concerns of Section 12.1C. In particular, consider the professor of Example 12.6 and assume that inflation will run at an average rate of 3.5% until his retirement in the year 2008. Refer to the computations of Example 12.8 and observe that the professor will pay $2.41 in September 2008 for what he paid $1.53 in September 1995. Since $\frac{1.53}{2.41} = 0.63$, and hence $\frac{0.63}{1.00} = \frac{1.53}{2.41}$, it follows that one 2008 dollar will have the same buying power that 63 cents had in 1995. Therefore, the payment of $6,000 that the professor will be receiving in September 2008 will be equivalent to ($6,000)(0.63) = $3700 in 1995 dollars. So for our professor, inflation—even at the modest rate of 3.5%—is a big problem. For the Chrysler Corporation, on the other hand, inflation is no problem at all. On the contrary, it will pay back the $245,000,000 that it borrowed in 1987 in much cheaper 1999 dollars.

Example 12.9. Consider an investment of $1,000. Assume that it earns interest at an annual rate of 0.06 compounded continuously. What will the investment be worth after 3 years, if an inflation rate of 3.5% is taken into consideration?

The interest rate that a bank offers is the so-called *nominal interest rate*. The *real interest rate* is equal to the nominal interest rate minus the expected rate of inflation. In the case under consideration, the real

interest rate is $0.06 - 0.035 = 0.025$. By the formula $A(t) = A_0 e^{rt}$ from Section 12.1A, we see that the inflation-adjusted value of the investment will be

$$A(3) = 1000e^{(0.025)3} = 1000e^{0.075} = \$1077.88.$$

This is much less than the $1197.22 that the nominal rate provides. Refer to Example 12.2.

12.3 Supply and Demand in a Market

$\boxed{A}$ *market* for a product consists of a group of producers of the product, a group of potential consumers, and the buying and selling interaction between the two groups. Many different markets for different products can be considered and analyzed. For instance, there is the world market for oil, the U.S. market for automobiles, the Dutch market for tulips, the Japanese market for textiles, and so forth.

The concepts of demand and supply are at the very core of economics. One can in fact regard demand to be the lifeblood of any business. If there is no demand for its products or services, a firm will have to close down. The *demand* for a product or service is the quantity that the consumers in a market are able (and willing) to buy in a given period of time at various prices. This demand depends on many factors, or variables. The price of the product is the most immediate, but the demand will also depend on the prices of similar and related products; the consumers' level of disposable income; their tastes, preferences and knowledge of the product (as influenced by advertising); taxes; and so on. Similarly, the *supply* for a product (or service) is the quantity that the producers in a market are able (and willing) to sell in the given period of time at various prices. The supply will also depend on many factors, such as the price and availability of raw materials, the efficiency of the production process, and so on.

When analyzing any aspect of a market, it is important to draw a distinction between the short run and long run. While *short run* generally refers to a relatively brief time frame (say, a few months) and *long run* to a lengthier one (say, a number of years), the essential distinction revolves around the variables that impact the particular analysis. For instance, in the short run, an increase in the price of gasoline may have only a small effect on its demand. While people might drive less, they are not likely to change to a smaller, more fuel-efficient car immediately. In the longer run, however, they may do both and the overall demand might decrease appreciably. In the case of a price increase in the automobile industry, many consumers are likely to delay purchasing a new car initially, so that the annual demand for new cars may drop. In the long run, however, the unreliability of older cars and the availability of public transportation may also become factors, and the annual demand could rebound. In general, the study of the short run is easier to undertake, since fewer variables need to be considered. Put another way, the fact that "longer range" variables can be discounted simplifies the analysis.

The discussion that follows will focus on the short run and in particular on the dependence of both supply and demand on price alone. So we will consider a situation in which all other factors that influence supply and demand are constant.

A. Supply and Demand Functions.   For a given market, a given price p of the product in dollars per unit will generate a certain demand $D(p)$ in numbers of units in a given period (typically a year, a quarter of a year, or a month) from the buyers in the market. In the same way, p will also generate a certain supply $S(p)$ from the suppliers of the market in numbers of units (in a given period).

The *demand curve* D tells us how many units of the product consumers are willing to buy at various prices p per unit. It slopes downward because consumers will generally buy less if the price is higher. The *supply curve* S tells us how many units of the product producers are willing to sell at various prices. This curve slopes upward because producers will generally produce more when the price is higher. For

example, a higher price may enable existing firms to expand production by hiring extra workers or by having existing workers work overtime. A higher price may also attract new firms that might have found entry into the market uneconomical at a lower price. Figure 12.4 provides the general shapes of the supply and demand curves.

The *market-equilibrium price* for the product is that price at which the quantity demanded and quantity supplied are equal. So this is the price p^* at which $D(p^*) = S(p^*)$. The *market mechanism* is the tendency in a free market for the price to change until the market *clears*, in other words, until the quantity supplied and quantity demanded are equal. At this point there is neither a shortage nor excess supply, and there is no pressure on the price to change further. Supply and demand might not always be in equilibrium, but the general tendency is for markets to clear.

To understand why markets tend to clear, suppose that the price is initially at p_1 with $p_1 > p^*$. This means,[4] see Figure 12.5, that $S(p_1) > D(p_1)$. So at the price p_1, an excess supply of $S(p_1) - D(p_1)$ units will accumulate during the period. To prevent this, and the storage and handling costs that it implies, producers will tend to lower their prices. The price will then fall in the direction of p^*. This trend will continue until p^* is reached. The opposite will happen if the price is initially at p_2 with $p_2 < p^*$. In this situation, again see Figure 12.5, $D(p_2) > S(p_2)$. A shortage of $D(p_2) - S(p_2)$ will develop because consumers will be unable to purchase all they would like. This will put upward pressure on the price, leading to an increase in the direction of p^*, and this will continue until p^* is attained.

A market is *stable* if there is no significant shift in either the supply or the demand of the good. A significant shift in the supply or the demand (or both) can destabilize the market. In such a situation, the earlier equilibrium price may no longer be the equilibrium price for new supply and demand conditions, and market forces will operate to create a new equilibrium price p^*. When condi-

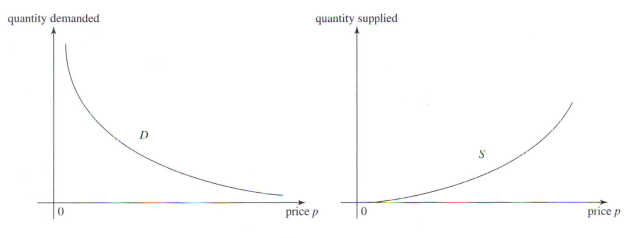

Figure 12.4

tions change suddenly, it may take some time for a new equilibrium price to be achieved. The *actual price* is the price that actually prevails. In a situation where the market is fairly stable, this should, in view of the market dynamics just described, be approximately equal to the equilibrium price.

Notice that the relationship between supply, demand, and price is subtle. Supply and demand curves are determined by an assessment of the supply and the demand at the various possible prices. In turn, the supply and demand curves together determine the market equilibrium price p^*, which is related to the actual price. See Figure 12.5. A shift in the demand or the supply will lead to a new market equilibrium price p^* and hence to a new actual price.

A market is said to be *competitive* if an individual producer or buyer has in essence no ability to influence the market price for the good. Each producer in a competitive market sells only a small fraction of the entire supply of the market. So the output of a single firm will have no effect on the market price. A producer in a competitive market will set the price of its product at (or close to) the price p^* that the market has determined. Such a producer is therefore a *price taker*; the producer must "take" the market price. Not all markets are competitive. For example, if a market

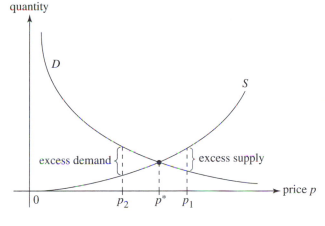

Figure 12.5

is supplied by a single producer, this producer has total control. Such a producer can decide on the price or the supply unilaterally and is said to have a *monopoly* on the market. A *cartel* is an organized group of producers that together control a significant portion of the supply. By acting together they can influence both the price and the supply. Some cartels have had a major impact on markets. We will see shortly that the oil cartel OPEC was able to drive up the price of crude oil sharply by cutting its production. Monopolies and cartels can also exist on the demand side.

As already pointed out, a change in the price of a product will generally lead to a change in the quantity demanded and in the quantity supplied. It will be of interest to both a supplier and buyer in the market to know *how much* the supply or demand will change for a given change in the price. In other words, they will want to know how sensitive the supply and the demand are to a given change in the price. For example, if the price of a Ford Taurus is increased by 10%, by how much will the demand drop? How much will the supply of gasoline increase if the price increases by 5%? The concept of *elasticity* clarifies such questions.

Elasticity measures the sensitivity of one variable relative to another. For example, the *price elasticity of demand* measures the sensitivity of the quantity demanded to change in price. Let's be more precise. Suppose that p is the price of the product and $D(p)$ the corresponding demand. Now suppose that the price changes by Δp to $p + \Delta p$. The corresponding change in the demand is $\Delta D = D(p + \Delta p) - D(p)$. The percentage change in the price is $\%\Delta p = \frac{\Delta p}{p}$, and the corresponding percentage change in the demand is $\%\Delta D = \frac{\Delta D}{D(p)}$. The *price elasticity of demand* is the ratio

$$E = \frac{\%\Delta D}{\%\Delta p} = \frac{\frac{\Delta D}{D(p)}}{\frac{\Delta p}{p}} = \frac{p}{D(p)}\frac{\Delta D}{\Delta p}.$$

Note that E depends on both the price p and on the change in price Δp. Since $\frac{\Delta D}{D(p)} = E \cdot \frac{\Delta p}{p}$, observe that if E is small, then a small change Δp in price will result in a very small ΔD and hence have relatively little effect on the demand. On the other hand, if E is large, then ΔD will be larger, and the effect on the demand can be significant. In rewritten form,

$$E = \frac{p}{D(p)}\frac{D(p + \Delta p) - D(p)}{\Delta p}.$$

The *point price elasticity of demand* $e_D(p)$ is the answer to the question: What happens to E when Δp is pushed to 0? Intuitively, $e_D(p)$ is equal to the ratio $\frac{\%\Delta D}{\%\Delta p}$ for a very small Δp. More precisely (assuming

that the function $D(p)$ is differentiable),

$$e_D(p) = \lim_{\Delta p \to 0} E = \lim_{\Delta p \to 0} \frac{p}{D(p)}\frac{D(p + \Delta p) - D(p)}{\Delta p}$$

$$= \frac{p}{D(p)} \cdot \lim_{\Delta p \to 0} \frac{D(p + \Delta p) - D(p)}{\Delta p}$$

$$= \frac{p}{D(p)}D'(p) = p\frac{D'(p)}{D(p)}.$$

So the point price elasticity of demand $e_D(p)$ is equal to the price p times the ratio of the rate of change of the demand $D'(p)$ to the demand $D(p)$.

In the typical situation the demand curve is decreasing (see Figure 12.4). So $D'(p) \le 0$ and hence $e_D(p) \le 0$ for all p. If $|e_D(p)| < 1$, then the demand is not very sensitive to a change in price from p. In this case, the demand is said to be *price inelastic* (think insensitive). If $|e_D(p)| \ge 1$, then the demand will be sensitive to change in price from p. In this case, the demand is said to be *price elastic* (think sensitive). It is possible for the demand to be very sensitive to price change at one price and relatively insensitive at another.

Suppose that $e_D(p) = p\frac{D'(p)}{D(p)}$ is equal to a constant k over some price range $p_1 \le p \le p_2$. At any price p in this range, $\frac{D'(p)}{D(p)} = \frac{k}{p} = kp^{-1}$. The derivative of $k \ln p$ is equal to $k \cdot \frac{1}{p} = kp^{-1}$, and the derivative of $\ln D(p)$ is $\frac{D'(p)}{D(p)}$. (Review Section 10.3 if necessary.) Because these derivatives are equal, the two functions $\ln D(p)$ and $k \ln p$ differ by a constant. So

$$\ln D(p) = k \ln p + C$$

for some constant C. A look at the graph of the natural logarithm function shows that $C = \ln a$ for some constant a. Note that

$$\ln D(p) = k \ln p + \ln a = \ln p^k + \ln a = \ln(ap)^k.$$

Therefore, $\ln D(p) = \ln(ap)^k$. It follows that

$$D(p) = ap^k$$

over the price range $p_1 \le p \le p_2$. For the sake of illustration, assume that a is equal to 1 (so $C = \ln 1 = 0$) and that $k = -\frac{1}{2}, -1$, or -2. The case $e_D(p) = k =$

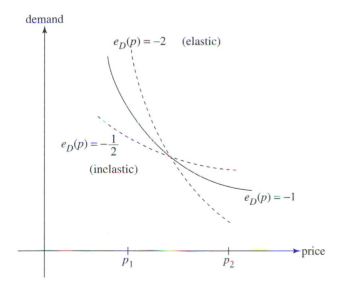

Figure 12.6

$-\frac{1}{2}$ is inelastic and corresponds to the demand function $D(p) = p^{-\frac{1}{2}}$. The cases $e_D(p) = -1$ and $e_D(p) = -2$ are elastic and correspond to $D(p) = p^{-1}$ and $D(p) = p^{-2}$. The three demand curves are sketched in Figure 12.6. In the inelastic case, the demand drops relatively slowly with an increase in price. In the two elastic cases the drop in demand is much more rapid.

The *point price elasticity of supply* is defined in a completely similar way by $e_S(p) = p\frac{S'(p)}{S(p)}$. It measures the sensitivity of the supply to change in price. In the typical case, the supply increases with price (see Figure 12.4), so that $S'(p) \geq 0$ and hence $e_S(p) \geq 0$.

There are many situations in economics where it is of interest to know how sensitive one variable is relative to a change in another and where the concept of elasticity is used. For example, consider the following questions. To what extent are bank deposits increased by a small increase in the interest rate? To what degree is the demand for a certain upscale automobile decreased by a decrease in the consumers' disposable income?

B. OPEC and the Price of Oil. In 1974 the world's oil market was destabilized by the Organization of Petroleum Exporting Countries, or OPEC. This cartel of primarily Middle Eastern oil-producing countries (Saudi Arabia, Kuwait, Iraq, Iran, the Arab Emirates, Oman, Venezuela,...) succeeded in pushing up the price of crude oil dramatically by collectively reducing production. In 1973 the world's demand and supply were both at 18 billion barrels (b.b.) per year at a market equilibrium price of $4 per barrel. OPEC supplied 12 b.b. per year and the competitive market (the United States, Norway, Mexico, ...) the other 6 b.b. per year. Suddenly, in 1974, OPEC cut its supply from 12 b.b. per year to 9 b.b. per year. The discussion[5] that follows analyzes the supply and demand dynamics of the world's oil market before and after the OPEC cuts. The dollars referred to are 1974 dollars. The CPI was 50.0 in August 1974 and 152.9 in August 1995. Therefore, a 1974 dollar was worth about three times a 1995 dollar.

We will suppose that the world's short-run demand D for oil in b.b. per year for the years 1973–1974 was a linear function of the price p. This means that $D(p) = a + bp$ for some constants a and b. Since $D'(p) = b$, the price elasticity of demand was

$$e_D(p) = p\frac{D'(p)}{D(p)} = \frac{bp}{a+bp}.$$

It was known that at a price of $4 per barrel, the world's demand for oil was not very sensitive to price change. More precisely, it had been estimated that the price elasticity for the demand at that price was a very small $e_D(4) = -0.05$. Recall that $D(4) = 18$. Inserting this information into our equations, we get

$$D(4) = a + 4b = 18$$

and

$$e_D(4) = 4\frac{D'(4)}{D(4)} = \frac{4b}{a+4b} = -0.05.$$

It follows that $\frac{4b}{a+4b} = \frac{4b}{18} = -0.05$. So $b = -(0.05)\frac{18}{4} = -(0.05)(4.5) = -0.225$. But $18 = a + 4b = a - 4(0.225) = a - 0.9$. Hence $a = 18.9$. Therefore, the world's short-run demand for oil was given by the linear function

$$D(p) = 18.9 - 0.225p$$

in b.b. per year.

Our focus next turns to the supply generated by the competitive market. We will assume that the short-run supply S_{cm} from the competitive market in the years 1973–74 was a linear function of p. So $S_{cm}(p) = c + dp$ for some constants c and d. Since $S'_{cm}(p) = d$, the price elasticity of the competitive supply was $p \frac{S'_{cm}(p)}{S_{cm}(p)} = \frac{dp}{c+dp}$. It was estimated that the price elasticity of the competitive supply at a price of \$4 per barrel was 0.10. Recall that $S_{cm}(4) = 6$. By subsituting $p = 4$ into the equations, we see that

$$S_{cm}(4) = c + 4d = 6$$

and

$$4 \frac{S'_{cm}(4)}{S_{cm}(4)} = \frac{4d}{c+4d} = \frac{4d}{6} = 0.10.$$

It follows that $d = (0.10)\frac{6}{4} = 0.15$ and $c = 6 - 4d = 6 - 4(0.15) = 6 - 0.6 = 5.4$. Therefore,

$$S_{cm}(p) = 5.4 + 0.15p$$

b.b. per year.

Adding the 12 b.b. per year that OPEC produced in 1973, we see that the world's short-run supply of oil was

$$S_1(p) = 12 + S_{cm}(p) = 17.4 + 0.15p$$

b.b. per year in 1973. The equality $S_1(4) = 17.4 + 0.6 = 18 = D(4)$ reflects the fact that the equilibrium price was \$4 per barrel.

When OPEC decreased its production by 3 b.b. per year in 1974, the world's supply of oil suddenly shifted to

$$S_2(p) = S_1(p) - 3 = 17.4 + 0.15p - 3 = 14.4 + 0.15p$$

b.b. per year. This destabilized the market, and market forces took over to determine a new equilibrium price p^*. What was this new equilibrium price? To find out, we need to set $S_2(p^*) = D(p^*)$ and solve for p^*. Doing

so, we obtain

$$14.4 + 0.15p^* = S_2(p^*) = D(p^*) = 18.9 - 0.225p^*.$$

Therefore, $0.375p^* = 4.5$. So $p^* = \frac{4.5}{0.375} = 12$ dollars per barrel!

This is in fact exactly what happened! The move by OPEC to cut its supply from 12 b.b. per year to 9 b.b. per year brought a price increase from 4 to 12 dollars per barrel in the short run. Figure 12.7 shows a graphic of the supply shift and its consequence for the price.

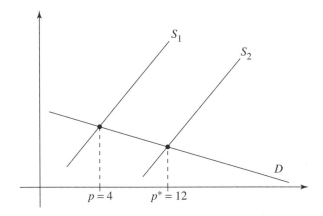

Figure 12.7

Notice that in 1973 at a production rate of 12 b.b. per year at \$4 per barrel, OPEC was earning revenues at a rate of $12 \times 4 = 48$ billion dollars per year. After the jump in the price to \$12 in 1974, it was earning revenues at rate of $9 \times 12 = 108$ billion dollars per year. In spite of the 3 b.b. cut in production, OPEC revenues soared. OPEC used a similar strategy in 1979–80. Because so much depends on petroleum and its products, the increases in the price of crude oil had a very negative effect on the world's economy. However, beginning in the early 1980s, oil conservation measures began to reduce the world's demand for oil. There were also increases on the supply side. These were brought about by an expansion of non-OPEC production and the fact that some OPEC countries refused to abide by the supply quotas set by the cartel. The fact is that OPEC has not been able to exert any significant pressure on oil prices since 1979–80. (An analysis of gasoline prices in the U.S. for the period 1962–87 confirms this. Refer to Exercises 12C.)

As we have just seen, the assumption that demand and supply functions are linear can provide important insights. However, it is frequently too simplistic. Forecasting market demand and supply, as well as the share that a particular firm can expect to fill, is usually a matter of considerable complexity. Every major corporation has a staff that is fully engaged in this process. Such staffs (often with the support of econometric forecasting services) will use wide-ranging statistical analyses to predict demand.[6] (Typically, these will make use of the gross national product, the unemployment rate, the government's monetary and regulatory policy, demographic changes in a region, the strengths of competitors, and so on.)

This process can be as complicated as a market analyst decides it should be. For example, in a study of the U.S. demand for gasoline undertaken in the 1980s, it was decided that the important variables affecting the demand are the pump price RP_R per gallon of gasoline, the consumer price index for public transportation RP_T, the number N of registered passenger cars, the average number of miles per gallon MPG, and, finally, the real disposable income RY_D (this is the income of a household available for spending, i.e., after all taxes have been paid). The variables RP_R, RP_T, and RY_D were adjusted for inflation. The demand for gasoline D for a given year (in millions of gallons) was then assumed to be given by an equation of the form

$$D = a_0 + a_1 RP_R + a_2 RP_T + a_3 N + a_4 MPG + a_5 RY_D,$$

where a_0, a_1, a_2, a_3, a_4, and a_5 are constants. These constants were then determined by a statistical analysis of data from the years 1962 to 1987. This resulted in the demand equation

$$D = 97868.796 - 416.025 RP_R + 6.914 RP_T + 0.714 N$$

$$- 6032.863 MPG - 1.312 RY_D.$$

See the Exercises 12G for more details about this study.

12.4 **A Firm's Cost of Production**

Suppose that firm has undertaken an analysis of the market and come to the conclusion that there is ample additional demand for one of its products. The market price for the product has been rising steadily, and the firm has no reason to expect that its competitors will be able to increase the supply of similar products substantially. The firm has tentatively decided to expand its operations and to increase the output of the product. Before it can determine whether this will be profitable, however, it needs to estimate the effect of the increased output on production costs. Such estimates begin with a careful collection of relevant data of previous production figures of the firm and the associated costs, and (when available) similar data for other firms. The cost of production consists of the sum of *labor costs* and *capital costs*. Labor costs will include those for the workers on the production line, the management and staff, the maintenance crew, and so on. Capital costs include the cost of the equipment, tools, and machinery needed for production; the various buildings that house them; the support facilities; the raw materials, any preassembled components (for instance, the batteries, brake systems, radios, and tires that Ford receives from suppliers); insurance and legal fees; and so on. The determination of the cost of a particular product is often a delicate task, especially when the firm is a conglomerate that has many product lines. Consider, for instance, the problem of allocating the appropriate proportion of maintenance and other general plant costs to a specific product.

A. Cost Functions. Suppose that the difficulties just mentioned can be dealt with and that the firm's projected output x of a particular product in units per year (or quarter of a year, or month, or day) can be correlated with the total cost $C(x)$ in dollars (or

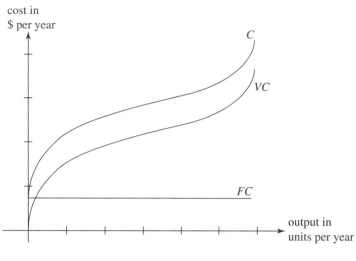

Figure 12.8

thousands, tens of thousands, or millions) of producing these x units. If $C(x)$ is understood for many different possible levels of output x, then cost increases brought about by a future increase of output can be estimated.

The total cost $C(x)$ consists of two components: the *fixed cost* and the *variable cost*. To understand the fixed cost, suppose, for a moment, that our firm has shut down its entire production and is only maintaining its facilities. The proportion of the cost for this "minimal operating state" that can be assigned to the product in question is part of the fixed cost of the product. This will include expenditures for plant maintenance, a minimal number of employees, and some insurance and legal fees. An appropriate portion of the capital investment expenditures in the plant is also part of the fixed cost. The focus in this section will be on the short run, and we will therefore take the fixed cost FC to be constant relative to the output x. The variable cost $VC(x)$ consists of all the costs related to the production of the x units. It thus depends on the number of units x produced per year. The variable cost includes most of the costs for labor, the maintainance of the production line, the raw materials, and components from suppliers. The total cost $C(x)$ of production at an output of x units of the product per year is the sum

$$C(x) = VC(x) + FC.$$

For a relatively small output per year, the variable cost per unit will generally be relatively high. As production increases, greater efficiency will often be possible, and the variable cost per unit will decrease. In the short run, this trend will generally continue until the production process becomes overextended and the facilities are no longer adequate. At this point, inefficiencies will begin to appear. In summary, the variable cost $VC(x)$ will typically rise rapidly for small x, then begin to level off, and again rise significantly when x becomes large. This means that the graph of a typical cost function $VC(x)$ will have the general shape given in Figure 12.8. Since $C(x) = VC(x) + FC$, the graph of $C(x)$ has the same shape as that of $VC(x)$.

Several related functions are important in the analysis of cost. The *marginal cost* is the derivative $MC(x) = C'(x) = VC'(x)$ of the total cost or the variable cost.

$$MC(x) = C'(x) = \lim_{\Delta x \to 0} \frac{C(x + \Delta x) - C(x)}{\Delta x}.$$

Since $MC(x) \approx \frac{C(x + \Delta x) - C(x)}{\Delta x}$ for a small Δx, think of $MC(x)$ as the cost of producing one additional unit, given that the output is at x per year. So

$$MC(x) \approx \frac{C(x+1) - C(x)}{1} = C(x+1) - C(x).$$

Taking $\Delta x = -1$, we see similarly that $MC(x) \approx \frac{C(x-1)-C(x)}{-1} = C(x) - C(x-1)$. So $MC(x)$ can also be thought of as the reduction in the cost brought about by cutting production from x to $x-1$ units per year.

Since the focus of the firm is on the cost of the units produced in the short run, it will be interested in the *average variable cost*

$$AVC(x) = \frac{VC(x)}{x}.$$

This is the average cost of producing a single unit at an output of x units per year. The firm will produce the product efficiently if it operates at an output x at which the average cost per unit $AVC(x)$ is small. Indeed, other things being equal (such as the quality of the product) the smaller the $AVC(x)$, the better.

Using the quotient rule,

$$\frac{d}{dx}AVC(x) = \frac{d}{dx}\frac{VC(x)}{x} = \frac{VC'(x)x - VC(x)}{x^2}$$
$$= \frac{MC(x)x - VC(x)}{x^2}.$$

In order to analyze this equation, we need to recall a basic fact (from Section 8.7): When the derivative of a function is negative, the function is decreasing, and when it is positive, the function is increasing.

A look at the equation just derived shows that if $MC(x)x - VC(x) < 0$, then $\frac{d}{dx}AVC(x)$ is negative. It follows that if $AVC(x) = \frac{VC(x)}{x} > MC(x)$, then $AVC(x)$ is decreasing. This means that if the firm operates at an ouput x for which

$$AVC(x) > MC(x)$$

then an increase (possibly only a small increase) in the production will decrease the average variable cost of the units produced. Let's look at this another way. Suppose that the output of the firm stands at x units per year with an average cost of $AVC(x)$ per unit. If the output is increased by one unit per year and if the cost $MC(x)$ of this additional unit is less than $AVC(x)$,

then the average cost $AVC(x+1)$ per unit for the $x+1$ units will be less than the average cost $AVC(x)$ per unit for the x units produced the year before. From the point of view of the cost of production, the suggestion is that output should be increased.

If $MC(x)x - VC(x) > 0$, on the other hand, then $\frac{d}{dx}AVC(x) > 0$. Therefore, if $AVC(x) = \frac{VC(x)}{x} < MC(x)$, then $AVC(x)$ is increasing. In this case, an increase in output will increase the average cost per unit. This suggests that the firm should lower its output.

At an output x at which the average cost per unit is a minimum, we must have $\frac{d}{dx}AVC(x) = 0$. This means that $MC(x)x - VC(x) = 0$, and therefore that

$$AVC(x) = \frac{VC(x)}{x} = MC(x).$$

Expressed in words, this says that the average cost per unit is a minimum at the output for which the average cost is equal to the marginal cost.

The firm may also want to take its fixed costs into consideration. In this case it will focus on the *average total cost* per unit

$$AC(x) = \frac{C(x)}{x}$$

and the average *fixed cost* per unit $AFC(x) = \frac{FC}{x}$. Because $C(x) = VC(x) + FC$,

$$AC(x) = AVC(x) + AFC(x).$$

An analysis of $\frac{d}{dx}AC(x)$ shows that the average total cost is a minimum when it is equal to the marginal cost. See Exercises 12G.

Typically, the marginal cost MC, the average variable cost AVC, and the average total cost AC are all functions that decrease over some range of output, reach a minimum, and then increase as output expands. These functions are therefore referred to as *U-shaped* cost functions. Figure 12.9 provides an illustration. Note that the average fixed cost $AFC(x) = \frac{FC}{x}$ will decline throughout.

A few years ago, industry analysts argued that the General Motors Corporation was too big.[7] In the late 1980s this company was almost twice as big as Ford and more than four times as big as Chrysler. Its huge size had created management inefficiencies that even the modernization and restructuring of the early 1980s could not resolve. For instance, Ford produced about 50% more cars per employee than GM. Industry estimates suggest that for roughly equivalent cars, Ford's costs were the lowest, followed by GM and Chrysler. While GM earned profits of 3.5% on its total sales in 1987 (only a fraction of this was from sales of automobiles), Ford earned 6.5% and Chrysler 3.9%. In reference to the average cost curve, this suggests that the three companies were positioned as shown in Figure 12.10. The figure indicates that in the late 1980s, Ford operated at near minimal average cost, Chrysler could have decreased its average cost by producing more, and GM's very large output caused it to operate inefficiently and at a high average cost.

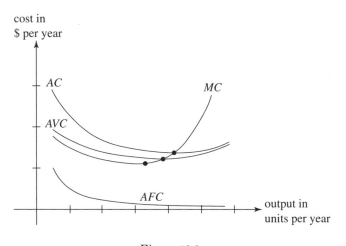

Figure 12.9

The approach discussed in this section can be used to provide a cost analysis for a single firm or an entire industry. For this to be possible, however, an explicit cost function must be available for the analysis. But how is such a cost function obtained? This is done as follows: Collect all the relevant cost data for the particular firm or industry. Then find a function whose graph fits the data that have been collected. This is the cost function that is required! The better and more

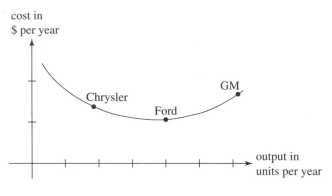

Figure 12.10

complete the data, the better the cost function will represent what is going on, and the more useful the analysis will be. Note that the strategy of moving from data to function was implicit in the supply and demand study of Section 12.3B.

The cost data for the electric utility industry are in the public domain and easily obtained. We will now turn to this industry to illustrate what has just been described. Table 12.1 lists the 1984 output and cost data[8] for a number of electric utility companies in New England. All the plants that are considered are powered by fossil fuel (either coal, oil, or gas). The output is given in millions of kilowatt hours (kwh) of electricity.[9] The variable cost of production is the sum of the labor and fuel costs. The cost is listed in thousands of dollars. The table determines the following (output, variable cost) data in order of increasing output:

(25.2, 1,804), (117.6, 10,051), (191.8, 15,507), (284.3, 16,518), (1880.9, 94,940), (2029.5, 119,165), (2615.9, 129,481), (2635.6, 145,237), (3512.9, 179,717), (3753.5, 164,103), (4147.7, 213,530), (9203.2, 272,272).

Taking output on the horizontal axis and variable cost on the vertical axis gives us the following points in the *x-y* plane. See Figure 12.11. It provides an overview of the variable cost information of Table 12.1.

Table 12.1

State	Plant	Output	Labor Expenditures	Fuel Expenditures	Variable Cost
Connecticut	Norwalk Harbor	1880.9	6,350	88,590	94,940
	Bridgeport Harbor	2635.6	13,431	131,806	145,237
	New Haven Harbor	2615.9	6,239	123,242	129,481
Maine	Graham Station	25.2	717	1,087	1,804
	Walter F. Wyman	2029.5	11,688	107,477	119,165
Massachusetts	Mystic	4147.7	23,056	190,474	213,530
	New Boston	3512.9	20,744	158,973	179,717
	Kendall	191.8	700	14,807	15,507
	Cannon Street	117.6	2,696	7,355	10,051
	Brayton Point	9203.2	21,922	250,350	272,272
	Salem Harbor	3753.5	22,735	141,368	164,103
	B. F. Cleary	284.3	3,063	13,455	16,518
New Hampshire	Merrimack	3272.7	8,598	65,390	73,988

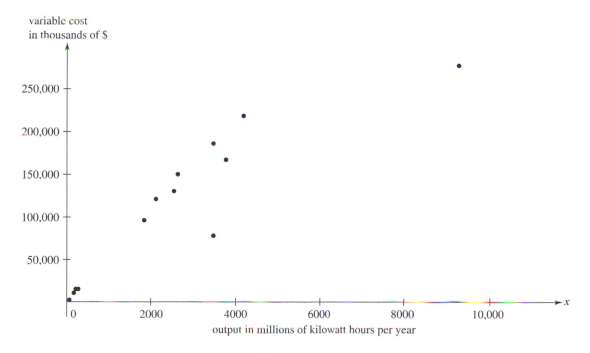

Figure 12.11

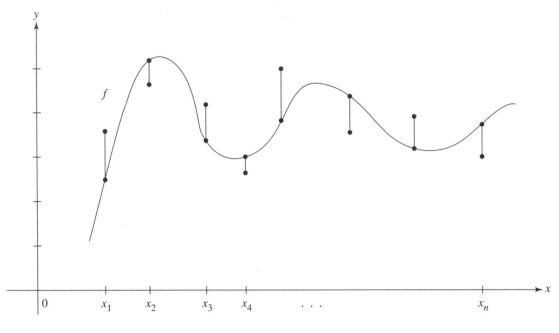

Figure 12.12

As was already pointed out, what is required now is a function whose graph "fits" the data points of Figure 12.11 at least in an approximate way. We will see that the analysis of such a function will provide us with important insight into the cost dynamics of the New England electric utility industry.

A comparison of the shape of the variable cost curve of Figure 12.8 with that of the graph of the function $f(x) = x^3$ suggests the following strategy: Start with a general cubic cost function

$$VC(x) = ax^3 + bx^2 + cx + d$$

where $a, b, c,$ and d are constants, and then determine the precise $a, b, c,$ and d that make the graph of $y = VC(x)$ fit the points of Figure 12.11 as tightly as possible.

The important two-fold problem that now remains is this: How is the fit of a curve to certain data points to be measured, and how does one go about finding the constants $a, b, c,$ and d such that the function $f(x) = ax^3 + bx^2 + cx + d$ fits the data points best?

B. The Method of Least Squares. Let's first consider the matter of fit. Fix any set of n points

$$(x_1, y_1), (x_2, y_2), \ldots, (x_n, y_n)$$

in the plane with the property that $x_1 < x_2 < \cdots < x_n$. Suppose a function f is defined on an interval containing $x_1, \ldots, x_n$ and

$$f(x_1) \approx y_1, \ f(x_2) \approx y_2, \ldots, f(x_n) \approx y_n.$$

So the graph of f "fits" the given points to some extent. What criterion can be used to decide how good the fit is? By what test can we say that one function provides a better fit than another? One of the most common and meaningful tests is the *least squares* criterion. The *error* of f is the number

$$E(f) = (y_1 - f(x_1))^2 + (y_2 - f(x_2))^2 + \cdots + (y_n - f(x_n))^2.$$

Refer to Figure 12.12. Of the two points over x_1, the one on the curve has coordinates $(x_1, f(x_1))$ and the other is the given point (x_1, y_1). The length of the segment joining them is $|y_1 - f(x_1)|$. Of the two points over x_2, the one on the curve has coordinates $(x_2, f(x_2))$

and the other is the given point (x_2, y_2). The length of the segment joining them is $|y_2 - f(x_2)|$. Continue the pattern and observe that the error of f is the sum of the squares of the lengths of all the vertical segments. The length of the segment measures the discrepancy between a data point and the curve. The fact that the lengths are squared magnifies the impact of a large discrepancy on the error $E(f)$.

It seems reasonable to stipulate that of two functions, it is the one with the smaller error that fits the data better. In reference to the question raised at the end of Section 12.4A, it remains to come up with a method for determining a cubic polynomial with the smallest possible error.

Let's look at a simple situation. Take the data points $(1, 11), (2, 3), (3, 9)$, and $(4, 5)$ and the cubic polynomial $f_1(x) = -4x^3 + 32x^2 - 70x + 52$. By substituting, we get

$$f_1(1) = -4 + 32 - 70 + 52 = 10$$

$$f_1(2) = -32 + 128 - 140 + 52 = 8$$

$$f_1(3) = -108 + 288 - 210 + 52 = 22$$

$$f_1(4) = -256 + 512 - 280 + 52 = 28$$

The fit gets progressively worse: 10 is close to 11, 3 is not too far from 8, but 9 is far from 22, and 28 even farther from 5. The error of f_1 is

$$E(f_1) = (11 - f_1(1))^2 + (3 - f_1(2))^2$$

$$+ (9 - f_1(3))^2 + (5 - f_1(4))^2$$

$$= 1^2 + (-5)^2 + (-13)^2 + (-23)^2$$

$$= 1 + 25 + 169 + 529 = 724.$$

As expected, the error is large. Notice that the large discrepancy of $|23|$ between $f_1(4) = 28$ and 5 provides by far the largest contribution to the error.

Let's do the same thing for the polynomial $f_2(x) = -4x^3 + 31x^2 - 72x + 56$. This time,

$$f_2(1) = -4 + 31 - 72 + 56 = 11$$

$$f_2(2) = -32 + 124 - 144 + 56 = 4$$

$$f_2(3) = -108 + 279 - 216 + 56 = 11$$

$$f_2(4) = -256 + 496 - 288 + 56 = 8$$

Observe that the fit is good: 11, 4, 11, and 8, are respectively close (or equal) to $11, 3, 9$, and 5. This time the error

$$E(f_2) = (11 - 11)^2 + (3 - 4)^2 + (9 - 11)^2 + (5 - 8)^2$$

$$= 0 + 1 + 4 + 9 = 14$$

is relatively small. A comparison of the coefficients shows that the polynomials

$$f_1(x) = -4x^3 + 32x^2 - 70x + 52$$

and

$$f_2(x) = -4x^3 + 31x^2 - 72x + 56$$

are not too different. But the difference in the errors is substantial. Is there a cubic polynomial that gives a smaller error? If so, how can it be found?

It turns out that for any collection of data points there is a cubic polynomial $f(x) = ax^3 + bx^2 + cx + d$ with the smallest possible error (the possibility that a, b, c, and/or d are zero must be allowed). If there are at least four data points, there is exactly one cubic polynomial that provides this smallest error. To find this polynomial is simple in principle: determine the constants a, b, c, and d that minimize the error $E(f)$. To carry this out involves elements from more advanced mathematics. (The methods of calculus of several variables are used to set up a system of four linear equations in the unknowns a, b, c, and d.) The details will not concern us, however, because there are computer programs that find the cubic polynomial with the smallest error at the touch of a button (well, a few buttons). The software packages *Maple* and *Mathematica* contain such programs.

Let's see what cubic polynomial *Maple* provides for the data points $(1, 11), (2, 3), (3, 9)$, and $(4, 5)$. By consulting a *Maple* manual about its "Least Squares" application, you will be able to generate the following on your computer screen:

```
>?leastsquares
```

```
> with(stats):
```

```
> xval := [1,2,3,4];
```
$$xval := [1,2,3,4]$$

```
> yval := [11,3,9,5];
```
$$yval := [11,3,9,5]$$

```
> p1 := a*x^3 + b*x^2 + c*x + d;
```
$$p1 := ax^3 + bx^2 + cx + d$$

```
> eq_fit := fit[leastsquare[[x,y],
    y = p1,{a,b,c,d}]]([xval, yval]);
```
$$eq\_fit := y = -4x^3 + 31x^2 - 73x + 57$$

The symbols $>$ are prompts that the monitor generates. Each of these is followed by the input that you must supply. The input xval := [1,2,3,4] comes from the x-coordinates of the data points $(1,11)$, $(2,3)$, $(3,9)$, and $(4,5)$. The input yval := [11,3,9,5] comes from the y-coordinates of the data points. A semicolon completes an input. After each input push the Enter key to trigger the response of the computer. It responds and draws a horizontal line. The last response of the computer lists

$$f(x) = -4x^3 + 31x^2 - 73x + 57.$$

This is the cubic polynomial with the smallest error! What is this error? By substituting, we see that

$$f(1) = -4 + 31 - 73 + 57 = 11$$

$$f(2) = -32 + 124 - 146 + 57 = 3$$

$$f(3) = -108 + 279 - 219 + 57 = 9$$

$$f(4) = -256 + 496 - 292 + 57 = 5$$

Notice that every y-coordinate is matched, and hence that the error

$$E(f) = (11 - f(1))^2 + (3 - f(2))^2$$
$$+ (9 - f(3))^2 + (5 - f(4))^2$$
$$= 0 + 0 + 0 + 0 = 0$$

is zero!!

The *Maple* program just described provides the cubic polynomial with the smallest possible error for any number of data points. Be aware that in general, this error is not necessarily equal to zero. In other situations it may be required that a linear or quadratic polynomial (or a polynomial of degree higher than three) be fit to certain data points. The linear case is relevant in the determination of the disintegration constant of a radioactive element and in the context of the logistics model. (Refer to Section 11.1B, especially Figure 11.1, and to Section 11.5C, in particular Figure 11.7.) The software packages *Maple* and *Mathematica* do this as well. (For instance, in the context of the illustration just given, simply enter p1:=a*x^2+b*x+c instead of p1:=a*x^3+b*x^2+c*x+d to get the quadratic polynomial with the smallest error.)

C. Cost Analysis for Electric Companies.

Let's return to the electric utility industry of New England. As already mentioned, what is needed is a cubic polynomial that fits the cost data of Table 12.1. The *Maple* program just described will show that

$$-0.0000001x^3 - 0.001425x^2 + 51.000703x + 4159.448062$$

is the polynomial that fits these data best. Therefore,

$$VC(x) = -0.0000001x^3 - 0.001425x^2 + 51.000703x$$
$$+ 4159.448062$$

is the cost function that needs to be analyzed. The variable cost $VC(x)$ is given in thousands of dollars at an output of x in millions of kwh. The graph of VC is

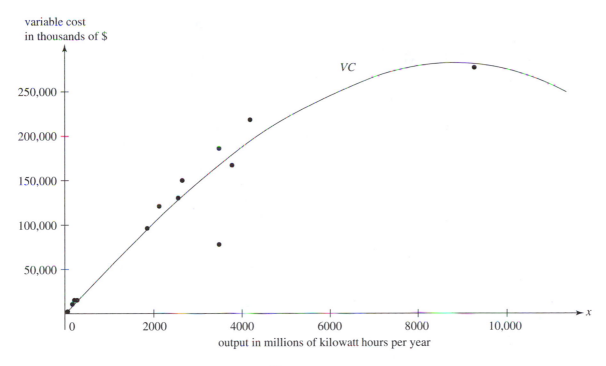

variable cost
in thousands of $

250,000

200,000

150,000

100,000

50,000

VC

0   2000   4000   6000   8000   10,000   x

output in millions of kilowatt hours per year

Figure 12.13

shown in Figure 12.13 along with the data points from Figure 12.11.

The average variable total cost AVC and the marginal cost MC are both readily computed:

$$AVC(x) = \frac{VC(x)}{x}$$

$$= -0.0000001x^2 - 0.00143x + 51.0007$$

$$+ 4159.45x^{-1}$$

and

$$MC(x) = VC'(x)$$

$$= -0.0000003x^2 - 0.00286x + 51.0007.$$

The graphs of the functions AVC and MC are sketched in Figure 12.14. Recall that cost is given in units of 1000 dollars and output in millions of kwh. So the average variable cost and the marginal cost are both in 1000 dollars per million kwh. Canceling the factor

1000 from this ratio shows that these costs are also given in dollars per 1000 kwh. This is the unit used in Figure 12.14.

What conclusions can be drawn from our analysis for particular electric utility plants and for the entire industry represented by the thirteen plants? Let's compare the actual average variable costs with the average costs predicted by the analysis. Note that this analysis was achieved by gathering together the data from all the plants. The actual average is computed from Table 12.1 by dividing the variable cost by the output x, and the predicted average is supplied by substituting x into the function AVC. Both are given in dollars per 1000 kilowatt hours. The Norwalk Harbor plant, with $50.48 (actual) and $50.17 (predicted) and the Brayton Point plant, with $29.58 (actual) and $29.82 (predicted), perform close to prediction. The Graham Station plant with $71.59 (actual) and $216.02 (predicted), and the Merrimack plant, with $22.61 (actual) and $46.52 (predicted), both produce much more cheaply at their levels of output than expected. They get high marks. On the other hand, both the Bridgeport Harbor plant, with $55.11 (actual) and $48.12 (predicted), and the Walter F. Wyman plant, with $58.72 (actual) and $49.74 (predicted), appear to be producing at too great a cost.

By an easy computation, $AVC(x) - MC(x)$

$$= (-0.0000001x^2 - 0.00143x + 51.0007 + 4159.45x^{-1})$$

$$- (-0.0000003x^2 - 0.00286x + 51.0007)$$

$$= 0.0000002x^2 + 0.00143x + 4,159.45x^{-1}.$$

It follows that $AVC(x) - MC(x) > 0$ for all x, and hence $AVC(x) > MC(x)$ for all x. Therefore, by the discussion in Section 12.4A, the average variable cost AVC is decreasing for any increase in output. This means that the New England utility plants in the study should be able to produce electricity more cheaply by increasing production. This suggests that New England would benefit from cheaper electric power if there were fewer plants with greater outputs. (Note, however, that we have assessed the New England utility industry in the short run only and have not taken into account the fact that larger plants would have greater fixed costs.)

What changes have occurred in the New England electric utility industry since 1984, the year that the data were collected? Did any of the fossil fuel plants listed increase their output? A comparison of the output figures of 1984 with those from 1994 shows that contrary to the recommendation of our analysis, the outputs of these plants dropped. In fact, they dropped dramatically: Norwalk Harbor from 1,880.9 to 946.1 million kwh; Bridgeport Harbor from 2,635.6 to 2,336.9; New Haven Harbor from 2,615.9 to 1,237.4; Walter F. Wyman from 2,029.5 to 695.3; Mystic from 4,147.7 to 2,434.6; New Boston from 3,512.9 to 3,136.3; Kendall from 191.8 to 112.4; Brayton Point from 9,203.2 to 8,171.6; Salem Harbor from 3,753.5 to 2,706.0; F. F. Cleary from 284.3 to 83.0; and Merrimack from 3,272.7 to 2,717.6. What had happened? To understand what happened we must look at the entire New England electric utility industry. The total output[10] of all the fossil fuel electric power plants (all those powered by coal, gas, or oil) of New England stood at 52,168 million kwh in 1984. (A simple addition shows that the plants of Table 12.1 provided 33,671 million kwh of this.) Adding the supply of 23,786 million kwh that was generated by the nuclear power plants of New England and the 4,778 million kwh that came from other sources (such as hydroelectric, wind, or solar power), we see that the entire production of electric power in New England in 1984 came to 80,731 million kwh. What was the picture in 1994? The total production, at 80,934 million kwh, remained the same! However, the component delivered by the fossil fuel plants dropped by about 17,000 million khw to 35,128 million kwh. The oil-powered plants accounted for most, namely 15,009 million kwh, of this drop. This slack was picked up by the nuclear plants, whose output rose by 17,384 million kwh. Why the shift from oil to nuclear? This shift was driven in large measure by the increases and uncertainties in the price of oil brought about by the OPEC supply manipulations discussed in Section 12.3B.

12.5 Price, Revenue, and Profit

In a perfectly competitive market, the price of a product is determined by market forces. In practice, markets are competitive, but never perfectly competitive. So a firm does have at least some control over the price that it sets for its product. Suppose that a firm has done an intensive study of the relationship between the price and the demand for one of its goods. Suppose it has also done a careful study of the connection between output and total cost. With this information in hand, the firm can now begin to turn its attention to the "bottom line." What price should it charge for the product, and how many units should it produce? The number of units produced will typically exceed the number of units sold, so that the firm will maintain an inventory of unsold units of the product.

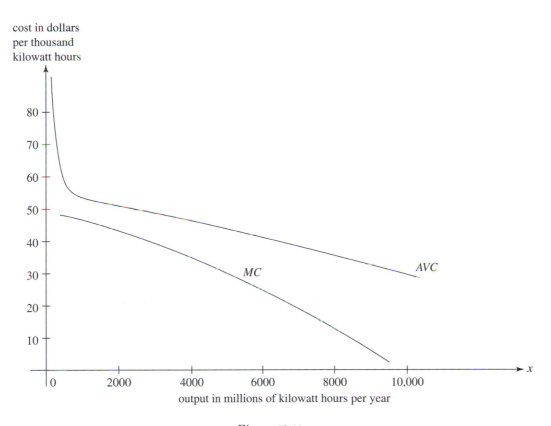

Figure 12.14

In the analysis that follows, we will ignore inventory and assume that output and sales coincide.

A. Revenue Functions. The analysis of the total cost should result in an understanding of the average cost $AC(x)$ per unit at various outputs of x units per year (or quarter of a year, or month, or day). This analysis should include not only the costs associated with production, but also the costs connected with sales and marketing (such as those for advertising, promotions, transportation, and training of the sales force). The next step is a decision about the price. A widely used practice is *cost-plus pricing*. The price is determined by cost plus a markup. More precisely, the price $p(x)$ per unit is set at

$$p(x) = AC(x) + \text{ markup}.$$

The price $p(x)$ that is decided upon will influence the demand for the product and hence the output x. It follows that the decision about the markup should take into account not only the prices that competitors are charging for similar products, but demand projections as well.

The *total revenue* that the firm takes in from the sale of x units per year at a price of $p(x)$ per unit is equal to

$$R(x) = p(x)x$$

dollars (or thousands, millions, or billions of dollars) per year.

Example 12.10. Refer back to the study of the impact of the OPEC cartel on the world's oil market. At an output of 12 billion barrels per year at $4 per bar-

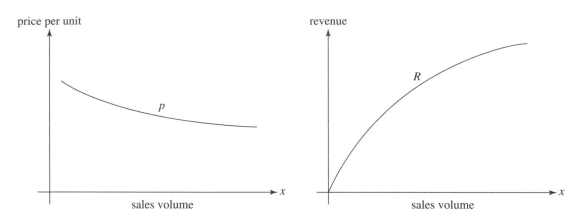

Figure 12.15

rel, OPEC earned a revenue of $R(12) = p(12)12 = 4 \cdot 12 = 48$ billion dollars per year. After it cut its production to 9 billion barrels per year, the price rose to $12 per barrel and OPEC's revenues soared to $R(9) = p(9)9 = 12 \cdot 9 = 108$ billion dollars per year.

Typical price and total revenue graphs are sketched in Figure 12.15. Both curves reflect general trends. An increase in the production x brings a decrease in the average cost $AC(x)$ per unit, and hence (at a fixed markup) a decrease in the price

$$p(x) = AC(x) + \text{markup}$$

per unit. Since $p(x)$ decreases with an increase in sales, the increases in revenue $R(x)$ will typically level off. The derivative

$$R'(x) = \lim_{\Delta x \to 0} \frac{R(x + \Delta x) - R(x)}{\Delta x}$$

of the revenue function $R(x)$ is the *marginal revenue* and is written $MR(x)$. Applying the product rule to $R(x) = p(x) \cdot x$ shows that

$$MR(x) = p'(x) \cdot x + p(x).$$

Suppose that our firm operates in a very competitive and also stable market. The supply provided by the firm has no influence on the market as a whole, and the firm sets the price $p(x)$ equal to the market-equilibrium price p^*. So $p(x) = p^*$ is constant. In this case, the revenue function $R(x) = p^*x$ is linear, and the marginal revenue $MR = R'(x)$ is equal to the market-equilibrium price p^*.

B. Maximizing Profit.   Suppose that the firm has done a careful analysis of the demand for one of its products, the costs of production, and various pricing strategies. It can now turn its attention to the matter of profit, and in particular the question as to the sales volume that will maximize it.

The profit from the manufacture and sale of the product is obtained by subtracting the total cost from the total revenue. More precisely, if x is the number of units sold per year (or quarter, or month), then the profit $\Pi(x)$ derived from the sale of these x units is given by

$$\Pi(x) = R(x) - C(x).$$

We will now suppose that the firm is interested in maximizing its profit. This is certainly a common strategy, although not the only one. (For instance, in the short run the executives of a firm might decide to maximize total revenue so as to achieve growth.) Therefore, the firm is interested in operating at a sales level x for which the profit $\Pi(x) = R(x) - C(x)$ is largest. This strategy can be reformulated as follows. Taking derivatives, we get

$$\Pi'(x) = R'(x) - C'(x) = MR(x) - MC(x).$$

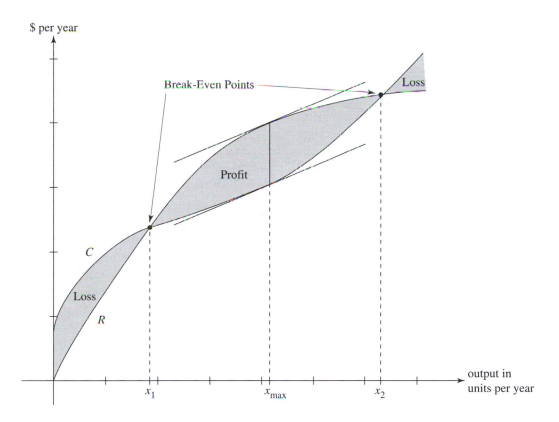

Figure 12.16

If the profit function $\Pi(x)$ is to be a maximum at x, then its derivative $\Pi'(x)$ must be equal to 0. Therefore, a sales level x_{max} that maximizes profits satisfies

$$\boxed{MR(x_{max}) = MC(x_{max})}$$

Expressed in words, this says that at a sales level that maximizes profit, the marginal revenue must be equal to the marginal cost.

Figure 12.16 illustrates the typical situation. It shows a firm's cost and revenue curves. Consider the curves in the direction of increasing output. For low levels of output x, the profit is negative because the total revenue is insufficient to cover the fixed and variable costs. In this first "Loss" region, the total cost $C(x)$ exceeds the total revenue $R(x)$. As output increases, the revenue catches up with the cost. At the break-even output of $x = x_1$ units, the total revenue is equal to the total cost. As output increases further, the revenues generated by the product exceed the costs of production, and the firm operates in "the black." It is now operating in the "Profit" region from x_1 to x_2 in which the total revenue $R(x)$ exceeds the total cost $C(x)$. The profit-maximizing output is x_{max}. This is the sales volume for which the profit is at a maximum. The vertical segment indicates the difference between the total revenue and the total cost. It therefore measures the maximum profit. Since $MR(x_{max}) = MC(x_{max})$, the slopes of the total revenue and cost functions are now equal. As output increases further, the profit begins to decline. This reflects the increase in the total cost of production. After the second break-even output x_2, these costs exceed revenues, and the firm is operating in another "Loss" region.

Since

$$\Pi(x) = R(x) - C(x) = p(x)x - C(x),$$

the firm will turn a profit as long as $p(x)x > C(x)$. This condition is the same as $p(x) > \frac{C(x)}{x} = AC(x)$. So the firm will turn a profit as long as it operates at an output at which the price of the product is above the average total cost.

A firm may decide to operate at a loss in the short run, if the expectation is that it will turn a profit in the future. For example, a demand projection might suggest that the firm will be able to raise its prices, or a cost analysis might point to an anticipated drop in production costs. A firm has the following choice if it decides to operate at a loss: It can either shut down its production temporarily, or it can produce at least some output. In the first case, the total cost is reduced to the fixed cost and is therefore minimal. In the second case, the total cost is not minimal, but there is some revenue coming in. What should the decision be?

The firm will likely choose the more profitable—actually less unprofitable—of the two alternatives. Observe that

$$\Pi(x) = R(x) - C(x) = R(x) - VC(x) - FC$$
$$= \big(R(x) - VC(x)\big) - FC.$$

If $R(x) - VC(x)$ is negative, then this term contributes to the loss. Since $R(x) = p(x)x$, this condition is $p(x)x - VC(x) < 0$, or

$$p(x) \le \frac{VC(x)}{x} = AVC(x).$$

It follows that the firm should shut down its production if it cannot operate at a production level at which the price of the product is above the average variable cost. If $p(x) > AVC(x)$, then $R(x) - VC(x) > 0$, and production should continue.

C. Profit-Maximizing for a Refinery.

Let's consider a firm that operates in a competitive market. So the price of its product is the market-equilibrium price p^* and the total revenue is $R(x) = p^*x$. Let's suppose that the firm wants to maximize profit. It therefore wants to operate at an output x for which

$$MC(x) = MR(x) = p^*.$$

Since the price p^* per unit is fixed, this equation provides a strategy for deciding what this output should be. The marginal cost is the key factor.

The short-run marginal cost function MC for a certain oil refinery[11] increases with output as shown in Figure 12.17. The shape of its graph is explained as follows: The refinery uses different processing units to manufacture gasoline from crude oil. It produces gasoline relatively inexpensively from light crude by a process that takes place in a "thermal cracker." Once these capacities are exhausted, however, additional gasoline can be produced (from light or heavy crude oil) only by more expensive methods. Note from the figure that the first capacity constraint becomes effective when production reaches about 9,700 barrels per day, at which point the marginal cost jumps suddenly. Another constraint sets in at around 10,700 barrels per day.

Our mathematical analysis will assume, now more precisely, that the marginal cost satisfies the data of Table 12.2. In view of Figure 12.17, we will also suppose that the marginal cost is a linear function in the two ranges $8{,}000 \le x \le 9{,}700$ and $10{,}700 \le x \le 12{,}000$, and that it is a quadratic function in the range $9{,}701 \le x \le 10{,}700$ whose graph at 9701 has zero slope. Finally, we will assume that $x = 8000$ barrels per day is the output at which the marginal and average total costs coincide. This means (refer to Figure 12.9 in Section 12.4A) that \$23.65 is the minimum average cost per barrel.

Table 12.2					
Output x	8,000	9,700	9,701	10,700	10,850
$MC(x)$	23.65	23.80	25.00	25.60	27.00

We will use this information to determine the marginal cost function completely. When MC is linear,

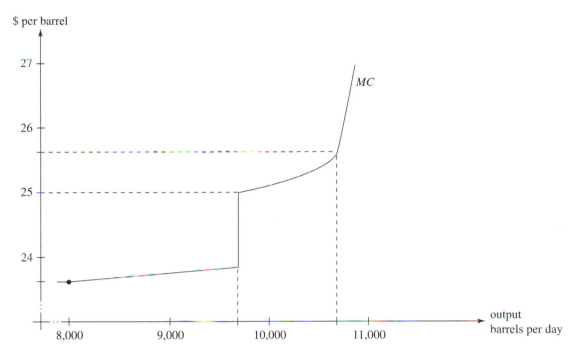

Figure 12.17

it is given by an equation of the form

$$MC(x) = mx + b.$$

Since $MC(8,000) = 23.65$ and $MC(9,700) = 23.80$, we get

$$23.80 = MC(9,700) = 9700m + b,$$

and

$$23.65 = MC(8,000) = 8000m + b.$$

By a subtraction, $0.15 = 1700m$, so $m = 0.00009$. Since $8000(0.00009) + b = 23.65$, $b = 23.65 - 0.72 = 22.93$. We have now shown that

$$MC(x) = 0.00009x + 22.93 \text{ for } 8,000 \leq x \leq 9,700.$$

Since $MC(10,700) = 25.60$ and $MC(10,850) = 27.00$, we get in exactly the same way that

$$MC(x) = 0.0093x - 73.91 \text{ for } 10,700 \leq x \leq 12,000.$$

It remains to determine the marginal cost in the range $9,701 \leq x \leq 10,700$. Here $MC(x)$ has the the

form

$$MC(x) = ax^2 + bx + c$$

with $MC'(9,701) = 0$. We will compute with 9,700 instead of 9,701 since this change has negligible impact. Using the values $MC(9,700) = 25.00$ and $MC(10,700) = 25.60$, we get the equations

$$25.60 = 10,700^2 a + 10,700b + c$$

and

$$25.00 = 9,700^2 a + 9700b + c.$$

By subtracting, $(10,700^2 - 9,700^2)a + 1000b = 0.6$, and therefore,

$$20,400,000a + 1000b = 0.6.$$

Since $MC'(x) = 2ax + b$ and $MC'(9,700) = 0$, we get $19,400a + b = 2a(9,700) + b = 0$, and hence $b = -19,400a$. Another substitution gives us

$$20,400,000a - 19,400,000a = 0.6,$$

and thus $10^6 a = 0.6$. Therefore,

$$a = 0.6 \times 10^{-6} = 0.0000006,$$

$$b = -19,400a = -0.0116,$$

$$c = 25.00 - 9,700^2 a - 9,700b$$

$$= 25.00 - 56.45 + 112.52 = 81.07.$$

So in the range $9,701 < x \leq 10,700$,

$$MC(x) = 0.0000006x^2 - 0.0116x + 81.07.$$

We have now determined that the marginal cost function is

$$MC(x) = \begin{cases} 0.00009x + 22.93, \\ \quad \text{for } 8,000 \leq x \leq 9,700, \\ 0.0000006x^2 - 0.0116x + 81.07, \\ \quad \text{for } 9,700 < x \leq 10,700, \\ 0.0093x - 73.91, \\ \quad \text{for } 10,700 \leq x \leq 12,000. \end{cases}$$

In view of the discussion at the beginning of this section, the profit-maximizing output x for the refinery can be obtained by setting $MC(x)$ equal to the market price of gasoline p^* and solving for x.

Consider, for example, a market price of $p^* = \$23.70$ per barrel. As just asserted, the profit-maximizing output at that price is found by setting $MC(x) = 23.70$ and solving for x. Refer to Figure 12.17 and notice that the output x must fall somewhere between 8,000 and 9,700 barrels per day. So the marginal cost is given by $MC(x) = 0.00009x + 22.93$. Setting $0.00009x + 22.93 = 23.70$, we get $0.00009x = 0.77$. It follows that the profit-maximizing output at $23.70 per barrel is

$$x = 8,556 \text{ barrels per day.}$$

The immediate question now is this: What is the firm's profit at this output? To compute it, we need to know both the total revenue and the total cost at $x = 8,556$. The total revenue is

$$R(8,556) = (23.70)(8,556) = \$202,780.$$

The total cost is more subtle. Begin by antidifferentiating both sides of the equation $C'(x) = MC(x) = 0.00009x + 22.93$ to get

$$C(x) = 0.000045x^2 + 22.93x + k$$

for some constant k. Next, recall that the average total cost at $x = 8,000$ is \$23.65. It follows that $C(8,000) = (23.65)(8,000) = 189,200$. This tells us that

$$k = 189,200 - (0.000045)(8,000)^2 - 22.93(8,000)$$

$$= 189,200 - 2,880 - 183,440$$

$$= 2,880.$$

So $C(x) = 0.000045x^2 + 22.93x + 2,280$, and hence

$$C(8,556) = (0.000045)(8,556)^2 + (22.93)(8,556) + 2,280$$

$$= 3,294 + 196,189 + 2,280$$

$$= \$201,763.$$

So the profit in question is

$$\Pi(8,556) = R(8,556) - C(8,556)$$

$$= 202,780 - 201,763$$

$$= \$1,017 \text{ per day.}$$

So the profit is small, and the firm is operating just barely beyond the break-even point. This is not surprising, because the market price of \$23.70 per barrel is only a little higher than the minimum average cost of \$23.65 per barrel.

Suppose now that the market price has risen to $p^* = \$25.40$ per barrel. What is the profit-maximizing output now? Another look at Figure 12.17 shows that for $M(x) = \$25.40$ to hold, the output x must fall to between 9,701 and 10,700 barrels per day. So this time the equation $MC(x) = 0.0000006x^2 - 0.0116x + 81.07$ applies. After setting

$$0.0000006x^2 - 0.0116x + 81.07 = 25.40,$$

we get $0.0000006x^2 - 0.0116x + 55.67 = 0$. by the quadratic formula,

$$x = \frac{0.0116 \pm \sqrt{0.0116^2 - 4(0.0000006)(55.67)}}{2(0.0000006)}$$

$$= \frac{0.0116 \pm \sqrt{0.000135 - 0.000134}}{0.0000012}$$

$$= \frac{0.0116 \pm 0.001}{0.0000012}$$

$$= \frac{0.0106}{0.0000012} = 8,833 \ \text{ or } \ \frac{0.0126}{0.0000012} = 10,500.$$

Since the output x needs to be between 9,701 and 10,700 barrels per day, the profit-maximizing output at a price of $25.40 per barrel is

$$10,500 \text{ barrels per day.}$$

What about the profit in this case? Again, the total revenue is easy. It is

$$R(10,500) = (25.40)(10,500) = \$266,700.$$

The computation of the total cost proceeds as before. By antidifferentating the equation $C'(x) = MC(x) = 0.0000006x^2 - 0.0116x + 81.07$, we get

$$C(x) = 0.0000002x^3 - 0.0058x^2 + 81.07x + k$$

for some constant k. Recall that

$$C(x) = 0.000045x^2 + 22.93x + 2,280$$

in the range $8,000 \le x \le 9,700$. It follows that $C(9,700) = 4,234 + 222,421 + 2,280 = 229,000$. This tells us that

$$k = 229,000 - (0.0000002)(9,700)^3$$

$$+ (0.0058)(9,700)^2 - (81.07)(9,700)$$

$$= 229,000 - 182,535 + 545,722 - 786,379$$

$$= -194,192.$$

So $C(x) = 0.0000002x^3 - 0.0058x^2 + 81.07x - 194,192$. Therefore,

$$C(10,500) = (0.0000002)(10,500)^3 - (0.0058)(10,500)^2$$

$$+ (81.07)(10,500) - 194,192$$

$$= 231,525 - 639,450$$

$$+ 851,235 - 194,192$$

$$= \$249,118.$$

This time the profit is

$$\Pi(10,500) = R(10,500) - C(10,500)$$

$$= 266,700 - 249,188$$

$$= \$17,582.$$

This daily profit, when projected over an entire year, is $(17,582)(365) = \$6,417,430$. Now the refinery is showing a healthy profit.

The message of this section has been that marginal cost analysis, in particular the use of the equation $MC(x) = p^*$, can provide important information about how to operate a firm. It must be stressed, however, that the usefulness of the estimates that are obtained is in direct proportion to the accuracy of the cost data that underlie them.

12.6 **Consumer Surplus**

Consider the market for a particular product. Suppose that market dynamics have resulted in a market-equilibrium price of p^* for the product and that x^* units per year (quarter, month, or day) are purchased at that price. Consider the prices

$$p_1 > p_2 > p_3 > p_4 > \cdots > p_{n-1} > p^*$$

for the product from the highest possible price p_1 at which there is at least some demand on down to the actual market price p^*. Suppose that the differences between these prices are all very small, say, one penny.

Group the x^* units of the product that were purchased as follows: A first group consisting of the small number of all the units, let's label it by Δx_1, that were purchased by consumers who would actually have bought the product at the highest price p_1; a second

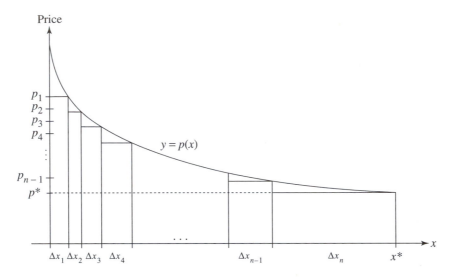

Figure 12.18

small group of Δx_2 units that were purchased by consumers who would not have paid p_1 for it, but who would have paid a penny less at p_2; a third group of Δx_3 units purchased by consumers who would have paid p_3 but not p_2; and so on, down to a last group consisting of the Δx_n units bought by those that would not have paid any more than the p^* at which they made their purchase.

Consider the function $p(x)$ that interrelates price p and demand x. This is the function obtained by solving the demand function $x = D(p)$ of Section 12.3 for p. So $p(x)$ is the price that generates a demand of x. Since $x^* = D(p^*)$, we know that $p(x^*) = p^*$. Figure 12.18 illustrates the grouping of the x^* purchases just discussed in terms of the graph of the function $p(x)$. Notice that

$$\Delta x_1 + \Delta x_2 + \cdots + \Delta x_n = x^*.$$

Now let $x_1, x_2, x_3, \ldots, x_{n-1}, x^*$ be the various demands that are generated by the prices p_1, p_2, $p_3, \ldots, p_{n-1}, p^*$. Thus

$$p(x_1) = x_1,\ p(x_2) = x_2,\ p(x_3) = x_3, \ldots,$$

$$p(x_{n-1}) = p_{n-1},\ p(x^*) = p^*.$$

Since the prices drop, the demands increase. So

$$x_1 \leq x_2 \leq x_3 \leq \cdots \leq x^*.$$

Notice that x_1 is the demand at the (highest) price. So $p_1 = \Delta x_1$. Consider the difference $x_2 - x_1$. A moment's reflection shows that this is the number of units that consumers would have been willing to pay a price of p_2 for but not p_1. So $x_2 - x_1 = \Delta x_2$. In the same way, $x_3 - x_2 = \Delta x_3$, and so on.

Observe that the buyers of the first group of Δx_1 units would have collectively been willing to pay a total of $p_1 \Delta x_1 = p(x_1) \Delta x_1$ for these units. Since they actually paid $p^* \Delta x_1$, they benefited implicitly from a saving of

$$p(x_1) \Delta x_1 - p^* \Delta x_1 = \big(p(x_1) - p^*\big) \Delta x_1.$$

This saving is the *consumer surplus* for these buyers. In the same way, the purchase of the second group of units resulted in a consumer surplus of

$$p(x_2) \Delta x_2 - p^* \Delta x_2 = \big(p(x_2) - p^*\big) \Delta x_2,$$

and so on. The *total consumer surplus* is equal to

$$\big(p(x_1) - p^*\big) \Delta x_1 + \big(p(x_2) - p^*\big) \Delta x_2 + \cdots$$

$$+ \big(p(x_{n-1}) - p^*\big) \Delta x_{n-1} + \big(p(x_n) - p^*\big) \Delta x_n$$

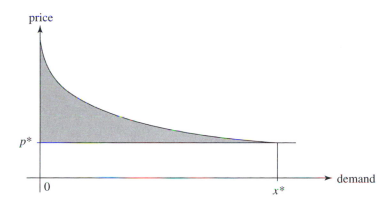

Figure 12.19

$$= p(x_1)\Delta x_1 + p(x_2)\Delta x_2 + \cdots + p(x_n)\Delta x_n$$

$$- p^*(\Delta x_1 + \Delta x_2 + \cdots + \Delta x_n)$$

$$= p(x_1)\Delta x_1 + p(x_2)\Delta x_2 + \cdots + p(x_n)\Delta x_n$$

$$- p^*x^*.$$

This is the amount that was, in effect, saved by all the purchasers of the product.

Denote the typical group of units Δx_i by dx. Now review the basics about the definite integral from Sections 5.3 and 5.5B. Since $\Delta x_1, \Delta x_2, \ldots, \Delta x_{n-1}$, and Δx_n are all small, your review will tell you that the sum

$$p(x_1)\Delta x_1 + p(x_2)\Delta x_2 + \cdots + p(x_n)\Delta x_n$$

is equal to the definite integral

$$\int_0^{x^*} p(x)dx.$$

This sum is equal to the area under the graph of the function $p(x)$ over the interval $0 \leq x \leq x^*$. Notice therefore that the total consumer surplus is equal to

$$\int_0^{x^*} p(x)dx - p^*x^*.$$

Viewed geometrically, this is equal to the area of the shaded region of Figure 12.19.

To illustrate the importance of the concept of consumer surplus, let's return to Section 12.3B and

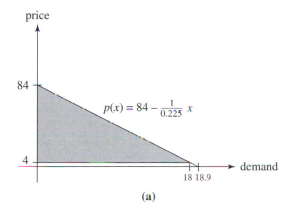

(a)

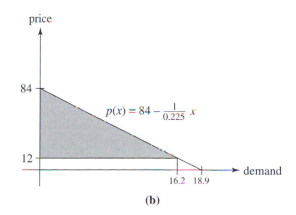

(b)

Figure 12.20

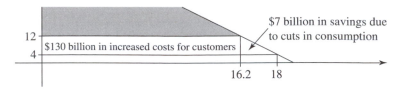

Figure 12.21

the world's oil market at the time of the OPEC supply cuts in 1974. The world's short-run demand for oil was $x = D(p) = 18.9 - 0.225p$. Solving $x = 18.9 - 0.225p$ for p, we get $0.225p = 18.9 - x$, or $p = 84 - \frac{1}{0.225}x$. Therefore,

$$p(x) = 84 - \frac{1}{0.225}x.$$

Before the OPEC cuts, the demand stood at $x^* = 18$ billion barrels per year at the market-equilibrium price of $p(18) = 84 - \frac{1}{0.225}(18) = 84 - 80 = 4$ dollars per barrel. The consumer surplus is represented by the area of the shaded triangular region shown in Figure 12.20(a). So it was equal to $\frac{1}{2}(18)(84 - 4) = 720$ billion dollars.

The OPEC supply cuts in 1974 destabilized the oil market, creating the new market-equilibrium price of \$12 per barrel. The new demand was $x^* = D(12) = 18.9 - (0.225)12 = 16.2$ b.b. per year. The corresponding consumer surplus is the area of the shaded triangle in Figure 12.20(b). It was therefore equal to $\frac{1}{2}(16.2)(84 - 12) = 583$ billion dollars.

The consumer surplus had dropped by $720 - 583 = 137$ billion dollars. This difference consists of two parts. See Figure 12.21. After the cuts, the consumers paid 8 dollars per barrel more for the 16.2 b.b. per year that they purchased. This represented a total loss to the consumers of $8(16.2) = 130$ billion dollars. The remaining drop of $\frac{1}{2}(18 - 16.2)(12 - 4) = (0.9)8 = 7$ billion dollars in the consumer surplus is explained by the reduced consumption.

The concept of consumer surplus, initially abstract, has now explained concretely the reallocation of large amounts of money. A similar surplus anal-

ysis applies to the producers of a product. This is undertaken in Exercise 12K.

12.7 Postscript

This chapter has given an illustration of some of the mathematical considerations used in finance and economics. The 1940s and 1950s saw a dramatic increase in the use of mathematics in these disciplines. A global view of an economy that takes into account the large number of commodities, the equally large number of prices, the many agents in the economic realm and their interactions, requires mathematical approaches. Some of these take a comprehensive view of any economic system and bring insights and solutions to several problems. For instance, pricing strategies can contribute to achieving an efficient use of resources, to equalize supply and demand for commodities, and also to prevent the formation of destabilizing conditions. Sophisticated mathematical theories, such as convex analysis and fixed-point theory can inform the economist as to how this might be done. Competition is perfect when every agent's influence on the outcome of economic activity is insignificant. However, the totality of their influence on the outcome is significant. The problem of adding myriads of negligible quantities so as to obtain a finite sum belongs to the domain of the calculus of the definite integral. In this respect the application of integration theory to the study of economic competition is entirely natural. In summary, mathematics provides the economist with a language and a method that permit an effective study of economic systems of forbidding complexity. This influence has transformed economics during the past five decades.

Exercises
12A. Basic Interest

1. An amount of $5,000 is invested in an account that pays an annual interest rate of $r = 0.08$. What will the investment be worth after 7 years, if interest is

   i.   compounded annually?
   ii. compounded semiannually?
   iii. compounded quarterly?
   iv. compounded monthly?
   v.   compounded daily?
   vi. compounded continuously?

2. An amount of A_0 dollars is deposited in an account that pays interest at an annual rate of 0.07 compounded continuously. In how many years will the account double to $2A_0$ dollars?

3. Show that the amount A_0 dollars will double in $t = \frac{69.3}{i}$ years at an annual interest of $i\%$ compounded continuously.

The number 69.3 is not convenient for mental arithmetic, and it is common to use 70 instead. The "Rule of 70" asserts that money invested at $i\%$ will double in about $\frac{70}{i}$ years. For $i = 6$ for instance, $\frac{70}{6} = 11\frac{4}{6} = 11\frac{2}{3}$. So at 6%, it will take about $11\frac{2}{3}$ years for the money in an account to double. The precise length of time is $\frac{69.3}{6} = 11.55$ years. The Rule of 70 is based on the conclusion of Exercise 3 and therefore applies to accounts for which interest is compounded continuously. We have already observed, however, that the frequency with which interest is compounded has little effect on the value of the account. See Example 12.1. So the Rule of 70 is a good rule of thumb.

4. Use the Rule of 70 to approximate the annual interest in percent (compounded continuously) that is necessary to double an initial investment in 8 years.

5. Suppose that an investment plan is set up with a bank that calls for a monthly investment of $250. Suppose that the plan earns interest at an annual rate of $r = 0.07$ compounded monthly. How much money will be in the account 10 years after it is started? How much money will it have after 15 years? How much money will this account have after 10 years if interest is paid at an annual rate of $r = 0.08$? How much will it have after 15 years at $r = 0.08$?

6. Your aunt has been paying $20 per month into an account earning interest at an annual rate of 0.06 starting on the day you were born. She will give you all the money that has accumulated in this account on your 21st birthday. How much will you get?

7. Given an annual interest rate of $r = 0.04$, compute the present value of $B = \$100$ in 6 months, in 24 months, and in 60 months. Compare the results with similar computations in Section 12.1C.

12B. Annuities, Mortgages, and Bonds

8. A younger colleague of the professor whom we met in Section 12.1C turned 35 in May of 1994 and has (since that time) been paying $1,000 per month into an account that pays at an annual rate of $r = 0.06$ compounded monthly. How much money will she have in this account when she retires at the age of 70 in May of the year 2029? She has decided to use this account to provide a monthly annuity payment for the first 15 years of her retirement. What monthly payments will she receive if it pays interest at an annual rate of $r = 0.08$ compounded monthly?

9. A professor turned 30 in June of 1994 and has (since that time) been paying $800 per month into an account that pays at an annual rate of $r = 0.04$ compounded monthly. How much money will she have in this account when she retires at the age of 68 in June of the year 2032? She has decided to use this account to provide a monthly annuity payment for the first 10 years of her retirement. What monthly payments will she receive from this account if it pays interest at an annual rate of $r = 0.05$ compounded monthly?

10. Arrangements have been made with a bank for a mortgage of $100,000 at an annual interest of 7% compounded monthly. Suppose that it is taken out now and that it is to be paid back in monthly payments over the next 30 years. What will the monthly payments be? [Hint: Let B be the monthly payment. From the bank's point of view the repayment can be viewed as a monthly annuity that pays it B dollars per month for $30 \times 12 = 360$ months. If the bank is willing to exchange the loan for this stream of payments, then this loan of $100,000 must be the present value PV of this income stream.]

11. Suppose that the 30 year mortagage of $100,000 of Exercise 10 is to be paid back at the annual interest of 9% (still compounded monthly). What will the monthly payments be then?

The companies AT&T, IBM, and Texaco have all issued $1000 bonds. AT&T issued 250,000 bonds on August 1, 1967; IBM issued 1,250,000 bonds on June 15, 1993; and Texaco issued 200,000 bonds on February 15, 1991. The coupon payments occur on February 1 and August 1 for the AT&T bond, on June 15 and December 15 for the IBM bond, and on February 15 and August 15 for the Texaco bond. The bond page of February 27, 1996, provides the following information about the three bonds.

AT&T 6s00	6.0	203	$100\frac{1}{8}$	$-\frac{1}{4}$
IBM $6\frac{3}{8}00$	6.3	125	102	...
Texaco $8\frac{1}{2}03$	7.6	5	112	-3

This table lists the amount of the semiannual coupon payment, the year of maturity of the bond, the number of bonds traded on February 27, 1996, the closing price, and the change in the closing price from the day before.

12. Consider each of the bonds on Feburary 27, 1996. How many coupon payments remain? Write a formula for the present value of one bond on February 27 in terms of the interest rate r. Show that the closing price on February 27, 1996, is based on an interest rate of about 0.06.

13. Suppose that on February 27, 1996, you purchased a certificate of deposit from a bank for $1120 that will earn interest at a guaranteed annual rate of 0.06 compounded monthly. How much will this investment be worth five years later on February 27 of the year 2001? On February 27, 1996, you also purchased one Texaco bond at the closing price of $1120. Beginning on August 15, 1996, you deposit the coupon payments from the bond on a semiannual basis into an investment plan that pays interest at an annual rate of 0.06 compunded semiannually. How much will this investment plan be worth on February 27, 2001? What will the value of the Texaco bond be on this day, under the assumption that its trading price is computed with the interest rate of 0.04? Will the vertificate of deposit have been the better investment or the bond.

12C. The Consumer Price Index

14. The CPI (base year 1983) was 31.6 in July 1965 and 152.5 in July 1995. What was the average rate of inflation over this 30 year period? The presidency of Jimmy Carter (1977 to 1981) was plagued by high inflation (brought about to a large extent by increases in the price of oil). Estimate the average rate of inflation for the 2 year period from December 1978 to December 1980. Use the fact that the CPI was 67.7 in December 1978 and 86.3 in December 1980.

15. The CPI was 153.2 in September 1995. Estimate the CPI in September 2000, assuming an average rate of inflation of 4% for this 5 year period. Suppose that this average rate of inflation persists until the year 2020. You will be sending your daughter to a private university in the fall of the year 2020. Assuming that the cost of an education at this university in 1995 was $25,000 per year, what would you expect to be paying then?

Recall that the current CPI was equal to 100 in August 1983. Since 1983 is its year of reference, let's label it by CPI_{83}. We saw in Section 12.2 that the CPI_{83} has increased steadily over the years. In particular, its value in 1995 was 152.9. Before 1983, a different consumer price index was in use. Since it was equal to 100 in July 1967, let's designate it by CPI_{67}. By August 1983, the CPI_{67} had risen to about 300. At this time the CPI_{83} was created and given the value 100. Notice that CPI_{83} is approximately $\frac{1}{3}CPI_{67}$. The U.S. Department of Labor gives the much more accurate formula

$$CPI_{83} = (0.3338279)CPI_{67}.$$

With this formula, the current CPI_{83} can be made retroactive to years prior to 1983. For example, the CPI_{83} for August 1967 is $(0.3338279)(100) = 33.38$. In this way, the CPI_{83} can be computed "backwards." This makes it possible to study inflation-adjusted price trends for consumer products over long periods of time.

16. Recall from Section 9.2 that the George Washington Bridge was constructed over the period 1927–1931 at a cost of 60 million dollars. Use the fact that the CPI_{83} is 16.7 for 1930 to show that this bridge would have cost about $550 million in 1995.

Incidentally, the CPI_{67} is still around. Why? Because many studies of price trends that go back to years before 1983 were undertaken with this index. Consider, for example, the behavior of the price of gasoline in the U.S. over the years. How was it affected by OPEC's manipulation of the world's oil market in the 1970s? What where the price trends before, and again thereafter? The data in Table 12.3 can be used to analyze these questions. They are taken from a study published by the American Petroleum Institute in 1989. The table lists the price of a gallon of gasoline in cents for the years from 1962 to 1987. The price listed below the given year is the average retail price for that gallon in terms of the dollar of that year. The price labeled "real" is the inflation-adjusted price in 1967 dollars. What does the table tell us? Notice the dramatic rise in the real price from 1973

Table 12.3

	1962	1963	1964	1965	1966	1967	1968	1969	1970	1971	1972	1973	1974
	30.64	30.42	30.35	31.15	32.08	33.16	33.71	34.84	35.69	36.43	36.13	38.82	52.41
real	33.82	33.17	32.67	32.96	33.00	33.16	32.35	31.73	30.69	30.03	28.83	29.17	35.48
	1975	1976	1977	1978	1979	1980	1981	1982	1983	1984	1985	1986	1987
	57.22	59.47	63.07	65.71	87.79	119.1	131.1	122.2	115.7	112.9	111.5	85.7	89.7
real	35.50	34.88	34.75	33.63	40.38	48.26	48.13	42.27	38.77	36.29	34.61	26.10	26.35

to 1974 and again from 1978 to 1980. The table shows that the real price declined in the years before 1973. After 1980 it shows a considerable real price decrease. In fact, observe that in real terms, gasoline was cheaper in 1987 than in 1962.

17. Compute the percentage of the increase in the real price of gasoline from 1973 to 1974, and again from 1979 to 1980. Compute the total decline in the real price of gasoline from 1981 to 1987. Compute the percentage of this decline and the yearly average of this percentage.

12D. Supply and Demand

18. The short-run supply and demand functions for a given product in a given market are both linear. At a price of 7 dollars, the demand and supply are both 220,000 units per month. At this price the price elasticity of supply is 0.20 and the price elasticity of demand is −0.12. Determine the supply and demand functions for the product. A group of suppliers gets together and (for the purpose of driving up the price) decides to cut production by 40,000 units. Assuming that the short-run demand situation remains the same, estimate the new equilibrium price that market forces will determine.

19. Consider the short-run demand and competitive supply functions for the world's oil determined in Section 12.3B. Suppose that OPEC had cut its supply to 10 b.b. per year instead of 9 b.b. per year. What would the new market-equilibrium price p^* have been in this case?

20. Continue to consider the world's oil market in the years 1973–1974 and recall in particular that the price of oil jumped to $12 per barrel after the OPEC supply cuts. Suppose that the supply curves S_1 and S_2

are exactly as described in Section 12.3B, but assume that the price elasticity of demand is −0.08 (instead of −0.05) at the price of $4 per barrel (and a demand of 18 b.b. per year). Notice that while the supply situation is the same, the demand is now more sensitive to price change. Consider the supply and demand curves after the OPEC cuts and refer to Figure 12.5. Would you expect the market-equilibrium price p^* to be more than $12 or less? Develop a linear function D for the world's demand based on the data just given. Use it to determine both the market-equilibrium price p^* and the demand for oil at that price.

12E. Least-Square Fit

In Exercises 21 and 22 compute the error E for each cubic polynomial relative to the data points listed. In each exercise, the polynomial of smallest error is in fact the cubic polynomial that fits the data best.

21. Consider the data points $(1, 7)$, $(2, 5)$, $(3, 9)$, $(4, 11)$ and the cubic polynomials

   i. $f(x) = -x^3 + 10x^2 - 25x + 22$
   ii. $g(x) = -x^3 + 11x^2 - 26x + 23$
   iii. $h(x) = -\frac{4}{3}x^3 + 11x^2 - \frac{77}{3}x + 23$

22. Consider the data points $(2, 3)$, $(5, 6)$, $(7, 2)$, $(9, 10)$ and the cubic polynomials

   i. $f(x) = \frac{1}{3}x^3 - 5x^2 + 23x - \frac{79}{3}$
   ii. $g(x) = \frac{3}{10}x^3 - \frac{24}{5}x^2 + \frac{229}{10}x - 26$

12F. The Hy-Tech Toy Company

This company, located in Hong Kong, is a manufacturer of innovative electronic recreational products for both children and adults. The company produces a variety of

Table 12.4

Sales	Fixed Cost	Variable Cost	Cost
250	50	100	150
275	55	118	173
300	60	150	210
280	60	112	172
315	65	142	207

* in thousands of dollars

video games and battery-operated remote-controlled toy automobiles, boats, and airplanes in a number of price ranges. Table 12.4 shows the following sales and cost records[12] for a recent five-year period. All the entries represent thousands of dollars. Because of the variety of products and range of prices for individual products, output is being measured by the dollar value of sales. In other words, the output x is taken to be equal to the revenue $R(x)$. The company assumes that a linear cost function will provide a reasonable estimate of costs in the short run. A least-squares analysis provides the total cost function

$$C(x) = 0.985x - 97.231$$

where 91.6 percent of total cost is accounted for by production and sales. Since the sales and cost data are both in thousands of dollars, this cost equation may be interpreted as saying the firm's total cost will be 98.5 percent of sales minus $97,231.

23. Compute the profit of the company as a function of the output x. What recommendation to the company does this analysis suggest?

12G. Economies and Diseconomies of Scale

Let x be a firm's output in units per year and let $C(x)$ be the associated total cost. Let $MC(x)$ be the marginal cost and $AC(x)$ the average total cost.

24. Differentiate the equation $AC(x) = \frac{C(x)}{x}$. What relationship between $MC(x)$ and $AC(x)$ tells us that $AC(x)$ is a decreasing function? What relationship tells us that it is an increasing function? Why is the average total cost a minimum at an output at which the marginal and average total costs coincide?

Define the *output elasticity of cost* by $e_C(x) = x\frac{C'(x)}{C(x)}$. If $e_C(x) < 1$, then the firm is said to enjoy *economies of scale*, and if $e_C(x) > 1$, then the firm is said to suffer from *diseconomies of scale*.

25. Show that $e_C(x) = \frac{MC(x)}{AC(x)}$. Explain why a firm should consider increasing its output if it enjoys economies of scale at an output of x and decreasing its output if it suffers from diseconomies of scale at an output of x.

12H. More About the Electric Utility Industry

Exercises 26 and 27 make use of information in Section 12.4.

26. Use Table 12.1 to check the following values for the actual average variable cost in dollars per million kilowatt hours: Brayton Point $29.58, Merrimack $22.61, and Walter F. Wyman $58.72. Now consider the variable cost function $AVC(x)$ that was derived from the cost data of all the plants. Check that this function gives the following average variable costs: Brayton Point $29.82, Merrimack $46.52, and Walter F. Wyman $49.74. Notice that the Walter F. Wyman plant produces electricity more expensively than the industry-wide cost data suggest, that the Merrimack plant produces it more cheaply, and that the Brayton Point plant operates at the expected cost.

27. Take the fixed cost of all of the plants listed in Table 12.1 to be $FC = 42,000$ in thousands of dollars per year. (This is the average fixed cost for these plants.) Verify that all of the plants in Table 12.1 enjoy economies of scale. [Hint: Show that $AC(x) - MC(x) > 0$ for all x.]

12I. Cost and Profit

Exercises 28–30 deal with the refinery studied in Section 12.5C. They make use of the information developed there.

28. What is the profit-maximizing output of the refinery at a market price of $26.50 per barrel, and what is the refinery's profit at that output?

29. Show that the refinery suffers from diseconomies of scale at any output x in the range $8,000 \leq x \leq 9,700$. [Hint: Use calculus to show that $MC(x) - AC(x)$ is an increasing function for all x in $8,000 \leq x \leq 9,700$ and deduce that $MC(x) - AC(x) > 0$ for all such x.] The strategy suggested in the hint can be used to show that the refinery also suffers from diseconomies of scale in the output range $9,700 < x \leq 10,700$.

30. The refinery earned a much greater profit at an output of 10,500 barrels per day than it did at 8,556 barrels per day. Why does this not contradict the conclusions of Exercise 29?

31. Assume that the market for one of the products of a firm is both competitive and stable, and that the firm sells everything it produces. Suppose that the market price for the product is $5760 per unit. It is known that the firm's marginal cost function for the product is given by $MC(x) = 0.00003x^2 - 0.48x + 2500$, where x is the number of units produced per quarter. Determine the production level at which the firm's quarterly profit is a maximum. It is known that the total cost for producing 15,000 units is $36.3 million dollars. What is the average cost per unit at that production level? Determine the cost function $C(x)$ and the profit function $\Pi(x)$. What is the maximal profit per quarter that the firm can earn?

12J.   A Cost Function for Freight Trucking

The regulation and deregulation of industries by the government are both important facts of economic life. In the 1980s, a number of industries saw a major relaxation of government controls as part of a trend to let market forces determine price levels, the number of competitors, and the allocation of services. This was particularly evident in the banking and transportation industries. A number of high-cost firms, formerly protected under regulation from potential competition, now faced new, more cost-efficient competition. Without significant cost reduction, these firms looked at the prospect of losses in the short run and going out of business in the long run. Consequently, an accurate assessment of cost conditions was of particular importance in the post-deregulation market.

A financial analyst interested in assessing the short-run relationship between output and cost for the general freight trucking industry in the deregulated period tabulated the following data for twelve trucking firms.[13] See Table 12.5. The variable cost listed is the total operating expense in thousands of dollars. Output is given in thousands of ton-miles.

32. Show that the average variable costs for the twelve firms (in the order in which they are listed in the table) are 0.509, 0.519, 0.303, 0.396, 0.517, 0.319, 0.293, 0.285, 0.381, 0.152, 0.277, and 0.341.

Consider the twelve data points (output, average variable cost) that Exercise 32 provides:

Table 12.5

Firm	Variable Cost	Output
1	102,813	201,953
2	196,121	377,940
3	296,416	979,267
4	226,356	571,714
5	176,163	340,608
6	450,666	1,413,807
7	607,082	2,071,861
8	624,680	2,195,352
9	636,133	1,670,195
10	222,885	1,466,024
11	378,446	1,367,596
12	579,696	1,701,125

Variable cost = total operating expenses, listed in thousands of dollars. Output listed in thousands of ton-miles.

$(102{,}813, 0.509)$, $(176{,}163, 0.517)$, $(196{,}121, 0.519)$, $(222{,}885, 0.152)$, $(226{,}356, 0.396)$, $(296{,}416, 0.303)$, $(378{,}446, 0.277)$, $(450{,}666, 0.319)$, $(579{,}696, 0.341)$, $(607{,}082, 0.293)$, $(624{,}680, 0.285)$, and $(636{,}133, 0.381)$.

A least-squares analysis shows that the polynomial

$$AVC(x) = 0.22083 \times 10^{-11}x^2 - 0.19739 \times 10^{-5}x + 0.70831$$

is the quadratic polynomial with the smallest least-squares error E.

33. Compare the actual average variable cost in Exercise 32 with that derived from the function AVC for firms 1, 9, and 11. How good is the fit at these points?

34. Compute the derivative of $AVC(x)$. Is $AVC(x)$ increasing or decreasing over the output range $100{,}000 \le x \le 1{,}000{,}000$? Determine the level of output that minimizes the average cost function AVC. What are the conclusions for the trucking firms being analyzed?

Suppose that $(703{,}649, 0.237)$, $(733{,}902, 0.190)$, and $(756{,}842, 0.115)$ are the respective (output, average variable cost) data for three additional firms. If these are combined with those for the earlier twelve firms, the least-squares analysis provides the average cost function

$$AVC(x) = -0.16959 \times 10^{-13}x^2 - 0.37410 \times 10^{-6}x + 0.48741.$$

35. Repeat the analysis of Exercise 34 for this expanded list of trucking firms.

12K.   Producer Surplus

Consider the market for a certain product. A producer who is willing to sell the product above the market-equilibrium price p^* will have the benefit of a savings (at least implicitly). These savings added over all the producers of the market are the *total producer surplus*. Let $x = S(p)$ be the supply that a price p determines in the market. Refer to Section 12.3A. Solve for p in terms of x to get the price $p(x)$ that generates a supply of x.

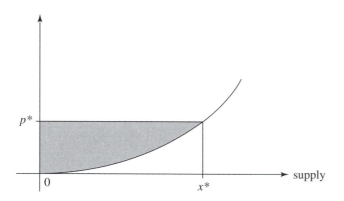

Figure 12.22

36. Study the analysis of consumer surplus in Section 12.6. Explain by drawing a diagram analogous to Figure 12.18, that the total producer surplus is equal to the area shown in Figure 12.22, where the graph is that of the function $p(x)$ just defined. (Why is this function different from the function $p(x)$ considered in Section 12.6?)

37. Express the total producer surplus in terms of a definite integral.

38. Recall the following from Section 12.3B. Before the OPEC supply cuts in 1974, the short-run supply function for the world's oil was given by $x = 17.4 + 0.15p$ b.b. per year, the market price was $p^* = 4$ dollars per barrel, and the supply $x^* = 18$ b.b. per year. Determine the associated function $p(x)$ for all $x \geq 0$.

Check that when Figure 12.22 is adapted to reflect the specifics of this situation, Figure 12.23 is obtained. Use this figure to show that the total producer surplus was 70.88 billion dollars.

39. Recall from Section 12.3B that after the OPEC cuts, the short-run supply function for oil was given by $x = 14.4 + 0.15p$ in b.b. per year. Recall also that the market price rose to $p^* = 12$ dollars per barrel at a supply of $x^* = 14.4 + 0.15(12) = 16.2$ b.b. per year. Repeat Exercise 38 in this case. Analyze the change in the producer surplus in terms of a figure analogous to Figure 12.21 (which does a similar thing for consumer surplus).

12L.   An Excerpt from Cournot's Mathematics of Value and Demand

Let p be the price per unit for some product and let $D(p)$ be the demand for the product in units per year. Cournot remarks[14] that "since the function $D(p)$ is continuous, the function $pD(p)$, which expresses the total value of the quantity annually sold, must be continuous also." Cournot goes on to point out "that it is theoretically always possible to assign to the symbol p a value so small that the value $pD(p)$ will vary imperceptibly from zero. The function disappears also when p becomes infinite, or, in other words, theoretically a value can always be assigned to p so great that the demand for the article and the production of it would cease. Since the function $pD(p)$ will at first increase and then decrease, there is therefore a value of p which makes this function a maximum, and which is given by the equation

$$D(p) + pD'(p) = 0$$

in which . . . ." Cournot lets q be such a value and considers a typical graph for D. In Figure 12.24, n is the point $(q, D(q))$, and the dotted line through n is the tangent to the graph of D at n. Note that the function $pD(p)$ that Cournot is considering is the revenue as a function of p and that $(pD(p))' = D(p) + pD'(p)$ is its derivative.

40. Verify Cournot's observation that q is the value of p for which the area of the rectangle with base Oq and height qn is a maximum and the triangle Ont is isosceles.

Cournot goes on to point out that it may be impossible to determine the function $D(p)$ empirically for each product. He does, however, suggest the possibility of an approximate determination of the value that renders the product $pD(p)$ a maximum and continues, "But even if it were impossible to obtain from statistics the value of p which should render the product $pD(p)$ a maximum, it would be easy to learn for

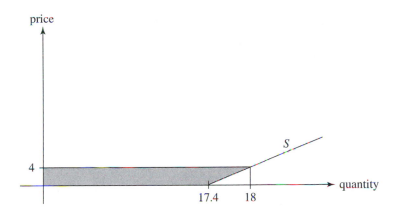

Figure 12.23

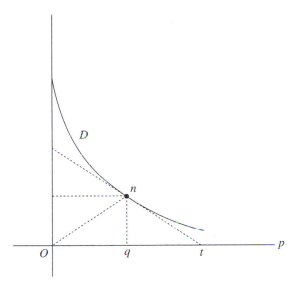

Figure 12.24

all articles to which the attempt has been made to extend commercial statistics, whether current prices are above or below this value. Suppose that when the price becomes $p + \Delta p$, the annual consumption as shown by statistics, such as customhouse records, becomes $D - \Delta D$. According as

$$\frac{\Delta D}{\Delta p} < \text{or} > \frac{D}{p}$$

the increase in price Δp will increase or diminish the product $pD(p)$; and consequently, it will be known whether the two values p and $p + \Delta p$ (assuming Δp to be a small fraction of p) fall above or below the value which makes the product un-

der consideration a maximum. Commercial statistics should therefore be required to separate articles of high economic importance into two categories, according as their current prices are above or below the value which makes a maximum of $pD(p)$." It seems clear that Cournot believed that his calculus of economics has potential practical value.

12M.   The Lerner Index

Suppose that a firm is operating in a market that is not completely competitive. So it has some ability to set the price p for its product. Suppose that the price p of the firm's product generates a demand of $x = D(p)$ in units per year. This D denotes the demand for the product of the firm and not the demand of the entire market (as it did in Section 12.3A and in Exercises 12L). The price elasticity of demand (again for the firm and not for the entire market) is $e(p) = p\frac{D'(p)}{D(p)}$. The ratio $\frac{-1}{e(p)} = i(p)$ is known as the *Lerner index* of the firm. The composite function $MC(D(p))$ is the marginal cost at the demand $D(p)$.

41.   Show that the profit as a function of p is given by

$$\Pi(p) = \Pi(D(p)) = p \cdot D(p) - C(D(p)).$$

[Hint: Start with Π expressed as a function of x.]

42.   Assume that p is a price that maximizes the profit.

    i.   Show that $-\frac{D(p)}{D'(p)} = p - MC(D(p))$.

    ii.   Deduce from (i) that the Lerner index is $i(p) = \frac{p - MC(D(p))}{p}$.

    iii.   Show $i(p^*) = 0$ in the competitive situation where p^* is the equilibrium price.

iv. Show that $p = \frac{MC(D(p))}{1 - i(p)}$.

Suppose that the firm has good statistical information about its Lerner index $i(p)$ and also about the marginal cost for various demands $D(p)$. It thus has information about the price $p = \frac{MC(D(p))}{1 - i(p)}$ at which its profit is a maximum. Therefore, the Lerner index can be used as a rule of thumb for setting the price.

12N. The Demand for Domestic Automobiles

Empirical estimates of demand conditions in the domestic automobile market have become an increasingly important subject of economic analysis. In the 1970s and 1980s American automobile manufacturers were struggling with problems ranging from poor productivity performance to intense foreign price competition. In addition to these firm-specific problems, the industry itself had to adjust to other pressures associated with a changing economic environment. Automobile sales are not only affected by the price of cars, but also by such factors as the consumers' disposable income and interest rates. Consequently, changes in these economic variables can have a negative effect on demand industry-wide, and therefore on the level of production of the individual firm. Table 12.6 provides[15] information about the domestic demand for new automobiles, as well as the variables that affect it, for the years 1970 to 1987. In Table 12.6,

D is the annual demand for new domestic automobiles.

p is a price index for new cars in constant dollars.

i is the prime rate of interest.

RY_D is the real (inflation adjusted) disposable income per family in 1988 dollars.

One essential issue is the following. Assuming that the nature of the quantitative relationships between demand, price, disposable income, and interest rate persists as in the table, is it possible to predict the demand in future years? We will assume that the dependence of D on the parameters p, RY_D, and i is given by an equation of the form

$$D = a_0 p^{a_1} RY_D^{a_2} i^{a_3}.$$

43. Which of the parameters a_0, a_1, and a_3 would you expect to be positive and which negative?

Taking the natural logarithm of the demand equation gives us

$$\ln D = \ln a_0 + a_1 \ln p + a_2 \ln RY_D + a_3 \ln i.$$

Table 12.6

Year	D	p	RY_D	i
1970	7,115,274	107.6	1668.1	7.91
1971	8,676,408	112.0	1728.4	5.72
1972	9,321,305	111.0	1797.4	5.25
1973	9,618,508	111.1	1916.3	8.03
1974	7,448,339	117.5	1896.6	10.81
1975	7,049,843	127.6	1931.7	7.86
1976	8,606,856	135.7	2001.0	6.84
1977	9,104,932	142.9	2066.6	6.83
1978	9,304,247	153.8	2167.4	9.06
1979	8,316,018	166.0	2212.6	12.67
1980	6,578,359	179.3	2214.3	15.27
1981	6,206,688	190.2	2248.6	18.87
1982	5,756,614	197.6	2261.5	14.86
1983	6,795,226	202.6	2331.9	10.79
1984	7,951,786	208.5	2469.8	12.04
1985	8,204,694	215.2	2542.2	9.93
1986	8,222,475	224.4	2645.1	8.33
1987	7,080,889	232.5	2676.1	8.22

A least-squares analysis using the data of the table shows that $\ln a_0 = 5.5527$, $a_1 = -1.184$, $a_2 = 2.183$, $a_3 = -0.191$. This provides the demand equation

$$D = 257.934 p^{-1.184} RY_D^{2.183} i^{-0.191}.$$

44. Compare the demand given by Table 12.6 with that determined by the equation for the years 1973, 1975, 1978, and 1982.

45. Treat RY_D and i as constants and show that the price elasticity of demand is the constant -1.184. Let $x = i$ vary, treat p and RY_D as constants, and show that the interest elasticity of the demand is -0.191. Is the demand more sensitive to change in price or change in the interest rate? Is your answer corroborated by the data of Table 12.6?

46. What parameters in addition to p, RY_D, and i might this study have included?

Table 12.7

Year	D	P_S	P_R	CPI	RP_T	N	MPG	RY_D
1962	43,771	20.36	30.64	90.6	87.4	66,638	14.37	2,148
1963	45,246	20.11	30.42	91.7	88.5	69,842	14.26	2,185
1964	47,567	19.98	30.35	92.9	90.1	72,969	14.25	2,304
1965	50,275	20.70	31.15	94.5	91.9	76,634	14.15	2,407
1966	53,312	21.57	32.08	97.2	95.2	80,106	14.10	2,494
1967	55,110	22.55	33.16	100.0	100.0	82,367	14.05	2,574
1968	58,524	22.93	33.71	104.2	104.6	85,793	13.91	2,647
1969	62,448	23.85	34.84	109.8	112.7	89,156	13.75	2,703
1970	65,784	24.55	35.69	116.3	128.5	92,059	13.70	2,786
1971	69,514	25.20	36.43	121.3	137.7	96,144	13.73	2,851
1972	73,463	24.46	36.13	125.3	143.4	100,658	13.67	2,933
1973	78,011	26.88	38.82	133.1	144.8	106,119	13.29	3,097
1974	74,217	40.41	52.41	147.7	148.0	109,823	13.65	3,038
1975	76,457	45.44	57.22	161.2	158.6	111,679	13.74	3,064
1976	78,847	47.44	59.47	170.5	174.2	115,170	13.93	3,143
1977	80,677	50.70	63.07	181.5	182.4	118,711	14.15	3,214
1978	83,233	53.09	65.71	195.4	187.8	121,717	14.26	3,335
1979	80,233	74.33	87.79	217.4	200.3	125,750	14.49	3,367
1980	73,375	104.73	119.1	246.8	251.6	127,448	15.32	3,330
1981	71,718	112.75	131.1	272.4	312.0	129,123	15.68	3,347
1982	72,848	102.65	122.2	289.1	346.0	129,500	16.36	3,331
1983	73,156	95.36	115.7	298.4	362.6	131,723	16.81	3,402
1984	71,180	91.46	112.9	311.1	385.2	133,751	17.80	3,569
1985	69,450	89.64	111.5	322.2	402.8	137,308	18.28	3,639
1986	71,404	63.63	85.7	328.4	426.4	140,693	18.35	3,750
1987	70,984	66.33	89.7	340.4	441.4	142,209	19.26	3,760

12O.   The Demand for Gasoline

The United States has experienced a persistent budget deficit for many years.[16] As one possible (and partial) solution to this deficit problem, the government at one time considered raising the gasoline tax. A preliminary estimate of the government's annual revenue gain was approximately $1 billion for every additional penny tax per gallon of gasoline. In addition to the revenue benefits, a higher gasoline tax was expected to generate several positive effects: The resulting higher effective price of gasoline would encourage more efficient driving, which would reduce pollution, congestion on urban roads and highways, and traffic fatalities. Moreover, the expected decline in gasoline consumption resulting from the increased tax would, in turn, lower U.S. dependence on the OPEC cartel and its manipulation of oil prices.

Before any decision about a tax is possible, an economic study of short-run consumer demand for gasoline is essential. The analysis that follows shows how far-reaching and complicated such a study can be.

Let D be annual quantity in millions of gallons of gasoline consumed per year. The quantifiable factors relevant to the demand of gasoline were taken to be the following:

P_S the service station price per gallon.

P_R the retail price per gallon.

$P_R - P_S$ the combined state and federal tax (note that the impact of this tax is the focus of the study).

CPI   the Consumer Price Index (this is the CPI reset to 100 in 1967).

RP_T the inflation adjusted price of public transportation.

N the total number of registered vehicles N.

MPG the miles per gallon of an average vehicle.

RY_D disposable income.

Let RP_R be the inflation adjusted retail price per gallon in 1967 dollars (refer to Table 12.3).

In an effort to simulate the demand for gasoline mathematically, the study assumed that the the domestic demand D for gasoline depends on the parameters RP_R, RP_T, N, MPG, and RY_D in the following way:

$$D = a_0 + a_1 RP_R + a_2 RP_T + a_3 N + a_4 MPG + a_5 RY_D,$$

where $a_0, a_1, a_2, a_3, a_4,$ and a_5 are constants.

47. Would you expect a_1 to be positive or negative? What about $a_2, a_3, a_4,$ and a_5?

The constants $a_0, a_1, a_2, a_3, a_4,$ and a_5 were determined by a least-squares analysis using the data of Table 12.7. This analysis provided the demand equation

$$D = 97868.796 - 416.025 RP_R + 6.914 RP_T$$

$$+ 0.714N - 6032.863 MPG - 1.312 RY_D.$$

48. Was it your expectation that a_5 would be negative? What conclusion can be drawn about the influence of disposable income on the demand for gasoline?

49. Compute the demand for gasoline from the equation for the years 1973 and 1974. How do these values compare with those listed in Table 12.7? Does the equation predict the sharp drop in the demand that actually occurred? (Incidentally, this drop in the demand is attributable to the OPEC price increase that we studied in Section 12.3B.)

50. What information does Table 12.7 provide about the tax (both state and federal) per gallon of gasoline for the year 1987? Convert this tax to 1967 dollars. What was the total tax revenue on the sale of gasoline for 1987 (in 1967 dollars).

51. Suppose that for the year 1988 all parameters remained at 1987 levels, except that the taxes on one gallon of gasoline were raised by a total of 5 cents (in 1967 dollars). Use the demand equation under these assumptions to estimate the demand for gasoline for 1988. Also estimate the tax revenue on the sale of gasoline for 1988 (in 1967 dollars). Compare this with the 1987 tax revenue computed in Exercise 50.

12P. The Impact of a Tax

When considered together, the concepts of consumer and producer surplus can be used to analyze the effect of a tax. This is done as follows. Suppose that at a price p^*, the supply and demand of a product in a certain market are in equilibrium. So $x^* = S(p^*) = D(p^*)$. At some point, the government decides to slap a tax t on the product.

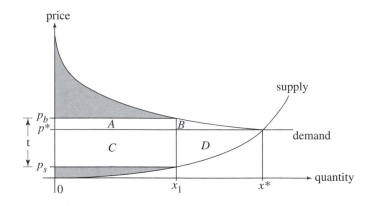

Figure 12.25

Now there are two prices. There is the price p_b that the buyer pays and the price p_s that the producer receives. The difference $p_b - p_s$ is the amount of the tax t. After some initial imbalance between demand and supply brought about by the new pricing situation, a new equilibrium emerges. Both the supply and demand will become equal to some quantity x_1 units per year. The price p_b will generate x_1 as demand, and the price p_s will generate x_1 as supply. This new equilibrium keeps some buyers and sellers out of the market. The buyers that are kept out of the market are those that are not willing to buy at p_b but would be willing to buy at a price p in the range $p^* \leq p < p_b$. The suppliers that are kept out of the market are those that will not supply at p_s but would supply at a price p in the range $p_s < p \leq p^*$. See Figure 12.25. The figure shows the new equilibrium as well as the old. The numbers A, B, C, and D are the areas of the indicated regions. The total tax revenue taken in by the government is $tx_1 = A + C$. Who pays for the tax? Refer to Figure 12.19 and compare the consumer surplus at the original demand x^* with the consumer surplus at the new demand x_1. Notice that the consumers suffer a loss of $A + B$ in their surplus (this involves the consumers that are kept out of the market). Now refer to Figure 12.22 of Exercises 12K. A comparison of the producer surplus at the original demand x^* with the producer surplus at the new demand x_1 shows that the producers suffer a loss of $C + D$ in their surplus (this involves the producers that are kept out of the market). The bottom line is that the consumers pay A of the tax and the producers C. The losses B for the consumer and D for the producer are called *deadweight* losses caused by the inefficiency that the tax introduces.

Notes

[1] Be careful to avoid roundoff errors when using your calculator. The standard practice is not to round off until a calculation is complete.

[2] The Bureau of Labor Statistics of the U.S. Department of Labor resets the index at 100 periodically. See Exercise 12C.

[3] It should be apparent that the assumption that $p(t)$ is differentiable cannot be justified in practical terms, if only because the statisticians in the Department of Labor compute the CPI for each month, and certainly not for any time $t + \Delta t$. However, as the discussion of this section will show, this assumption provides us with important insights into the way the CPI operates.

[4] In Figures 12.4 and 12.5 the price is placed on the horizontal axis and quantity (either demanded or supplied) on the vertical axis. This is how it was done by Cournot (1801–1877). The current practice in economics of placing price on the vertical axis and quantity on the horizontal dates back to Alfred Marshall (1842–1924).

[5] This study is adapted from a discussion in R. S. Pindyck and D. A. Rubinfeld, *Microeconomics*, Prentice Hall, Englewood Cliffs, New Jersey, Third Edition, 1995.

[6] For an overview of the forecasting methods at General Motors, see the essay by Richard DeRoeck, pages 309–311 in *Managerial Economics* by Keating and Wilson, Harcourt Brace Jovanovich, 1992.

[7] See pages 473–74 in Keating and Wilson, *Managerial Economics*, Harcourt Brace Jovanovich, 1992.

[8] The data in Table 12.1 is taken from Evan Douglas, *Managerial Economics: Analysis and Strategy*, 4th edition, Prentice Hall, Englewood Cliffs, 1979.

[9] A watt is a unit of power defined to be the energy production of one joule per second (refer to Exercises 11.B). One kilowatt is a thousand watts. If power is produced at a constant rate of one kilowatt in one hour, the amount of energy produced is called one kilowatt hour (kwh). Therefore,

$$1 \text{ kwh} = 1000 \text{ joules/sec} \times 3600 \text{ seconds}.$$

Hence 1 kwh $= 3.6 \times 10^6$ joules. The Brayton Point plant produces roughly 10,000 million kwh per year, more than any other plant in Table 12.1. It therefore produces $(10{,}000 \times 10^6)(3.6 \times 10^6) = 3.6 \times 10^{16}$ joules per year. Refer to Exercises 11D, and observe that this is about the same as the fission energy in 1000 pounds of uranium-235.

[10] All the data that follow were provided by the Energy Information Administration of the U.S. Department of Energy.

[11]This study is based on James M. Griffin, "The Process Analysis Alternative to Statistical Cost Functions: An Application to Petroleum Refining," *American Economic Review* 62 (1972), 46–56, as modified in *Microeconomics*, by R. S. Pindyck and D. A. Rubinfeld.

[12]This study is adapted from K. K. Seo, *Managerial Economics*, 7th edition, Irwin, Homewood, Ill., 1991.

[13]Study adapted from Evan Douglas, *Managerial Economics, Analysis and Strategy*, 4th edition, Prentice Hall, Englewood Cliffs, N.J., 1979.

[14]Taken from pages 1213–1215 of volume 2 of James R. Newman, *The World of Mathematics*, Simon and Schuster, New York, 1956.

[15]Study adapted from Evan Douglas, *Managerial Economics, Analysis and Strategy*, 4th edition, Prentice Hall, Englewood Cliffs, N.J., 1979. The data in Table 12.6 were taken from the *Economic Report of the President*, 1988, and various issues of *Automotive News, Market Data Book*.

[16]Study adapted from Evan Douglas, *Managerial Economics, Analysis and Strategy*, 4th edition, Prentice Hall, Englewood Cliffs, N.J., 1979. The data in Table 12.7 were taken from the *Basic Petroleum Data Book*, published by the American Petroleum Institute in 1989, and from the *Economic Report of the President*, 1988.

Integral Calculus: Meaning and Method

It was one of the central aims of several of the previous chapters to make the following point: Calculus infuses a number of areas of study (ranging from engineering, to biology, to finance) with important, indeed crucial, information. A review of this information shows that most of it was supplied by just a few mathematical elements: The properties of certain basic functions and their derivatives.

The time has come to introduce additional mathematical methods and techniques. Because these involve the integral either directly or indirectly, we will begin with another look at the definite integral and the Fundamental Theorem of Calculus. We have already seen in Sections 5.3–5.5 that this fundamental theorem consists in an equality. On the one side is a sum consisting of a huge number of very small terms: the definite integral of a function. On the other there is the value of this sum: a number evaluated from the antiderivative of the function. As is the case with any significant equality, both sides are important: The sum has many different meanings in a variety of disciplines, and the antiderivative provides one of the methods with which this sum can be computed. Both sides of the equation—the meanings as well as the methods—are the concern of integral calculus and will receive attention in this chapter.

Much of the mathematics that has been developed in this book involves the Cartesian coordinate system in a most fundamental way. But there are situations, namely those that feature a "central point" (such as the Sun in the solar system), that are analyzed more incisively by the *polar coordinate system*. The polar coordinate system and its integral calculus is the second topic of this chapter. It is followed by an introduction to the theory of differential equations. This far-reaching theory includes the methods of finding antiderivatives of functions as a very special case.

13.1 Riemann Sums and the Definite Integral

The theory of Leibniz and Newton provides a workable calculus, as well as an intuitive understanding of its theoretical aspects. This was the lesson of Chapters 5 and 6. But as was already observed, some elements of this theory do not measure up to the modern standards of rigor and precision. This section will develop the definite integral afresh. It will show what refinements have to be made in the work of Leibniz and Newton. This part of the chapter is analogous to Chapter 8, which does the same for the derivative.

A. A Return to the Approach of Leibniz. Let's begin by recalling Leibniz's approach to the definite integral from Section 5.3.

Let f be a function that is continuous over a closed interval $[a, b]$. Place a lot of points on the segment $[a, b]$ each very close to the next. Let x be any one of them and let dx be the distance to the next one. Draw the rectangle with height $f(x)$ and base dx. Because dx is extremely small, the area $f(x)\,dx$ of the rectangle is very small. Figure 13.1 shows a larger version of a typical rectangle. Consider all rectangles obtained in this way from the points chosen. Add up all their areas, in other words, all the values $f(x)\,dx$, and denote the sum by

$$\int_a^b f(x)\,dx.$$

This is a sum that consists of a large number of very small terms. Its value approximates the area under the graph from a to b in the same way that, say, 5.66666 approximates $5\frac{2}{3}$.

There is something not entirely satisfactory with the procedure just described, and it is this: Suppose that it is carried out twice. The first time choose, say, 1 million points, pack them in tightly between a and b with one very close to the next, and form the sum $\int_a^b f(x)\,dx$. Then do the same thing again, this time with, say, 1 billion points, to get the sum $\int_a^b f(x)\,d(x)$ once more. Each sum is an approximation for the area

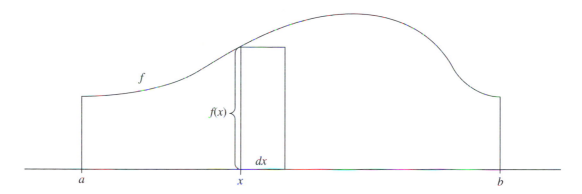

Figure 13.1

under the graph of f and above the x-axis. However, one of these approximations (the second) is likely to be much better than the other (in the same way that 5.66666666 is a much better approximation for $5\frac{2}{3}$ than 5.666). One approximation may fall within the tolerances that a particular computation might require and the other not. The ambiguity can be rectified by "going to the limit." This is a refinement that comes at a price: There is a considerable increase in the level of complication, particularly notational complication.

We will continue to assume that $y = f(x)$ is a continuous function defined on a closed interval $[a, b]$. It will not be necessary for the graph of f to lie above the x-axis.

1. Take a *partition* P of the interval $[a, b]$. This is a division of $[a, b]$ into subintervals given by a collection of numbers

    $$a = x_0 < x_1 < x_2 < x_3 < \cdots < x_{i-1} < x_i < \cdots$$

    $$< x_{n-1} < x_n = b$$

    between a and b, where n can be any positive integer. A typical division of $[a, b]$ is shown in Figure 13.2. For example, the numbers

    $$2 < 2.5 < 2.9 < 3.2 < 3.8 < 4.1 < 4.3 < 4.8 < 5$$

    determine a partition of $[2, 5]$ with $n = 8$. Now set

    $$\Delta x_1 = x_1 - x_0, \ \Delta x_2 = x_2 - x_1,$$

    $$\Delta x_3 = x_3 - x_2, \ldots,$$

    $$\Delta x_i = x_i - x_{i-1}, \ldots, \Delta x_n = x_n - x_{n-1}$$

    and observe that Δx_1 is the length of the first subinterval $[x_0, x_1]$, Δx_2 is the length of $[x_1, x_2]$, ..., Δx_i that of the ith subinterval $[x_{i-1}, x_i]$, and so on.

    The *norm* of the partition P, written $\|P\|$, is defined to be the largest of all the Δx_i. For instance, the norm of the partition of $[2, 5]$ given earlier is 0.6.

2. Next pick a point—this can be any point—p_1 in the first subinterval, any p_2 in the second, any p_3 in the third, ..., any p_i in the ith, and so on, as indicated in Figure 13.3. (Freedom in the choice of the points p_i can be important. The application given in Section 13.2 illustrates this.) Now form the sum

    $$f(p_1)\Delta x_1 + f(p_2)\Delta x_2 + \cdots + f(p_n)\Delta x_n.$$

    This sum is called a *Riemann sum*. In "sigma" notation (refer to Section 3.3) it can be rewritten as

    $$\sum_{i=1}^{n} f(p_i)\Delta x_i.$$

3. Repeat the construction of (**2**) again and again (as a thought experiment) with partitions of successively smaller and smaller norm. Each step produces a Riemann sum $\sum_{i=1}^{n} f(p_i)\Delta x_i$. Each sum

Figure 13.2

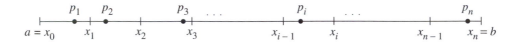

Figure 13.3

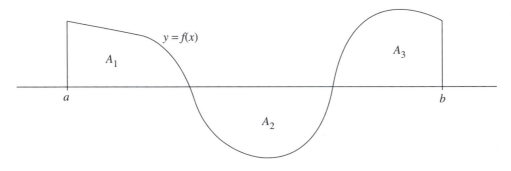

Figure 13.4

is a number. The progression of numbers will close in on a limit, and this limit is denoted by

$$\int_a^b f(x)\,dx.$$

It is called the *definite integral of the function f* from *a* to *b*. The equation

$$\lim_{\|P\|\to 0}\sum_{i=1}^n f(p_i)\Delta x_i = \int_a^b f(x)\,dx$$

is the mathematical shorthand that summarizes step (**3**).

This construction can be visualized and made plausible by considering areas. Figure 13.4 shows the graph of a continuous function f. The areas of the region formed by the graph and the x-axis are designated by A_1, A_2, and A_3. Figure 13.5 includes the partition from step (**1**). It also illustrates step (**2**). If

$f(p_i) \geq 0$, then $f(p_i)\Delta x_i$ is the area of a rectangle that lies above the x-axis. If $f(p_i) < 0$, then the rectangle is below the x-axis and $f(p_i)\Delta x_i$ is equal to minus its area. If the norm $\|P\|$ is small, then all the Δx_i will be small, and it follows that

$$\sum_{i=1}^n f(p_i)\Delta x_i \approx A_1 - A_2 + A_3.$$

Step (**3**) provides a progression of numbers

$$\sum_{i=1}^n f(p_i)\Delta x_i,$$

one for each partition. Each is an approximation of $A_1 - A_2 + A_3$. As the norms $\|P\|$ become smaller, these numbers get progressively closer to $A_1 - A_2 + A_3$. A comparison of the highlighted regions of Figures 13.5 and 13.6 gives an indication of this. The limit of step (**3**) is the number $A_1 - A_2 + A_3$.

In this way, the construction of the limit $\int_a^b f(x)\,dx$ in steps (**1**)–(**3**) can be visualized in terms of areas

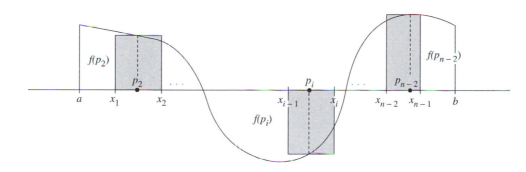

Figure 13.5

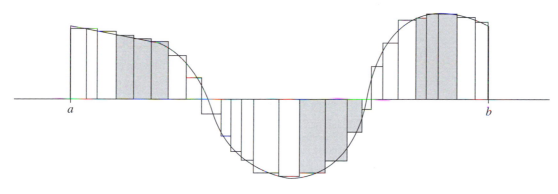

Figure 13.6

under the graph of f. This interpretation of $\int_a^b f(x)\,dx$ is always valid: For any continuous function $y = f(x)$ defined on a closed interval $[a, b]$,

$$\int_a^b f(x)\,dx = \text{the sum of areas above the axis (and below the graph) } minus \text{ the sum of the areas below the axis (and above the graph).}$$

We saw in Sections 5.6B and 5.6C that $\int_a^b f(x)\,dx$ can also represent volumes and lengths of arcs. We will see in this chapter and the next that it has a number of other interpretations as well.

Does the progression of numbers that step **(3)** produces close in on a limit? If yes, is this limit independent of the way the partitions are chosen? In other words, has the ambiguity that we mentioned at the be-

ginning of this section been completely removed? The answer to all of these questions is yes. But the mathematical proofs that this is so are very demanding. This was the accomplishment of the mathematicians of the 19th century. The work of the Frenchman Augustin Louis Cauchy in the 1820s and that of the Germans Lejeune Dirichlet and Georg Bernhard Riemann later in the century deserves to be singled out.

B. A Return to the Approach of Newton. We will now have a second look at Newton's approach to area. See Section 6.2.

As before, let f be a function that is continuous on a closed interval $[a, b]$. Let x denote a point on the axis between a and b. Figure 13.7 illustrates a typical situation. In order to avoid notational ambiguities (between x as point and x as variable), we will now let t denote the variable of the horizontal axis. Making use

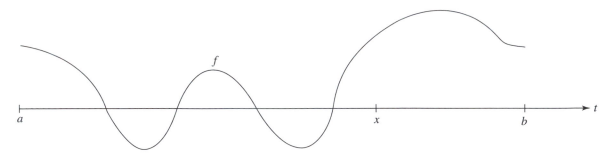

Figure 13.7

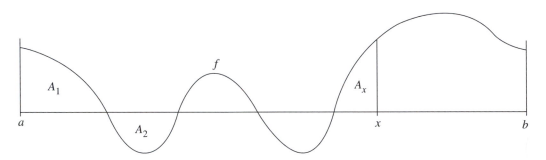

Figure 13.8

of the construction (**1**) – (**3**) of Section 13.1A, define $A(x)$ by

$$A(x) = \int_a^x f(t)\, dt.$$

Because this can be done for any x between a and b, this rule defines a function $A(x)$ with domain $[a, b]$. By the discussion in Section 13.1A,

$A(x)$ = the sum of areas above the axis (and below the graph) *minus* the sum of the areas below the axis (and above the graph) from a to x.

In the typical situation that Figure 13.8 illustrates, $A(x) = A_1 - A_2 + \cdots + A_x$. We will see next that the function $A(x)$ has the following important property: It is differentiable on $[a, b]$ and

$$A'(x) = f(x)$$

for all x in $[a, b]$. In other words, the derivative of $A(x)$ is the function $f(x)$ that we started with. The

proof of this assertion does not rely in any way on the interpretation of $A(x)$ as area. The connection with area makes the argument more transparent, however, and we will make use of it.

By the definition of the derivative, $A'(x) = \lim_{\Delta x \to 0} \frac{A(x+\Delta x)-A(x)}{\Delta x}$. Therefore, it must be verified that for any x between a and b

$$\lim_{\Delta x \to 0} \frac{A(x + \Delta x) - A(x)}{\Delta x} = f(x).$$

Take any x and consider it fixed (in the argument that follows). For any Δx observe that

$$A(x + \Delta x) - A(x) = (A_1 - A_2 + \cdots + A_{x+\Delta x})$$
$$- (A_1 - A_2 + \cdots + A_x)$$
$$= A_{x+\Delta x} - A_x.$$

Notice that Δx must be appropriately small (for instance $x + \Delta x \leq b$), but this is no problem since the focus is on what happens when $\Delta x \to 0$. We will continue by assuming that Δx is positive (although this is not essential). This is the case that is illustrated in Figure 13.9. Since it is continuous on the interval $[x, x + \Delta x]$, the function f achieves a minimum value m and a maximum value M on this closed interval. It follows that

$$m \cdot \Delta x \le A(x + \Delta x) - A(x) \le M \cdot \Delta x.$$

Therefore,

$$m \le \frac{A(x + \Delta x) - A(x)}{\Delta x} \le M.$$

Let $v = \frac{A(x+\Delta x) - A(x)}{\Delta x}$. By the Intermediate Value Theorem of Section 8.2, there is a number u with $x \le u \le x + \Delta x$ such that $f(u) = v = \frac{A(x+\Delta x)-A(x)}{\Delta x}$. So $A(x + \Delta x) - A(x) = f(u)\Delta x$. Refer to Figure 13.10. Now push $\Delta x \to 0$. Since $u \to x$ in the process and f is continuous at x, it follows that

$$A'(x) = \lim_{\Delta x \to 0} \frac{A(x + \Delta x) - A(x)}{\Delta x} = \lim_{\Delta x \to 0} \frac{f(u)\Delta x}{\Delta x}$$

$$= \lim_{\Delta x \to 0} f(u) = f(x).$$

Therefore, as asserted,

$$A'(x) = \frac{d}{dx}\left(\int_a^x f(t)\, dt \right) = f(x)$$

Now let F be any antiderivative of f. Since $A'(x) = f(x) = F'(x)$, we see that

$$\frac{d}{dx}\big(A(x) - F(x)\big) = A'(x) - F'(x) = 0.$$

As a consequence, $A(x) - F(x)$ is equal to a constant C. Evaluating $A(x) = F(x) + C$ at $x = a$ gives us $0 = A(a) = F(a) + C$. So $C = -F(a)$, and $A(x) = F(x) - F(a)$. Finally, take $x = b$ to get

$$A(b) = \int_a^b f(t)\, dt = F(b) - F(a).$$

The number $\int_a^b f(t)\, dt$ does not depend on the notation t for the variable of the horizontal axis. The fact that this number is equal to the "sum of the areas above minus the sum of the areas below" means that the symbol that is used for the variable is irrelevant. So we can now replace t by x and write this equation as

$$\int_a^b f(x)\, dx = F(x) \Big|_a^b = F(b) - F(a)$$

Example 13.1. Consider $\int_{-\pi}^{3\pi} \sin x\, dx$ (See Figure 13.11). By considering the "areas above minus the areas below" we see that $\int_{-\pi}^{3\pi} \sin x\, dx = 0$. What about the answer using an antiderivative? Let $F(x) = -\cos x$. Since $F'(x) = -(-\sin x) = \sin x$, $F(x)$ is an antiderivative of $f(x) = \sin x$. Therefore,

$$\int_{-\pi}^{3\pi} \sin x\, dx = F(3\pi) - F(\pi) = -\cos 3\pi + \cos(-\pi)$$

$$= -\cos \pi + \cos(-\pi) = 1 - 1 = 0,$$

as expected.

Example 13.2. The area under a single loop of the graph of $f(x) = \sin x$ is computed in the same way:

$$\int_0^{\pi} \sin x\, dx = F(\pi) - F(0) = -\cos \pi + \cos 0$$

$$= 1 + 1 = 2.$$

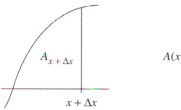
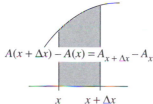

Figure 13.9

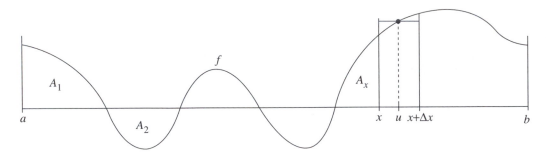

Figure 13.10

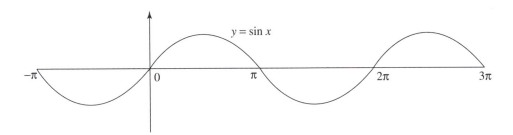

Figure 13.11

C. The Fundamental Equality.

Let's review what has been done. Let $y = f(x)$ be any function that is continuous on a closed interval $[a, b]$. Take any partition P of $[a, b]$, select points p_i in the various subintervals,

and form the number $\sum_{i=1}^{n} f(p_i) \Delta x_i$. Repeat this with finer and finer partitions to get the limiting number

$$\lim_{\|P\| \to 0} \sum_{i=1}^{n} f(p_i) \Delta x_i = \int_{a}^{b} f(x)\, dx.$$

This number is the *definite integral of f from a to b*. It can be computed with the equality

$$\int_{a}^{b} f(x)\, dx = F(b) - F(a)$$

where F is any antiderivative f. This is the Fundamental Equality of calculus.

The following message has been one of the primary aims of this book: The basic concepts of calculus give us a way of thinking about some fundamental physical phenomena. And when the connection between these concepts and these phenomena are combined with the computational methods of calculus, a wealth of quantitative information about our world emerges. Several chapters have already made this point, relying primarily on the derivative to do so.

So the important question now is this. Is the integral another basic concept that illustrates this message? In view of what was just said, this question has two parts:

A. What meaningful physical interpretations does the number $\int_{a}^{b} f(x)\, dx$ have?

B. Can the Fundamental Equality be used to compute this number effectively?

Two answers to question (**A**) were already given in Sections 5.6B and 5.6C. We saw there that

$$\int_a^b \pi f(x)^2 \, dx$$

is the volume V obtained by rotating the region under the graph of f from a to b one revolution about the x-axis (if $f(x) \geq 0$ for $a \leq x \leq b$), and that

$$\int_a^b \sqrt{1 + (f'(x))^2} \, dx$$

the length of the graph of f from the point $(a, f(a))$ to the point $(b, f(b))$.

Example 13.3. The definite integral $\int_0^\pi \pi \sin^2 x \, dx$ is equal to the volume obtained by rotating the graph of $f(x) = \sin x$, $0 \leq x \leq \pi$, one revolution about the x-axis. Because $\pi \sin^2 x \geq 0$, observe that the graph of the function $\pi \sin^2 x$ lies above the x-axis. Therefore, this integral is also equal to the area under the graph of $\pi \sin^2 x$ from 0 to π.

Example 13.4. What interpretation does

$$\int_0^\pi \sqrt{1 + \cos^2 x} \, dx$$

have? Again let $f(x) = \sin x$ and recall from Section 8.6 that $f'(x) = \cos x$. It follows that this definite integral is equal to the length of the graph of the sine from $(0, 0)$ to $(0, \pi)$. Observe that it is also equal to the area under the graph of the function $\sqrt{1 + \cos^2 x}$ from 0 to π.

We will see in Sections 13.2 and 13.4B that the definite integral also represents areas of surfaces and pie-shaped regions. Additional interpretations of $\int_a^b f(x) \, dx$ come from the context of the function f. For example, when f represents force, then the definite integral represents energy as well as momentum. The definite integral is therefore basic to the physical sciences and many disciplines of engineering. These and other connections will be taken up in Chapter 14.

What about question (**B**)? This has two answers. There is a "bag of tricks" that can be used to determine antiderivatives of functions, and thus to compute definite integrals. There are also—and this is ultimately more relevant for applications—a number of approximation methods that can be used for the same purpose. The method of power series, already encountered in Section 6.3, is one of these. Another consists of techniques for approximating areas in the plane and relies on the fact (already pointed out) that every definite integral can be interpreted in terms of area.

Questions (**A**) and (**B**) are usually difficult to answer. The next section will show how difficult the answer to question (**A**) can be.

13.2 The Definite Integral as Surface Area

This section turns to the computation of the area of a surface obtained by revolving the graph of a function. Let $y = f(x)$ be a differentiable function defined on a closed interval $[a, b]$. Assume that $f(x) \geq 0$ for all x in $[a, b]$.

Consider the graph of f and rotate it about the x-axis as shown in Figure 13.12. This rotation produces the "vase" shown in Figure 13.13. This vase consists of two circular regions, one at the top (the point a is the center of this circle) and one at the base (its center is b), and the curved surface between them. Because the circular areas at the top and bottom of the vase are easily computed (with the formula πr^2), we will concentrate on the curved surface A of the vase.

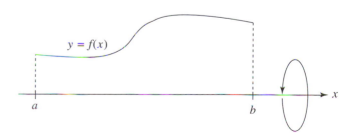

Figure 13.12

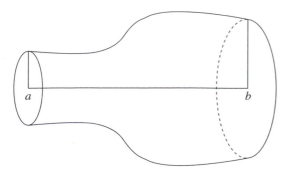

Figure 13.13

Figure 13.14

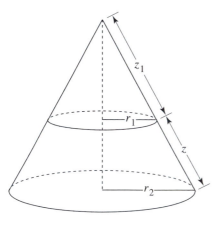

Figure 13.15

Begin by recalling the fact that the area of the surface of a cone (excluding the circular base) of radius r and slant height s is πrs. See Figure 13.14. This fact was verified in Example 8.41 of Section 8.8. Slice this cone into two pieces with a cut that is parallel to the base. Refer to Figure 13.15 and note that $r = r_2$ and $s = z_1 + z$. Remove the smaller upper cone and consider the surface area C of the piece that is left. (The two circular areas at the top and bottom are not included.) See Figure 13.16. Because this piece is the difference between two cones, it follows that

$$C = \pi r_2 s - \pi r_1 z_1 = \pi r_2(z_1 + z) - \pi r_1 z_1$$

$$= \pi r_2 z_1 + \pi r_2 z - \pi r_1 z_1 = \pi(r_2 z_1 + r_2 z - r_1 z_1).$$

By similar triangles, $\frac{z_1}{r_1} = \frac{z_1 + z}{r_2}$, and therefore, $r_2 z_1 = r_1(z_1 + z) = r_1 z_1 + r_1 z$. Popping this into the last expression for C gives us

$$C = \pi(r_1 z_1 + r_1 z + r_2 z - r_1 z_1)$$

$$= \pi(r_1 z + r_2 z) = \pi(r_1 + r_2)z.$$

To obtain an understanding of the surface area A of the vase, we turn to the "strategy of the definite integral," in other words, to steps **(1)**–**(3)** of Section 13.1A.

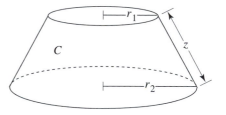

Figure 13.16

1. Start with a partition P

$$a = x_0 < x_1 < x_2 < x_3 < \cdots$$

$$< x_{i-1} < x_i < \cdots < x_{n-1} < x_n = b$$

of the interval $[a, b]$. Through each partition point draw in a vertical segment up the graph of f

as shown in Figure 13.17. Rotating this diagram one revolution about the x-axis produces Figure 13.18. Notice that the curved surface A under study has been cut up into n sections. Label the areas of these sections respectively $A_1, \ldots, A_i, \ldots, A_n$ and notice that

$$A = A_i + \cdots + A_i + \cdots + A_n.$$

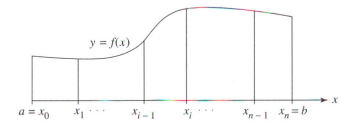

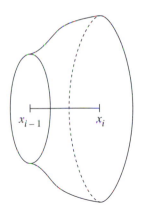 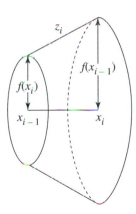

Figure 13.19

Figure 13.17

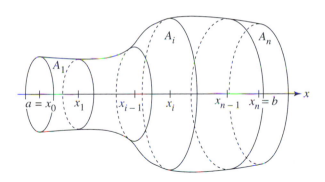

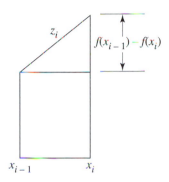

Figure 13.20

Figure 13.18

2. We now assume that the partition is fine and focus on a typical section, the ith section, with area A_i. Observe that this section is approximated by the surface of the conical piece shown in Figure 13.19. Taking $r_1 = f(x_i)$, $r_2 = f(x_{i-1})$, and $z = z_i$ in the formula for C, we see that

$$A_i \approx \pi(r_1 + r_2)z = \pi\big(f(x_i) + f(x_{i-1})\big)z_i.$$

Let $\Delta x_i = x_i - x_{i-1}$ and apply Pythagoras's theorem to the triangle in Figure 13.20, to get that

$$z_i = \sqrt{\Delta x_i^2 + \big(f(x_{i-1}) - f(x_i)\big)^2}.$$

The next step is an application of the Mean Value Theorem of Chapter 8.5B to the function f and the closed interval $[x_{i-1}, x_i]$. Note that this theorem provides a number p_i with $x_{i-1} \le p_i \le x_i$ such that

$$f'(p_i) = \frac{f(x_i) - f(x_{i-1})}{x_i - x_{i-1}}.$$

Because $f(x_i) - f(x_{i-1}) = f'(p_i)(x_i - x_{i-1}) = f'(p_i)\Delta x_i$, it follows that

$$z_i = \sqrt{\Delta x_i^2 + \big(f'(p_i)\Delta x_i\big)^2} = \sqrt{1 + f'(p_i)^2}\,\Delta x_i.$$

Since the partition is fine, x_{i-1} and x_i are close together. So $x_{i-1} \approx p_i$ and $x_i \approx p_i$. Recall that we are assuming that f is a differentiable function. Therefore, f is continuous by remarks in Section

8.5B. It follows from the continuity criterion of Section 8.2 that $f(x_{i-1}) \approx f(p_i)$ and $f(x_i) \approx f(p_i)$. Inserting these approximations together with $z_i = \sqrt{1+f'(p_i)^2}\Delta x_i$ into the earlier approximation for A_i, we get

$$A_i \approx \pi\big(f(x_i)+f(x_{i-1})\big)z_i \approx 2\pi f(p_i)\sqrt{1+f'(p_i)^2}\Delta x_i.$$

Doing this for $i = 1,\ldots,n$, we get

$$A = \sum_{i=1}^{n} A_i \approx \sum_{i=1}^{n} 2\pi f(p_i)\sqrt{1+f'(p_i)^2}\Delta x_i.$$

3.   Repeat this construction with finer and finer partitions. Each repetition gives us a number of the form $\sum_{i=1}^{n} 2\pi f(p_i)\sqrt{1+f'(p_i)^2}\Delta x_i$ that approximates the area A. As the partitions get finer, these approximations get tighter. So in terms of the limit notation of Section 13.1A,

$$A = \lim_{\|P\|\to 0} \sum_{i=1}^{n} 2\pi f(p_i)\sqrt{1+f'(p_i)^2}\Delta x_i.$$

But this is exactly how $\int_a^b 2\pi f(x)\sqrt{1+f'(x)^2}\,dx$ is defined. Therefore,

$$A = \int_a^b 2\pi f(x)\sqrt{1+f'(x)^2}\,dx$$

What has our complicated discussion accomplished? The area A of the curved surface obtained by revolving the graph of the function f has been expressed as a definite integral. What was the essence of the argument? The curved surface was sliced into small sections, and the area of each section was approximated. The sum of the areas of the section gave an approximation of A. This approximation was "tightened up" by taking finer and finer slices.

A combination of the formula that was just derived with the Fundamental Equality of Section 13.1C gives us a strategy for computing A.

Example 13.5. Consider the circle of radius r centered at the origin. Its equation is $x^2 + y^2 = r^2$. Note that the graph of the function $y = f(x) = \sqrt{r^2 - x^2} =$

$(r^2 - x^2)^{1/2}$ with domain $[-r, r]$ is the upper semicircle shown in Figure 13.21. The surface obtained by revolving this semicircle around the x-axis is a sphere of radius r. It follows that the surface area of the sphere of radius r can be expressed as the definite integral

$$\int_{-r}^{r} 2\pi f(x)\sqrt{1+f'(x)^2}\,dx$$

where $f(x) = (r^2 - x^2)^{1/2}$. Let's see if we can compute it. By the chain rule,

$$f'(x) = \frac{1}{2}(r^2 - x^2)^{-1/2}(-2x)$$

$$= \frac{-x}{(r^2 - x^2)^{1/2}}.$$

So $f'(x)^2 = \frac{x^2}{r^2-x^2}$, and hence,

$$\sqrt{1+f'(x)^2} = \sqrt{1+\frac{x^2}{r^2-x^2}} = \sqrt{\frac{r^2-x^2+x^2}{r^2-x^2}}$$

$$= \frac{r}{(r^2-x^2)^{1/2}}.$$

So the surface area of a sphere of radius r is equal to

$$\int_{-r}^{r} 2\pi f(x)\sqrt{1+f'(x)^2}\,dx$$

$$= \int_{-r}^{r} 2\pi(r^2-x^2)^{1/2}\frac{r}{(r^2-x^2)^{1/2}}\,dx$$

$$= \int_{-r}^{r} 2\pi r\,dx.$$

Because $2\pi rx$ is an antiderivative of $2\pi r$, we get

$$\int_{-r}^{r} 2\pi r\,dx = 2\pi rx\,\Big|_{-r}^{r} = 2\pi r^2 - (-2\pi r^2) = 4\pi r^2.$$

We have shown that the surface area of a sphere of radius r is $4\pi r^2$.

Example 13.6. Instead of rotating the full semicircle of radius r, rotate only the part between a and b. By the argument of Example 13.5, the area of the section of the surface of the sphere shown in Figure 13.22 is

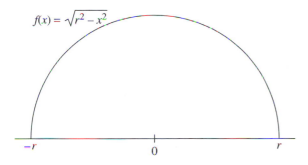

$$f(x) = \sqrt{r^2 - x^2}$$

Figure 13.21

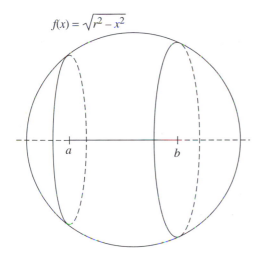

$$f(x) = \sqrt{r^2 - x^2}$$

Figure 13.22

equal to

$$\int_a^b 2\pi r \, dx = 2\pi r x \Big|_a^b = 2\pi r b - 2\pi r a = 2\pi r (b - a).$$

Notice that for a sphere of given radius r, the area of this surface depends only on the width $b - a$ of the section and not on its location on the sphere. This can be rephrased more graphically this way. Suppose you have a loaf of bread in the shape of a sphere. Cut any two slices from the loaf (making sure that the cuts you have made are parallel to each other). If the slices have the same thickness, then they have the same amount of crust, no matter which part of the loaf they are taken from!

Example 13.7. Consider the area A of the surface obtained by revolving the graph of $f(x) = \sqrt{x}, 3 \le x \le 5$, one revolution around the x-axis. See Figure 13.23. The area A of this surface is equal to

$$A = \int_3^5 2\pi f(x) \sqrt{1 + f'(x)^2} \, dx$$

where $f(x) = \sqrt{x} = x^{1/2}$. Because $f'(x) = \frac{1}{2} x^{-1/2}$, we get that

$$A = \int_3^5 2\pi \sqrt{x} \sqrt{1 + \frac{1}{4} x^{-1}} \, dx = \int_3^5 2\pi \sqrt{x + \frac{1}{4}} \, dx.$$

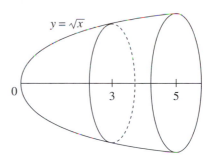

$$y = \sqrt{x}$$

Figure 13.23

Example 13.8. Consider the area of the surface obtained by revolving the graph of $f(x) = \sin x$, $0 \le x \le \pi$, around the x-axis. Since $f'(x) = \cos x$, this area is

$$A = \int_0^\pi 2\pi \sin x \sqrt{1 + \cos^2 x} \, dx.$$

Examples 13.7 and 13.8 have made a start towards the computations of the two surface areas. But the two definite integrals still need to be solved. Can we find antiderivatives of the functions $\sqrt{x + \frac{1}{4}}$ and $\sin x \sqrt{1 + \cos^2 x}$? How does one solve definite integrals? Are there general methods for finding antiderivatives of functions?

13.3 Methods of Integration

Recall from Section 10.5 that the "general" antiderivative is called indefinite integral. The title *Methods of Integration* of this section refers to the bag of various tricks for finding indefinite integrals of functions. The essential strategy common to all the tricks concentrates on simplifying or recasting the *integrand*—this is the function on the right of the integral sign—in such a way that its indefinite integral can be found by using formulas already established, for example, those of Section 10.5. Some of these tricks work in isolated instances only, while others deserve to be called methods.

Let's start off with the trick that solves the indefinite integral

$$\int \sec x \, dx.$$

Begin with the observation that

$$\sec x = \sec x \frac{\sec x + \tan x}{\sec x + \tan x} = \frac{\sec^2 x + \sec x \tan x}{\sec x + \tan x}.$$

A look at Examples 8.31 and 8.32 in Section 8.6 confirms that $\frac{d}{dx} \sec x = (\sec x)(\tan x)$ and $\frac{d}{dx} \tan x = \sec^2 x$. Next, take $g(x) = \sec x + \tan x$ and notice that $\sec x = \frac{g'(x)}{g(x)}$. The use of the formula $\frac{d}{dx} \ln g(x) = \frac{g'(x)}{g(x)}$ from Section 10.3 shows us that

$$\frac{d}{dx} \ln(\sec x + \tan x) = \frac{d}{dx} \ln g(x) = \frac{g'(x)}{g(x)} = \sec x.$$

So $\ln(\sec x + \tan x)$ is an antiderivative of $\sec x$ and we have verified that

$$\int \sec x \, dx = \ln(\sec x + \tan x) + C.$$

The reference to methods of integration as a "bag of tricks" should now have at least some credibility.

We saw another method of integration in action in Section 11.5B. In the context of the logistics model, the solution

$$\int \frac{M}{y(M - y)} \, dy = \ln y - \ln(M - y) + C$$

was obtained by "reversing" common denominators. Let's recall how this was done. Let A and B be any constants and consider the expression $\frac{A}{y} + \frac{B}{M-y}$. Take common denominators to get

$$\frac{A}{y} + \frac{B}{M - y} = \frac{A(M - y) + By}{y(M - y)} = \frac{AM + (B - A)y}{y(M - y)}.$$

In order to get the right side to have the form $\frac{M}{y(M-y)}$ we now set $A = 1$ and $B - A = 0$. Because $B = A = 1$, we see that $\frac{M}{y(M-y)} = \frac{1}{y} + \frac{1}{M-y}$. Therefore,

$$\int \frac{M}{y(M - y)} \, dy = \int \frac{1}{y} \, dy + \int \frac{1}{M - y} \, dy.$$

Taking derivatives confirms that the two integrals (or antiderivatives) on the right are

$$\int \frac{1}{y} \, dy = \ln y + C_1$$

and

$$\int \frac{1}{M - y} \, dy = -\ln(M - y) + C_2.$$

By combining the constants into the single constant C, we get

$$\int \frac{M}{y(M - y)} \, dy = \ln y - \ln(M - y) + C.$$

Both of the two tricks just illustrated make the following point: It is sometimes possible to rewrite the integrand algebraically in such a way that its antiderivative is readily found. This is the essence of the method that we will consider next.

A. The Substitution Method. Consider two functions f and g. Let F be an antiderivative of f, and form the composite function $F(g(x))$. By the chain rule,

$$\frac{d}{dx} F(g(x)) = F'(g(x)) \cdot g'(x) = f(g(x)) \cdot g'(x).$$

So $F(g(x))$ is an antiderivative of the function $f(g(x)) \cdot g'(x)$. It follows that

$$\boxed{\int f(g(x)) \cdot g'(x) \, dx = F(g(x)) + C}$$

An abbreviated version of this formula is obtained by letting $u = g(x)$. Then $\frac{du}{dx} = g'(x)$. So $du = g'(x)\,dx$, and hence

$$\int f(u)\,du = F(u) + C.$$

In this way, a carefully chosen substitution $u = g(x)$ can often transform a complicated integral into one that is readily solved. Examples 13.9–13.12 will illustrate this.

Example 13.9. Compute $\int (x-5)^8\,dx$. One way to do this is to multiply the integrand out and then to antidifferentiate term by term. But a substitution provides a much less tedious solution. Let $u = x - 5$. Since $\frac{du}{dx} = 1, du = dx$. By substituting, $\int (x-5)^8\,dx = \int u^8\,du$. But $\int u^8\,du = \frac{1}{9}u^9 + C$. Since $u = x-5$, we get

$$\int (x-5)^8\,dx = \frac{1}{9}(x-5)^9 + C.$$

Example 13.10. Compute $\int x\sqrt{x+3}\,dx$. Let $u = x + 3$. So $\frac{du}{dx} = 1$, and hence $du = dx$. Since $x = u - 3$, $\int x\sqrt{x+3}\,dx$ can be rewritten as $\int (u-3)u^{1/2}\,du$. Has anything meaningful been accomplished? Noting that $(u-3)u^{1/2} = u^{3/2} - 3u^{1/2}$, we get that

$$\int x\sqrt{x+3}\,dx = \int (u^{3/2} - 3u^{1/2})\,du$$

$$= \int u^{3/2}\,du - \int 3u^{1/2}\,du.$$

Taking antiderivatives term by term, we now get

$$\int x\sqrt{x+3}\,dx = \frac{2}{5}u^{5/2} - 3 \cdot \frac{2}{3}u^{3/2} + C.$$

The remaining thing to do is to state the answer in terms of the original variable x. Because $u = x + 3$ this is

$$\int x\sqrt{x+3}\,dx = \frac{2}{5}(x+3)^{5/2} - 2(x+3)^{3/2} + C.$$

Example 13.11. Solve $\int \frac{3(\ln t)^2}{t}\,dt$. Let $u = \ln t$. Then $\frac{du}{dt} = \frac{1}{t}$, and hence $du = \frac{1}{t}\,dt$. So

$$\int \frac{3(\ln t)^2}{t}\,dt = \int 3u^2\,du = u^3 + C = (\ln t)^3 + C.$$

Example 13.12. Compute $\int \frac{\sin\theta}{1+\cos^2\theta}\,d\theta$. Let $u = \cos\theta$. Since $\frac{du}{d\theta} = -\sin\theta, du = -\sin\theta\,d\theta$. So

$$\int \frac{\sin\theta}{1+\cos^2\theta}\,d\theta = \int \frac{-1}{1+u^2}\,du.$$

Recall from Section 10.5 that

$$\int \frac{1}{x^2+1}\,dx = \tan^{-1}x + \text{ constant.}$$

Therefore, $\int \frac{-1}{1+u^2}\,du = -\tan^{-1}u + C$, and hence

$$\int \frac{\sin\theta}{1+\cos^2\theta}\,d\theta = -\tan^{-1}(\cos\theta) + C.$$

The strategy of the substitution method should begin to become apparent: Look at the integrand and pick out a function $u = g(x)$ whose derivative $\frac{du}{dx} = g'(x)$ is in essence also present in the integrand; try to rewrite the integral completely in terms of the variable u; and hope that this new integral is manageable. Unfortunately, success is not guaranteed. It may well be that the integral cannot be solved by this method (or any other).

Example 13.13. In the three cases

$$\int \frac{5x^6}{12x^7+19}\,dx$$

$$\int \frac{\sec^2\varphi}{\tan^2\varphi+1}\,d\varphi$$

$$\int \sqrt{e^z+1}e^{2z}\,dz$$

the promising choices for u are respectively $u = 12x^7 + 19$, $u = \tan\varphi$, and $u = e^z + 1$. Make use of them and rewrite each of the integrals in terms of u.

Consider the definite integral $\int_{-2}^{5} x\sqrt{x+3}\,dx$. Example 13.10 tells us that the function

$$\frac{2}{5}(x+3)^{\frac{5}{2}} - 2(x+3)^{\frac{3}{2}}$$

is an antiderivative of $x\sqrt{x+3}$. So this definite integral is easily computed:

$$\int_{-2}^{5} x\sqrt{x+3}\,dx = \left(\frac{2}{5}(x+3)^{5/2} - 2(x+3)^{3/2}\right)\Big|_{-2}^{5}$$

$$= \left(\frac{2}{5}(8)^{5/2} - 2(8)^{3/2}\right) - \left(\frac{2}{5} - 2\right)$$

$$\approx 28.75.$$

There is a more efficient way of doing this. The idea is to use the substitution not only to transform $x\sqrt{x+3}\,dx$ but also the limits 5 and -2. Refer back to Example 13.10. Substituting $x = 5$ and $x = -2$ into $u = x + 3$ gives us $u = 8$ and $u = 1$, and therefore,

$$\int_{-2}^{5} x\sqrt{x+3}\,dx = \int_{1}^{8} (u-3)u^{\frac{1}{2}}\,du$$

$$= \left(\frac{2}{5}u^{\frac{5}{2}} - 3 \cdot \frac{2}{3}u^{\frac{3}{2}}\right)\Big|_{1}^{8}.$$

Notice that the answer is the same as before.

This procedure works more generally. It is summarized by the formula

$$\boxed{\int_{a}^{b} f(g(x))g'(x)\,dx = \int_{g(a)}^{g(b)} f(u)\,du}$$

where $u = g(x)$.

We can now return to the unfinished business of computing the area of the parabolic surface of Example 13.7.

Example 13.14. To compute $\int_{3}^{5} 2\pi\sqrt{x+\frac{1}{4}}\,dx$, we let $u = x + \frac{1}{4}$. Note that $du = dx$, and that $u = \frac{13}{4}$ when

$x = 3$, and $u = \frac{21}{4}$ when $x = 5$. Therefore,

$$\int_{3}^{5} 2\pi\sqrt{x+\frac{1}{4}}\,dx = \int_{13/4}^{21/4} 2\pi u^{\frac{1}{2}}\,du = \frac{4}{3}\pi u^{\frac{3}{2}}\Big|_{13/4}^{21/4}$$

$$= \frac{4}{3}\pi\left(\frac{21}{4}\right)^{3/2} - \frac{4}{3}\pi\left(\frac{13}{4}\right)^{3/2}$$

$$\approx 50.39 - 24.54 = 25.85.$$

Another integral that remains to be solved is the integral

$$\int_{0}^{\pi} \sin x\sqrt{1 + \cos^2 x}\,dx$$

that arose in Example 13.8. Consider the substitution $u = \cos x$. Since $\frac{du}{dx} = -\sin x$, we see that $\sin x\,dx = -du$. Therefore,

$$\int \sin x\sqrt{1 + \cos^2 x}\,dx = \int -\sqrt{1+u^2}\,du.$$

This leaves us with the integral $\int \sqrt{1+u^2}\,du$. But now what?

B. Trigonometric Substitutions. There is another substitution strategy—the method of trigonometric substitutions—where the substitution does not come from the integrand directly.

Since it is best described by an example, we will use this method to analyze the integral

$$\int \frac{x^3}{\sqrt{5-x^2}}\,dx.$$

Since $\sin^2\theta + \cos^2\theta = 1$, $1 - \sin^2\theta = \cos^2\theta$, and hence $5 - 5\sin^2\theta = 5\cos^2\theta$. Now let $x = \sqrt{5}\sin\theta$. Let's see how this substitution transforms the integral. Note that

$$\sqrt{5-x^2} = \sqrt{5 - 5\sin^2\theta} = \sqrt{5\cos^2\theta} = \sqrt{5}\cos\theta$$

and $\frac{dx}{d\theta} = \sqrt{5}\cos\theta$. So $dx = \sqrt{5}\cos\theta\,d\theta$, and hence

$$\int \frac{x^3}{\sqrt{5-x^2}}\,dx = \int \frac{5\sqrt{5}\sin^3\theta}{\sqrt{5}\cos\theta}\sqrt{5}\cos\theta\,d\theta$$

$$= 5\sqrt{5}\int \sin^3\theta\,d\theta.$$

The substitution has transformed the original integral into a so-called trigonometric integral. There is a bag of tricks to handle these, but in the present case it is simple. Rewrite $\sin^3 \theta$ as

$$(\sin \theta)(\sin^2 \theta) = \sin \theta (1 - \cos^2 \theta)$$

and let $u = \cos \theta$. So $du = - \sin \theta \, d\theta$, and

$$5\sqrt{5} \int \sin^3 \theta \, d\theta = 5\sqrt{5} \int \sin \theta (1 - \cos^2 \theta) \, d\theta$$

$$= -5\sqrt{5} \int (1 - u^2) \, du$$

$$= -5\sqrt{5} \left(u - \frac{1}{3} u^3 \right) + C$$

$$= -5\sqrt{5} \left(\cos \theta - \frac{1}{3} \cos^3 \theta \right) + C.$$

Recall, finally, that $\sqrt{5 - x^2} = \sqrt{5} \cos \theta$. So $\cos \theta = \frac{\sqrt{5-x^2}}{\sqrt{5}} = \frac{1}{\sqrt{5}}(5 - x^2)^{\frac{1}{2}}$. Putting everything together, we get

$$\int \frac{x^3}{\sqrt{5 - x^2}} \, dx = 5\sqrt{5} \int \sin^3 \theta \, d\theta$$

$$= -5\sqrt{5} \left(\cos \theta - \frac{1}{3} \cos^3 \theta \right) + C$$

$$= -5(5 - x^2)^{1/2} + \frac{1}{3}(5 - x^2)^{3/2} + C.$$

The integral $\int \sqrt{1 + u^2} \, du$ can be handled in a similar way. How? Looking back, notice that the key to the solution of $\int \frac{x^3}{\sqrt{5-x^2}} \, dx$ was the fact that the formula $1 - \sin^2 \theta = \cos^2 \theta$ turns the difference of the two squares $5 - x^2 = (\sqrt{5})^2 - x^2$ into a square, thereby simplifying $\sqrt{5 - x^2}$. Now it is $1 + u^2$ that must be made into a square. This is accomplished by the substitution that the formula $1 + \tan^2 \theta = \sec^2 \theta$ suggests.

C. Integration by Parts. Any formula for a derivative tells us, "this" is the derivative of "that." So it also tells us, "that" is an antiderivative of "this." In other words, every formula for a derivative gives rise to an integration formula. The product rule provides an important illustration of this. Recall that if $f(x)$ and $g(x)$ are differentiable functions of x, then

$$(f(x) \cdot g(x))' = f'(x) \cdot g(x) + f(x) \cdot g'(x)$$

and hence

$$f(x) \cdot g'(x) = (f(x) \cdot g(x))' - f'(x) \cdot g(x).$$

Suppose that $G(x)$ is an antiderivative of the function $f'(x) \cdot g(x)$. Then this last equality can be rewritten as

$$f(x) \cdot g'(x) = (f(x) \cdot g(x))' - G'(x) = (f(x) \cdot g(x) - G(x))'.$$

In particular, $f(x) \cdot g(x) - G(x)$ is an antiderivative of $f(x) \cdot g'(x)$. Therefore, in order to find an antiderivative of $f(x) \cdot g'(x)$ it suffices to find one for $f'(x) \cdot g(x)$ and to subtract it from $f(x) \cdot g(x)$. This observation can be expressed in terms of the formula

$$\int f(x) \cdot g'(x) \, dx = f(x) \cdot g(x) - \int f'(x) \cdot g(x) \, dx.$$

This formula is used in the following rewritten form. Let $u = f(x)$ and $v = g(x)$. Since $\frac{du}{dx} = f'(x)$ and $\frac{dv}{dx} = g'(x)$, we see that $du = f'(x) \, dx$ and $dv = g'(x) \, dx$. So the formula becomes

$$\boxed{\int u \, dv = u \cdot v - \int v \, du}$$

This is the *Integration by Parts* formula.

Examples 13.15, 13.16, and 13.17 will illustrate how this formula is used. The basic idea is the same as before: Try to convert the original integral into one that can be solved.

Example 13.15. Consider $\int \ln x \, dx$. To use the formula, the integrand $\ln x \, dx$ has to be written in the form $u \, dv$. One thing we can do is to let $u = \ln x$ and $dv = dx$. In order to apply

$$\int u \, dv = u \cdot v - \int v \, du$$

v and du must be found. This is not hard: The function $v = x$ satisfies the equation $dv = dx$; and in view of the fact that $\frac{du}{dx} = \frac{1}{x}$, we see that $du = \frac{1}{x}\,dx$. In summary,

$$u = \ln x \qquad dv = dx$$

$$du = \frac{1}{x}\,dx \qquad v = x$$

By substituting into the formula,

$$\int \ln x\,dx = x\ln x - \int x\frac{1}{x}\,dx = x\ln x - \int 1\,dx.$$

The "unknown" integral $\int \ln x\,dx$ has therefore been reduced—and this is the point—to the manageable (in this case very easy) integral $\int 1\,dx$. Since this is equal to $x + \text{constant}$, we get

$$\int \ln x\,dx = x\ln x - x + C.$$

Example 13.16. Try integration by parts on

$$\int x\cos x\,dx.$$

Let's try $u = \cos x$ and $dv = x\,dx$. So $du = -\sin x\,dx$. Because $\frac{dv}{dx} = x$, we take $v = \frac{1}{2}x^2$. So

$$u = \cos x \qquad dv = x\,dx$$

$$du = -\sin x\,dx \qquad v = \frac{1}{2}x^2$$

By substituting into the integration by parts formula, we get

$$\int x\cos x\,dx = \frac{1}{2}x^2\cos x - \int\left(-\frac{1}{2}x^2\sin x\right)dx.$$

This formula (while it is perfectly valid) fails in an essential point: The integral $\int x\cos x\,dx$ that we set out to solve has been converted to the more complicated integral $\int \frac{1}{2}x^2\sin x\,dx$.

So let's start over and try $u = x$ and $dv = \cos x\,dx$ instead. Now,

$$u = x \qquad dv = \cos x\,dx$$

$$du = dx \qquad v = \sin x$$

Substituting into $\int u\,dv = u\cdot v - \int v\,du$ this time, we get

$$\int x\cos x\,dx = x\sin x - \int \sin x\,dx.$$

Since $\int \sin x\,dx = -\cos x + \text{constant}$, we have now have the solution:

$$\int x\cos x\,dx = x\sin x + \cos x + C.$$

Example 13.17. Try integration by parts on

$$\int \sec^3 x\,dx.$$

Let $u = \sec x$ and $dv = \sec^2 x\,dx$. So

$$u = \sec x \qquad dv = \sec^2 x\,dx$$

$$du = (\sec x)(\tan x)\,dx \qquad v = \tan x$$

By the integration by parts formula, we get

$$\int \sec^3 x\,dx = (\sec x)(\tan x) - \int (\sec x)(\tan^2 x)\,dx.$$

Because $\tan^2 x = \sec^2 x - 1$, we now see that

$$\int \sec^3 x\,dx = (\sec x)(\tan x) - \int (\sec x)(\sec^2 x)\,dx$$

$$+ \int \sec x\,dx$$

$$= (\sec x)(\tan x) - \int \sec^3 x\,dx$$

$$+ \int \sec x\,dx.$$

Was all this of help? We saw earlier that

$$\int \sec x\,dx = \ln(\sec x + \tan x) + \text{constant}.$$

But the presence of $\int \sec^3 x\,dx$ suggests that we are back where we started. However, a more careful look at the equation that we derived shows that we can bring $-\int \sec^3 x\,dx$ from the left to the right to get

$$2\int \sec^3 x\,dx$$

$$= (\sec x)(\tan x) + \int \sec x\,dx$$

$$= (\sec x)(\tan x) + \ln(\sec x + \tan x) + \text{constant}.$$

We have therefore discovered that

$$\int \sec^3 x \, dx$$

$$= \frac{1}{2}(\sec x)(\tan x) + \frac{1}{2}\ln(\sec x + \tan x) + C.$$

Our survey of integration is by no means complete, but it has described and illustrated the most important methods.

13.4 **Polar Coordinates**

T he fact that the rectangular, or Cartesian, coordinate system has been crucial to virtually all the topics of this book should need no elaboration. It is clear that much of the mathematics that has been presented is ultimately based on the interplay between geometry and algebra that this coordinate system makes possible. However, this system is not the only one that makes this connection. We will now discuss a coordinate system—the polar coordinate system introduced by Isaac Newton—that provides a better framework for certain mathematical investigations than the Cartesian system. It turns out that this system is tailor-made for the mathematical description of the orbits of the planets about the Sun. In this section, we will study the basic properties of polar coordinates. The connection with the solar system will be taken up in Chapter 14.

Start with a plane and fix a point in it. Call it the *origin* or *pole* and designate it by O. Next, take a straight line—actually a "half-line"—that emanates from the origin. This is called the *polar axis*. Finally, fix a unit of length. Our *polar coordinate system* is complete.

The polar axis is customarily drawn horizontally and to the right of the origin.

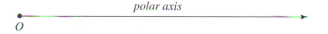

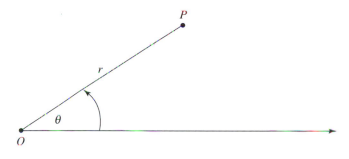

Figure 13.24

Let P be any point in the plane and draw the segment OP. Let r be the length of this segment and let θ be the angle in radians that it makes with the polar axis. See Figure 13.24. As in Section 4.4, angles measured off in the counterclockwise direction are positive, and those measured in the clockwise direction are negative. In this way, the point P is determined by the ordered pair (r, θ) of real numbers. The numbers r and θ are called the *polar coordinates* of P. This procedure can be reversed. Namely, to a pair of real numbers (r, θ) there corresponds a point in the plane. This is described by Figure 13.24. The understanding is that the point that corresponds to a pair (r, θ) with a negative r is given as follows: take the ray that θ determines and then mark off the distance $|r|$ in the opposite direction (in other words, along the ray $\theta + \pi$). Figure 13.25 provides some examples.

One difference between the Cartesian coordinate system and the polar coordinate system is the fact that a point P can be represented in many (indeed infinitely many) ways as a pair (r, θ). Figure 13.26 shows how four different pairs of coordinates all determine the same point.

Refer to back to Section 4.7 and notice that the analysis of Kepler's investigations made use of polar coordinates. (Just replace α by θ in Figure 4.33.)

A. Graphing Polar Equations. Suppose that the plane is equipped with a polar coordinate system. The *graph* of an equation in the variables r and θ is the set of all points (r, θ) in the plane whose coordinates satisfy it.

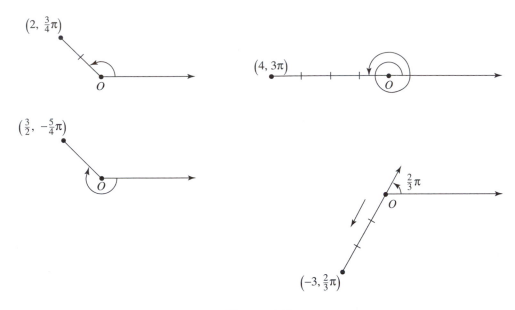

Figure 13.25

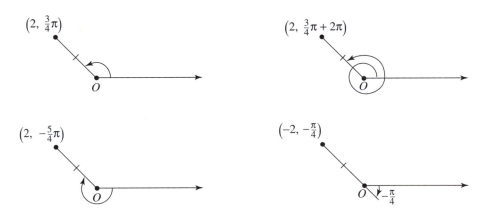

Figure 13.26

Example 13.18. The graph of the equation $\theta = \frac{5\pi}{4}$ is the set of all (r, θ) with $\theta = \frac{5\pi}{4}$. This set of points consists of the line determined by the ray $\theta = \frac{5\pi}{4}$. Refer to Figure 13.27.

Example 13.19. The graph of the equation $r = 5$ is the set of all (r, θ) with $r = 5$. Its graph is the circle of radius 5 with center the origin. See Figure 13.28.

Consider a function $r = f(\theta)$. Since it is an equation in the variable r and θ it has a graph in the polar plane.

Example 13.20. Graph the function $r = f(\theta) = \theta$, $\theta \geq 0$. Note that $(0, 0)$ is on the graph. As θ increases, the ray that it determines rotates counter-clockwise. In the process the corresponding r (equal to θ) increases. In other words, the point (r, θ) slides up on the the ray as it rotates counterclockwise. The curve that is generated is a spiral. It is known as the *Spiral of Archimedes*. It is sketched in Figure 13.29. (What, if anything, would be added to the graph if negative θ were allowed?)

graph consists of a loop. Could it be a circle? How can we be sure?

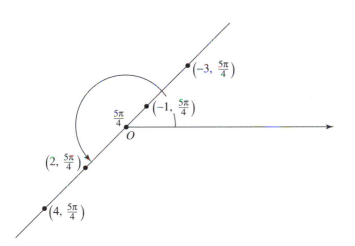

Figure 13.27

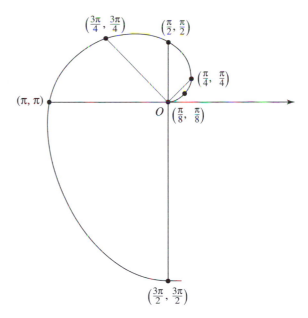

Figure 13.29

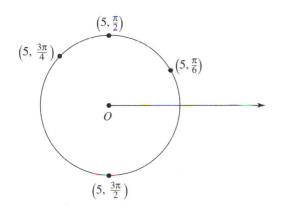

Figure 13.28

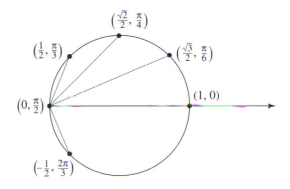

Figure 13.30

Example 13.21. What is the graph of $r = f(\theta) = \cos\theta$? Plot the points corresponding to $\theta = 0, \frac{\pi}{6}, \frac{\pi}{4}, \frac{\pi}{3}, \frac{\pi}{2}$, and $\frac{2\pi}{3}$, and check that you get the pattern shown in Figure 13.30. Let's consider $\theta = \pi$ next. Since $\cos\pi = -1$, the corresponding point is $(-1, \pi)$. Note that this represents the same point as $(1, 0)$. It seems that the

To answer this question, we will consider a Cartesian and polar coordinate system simultaneously. Fix a Cartesian coordinate system, and take the origin along with the positive x axis as the polar coordinate system. Now let P be a point. Let (x, y) be the Cartesian coordinates of P and (r, θ) the polar coordinates of P. See Figure 13.31. By the Pythagorean theorem and the definitions of the sine, cosine, and tangent, we get the following *transformation equations*. They can be used to "transform" a polar equation into a Cartesian equation and conversely:

$$r^2 = x^2 + y^2 \quad x = r \cos \theta \quad y = r \sin \theta \quad \tan \theta = \frac{y}{x}$$

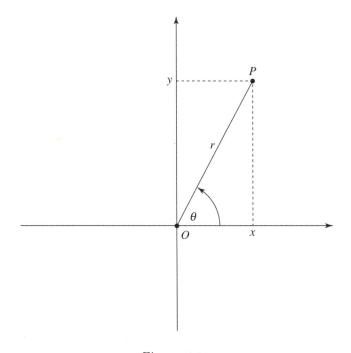

Figure 13.31

These relationships hold regardless of the location of the point and regardless of the choice of its polar coordinates. For instance, the fact that the polar coordinates $(-r, \theta + \pi)$ represent the same point as (r, θ) will have to mean that its Cartesian coordinates are also the same. Since $\sin(\theta + \pi) = -\sin \theta$ and $\cos(\theta + \pi) = -\cos \theta$ (see Example 4.11 of Section 4.4), it is indeed true that

$$x = r \cos \theta = -r \cos(\theta + \pi) \quad \text{and}$$

$$y = r \sin \theta = -r \sin(\theta + \pi).$$

Now to return to the equation $r = \cos \theta$. What is its Cartesian equivalent? Noting that $r = \pm\sqrt{x^2 + y^2}$, we see that

$$\pm\sqrt{x^2 + y^2} = r = \cos \theta = \frac{x}{r} = \frac{x}{\pm\sqrt{x^2 + y^2}}.$$

By multiplying through by $\pm\sqrt{x^2 + y^2}$, we get $x^2 + y^2 = x$ and hence $x^2 - x + y^2 = 0$. After completing the square, $x^2 - x + \left(\frac{1}{2}\right)^2 + y^2 = \left(\frac{1}{2}\right)^2$ and hence $\left(x - \frac{1}{2}\right)^2 + y^2 = \left(\frac{1}{2}\right)^2$. This computation has shown that if the polar coordinates (r, θ) of a point satisfy $r = \cos \theta$, then its Cartesian coordinates (x, y) satisfy $\left(x - \frac{1}{2}\right)^2 + y^2 = \left(\frac{1}{2}\right)^2$. This means that any point on the graph of $r = \cos \theta$ must lie on the circle with center $\left(\frac{1}{2}, 0\right)$ and radius $\frac{1}{2}$. This argument can be continued to show that the graph of $r = \cos \theta$ and this circle actually coincide.

Example 13.22. Sketch the graph of $r = 1 + \cos \theta$. To understand the relationship between r and θ, we begin

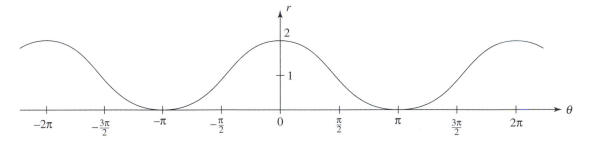

Figure 13.32

by regarding $r = 1 + \cos\theta$ as an equation in Cartesian coordinates. Its graph is obtained by "adding 1" to the graph of the cosine. See Figure 13.32. Turning to polar coordinates, we now see the following: As the ray determined by θ rotates from 0 to $\frac{\pi}{2}$, the r coordinate of the point (r, θ) slides from 2 to 1; as the ray rotates from $\frac{\pi}{2}$ to π, the r coordinate slides from 1 to 0; as θ rotates from π to $\frac{3\pi}{2}$, it slides from 0 to 1; and finally, as the ray given by θ rotates from $\frac{3\pi}{2}$ to 2π, r slides from 1 to 2. Converting this information into a polar graph provides Figure 13.33. (What happens for negative θ?) Since the graph resembles a heart, it is called a *cardioid*.

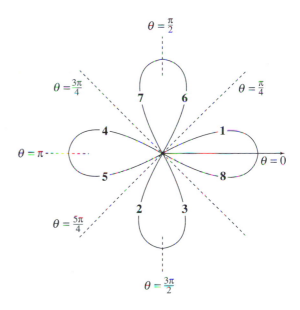

Figure 13.35

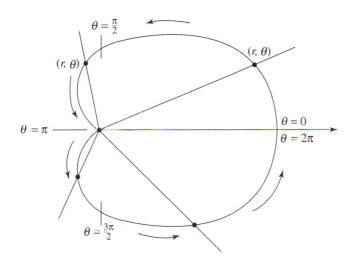

Figure 13.33

Example 13.23. Sketch the graph of $r = \cos 2\theta$. The strategy is the same as that of the previous example. Start with the graph of $r = \cos 2\theta$ in Cartesian coordinates. Since $\cos 2\theta = \cos(2\theta + 2\pi) = \cos 2(\theta + \pi)$, this graph has period π. It is the graph of $\cos\theta$ compressed by a factor of 2. See Figure 13.34. Now let (r, θ) be a point on the polar graph of $r = \cos\theta$ and consider Figures 13.34 and 13.35 together. Observe the following: As the ray θ rotates from 0 to $\frac{\pi}{4}$, r slides from 1 to 0 and (r, θ) traces out the arc labeled **1**; as θ rotates from $\frac{\pi}{4}$ to $\frac{\pi}{2}$, r moves from 0 to -1 and (since r is negative) (r, θ) traces out the arc labeled **2**; as θ rotates from $\frac{\pi}{2}$

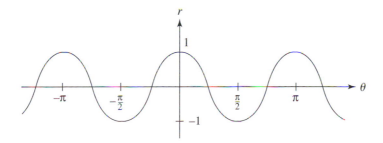

Figure 13.34

to $\frac{3\pi}{4}$, r slides from -1 to 0, so that (r, θ) traces out the arc labeled **3**; as θ rotates from $\frac{3\pi}{4}$ to π, r slides from 0 to 1, and (r, θ) traces out the arc labeled **4**. Continuing in this way, we get the graph depicted in Figure 13.35. It is a *four-leaved rose*.

With the application of polar coordinates to the solar system in mind, we now turn to an analysis of the graph of

$$r = \frac{L}{2(1 + \varepsilon \cos \theta)}$$

where $\varepsilon \geq 0$ and $L > 0$ are constants. It turns out (refer to Figure 13.36) that

1. The graph is a conic section, in other words, an ellipse, a parabola, or a hyperbola.
2. One of the focal points of the conic section (there is only one in the case of the parabola) falls on the origin.
3. The axis of the conic section (this is the axis if it is a parabola, or the line determined by the two focal points if not) falls on the line determined by the polar axis.
4. The constant L is the latus rectum of the conic section.
5. The constant ε determines the extent to which the conic section curves back towards its axis: If $0 \leq \varepsilon < 1$, the conic section is an ellipse (and ε is its astronomical eccentricity); if $\varepsilon = 1$, it is a parabola; and if $\varepsilon > 1$, it is a hyperbola.

These assertions are best verified by transforming $r = \frac{L}{2(1+\varepsilon \cos \theta)}$ into an equation in Cartesian coordinates (with the transformation equations) and by recalling what we have already learned about conic sections (for example in Sections 3.1, 4.3, and 4.5).

We begin by rewriting the equation. Because $2r(1 + \varepsilon \cos \theta) = L$, we get $2r + 2r\varepsilon \cos \theta = L$ and hence $2r = L - 2r\varepsilon \cos \theta$. So $4r^2 = (L - 2r\varepsilon \cos \theta)^2$. By inserting the transformation equations and squaring, we get the Cartesian equation (as before, the Cartesian coordinate system has been chosen so that the

positive x-axis coincides with the polar axis)

$$4x^2 + 4y^2 = (L - 2\varepsilon x)^2 = L^2 - 4\varepsilon Lx + 4\varepsilon^2 x^2.$$

Bringing all the variables to the left, we see that $4x^2 - 4\varepsilon^2 x^2 + 4\varepsilon Lx + 4y^2 = L^2$, and after factoring out x^2 that

$$(*) \qquad 4(1 - \varepsilon^2)x^2 + 4\varepsilon Lx + 4y^2 = L^2.$$

Assume for a moment that $\varepsilon = 1$. In this case the equation reduces to $4Lx + 4y^2 = L^2$ or, after dividing by $4L$, to $x = -\frac{1}{L}y^2 + \frac{L}{4}$. This (by a remark towards the end of Section 4.3) is a parabola that has the general form of Figure 13.36. The verification of points (**2**)–(**4**) in this case is left to the Exercises 13L.

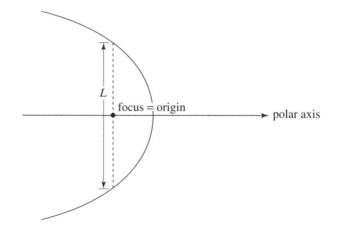

Figure 13.36

Suppose from now on that $\varepsilon \neq 1$. So $\varepsilon^2 \neq 1$, because $\varepsilon \geq 0$. By dividing both sides of equation $(*)$ by $4(1 - \varepsilon^2)$, we get

$$x^2 + \frac{\varepsilon L}{1 - \varepsilon^2}x + \frac{y^2}{1 - \varepsilon^2} = \frac{L^2}{4(1 - \varepsilon^2)}.$$

Because,

$$x^2 + \frac{\varepsilon L}{1 - \varepsilon^2}x + \left(\frac{\varepsilon L}{2(1 - \varepsilon^2)}\right)^2 + \frac{y^2}{1 - \varepsilon^2}$$

$$= \frac{L^2}{4(1 - \varepsilon^2)} + \left(\frac{\varepsilon L}{2(1 - \varepsilon^2)}\right)^2$$

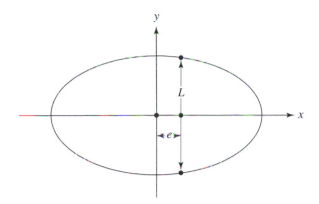 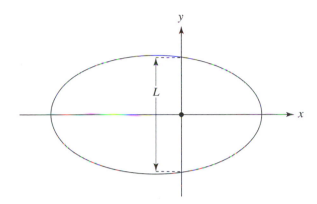

Figure 13.37

we get by completing the square,

$$\left(x + \frac{\varepsilon L}{2(1 - \varepsilon^2)}\right)^2 + \frac{y^2}{1 - \varepsilon^2}$$

$$= \frac{L^2}{4(1 - \varepsilon^2)} + \left(\frac{\varepsilon L}{2(1 - \varepsilon^2)}\right)^2$$

$$= \frac{L^2(1 - \varepsilon^2) + \varepsilon^2 L^2}{4(1 - \varepsilon^2)^2}.$$

Therefore,

(**) $$\left(x + \frac{\varepsilon L}{2(1 - \varepsilon^2)}\right)^2 + \frac{y^2}{1 - \varepsilon^2} = \frac{L^2}{4(1 - \varepsilon^2)^2}.$$

Turning to the case of the ellipse, we will now assume that $0 \le \varepsilon < 1$. So $0 < 1 - \varepsilon^2 \le 1$. Let $a = \frac{L}{2(1 - \varepsilon^2)}$ and $b = a\sqrt{1 - \varepsilon^2}$. Notice that $a > 0$, $b > 0$, and $a \ge b$. In addition, $b^2 = a^2(1 - \varepsilon^2)$ and $a^2 = \frac{L^2}{4(1 - \varepsilon^2)^2}$. The equation (**) can now be written as $(x + \varepsilon a)^2 + \frac{a^2 y^2}{b^2} = a^2$. Put $e = \varepsilon a$ and divide both sides by a^2 to get

$$\frac{(x + e)^2}{a^2} + \frac{y^2}{b^2} = 1.$$

Refer to "Symmetry and Shifting" in Section 10.8 for the fact that the graph of this equation is obtained by shifting the graph of the equation $\frac{x^2}{a^2} + \frac{y^2}{b^2} = 1$ to the left by e units. This last equation is the standard equation of the ellipse. Its graph is shown in Figure 4.28 of Section 4.5. It remains to turn to some basic facts

from Section 4.5. Because $b^2 = a^2 - a^2\varepsilon^2 = a^2 - e^2$, it follows that e is the eccentricity. So the shift places the right focus of this ellipse at the origin. In Figure 13.37 the standard ellipse is shown on the left and the shifted version on the right. The graph on the right is the graph of $\frac{(x+e)}{a^2} + \frac{y^2}{b^2} = 1$ and hence that of the polar equation $r = \frac{L}{2(1+\varepsilon\cos\theta)}$. Notice that it is positioned as indicated in Figure 13.36. A review of what has been done shows that points (**2**) and (**3**) have now been verified. To verify (**4**), we need to set $x = 0$ in the equation $\frac{(x+e)^2}{a^2} + \frac{y^2}{b^2} = 1$ and solve for y. Doing so, we get $\frac{y^2}{b^2} = 1 - \frac{e^2}{a^2} = \frac{a^2 - e^2}{a^2} = \frac{b^2}{a^2}$. Hence $y^2 = \frac{b^4}{a^2}$ and $y = \pm\frac{b^2}{a}$. Therefore, the latus rectum of the ellipse is

$$2\frac{b^2}{a} = 2\frac{a^2 - a^2\varepsilon^2}{a} = 2a(1 - \varepsilon^2)$$

$$= 2\frac{L}{2(1 - \varepsilon^2)}(1 - \varepsilon^2) = L$$

as required. Because $\varepsilon = \frac{e}{a}$, ε is the astronomical eccentricity of the ellipse as stated in (**5**). Our discussion of the case $0 \le \varepsilon < 1$ is now complete. The case $\varepsilon > 1$ (of the hyperbola) is similar. It too starts with equation (**). It is discussed in Exercises 13L.

Example 13.24. What is the graph of the equation $r = \frac{5}{2 + \cos\theta}$? Putting this equation into the form that we have studied, we get $r = \frac{5}{2(1 + \frac{1}{2}\cos\theta)}$. So $\varepsilon = \frac{1}{2}$ and $L = 5$. Because $0 \le \varepsilon < 1$, the graph is an ellipse.

Using information already developed, we see that this ellipse has

$$a = \frac{L}{2(1 - \varepsilon^2)} = \frac{5}{2\left(1 - \frac{1}{4}\right)} = \frac{5}{\frac{3}{2}} = \frac{10}{3},$$

$$b = a\sqrt{1 - \varepsilon^2} = \frac{10}{3}\sqrt{1 - \frac{1}{4}} = \frac{10}{3}\frac{\sqrt{3}}{2} = \frac{5}{3}\sqrt{3}.$$

Also, $e = \varepsilon a = \frac{1}{2} \cdot \frac{10}{3} = \frac{5}{3}$. It follows that we are dealing with the standard ellipse

$$\frac{x^2}{\left(\frac{10}{3}\right)^2} + \frac{y^2}{\left(\frac{5}{3}\sqrt{3}\right)^2} = 1$$

shifted $\frac{5}{3}$ units to the left.

Arguments virtually identical to those described show that the graph of any of the equations

$$r = \frac{L}{2(1 - \varepsilon\cos\theta)}, \quad r = \frac{L}{2(1 + \varepsilon\sin\theta)},$$

and

$$r = \frac{L}{2(1 - \varepsilon\sin\theta)}$$

is a parabola if $\varepsilon = 1$, an ellipse if $0 \le \varepsilon < 1$, and a hyperbola if $\varepsilon > 1$. In each case L is the latus rectum. The graphs are rotated versions of those already discussed.

B. Areas in Polar Coordinates.

We saw in Section 4.7 and again in Section 7.3 that Kepler's "equal

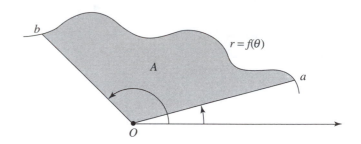

Figure 13.38

areas in equal times" law (see Section 4.1) is an essential element of the theory that explains the workings of the solar system. This section will study the areas of regions determined by graphs of functions in polar coordinates. The analysis of the solar system using polar coordinates will make important use of this study. See Chapter 14.

Let $r = f(\theta)$ be a continuous function defined on a closed interval $a \le \theta \le b$. We will assume that $f(\theta) \ge 0$, and (in order to avoid overlap) that $b - a \le 2\pi$. The graph of a typical situation is shown in Figure 13.38. Our concern is the area A that the graph determines. We will use the method of Riemann sums. See Section 13.1A.

Consider a partition P of the closed interval $[a, b]$ and let it be given by the real numbers

$$a = \theta_0 < \theta_1 < \cdots < \theta_{i-1} < \theta_i < \cdots < \theta_{n-1} < \theta_n = b.$$

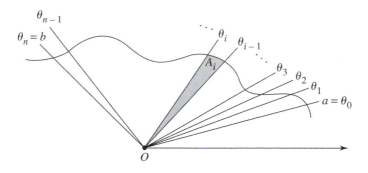

Figure 13.39

The rays given by the angles $\theta_1, \ldots, \theta_{n-1}$ divide the area A into n pie-shaped regions of areas $A_1, \ldots, A_i, \ldots, A_n$, respectively. See Figure 13.39.

Next, select angles $\varphi_1, \ldots, \varphi_i, \ldots, \varphi_n$ with

$$\theta_0 \leq \varphi_1 \leq \theta_1, \theta_1 \leq \varphi_2 \leq \theta_2, \ldots, \theta_{i-1} \leq \varphi_i \leq \theta_i, \ldots .$$

Let $f(\varphi_i) = r_i$, and consider the point (r_i, φ_i) on the graph of f. Now draw a circular arc centered at the origin and through (r_i, φ_i). Consider the circular sector determined by this arc and the lines $\theta = \theta_{i-1}$ and $\theta = \theta_i$. Figure 13.40 shows the circular sector, as well as the relevant part of the graph of the function f. Notice that A_i is approximated by the area of this sector. Let $\Delta\theta_i$ be the angle $\Delta\theta_i = \theta_i = \theta_{i-1}$. Since r_i is the radius of the circular arc, it follows that the area of this circular sector (see Section 3.2) is

$$\frac{1}{2}r_i^2\Delta\theta_i = \frac{1}{2}f(\varphi_i)^2\Delta\theta_i.$$

Therefore, $A_i \approx \frac{1}{2}f(\varphi_i)^2\Delta\theta_i$. By doing this for $i = 1, 2, \ldots, n$, we find that the area A that is being studied is approximated by

$$A = A_1 + \cdots + A_i + \cdots + A_n$$

$$\approx \frac{1}{2}f(\varphi_1)^2\Delta\theta_1 + \cdots + \frac{1}{2}f(\varphi_i)^2\Delta\theta_i +$$

$$\cdots + \frac{1}{2}f(\varphi_n)^2\Delta\theta_n.$$

So $A \approx \sum_{i=1}^{n} \frac{1}{2}f(\varphi_i)^2\Delta\theta_i$.

Repeat this construction with finer and finer partitions P of $[a, b]$. Each repetition gives us a number of the form $\sum_{i=1}^{n} \frac{1}{2}f(\varphi_i)^2\Delta\theta_i$ that approximates the area A. As the partitions get finer, these approximations get tighter. So in terms of the limit notation of Section 13.1A,

$$A = \lim_{\|P\| \to 0} \sum_{i=1}^{n} \frac{1}{2}f(\varphi_i)^2\Delta\theta_i.$$

On the other hand (because φ_i plays the role of p_i and $\Delta\theta_i$ that of Δx_i), this limit is precisely how $\int_a^b \frac{1}{2}f(\theta)^2\, d\theta$ is defined. (Refer to Section 13.1A once

more.) Therefore, it follows that

$$A = \int_a^b \frac{1}{2}f(\theta)^2\, d\theta$$

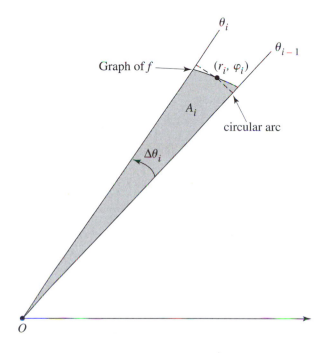

Figure 13.40

Our first example tests this formula in a situation where the result can be determined by other means.

Example 13.25. Consider the area of the region bounded by the graph of the function $f(\theta) = \cos\theta$ and the rays $\theta = 0$ and $\theta = \frac{\pi}{4}$. By Example 13.21, the part of the graph of this function that is relevant is the semicircle shown in Figure 13.41. Notice that the area in question consists of a quarter circle of radius $\frac{1}{2}$ and a triangle of height and base both $\frac{1}{2}$. This area is therefore equal to $\frac{1}{4}\pi\left(\frac{1}{2}\right)^2 + \frac{1}{8} = \frac{\pi}{16} + \frac{1}{8}$. Does the formula

$$A = \int_0^{\pi/4} \frac{1}{2}\cos^2\theta\, d\theta$$

provide the same result? The half-angle formula $\cos^2\theta = \frac{1+\cos 2\theta}{2}$ is a consequence of the addition formula for the cosine. (See Exercises 2C.) Using it shows that

$$A = \int_0^{\pi/4} \frac{1}{2}\cos^2\theta\,d\theta = \int_0^{\pi/4} \frac{1}{4}(1+\cos 2\theta)\,d\theta$$

$$= \left(\frac{1}{4}\theta + \frac{1}{8}\sin 2\theta\right)\Big|_0^{\pi/4}$$

$$= \left(\frac{\pi}{16} + \frac{1}{8}\right) - 0 = \frac{\pi}{16} + \frac{1}{8}.$$

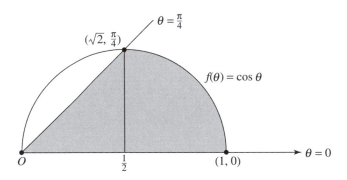

Figure 13.41

So the formula gives the same answer. We will next illustrate the formula for various regions determined by the spiral of Archimedes. See Example 13.20.

Example 13.26. Since $r = f(\theta) = \theta$, the area formula specializes to

$$A = \int_a^b \frac{1}{2}\theta^2\,d\theta = \frac{1}{6}\theta^3\,\Big|_a^b = \frac{1}{6}(b^3 - a^3).$$

Refer to Figure 13.29. Let A_1 be the area bounded by the spiral and the rays $\theta = 0$ and $\theta = \frac{\pi}{2}$; let A_2 be the area bounded by the spiral and the rays $\theta = \frac{\pi}{2}$ and $\theta = \pi$; let A_3 be that determined by the spiral and the rays $\theta = \pi$ and $\theta = \frac{3\pi}{2}$; and let A_4 be the area bounded

by the spiral and the rays $\theta = \frac{3\pi}{2}$ and $\theta = 2\pi$. Plugging these data into the formula above, we get

$$A_1 = \frac{1}{6}\left(\left(\frac{\pi}{2}\right)^3 - 0\right) = \frac{\pi^3}{48}$$

$$A_2 = \frac{1}{6}\left(\pi^3 - \left(\frac{\pi}{2}\right)^3\right) = \frac{1}{6}\left(\frac{7\pi^3}{8}\right) = \frac{7\pi^3}{48}$$

$$A_3 = \frac{1}{6}\left(\left(\frac{3\pi}{2}\right)^3 - \pi^3\right) = \frac{1}{6}\left(\frac{19\pi^3}{8}\right) = \frac{19\pi^3}{48}.$$

Guess what A_4 is equal to and then confirm your guess.

13.5 **Differential Equations**

This section gives a short introduction to the theory of differential equations. This theory has a wide range of applications. Any problem that calls for the mathematical simulation of a physical process that unfolds smoothly falls within its domain. For example, the mathematical elements that underlie Chapter 11 are aspects of this theory.

A *differential equation* is an equation involving one or more derivatives of an unknown function. For example,

$$f'(x) - f(x)^2 - 1 = 0$$

and

$$(1-t)f''(t) + tf'(t) - f(t) = 0$$

are differential equations. They can also be written as

$$\frac{dy}{dx} - y^2 - 1 = 0$$

and

$$(1-t)\frac{d^2z}{dt^2} + t\frac{dz}{dt} - z = 0$$

where y is understood to be a function of x in the first case and z a function of t in the second.

From the point of view of mathematics, the equations $x^2 - 3x + 5 = 0$ and $\varphi^2 - 3\varphi + 5 = 0$ are identical. The notation for the variables is also irrelevant in the context of differential equations. For example, the differential equations $\frac{dx}{dt} - x^2 - 1 = 0$ and $\frac{d\theta}{du} - \theta^2 - 1 = 0$ are both the same as $\frac{dy}{dx} - y^2 - 1 = 0$.

Let f be a function that can be differentiated again and again. The successive derivatives, f', $(f')'$, $((f')')'$, ... are called the first, second, third, ... derivatives of f. The word *order* is also used. So $(f')'$ is a derivative of order two, $((f')')'$ has order three, and so on. The *order* of a differential equation is the order of the highest derivative that occurs in the equation. For instance, $\frac{dy}{dx} - y^2 - 1 = 0$ is a *first-order* differential equation, and $(1-t)\frac{d^2z}{dt^2} + t\frac{dz}{dt} - z = 0$ is a *second-order* differential equation.

A *solution* of a differential equation is a function that satisfies the equation. A *solution* of a differential equation *on an interval I* is a function that satisfies the equation for all real numbers in I. It is understood that the function, as well as all the derivatives of the function up to the order of the equation, exist for all real numbers in I.

Example 13.27. Consider the differential equation $\frac{dy}{dx} - y^2 - 1 = 0$. Let $y = \tan x$. Then $\frac{dy}{dx} = \sec^2 x$. Because $\sec^2 x = \tan^2 x + 1$, it follows that $\frac{dy}{dx} - y^2 - 1 = 0$. So $y = \tan x$ is a solution of this equation on the interval $(-\infty, \infty)$. Not so fast! A look at Example 10.30 in Section 10.8 shows that $y = \tan x$ is not defined for $x = \pm\frac{\pi}{2}, \pm\frac{3\pi}{2}, \pm\frac{5\pi}{2}, \ldots$. Therefore, $y = \tan x$ cannot be a solution of $\frac{dy}{dx} - y^2 - 1 = 0$ on the interval $(-\infty, \infty)$. But $y = \tan x$ is differentiable for any x with $-\frac{\pi}{2} < x < \frac{\pi}{2}$. Hence it is a solution of the differential equation on $I = \left(-\frac{\pi}{2}, \frac{\pi}{2}\right)$.

Example 13.28. Let $f(t) = e^t + t$. Since $f'(t) = e^t + 1$ and $f''(t) = e^t$, notice that

$$(1-t)f''(t) + tf'(t) - f(t)$$

$$= (1-t)e^t + t(e^t + 1) - e^t - t = 0.$$

Because the derivatives $f'(t) = e^t + 1$ and $f''(t) = e^t$ both exist for all t, it follows that $f(t) = e^t + t$ is a solution of $(1-t)f''(t) + tf'(t) - f(t) = 0$ on the interval $(-\infty, \infty)$.

Given a function and a differential equation, it is usually a routine matter (as we have just seen) to check whether the function is a solution to the equation or not. However, if only an equation is given, then the problem of finding a solution is often very challenging! The theory of differential equations is a response to this challenge. Notice that this theory includes the methods of integration as a special case, because the problem of finding an antiderivative of a function $g(x)$ is nothing else than the problem of finding a solution of the differential equation $\frac{dy}{dx} = g(x)$.

Differential equations have already arisen in the analysis of a number of different questions in this book. A careful review of these analyses shows that they follow essentially the same strategy. Each has as its starting point a fundamental insight that is formulated in terms of a differential equation. The differential equation is then subjected to a mathematical analysis and solved. This solution of the differential equation is in each case an important component of the answer to the question.

For example, the basic insight (which can be attributed to Galileo) in the study of the projectile was the differential equations

$$x''(t) = 0 \quad \text{and} \quad y''(t) = -g.$$

Their solutions—refer back to Section 6.5—form the basis of the entire theory. In Section 9.2, the differential equation

$$f'(x) = \frac{w}{T_0}x$$

was an important initial element in the study of the suspension bridge. Its solution informed us that the main cable over the center span has the shape of a parabola. The basic insight in the analysis of the experiment of Galileo in Section 9.3C was the differential equation

$$p''(t) = a(t) = \frac{5g}{7}\sin\beta.$$

Rutherford's experiments with radioactivity (see Section 11.1B) led directly to the differential equation

$$y'(t) = y'(0)e^{-\lambda t}.$$

Its solution $y(t) = y_0 e^{-\lambda t}$ was basic to the understanding of radioactivity. The starting point of Verhulst's logistics model of Section 11.5B was the differential equation $\frac{y'(t)}{y(t)} = \mu - ky(t)$, or in rewritten form,

$$y'(t) = \mu y(t) - ky(t)^2.$$

Its solution resulted in mathematical simulations of the world's population and populations of bacteria.

Many problems arising in the sciences and engineering are attacked with this approach. The differential equations that arise are generally as complicated as the problem that is being considered, and their solutions often require considerable effort. The rest of this section will concern itself with some abstract differential equations and their solutions.

There is a general class of first-order differential equations that can be solved, at least in the sense that they can be reduced to the problem of finding antiderivatives of functions. They are the equations of the form

$$\frac{dy}{dx} = g(x) \cdot h(y)$$

for some functions $g(x)$ of x and $h(y)$ of y. For example, the equation $\frac{dy}{dx} = ky$, where k is a constant, has this form; just take $g(x) = k$ and $f(y) = y$. So does the equation $\frac{dy}{dx} - y^2 - 1 = 0$ of Example 13.27; just take $g(x) = 1$ and $f(y) = y^2 + 1$.

Let $\frac{dy}{dx} = g(x) \cdot h(y)$ be any differential equation of this form. Assume that $y = f(x)$ is a solution. So $f'(x) = g(x) \cdot h(y)$. Taking $k(y) = \frac{1}{h(y)}$, we have $k(y)f'(x) = g(x)$. Now let $K(y)$ be an antiderivative of $k(y)$. The function $K(y) = K(f(x))$ is also a function of x. Differentiating it with the chain rule gives us $\big(K(f(x))\big)' = K'(y) \cdot f'(x)$. Therefore,

$$\big(K(f(x))\big)' = K'(y) \cdot f'(x) = k(y) \cdot f'(x) = g(x).$$

So $K(f(x))$ is an antiderivative of $g(x)$. If $G(x)$ is an antiderivative of $g(x)$, then $K(f(x))$ and $G(x)$ differ by

a constant, and therefore,

$$(*) \qquad K(y) = K(f(x)) = G(x) + \text{constant}.$$

We have shown that any solution $y = f(x)$ of the equation $\frac{dy}{dx} = g(x) \cdot h(y)$ also satisfies equation $(*)$. The point is this: Find antiderivatives $K(y)$ of the function $k(y) = \frac{1}{h(y)}$ and $G(x)$ of the function $g(x)$ and then find a function $y = f(x)$ that satisfies equation $(*)$. Differentiating equation $(*)$ with respect to x (using the chain rule) shows that $k(y) \cdot \frac{dy}{dx} = k(y) \cdot f'(x) = g(x)$. So $f'(x) = \frac{g(x)}{k(y)} = g(x) \cdot h(y)$, and therefore $y = f(x)$ is a solution of $\frac{dy}{dx} = g(x) \cdot h(y)$.

More briefly formulated, the strategy is this: Separate the variables of the differential equation $\frac{dy}{dx} = g(x) \cdot h(y)$ by rewriting it as $\frac{1}{h(y)} dy = g(x)\, dx$ and consider the indefinite integrals

$$\int \frac{1}{h(y)}\, dy = \int g(x)\, dx.$$

If both of them can be solved, then an equation in the variables y and x is obtained. If this equation can be solved for y in terms of x, then a function $y = f(x)$ of $(*)$ is determined, and it is a solution of the differential equation $\frac{dy}{dx} = g(x) \cdot h(y)$.

Example 13.29. The equation $\frac{dy}{dt} = \mu y$ that arose in the context of nuclear decay is a separable equation with $g(t) = \mu$ and $h(y) = y$. According to the method just described, we get

$$\int \frac{dy}{y} = \int \mu\, dt.$$

After taking antiderivatives, $\ln y = \mu t + C$. This is the equation $(*)$ that must be solved for y. Doing so, we find that $y = e^{\ln y} = e^{\mu t + C} = e^C e^{\mu t}$. By substituting $t = 0$, we see that $e^C = y(0)$. Observe that this is the solution obtained in Section 11.5A.

Example 13.30. Return to the differential equation $\frac{dy}{dx} - y^2 - 1 = 0$ of Example 13.27. Because $\frac{dy}{dx} = y^2 + 1$, we get after separating variables that

$$\int \frac{dy}{y^2 + 1} = \int dx.$$

Recall from Section 10.5 that

$$\int \frac{dy}{y^2 + 1} = \tan^{-1} y + \text{constant}.$$

By antidifferentiating the other side, we see that $\tan^{-1} y = x + C$. This is the equation that must be solved for y. Taking the tangent of both sides, we find that $y = \tan(\tan^{-1} y) = \tan(x + C)$. Therefore, the functions $\tan x$, $\tan(x + 1)$, $\tan(x + 2)$, $\tan(x + \pi)$, and so on are all solutions of the equation $\frac{dy}{dx} - y^2 - 1 = 0$. (Check this for $\tan(x + 2)$.)

Example 13.31. Consider the equation $\frac{dy}{dx} - x^2 y^2 = 0$. By separating variables, $y^{-2} dy = x^2 dx$. So

$$\int y^{-2} \, dy = \int x^2 \, dx.$$

Taking antiderivatives of both sides, we get $-y^{-1} = \frac{1}{3}x^3 + C$. So $y = -\left(\frac{1}{3}x^3 + C\right)^{-1}$. Is every function $y = f(x)$ of this form a solution of $\frac{dy}{dx} = x^2 y^2$? Yes! By the chain rule,

$$\frac{dy}{dx} = f'(x) = \left(\frac{1}{3}x^3 + C\right)^{-2} (x^2) = x^2 y^2.$$

Differential equations of the form $\frac{dy}{dx} = g(x) \cdot h(y)$ are known as *separable differential equations*. Many first-order differential equations are separable. The differential equation $\frac{dy}{dt} = \mu y \left(1 - \frac{y}{M}\right)$ of the logistics model is another example (with $g(t) = \mu$ and $h(y) = y \left(1 - \frac{y}{M}\right)$). This equation was solved by the method just described (see Section 11.5B for the details). Another class of first-order differential equations, while not explicitly separable, can be reduced to separable equations by algebraic tricks. Examples of such equations will be studied in Exercise 13Q. It is important to note, however, that important classes of first-order differential equations are not reducible to separable equations.

The *general solution* of a differential equation is a solution that contains one or more constants and satisfies the following requirement: Every solution of the differential equation can be obtained by taking specific values for these constants.

Example 13.32. Find a solution $y = f(x)$ of the equation $\frac{dy}{dx} - x^2 y^2 = 0$ that satisfies $f(0) = 8$. By Example 13.31, the general solution is $f(x) = -\left(\frac{1}{3}x^3 + C\right)^{-1}$. Since we need to have $f(0) = 8$, we set $f(0) = 8$ and solve for C. Doing this, we get $-\left(\frac{1}{3}(0)^3 + C\right)^{-1} = -C^{-1} = 8$, and hence that $C = -\frac{1}{8}$. So the solution that we are looking for is $f(x) = -\left(\frac{1}{3}x^3 - \frac{1}{8}\right)^{-1}$.

Example 13.33. Find a solution $y = g(x)$ of the differential equation $\frac{dy}{dx} - y^2 - 1 = 0$ that satisfies $g(5) = 1$. By Example 13.30, the general solution is $g(x) = \tan(x + C)$. Setting $g(5) = 1$, we get $\tan(5 + C) = 1$. This is satisfied when $5 + C = \frac{\pi}{4}$, or $C = \frac{\pi}{4} - 5$. So $g(x) = \tan\left(x + \frac{\pi}{4} - 5\right)$ solves the problem.

A condition such as $f(0) = 8$ or $g(5) = 1$ is called an *initial condition* or *boundary condition* of a differential equation.

We conclude our discussion of differential equations by considering the second-order equation

$$\frac{d^2 y}{dx^2} + \frac{k}{m} y = 0$$

where k and m are positive constants. It arises in the study of planetary motion (see Exercises 14I) and oscillations (see Exercises 14M).

We begin by checking that the function $f(x) = A\left(\sin \sqrt{\frac{k}{m}} x\right)$, with A a constant, is a solution. Differentiations using the chain rule give us

$$f'(x) = A\left(\cos \sqrt{\frac{k}{m}} x\right) \sqrt{\frac{k}{m}} = A\sqrt{\frac{k}{m}} \left(\cos \sqrt{\frac{k}{m}} x\right)$$

$$f''(x) = -A\sqrt{\frac{k}{m}} \left(\sin \sqrt{\frac{k}{m}} x\right) \sqrt{\frac{k}{m}}$$

$$= -A\frac{k}{m} \left(\sin \sqrt{\frac{k}{m}} x\right).$$

It follows that $f''(x) + \frac{k}{m} f(x) = 0$ and hence that $y = f(x)$ is a solution of $\frac{d^2 y}{dx^2} + \frac{k}{m} y = 0$. A similar calculation shows that $g(x) = B\left(\cos \sqrt{\frac{k}{m}} x\right)$, with B any constant, is another solution. Observe next that

$$\frac{d^2}{dx^2}(f(x) + g(x)) + \frac{k}{m}(f(x) + g(x))$$

$$= \frac{d^2}{dx^2}f(x) + \frac{d^2}{dx^2}g(x) + \frac{k}{m}f(x) + \frac{k}{m}g(x) = 0.$$

Therefore, $A\left(\sin\sqrt{\frac{k}{m}}x\right) + B\left(\cos\sqrt{\frac{k}{m}}x\right)$ is a solution of $\frac{d^2y}{dx^2} + \frac{k}{m}y = 0$ for any constants A and B. In more advanced courses, it is shown that

$$A\left(\sin\sqrt{\frac{k}{m}}x\right) + B\left(\cos\sqrt{\frac{k}{m}}x\right)$$

is the general solution of $\frac{d^2y}{dx^2} + \frac{k}{m}y = 0$.

Example 13.34. Find a solution $y = h(x)$ of

$$\frac{d^2y}{dx^2} + \frac{k}{m}y = 0$$

that satisfies the boundary conditions $h(0) = 5$ and $h'(0) = 7$. Start with the general solution

$$h(x) = A\left(\sin\sqrt{\frac{k}{m}}x\right) + B\left(\cos\sqrt{\frac{k}{m}}x\right).$$

Taking $x = 0$, we get $5 = A\sin 0 + B\cos 0 = B$. Since

$$h'(x) = A\left(\cos\sqrt{\frac{k}{m}}x\right)\sqrt{\frac{k}{m}} - B\left(\sin\sqrt{\frac{k}{m}}x\right)\sqrt{\frac{k}{m}},$$

$7 = h'(0) = A\left(\cos\sqrt{\frac{k}{m}}0\right)\sqrt{\frac{k}{m}} = A\sqrt{\frac{k}{m}}$. So $A = 7\sqrt{\frac{m}{k}}$.
Therefore, the solution we are looking for is $h(x) =$
$7\sqrt{\frac{m}{k}}\left(\sin\sqrt{\frac{k}{m}}x\right) + 5\left(\cos\sqrt{\frac{k}{m}}x\right)$.

13.6 Postscript

The concepts and methods discussed in this chapter have a wide range of application in science and engineering. It is the goal of the upcoming chapter to give examples of such applications in a number of disciplines. Because they have the definite or indefinite integral at their core, these applications depend on the solution of an integral or a differential equation. In some of these applications the solutions can be achieved explicitly, but in others this is impossible (the general solution of the rocket equation is an example). The problem is that the solutions of many integrals and differential equations *cannot* be found explicitly! The reason for this is "existential." Let's call a function *elementary* if it can be put together by the operations of addition, subtraction, multiplication, taking powers and roots, and composition of functions (as many times and in any combination that one likes) by using the following basic functions: $f(x) =$ a constant, $f(x) = x$, any trigonometric function, any exponential function, and all the inverses of such functions (for example, logs and inverse trig functions). For instance, the function

$$f(x) = 5^{\sin(x^3 - x^{-5/2})} + \sqrt{\log_{57}\left(\cos\left(\frac{1}{7x^{-4} + \tan^{-1}x^5}\right)\right)}$$

is an elementary function. All the functions that we have met in this text (and many more) are elementary. Now let $y = f(x)$ be a "random" elementary function that is continuous on a closed interval $[a, b]$ and consider the integral $\int_a^b f(x)\,dx$. The fact of the matter is that the antiderivative of f—which is crucial in the solution of the integral—may *not* be an elementary function! A similar difficulty arises with differential equations. How then, can such integrals or differential equations be solved?

There are two general strategies. One of them relies on the fact—see Examples 13.3 and 13.4 of Section 13.1C for instance—that every definite integral $\int_a^b f(x)\,dx$ can be interpreted as the area of some region in the plane. It follows that definite integrals can be evaluated (or approximated) if such areas can be evaluated (or approximated). There are several geometric methods that do this accurately and effectively. The other strategy involves power series. See Section 6.3 for an introduction to this topic. It turns out that integrals and differential equations that cannot be solved with elementary functions can often be solved with functions that are given as power series.

The two strategies just discussed become especially powerful in combination with computers. Graphs of functions, derivatives, solutions to integrals and differential equations, and a large variety of other mathematical procedures—especially those that involve lots of numerical data—can be carried out with powerful software tools at the touch of a few buttons! *Maple*, *Mathematica*, *Macsyma*, and *Matlab* are examples of such software packages. Thick manuals say exactly which series of buttons a particular problem requires. We have already seen an example of this. Section 12.4B discussed how *Maple* can be used to fit curves to data points with the method of least-squares approximation. The speed with which such computer tools operate and the large quantities of data that they can keep track of have expanded the horizon of the applications of mathematics considerably. For example, the mathematical methods discussed in this book can be used to create computer models that simulate the planets orbiting the Sun, the motion of projectiles, the statics and dynamics of suspension bridges, the movement of the Earth's tectonic plates, the growth of populations, fermentation processes, the formation of the universe (!), and so on.

Exercises
13A. Riemann Sums

In each of Exercises 1 to 4 a function and an interval of definition are given. Also, a partition of the interval is specified, as well as a point in each of the subintervals that the partition determines. In each case, sketch the graph of f and the rectangles that this information provides and compute the Riemann sum.

1. $f(x) = 25 - x^2$, $[0,5]$, $\{0,2,3,4,5\}$, p_i = right endpoint.

2. $f(x) = 16 - x^2$, $[0,4]$, $\{0,1,2,3,4\}$, p_i = midpoint.

3. $f(x) = x^3 + 2$, $[-1,2]$, $\{-1,-0.5,0,0.5,1.0,1.5,2\}$, p_i = left endpoint.

4. $f(x) = \frac{1}{x+1}$, $[0,2]$, $\{0,0.5,1.0,1.5,2\}$, $p_1 = 0.25$, $p_2 = 1$, $p_3 = 1.25$, $p_4 = 2$.

13B. Pushing a Riemann Sum to the Limit

Exercises 5–7 are a "walk-through" of the limit procedure (1)–(3) of Section 13.1A in a concrete case. The notation will be as in Section 13.1A. In Exercises 5 and 6 use the sum of squares formula

$$\sum_{i=1}^{n} i^2 = 1^2 + 2^2 + \cdots + n^2 = \frac{n(n+1)(2n+1)}{6}.$$

5. Consider the function $f(x) = 16 - x^2$, $0 \le x \le 3$. Let P_7 be the partition obtained by dividing the interval $0 \le x \le 3$ into 7 equal subintervals. So the norm $\|P_7\| = \frac{3}{7}$, and $\Delta x_i = \frac{3}{7}$ for all i. Let p_i be the right endpoint of the ith subinterval.

    i.   Check that $p_1 = 1 \cdot \frac{3}{7}$, $p_2 = 2 \cdot \frac{3}{7}$, $p_3 = 3 \cdot \frac{3}{7}$, $p_4 = 4 \cdot \frac{3}{7}$, $p_5 = 5 \cdot \frac{3}{7}$, $p_6 = 6 \cdot \frac{3}{7}$, $p_7 = 7 \cdot \frac{3}{7}$.
    ii. Write out the sum $\sum_{i=1}^{7} f(p_i)\Delta x_i$.
    iii. Compute this sum by using the sum of squares formula.

6. Continue to consider the function $f(x) = 16 - x^2$, $0 \le x \le 3$. Let P_n be the partition obtained by dividing the interval $0 \le x \le 3$ into n equal subintervals. So the norm $\|P_n\| = \frac{3}{n}$ and $\Delta x_i = \frac{3}{n}$ for all i. Again let p_i be the right endpoint of the ith subinterval.

    i.   What are the numbers p_1, p_2, p_i, and p_n equal to?
    ii. Show that $f(p_i) = 16 - i^2 \left(\frac{3}{n}\right)^2$.
    iii. Use the sum of squares formula to conclude that

$$\sum_{i=1}^{n} f(p_i)\Delta x_i = \left[16n - \left(\frac{3}{n}\right)^2 \frac{n(n+1)(2n+1)}{6}\right]\left(\frac{3}{n}\right)$$

$$= 48 - \frac{9}{2}\frac{(n+1)(2n+1)}{n^2}$$

$$= 48 - \frac{9}{2}\left(2 + \frac{3}{n} + \frac{1}{n^2}\right).$$

7. Continue to consider the function $f(x) = 16 - x^2$, $0 \le x \le 3$, and the partition P_n.

    i.   What happens to the norm $\|P_n\|$ as n is pushed to infinity?
    ii. Why is

$$\lim_{\|P\| \to 0} \left(\sum_{i=1}^{n} f(p_i)\Delta x_i\right)$$

$$= \lim_{n \to \infty} \left[48 - \frac{9}{2} \left(2 + \frac{3}{n} + \frac{1}{n^2} \right) \right] ?$$

iii. Use (ii) to conclude that $\int_0^3 (16 - x^2) \, dx = 37$. Then confirm this by using the fundamental theorem of calculus.

13C. Applying the Fundamental Theorem of Calculus

Use the fundamental theorem of calculus to compute the following definite integrals. (You will need to make use of formulas and examples from Sections 8.6, 10.1, 10.2, 10.3, and 10.5.)

8. $\int_0^{\pi/2} \sin x \, dx$

9. $\int_0^{\pi/4} \tan x \, dx$

10. $\int_0^{\pi/4} \sec^2 x \, dx$

11. $\int_{\ln 2}^{\ln 5} e^x \, dx$

12. $\int_3^7 \frac{1}{x} \, dx$

13. $\int_1^4 \frac{\sqrt{25 - x^2}}{x} \, dx$

Let a and b be numbers with $a < b$ and consider the definite integral

$$\int_a^b f(x) \, dx$$

of some function $f(x)$. Suppose that the roles of a and b are reversed, and consider

$$\int_b^a f(x) \, dx$$

Since $b > a$, we have not as yet considered such a situation. Can this definite integral be given a meaningful interpretation? Let F be an antiderivative of f. If the fundamental theorem of calculus is to apply, then we need to have

$$\int_b^a f(x) \, dx = F(a) - F(b) = -(F(b) - F(a)) = - \int_a^b f(x) \, dx.$$

We therefore define

$$\int_b^a f(x) \, dx = - \int_a^b f(x) \, dx$$

in any situation where $a < b$.

Compute the following definite integrals.

14. $\int_{20}^6 (2x - 7) \, dx$

15. $\int_8^1 \left(x^{\frac{3}{2}} + 4x^{\frac{1}{2}} - \pi \right) dx$

16. $\int_\pi^0 (\cos x - 8x^2) \, dx$

In Exercises 17 to 19 find the derivative of each function.

17. $F(x) = \int_0^x t(1 + t^3)^9 \, dt$

18. $F(x) = \int_4^x \sin t (\cos t - 4t^2) \, dt$

19. $G(x) = \int_0^{3x^2} (\sin t + 2t^{-3}) \, dt$. [Hint: Use the chain rule.]

13D. The Substitution Method

In Exercises 20 to 27 solve the integral by making the indicated substitution.

20. $\int \sqrt{4x - 5} \, dx$; $u = 4x - 5$.

21. $\int 10x(1 - 5x^2)^{\frac{2}{3}} \, dx$; $u = 1 - 5x^2$.

22. $\int x \cos x^2 \, dx$; $u = x^2$.

23. $\int \sin^3 t \cos t \, dt$; $u = \sin t$

24. $\int (x - 1)\sqrt{x + 1} \, dx$; $u = x + 1$.

25. $\int x^2 \sqrt{x + 3} \, dx$; $u = x + 3$.

26. $\int \frac{x^2}{(x-2)^3} \, dx$; $u = x - 2$.

27. $\int \frac{\sec^2 \varphi}{\tan^2 \varphi + 1} \, d\varphi$; $u = \tan \varphi$.

In Exercises 28 to 35 solve the given integral by making an appropriate substitution.

28. $\int \frac{5x^6}{12x^7 + 19} \, dx$

29. $\int (1 + 4x)\sqrt{1 + 2x + 4x^2} \, dx$

30. $\int \sin^6 t \cos t \, dt$

31. $\int \sqrt{e^z + 1} \, e^{2z} \, dz$

32. $\int_1^8 x^{-2/3} \sqrt{1 + 4x^{1/3}} \, dx$

33. $\int_0^3 \frac{dx}{\sqrt[3]{(1 + 2x)^2}}$

34. $\int_0^1 t^9 \tan(t^{10}) \, dt$

35. $\int_3^1 \frac{(\ln x)^2}{x} \, dx$

13E. Computations of Areas

In Exercises 36 to 42 compute the area of the region between the graph of f and the x axis over the given interval.

36. $f(x) = x^4$, $-1 \le x \le 1$.

37. $f(x) = \frac{1}{x^2}$, $-2 \le x \le -1$.

38. $f(x) = x^{\frac{1}{2}}$, $1 \le x \le 4$.

39. $f(x) = x^{\frac{1}{3}}$, $-3 \le x \le 8$.

40. $f(x) = \sec x$, $0 \le x \le \frac{\pi}{4}$.

41. $f(x) = \sqrt{x + 1}$, $0 \le x \le 3$.

42. $f(x) = \frac{x}{(x^2 + 1)^2}$, $-1 \le x \le 2$.

In Exercises 43 to 45 make a rough sketch of the graph of the integrand over the interval in question. Point out what area (or difference between areas) the integral represents. Evaluate the integral.

43. $\int_1^3 \frac{1}{x}\, dx$

44. $\int_2^5 \frac{1}{x-1}\, dx$

45. $\int_{-4}^5 \frac{x}{x^2+4}\, dx$

13F. Integration by Parts

In Exercises 46 to 51 evaluate the integral by parts by taking the suggested u and dv.

46. $\int x \sin x\, dx$; $u = x$ and $dv = \sin x\, dx$.

47. $\int_0^{\pi/2} (x + x \cos x)\, dx$; refer to the suggestion for Exercise 46.

48. $\int x \ln x\, dx$; $u = \ln x$ and $dv = x\, dx$.

49. $\int x^2 \ln x\, dx$; try something similar to the suggestion in Exercise 48.

50. $\int_0^1 xe^{5x}\, dx$; $u = x$ and $dv = e^{5x}\, dx$.

51. $\int x^2 e^{4x}\, dx$; does $u = x^2$ and $dv = e^{5x}\, dx$ accomplish anything?

52. Express $\int e^x \sin x\, dx$ in terms of $\int e^x \cos x\, dx$.

53. Express $\int e^x \cos x\, dx$ in terms of $\int e^x \sin x\, dx$.

54. Combine the results of Exercises 52 and 53 to evaluate both $\int e^x \cos x\, dx$ and $\int e^x \sin x\, dx$.

In Exercises 55 and 56 make a substitution first and then use integration by parts to evaluate the integral.

55. $\int_0^1 \ln(x + 1)\, dx$

56. $\int \cos \sqrt{t}\, dt$

13G. Trigonometric Substitution

Use the method of trigonometric substitution and the hints that are supplied to solve the integrals in Exercises 57 to 60.

57. $\int_0^{1/2} \sqrt{1 - 4x^2}\, dx$; use $x = \frac{1}{2} \sin \theta$ and $\cos^2 \theta = \frac{1 + \cos 2\theta}{2}$.

58. $\int_{-1}^1 \frac{1}{(x^2 + 1)^{\frac{1}{2}}}\, dx$; take $x = \tan \theta$.

59. $\int \frac{x^2}{\sqrt{x^2 - 9}}\, dx$; take $x = 3 \sec \theta$ to reduce the integral to $\int \sec^3 \theta\, d\theta$.

60. With what substitution would you try to solve the indefinite integral $\int \frac{\sqrt{a^2 - x^2}}{x}\, dx$. This integral arises in the analysis of the tractrix. Refer to Section 10.4.

13H. Surface Area

In Exercises 61 to 63 find the surface area of the surface generated by revolving the graph of f on the given interval about the x axis.

61. $f(x) = x^3$, $0 \le x \le 5$.

62. $f(x) = e^{-x}$, $0 \le x \le 1$.

63. $f(x) = \sqrt{4 - x^2}$, $-\frac{1}{2} \le x \le \frac{3}{2}$.

13I. Up to the Gills

Figure 13.42 shows the structure of the respiratory tissue in the gills of a fish.[1] Blood rushes through the chamber and past the pillar cells (these are the stems in the figure). As it does so, the enzymes (these are microscopic organic agents that facilitate chemical reactions) that line the surface activate or inactivate hormones in the blood. Hormones are biochemical messengers that regulate virtually all biological function. Therefore, this hormonal activity drives the behavior of the fish. The rate at which the organs respond to the hormones is proportional to the rate at which the hormones are activated (or inactivated). This in turn is proportional to the number of enzymes on the surface of the inner membrane, and this number, finally, is proportional to the surface area of the membrane. This surface area is therefore important in investigations into the physiology of fish. But how is this area to be estimated? To produce such an estimate, let's start with an idealized version of Figure 13.42. This is sketched in Figure 13.43. It shows a frontal cross-section of an abstract (straightened out) model of the tissue. It includes two of the pillar cells, which are repeated in perspective below the cross-section. The dimensions that are indicated are based on observations with an electron microscope. The unit μm is the micrometer, where $1\ \mu m = 10^{-6}$ meters. So these passages are very tiny. (As a comparison, recall from Section 11.6 that an $E.\ coli$ bacterium has a size of $3\ \mu$m and that an average human hair is about 100μm thick.) The most challenging part of the computation of the surface area of the membrane is that of the curved surface of the top and base of the pillar cells. Place the pillar cell (the abstract model) on its side and observe that it is generated by the rotation shown in Figure 13.44. Concentrate on one of the circular parts of the rotated arc and observe that the curved part at either the top or the base of a pillar cell is generated by the rotation of a quarter circle of radius $2\ \mu$m. Some thought about what is going on (and a look at the dimensions given in Figure 13.43) shows that the curved part of the top (or base) of the pillar is the surface of revolution obtained by rotating one quarter of the circle with center $(0, 5.5)$ and radius 2 about the x-axis. See Figure 13.45. We are now in a position to compute the area of this surface.

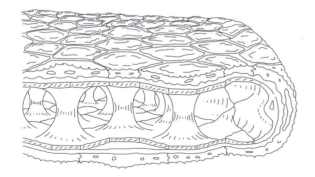

Figure 13.42

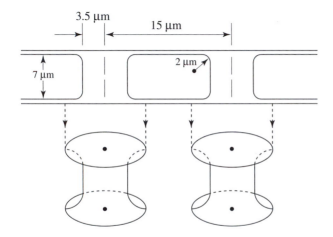

Figure 13.43

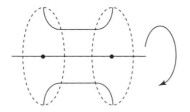

Figure 13.44

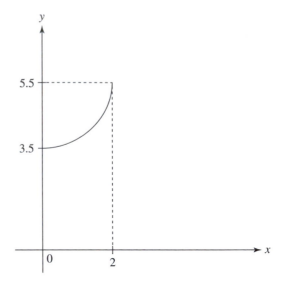

Figure 13.45

66. Solve the integral and show that the area of this surface of revolution is 83.43 μm^2.

13J. Points in the Polar Plane

67. Find the Cartesian coordinates of the following points, which are in polar coordinates.

    i. $\left(3, \frac{\pi}{4}\right)$
    ii. $\left(-2, -\frac{\pi}{6}\right)$
    iii. $\left(3, \frac{7\pi}{3}\right)$
    iv. $(5, 0)$
    v. $\left(-2, \frac{\pi}{2}\right)$
    vi. $\left(-2, \frac{3\pi}{2}\right)$
    vii. $\left(4, \frac{3\pi}{2}\right)$
    viii. $\left(0, \frac{6\pi}{7}\right)$
    ix. $\left(-1, -\frac{23\pi}{3}\right)$
    x. $\left(-1, -\frac{23\pi}{3}\right)$
    xi. $\left(1, \frac{3\pi}{2}\right)$
    xii. $\left(3, -\frac{5\pi}{6}\right)$

68. Find all sets of polar coordinates for each of the following points, which are in Cartesian coordinates.

    i. $(3, 3)$
    ii. $(4, -4)$
    iii. $(0, 5)$
    iv. $(-4, 0)$
    v. $(3, 3\sqrt{3})$

64. Why is the function $f(x) = 5.5 - \sqrt{4 - x^2}$ (and not $g(x) = 5.5 + \sqrt{4 - x^2}$) with $0 \le x \le 2$ the function whose graph is the quarter circle of Figure 13.45?

65. Express the area of the surface obtained by rotating this quarter circle one revolution about the x-axis as a definite integral.

    vi. $\left(-\frac{1}{3},\frac{\sqrt{3}}{3}\right)$
   vii. $(-3,\sqrt{3})$
 viii. $(-2\sqrt{3},2)$
    ix. $(0,0)$
     x. $(-5\sqrt{3},-5)$

13K. Equations in Polar Coordinates

In Exercises 69 to 71 rewrite the equation in polar coordinates. Express the answer in the form $r = f(\theta)$ whenever possible.

69. $2x + 3y = 4$

70. $9x^2 + y^2 = 4y$

71. $x^2 + y^2 = x(x^2 - 3y^2)$

In Exercises 72 to 75 rewrite the polar equation as an equation in Cartesian coordinates.

72. $r = 5$

73. $r = 3\cos\theta$

74. $\tan\theta = 6$

75. $r = 2\sin\theta\tan\theta$

In Exercises 76 to 83 sketch the graph of the equation. For a complete understanding of the graph, it may be necessary to convert the equation to rectangular coordinates.

76. $r = 6$

77. $\theta = -\frac{8\pi}{6}$

78. $r = \sin\theta$

79. $r(\sin\theta + \cos\theta) = 1$

80. $r = 2\cos 2\theta$. (The *four-leaved rose.*)

81. $r = -4\sin 3\theta$. (The *three-leaved rose.*)

82. $r^2 = 9\sin 2\theta$. (The *lemniscate of Bernoulli.* The word *lemniscate* comes from the Greek *lemniscos* meaning *ribbon.*)

83. $r = 2 - \cos\theta$. (The *limaçon of Pascal. Limaçon* is French for *snail.*)

13L. Parabolas and Hyperbolas in Polar Coordinates

Return to the equation (*) $4(1 - \varepsilon^2)x^2 + 4\varepsilon Lx + 4y^2 = L^2$ of Section 13.4. The case of the ellipse (this is where $0 \le \varepsilon < 1$) was already analyzed. In particular, it was shown that (**2**)–(**4**) hold.

84. Consider the case $\varepsilon = 1$. Check that equation (*) can be rewritten as $x = -\frac{1}{L}y^2 + \frac{L}{4}$. Make a rough sketch of the graph of this parabola. Draw in the focus F

and the directrix D in a general way. (You may need to review some of the facts from Section 3.1.) You may assume that F lies on the x-axis and that D is perpendicular to the x-axis. (This can be verified by an analysis similar to that in Section 4.3.) It follows that $F = (a,0)$ for some constant a, and that D has an equation of the form $x = d$ for some constant d. Note that the parabola crosses the x-axis at the point $V = (\frac{L}{4},0)$. Observe that $a < \frac{L}{4} < d$. Use the fact that V is on the parabola to show that $d+a = \frac{L}{2}$. Let (a,b) be the point on the parabola above the focus and show that $b = d-a = \frac{L}{2} - 2a$. Plug (a,b) into the equation of the parabola and conclude that a must be zero and that L is the latus rectum. Points (**2**) and (**4**) have been verified. Why does (**3**) hold as well?

85. Assume $\varepsilon > 1$. Note that $\varepsilon^2 - 1 > 0$. Put $a = \frac{L}{2(\varepsilon^2-1)}$ and $b = a\sqrt{\varepsilon^2 - 1}$. Both a and b are positive constants. Refer back to Section 13.4 and in particular to the rewritten version (**) of equation (*). After it is multiplied through by $\frac{1}{a^2} = \frac{4(\varepsilon^2-1)^2}{L^2}$, equation (**) becomes

$$\frac{(x-\varepsilon a)^2}{a^2} - \frac{y^2}{b^2} = 1.$$

An analysis similar to that of Section 4.6 (after a review of the basics from Section 3.1) can be used to show that the graph of this equation is a hyperbola of the form sketched in Figure 13.46. By arguing as in

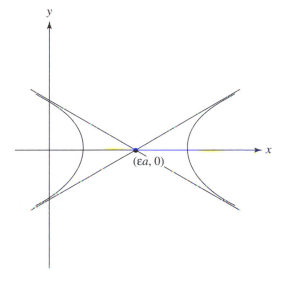

Figure 13.46

the case of the ellipse, points (**2**)–(**4**) can be verified as well.

13M.   Areas of Regions Given in Polar Coordinates

In Exercises 86 to 89 find the area of the region bounded by the graphs of the given equations.

86. $r = 4$

87. $r = 3 \sin \theta$

88. $r = 3 \sin \theta$ with $0 \leq \theta \leq \frac{\pi}{3}$, and the line $\theta = \frac{\pi}{3}$.

89. $r = 2 + 2 \cos \theta$ (the *cardioid*).

13N.   Differential Equations

In Exercises 90 to 93 verify that the function y satisfies the given differential equation.

90. $\frac{dy}{dx} = e^{3x}; y = \frac{1}{3} e^{3x}$.

91. $2\frac{dy}{dx} - y^2 = 1; y = \tan x + \sec x$ for $0 < x < \frac{\pi}{2}$.

92. $\frac{d^2 y}{dx^2} + 4y = 0; y = \sin 2x - \cos 2x$.

93. $\frac{d^2 y}{dx^2} + 4\frac{dy}{dx} + 4y = 0; y = xe^{-2x}$.

13O.   The Method of Separation of Variables

In Exercises 94 to 99 find the general solution of the differential equation. Then find the particular solution that satisfies the given initial condition.

94. $\frac{dy}{dx} = \frac{x}{y}; y(0) = 4$.

95. $\frac{dy}{dx} = xy; y(1) = 3$.

96. $\frac{dy}{dx} = \frac{y}{x}; y(2) = 0$.

97. $(y^2 - 3)\frac{dy}{dt} = 1; y(0) = 3$. [Hint: use the quadratic formula to solve for y in terms of t.]

98. $\frac{dy}{dx} = \frac{\sin x - \cos x}{1+y}; y(\pi) = 0$.

99. $(\ln y)^2 \frac{dy}{dx} = x^2 y; y(2) = 1$. [Hint: after separating variables, integrate by substitution.]

13P.   Chemical Reactions

Suppose that two chemical substances in a solution react together to form a compound. If the reaction occurs by means of the collision and interaction of the molecules of the two substances, then the rate of formation of the compound should be proportional to the number of collisions per unit time. This in turn should be proportional to the product of the amounts of the substances yet untransformed. Let's be more precise. Suppose that there are a molecules of the one chemical and b molecules of the other when the reaction starts at time $t = 0$. Suppose that $y(t)$ is the number of molecules of the compound at any time $t \geq 0$. One molecule of the compound is produced when one molecule from the one chemical combines with one molecule from the other. It follows that at time $t \geq 0$, there are $a - y(t)$ molecules of the one substance and $b - y(t)$ of the other as yet untransformed. In view of the earlier assertion about the rate of formation of the compound, we see that the function $y(t)$ satisfies the differential equation

$$\frac{dy}{dt} = k(a - y)(b - y)$$

where k is a constant.

100.   Solve this differential equation. [Hint: Consider the two cases $a = b$ and $a \neq b$ separately.]

A chemical reaction that proceeds as just described is called a *second-order reaction*.

13Q.   First-Order Linear Differential Equations

Consider any differential equation of the form

(1) $\qquad a(t)y' + b(t)y = c(t)$

where $a(t)$, $b(t)$, and $c(t)$ are continuous functions of t. Dividing through by $a(t)$ and letting $p(t) = \frac{b(t)}{a(t)}$ and $q(t) = \frac{c(t)}{a(t)}$, we can rewrite the equation as

(2) $\qquad y' + p(t)y = q(t)$.

The solution of this equation can be reduced to an integration as follows: Let $P(t)$ be an antiderivative of $p(t)$ and let $y(t)$ be any (differentiable) function. By applying the product rule and the chain rule,

$$\frac{d}{dt}\left(y(t)e^{P(t)} \right) = y'(t)e^{P(t)} + y(t)e^{P(t)}P'(t)$$

$$= y'(t)e^{P(t)} + y(t)e^{P(t)}p(t)$$

$$= e^{P(t)}[y'(t) + y(t)p(t)].$$

Observe that the term in the brackets is the same as that on the left side of equation (2). This suggests that it might be profitable to multiply equation (2) by $e^{P(t)}$. Doing this, we get

$$e^{P(t)}[y'(t) + y(t)p(t)] = q(t)e^{P(t)}.$$

Combining this with the previous equation, we get

$$\frac{d}{dt}\left(y(t)e^{P(t)} \right) = q(t)e^{P(t)}.$$

Taking antiderivatives of both sides of this last equation and combining the constants of integration on the right, we get $y(t)e^{P(t)} = \int q(t)e^{P(t)} \, dt + C$. Therefore,

$$(3) \qquad y(t) = e^{-P(t)} \left[\int q(t)e^{P(t)} \, dt + C \right].$$

This formula provides solutions for equation (2), provided that the integral $\int q(t)e^{P(t)} \, dt$ can be solved.

101. Consider the differential equation $y' + y = t$. Does it fit into the scheme just discussed? What are $q(t)$ and $P(t)$ equal to? Find the general solution of the equation. Find the particular solution $y = f(t)$ that satisfies $f(0) = 2$.

102. Find the particular solution $y(t)$ of the differential equation $y' + 2ty = 2t$ that satisfies $y(0) = 0$.

103. Find the particular solution $y(t)$ of the equation $y' + \frac{1}{t}y = \cos t$ that satisfies $y(\pi) = 0$.

104. Solve the differential equation $x^2 \frac{dy}{dx} - x + 1 = 0$ with the boundary condition $y(1) = 3$.

13R. Application to Radioactive Decay

As was pointed out in Section 11.1A, most radioactive decay occurs in sequences. Indeed, only twelve of the radioactive substances that occur in nature (this includes rubidium-87, potassium-40, and carbon-14) are not parts of radioactive series. For example, uranium-238 decays in a sequence to produce fourteen radioactive elements, one decaying to the next, until the process ends with the stable isotope lead-206. Consider a sample of pure uranium-238 at time $t = 0$. Eventually, this sample will contain isotopes of sixteen different elements: uranium-238, the fourteen radioactive decay products, and lead-206. Given all the disintegration constants, it is possible to compute the percentages of each element in the sample at any time $t \geq 0$. The discussion below considers the mathematics of the typical step in the sequence.

Suppose that a radioactive parent element decays to a radioactive daughter. Let λ_1 be the disintegration constant of the parent and let λ_2 be that of the daughter. Suppose that a sample contains x_0 atoms of the parent element at time $t = 0$ and $x(t)$ atoms at any time $t \geq 0$. By Section 11.1B, $x'(t) = -\lambda_1 x(t)$ and $x(t) = x_0 e^{-\lambda_1 t}$ for any $t \geq 0$. By combining equations, $x'(t) = -\lambda_1 x_0 e^{-\lambda_1 t}$ for any $t \geq 0$. Now let $y(t)$ be the number of daughter atoms in the sample at any time t. The rate at which $y(t)$ changes is influenced by two factors:

i. On the one hand, since its decay constant is λ_2, it decreases at the rate $\lambda_2 y(t)$.

ii. On the other hand, it increases at the same rate that $x(t)$ decreases. So this rate of increase is $\lambda_1 x_0 e^{-\lambda_1 t}$.

It follows that

$$y'(t) = x_0 \lambda_1 e^{-\lambda_1 t} - \lambda_2 y(t).$$

This can be rewritten as $y' + p(t)y = q(t)$, where $p(t) = \lambda_2$ and $q(t) = x_0 \lambda_1 e^{-\lambda_1 t}$. Since $P(t) = \lambda_2 t$ is an antiderivative of $p(t) = \lambda_2$, we get by equation (3) of Exercise 13Q that

$$y(t) = e^{-\lambda_2 t} \left[\int x_0 \lambda_1 e^{-\lambda_1 t} e^{\lambda_2 t} \, dt + C \right]$$

$$= e^{-\lambda_2 t} \left[\int x_0 \lambda_1 e^{(\lambda_2 - \lambda_1)t} \, dt + C \right]$$

$$= e^{-\lambda_2 t} \left[\frac{x_0 \lambda_1}{(\lambda_2 - \lambda_1)} e^{(\lambda_2 - \lambda_1)t} + C \right].$$

Put $y(0) = y_0$. Then $y_0 = 1 \cdot \left[\frac{x_0 \lambda_1}{(\lambda_2 - \lambda_1)} \cdot 1 + C \right]$. So $C = y_0 - \frac{x_0 \lambda_1}{(\lambda_2 - \lambda_1)}$. By a substitution,

$$y(t) = \frac{x_0 \lambda_1}{(\lambda_2 - \lambda_1)} e^{-\lambda_1 t} - \frac{x_0 \lambda_1}{(\lambda_2 - \lambda_1)} e^{-\lambda_2 t} + y_0 e^{-\lambda_2 t}$$

$$= \frac{x_0 \lambda_1}{(\lambda_2 - \lambda_1)} (e^{-\lambda_1 t} - e^{-\lambda_2 t}) + y_0 e^{-\lambda_2 t}.$$

The determination of $y(t)$ has been achieved.

For the exercises below, recall that 99.28% of naturally occurring uranium consists of the isotope $^{238}_{92}\text{U}$ and that $^{235}_{92}\text{U}$ with 0.715% comprises most of the rest. Also, take the half-lives of $^{238}_{92}\text{U}$ and $^{235}_{92}\text{U}$ to be 4.5×10^9 and 7.1×10^8 years respectively.

105. Consider one gram of naturally occurring uranium at time $t = 0$. Given that the radioactive daughter $^{234}_{90}\text{Th}$ of $^{238}_{92}\text{U}$ has a half-life of 24.1 days, determine the amount of this thorium isotope in the sample at any time t. How much $^{234}_{90}\text{Th}$ is there after 1 billion years?

106. Given that the radioactive daughter $^{231}_{90}\text{Th}$ of $^{235}_{92}\text{U}$ has a half-life of 25.6 hours, determine the amount of this thorium isotope in the sample of Exercise 105 at any time t. How much $^{234}_{90}\text{Th}$ is there after 500 million years?

Notes

[1]Taken from K. Olson, Hormone metabolism by the fish gill, *Comparative Biochemistry and Physiology*, Vol. 119A, 1998, 55–65.

14

Integral Calculus and the Action of Forces

This entire chapter is devoted to the mathematics of force and its consequences. It makes use of all the concepts and methods of integral calculus that were developed in Chapter 13. After everything in this chapter is said and done, there is one point that will need no further elaboration: Integral calculus lies at the very core of our understanding of the physical universe. We begin with the mathematics of work and energy and its application to an analysis of the propulsion of a bullet in the barrel of a gun. This is followed by a development of the calculus of momentum. It is illustrated and applied in a study of the thrust of a rocket engine and the dynamics of a rocket in flight. Another section makes the point that integral calculus is essential to an understanding of the most basic properties of gravitational force. The chapter then turns to the laws of Kepler and the inverse square law. We will see that calculus in polar coordinates provides a much better framework for the analysis of these fundamental concerns than the original methods of Newton (as discussed in Chapter 7). The final section of the chapter applies integral calculus in an investigation into the structure and age of the universe.

14.1 **Work and Energy**

Before we start, we need to review some basic matters. In American units, distance is expressed in feet, force in pounds, and mass in slugs. In the M.K.S. system, distance is in meters, mass in kilograms, and force in newtons. (This was already discussed in Exercises 7E.) The American units and M.K.S. units are related as follows: 1 meter = 3.28 feet, 1 newton = 0.22 pounds, and 1 slug = 14.59 kilograms. The weight of an object is understood to be the magnitude of the gravitational force that acts on it when the object is on the surface of the Earth. By Newton's second law, weight = $m \times g$, where m is the mass of the object and g is the gravitational constant. If we combine Newton's second law with his law of universal gravitation (review Sections 7.5 and 7.6

if needed), we see that the weight mg of an object of mass m is also equal to $\frac{GmM}{R^2}$, where M is the mass of Earth, G is the universal gravitational constant, and R is the radius of the Earth. Canceling m tells us that

$$g = \frac{GM}{R^2}.$$

Because the Earth has the shape of a sphere that is flattened at the poles, R must more accurately be taken as the distance from the center of the Earth to its surface. In particular, R is slightly larger at the equator than at the poles. So g is slightly smaller at the equator than at the poles. At a latitude of 40°— the cities of San Francisco, New York, Madrid, Rome, Athens, Ankara, Beijing, and Tokyo all have approximately this latitude—g is close to 32.17 feet/sec^2 or 9.81 meters/sec^2. This is the value that we will use throughout the chapter. It corresponds to the distance $R = 3960$ miles, or 6.373×10^6 meters.

The equation weight = mass × g converts units of mass to units of force, and the other way around. For example, a mass of 1 slug weighs

$$\text{mass} \times g = 1 \text{ slug} \times 32.17 \text{ feet/sec}^2 = 32.17 \text{ pounds,}$$

and a mass of 1 kilogram weighs

$$\text{mass} \times g = 1 \text{ kilogram} \times 9.81 \text{ meters/sec}^2$$

$$= 9.81 \text{ kilogram-(meters/sec}^2)$$

$$= 9.81 \text{ newtons.}$$

Two of the consequences of the action of a force are closely related: work and energy. This section introduces these concepts and develops the mathematical connection between them.

A. Variable Force and Work. Suppose that an object moves along a straight line and that a constant force acts on the object along the line of its motion. If the object moves a distance d and the magnitude of the force is f, then the work that the force does on the object is defined to be the product $f \times d$.

We now formulate this more precisely. Suppose the object moves along a coordinate axis from a to b as

Figure 14.1

indicated in Figure 14.1. Let a constant force act on the object along the axis and let c be the magnitude of the force. So $c \geq 0$. If the force acts in the positive direction, designate it by $f = c$. If the force is in the negative direction, designate it by $f = -c$. In the situation of the figure the force f is positive if it acts to the right and negative if it acts to the left. If f is positive, the object will speed up as it moves from a to b, and if f is negative, the object will slow down as it moves from a to b. The magnitude of the force is $|f|$ in either case. The *work* W done on the object is defined to be the product

$$W = f(b - a).$$

Notice that W is positive if f is positive and negative if f is negative.

Because work = force × distance, work is given in foot-pounds in the American system and in newton-meters in M.K.S. The newton-meter is called *joule*. The connection between these units of work is

1 joule = 1 newton-meter = $(0.22)(3.28)$ foot-pounds

= 0.72 foot-pounds.

Example 14.1. Suppose that you have pushed a shopping cart down an aisle of a market with a force of 1.5 pounds for a distance of 20 feet. The work done on the cart was $1.5 \times 20 = 30$ foot-pounds. After having loaded some heavy items into the cart you then push the cart for another 25 feet with a force of 2.0 pounds. How much work was done on the cart this time?

Example 14.2. An Olympic weightlifter is going for gold. He accelerates a 535 pound barbell upward from the ground with a force of 575 pounds and sustains this force for the first $\frac{1}{2}$ foot of the lift. The work done on the barbell is $575 \times 0.5 = 287.5$ foot-pounds.

Example 14.3. An object of mass m kilograms is dropped from a height of h meters and falls to the ground. Observe that the weight of the object is $f = mg$ newtons. Because this is equal to the gravitational force acting on the object, the work done by gravity on the object is $fh = mgh$ newton-meters. This equality is also valid when m is given in slugs, $f = mg$ in pounds, and h in feet.

Forces generally vary in magnitude and are often constant only for a part of the range over which they act. Consider the situation of the shopping cart, for instance. You will stop the cart often to load something. When you then begin to move the cart forward again, you must not only overcome friction, but you must also accelerate the cart in order to give it the velocity you wish to have. Once it has this velocity, you need to push just hard enough to balance the frictional force on the cart. This will keep it moving at the desired velocity. In order to slow the cart down in preparation for the next stop, you will have to decrease the force on the cart. In this way, the magnitude of the force with which you push on the cart will vary throughout your visit to the store.

Turning to the weightlifter, observe that he will not apply the force of 575 pounds for long. After all, he wishes to bring the barbell to a stop chest high and keep it there (at least for a while). To keep it there, he has to balance the gravitational force of 535 pounds that acts on the barbell. So the force he applies to the barbell is 575 pounds at the beginning of the lift and 535 pounds at the end. It follows that the force that is applied to the barbell varies.

What about the force of gravity on a falling object? Because it is equal to the weight of the object, this force does seem to be constant. Or is it? Consider an object in free fall as in Figure 14.2(a). Now "step back" and consider what is happening from a greater distance. In Figure 14.2(b) the object is shown at a height h above the surface of the Earth, with C the center of the Earth and R the distance from C to the Earth's surface. According to Newton's law of universal gravitation, the gravitational force on the object is given by

$$f = G\frac{mM}{r^2}$$

where M is the mass of the Earth, G is the universal gravitational constant, and r is the distance of the object to the center of the Earth. A look at Figure 14.2(b) shows that r varies from a maximum of $h + R$ to a minimum of R. So the gravitational force on the object varies from a minimum of $\frac{GmM}{(h+R)^2}$ at the beginning of the fall to a maximum of $\frac{GmM}{R^2}$ at the end (the moment of impact). Therefore, this force varies. Of course, if h is very small relative to R, then $R + h \approx R$. In this case, the assumption that the gravitational force is equal to the weight $mg = \frac{GmM}{R^2}$ of the object does approximate the actual situation well.

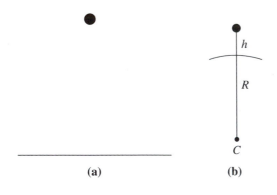

(a)   (b)

Figure 14.2

The question that will now concern us is this. If an object moves along a straight line and a variable force acts on it along the line of its motion, then what is the work done on the object? The answer consists in a "classical" application of the strategy of the definite integral as it was developed in Section 13.1.

Suppose that the object moves from a to b on a coordinatized line. Let the force be given by a function $f(x)$ for $a \le x \le b$. See Figure 14.3. If $f(x) > 0$, this means that the force acts from left to the right at x with magnitude $f(x)$; if $f(x) < 0$, this means that it acts from right to left with a magnitude of $-f(x)$; if $f(x) = 0$, then the force has zero magnitude at x. What can we say about the work done on the object? The discussion of the falling object suggests a strategy: Break up the distance from a to b through which the object moves into very small intervals; assume that the force is constant over each one of them; compute the work done on the object over each of these small intervals; and finally, add up all these little "bits" of work. The strategy is simple in principle, but it is complicated to "write down" with precision.

To break up the distance through which the object moves, take a partition P of $[a, b]$ into n subintervals. Let ΔW_1 be the work done on the object over the first subinterval, ΔW_2 the work done over the second,

$$\overset{\Delta W_1 \quad \Delta W_2 \quad \Delta W_3 \qquad\qquad \Delta W_i \qquad\qquad \Delta W_n}{\underset{a \hphantom{xxxxxxxxxxxxxxxxxxxxxxxxxxxxxxxxxx} b}{\rule{0pt}{0pt}}}$$

ΔW_3 that over the third, and so on; finally, let ΔW_n be the work done over the last subinterval. So the total work W done on the object as it moves from a to b is

$$W = \Delta W_1 + \Delta W_2 + \Delta W_3 + \cdots + \Delta W_n$$

$$= \sum_{i=1}^{n} \Delta W_i.$$

Let Δx_1 be the length of the first subinterval, Δx_2 the length of the second, Δx_3 that of the third, and so on, and Δx_n that of the last. Pick a point p_1 in the first subinterval, p_2 in the second, ..., and p_n in the last. Assume now that the norm $\|P\|$ of the partition is very small.

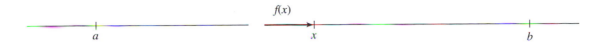

$$f(x)$$

a x b

Figure 14.3

p_1 p_2 p_3 $\cdots$ p_i $\cdots$ p_n

a Δx_1 Δx_2 Δx_3 Δx_i Δx_n b

This means that all the distances Δx_i are very small. As in the example of the object in free fall, we can therefore (as an approximation) take the force to be a constant over each of the subintervals. So the force is $f(p_1)$ over the first interval, $f(p_2)$ over the second, $f(p_3)$ over the third, etc. It follows that

$$\Delta W_1 \approx f(p_1)\Delta x_1, \ \Delta W_2 \approx f(p_2)\Delta x_2, \ldots$$

$$\Delta W_i \approx f(p_i)\Delta x_i, \ldots, \ \Delta W_n \approx f(p_n)\Delta x_n.$$

Therefore,

$$W \approx f(p_1)\Delta x_i + f(p_2)\Delta x_2 + \cdots + f(p_n)\Delta x_n$$

$$= \sum_{i=1}^{n} f(p_i)\Delta x_i.$$

The smaller a subinterval is, the less variation there will be in the force as it acts through it, so the more accurate it is to say that the force is approximated by a constant. It follows that if the partitions are taken to be finer and finer, the approximation for W becomes tighter and tighter. Therefore, by the discussion in Section 13.1,

$$W = \lim_{\|P\| \to 0} \sum_{i=1}^{n} f(p_i)\Delta x_i.$$

This last step requires the assumption that the function $y = f(x)$ is continuous over the interval $[a, b]$. So if the force function f is continuous, then by the definition of the definite integral in Section 13.1C, the work W done on the object as it moves from a to b is equal to

$$W = \int_a^b f(x)\, dx$$

What happens if we apply the formula $W = \int_a^b f(x)\, dx$ in a situation where the force $f(x)$ is constant, say, equal to c? Because

$$W = \int_a^b f(x)\, dx = cx \Big|_a^b = c(b - a)$$

we get the expected result: The work W done on the object is the product of the force c times the distance $b - a$ that the object has moved.

Let's return to the shopping cart of Example 14.1 to illustrate the equation that was just derived. Recall that you pushed the cart with a force of 1.5 pounds from the beginning of the aisle for a distance of 20 feet. Assume, for the sake of argument, that you then wanted to stop at a certain location in the aisle. So you decreased the force gradually, and after 2 more feet the force with which you pushed was zero. You were then 22 feet down the aisle. Since the cart had a certain velocity at this point, it continued to roll. After 1 additional foot it rolled to a stop, 23 feet down the aisle (at the location you had in mind). You then loaded on some items. To get the cart moving again you applied a force of 2.5 pounds. As it picked up speed, you decreased this force gradually, so that it was down to 2 pounds 25 feet down the aisle. The cart now had the speed you wanted. You continued pushing with a force of 2 pounds. This force balanced the frictional forces inhibiting the motion of the cart and kept it moving down the aisle with a constant speed for another 25 feet. Figure 14.4 is a sketch of the graph of the force function $f(x)$ over the interval $[0, 50]$. Because the force is always in the direction of the motion (or zero), $f(x) \geq 0$ for all x in this interval. Over the intervals $20 \leq x \leq 22$ and $23 \leq x \leq 25$, the graph is understood to be given by line segments. What is the work W done on the cart?

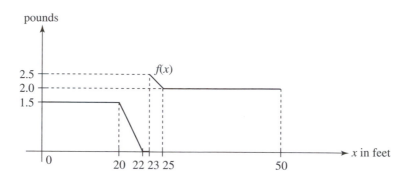

Figure 14.4

By the equation derived earlier,

$$W = \int_0^{50} f(x)\, dx.$$

Because $f(x) \geq 0$, this integral is also equal to the area under the graph of f. By a look at Figure 14.4, this area is equal to: $(1.5)(20) = 30$ over $0 \leq x \leq 20$; a triangle of area $\frac{1}{2}(1.5)(2) = 1.5$ over $20 \leq x \leq 22$; zero over $22 \leq x \leq 23$; the area $\frac{1}{2}(0.5)(2) + (2)(2) = 4.5$ over $23 \leq x \leq 25$; and $(2.0)(25) = 50$ over $25 \leq x \leq 50$. Therefore,

$$\int_0^{50} f(x)\, dx = 30 + 1.5 + 0 + 4.5 + 50 = 86 \text{ foot-pounds.}$$

For another illustration, let's return to the weightlifter and his barbell. We will suppose that the circular weights at the two ends of the bar have a diameter of $1\frac{1}{2}$ feet. (A diameter of 18 inches is standard for such weights.) So at the start of the lift the bar is at rest at a height $\frac{3}{4}$ of a foot above the floor. The 575 pounds of force that the weightlifter applies for the first $\frac{1}{2}$ foot of the lift exceeds the 535 pound weight of the barbell by $575 - 535 = 40$ pounds. This net upward force of 40 pounds accelerates the barbell upward. At some point during the lift the barbell must begin to decelerate, or else it would not come to a stop at the weightlifter's chest. Because force = mass × acceleration and a deceleration is a negative acceleration, this can only happen if the net force on the barbell is negative, in other words, if it is down-

ward. Notice, therefore, that during this phase of the lift, the weightlifter applies a force that is less than the 535 pounds of the barbell. Finally, at the end of the lift, when he holds the barbell chest-high at, say, 5 feet above the ground, he must apply a force equal to the weight of 535 pounds of the barbell. At this point the barbell is at rest, and the net force on it is zero.

Let the net force on the barbell be given by the function $n(x)$, with $\frac{3}{4} \leq x \leq 5$. So $n(x) = 40$ at the start of the lift, after some point $n(x)$ becomes negative, and finally, $n(5) = 0$. Figure 14.5 provides a sketch of the general shape of the graph of $n(x)$. The letters A and B refer to the areas of the two regions. The formula already developed tells us that

$$W = \int_{3/4}^{5} n(x)\, dx.$$

Combining this with the fact that any definite integral is equal to "areas above" minus "areas below" tells us that the work done on the barbell by the net force is $W = A - B$.

Let's turn to the falling object of Example 14.3 next. Place a coordinatized axis as shown in Figure 14.6. Consider the falling object at a typical position x between 0 and h. Because its distance above the ground is $h - x$, its distance from the center of the Earth is $R + (h - x)$. The force of gravity on the object at this point is $G\frac{mM}{(R+h-x)^2}$. It follows that the work done

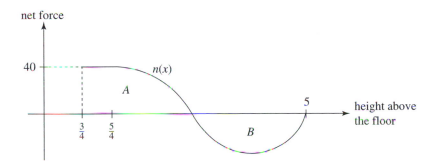

Figure 14.5

on the object by gravity is

$$W = \int_0^h G \frac{mM}{(R+h-x)^2}\, dx.$$

This integral can be solved by a substitution. Let $u = R + h - x$. So $dx = -du$ and hence

$$\int G \frac{mM}{(R+h-x)^2}\, dx = \int -GmMu^{-2}\, du$$

$$= GmMu^{-1} + C$$

$$= \frac{GmM}{R+h-x} + C.$$

Therefore,

$$W = \int_0^h G \frac{mM}{(R+h-x)^2}\, dx = \frac{GmM}{(R+h-x)} \Big|_0^h$$

$$= \frac{GmM}{R} - \frac{GmM}{(R+h)} = \frac{GmM(h+R) - GmMR}{R(h+R)}$$

$$= \frac{GmMh}{R(h+R)} = \frac{GM}{R^2} \cdot mh \cdot \frac{R}{h+R} = mgh \cdot \frac{R}{h+R}$$

$$= mgh \left(\frac{1}{1 + \frac{h}{R}} \right).$$

In Example 14.3 the gravitational force was assumed to be equal to the weight of the object throughout the fall. As a consequence, the work done by gravity on the falling object was computed to be mgh. The more accurate computation just undertaken shows that the work W done on the object is obtained

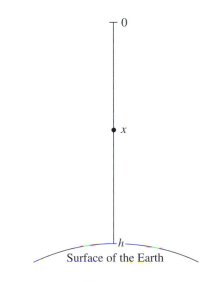

Figure 14.6

by multiplying mgh by the factor $\frac{1}{1+\frac{h}{R}}$. If h is small relative to R, that is, if the fall of the object occurs near the Earth's surface, then $\frac{h}{R} \approx 0$ and therefore $\frac{1}{1+\frac{h}{R}} \approx 1$. In such a case $W \approx mgh$.

Suppose for example that the object is dropped from the Leaning Tower of Pisa. The Leaning Tower has a height of $h = 177$ feet. Taking the distance R from the surface to the center of the Earth to be equal to 3960 miles and hence 3960×5280 feet, we get that $\frac{1}{1+\frac{h}{R}}$ is equal to $\frac{1}{1+\frac{177}{3960\times 5280}} = 0.99999$. It follows that the difference between mgh and the more accurate value $W = 0.99999mgh$ is negligible.

Let's illustrate the difference in another situation. Suppose that a truck with a mass of 7000 kilograms (this corresponds to about 15 tons) becomes dislodged from the hold of a cargo plane that flies at a height of 10,000 meters (about 32,800 feet) and falls to the Earth. Example 14.3 provides the estimate

$$mgh = (9.81)(7000)(10,000)$$

$$= 686.70 \times 10^6 \text{ newton-meters}$$

for the work done by gravity on the truck during its fall. Because $R = 6.373 \times 10^6$ meters,

$$\frac{1}{1 + \frac{h}{R}} = \frac{1}{1 + \frac{1.0 \times 10^4}{6.373 \times 10^6}} = \frac{1}{1.0016} = 0.9984.$$

So the more accurate value for the work done by gravity is

$$(686.70 \times 10^6)(0.9984) = 685.60 \times 10^6 \text{ newton-meters}.$$

This time the difference is discernible.

B. Kinetic and Potential Energy. The *kinetic energy* of a moving object with mass m and velocity v is defined to be $\frac{1}{2}mv^2$.

Example 14.4. Suppose an automobile of mass 900 kilograms is moving at a constant speed of 25 meters per second (about 55 miles per hour). Its kinetic energy is $\frac{1}{2}mv^2 = \frac{1}{2}(900)(25)^2 = (450)(625) = 281,250 \text{ kilogram-(meters/sec)}^2$.

Example 14.5. Suppose that a shopping cart, together with the items in it, weighs 120 pounds and that it is being pushed at a constant velocity of 1.5 feet per second. What is its kinetic energy? To answer this question, we need to know the mass m of the cart. Because weight $= g \times$ mass, we see that $m = \frac{120}{32.17} = 3.73$ slugs. So the kinetic energy of the cart is $\frac{1}{2}mv^2 = \frac{1}{2}(3.73)(1.5)^2 = 4.20 \text{ slug-(feet/sec)}^2$.

A look at the units involved tells us that

$$1 \text{ kilogram-(meters/sec)}^2$$

$$= 1 \text{ kilogram-(meters/sec}^2)\text{-meter}$$

$$= 1 \text{ newton-meter} = 1 \text{ joule}$$

in the M.K.S. system, and that

$$1 \text{ slug-(feet/sec)}^2 = 1 \text{ slug-(feet/sec}^2)\text{-feet}$$

$$= 1 \text{ pound-feet} = 1 \text{ foot-pound}$$

in the American system. Observe therefore, that work and kinetic energy are expressed in the same units! This suggests a connection between work and energy. Indeed, the action of a force on an object will bring about a change in its velocity and hence a change in its kinetic energy. But what is the exact mathematical relationship?

To answer this question, let's consider an object of mass m moving along a coordinatized line from a to b. Let the variable force that acts on the object along the line of its motion be given by the function $f(x)$ for $a \leq x \leq b$. Take a stopwatch and time what is happening. Suppose that the object is at a at time t_0 and at b at time t_1. Let its position at any time t with $t_0 \leq t \leq t_1$ be $x = x(t)$. See Figure 14.7. The velocity $v(t)$ of the object at any time t is the derivative $v(t) = x'(t)$, and its acceleration is $a(t) = v'(t) = x''(t)$. The force is also a function of time. This function is the composite $f(x(t))$ of the two functions $f(x)$ and $x(t)$. Note that by Newton's second law, $f(x(t)) = ma(t)$ at any time t with $t_0 \leq t \leq t_1$.

Consider the work

$$W = \int_a^b f(x)\, dx$$

done on the object. Refer back to Section 13.3A and observe that the substitution $x = x(t)$ converts this definite integral to

$$W = \int_a^b f(x)\, dx = \int_{t_0}^{t_1} f(x(t))x'(t)\, dt.$$

The kinetic energy of the object at any time t is given by $\frac{1}{2}mv(t)^2$. Differentiating this function by the chain rule, we get

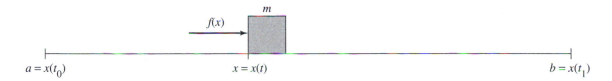

Figure 14.7

$$\frac{d}{dt}\left(\frac{1}{2}mv(t)^2\right) = mv(t) \cdot v'(t) = ma(t)v(t)$$

$$= f(x(t))x'(t).$$

So the kinetic energy function $\frac{1}{2}mv(t)^2$ is an antiderivative of $f(x(t))x'(t)$. Therefore, by the Fundamental Theorem of Calculus (see Section 13.1C),

$$W = \int_{t_0}^{t_1} f(x(t))x'(t)\,dt = \frac{1}{2}mv(t_1)^2 - \frac{1}{2}mv(t_0)^2$$

This equality provides us with the precise connection between kinetic energy and force that we were looking for. If an object is moving along a coordinatized line and a variable force acts on it along the line of its motion, then the work done by the force as the object moves from a to b is equal to

the kinetic energy of the object at b
minus the kinetic energy of the object at a.

In other words, the work done by the force on the object is equal to the change that it effects in the kinetic energy of the object. If the object starts from rest, then $v(t_0) = 0$ and $\frac{1}{2}mv(t_0)^2 = 0$. In this case, $W = \frac{1}{2}mv(t_1)^2$. So the work done on the object is equal to the kinetic energy that the object picks up during its motion.

Notice that the Fundamental Theorem of Calculus has just come to life as a link between two important concepts of physics. In particular, it is not just a computational technique that allows us to sum up lots of very small numbers.

Let's return once more to the weightlifter of Section 14.1A. Let $n(x)$ be the net force on the barbell at a height x above the floor. A look back at Figure 14.5 shows that

$$\int_{3/4}^{5/4} n(x)\,dx = (40)(0.5) = 20.$$

By the facts just developed, this amount of work is related to the velocity of the moving barbell by

$$\int_{3/4}^{5/4} n(x)\,dx = \frac{1}{2}m \cdot \left(\text{velocity at } x = \frac{5}{4}\right)^2$$
$$- \frac{1}{2}m \cdot \left(\text{velocity at } x = \frac{3}{4}\right)^2$$
$$= \frac{1}{2}m \cdot \left(\text{velocity at } x = \frac{5}{4}\right)^2 - 0.$$

It follows that $\frac{1}{2}m \cdot \left(\text{velocity at } x = \frac{5}{4}\right)^2 = 20$. So $\left(\text{velocity at } x = \frac{5}{4}\right)^2 = \frac{40}{m}$, and hence the velocity of the barbell at $x = \frac{5}{4}$ is equal to $\sqrt{\frac{40}{m}}$. What is the mass m of the barbell? Because the barbell weighs 535 pounds, we know that $mg = 535$, and hence that $m = \frac{535}{32.17} = 16.63$ slugs. Therefore, the velocity at $x = \frac{5}{4}$ is equal to $\sqrt{\frac{40}{16.63}} = 1.55$ feet per second.

What is the work done by the net force on the barbell from $x = \frac{3}{4}$ to $x = 5$? This is easy! After all, the barbell is at rest both at the beginning and the end of the lift. So its kinetic energy is zero both at the beginning and the end of the lift. It follows that the work done is zero.

So far, we have considered the work done on the barbell by the net force. But what is the work W done by the force that the weightlifter applies? Let $f(x)$ be the force that the weightlifter applies at a height x above the floor. Since $f(x) - 535 = n(x)$ is the net force on the barbell, we see that $f(x) = n(x) + 535$. Therefore

$$W = \int_{3/4}^{5} f(x)\, dx = \int_{3/4}^{5} n(x)\, dx + \int_{3/4}^{5} 535\, dx.$$

Since we have just seen that the work done by the net force is zero, we know that

$$\int_{3/4}^{5} n(x)\, dx = 0$$

and therefore that

$$W = 0 + (535x) \Big|_{3/4}^{5} = (535)\left(5 - \frac{3}{4}\right) = (535)(4.25)$$

$$= 2273.75 \text{ foot-pounds.}$$

Notice that the particulars of the force $f(x)$ applied by the weightlifter are irrelevant. No matter how the weightlifter does it, the work is equal to the 535 pound weight of the barbell times the distance of $5 - \frac{3}{4} = 4.25$ feet it was lifted.

Consider a force and a region in a plane or in space. Suppose that the force has the ability to act at every point in the region and that both the magnitude and the direction are determined by the location of the point. We will call such a region a *force field*. Think of an object of mass m in the vicinity of the Earth. If the object is a distance r from the center of the Earth, then gravity will act in the direction of the center with a magnitude of $\frac{GmM}{r^2}$, where M is the mass of the Earth. If m is fixed, then not only the direction, but also the magnitude of the force is determined by the location of the point. So this "gravitational force field" meets our requirements. Figure 14.8(a) shows this force field from "afar" and Figure 14.8(b) depicts it nearer to the surface of the Earth. Figure 14.8(c) is another (hypothetical) example of a force field.

Consider a force field and an object of mass m in it. Assume that the initial velocity of the object is $v_0 = 0$. To focus the discussion, suppose that we are dealing with one of the force fields depicted in Figure 14.8. A look at the flow of the vectors in any of them shows

that the object will move in a straight line. In case (a) it will move towards the center of the black disc, in case (b) towards the bottom, and in case (c) towards the top. Place a coordinate axis along the path of the object. Let the origin coincide with its initial position and let b be the farthest possible point in the region that the object can reach. (In Figure 14.8(b), b would be at the bottom.) Let $f(x)$ be the force at any point x between 0 and b. The force field propels the object from 0 towards b. Figure 14.9 shows it at a typical position c. In view of our earlier analysis, the kinetic energy of the object at c is

$$\int_{0}^{c} f(x)\, dx = \frac{1}{2}m(v_c)^2 - \frac{1}{2}m(v_0)^2 = \frac{1}{2}m(v_c)^2$$

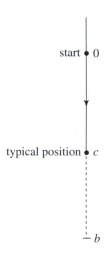

Figure 14.9

where v_c is the velocity of the object at c. The object has the potential of gaining additional kinetic energy. If the object were to continue to the farthest point b and reach it with a velocity v_b, then it would have a kinetic energy of $\frac{1}{2}m(v_b)^2$ at that point. Therefore, there is the potential of a gain of

$$\int_{c}^{b} f(x)\, dx = \frac{1}{2}m(v_b)^2 - \frac{1}{2}m(v_c)^2$$

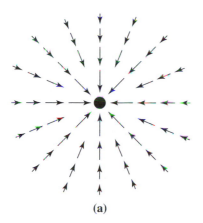 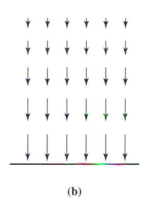 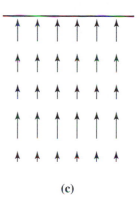

(a)   (b)   (c)

Figure 14.8

in kinetic energy. This difference is the object's *potential energy* at the point c. Notice that

kinetic energy + potential energy

$$= \frac{1}{2}m(v_c)^2 + \left(\frac{1}{2}m(v_b)^2 - \frac{1}{2}m(v_c)^2\right) = \frac{1}{2}m(v_b)^2.$$

Observe that the term $\frac{1}{2}m(v_b)^2$ does not depend on the specific point c. It follows therefore that the kinetic energy plus the potential energy of the object is the same no matter where the object is along its path. We have discovered a specific instance of the *law of conservation of energy*. Because $\int_0^b f(x)\,dx = \frac{1}{2}m(v_b)^2$, it can be expressed (in the situation we are discussing) as the equality

$$\int_0^b f(x)\,dx = \int_0^c f(x)\,dx + \int_c^b f(x)\,dx$$

of definite integrals.

The law of conservation of energy is one of the cornerstones of physics. In most situations it must be expanded to include not only "mechanical energy" (this is the potential and kinetic energy related to motion) of the object, but also any "internal energy" of the object. In other words, energy is not conserved unless internal energy is included. The phenomenon of friction is an example of a situation where mechanical energy of an object is transformed into internal

energy of the object, namely heat. There are many processes in which mechanical, heat, chemical, and nuclear energy are exchanged. We will encounter one such process in a moment: The explosive flash in the chamber of a gun that drives a bullet through the barrel.

14.2 Interior Ballistics

T his section will apply what we have learned to the study of the forces inside the barrel of a gun. The emphasis will be on the following questions: Can these forces be analyzed quantitatively? Exactly how is a bullet or shell propelled forward in the barrel? Can the muzzle velocity be computed? We will first study the force generated by a steel spring. This is done primarily for two reasons. One is that a spring-based mechanism is an important element in the triggering device of a gun (and in other mechanical devices, including the suspension systems of automobiles). The other is that such a study serves as an elementary introduction to the more difficult analysis of the force on a bullet.

A. Analysis of Springs. Consider a coil spring made of an appropriately resilient metal such as steel. The length of the spring, when it is not subject to an

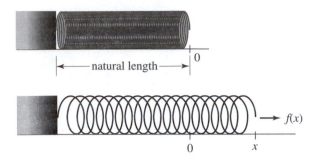

Figure 14.10

external force, is its *natural length*. Put the spring on a horizontal plane and anchor one end into place. Put an *x*-axis along the length of the spring in such a way that when the spring is at its natural length, its free end falls on the origin. Now stretch the spring by applying a force at the free end. Figure 14.10 shows what is happening. *Hooke's Law* for springs asserts that the force $f(x)$ necessary to keep the spring stretched x units beyond its natural length is proportional to x. More precisely, there is a constant k—it is called the *spring constant* of the spring—such that

$$f(x) = kx$$

for any displacement x of the spring beyond its natural length. This equation also holds when the spring is compressed. In this case, both x and the force $f(x)$ are negative. The spring constant is given in pounds/feet in American units and newtons/meter in M.K.S. It depends on the properties of the given spring. A rigid spring has a large spring constant, and a more flexible spring has a smaller spring constant. (Why?) When stretched or compressed too far, a spring may deform and no longer perform as just described. We will assume that the springs we are considering are not stretched (or compressed) beyond such a critical length.

Example 14.6. Suppose a spring has a natural length of 30 centimeters. Determine its spring constant, knowing that it takes a force of 1.00 newtons to keep it stretched at a length of 34 centimeters.

Since the force is given in newtons, the operational unit of length is the meter. So take the natural length of the spring as 0.30 meters and the length in its stretched state as 0.34 meters. Since the force of 1.00 newtons displaces the spring 0.04 meters beyond its natural length, we find from Hooke's Law that $1.00 = k(0.04)$. So $k = \frac{1.00}{0.04} = 25$ newtons/meter.

Example 14.7. Compute the work W done in stretching the spring of the previous example from its natural length to 4 centimeters beyond its natural length.

Specializing the formula $W = \int_a^b f(x)\, dx$ to the current situation, we find that

$$W = \int_0^{0.04} f(x)\, dx = \int_0^{0.04} kx\, dx = \int_0^{0.04} 25x\, dx$$

$$= \frac{25}{2}x^2 \Big|_0^{0.04} = \frac{25}{2}(0.04)^2 = 0.02 \text{ newton-meters}$$

$$= 0.02 \text{ joules.}$$

We will use springs to illustrate the connection between work and kinetic energy. Compress the spring that we are discussing by 3 centimeters and hold it in this compressed state. Place an ice cube in front of it, and release the spring. The ice cube, propelled by the force of the spring, will accelerate and pick up velocity. We will assume that it glides frictionlessly. The ice cube will be pushed until the spring reaches its natural length, and at that point it will lose contact with the spring. The spring constant $k = 25$ newton/meter determines the force that the spring exerts, and the compression of 0.03 meters gives us the distance through which it acts. If the mass of the ice cube is known, then we should be able to determine the motion of the ice cube, and in particular its final velocity. But how?

Let's start by computing the work that the spring does on the ice cube. Place a coordinate axis as in Figure 14.10. In particular, the right end of the spring is at the origin when the spring is at its natural length. Note that when the spring is compressed by 0.03 meters, its right end has coordinate −0.03. Figure

14.11 shows the ice cube in typical position, some time after its release. Observe that x is negative and hence that the spring is displaced $-x$ units beyond its natural length. The work done on the ice cube by the spring is

$$W = \int_{-0.03}^{0} f(x)\,dx = \int_{-0.03}^{0} -25x\,dx = -\frac{25}{2}x^2 \Big|_{-0.03}^{0}$$

$$= 0 - \left(-\frac{25}{2}(-0.03)^2\right) = \frac{25}{2}(0.03)^2$$

$$\approx 0.011 \text{ joules.}$$

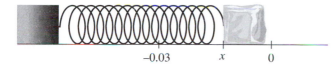

Figure 14.11

The coordinate axis could have been chosen differently. Figure 14.12 shows another way. Now the right end of the spring is at the point 0.03, when the spring is at its natural length. When the ice cube is at a typical position x, the spring is displaced by $0.03 - x$ meters, and the force that it exerts on the ice cube is $25(0.03 - x)$. So the work done on the ice cube is

$$W = \int_{0}^{0.03} f(x)\,dx = \int_{0}^{0.03} 25(0.03 - x)\,dx$$

$$= \int_{0}^{0.03} (0.75 - 25x)\,dx$$

$$= \left(0.75x - \frac{25}{2}x^2\right)\Big|_{0}^{0.03} = 0.75(0.03) - \frac{25}{2}(0.03)^2$$

$$= 25(0.03)^2 - \frac{25}{2}(0.03)^2 = \frac{25}{2}(0.03)^2$$

$$\approx 0.011 \text{ joules.}$$

As expected, the answer is the same as before. A comparison of the two definite integrals shows that the second integral is reduced to the first by the substitution $u = 0.03 - x$.

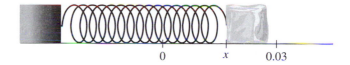

Figure 14.12

Take a stopwatch and time this "event." Let t_0 be the instant the compressed spring is released. The velocity of the ice cube at this time is $v(t_0) = 0$. Let t_1 be the instant that the spring and the ice cube part company, and let $v(t_1)$ be the velocity of the ice cube at that time. The connection between work and kinetic energy developed in Section 14.1B tells us that

$$\frac{25}{2}(0.03)^2 = W = \frac{1}{2}mv(t_1)^2 - \frac{1}{2}mv(t_0)^2 = \frac{1}{2}mv(t_1)^2$$

where m is the mass of the ice cube. Assume that the mass of the ice cube is 20 grams = 0.02 kilograms. So

$$25(0.03)^2 = 0.02v(t_1)^2.$$

Therefore, $v(t_1)^2 = 1.125$, and hence the final velocity of the ice cube is $v(t_1) \approx 1.06$ meters/sec. See Exercises 14E for more precise information about the motion of this ice cube.

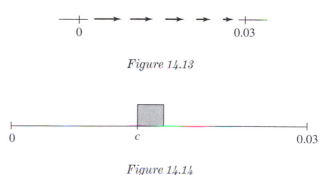

Figure 14.13

Figure 14.14

Refer to Figure 14.12 and observe that the spring can exert a force at any point of the interval $[0, 0.03]$. So the spring determines a force field along this interval. See Figure 14.13. Consider the ice cube in a typical position c as shown in Figure 14.14. We know

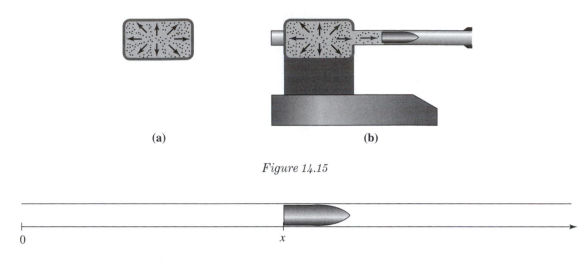

Figure 14.15

Figure 14.16

that its kinetic energy at c is equal to the work

$$\int_0^c f(x)\,dx = \int_0^c 25(0.03 - x)\,dx$$

$$= \left(0.75x - \frac{25}{2}x^2\right)\bigg|_0^c = 0.75c - \frac{25}{2}c^2$$

done on the ice cube. The potential energy at c is the kinetic energy that the ice cube will gain during its motion from c to 0.03. This is equal to the work

$$\int_c^{0.03} f(x)\,dx = \int_c^{0.03} 25(0.03 - x)\,dx$$

$$= \left(0.75x - \frac{25}{2}x^2\right)\bigg|_c^{0.03}$$

$$= 0.75(0.03) - \frac{25}{2}(0.03)^2 - \left(0.75c - \frac{25}{2}c^2\right).$$

Notice that the kinetic plus the potential energy of the ice cube is the same no matter where c is taken. This illustrates the principle of conservation of energy.

We have assumed that the ice cube glides in a frictionless way. A more exacting analysis would have to consider friction and the conversion of mechanical energy to heat.

B. The Force in a Gun Barrel.   Suppose that an explosion occurs inside a closed chamber. The forces that are released will push against all surfaces of the chamber. If one of the walls of the chamber is free to move, it will be "blown away." This, in principle, is the situation inside the chamber of a gun, where it is the shell or bullet that can move freely. The force of the explosion will drive the shell or bullet down and through the barrel. See Figure 14.15. The situation is entirely analogous to that of the ice cube considered earlier. The discipline that studies what happens inside the barrel of a gun is known as *interior ballistics*. (Once the shell leaves the muzzle, it becomes the focus of study of *exterior ballistics*.)

Suppose that a gun is fired at time $t = 0$. Place a coordinate system as shown in Figure 14.16 and let x be a typical position of the shell inside the barrel. A formula of historical and practical interest in interior ballistics is the formula of Le Duc. It says that the velocity $v(x)$ and the position x of the shell are related by a formula of the form

$$v(x) = \frac{rx}{s + x}$$

where r and s are positive constants.[1] These constants depend on the specifics of the situation: properties of the barrel, the shell, the explosive force, and so on.

Because the position x of the shell in the barrel is a function of time $t \geq 0$, we can put $x = x(t)$. The velocity of the shell as a function of time is the composite function $v(t) = v(x(t))$. But we also know that $v(t) = x'(t)$. It follows that the formula of Le Duc can be rewritten as

$$x'(t) = v(x(t)) = \frac{rx(t)}{s + x(t)}.$$

The acceleration of the shell at any time $t \geq 0$ is $a(t) = x''(t)$. Let's compute $x''(t)$ by differentiating $x'(t)$. By the formula of Le Duc and the quotient rule,

$$x''(t) = \frac{d}{dt}\left(x'(t)\right) = \frac{d}{dt}\left(\frac{rx(t)}{s + x(t)}\right)$$

$$= \frac{rx'(t)(s + x(t)) - rx(t)x'(t)}{(s + x(t))^2} = \frac{rsx'(t)}{(s + x(t))^2}.$$

Inserting the formula of Le Duc, we get

$$x''(t) = \frac{rs\frac{rx(t)}{s+x(t)}}{(s + x(t))^2},$$

so that after an algebraic simplification,

$$x''(t) = \frac{r^2sx(t)}{(s + x(t))^3}.$$

This is the acceleration of the shell produced by the net force on it. This net force on the shell is the force generated by the explosion of the powder minus the frictional forces (that are negligible in comparison). Let m be the mass of the shell. By Newton's second law, the net force on the shell is equal to

$$F(t) = ma(t) = mx''(t)$$

at any time $t \geq 0$. Putting in the expression for $x''(t)$ just derived, we get that

$$F(t) = \frac{mr^2sx(t)}{(s + x(t))^3}.$$

Notice that as a function of its position x in the barrel, the force on the shell is given by

$$f(x) = \frac{mr^2sx}{(s + x)^3}.$$

Let's analyze the force function $f(x)$. By the quotient rule,

$$f'(x) = \frac{mr^2s \cdot (s + x)^3 - mr^2sx \cdot 3(s + x)^2}{(s + x)^6}$$

$$= \frac{mr^2s \cdot (s + x)^2(s + x - 3x)}{(s + x)^6}$$

$$= \frac{mr^2s(s - 2x)}{(s + x)^4}.$$

So $f'(x) = 0$ when $x = \frac{s}{2}$. Observe that if $x < \frac{s}{2}$, i.e., if $2x < s$, then $f'(x) > 0$; and if $x > \frac{s}{2}$, then $f'(x) < 0$. It follows that f increases over the interval $[0, \frac{s}{2}]$, has an absolute maximum at $\frac{s}{2}$, and decreases thereafter. (Review the discussion in Section 8.7 if necessary.)

Using the quotient rule once more shows us that the second derivative of f is

$$f''(x) = \frac{-2mr^2s \cdot (s + x)^4 - mr^2s(s - 2x) \cdot 4(s + x)^3}{(s + x)^8}$$

$$= \frac{-2mr^2s \cdot (s + x) - 4mr^2s(s - 2x)}{(s + x)^5}$$

$$= \frac{6mr^2sx - 6mr^2s^2}{(s + x)^5}$$

$$= \frac{6mr^2s(x - s)}{(s + x)^5}.$$

So $f''(x) = 0$ when $x = s$. Observe that $f''(x) < 0$ when $x < s$, and $f''(x) > 0$ when $x > s$. It follows that the graph of f is concave down over $[0, s]$, has a point of inflection at s, and is concave up thereafter. (If necessary, refer to Section 10.6 for the connection between the second derivative and concavity.)

The graph of f will depend on the constants r and s. Its general shape is shown in Figure 14.17. The point of inflection is highlighted. The domain of f is the interval $[0, b]$, where b is the length of the barrel of the gun.

Let's conclude our analysis with a focus on the pressure that is generated inside the barrel. This is a concern of obvious importance in the design of the gun. The inside of the barrel is a cylinder. Let k be the

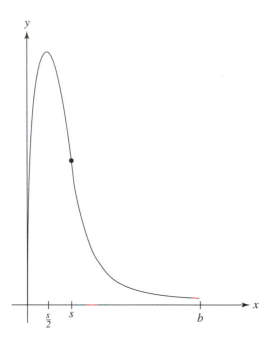

Figure 14.17

diameter of the circular cross-section. This diameter is known as the *caliber* of the gun. The area of the cross-section is $\pi\left(\frac{k}{2}\right)^2 = \frac{\pi k^2}{4}$. Now let $p(x)$ be the pressure in the barrel at x. (Recall from Figure 14.16 that x is the location of the shell.) Because force = pressure × area, we see that $f(x) = \frac{\pi k^2}{4}p(x)$. Therefore,

$$p(x) = \frac{4}{\pi k^2}f(x).$$

So the pressure curve has the same general shape as the force curve. In particular, the pressure in the barrel (at the location of the shell) rises very rapidly and reaches a maximum at $\frac{s}{2}$. It then drops off, rapidly at first, and then more gradually.

C. The Springfield Rifle.
Let's load a rifle by placing a cartridge into what is called the chamber. A typical cartridge is shown below. It consists not only of a bullet (shown in black) but also of a cylinder that contains an explosive powder.

The action of pulling the trigger pushes a spring-driven pin into the cartridge. This detonates a small charge, which brings the powder in the cylinder to explosion. The explosion separates the bullet from the rest of the cartridge and sends it hurtling down the barrel. See Figure 14.18. Any rifle is *rifled*. This means that the inside of the barrel has grooves that turn like the threading of a screw. These grooves will cause the bullet to rotate rapidly as it rushes through the barrel. This rotation gives the bullet greater stability in flight and the weapon greater accuracy.

We will now consider the famous Springfield rifle. This five-shot repeating rifle (five bullets can be fired in rapid succession) was closely modeled after a rifle used by the German military. The Springfield rifle was first issued in 1904 and remained in the U.S. arsenal until 1938. It was one of the most reliable and accurate firearms in military history. A version of it is a popular hunting rifle today.

The Springfield rifle has a caliber of $k = 0.025$ feet (or about 0.30 inches) and a barrel length of 2 feet. The barrel has four grooves that make one complete turn every 0.833 feet (or 10 inches). The traditional measure of the weight of a bullet is the *grain*, where 7000 grains = 1 pound. The weight of a bullet fired by the Springfield rifle is 165 grains, or $\frac{165}{7000}$ pounds. So its mass is $m = \frac{165}{7000} \cdot \frac{1}{32.17} = 0.000733$ slugs.

Carefully devised experiments[2] have verified that the explicit version

$$v(x) = \frac{3716x}{0.65 + x}$$

of the formula of Le Duc is valid for the Springfield rifle. The distance x (as indicated in Figure 14.16) is given in feet, and the velocity $v(x)$ is in feet per second. In reference to the general case of the formula, note that $r = 3716$ and $s = 0.650 = 6.50 \times 10^{-1}$.

Because the length of the barrel corresponds to $x = 2$, we see that the *muzzle velocity* (this is the

Figure 14.18

velocity of the bullet at the point of exit from the barrel) is equal to

$$v(2) = \frac{(3716)(2)}{0.65 + 2.00} = \frac{(3716)(2)}{2.65}$$

$$\approx 2800 \text{ feet per second.}$$

This agrees with the official[3] muzzle velocity of 2800 feet per second.

Let's consider some of the information that the discussion of Section 14.2B provides. The force on the bullet at any point x in the barrel is

$$f(x) = \frac{mr^2 sx}{(s+x)^3} = \frac{(7.33 \times 10^{-4})(3716^2)(6.5 \times 10^{-1})x}{(0.65 + x)^3}$$

$$\approx \frac{6580x}{(0.65 + x)^3}.$$

A look at the units that we are using tells us that this is in pounds. The force on the bullet in the barrel reaches its maximum when $x = \frac{s}{2} = 0.325$. This maximum force is

$$f(0.325) \approx \frac{6580(0.325)}{(0.650 + 0.325)^3} \approx 2310 \text{ pounds.}$$

The graph of the force function is sketched in Figure 14.19. Because the pressure in the barrel is equal to $p(x) = \frac{4}{\pi k^2}f(x) \approx \frac{4}{(3.14)(2.50 \times 10^{-2})^2}f(x) \approx 2040f(x)$, we get that the maximum pressure is an incredible 4,710,000 pounds per square foot!

Let's use the Springfield rifle to illustrate the connection between work and kinetic energy. The kinetic energy of the bullet when it leaves the barrel is

$$\frac{1}{2}mv(2)^2 \approx \frac{1}{2}(7.33 \times 10^{-4})(2800)^2 \approx 2870 \text{ foot-pounds.}$$

Because $f(x) \approx \frac{6580x}{(0.65+x)^3}$, the work done on the bullet during its motion through the barrel is approximated

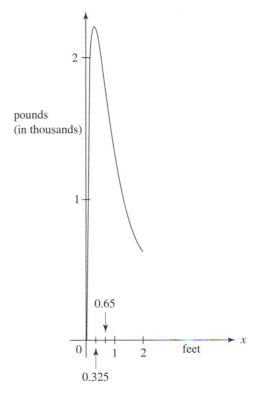

Figure 14.19

by the definite integral

$$\int_0^2 \frac{6580x}{(0.65 + x)^3}\, dx.$$

This integral is easy to evaluate by the method of substitution. Let $u = 0.65 + x$. Then $du = dx$. Since $u = 0.65$ when $x = 0$, and $u = 2.65$ when $x = 2$,

$$\int_0^2 \frac{6580x}{(0.65 + x)^3}\, dx = \int_{0.65}^{2.65} \frac{6580(u - 0.65)}{u^3}\, du$$

$$= 6580 \int_{0.65}^{2.65} (u^{-2} - 0.65u^{-3})\, du$$

$$= 6580\left(-u^{-1} + \frac{0.65}{2}u^{-2}\right)\Big|_{0.65}^{2.65}$$

$$= 6580\left(-\frac{1}{2.65} + \frac{0.65}{2(2.65)^2} + \frac{1}{0.65} - \frac{0.65}{2(0.65)^2}\right)$$

$$\approx 6580(0.438)$$

$$\approx 2880 \text{ foot-pounds.}$$

According to the theory of Section 14.1B, the work done on the bullet should be equal to the kinetic energy of the bullet when it leaves the barrel. Why the discrepancy? The short answer is roundoff error! These are the errors that "creep" into a calculation as a result of rounding off. The more detailed and definitive answer that follows addresses a problem that was already encountered. For example, in Section 9.2 the maximal tension in one of the main cables of the George Washington Bridge was computed to be

$$(20{,}562{,}500)(2.85659) \approx 58{,}700{,}000 \text{ pounds.}$$

The implicit understanding is that only the three numbers, or "figures," 5, 8, and 7 of the answer 58,700,000 are "significant" and that the zeros are meaningless as far as accuracy is concerned. Why? Simply because the data $w = 11{,}750$, $d = 1750$, and $s = 327$ on which this computation was based is unlikely to be reliable beyond three significant figures. The roundoff strategy in the calculations for the Springfield rifle was based on similar assumptions about the data for this rifle. For example, in view of the fact that 0.01 feet is approximately equal to $\frac{1}{8}$ inches, it was reasonable to assume that the barrel length is $2 = 2.00$ feet where all three significant figures, 2, 0, and 0, are accurate. A similar assumption is warranted about the three significant figures 7, 3, and 3, in the expression for the mass $m = 0.000733$ slugs of the bullet. However, there is no evidence that the constant 0.65 in Le Duc's formula is accurate beyond its two significant figures (even though a zero was formally added to the decimal expansion). These considerations suggest that the calculations should be rounded off at the third significant figure and that any accuracy beyond that is meaningless. This is the strategy that was used. In a more precise analysis, each of the relevant constants would be supplied with an error estimate — for example, an error estimate of ± 0.002 feet for the barrel length $b = 2$ feet would mean that b falls into the interval $1.998 \leq b \leq 2.002$. The error estimates would then be carried along in the calculations and result in "confidence interval" for the solution. Refer to Examples 11.9 and 11.12 for simple illustrations of this approach. The discrepancy between the numbers 2880 and 2870 means that both 2870 and 2880 foot-pounds fall into the confidence interval of the solution to the work/kinetic-energy problem considered earlier. The accuracy of the numerical results about the Springfield rifle are also affected by the fact that the formula of Le Duc is only an approximation for the velocity. Ideally, this formula should also come with an error estimate.

14.3 **Momentum**

Momentum is a concept of fundamental importance in the analysis of motion. The *momentum* of an object of mass m moving with velocity v is defined to be the product $m \times v$. In the sense that it measures the "quantity" of motion, the concept of momentum is similar to that of kinetic energy.

Let's return to the examples of Section 14.1B. The shopping cart has mass $m = 3.73$ slugs. When it is moving with a velocity $v = 1.5$ feet per second, its momentum is equal to

$$mv = (3.73)(1.5) = 5.56 \text{ slug-feet/sec.}$$

On a larger scale, consider the automobile. It has a mass of $m = 900$ kilograms and it is moving at a velocity of $v = 25$ meters per second. So its momentum is

$$mv = (900)(25) = 22{,}500 \text{ kilograms-meters/sec.}$$

Recalling that 1 slug = 14.59 kilograms and 1 meter = 3.28 feet, this is equal to $\frac{(22{,}500)(3.28)}{14.59} = 5058.26$ slug-feet/sec. As expected, its momentum is considerably larger than that of the shopping cart.

Figure 14.20

The barbell has mass $m = \frac{535}{32.17} = 16.63$ slugs. When it is $\frac{5}{4}$ feet off the ground, its velocity is 1.55 feet per second. So it has a momentum of

$$mv = (16.63)(1.55) = 25.78 \text{ slug-feet/sec}$$

at that point. We saw in Section 14.2C that a bullet of the Springfield rifle has mass $m = 0.000733$ slugs and that it has a muzzle velocity of $v = 2800$ feet per second. Therefore, it has a momentum

$$mv = (0.000733)(2800) = 2.05 \text{ slug-feet/sec}$$

as it emerges from the barrel.

A. Force as Function of Time. When a force acts on an object, it will effect a change in its velocity and therefore also in its momentum. Can this change be computed? More fundamentally, what is the precise connection between force and momentum? The key to the answers to these questions (as in the case of the connection between force and energy) lies in the consideration of force as a function of time.

Consider an object moving along a coordinatized line and suppose that a variable force acts on it along the line of its motion. Suppose that the object moves from a to b on the line. Let the object have mass m and let the force be given by the function $f(x)$ for any x with $a \leq x \leq b$. Suppose that the object is at a at time t_0 and at b at time t_1. Let its position at any time t with $t_0 \leq t \leq t_1$ be $x = x(t)$. See Figure 14.20. The velocity of the object at any time t is $v(t) = x'(t)$, and its acceleration is $a(t) = v'(t) = x''(t)$. Observe that

$$ma(t) = mv'(t) = \frac{d}{dt}(mv(t)).$$

The force as a function of time is the composite $F(t) = f(x(t))$. By Newton's second law, $F(t) = ma(t)$, and

hence

$$F(t) = \frac{d}{dt}(mv(t)).$$

So the force on the object at any time t is equal to the rate at which the momentum is changing at that time. Note that the momentum $mv(t)$ is an antiderivative of the force function $F(t)$. So by the Fundamental Theorem of Calculus,

$$\int_{t_0}^{t_1} F(t)\,dt = mv(t)\Big|_{t_0}^{t_1} = mv(t_1) - mv(t_0)$$

The integral $\int_{t_0}^{t_1} F(t)\,dt$ is called the *impulse of the force* over the time interval $[t_0, t_1]$.

The analysis carried out in Section 14.1B showed us that the change in the kinetic energy of an object brought about by a force acting over a *distance* is equal to the work done on the object over this distance. The equality that we have just derived says that the change in the momentum of an object caused by a force acting over an interval of *time* is equal to the impulse of the force over this time interval. In particular, the relationship between momentum and impulse is analogous to the relationship between energy and work.

Example 14.8. Consider the force on a bullet in the barrel of the Springfield rifle from the time t_0 that the trigger is pulled to the time t_1 at which it leaves the barrel. Since the velocity of the bullet at time t_0 is 0, it follows that the impulse of the force over $[t_0, t_1]$ is equal to the momentum $mv(t_1) = (0.000733)(2800) = 2.05$ slug-feet per second that the bullet has when it leaves the barrel.

Suppose that the force F acting on the object is constant. Because Ft is an antiderivative of F, the impulse

$$\int_{t_0}^{t_1} F(t)\, dt = Ft \, \Big|_{t_0}^{t_1} = F(t_1 - t_0)$$

is equal to the product of F times the length of time F acts on the object. Therefore, in the situation of a constant force F, the impulse-momentum relationship is

$$mv(t_1) - mv(t_0) = F(t_1 - t_0).$$

Let's illustrate this equation in the context of the weightlifter and his barbell. (See Section 14.1B.) We will concentrate on the first phase of the lift, during which the weightlifter raises the barbell from $\frac{3}{4}$ feet to $\frac{5}{4}$ feet off the ground.

Set your stopwatch and let time $t_0 = 0$ be the beginnning of the lift. At this moment, the barbell is at rest $\frac{3}{4}$ feet off the ground. So the momentum $mv(t_0)$ is zero. Let t_1 be the moment that the barbell is $\frac{5}{4}$ feet off the ground. We saw at the beginning of Section 14.3 that the momentum $mv(t_1)$ of the barbell is 25.78 slug-feet/sec. So the change of momentum $mv(t_1) - mv(t_0)$ that the barbell undergoes is

$$25.78 \text{ slug-feet/sec} = 25.78 \text{ (slug-feet/sec}^2)\text{-seconds}$$

$$= 25.78 \text{ pound-seconds.}$$

During this phase of the lift, a net force of $F = 575 - 535 = 40$ pounds is acting on the barbell. To compute the impulse $F(t_1 - t_0) = Ft_1 = 40t_1$, we still need to compute the time t_1. Let's consider what we know. We know that the mass of the barbell is 16.63 slugs. So if a is the acceleration of the barbell, then by Newton's second law, $40 = 16.63a$. Therefore, $a = 2.41$ feet/sec^2. Because the initial velocity of the barbell is $v(0) = 0$, its velocity is $v(t) = 2.41t$ feet/sec at any time $t \geq 0$ during this phase of the lift. (Why?) So the height $x(t)$ of the barbell above the ground at any time $t \geq 0$ is $x(t) = 1.20t^2 + x(0) = 1.20t^2 + \frac{3}{4}$. Because $x(t_1) = \frac{5}{4}$, it remains to solve the equation $1.20(t_1)^2 + \frac{3}{4} = \frac{5}{4}$ for t_1. Doing this, we get $(t_1)^2 = \frac{0.5}{1.20} = 0.42$ and hence

that $t_1 = 0.65$ seconds. Therefore, the impulse that the barbell receives is

$$F(t_1 - t_0) = Ft_1 = 40t_1 = (40)(0.65)$$

$$= 26 \text{ pound-seconds.}$$

But the change of momentum was equal to 25.78 pound-seconds. Why the difference? Again, it's simply a round-off error. See Exercise 18.

Once more, consider an object moving along a coordinatized line and suppose that a variable force acts on it along the line of its motion. Again regard the force as a function $F(t)$ of time. This time, however, suppose that the mass m of the object is changing. So $m = m(t)$ is also a function of time. It turns out—as in the case where the mass m is constant—that the force $F(t)$ on the object at any time t is equal to the rate at which the momentum is changing at that time. Restated in a formula, the assertion is that

$$F(t) = \frac{d}{dt}\left(m(t) \cdot v(t)\right).$$

In the case where $m = m(t)$ is a constant, this says that $F(t) = m\frac{d}{dt}v(t) = ma(t)$. Therefore, $F(t) = \frac{d}{dt}\left(m(t) \cdot v(t)\right)$ is Newton's second law formulated in a way that allows the mass of the object to vary.

Let's consider a concrete example of such a situation. Suppose that at time $t = 0$ a chute opens and begins to deposit sand onto a conveyor belt at a rate of 3 pounds per second. The conveyor belt moves at a rate of $\frac{1}{2}$ foot per second and transports the sand to a location 20 feet away. See Figure 14.21. If $x(t)$ is the position of the leading edge of the sand, then $x'(t) = v(t) = 0.5$. It follows that $x(t) = (0.5)t$, and in particular, that $x(40) = 20$. So at $t = 40$ seconds, the sand will begin to drop off the belt. Let $w(t)$ and $m(t)$ be the weight and mass, respectively, of the sand on the conveyor belt at any time $t \geq 0$. We know that $w(t) = m(t) \cdot g$. Consider the time interval $0 \leq t \leq 40$. During this interval, sand is added to the belt at a rate of 3 pounds/sec. It follows that

$$3 = w'(t) = g \cdot m'(t) = 32.17m'(t).$$

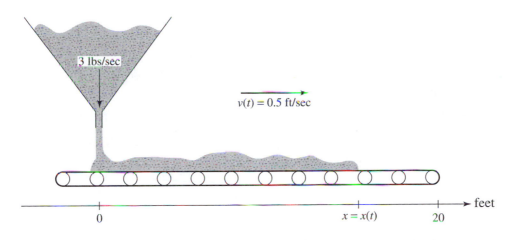

Figure 14.21

So $m'(t) = 0.093$. Let $F(t)$ be the force that is required to move the sand. So $F(t)$ is the force produced by the motor driving the belt minus all frictional forces. By Newton's second law,

$$F(t) = \frac{d}{dt}(m(t) \cdot v(t)) = \frac{d}{dt}((0.5)m(t))$$

$$= (0.5)m'(t) = (0.5)(0.093) = 0.046 \text{ pounds.}$$

So the force is constant even though the mass that it is moving is increasing. If this seems surprising, think about the fact that when a force is applied to a constant mass it produces an increase in velocity. At any $t \geq 40$, the sand will drop from the belt at the same rate with which it is deposited, so that the mass $m(t)$ of the sand on the belt is constant. So $m'(t) = 0$, and hence $F(t) = 0$. Now the motor needs only to generate enough force to counterbalance friction. Assume that the chute is shut off at $t = 60$ seconds. For the next 40 seconds thereafter, i.e., for $60 \leq t \leq 100$, we see that $m'(t) = -0.093$. So by the calculation above,

$$F(t) = (0.5)(-0.093) = -0.046 \text{ pounds.}$$

This means that a breaking action is required if the conveyer belt is to continue to move forward with the speed of $\frac{1}{2}$ foot per second. Is this surprising?

B. Conservation of Momentum. Return to the force that propels a bullet inside the barrel of a rifle.

We saw that this force is a function $F(t)$ of time. If m_b is the mass of the bullet and $a_b(t)$ its acceleration, then by Newton's second law,

$$F(t) = m_b a_b(t).$$

Newton's third law asserts that every action has an equal and opposite reaction. In the case of the rifle, this says that corresponding to the force on the bullet, there is a force acting on the rifle. Because this force has the same magnitude and is opposite in direction, it must be equal to $-F(t)$. If m_r is the mass of the rifle and $a_r(t)$ is the acceleration of the rifle produced by this force, then (again by Newton's second law),

$$-F(t) = m_r a_r(t).$$

What we have observed is confirmed by Figure 14.15(b). Corresponding to the force that propels the bullet to the right, there is a force of equal magnitude that pushes the system to the left. So as the bullet is pushed forward, the rifle is pushed backward. This phenomenon is known as the *recoil* of the rifle.

Let $v_b(t)$ be the velocity of the bullet, and let $v_r(t)$ be the recoil velocity of the rifle. By differentiating, we get that

$$\frac{d}{dt}\left(m_b v_b(t) + m_r v_r(t)\right) = m_b a_b(t) + m_r a_r(t)$$

$$= F(t) - F(t) = 0.$$

Therefore,

$$m_b v_b(t) + m_r v_r(t) = \text{constant}.$$

So the total momentum of the rifle-bullet system is the same for any time $t \geq 0$. This assertion is a special case of the *law of conservation of momentum*.

Assume that the rifle is fired at time $t = 0$. We know that $v_b(0) = v_r(0) = 0$. So the total momentum $m_b v_b(t) + m_r v_r(t)$ of the rifle-bullet system is equal to the constant $m_b v_b(0) + m_r v_r(0) = 0$. It follows that

$$m_b v_b(t) = -m_r v_r(t).$$

Example 14.9. The Springfield rifle is known to weigh 8.5 pounds. Hence it has a mass of $m_r = \frac{8.5}{32.17} = 0.264$ slugs. Recall that a bullet for this rifle has a mass of $m_b = 0.000733$ slugs. At the instant a bullet exits the barrel it has a velocity $v_b = 2800$ feet per second (this is the muzzle velocity). So $m_b v_b = (0.000733)(2800) = 2.05$ slug-(feet/sec). At the same instant,

$$0.264 v_r = m_r v_r = -m_b v_b = -2.05.$$

Therefore, the recoil velocity of the rifle is $v_r = -7.77$ feet per second. The minus sign indicates that the direction of motion is opposite to that of the bullet.

The forces on a system of particles will, in general, consist of both *external* forces and *internal* forces. Consider a simple system consisting of two particles A and B. Suppose that B acts on A with a certain force F and that A acts on B with an equal and opposite force $-F$. These are forces internal to the system. Assume that the only other force, F_e, is external to the system and acts only on B. Begin to observe the system at time $t = 0$. Both the direction and magnitude of the forces F and F_e can vary with time $t \geq 0$. However, $-F$ remains equal and opposite to F for all $t \geq 0$. Let $\varphi(t)$ be the angle that F_e makes with F. Figure 14.22 shows what is going on. The masses of A and B can also vary. They are given by the functions $m_A(t)$ and $m_B(t)$ respectively.[4]

Consider the vector F_e at a fixed time t. Place a co-ordinate axis in such a way that the origin is at B and the positive direction coincides with the direction of

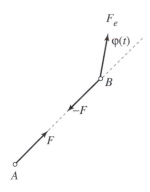

Figure 14.22

F_e. Let $F_e(t)$ be the magnitude of F_e and let $v_A(t)$ and $v_B(t)$ be the components of the velocities of the particles A and B in the direction of F_e. This is illustrated in Figure 14.23. Let $F(t)$ be the magnitude of F and check that the component of F in the direction of F_e is $F(t) \cos \varphi(t)$. (See Section 9.1B.) Have a careful look at Figures 14.22 and 14.23. Newton's second law applied in the direction of F_e first to the particle A and then to the particle B gives us that

$$F(t) \cos \varphi(t) = \frac{d}{dt}(m_A(t) v_A(t))$$

and

$$\left(F_e(t) - F(t) \cos \varphi(t)\right) = \frac{d}{dt}(m_B(t) v_B(t)).$$

By adding these two equations, we get

$$F_e(t) = \frac{d}{dt}(m_B(t) v_B(t)) + \frac{d}{dt}(m_A(t) v_A(t))$$

$$= \frac{d}{dt}(m_A(t) v_A(t) + m_B(t) v_B(t)).$$

So at any time t, the rate of change of the momentum of the system in the direction of the external force F_e is equal to the magnitude of F_e.

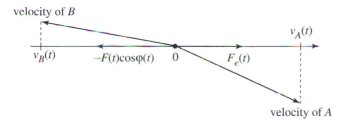

velocity of B

$v_B(t)$ $-F(t)\cos\varphi(t)$ 0 $F_e(t)$ $v_A(t)$

velocity of A

Figure 14.23

Now repeat the discussion just undertaken, but this time let F provide the direction of reference instead of F_e. Let $F(t)$ be the magnitude of F and let $v_A(t)$ and $v_B(t)$ be the components of the velocities of A and B in the direction of F. The component of F_e in the direction of F is $F_e(t)\cos\varphi(t)$. Take one more look at Figure 14.22. Newton's second law, this time applied in the direction of F, tells us that

$$F(t) = \frac{d}{dt}\left(m_A(t)v_A(t)\right)$$

and

$$\left(F_e(t)\cos\varphi(t) - F(t)\right) = \frac{d}{dt}\left(m_B(t)v_B(t)\right).$$

By an addition,

$$F_e(t)\cos\varphi(t) = \frac{d}{dt}\left(m_B(t)v_B(t)\right) + \frac{d}{dt}\left(m_A(t)v_A(t)\right)$$

$$= \frac{d}{dt}\left(m_A(t)v_A(t) + m_B(t)v_B(t)\right).$$

Thus the rate of change of the momentum of the system in the direction of F is equal to the magnitude of the component of the external force F_e in the direction of F.

This reasoning can be extended to a system of any number of particles and any number of external forces to show that the rate of change of the total momentum of the system in any direction is equal to the magnitude of the component of the resultant of the external forces in that direction. Suppose that the net external force F_e on a system is equal to zero. Then the rate of change in the total momentum of the system is zero, and it follows that its total momentum is

a constant. This is the *law of conservation of momentum*. It is valid from the submicroscopic world of the atom to the vast world of the universe. Along with the law of conservation of energy, the law of conservation of momentum is one of the most basic and important principles in all of physics.[5]

We close Section 14.3 with an application of both the law of conservation of momentum and the law of conservation of energy. The purpose of a *ballistic pendulum* is to measure muzzle velocities of guns. A ballistic pendulum consists of a block of some material of mass m suspended like a pendulum. See Figure 14.24. Let the block be at rest and fire a bullet with mass m_b and velocity v_b at the block. The bullet will strike the block, embed itself in it, and cause the block to swing. By knowing m_b, m, and the vertical displacement h of the block at the top of the swing, the velocity v_b of the bullet can be determined.

This is done as follows. After the bullet has left the barrel of the gun, the net external force acting on the block-bullet system in the horizontal direction is zero. (Gravity acts, of course, but it is counterbalanced by the suspension of the pendulum and has no effect on the horizontal motion of the bullet.) Therefore, the momentum of the block-bullet system in the horizontal direction is conserved. Before the bullet strikes it, the pendulum is at rest, and it follows that the momentum of the system is $m_b v_b$. See Figure 14.24(a). Now turn to Figure 14.24(b). The bullet has struck the block, entered it, and has come to rest within it. Suppose all this happens in an instant, and that the bullet imparts a certain initial velocity v to the block. The block with the bullet embedded within has a mass of $m_b + m$, so that the momentum of the system is now $(m_b + m)v$. By the law of conservation of momentum,

$$m_b v_b = (m_b + m)v.$$

The initial kinetic energy of the block is $\frac{1}{2}(m_b + m)v^2$. By the law of conservation of energy, this kinetic energy is converted to the potential energy that the block has at the time it reaches the top of its swing. This potential energy is equal to the work done by

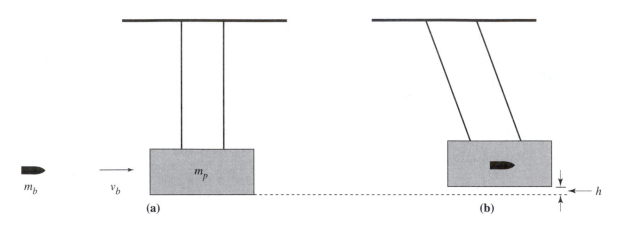

Figure 14.24

the force of gravity $(m_b + m)g$ in moving the block through the distance h (refer to Section 14.1B). It follows that

$$\frac{1}{2}(m_b + m)v^2 = (m_b + m)gh.$$

So $\frac{1}{2}v^2 = gh$, and $v = \sqrt{2gh}$. Inserting this into the equation $m_b v_b = (m_b + m)v$ and solving for v_b, we get

$$v_b = \sqrt{2gh}\left(\frac{m_b + m}{m_b}\right).$$

So by knowing the masses m_b and m and the displacement h, the velocity v_b of the bullet can be computed.

Example 14.10. Suppose that the Springfield rifle is fired at a ballistic pendulum that weighs exactly 20 pounds. Compute the maximum vertical displacement h of the block. The bullet has a mass of $m_b = 0.00073$ slugs and a muzzle velocity of $v_b = 2800$ feet per second. (See Section 14.2C.) Since 1 slug of mass weighs 32.17 pounds, the block has mass $m = \frac{20}{32.17} = 0.62$ slugs. Substituting this information into the formula $v_b = \sqrt{2gh}\left(\frac{m_b + m}{m_b}\right)$, we get

$$2800 = \sqrt{(2)(32.17)h}\left(\frac{0.00073 + 0.62}{0.00073}\right)$$

$$\approx 8.02\sqrt{h}\left(\frac{0.621}{0.00073}\right) \approx 6800\sqrt{h}.$$

It follows that $\sqrt{h} \approx 0.41$, and $h \approx 0.17$ feet.

14.4 **The Calculus of Rocket Propulsion**

The basic principle of rocket propulsion is simple. A rocket engine consists in essence of a chamber in which a volatile fuel mix is brought into a state of continued and controlled explosion. The gases produced by the combustion process are pushed out with high velocity in one direction and—by the law of conservation of momentum—the rocket is propelled forward in the other.

The basic analysis of rocket propulsion was undertaken in the last years of the 19th century by Konstantin Eduardovich Tsiolkovsky (1857–1935). This largely self-educated Russian mathematics and physics teacher was the first to suggest liquid propellants and multiple-stage rockets. He foresaw space exploration as an integral part of human development

(a)

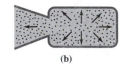
(b)

Figure 14.25

and wrote numerous scientific and popular articles to advance his ideas. His work remained largely unknown, and Tsiolkovsky received recognition only towards the end of his life. The second pioneer of space flight was the American Robert Hutchings Goddard (1882–1945). This professor of physics at Clark University described his early work in the remarkable paper *A Method for Reaching Extreme Altitudes*, published by the Smithsonian Institution in 1919. It discussed the basic differential equation of rocket propulsion and developed the solution. It describes Goddard's many tests and experiments with fuels, fuel tanks, fuel pumps, combustion chambers, and nozzles. He dedicated his entire life to the development of rockets and their components. With only meager resources at his disposal, he made enormous progress: He supplied the oxygen necessary for combustion in the form of pressurized liquid oxygen; he developed turbine-operated fuel pumps, used gyroscopes to stabilize flight, and installed parachutes to bring his rockets to a soft landing. Had his work received the broad-based attention and support that it deserved, there seems little doubt that he would have succeeded in developing an operational rocket. Let's interrupt our brief historical survey and turn to the basic principles of rocket engines.

A. Thrust.

Let's begin our analysis of a rocket engine by considering its most essential component. This is the chamber in which a fuel-oxygen mix is brought to a continuous explosion. The forces released by an explosion in a closed chamber will push against all surfaces of the chamber. See Figure 14.25(a). If one of the sides of the chamber is missing, then, obviously, there is nothing to push against on that side. For instance, in the diagram in Figure 14.25(b), there is nothing to oppose the force pushing against the wall on the right. Consequently, the chamber will be propelled to the right. At the same time, the particles produced by the explosion will be pushed out to the left. The rest of the rocket engine consists of fuel tanks and pumps. The essence of such a scheme is shown in Figure 14.26. Liquid oxygen and a fuel are both pumped from their

tanks into the combustion chamber and pressurized, and the mix is brought to ignition. Both the oxygen and the fuel are cooled to very low temperatures. Before it enters the combustion chamber, the fuel is forced through a system of pipes attached to both the combustion chamber and the nozzle to cool the engine. The engine, guidance system, and a payload are built into a streamlined cylindrical housing, and the rocket is complete. We have described the *liquid fuel rocket*. See Figure 14.27. There are also solid fuel rockets. The fuel and the combustion process are different, but the basic principles are the same.

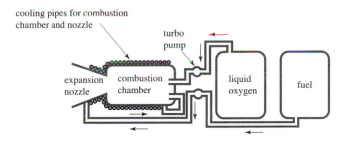

Figure 14.26

Let's begin our mathematical analysis with a static test of the engine. Bolt a rocket engine firmly into place vertically. Fire up the engine and allow it to reach a steady state at time $t = 0$. The exhaust gases will be pushed through the expansion nozzle with a constant velocity v_e. For any time $t \geq 0$, let $m_e(t)$ be the total mass of the gases that have been expelled from time $t = 0$ to time t. The total momentum of the exhaust gases at the point of exit from the nozzle (and before they are slowed down by their interaction with the air) from the time $t = 0$ to time t is equal to

$$v_e m_e(t).$$

Because the engine operates at a steady state, the rate $\frac{d}{dt} m_e(t)$ at which the mass of the exhaust gases is expelled is constant. Since "mass in" equals "mass out," the derivative $\frac{d}{dt} m_e(t)$ is also equal to the rate at which the fuel mix is consumed by the engine. (Total

Figure 14.27

mass is conserved in any combustion even though heat is produced.)

Consider the force that expels the gases. As already pointed out in Section 14.3B, the magnitude of this force is equal to the rate of change $\frac{d}{dt}\left(v_e m_e(t)\right)$ in the momentum of these gases. Call this force the *momentum component* of the thrust of the engine and denote its magnitude by T_{mo}. As just asserted,

$$T_{\mathrm{mo}} = \frac{d}{dt}\left(v_e m_e(t)\right) = v_e \frac{d}{dt} m_e(t).$$

Since the engine is firing in steady state, T_{mo} is constant. We will see in Exercise 34 that the formula $T_{\mathrm{mo}} = v_e \frac{d}{dt} m_e(t)$ for the momentum component of the thrust is also valid when the exhaust velocity v_e is allowed to vary with time.

Assume that the expansion nozzle (see Figure 14.26) has a circular cross-sectional area A at its exit. Let ρ_e be the density of the gas at the point of exit from the nozzle. The exhaust gases will disperse and form a cone as they are being expelled. In order to be able to compute the total mass of the gas that has been expelled from time 0 to time t, assume that the gas does not disperse but that it retains its density ρ_e. In this way, the exhaust gases can be considered to form a cylinder that has a circular base of area A. The length of the cylinder will be the exit velocity of the gas v_e times the elapsed time t. Therefore, the total volume of the gases expelled at time t is $A \cdot v_e t$. See Figure 14.28. Because mass = density × volume, it follows that the total mass of the gases expelled at time t is

$$m_e(t) = \rho_e A v_e t.$$

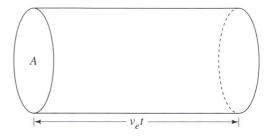

Figure 14.28

Therefore, the momentum of the exhaust gases at time t is

$$v_e m_e(t) = \rho_e A \cdot (v_e)^2 t.$$

The earlier expression for the momentum component of the thrust can now be written as

$$T_{\mathrm{mo}} = v_e \frac{d}{dt} m_e(t) = v_e \frac{d}{dt}\left(\rho_e A v_e t\right) = \rho_e A \cdot (v_e)^2.$$

There is another aspect to thrust. It is determined by the difference $p_e - p_0$ between the pressure of the exiting gases p_e and the external (atmospheric) pressure p_0 outside the rocket. The relationship

$$\text{force} = \text{area} \times \text{pressure}$$

shows that the engine produces an additional force of

$$A(p_e - p_0).$$

This *pressure component* of the thrust is denoted by T_{pr}. In the test that we are considering, the engine is bolted down and fired in steady state. So p_e and p_0 and therefore T_{pr} are all constants. The *thrust* T of the engine is defined to be the sum

$$T = T_{\mathrm{mo}} + T_{\mathrm{pr}} = v_e \frac{d}{dt} m_e(t) + A(p_e - p_0)$$

$$= \rho_e A \cdot (v_e)^2 + A(p_e - p_0).$$

The first fully operational rocket was developed for the German war effort in the 1940s. Known as the V-2, this rocket was 47 feet long and weighed 28,229 pounds fully fueled. Of this weight, 19,392 pounds was liquid oxygen and fuel (ethyl alcohol). The turbine-driven pumps pushed the fuel mix into the combustion chamber at a rate of 275 pounds per second. The velocity of the exhaust gases was 6560 feet per second, and the thrust generated by the engine was 59,500 pounds. The V-2 reached a maximum velocity of 5575 feet per second and had an operational range of 180 to 190 miles.[6] When fired vertically, it reached an altitude of 116 miles. The destructive power of the bomb that it carried terrorized London. The V-2, both the hardware and the scientists, fell into the hands of the allies at the end of the war and later became an important part of the American rocket program.

Consider a test of an engine of the V-2. Recall that it burns the fuel mix at a rate of 275 pounds per second. In units of mass, this corresponds to $\frac{275}{32.17} = 8.55$ slugs per second. It follows that $\frac{d}{dt} m_e(t) = 8.55$ slugs per second. Since the exhaust velocity of the gases is $v_e = 6560$ feet per second, the momentum component of the thrust is

$$T_{\mathrm{mo}} = v_e \frac{d}{dt} m_e(t) = (6560)(8.55) = 56{,}100 \text{ pounds.}$$

Recalling that the rocket's thrust T is 59,500 pounds, we see that the pressure component of the thrust is $T_{\mathrm{pr}} = T - T_{\mathrm{mo}} = 59{,}500 - 56{,}100 = 3{,}400$ pounds. Notice that T_{mo} is much the larger of the two components of the thrust. The radius of the expansion nozzle of the V-2 at the exit is 1.21 feet. So $A = \pi(1.21)^2 = 4.60$ square feet. It follows that the density ρ_e of the gas at the point of exit from the nozzle is

$$\rho_e = \frac{T_{\mathrm{mo}}}{A \cdot (v_e)^2} = \frac{56{,}100}{(4.60)(6560)^2} = 2.83 \times 10^{-4} \text{ slugs/ft}^3.$$

The atmospheric pressure at sea level is known to be $p_0 = 14.6$ pounds/inch^2, and therefore, $p_0 = (14.6)(12^2) = 2{,}102$ pounds/ft^2. Recall that $T_{\mathrm{pr}} = A(p_e - p_0)$, where p_e is the pressure of the exiting gases. Making the appropriate substitutions, we get

$$3{,}400 = 4.60(p_e - 2{,}102).$$

So $(p_e - 2{,}102) = 740$, and hence $p_e \approx 2{,}840$ pounds/ft^2.

B. The Rocket Equation.   The analysis of the motion of a rocket in flight is based on the considerations in Section 14.3B, and in particular the answer to two questions: At what rate does the momentum of the rocket/exhaust-gas system change, and what are the external forces on this system?

Begin to observe a rocket at time $t = 0$. This can be at the time of liftoff, or some time into the flight. At any time $t \geq 0$, let $m(t)$ be the mass of the rocket and let $v(t)$ be its velocity in the direction of its motion in reference to some observation point on the Earth. The mass and velocity at time $t = 0$ are $m(0) = m_0$ and $v(0) = v_0$ respectively. The momentum of the rocket at any time t is

$$m(t)v(t).$$

Now turn to the exhaust gases. Since the throttle setting of the engine can vary, we will assume that the gases exit the nozzle with a variable velocity of $-v_e = -v_e(t)$ relative to the rocket. So the gases move with a velocity of $v(t) - v_e(t)$ relative to the observation point. The total mass of the gases that have been expelled between time $t = 0$ and time t is equal to $m_e(t) = m_0 - m(t)$. Let $\mu(t)$ be the momentum of the exhaust gases expelled from time $t = 0$ to time t.

The total momentum of the rocket/exhaust-gas system is

$$M(t) = m(t)v(t) + \mu(t)$$

at any time $t \geq 0$. Fix a time t and let a very small additional amount of time Δt elapse. Observe that

$$M(t + \Delta t) - M(t)$$

$$= m(t + \Delta t)v(t + \Delta t) + \mu(t + \Delta t) - \big(m(t)v(t) + \mu(t)\big)$$

$$= m(t + \Delta t)v(t + \Delta t) - m(t)v(t) + \big(\mu(t + \Delta t) - \mu(t)\big).$$

The difference $\mu(t+\Delta t) - \mu(t)$ is the momentum of the gases expelled during time Δt. The mass of the gases expelled during this time is equal to

$$m_e(t + \Delta t) - m_e(t)$$

$$= \big(m_0 - m(t + \Delta t)\big) - \big(m_0 - m(t)\big)$$

$$= -\big(m(t + \Delta t) - m(t)\big).$$

Since Δt is very small, we can take the velocity of these gases during time Δt to be equal to their velocity $v(t) - v_e(t)$ at time t. Therefore, the change in the momentum of the exhaust gases during time Δt is

$$\mu(t + \Delta t) - \mu(t) = -\big(m(t + \Delta t) - m(t)\big)\big(v(t) - v_e(t)\big)$$

$$= \big(m(t + \Delta t) - m(t)\big)\big(v_e(t) - v(t)\big).$$

Inserting this into the expression for the change in the momentum $M(t)$ gives us

$$M(t + \Delta t) - M(t)$$

$$= m(t + \Delta t)v(t + \Delta t) - m(t)v(t)$$

$$+ \big(v_e(t) - v(t)\big)\big(m(t + \Delta t) - m(t)\big).$$

Dividing both sides of this equation by Δt, we get

$$\frac{M(t + \Delta t) - M(t)}{\Delta t}$$

$$= \frac{m(t + \Delta t)v(t + \Delta t) - m(t)v(t)}{\Delta t}$$

$$+ \frac{\big(v_e(t) - v(t)\big)\big(m(t + \Delta t) - m(t)\big)}{\Delta t}.$$

After pushing Δt to 0 we see that

$$\frac{d}{dt}M(t) = \frac{d}{dt}\big(m(t)v(t)\big) + \big(v_e(t) - v(t)\big)\frac{d}{dt}m(t).$$

By the product rule,

$$\frac{d}{dt}\big(m(t)v(t)\big) = \left(\frac{d}{dt}m(t)\right)v(t) + m(t)\frac{d}{dt}v(t).$$

After inserting this into the equation just derived, we get

$$\frac{d}{dt}M(t) = m(t)\frac{d}{dt}v(t) + v_e(t)\frac{d}{dt}m(t).$$

Because $m(t) = m_0 - m_e(t)$, it is clear that

$$\frac{d}{dt}m(t) = -\frac{d}{dt}m_e(t).$$

Therefore, the rate of change of the momentum of the rocket/exhaust-gas system is equal to

$$(*) \qquad \frac{d}{dt}M(t) = m(t)\frac{d}{dt}v(t) - v_e(t)\frac{d}{dt}m_e(t).$$

The first part of the analysis is complete. Our attention now turns to the study of the forces on this system. Consider the forces generated by the explosion in the combustion chamber. One of these, say F, propels the rocket forward, and the other is the "recoil" force $-F$ that pushes the gases backward. Notice that F is in the direction of flight. All other forces are external to the rocket/exhaust-gas mass system. What we need to do is to compute the component of the net external force F_e on the rocket in the direction of F. Why? Because by the results in Section 14.3B—compare Figures 14.22 and 14.29—this component of the net external force is equal to the rate of change of the momentum as given by equation $(*)$.

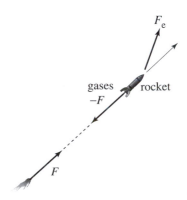

Figure 14.29

Consider a rocket in flight in the vicinity of, say, the Earth. What are the various external forces on the rocket. One of them is the pressure component T_{pr} of the thrust. Note that it acts in the direction of F. Recall that $T_{\mathrm{pr}} = A(p_e - p_0)$, where A is the cross-sectional area of the nozzle exit and p_e and p_0 are respectively the pressure of the exhaust gases and the external (atmospheric) pressure. Because both p_e and p_0 can vary in flight, $T_{\mathrm{pr}} = T_{\mathrm{pr}}(t)$ is a function of time. Another external force is the atmospheric resistance on the rocket. This too will vary with atmospheric pressure, and therefore with the altitude and the time t. Since it acts in a direction opposite to F, we will denote it by $-R(t)$. One more external force to consider is gravity. Refer to Figure 14.30. By Newton's universal law of gravitation, the gravitational force on the rocket is equal to

$$G(t) = G\frac{m(t)M_E}{r(t)^2}$$

where $m(t)$ is the mass of the rocket, M_E is the mass of the Earth, and $r(t)$ is the distance between the rocket and the center of the Earth. Let $\varphi(t)$ be the angle between the direction of flight and the line joining the rocket and the center of the Earth. Note that the component of gravity in the direction of F is $-G(t)(\cos\varphi(t))$. In summary, the magnitude of the net external force in the direction of the flight of the rocket is

$$T_{\mathrm{pr}}(t) - G(t)(\cos\varphi(t)) - R(t).$$

As already asserted, this is equal to the rate of change of the momentum of the rocket/exhaust-gas system as given by equation (∗). It follows that

$$T_{\mathrm{pr}}(t) - G(t)(\cos\varphi(t)) - R(t) = m(t)\frac{d}{dt}v(t) - v_e(t)\frac{d}{dt}m_e(t).$$

Refer to Section 14.4A, and notice that $v_e(t)\frac{d}{dt}m_e(t)$ is the momentum component T_{mo} of the thrust. Since the thrust T is equal to $T = T_{\mathrm{mo}} + T_{\mathrm{pr}}$ (as functions of time), we now see that $T(t) = T_{\mathrm{pr}}(t) + v_e(t)\frac{d}{dt}m_e(t)$ and

therefore that

$$m(t)\frac{d}{dt}v(t) = T(t) - G(t)\cos\varphi(t) - R(t)$$

This differential equation is the *rocket equation*. In words, it says that

the mass times the acceleration of the rocket

= thrust − gravity − drag.

The rocket equation with $\varphi = 0$ and hence $\cos\varphi = 1$ was known to Goddard and constitutes the important theoretical component in his investigations. The solution of the rocket equation is difficult and requires delicate numerical methods and computer analysis.

There is one important special case where the solution of the rocket equation is routine. Suppose that the engine is operating at constant throttle. We saw in Section 14.4A that in this case,

$$T = A\left(\rho_e(v_e)^2 + (p_e - p_0)\right)$$

where A is the cross-sectional area of the nozzle exit, ρ_e and p_e are the density and pressure of the exhaust gases (at the nozzle exit), and p_0 is the external atmosphere pressure. The fact that the engine operates at constant throttle implies that both ρ_e and p_e are constants. Suppose also that the rocket is in flight in a situation (say somewhere between the Earth and the Moon) where both gravity and atmospheric resistance are minimal and can be ignored. So $G(t) = 0$, $R(t) = 0$, and $p_0 = 0$. The rocket equation now has the much simpler form

$$m(t)\frac{d}{dt}v(t) = A\left(\rho_e(v_e)^2 + p_e\right).$$

The mass $m(t)$ of the rocket at any time t is equal to its mass m_0 at time $t = 0$ minus the total mass $m_e(t)$ of the gases expelled from time $t = 0$ to time t. But at constant throttle, the rate $\frac{d}{dt}m_e(t)$ of flow of the exhaust gases is constant and is equal to the rate r at which the fuel is consumed. So $\frac{d}{dt}m_e(t) = r$. Hence

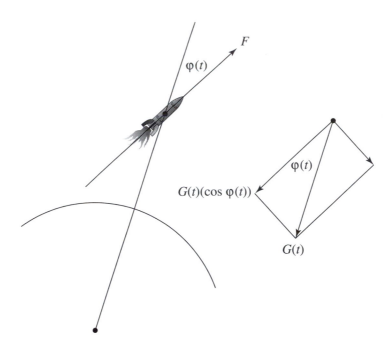

Figure 14.30

$m_e(t) = rt + C$. Since $m_e(0) = 0$, we see that $m_e(t) = rt$. It follows that $m(t) = m_0 - rt$ and therefore that

$$(m_0 - rt)\frac{dv}{dt} = A\left(\rho_e(v_e)^2 + p_e\right).$$

This is a separable differential equation that is easily solved. After separating the variables v and t, we get $dv = A(\rho_e(v_e)^2 + p_e)\frac{dt}{m_0 - rt}$. So

$$\int dv = A\left(\rho_e(v_e)^2 + p_e\right)\int \frac{dt}{m_0 - rt}.$$

The integral on the left is just the velocity v. To solve the one on the right, let $u = m_0 - rt$. So $du = -r\,dt$, and

$$\int \frac{dt}{m_0 - rt} = -\frac{1}{r}\int \frac{du}{u} = -\frac{1}{r}\ln u + \text{constant}$$

$$= -\frac{1}{r}\ln(m_0 - rt) + \text{constant}.$$

Therefore,

$$v(t) = -\frac{1}{r}A\left(\rho_e(v_e)^2 + p_e\right)\ln(m_0 - rt) + C.$$

At $t = 0$, the velocity is

$$v_0 = v(0) = -\frac{1}{r}A\left(\rho_e(v_e)^2 + p_e\right)\ln m_0 + C.$$

After solving for C and substituting, we get

$$v(t) = \frac{1}{r}A\left(\rho_e(v_e)^2 + p_e\right)\ln m_0 + v_0$$

$$- \frac{1}{r}A\left(\rho_e(v_e)^2 + p_e\right)\ln(m_0 - rt)$$

$$= \frac{1}{r}A\left(\rho_e(v_e)^2 + p_e\right)\left(\ln m_0 - \ln(m_0 - rt)\right) + v_0$$

$$= \frac{1}{r}A\left(\rho_e(v_e)^2 + p_e\right)\ln\left(\frac{m_0}{m_0 - rt}\right) + v_0.$$

This solution of the rocket equation gives us the velocity of the rocket at any time $t \geq 0$ in terms of the parameters.

Example 14.11. Suppose that a V-2 rocket is in flight at time $t = 0$ between the Moon and the Earth. Suppose that it has a velocity of $v_0 = 6000$ feet per second and a mass of $m_0 = 800$ slugs. This corresponds to a weight of $(800)(32.17) = 25{,}736$ pounds on the surface of the Earth. It has 197 slugs of fuel at time $t = 0$, and this is burning at a rate $r = 8.55$ slugs per second. What will the velocity of the rocket be at the time it runs out of fuel?

This question can be answered by inserting the data for the V-2 already provided in Section 14.4A into the velocity formula just derived. Note that the fuel will last for $t = \frac{197}{8.55} = 23.0$ seconds. Recall next that $\rho_e A(v_e)^2 = T_{\mathrm{mo}} = 56{,}100$ pounds, $A = 4.60$ ft^2, and $p_e \approx 2{,}840$ pounds/ft^2. So $\rho_e(v_e)^2 \approx 12{,}200$. Therefore, when the rocket runs out of fuel, it has a velocity of

$$v(t) = \frac{1}{r} A \left(\rho_e(v_e)^2 + p_e \right) \ln \left(\frac{m_0}{m_0 - rt} \right) + v_0$$

$$\approx \frac{1}{8.55} 4.60(12{,}200 + 2{,}840)$$

$$\times \ln \left(\frac{800}{800 - (8.55)(23.0)} \right) + 6000$$

$$\approx 8{,}090(\ln 1.33) + 6000$$

$$\approx 8{,}300 \text{ feet/sec.}$$

This example completes our mathematical analysis of rockets. The development of rockets (both the theory and the hardware) have become a central part of the world's scientific and technological agenda, especially in the context of space exploration and the placement of communication satellites in orbit around the Earth. This began in 1957 when the Russians launched the first artificial satellite, "Sputnik." This launch (and its military implications) initiated a "space race" between the United States and the Soviet Union. The high point of these efforts to "conquer space" was the development of the Saturn V rocket and the American Apollo Moon program. The Saturn V was a three-stage rocket that featured the most powerful rocket engines ever made. The first stage was propelled by five engines, each of which generated a thrust of 1,510,000 pounds. For

the Moon missions this rocket propelled a service module with three astronauts and an attached lunar landing vehicle into orbit around the Earth. These (each equipped with its separate engine and guidance system) continued together to the Moon. After establishing orbit around the Moon, the lunar lander separated from the service module and was flown by two of the astronauts to the Moon's surface. The service module with the third astronaut remained in orbit around the Moon. After the completion of the exploration of the Moon's surface, the lunar lander returned to the service vehicle, and the service vehicle returned the three astronauts to the vicinity of the Earth. At this point, the Apollo capsule that carried the astronauts separated from the service vehicle. The capsule reentered the Earth's atmosphere and landed by parachute in the Pacific. The first Moon landing occurred on July 20, 1969, during the Apollo 11 mission, and the last on December 20, 1972, during the Apollo 17 mission. They were incomparable feats of science, engineering, endurance, and courage.

14.5 Surprises about Gravity?

Some historians of science are of the opinion that one of the reasons why Newton delayed the publication of the *Principia* was a subtle difficulty involving the law of universal gravitation. The issue is this. Consider a situation where a larger body A of mass M exerts a gravitational force on a very small particle, or "point mass," P of mass m. The magnitude F of this force as given by Newton's universal law of gravitation is

$$(*) \qquad F = G\frac{mM}{d^2}$$

where d is the distance from P to the center of mass of A. Right? Surprise! This is false in general! Fortunately, the formula is correct in the most important situations: the case where the large attracting body

A is the Sun, the Earth, a planet, or, more generally, a sphere whose mass is distributed in a certain homogeneous way.

It is the purpose of this section to look at an example where the formula fails and then to verify that it is correct in the case of a sphere that is made up of homogeneous shells (think of the way an onion is layered). The basic strategies of integral calculus will be the key ingredient in this discussion.

A. Gravity and Shape.   To be able to analyze what is going on, consider the larger mass A as a sum total of very small masses or particles, say each of mass M_i. Focus on one of these small particles. It is true that the magnitude of the force of attraction between it and the particle P is given by Newton's law

$$G\frac{mM_i}{d^2}$$

where d is the distance between the two particles. Each one of the particles of the larger mass attracts P, and the gravitational force exerted by A is the resultant of all of these smaller gravitational forces. But as we will now see, this resultant force depends on the geometry of the object, and its magnitude is not necessarily given by Newton's law.

Start by taking A to be a thin homogeneous wire bent into a circle of radius R, and let the point mass P lie on the perpendicular to the circle through its center at a distance c from the center. Divide the circular wire into a larger number, say n, small equal pieces, each of mass M_i. So $M = nM_i$. Refer to Figure 14.31 and notice that the distance between each of the small pieces and P is equal to $\sqrt{c^2 + R^2}$. By Newton's law, the force with which each small piece pulls on P has magnitude

$$\frac{GmM_i}{(\sqrt{c^2 + R^2})^2} = \frac{GmM_i}{c^2 + R^2}.$$

Decompose this force into the two components shown in Figure 14.32. One component points from P to the center of the circle, and the other lies in the plane of the circle and points from the center to the small piece on it. Notice that the magnitude of the

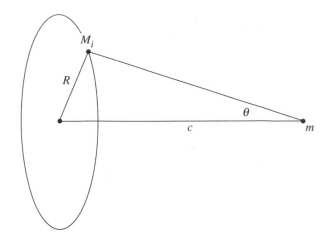

Figure 14.31

component that points to the center of the circle is equal to $F_i = \frac{GmM_i}{c^2+R^2}\cos\theta$. (See Section 9.1B.) Refer to Figure 14.31, to see that $\cos\theta = \frac{c}{\sqrt{c^2+R^2}}$. Therefore,

$$F_i = \frac{GmM_i}{c^2 + R^2} \cdot \frac{c}{\sqrt{c^2 + R^2}} = \frac{GmM_ic}{(c^2 + R^2)^{3/2}}.$$

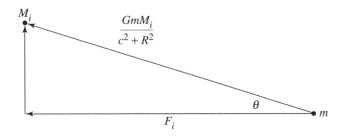

Figure 14.32

Let's now consider the resultant of the forces produced by all the n pieces on the circle. Consider first the sum of the components that lie in the plane of the circle. As we move completely around the circle from one small piece to the next, we see—refer to Figure 14.33—that these forces add to 0. Each one has a counterpart that pulls in the opposite direction. As to the n components pulling towards the center of the circle, note that they are all equal, and that their

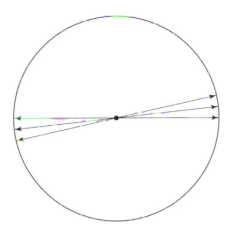

Figure 14.33

magnitudes add up to

$$\sum_{i=1}^{n} F_i = \sum_{i=1}^{n} \frac{GmM_ic}{(c^2 + R^2)^{3/2}} = n\frac{GmM_ic}{(c^2 + R^2)^{3/2}}$$

$$= \frac{Gm \cdot nM_ic}{(c^2 + R^2)^{3/2}} = \frac{GmMc}{(c^2 + R^2)^{3/2}}.$$

The argument just completed has informed us that the gravitational force with which the circular wire of mass M and radius R attracts the particle P of mass m is directed to the center of the circle, and that its magnitude is

$$F = \frac{GmMc}{(c^2 + R^2)^{3/2}}$$

where c is the distance from P to the center of the circle. Because the center of mass of the circular wire is the center of the circle, the gravitational force does point from P to the center of mass of the wire. However, its magnitude is *not* given by Newton's law $F = \frac{GmM}{c^2}$.

B. The Case of the Sphere. We continue to assume that P is a particle of mass m, but we turn now to the situation where the larger mass A is a homogeneous spherical shell of radius R and mass M. Suppose that P is outside the shell, a distance c from its center. We will analyze the gravitational force that the

shell exerts on the particle. Note that we are only considering the "skin" of the sphere and not the inside.

Begin by placing an x-axis through both the center of the sphere and the particle P in such a way that the origin falls on the center of the sphere and the particle has x-coordinate c on the positive x-axis. (Look ahead to Figure 14.35.) Now let

$$-R = x_0 < x_1 < x_2 < \cdots < x_{i-1}$$

$$< x_i < \cdots < x_{n-1} < x_n = R$$

be a partition of the interval $[-R, R]$ into n subintervals of equal lengths. So if Δx_1 is the length of the first subinterval, Δx_2 that of the second, $\ldots$, Δx_i that of the ith, $\ldots$, and Δx_n that of the last, then $\Delta x_1 = \Delta x_2 = \cdots = \Delta x_i = \cdots = \Delta x_n$. Since $[-R, R]$ has length $2R$, notice that $n\Delta x_i = 2R$. So the common length of each subinterval is $\Delta x_i = \frac{2R}{n}$ for all i. Suppose that there are many subintervals and hence that their common length is very small.

Take p_1 to be the midpoint of the first subinterval, p_2 that of the second, and so on as shown in the accompanying diagram.

$$\begin{array}{c}\underset{-R=x_0 \quad x_1}{\overset{p_1}{\bullet}} \underset{x_2}{\overset{p_2}{\bullet}} \underset{x_3}{\overset{p_3}{\bullet}} \cdots \underset{x_{i-1} \quad x_i}{\overset{p_i}{\bullet}} \cdots \underset{x_{n-1} \ x_n=R}{\overset{p_n}{\bullet}}\end{array}$$

All these points lie on the diameter of the sphere. Typical points are shown in Figure 14.34. Now cut the sphere with planes perpendicular to this diameter through the points

$$x_0 < x_1 < x_2 < x_3 < \cdots < x_{i-1} < x_i < \cdots < x_{n-1} < x_n.$$

This decomposes the sphere into n circular bands. (Again, note that we are only considering the surface and not the inside.) Figure 14.35 shows two of them. All these bands have the same thickness, and therefore—by Example 13.6 in Section 13.2—the same surface area. Therefore, each circular band has the same mass $M_i = \frac{M}{n}$. Because the bands are thin, think of each band as a circular wire. In particular, regard the ith band as a circular wire of mass M_i with center at p_i. It follows from the discussion of Section 14.5A that the gravitational force of the ith circular band on the mass m is directed to the center of the circle and has magnitude

$$F_i = \frac{GmM_i c_i}{(c_i^2 + R_i^2)^{3/2}},$$

where R_i is the radius of the circle and c_i is the distance from the center p_i of the circle to the mass m. Notice that c_i is equal to $c_i = c - p_i$, regardless of whether p_i is positive or negative.

Now put in the y-axis. Since the circle of Figure 14.34 has center 0 and radius R, its equation is $x^2 + y^2 =$

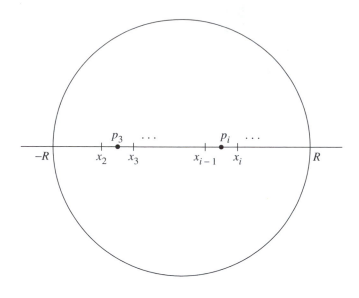

Figure 14.34

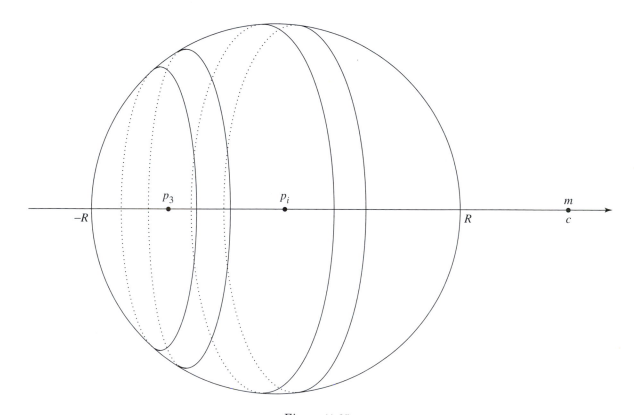

Figure 14.35

R^2, or $y^2 = R^2 - x^2$. So by a look at Figure 14.35 the radius R_i of the ith circular band satisfies $R_i^2 = R^2 - p_i^2$. So

$$c_i^2 + R_i^2 = (c - p_i)^2 + R^2 - p_i^2$$

$$= c^2 - 2cp_i + p_i^2 + R^2 - p_i^2 = c^2 - 2cp_i + R^2.$$

Since $\frac{1}{n} = \frac{\Delta x_i}{2R}$, each circular band has mass $M_i = \frac{M \Delta x_i}{2R}$. (Recall that M is the mass of the entire spherical shell.) Making the appropriate substitutions into the formula for F_i gives us

$$F_i = \frac{GmM}{2R} \left[\frac{c_i}{(c_i^2 + R_i^2)^{3/2}} \right] \Delta x_i$$

$$= \frac{GmM}{2R} \left[\frac{c - p_i}{(c^2 - 2cp_i + R^2)^{3/2}} \right] \Delta x_i.$$

A look at Figure 14.35 shows that the gravitational force of the entire spherical shell on the mass m is approximated by adding the effect of the gravitational forces determined by all the n circular bands. Therefore, this force points to the center of the sphere, and its magnitude F is approximately equal to

$$F \approx \sum_{i=1}^{n} F_i = \sum_{i=1}^{n} \frac{GmM}{2R} \left[\frac{c - p_i}{(c^2 - 2cp_i + R^2)^{3/2}} \right] \Delta x_i$$

$$= \frac{GmM}{2R} \sum_{i=1}^{n} \left[\frac{c - p_i}{(c^2 - 2cp_i + R^2)^{3/2}} \right] \Delta x_i.$$

Finer and finer partitions create more and more circular bands that are ever thinner and ever closer to being "circles with mass." So the accuracy of the approximation will be improved step by step. We can conclude (see Section 13.1A) that

$$F = \frac{GmM}{2R} \lim_{\|P\| \to 0} \sum_{i=1}^{n} \left[\frac{c - p_i}{(c^2 - 2cp_i + R^2)^{3/2}} \right] \Delta x_i.$$

Because

$$\sum_{i=1}^{n} \left[\frac{c - p_i}{(c^2 - 2cp_i + R^2)^{3/2}} \right] \Delta x_i$$

is a Riemann sum for the function

$$f(x) = \frac{c - x}{(c^2 - 2cx + R^2)^{3/2}}, \qquad -R \le x \le R,$$

it follows by the analysis of Section 13.1A that

$$\lim_{\|P\| \to 0} \sum_{i=1}^{n} \left[\frac{c - p_i}{(c^2 - 2cp_i + R^2)^{3/2}} \right] \Delta x_i$$

$$= \int_{-R}^{R} \frac{c - x}{(c^2 - 2cx + R^2)^{3/2}} \, dx.$$

To complete this investigation, it remains to consider

$$\int \frac{c - x}{(c^2 - 2cx + R^2)^{3/2}} \, dx.$$

This integral can be evaluated by the method of integration by parts as explained in Section 13.3C. Let

$$u = c - x \qquad \text{and} \qquad dv = \frac{dx}{(c^2 - 2cx + R^2)^{3/2}}.$$

Clearly, $du = -dx$. To find v, i.e., an antiderivative of dv, make the substitution $z = c^2 - 2cx + R^2$. Because $dz = -2c \, dx$, we see that $dv = -\frac{1}{2c} \frac{dz}{z^{3/2}} = -\frac{1}{2c} z^{-3/2} \, dz$. So $v = \frac{1}{c} z^{-1/2} = \frac{1}{c}(c^2 - 2cx + R^2)^{-1/2}$. By the integration by parts formula,

$$\int \frac{c - x}{(c^2 - 2cx + R^2)^{3/2}} \, dx = \int u \, dv = uv - \int v \, du$$

$$= \frac{1}{c}(c - x)(c^2 - 2cx + R^2)^{-1/2}$$

$$+ \int \frac{1}{c}(c^2 - 2cx + R^2)^{-1/2} \, dx.$$

This last integral falls to the method of substitution. With $z = c^2 - 2cx + R^2$, we get $dz = -2c \, dx$, and hence that

$$\int \frac{1}{c}(c^2 - 2cx + R^2)^{-1/2} \, dx = -\int \frac{1}{2c^2} z^{-1/2} \, dz$$

$$= -\frac{1}{c^2} z^{1/2} + C = -\frac{1}{c^2}(c^2 - 2cx + R^2)^{1/2} + C.$$

Putting all the pieces together, we can conclude that

$$\int \frac{c - x}{(c^2 - 2cx + R^2)^{3/2}}\, dx$$

$$= \frac{1}{c}(c - x)(c^2 - 2cx + R^2)^{-1/2}$$

$$- \frac{1}{c^2}(c^2 - 2cx + R^2)^{1/2} + C.$$

Because $c \geq R$, we get

$$(c^2 - 2cR + R^2)^{-1/2} = \left((c - R)^2\right)^{-1/2} = (c - R)^{-1}.$$

In a similar way,

$$(c - 2cR + R^2)^{1/2} = c - R,$$

$$(c^2 + 2cR + R^2)^{-1/2} = (c + R)^{-1},$$

and

$$(c^2 + 2cR + R)^{1/2} = c + R.$$

Substituting all this into the solution of the indefinite integral that was just derived, we get

$$\int_{-R}^{R} \frac{c - x}{(c^2 - 2cx + R^2)^{3/2}}\, dx$$

$$= \left(\frac{1}{c}(c - R)(c^2 - 2cR + R^2)^{-1/2}\right.$$

$$\left. - \frac{1}{c^2}(c^2 - 2cR + R^2)^{1/2}\right)$$

$$- \left(\frac{1}{c}(c + R)(c^2 + 2cR + R^2)^{-1/2}\right.$$

$$\left. - \frac{1}{c^2}(c^2 + 2cR + R^2)^{1/2}\right)$$

$$= \frac{1}{c}(c - R)(c - R)^{-1} - \frac{1}{c^2}(c - R)$$

$$- \frac{1}{c}(c + R)(c + R)^{-1} + \frac{1}{c^2}(c + R)$$

$$= \frac{1}{c^2}(R - c) + \frac{1}{c^2}(c + R) = \frac{2R}{c^2}.$$

Recalling that

$$F = \frac{GmM}{2R} \lim_{\|P\| \to 0} \sum_{i=1}^{n} \left[\frac{c - p_i}{(c^2 - 2cp_i + R^2)^{3/2}}\right] \Delta x_i,$$

we find after one last substitution that

$$F = \frac{GmM}{2R} \cdot \frac{2R}{c^2} = \frac{GmM}{c^2}.$$

We have arrived at "the bottom line" of our discussion and it is this. The gravitational force with which a homogenous spherical shell A attracts a particle P positioned outside it behaves in accordance with Newton's formula (∗): It is directed towards the center (which is also the center of mass) of the spherical shell A and its magnitude is equal to G times the product of the masses of P and A divided by the square of the distance from P to the center of A.

We conclude this section with some consequences of this fact. Consider a sphere of matter. Now it is not only the outer shell that we have in mind, but the inside as well. Assume that the matter of the sphere is arranged as follows: any two small particles have the same density if they are the same distance from the center of the sphere. The Sun, the planets, and the Moon satisfy this property, at least in an approximate way. Since such a sphere can be considered to be composed of many concentric homogeneous shells—think of an onion—and since each shell attracts a small particle P in accordance with Newton's formula (∗), the entire solid sphere attracts P according to this formula. Suppose, finally, that two spheres A and B of matter satisfy this "onion property." We know that every particle of B is attracted by A in accordance with formula (∗) where d is the distance from P to the center of A. From the point of view of B, therefore, the entire mass of A can be considered to be concentrated at the center of A. In other words, A can be considered as a point mass. But this point mass is attracted by B in accordance with formula (∗). Therefore, Newton's formula holds for the two spheres A and B: The gravitational force of attraction is equal to G times the product of their masses divided by the distance between their centers.

The problem of the gravitational attraction between two bodies in the solar system, as you have just seen, is a formidable problem. It must have challenged Newton and could have contributed to the delay in the publication of the *Principia*.

14.6 **Returning to Newton's Principia**

T urn to the various diagrams of Chapter 7 and compare them with the basic setup of the polar coordinate system as developed in Section 13.4. Such a comparison suggests that the matter of centripetal force and planetary motion is best pursued within the framework of such a coordinate system. This modern approach is the topic of this section. The approach is actually not that modern. It was first published by Euler in 1749 and appeared in textbooks a few decades later. It can be found, for example, in the interesting (and also demanding) two-volume work

J. M. F. Wright, *A Commentary on Newton's Principia*, London: printed for T. & J. Tegg, 73, Cheapside; and Richard Griffin & Co., Glasgow, 1833.

This textbook was "designed for the use of students at the universities" and was "dedicated to the tutors of the several Colleges at Cambridge."[7]

A. Forces and Polar Coordinates.   Suppose that an object P of nonzero mass m is in motion and subject to a single force that acts in the direction of a fixed point S. The motion of P takes place in a plane containing S. See Figure 14.36. If P were to be at S, then the force would have no defined direction, so we will rule this possibility out. This is also reasonable in practical terms; think of the case where P is a planet and S the center of the Sun. Place an x-y coordinate system so that S is at the origin. Refer to Figure 14.37. Let P be in typical position and let x and y be the coordinates of P. Since P moves, both x and y vary

with time. Therefore, $x = x(t)$ and $y = y(t)$ are both functions of time $t \geq 0$.

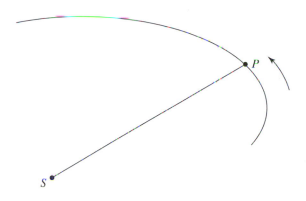

Figure 14.36

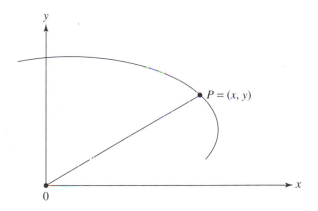

Figure 14.37

Let r be the distance from P to S. Because the positions of P and S cannot coincide, $r > 0$. Let F be the centripetal force and let F_x and F_y be the components of this force in the x- and y-directions respectively. Observe that F, F_x, F_y, and r are all functions of t. By the parallelogram law of forces and similar triangles,

$$\frac{F_x}{F} = \frac{x}{r} \quad \text{and} \quad \frac{F_y}{F} = \frac{y}{r}.$$

See Figure 14.38. As a consequence of Newton's second law, we get

$$F_x = -m\frac{d^2x}{dt^2} \quad \text{and} \quad F_y = -m\frac{d^2y}{dt^2}.$$

The minus signs are necessary, since the x-component of the force acts in the direction of negative x when x is positive and in the positive x-direction when x is negative, and analogously for y. Combining the equations just derived gives us

$$m\frac{d^2x}{dt^2} = -F\frac{x}{r} \quad \text{and} \quad m\frac{d^2y}{dt^2} = -F\frac{y}{r}.$$

It follows that

$$m\left(y\frac{d^2x}{dt^2} - x\frac{d^2y}{dt^2}\right) = my\frac{d^2x}{dt^2} - mx\frac{d^2y}{dt^2}$$

$$= -F\frac{yx}{r} + F\frac{xy}{r} = 0,$$

and hence that

$$(*) \qquad y\frac{d^2x}{dt^2} = x\frac{d^2y}{dt^2}.$$

Consider the difference $x \cdot \frac{dy}{dt} - y \cdot \frac{dx}{dt}$. By the product rule, the derivative of this term is

$$\frac{d}{dt}\left(x \cdot \frac{dy}{dt} - y \cdot \frac{dx}{dt}\right) = \frac{d}{dt}\left(x \cdot \frac{dy}{dt}\right) - \frac{d}{dt}\left(y \cdot \frac{dx}{dt}\right)$$

$$= \left(\frac{dx}{dt} \cdot \frac{dy}{dt} + x \cdot \frac{d^2y}{dt^2}\right) - \left(\frac{dy}{dt} \cdot \frac{dx}{dt} + y \cdot \frac{d^2x}{dt^2}\right)$$

$$= \frac{dx}{dt} \cdot \frac{dy}{dt} - \frac{dy}{dt} \cdot \frac{dx}{dt} + x \cdot \frac{d^2y}{dt^2} - y \cdot \frac{d^2x}{dt^2}.$$

In view of the equality $(*)$ already verified, it follows that $\frac{d}{dt}(y \cdot \frac{dx}{dt} - x \cdot \frac{dy}{dt}) = 0$. We can therefore conclude that

$$x \cdot \frac{dy}{dt} - y \cdot \frac{dx}{dt} = c,$$

where c is a constant. This fact contains essential information about the motion of P. To extract it we will now transfer our discussion into the context of polar coordinates.

Consider the positive x-axis as polar axis (so S is at the pole) and let (r, θ) be the polar coordinates of

P. See Figure 14.39. Both $r = r(t)$ and $\theta = \theta(t)$ are functions of t. Recall from Chapter 13.4A that

$$x = r\cos\theta \quad \text{and} \quad y = r\sin\theta.$$

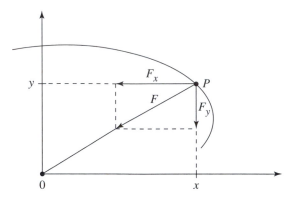

Figure 14.38

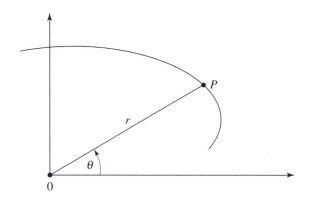

Figure 14.39

We will now rewrite the equality $x \cdot \frac{dy}{dt} - y \cdot \frac{dx}{dt} = c$ in terms of the variables r and θ. By the product rule,

$$\frac{dx}{dt} = \frac{dr}{dt}\cos\theta + r\frac{d}{dt}(\cos\theta).$$

Since $\theta = \theta(t)$ is a function of t, we get by the chain rule that $\frac{d}{dt}(\cos\theta) = -\sin\theta \cdot \frac{d\theta}{dt}$ and hence that

$$\frac{dx}{dt} = \frac{dr}{dt}\cos\theta - r\sin\theta \cdot \frac{d\theta}{dt}.$$

By the same argument,

$$\frac{dy}{dt} = \frac{dr}{dt}\sin\theta + r\cos\theta \cdot \frac{d\theta}{dt}.$$

Combining the equations $x = r\cos\theta$ and $y = r\sin\theta$ with the two equations just derived, we get

$$x \cdot \frac{dy}{dt} - y \cdot \frac{dx}{dt} = (r\cos\theta)\left(\frac{dr}{dt}\sin\theta + r\cos\theta \cdot \frac{d\theta}{dt}\right)$$

$$- (r\sin\theta)\left(\frac{dr}{dt}\cos\theta - r\sin\theta \cdot \frac{d\theta}{dt}\right)$$

$$= (r^2\cos^2\theta)\frac{d\theta}{dt} + (r^2\sin^2\theta)\frac{d\theta}{dt}$$

$$= r^2(\sin^2\theta + \cos^2\theta)\frac{d\theta}{dt}$$

$$= r^2\frac{d\theta}{dt}.$$

Observe that we have verified that

$$r^2\theta'(t) = r^2\frac{d\theta}{dt} = c.$$

This fact has an immediate payoff. Consider P at two different times in its orbit, say at time t_0 and then again at time t_1. Let A be the area that is swept out by the segment SP in the time interval $[t_0, t_1]$. Let $\theta(t_0) = a$ and $\theta(t_1) = b$. Refer to Figure 14.40. Recall from Section 13.4B that

$$A = \int_a^b \frac{1}{2}f(\theta)^2 d\theta$$

where $r = f(\theta)$ is the polar equation of the orbit. See Figure 14.39. Refer to the method of substitution of Section 13.3A. With the substitution $\theta = \theta(t)$ and $d\theta = \theta'(t)\,dt$, we get

$$A = \int_a^b \frac{1}{2}f(\theta)^2\,d\theta = \int_{t_0}^{t_1}\frac{1}{2}r^2\theta'(t)\,dt = \int_{t_0}^{t_1}\frac{1}{2}c\,dt$$

$$= \frac{1}{2}ct\Big|_{t_0}^{t_1} = \frac{1}{2}c(t_1 - t_0).$$

It follows that the area swept out by P depends only on the time span $t_1 - t_0$. Kepler's second law is an immediate consequence: The object P sweeps out equal areas in equal time intervals. Compare this verification of Kepler's second law with that of Newton as carried out in Section 7.1. Which is mathematically more compelling?

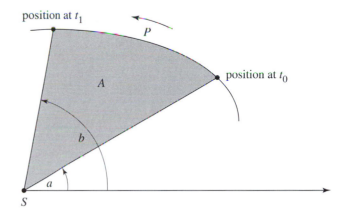

position at t_1

P

position at t_0

A

b

a

S

Figure 14.40

Let $t = t_1 - t_0$ be the time needed to sweep out the area A. Then

$$\frac{A}{t} = \frac{1}{2}c.$$

A look at the definition of Kepler's constant κ in Section 7.1 shows that $\frac{1}{2}c = \kappa$ and hence that $c = 2\kappa$.

B. The Inverse Square Law. We begin this discussion by converting the force equations

$$m\frac{d^2x}{dt^2} = -F\frac{x}{r} \quad \text{and} \quad m\frac{d^2y}{dt^2} = -F\frac{y}{r}$$

into the polar variables r and θ. (We will see that in fact r alone suffices.) From Section 14.6A, we know that $r^2\frac{d\theta}{dt} = 2\kappa$. Because this is a constant, we get that $0 = \frac{d}{dt}(r^2\frac{d\theta}{dt}) = 2r\frac{dr}{dt}\cdot\frac{d\theta}{dt} + r^2\cdot\frac{d^2\theta}{dt^2}$ and hence that

$$2\frac{dr}{dt}\cdot\frac{d\theta}{dt} + r\cdot\frac{d^2\theta}{dt^2} = 0.$$

Recall from Section 14.6A that

$$\frac{dx}{dt} = \frac{dr}{dt}\cos\theta - r\sin\theta \cdot \frac{d\theta}{dt}.$$

By differentiating this equation (use the product and chain rules several times), we get that

$$\frac{d^2x}{dt^2} = \left(\frac{d^2r}{dt^2}\cos\theta - \frac{dr}{dt}\sin\theta \cdot \frac{d\theta}{dt}\right) - \frac{dr}{dt}\sin\theta \cdot \frac{d\theta}{dt}$$

$$- r\left(\cos\theta \cdot \frac{d\theta}{dt} \cdot \frac{d\theta}{dt} + \sin\theta \cdot \frac{d^2\theta}{dt^2}\right)$$

$$= \frac{d^2r}{dt^2}\cos\theta - 2\frac{dr}{dt}\sin\theta \cdot \frac{d\theta}{dt} - r\cos\theta \cdot \left(\frac{d\theta}{dt}\right)^2$$

$$- r\sin\theta \cdot \frac{d^2\theta}{dt^2}$$

$$= \frac{d^2r}{dt^2}\cos\theta - r\cos\theta \cdot \left(\frac{d\theta}{dt}\right)^2$$

$$- \sin\theta\left(2\frac{dr}{dt} \cdot \frac{d\theta}{dt} + r \cdot \frac{d^2\theta}{dt^2}\right).$$

It follows that

$$\frac{d^2x}{dt^2} = \cos\theta\left[\frac{d^2r}{dt^2} - r \cdot \left(\frac{d\theta}{dt}\right)^2\right].$$

It can be shown in exactly the same way that

$$\frac{d^2y}{dt^2} = \sin\theta\left[\frac{d^2r}{dt^2} - r \cdot \left(\frac{d\theta}{dt}\right)^2\right].$$

In view of the equations $-rm\frac{d^2x}{dt^2} = Fx = Fr\cos\theta$ and $-rm\frac{d^2y}{dt^2} = Fy = Fr\sin\theta$, we now get that

$$F\cos\theta = m\cos\theta\left[r \cdot \left(\frac{d\theta}{dt}\right)^2 - \frac{d^2r}{dt^2}\right]$$

and

$$F\sin\theta = m\sin\theta\left[r \cdot \left(\frac{d\theta}{dt}\right)^2 - \frac{d^2r}{dt^2}\right].$$

A look at their graphs shows that $\sin\theta$ and $\cos\theta$ are never equal to 0 simultaneously. (Compare Figures

10.28 and 10.29 of Section 10.8.) So it follows by a cancellation that

$$\boxed{F = m\left[r \cdot \left(\frac{d\theta}{dt}\right)^2 - \frac{d^2r}{dt^2}\right]}$$

Now recall that $r^2\frac{d\theta}{dt} = 2\kappa$. Therefore, $\frac{d\theta}{dt} = \frac{2\kappa}{r^2}$, and hence

$$F = m\left(r \cdot \frac{4\kappa^2}{r^4} - \frac{d^2r}{dt^2}\right) = m\left(\frac{4\kappa^2}{r^3} - \frac{d^2r}{dt^2}\right).$$

Observe that our preliminary goal has been achieved. The centripetal force F has been expressed in terms of the distance function $r(t)$ and the constants m and κ.

We will now assume that the orbit of P—note that we have made no assumptions about the geometry of the orbit so far—is a conic section: either an ellipse, a parabola, or a hyperbola. A careful review of the analysis of the equation

$$r = \frac{L}{2(1 + \varepsilon\cos\theta)}$$

undertaken in Section 13.4A and Exercises 13L shows that it is possible to: i) choose ε and L, and ii) place the polar axis on the axis of the conic section with the pole (the position S to which the force is directed) at a focal point, in such a way that

$$r = \frac{L}{2(1 + \varepsilon\cos\theta)}$$

is the polar equation of the conic section. The parameter L is the latus rectum of the conic section. If the orbit is an ellipse, then ε is its astronomical eccentricity and falls in the range $0 \leq \varepsilon < 1$. If the orbit is a parabola, then $\varepsilon = 1$, and if it is a hyperbola, $\varepsilon > 1$.
Differentiate the equation

$$r = \frac{1}{2}L(1 + \varepsilon\cos\theta)^{-1}$$

(apply the chain rule) and use the fact that $\frac{d\theta}{dt} = \frac{2\kappa}{r^2}$ to get

$$\frac{dr}{dt} = -\frac{1}{2}L(1 + \varepsilon \cos\theta)^{-2}(-\varepsilon \sin\theta)\frac{d\theta}{dt}$$

$$= \frac{1}{2}L(1 + \varepsilon \cos\theta)^{-2}(\varepsilon \sin\theta)\frac{2\kappa}{r^2}.$$

Because $r^2 = \frac{1}{4}L^2(1 + \varepsilon \cos\theta)^{-2}$, we see that

$$L(1 + \varepsilon \cos\theta)^{-2} = \frac{4r^2}{L}.$$

So the above equation can be rewritten as

$$\frac{dr}{dt} = \frac{4r^2}{L}(\varepsilon \sin\theta)\frac{\kappa}{r^2} = \frac{4\varepsilon\kappa}{L}\sin\theta.$$

Applying the chain rule to $\frac{dr}{dt} = \frac{4\varepsilon\kappa}{L}\sin\theta$, we get

$$\frac{d^2r}{dt^2} = \frac{4\varepsilon\kappa}{L}(\cos\theta)\frac{d\theta}{dt} = \frac{4\varepsilon\kappa}{L}(\cos\theta)\frac{2\kappa}{r^2}$$

$$= \frac{8\varepsilon\kappa^2}{L}(\cos\theta)\frac{1}{r^2}.$$

Substituting this into the earlier formula for the centripetal force F, we obtain

$$F = m\left(\frac{4\kappa^2}{r^3} - \frac{d^2r}{dt^2}\right) = m\left(\frac{4\kappa^2}{r^3} - \frac{8\varepsilon\kappa^2}{L}(\cos\theta)\frac{1}{r^2}\right)$$

$$= 4m\kappa^2\left(\frac{1}{r} - \frac{2\varepsilon}{L}\cos\theta\right)\frac{1}{r^2}.$$

Because $r = \frac{L}{2(1 + \varepsilon \cos\theta)}$, we see that

$$\frac{1}{r} = \frac{2(1 + \varepsilon \cos\theta)}{L} = \frac{2}{L} + \frac{2\varepsilon}{L}\cos\theta$$

and therefore

$$F = 4m\kappa^2\left(\frac{2}{L}\right)\frac{1}{r^2}.$$

What has been accomplished? We have established that if an object that is propelled by the action of a centripetal force has an orbit that is either an ellipse, a parabola, or a hyperbola, then the magnitude of the force satisfies the inverse square law

$$F = \frac{8m\kappa^2}{L}\frac{1}{r^2}$$

where m is the mass of the object, κ is Kepler's constant and L the latus rectum of the orbit, and r is the distance from the object to the point of origin of the force.

It has become tradition in modern mathematics and certain areas of physics to follow an axiomatic approach. In other words, the definitive versions of theories[8] of mathematics and physics are often cast in the following way: Certain basic underlying laws or principles, referred to as axioms, are taken as starting point, and the remaining relevant propositions and principles of the theory are deduced from these by the force of logic and mathematics alone. This can be done with the theory of gravitation. Taking Newton's three basic laws of motion (see Chapter 7) and his law of universal gravitation $F = G\frac{m_1 \cdot m_2}{r^2}$ as axioms, Kepler's three laws of planetary motion can be derived mathematically. Exercises 14I will lead you through these derivations.

14.7 Hubble's Law and Einstein's Universe

From the beginning of the 18th century to the middle of the 19th, Newtonian science reigned supreme over the physical universe. But then, history began to repeat itself (ultimately more dramatically than ever). The problem of "discrepancy" between theory and observation that had brought about the downfall of the astronomy of the Greeks, and later that of Copernicus, was now beginning to threaten the theory of Newton. What was the problem? The location of Mercury's perihelion—this is the point in the orbit at which the planet is closest to the Sun—changes position from one orbit to the next. In other words, Mercury's ellipse undergoes a rotation. This change in position is predicted by Newton's theory. However, the precise amount of this change was discovered to be a little greater than predicted. This discovery was made by the Frenchman Leverrier in the middle of the 19th century. When the Newtonian theory could not explain this difference away, the stage was set for another revolution.

Newton's theory of gravitation explains planetary motion by asserting that the celestial bodies pull on each other across millions of miles of empty space with a force that is inversely proportional to the square of the distance between them. This stood as the ultimate explanation of why the planets move the way they do, until Albert Einstein, in 1916, laid the foundations of his General Theory of Relativity. Einstein took the point of view that heavenly bodies don't pull on each other across empty space, but rather that the masses of the objects in the universe cause space itself to be curved. The force of gravity disappeared and was replaced by the geometry of space itself. Matter curves space, and what we call gravitation is but the acceleration of objects as they glide along the undulations of space like a golf ball on a challenging putting green. In particular, the elliptical orbits of the planets are explained in this way. Einstein's theory makes precise mathematical sense of a dynamically evolving curved space-time determined by the stars and planets (indeed all of the mass) in the universe. While the basic concept is simple, the mathematics is formidable. It relies on the non-Euclidean geometries of Gauss and Riemann (that we have mentioned before in passing).

The theory was not only elegant, but it also held up against observations. When applied to the orbit of Mercury, it showed that the perihelion position would advance 574 seconds. It therefore accounted for the 43 seconds that had eluded Newton's astronomy.[9] This was the first experimental evidence that Newton's theory is only an approximation of Einstein's general relativity. General relativity also predicts that light rays are bent by gravitational fields. Observations of a total solar eclipse in 1919 dramatically confirmed that this occurred precisely as general relativity asserted. Einstein's theory had (quite literally) eclipsed Newton's.

When Einstein began to investigate the implications of his theory for the structure of the universe, he found something disturbing. The theory implied that the universe could not be stationary, or static, but that it was either expanding or contracting. This was a strange idea for which there was no observational evidence whatsoever at the time. There was evidence for the motion of stars, but none of it suggested a consistent pattern of expansion or contraction. Einstein reluctantly concluded that there must be something wrong with his theory and modified[10] its equations by including a term that he called the "cosmological constant." Its purpose was to make the radius of the universe hold steady with time.

In 1917, the very year that Einstein introduced his constant, the first published evidence appeared that suggested that the universe is indeed expanding. An analysis of the light coming from certain galaxies revealed a shift of the spectrum towards the red, or lower, frequencies, an indication of the fact that these stars are moving away from the Earth, indeed, away from our own Milky Way galaxy. Only the examination of the distance-velocity data for many galaxies could confirm a general trend. The missing element was a reliable way of measuring the distances to the stars. In 1923, the American astronomer Edwin Hubble made the discovery that certain pulsating stars, known as Cepheid variables, can be used as reliable indicators of astronomical distances. When Hubble was able to place the "Andromeda nebula" well beyond the bounds of our galaxy, the question as to whether the universe is expanding was effectively settled. In fact, what emerged was a quantitative connection, known as *Hubble's law*:

Consider any galaxy. Let d be its distance from the Milky Way, and let the v be the velocity with which it is moving away from the Milky Way. Then the ratio $\frac{v}{d}$ is the same constant, no matter which galaxy is being considered. This constant is now known as *Hubble's constant*. We will denote it by H_0.

In the years since Hubble's discoveries, thousands of observations have extended and refined the measurements and have confirmed the general correctness of this velocity-distance relation. The vast distances in question are measured using the *mega light year*, defined to be equal to one million light years. So 1000 mega light years is 1 billion light years. For example, the galaxy cluster Ursa Major is a distance $d = 978$ mega light years away and receding at a rate of $v = 15,000$ kilometers per second. The cluster Corona Borealis is $d = 1400$ mega light years distant and receding at $v = 22,000$ km/sec. The cluster Hydra is $d = 4000$ mega light years away and moving away at a clip of $v = 61,000$ km/sec. The respective velocity/distance ratios are $15,000/978 = 15.3$ for Ursa Major, $22,000/1400 = 15.7$ for Corona Borealis, and $61,000/4000 = 15.3$ for Hydra. These three values suggest that H_0 is around 15 km/sec per mega light year.

Figure 14.41 is a pictorial confirmation of Hubble's law: If the (velocity, distance) points are plotted for all galaxy clusters (for which these data are known), then these points all lie close to a line. The data points originally collected by Hubble all fall into the rectangle shown near the origin. Notice that H_0 is equal to the slope of this line. The definitive value of H_0 is a very difficult and elusive target. The most recent estimates (still based on the observation of Cepheid variables) make use of the more precise distance and velocity measurements that the Hubble Space Telescope makes possible. This unique telescope is in orbit around the Earth. Because it is positioned beyond the Earth's atmosphere, the light that it collects is undisturbed by atmospheric effects. It therefore provides more accurate measurements of intergalactic distances and velocities. These measurements place H_0 in the range of

$$14 \leq H_0 \leq 23$$

km/sec per mega light year. It is known today that Hubble's constant has in fact varied over time. The value H_0 that we have considered is today's value. Incidentally, the Hubble telescope has also shown that there are many more galaxies in the universe than had been previously believed. Recent images produced by the telescope have given indication that there are

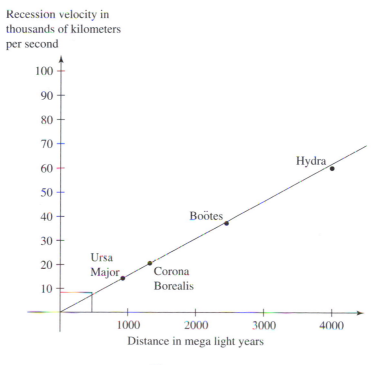

Figure 14.41

perhaps 50 billion galaxies, rather than the 10 billion estimated before.

Because there is nothing unique about the Milky Way galaxy (other than the fact that our own planet is a tiny speck in it), it must follow that the velocity-distance relationship just discussed holds from the vantage point of every galaxy. Therefore, the entire universe must be in a state of expansion. The "nonuniqueness" of the Milky Way also suggests that Hubble's law must hold for the entire universe. This observation has interesting consequences. Imagine, for example, a movie that features the universe as it has expanded over time. Now run the movie backwards and observe the galaxies and clusters of galaxies coming together, until you "see," at some time in the very distant past, all matter crowded together in an extremely dense state. You are now looking at the initial condition of the universe. From this initial state, the universe suddenly began its expansion. So by "playing" Hubble's expanding universe backwards, we can conclude that the universe began with a big explosion from a very dense state. This explanation is now known as the theory of the *Big Bang*.

Observations also indicate that the universe is homogeneous. In other words, the galaxies in the universe are distributed uniformly in all directions and at all distances. Modern astronomy has confirmed this so-called *cosmological principle*. Incidentally, the average density of the current universe is thought to be approximately 10^{-27} kg/m³. The discussion in Section 11.1A about the masses of the subatomic particles tells us that this corresponds to about 1 proton per cubic meter. So the universe is, on average, very empty.

The observations and principles that we have discussed have profound ramifications that are explored in the esoteric realm of mathematical astrophysics, also called cosmology. This discipline is as inaccessible to the nonspecialist as the distant galaxies themselves. However, some basic calculus does give us a glimpse at what is involved. This will be the purpose of the rest of this section.

Think of the universe as an evolving giant sphere and assume that its total mass M_u is a constant. (So we will ignore the exchange between mass and energy. See Exercises 11.B.) Let t denote elapsed time (in years, for example) since the Big Bang. Let $R = R(t)$ be the radius of the universe at any time $t \geq 0$. We will assume that the growth of the universe has occurred in a continuous and smooth flow and that it will continue to do so. In particular, we will assume that both its radius $R = R(t)$ and its rate of expansion $\frac{dR}{dt} = R'(t)$ are differentiable and hence continuous functions of time. (Review Sections 8.3 and 8.5B if necessary.) Consider an object of mass m "riding" at the outer edge of the universe, a distance $R(t)$ from its center. See Figure 14.42. (Think of the center of the universe at the origin of a coordinate axis and the object at $R(t)$ on the positive side of the axis.) Combining the cosmological principle with the conclusions of Section 14.5B, we see that the force of gravity on the object is

$$-\frac{GmM_u}{R(t)^2}.$$

(Gravity acts in the direction of the center of the universe and this is the negative direction.) By Newton's Second Law,

$$-\frac{GmM_u}{R(t)^2} = mR''(t).$$

It follows that $R''(t) = -\frac{GM_u}{R(t)^2}$. Multiplying both sides of this equation by the velocity $R'(t)$ of the expansion, we get

$$R''(t) \cdot R'(t) = -\frac{GM_u}{R(t)^2} \cdot R'(t).$$

By the chain rule, $\frac{1}{2}(R'(t))^2$ is an antiderivative of the left side. Notice that $\frac{GM_u}{R(t)} = GM_u \cdot R(t)^{-1}$ is an antiderivative of the right side. So the functions $\frac{1}{2}(R'(t))^2$ and $\frac{GM_u}{R(t)}$ have the same derivative. They therefore differ by a constant. Denoting this constant by E, we get that

$$(*) \qquad \frac{1}{2}\left(R'(t)\right)^2 - \frac{GM_u}{R(t)} = E.$$

It follows that $(R'(t))^2 = 2(E + \frac{GM_u}{R(t)})$. Note that $E + \frac{GM_u}{R(t)} \geq 0$. Since G and M_u are positive constants and $R(t)$ is positive (at least for $t > 0$), $\frac{GM_u}{R(t)}$ is also positive.

(A) Suppose that $E \geq 0$. We will see that under this assumption, the universe will continue to grow and that it will do so without bound. Observe first that if $R'(t)$ where to be equal to zero for some t, then $E + \frac{GM_u}{R(t)} = 0$. But this means that $E = -\frac{GM_u}{R(t)}$ is negative. Therefore, $R'(t)$ is never zero. Because the universe is currently expanding, $R'(t)$ is positive now. The fact that $R'(t)$ is a continuous function implies that $R'(t)$ will always be positive. By taking square roots, $R'(t) = \sqrt{2(E + \frac{GM_u}{R(t)})}$ for all $t > 0$. Assume, if possible, that the universe will not grow without bound. Then there is some positive constant L such that $R(t) \leq L$ for all $t \geq 0$. So $\frac{1}{R(t)} \geq \frac{1}{L}$ and hence

$$R'(t) = \sqrt{2\left(E + \frac{GM_u}{R(t)}\right)} \geq \sqrt{2\left(E + \frac{GM_u}{L}\right)}$$

for all $t > 0$. Therefore the derivative of $R(t)$ is greater than or equal to a fixed positive constant for all $t > 0$. This means (think about it), that $R(t)$ is an increasing function of t that will grow beyond any given bound, and hence also beyond L. The contradiction that we have arrived at implies that there is no positive constant L such that $R(t) \leq L$. It follows that

$$\lim_{t \to \infty} R(t) = +\infty.$$

So in the case $E \geq 0$, the universe will expand forever. This model is called the *Open Universe*. If t is pushed to infinity, then $\frac{GM_u}{R(t)}$ will diminish to zero. So the rate of expansion of the universe $R'(t) = \sqrt{2(E + \frac{GM_u}{R(t)})}$ will decrease and close in on $\sqrt{2E}$. This is the limiting rate of expansion of the open universe. In particular if $E = 0$, then the rate of expansion will slow to zero.

(B) Suppose that $E < 0$. In this case, $-E$ is a positive constant. Because $E + \frac{GM_u}{R(t)} \geq 0$ for all $t > 0$, we see that $\frac{GM_u}{R(t)} \geq -E$ for all $t > 0$. Thus, $\frac{R(t)}{GM_u} \leq \frac{1}{-E}$ and $R(t) \leq \frac{GM_u}{-E}$ for all $t > 0$. Therefore, the universe

will not expand beyond this fixed bound. What is in store for this *Closed Universe*? We will see that the expansion of the universe will stop, that it will begin to contract, and that its ultimate fate will be a devastating *Big Crunch*. The demonstration of this calamity follows. Observe first that $R''(t) = -\frac{GM_u}{R(t)^2}$ is negative for all $t > 0$. Therefore, $R(t)$ is concave down throughout. Note that $\frac{(GM_u)^2}{R(t)^2} \geq (-E)^2 = E^2$ and hence $-\frac{(GM_u)^2}{R(t)^2} \leq -E^2$ for all $t > 0$. Therefore,

$$R''(t) = -\frac{GM_u}{R(t)^2} = -\frac{(GM_u)^2}{R(t)^2} \cdot \frac{1}{GM_u} \leq \frac{-E^2}{GM_u}$$

for all $t > 0$. So $\frac{d}{dt}\left(R'(t) + \frac{E^2}{GM_u}t\right) = R''(t) + \frac{E^2}{GM_u} \leq 0$ for all $t > 0$. As a consequence, $R'(t) + \frac{E^2}{GM_u}t$ is a decreasing (or more precisely, nowhere increasing) function of time. Let t_0 be the time that has elapsed since the Big Bang. So t_0 is the current age of the universe and $C = R'(t_0) + \frac{E^2}{GM_u}t_0$ is today's value of this function. The fact that $R'(t) + \frac{E^2}{GM_u}t$ is a decreasing function implies that $R'(t) + \frac{E^2}{GM_u}t \leq C$ for all $t \geq t_0$. Thus,

$$R'(t) \leq C - \frac{E^2}{GM_u}t$$

for all $t \geq t_0$. It follows that after some time, $R'(t)$ will be negative. So after some time, $R(t)$ will be a decreasing function of time. The fact that $R(t)$ is also concave down, implies that $R(t)$ must eventually collapse to zero (review Section 10.6 if needed). So the universe will end in a Big Crunch.

We are now face to face with the crucial question: Is $E < 0$ or is $E \geq 0$? Will there be a Big Crunch or do we have an open, continuously expanding universe? Let's proceed to speculate.

Recall that Hubble's constant varies with time t. Accordingly, we will consider the function $H(t)$ defined by

$$H(t) = \frac{R'(t)}{R(t)}.$$

The constant H_0 is the value $H(t_0)$, where t_0 is the current age of the universe, in other words, the number

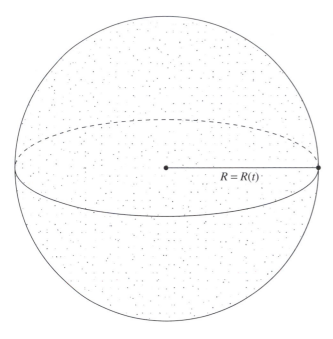

Figure 14.42

of years that have elapsed since the Big Bang. Now let ρ be the average density of the universe. Because M_u is constant and the universe is expanding, $\rho = \rho(t)$ is a decreasing function of time. Recall that mass = volume × density. So $M_u = (\frac{4}{3}\pi R(t)^3)\rho(t)$. By substituting this and $R'(t) = H(t)R(t)$ into equation (∗), we get

$$\frac{1}{2}H(t)^2 R(t)^2 - \frac{\frac{4}{3}\pi G \rho(t) R(t)^3}{R(t)} = E.$$

After multiplying through by $\frac{2}{R(t)^2}$, we obtain

$$H(t)^2 - \frac{8}{3}\pi G \rho(t) = \frac{2E}{R(t)^2}$$

This is *Friedmann's equation*. It is named after its discoverer, the brilliant Russian mathematical physicist Alexander Friedmann (1888–1925). Friedmann was born (and died) in Saint Petersburg and graduated from the city's university in 1910. During World War I (1914–1918) he worked for the Russian army as an aerologist. (Aerology concerns itself

with the observation of weather-related phenomena in the atmosphere by deploying balloons and airplanes.) Later, he was the director of a scientific observatory of the Saint Petersburg Academy of Sciences and taught mathematics and mechanics at several scientific institutes.

Plugging t_0 into Friedmann's equation gives us

(∗∗) $$(H_0)^2 - \frac{8}{3}\pi\rho(t_0)G = \frac{2E}{R(t_0)^2}.$$

Assuming for a moment that $E = 0$ and solving for $\rho(t_0)$, we get the *critical density*

$$\rho_{cr}(t_0) = \frac{3(H_0)^2}{8\pi G}$$

of the universe. Notice that the critical density is determined by Hubble's constant H_0. The ratio

$$\frac{\rho(t_0)}{\rho_{cr}(t_0)}$$

of the actual density divided by the critical density is denoted by Ω (the Greek letter omega). Observe that $\rho(t_0) = \Omega\rho_{cr}(t_0)$. Inserting this into (∗∗) gives us

$$\frac{2E}{R(t_0)^2} = (H_0)^2 - \frac{8}{3}\pi\rho(t_0)G = (H_0)^2 - \frac{8}{3}\pi\Omega\rho_{cr}(t_0)G$$

$$= (H_0)^2 - \frac{8}{3}\pi\Omega\frac{3(H_0)^2}{8\pi G}G = (H_0)^2 - \Omega(H_0)^2$$

$$= (H_0)^2(1 - \Omega).$$

It follows from the equality $\frac{2E}{R(t_0)^2} = (H_0)^2(1 - \Omega)$ that $E = 0$ precisely when $\Omega = 1$, $E > 0$ precisely when $\Omega < 1$, and $E < 0$ precisely when $\Omega > 1$. Refer back to discussion (A) and (B) about the constant E, and notice that the Open Universe corresponds to $\Omega \leq 1$ and the Big Crunch to $\Omega > 1$.

The fact that the constant Ω controls the fate of the universe has generated lots of interest in it. For a long time most cosmologists (the theoreticians and observers of the cosmos) believed that $\Omega \approx 1$. This is based on the following reasoning: The average density of the visible matter of the universe has been estimated to be about 10^{-27} kg/m^3. In addition, it is known that there are considerable amounts of "dark" matter. This is matter that is "invisible" in the sense that it emits neither light nor any other form of detectable radiation. For example, the fact that some stars are seen to be pulled by strong yet "invisible" forces suggests that these stars are in the gravitational field of massive invisible objects (the densest of which are called "black holes"). The conventional belief had been that the total amount of matter in the universe is about 10 times the visible amount. Inserting this into our discussion gives us that

$$\rho(t_0) = 10 \times 10^{-27} \text{ kg/m}^3 = 10^{-26} \text{ kg/m}^3.$$

On the other hand, measurements of Hubble's constant had provided the conclusion that this was also the estimate for the critical density $\rho_{cr}(t_0)$. Therefore, $\rho(t_0) \approx \rho_{cr}(t_0)$, and hence $\Omega \approx 1$.

This estimate is now being drawn into question. The Hubble Space Telescope has provided more accurate assessments of Hubble's constant and hence of the critical density $\rho_{cr}(t_0)$. Also, the existence of new dark matter in the form of "primordial" helium has recently been confirmed. In addition, innovative computer analyses have been undertaken. This is one of the most revealing ways, sometimes the only way, to test alternative mathematical models of the universe. Cosmologists feed known conditions about the early universe into the most powerful supercomputers. They then add conclusions contained in various mathematical models of the early universe and see by computer simulation what these might lead to in billions of years. This approach has suggested that $0.2 \leq \Omega \leq 0.3$ and that the universe is more "expansionist."

These investigations have a direct impact on the value of t_0, in other words, on the estimates of the age of the universe. To illustrate this connection, let's suppose that $\Omega = 1$. So $E = 0$, and hence equation $(*)$ reduces to $\frac{1}{2}(R'(t))^2 = \frac{GM_u}{R(t)}$. Therefore,

$$R'(t) = \sqrt{2GM_u}R(t)^{-\frac{1}{2}}.$$

So

$$R(t)^{\frac{1}{2}}R'(t) = \sqrt{2GM_u},$$

and after taking antiderivatives of both sides,

$$\frac{2}{3}R(t)^{\frac{3}{2}} = (2GM_u)^{\frac{1}{2}}t + \text{constant}.$$

Since $R(t)$ shrinks to 0 when t is pushed to 0, the constant is equal to 0. So,

$$\frac{2}{3}R(t)^{\frac{3}{2}} = (2GM_u)^{\frac{1}{2}}t.$$

After squaring both sides,

$$\frac{4}{9}R(t)^3 = 2GM_u t^2.$$

Because $M_u = \frac{4}{3}\pi\rho(t)R(t)^3$, we see that

$$\frac{4}{9}R(t)^3 = \frac{8}{3}\pi G\rho(t)R(t)^3 t^2,$$

and hence that $\frac{4}{9} = \frac{8}{3}\pi G\rho(t)t^2$. Since $E = 0$, $\frac{8}{3}\pi G\rho(t) = H(t)^2$ by Friedmann's equation. Therefore, $t^2 = \frac{4}{9}\frac{1}{H(t)^2}$ and hence

$$t = \frac{2}{3}\frac{1}{H(t)}.$$

Taking $t = t_0$ gives the value $t_0 = \frac{2}{3}\frac{1}{H_0}$. Now recall the estimate

$$14 \leq H_0 \leq 23$$

for Hubble's constant given in km/sec per mega light year. Since 1 light year is equal to 9.46×10^{12} km, 1 mega light year is 9.46×10^{18} km. Therefore,

$$1\frac{\text{kilometer}}{\text{second}} \cdot \frac{1}{\text{mega light year}} = \frac{1}{9.46 \times 10^{18}}\frac{1}{\text{sec}}$$

$$\approx 1 \times 10^{-19}\frac{1}{\text{sec}}$$

and hence

$$14 \times 10^{-19} \leq H_0 \leq 23 \times 10^{-19}$$

in $\frac{1}{\text{sec}}$. Inverting all three terms tells us that

$$43 \times 10^{16} \leq \frac{1}{H_0} \leq 72 \times 10^{16}$$

in seconds. After multiplying through by $\frac{2}{3}$ and using the equality $t_0 = \frac{2}{3}\frac{1}{H_0}$, we get

$$29 \times 10^{16} \leq t_0 \leq 48 \times 10^{16}$$

in seconds. Since 1 year has about 31.5×10^6 seconds, we finally get

$$9 \times 10^9 \leq t_0 \leq 15 \times 10^9$$

in years. So the assumption that $\Omega = 1$ has provided the estimate that the universe is between 9 and 15 billion years old. This is in agreement with the estimates of modern cosmology.[11]

14.8 Postscript

The journey that this book has undertaken has reached its end. It has discussed many subjects and engaged many problems. If it has a single message, it is this: mathematics, and especially calculus, is a powerful and elegant discipline that informs all sorts of different enterprises. The end of this volume is in fact only a beginning. A number of areas of mathematics were either only touched upon (for example, real analysis, differential equations, numerical methods, and probability theory), and many others—for reasons of the objectives and scope of this volume—were omitted altogether (for instance, calculus of many variables; partial differential equations; probabilistic analysis; statistics; and topology, the mathematics of shape). This is true to an even greater extent about the areas of science and engineering that this book has engaged. Our remarks have pointed to the exciting developments of the past and present. The future promises much more. As science continues its fact-finding mission into the relevant phenomena both "under the Sun" and beyond, mathematics will be there. Indeed, it will provide much of the language and many of the thought processes needed for the task.

Exercises
14A.   Force and Acceleration

The answers to the problems below involve basic facts such as $F = ma$ and the relationships between acceleration, velocity, and displacement developed in Section 6.4.

1. A person with a mass of 70 kilograms stands in an elevator. Determine the net force that is exerted on the person when:

    i.   the elevator has stopped.
    ii. the elevator goes up with a constant speed.
    iii. the elevator goes down with a constant speed.
    iv. the elevator accelerates upward at a rate of 2 meters/sec^2.
    v.   the elevator accelerates downward at a rate of 2 meters/sec^2.

2. An object is in free fall at time $t = 0$ at a height of y_0 meters (near the surface of the Earth) with a velocity of v_0 meters/sec. The object is massive enough and y_0 and v_0 are both small enough, so that air resistance is negligible. Determine at any time $t \geq 0$, the acceleration of the object, its velocity, as well as its height above the ground. [Hint: Start with the fact that the acceleration due to gravity is $a(t) = -9.81$ meters per second^2.]

3. An object with a mass of 2 kilograms falls on a pile of sand from a height of 6 meters. The object penetrates the sand a distance of 3 centimeters before coming to a stop. Compute the force that the sand exerts on the object under the assumption that this force is constant. [Hint: What is the velocity of the object when it first touches the sand?]

4. A 5 pound block of ice slides without friction down a ramp that is inclined at an angle of 30° with the horizontal. With what acceleration will the block slide down the ramp?

5. A car weighing 3000 pounds is moving on a horizontal street.

    i.   Suppose it has a speed of 60 miles per hour. When the brakes are applied to produce a constant deceleration, the car stops in 15 seconds. Determine the force that the brakes exert on the car. [Convert to feet by using the fact that 1 mile = 5280 feet.]
    ii. What constant force generated by the engine will increase the speed of the car from 10 miles per hour to 80 miles per hour in 10 seconds. [Hint:

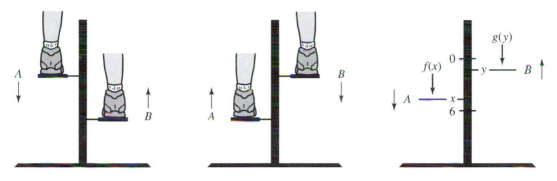

Figure 14.43

What constant acceleration will the car have to undergo?]

14B. Work and Energy

6. A car was pushed along a level road from a point A to a point B. The two points are 35 feet apart, and a force of 200 pounds was exerted on the car. Find the work done in pushing the car.

7. Compare the kinetic energy of a 12 gram bullet moving at 1000 meters per second with that of a 10 metric ton truck moving at 7 meters per second. [Hint: 1 metric ton = 1000 kilograms.]

8. A construction crane has its horizontal boom 90 feet above the ground. Its load-bearing cable starts its descent from there. It weighs 6 pounds per foot. The crane hoists a 500 pound concrete slab from the ground to a height of 70 feet. What is the work done on the slab? [Hint: The weight of the cable must be included. What is the downward force on the cable at a height of 90 feet above the ground when the slab is a height of x feet above the ground? Draw a sketch of what is going on.]

9. A Stairmaster is a gym apparatus that simulates a stair climb. The essence of the machine consists of two pedals that can move up and down vertically. The user stands on the two pedals. Figure 14.43 shows the backs of the feet of a woman using the equipment. If she puts more than one-half of her weight on one pedal, this pedal is pushed down. At the same time, the other pedal rises. When the woman shifts her weight, the situation is reversed. The two pedals together support the full weight of the woman. Figure 14.43 shows both situations. On the left, it is pedal A that is pushed down while pedal B rises. In the middle, pedal B supports more weight. So it is pushed down,

while A rises. The diagram on the right shows the typical position of the two pedals "in the abstract." A coordinate axis has been placed along the motion of the feet. Length is given in inches. The highest position reached by the two pedals is the origin, and 6 is the lowest. A typical situation is shown. Pedal A is at x, and pedal B is at y. The weight on A is $f(x)$, and the weight on B is $g(y)$. The pedals are always in "symmetric" position, meaning that the distance from the one pedal to the highest position is the same as that from the other pedal to the lowest position. Observe that the work done on pedal A (as this pedal moves from the highest to the lowest position) is $\int_0^6 f(x)\,dx$. Similarly, the work done on pedal B (as this pedal moves from its lowest to its highest position) is $\int_6^0 g(y)\,dy$.

i. Suppose that the woman weighs 120 pounds. Compute the work done on the two pedals as one pedal moves from the lowest position to the highest and the other from the highest to the lowest. In other words, compute the sum of the two integrals. [Hints: Express the integral of $g(y)$ in terms of the integral of $f(x)$. To do so, ask yourself the following questions: What are $f(x) + g(y)$ and $x + y$ equal to? In addition, make use of the discussion preceding Problem 14 in Exercises 13C.]

ii. Suppose that the woman completes 50 such "ups and downs" in 1 minute. What is the work done in 30 minutes? How many calories will she burn up during this time? (One calorie is equivalent to about 3.1 foot-pounds.)

iii. How many calories are burned up by a man weighing 180 pounds if he operates the apparatus in the same way as the woman did?

14C.   The Internal Combustion Engine

An engine of an automobile operates on the same basic principle as a jet or rocket engine. An engine consists typically of 4, 6, or 8 cylinders. Consider one of them. One side of the cylinder is open, and it is into that side that a piston is placed. See Figure 14.44. A fuel mix is injected into the cylinder from the carburetor (at this time the exhaust valve is closed) and compressed by the piston. A sparkplug ignites the fuel mix. The resulting explosion drives the piston downward. The same thing happens in all of the cylinders. Each piston pushes on a connecting rod, which, in turn, rotates the crankshaft. The explosions in the various cylinders occur in rapid-fire succession. They are carefully synchronized so that overall, a smooth force is produced. The rotating crankshaft is the source of power for the car. Only every second downward motion of the piston produces a force. This is the so-called *power stroke* of the piston. The other downward motion allows the fuel mix to be drawn into the cylinder. This is the *induction stroke* of the piston.

Now consider the engine of a particular six-cylinder automobile. Figure 14.45 shows one of its cylinders and pistons "in the abstract." (The cylinder is turned sideways, and the explosion pushes the piston to the right.) The range of motion of the piston is $0.40 \leq x \leq 4$ inches. Tests of the engine have determined that the pressure that the explosion generates in the cylinder is given by the function $p(x) = \frac{120}{x}$ pounds per square inch for $0.40 \leq x \leq 4$.

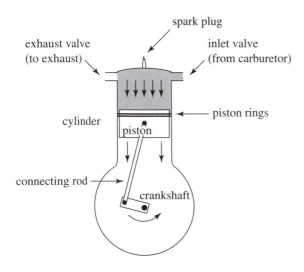

Figure 14.44

10. The inner radius of a cylinder is 2 inches. Show that the force driving a piston is given by $f(x) = Ap(x) = \frac{1507.96}{x}$ pounds, where A is the cross-sectional area of the cylinder.

11. Show that the work done on a piston by the explosive force in a cylinder is 289.35 foot-pounds.

12. Suppose that the six cylinders of the engine operate at 45 power strokes per second. Show that when the engine is operated in this way, it produces 142 horsepower. (1 horsepower = 550 foot-pounds per second.) Show that in this mode the crankshaft rotates at 5400 rpm (revolutions per minute).

14D.   Impulse and Momentum

13. Compare the momentum of a 12 gram bullet moving at 1000 meters per second with that of a 10 metric ton truck moving at 7 meters per second. Contrast this result with that of Exercise 7. [Hint: 1 metric ton = 1000 kilograms.]

14. A box sliding on the floor of a room is brought to rest by friction. What happens to the momentum of the box? Is the momentum conserved? If not, does this process contradict the principle of conservation of momentum?

15. A 5 kilogram object undergoes a change in velocity from 2 meters/sec to 8 meters/sec in 3 seconds. Compute the change in momentum. Assume that the acceleration of the object is constant during this increase in velocity. Compute the force that produces this acceleration.

16. A cart with a mass of 1.5 kilograms moves along a track at 0.20 meters/sec until it runs into a flexible bumper at the end of the track. What is the change in momentum and the average force exerted on the cart

    i. if it is brought to a complete stop in 0.1 seconds?
    ii. if, after 0.1 seconds, it has rebounded with a velocity of 0.10 meters/sec?

    [Hint: Take the average force to be the constant force that has the same effect as the actual force that is exerted on the cart.]

17. A baseball bat exerts an average force of 200 pounds on a baseball for a time of 0.01 seconds. What is the impulse of the force? A baseball weighs about 0.34 pounds. If it arrives at the plate at 90 miles per hour, with what velocity will it leave the bat?

18. Recall the study of the barbell in Section 14.3A. During the first phase of the lift, the 535 pound barbell is

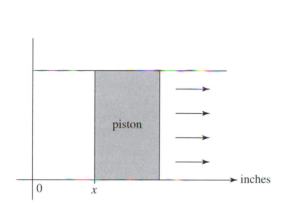

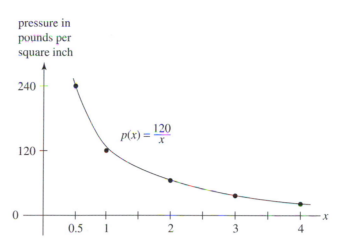

Figure 14.45

pulled up by a net force of 40 pounds. Show that the impulse that the barbell receives in the first phase of the lift and the change in momentum during this phase are both equal to $\frac{\sqrt{40} \cdot \sqrt{535}}{\sqrt{g}}$. (So the discrepancy between 25.78 and 26 pointed out in Section 14.3A is due to round-off error.)

19. The force of an explosion on the piston of an engine is due to the change in momentum of the gas particles. Consider an explosion that lasts 0.001 seconds. What must the speed of the molecules be if 0.6 grams of gas exerts a force of 1400 newtons on the piston.

14E.   More About Springs and Guns

20. Consider a spring of a natural length of 11 inches. A force of 9 pounds is required to keep the spring stretched at a length of 12 inches.

   i. Find the force necessary to keep the spring stretched at a length of 13 inches.
   ii. Find the work done in stretching the spring from its natural length to a length of 12 inches.
   iii. Find the work done in stretching the spring from a length of 12 inches to a length of 13 inches.

   The next three exercises analyze the motion of the ice cube of Section 14.2A. The operative situation is depicted in Figure 14.11. The units are M.K.S. Take out a stopwatch and release the spring at time $t_0 = 0$. The position of the ice cube is a function $x = x(t)$, and its velocity is the derivative $x'(t)$. Observe that $x(0) = -0.03$ and $x'(0) = 0$.

21. Show that the position function $x(t)$ satisfies the differential equation $x''(t) + \frac{k}{m}x(t) = 0$, where k is the spring constant and m the mass of the ice cube.

22. Apply the results of Section 13.5 to the conclusion of Exercise 21 to deduce that $x(t) = -0.03 \cos\left(\sqrt{\frac{k}{m}}\, t\right)$ and $x'(t) = 0.03\sqrt{\frac{k}{m}} \sin\left(\sqrt{\frac{k}{m}}\, t\right)$.

23. Recall that the spring constant is $k = 25$ newtons/meter and that the mass of the ice cube is $m = 0.02$ kilograms. Deduce that the left edge of the ice cube arrives at the origin when $t = \frac{\pi}{2\sqrt{1250}} = 0.04$ seconds with velocity of $(0.03)(\sqrt{1250}) \approx 1.06$ meters/sec.

24. A bullet of mass 0.05 kilograms leaves the muzzle of a gun of mass 4 kilograms at 800 meters/sec. What is the velocity of recoil of the gun?

25. A battle tank weighing 50 metric tons (recall that 1 metric ton = 1000 kilograms) fires a 25 kilogram shell at an angle of $10°$. If the shell leaves the barrel with a velocity of 1000 meters/sec, compute the horizontal component of the initial recoil velocity of the tank. The recoil pushes the tank 0.5 meters backward. Assume that the horizontal component of the frictional force that brings the tank to a stop is constant. What is this force equal to?

26. The time it takes for a shell to travel through the barrel is the *barrel time*. Let's denote it by t_1. One rule of thumb used to estimate the barrel time is $t_1 \approx \frac{3b}{2v}$, where b is the barrel length and v the muzzle velocity.

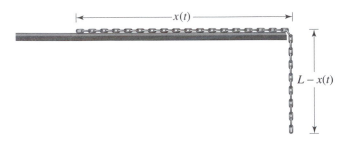

Figure 14.46

Use it to estimate the barrel time for the Springfield rifle.

27. Let $x > 0$ be any position of the shell in the barrel t units of time after the trigger is pulled. Show that $t = \frac{s}{r} \ln x + \frac{x}{r} + C$ for some constant C, where s and r are the constants in the formula of Le Duc in Section 14.2B. [Hint: Start with the formula of Le Duc and "separate variables."]

28. Refer to the analysis of the motion of the shell in Section 14.2B. Let $x(0) = 0$. Turn to the formulas for the velocity and the force and notice that $v(0) = 0$ and $F(0) = 0$. So the velocity of the bullet is zero and the force on it is zero. But doesn't this mean that the bullet will not move away from the origin? [Answer: Indeed it does!] So the formula of Le Duc requires the assumption that $x(0)$ is some small positive number a. How is this number a related to the equation in Exercise 27?

29. Suppose that an error was made and that the constants in Le Duc's formula for the Springfield rifle are 3720 and 0.64, respectively. Carry out the analysis of Section 14.2C again to compute the maximum force and pressure in the barrel, as well as the kinetic energy of the bullet when it leaves the muzzle. Compare these values with those obtained in Section 14.2C.

30. Suppose that the Springfield rifle is fired at a ballistic pendulum that weighs 20 pounds and consider the data provided in Example 14.10 of Section 14.3B. Now refer to Figure 14.24 and compare states (a) and (b) from the point of view of energy. Is energy conserved? In other words, is the kinetic energy of the bullet in state (a) equal to the potential energy of the block at the top of the swing?

14F. The Sliding Chain

31. A chain of length $L = 100$ centimeters and a mass of 1 gram per centimeter is laid out straight on a horizontal table that is perfectly polished and hence provides a frictionless surface. The chain is held at one end in such a way that 10 centimeters of the other end of the chain hang over the edge. At time $t = 0$, the chain is released. How long will it take for the chain to glide completely off the table? [Hint: Refer to Figure 14.46 and consider the rate of change of the momentum of the chain. Recall that 1 gram is $10^{-3} =$ kilograms.]

This problem is adapted from Question 447 in *Leybourn's Mathematical Questions*, Vol. II, J. Mawman, Ludgate St., London, 1817. They are "the mathematical questions proposed in the ladies' diary and their original answers, together with some new solutions from its commencement in the year 1704 to 1816." Question 447 was posed in 1757 and answered in 1758.

14G. More About Rockets

32. A rocket of mass 10^4 kilograms starts from rest and is propelled by a force of 2×10^5 newtons for 20 seconds. What is the final velocity of the rocket?

33. A fully fueled V-2 rocket is launched vertically from the surface of the Earth. It flies straight up until it runs out of fuel. Estimate its velocity at that time. [Hint: Ignore air resistance and use the data of Section 14.4A.]

34. Consider a test of a rocket engine with the rocket bolted down in vertical position (as discussed in Section 14.4A). Let the test start at time $t = 0$, but assume this time that the velocity of the exhaust gases is a function $v_e(t)$ of time. Show that the momentum component T_{mo} of the thrust is equal to $v_e(t) \frac{d}{dt} m_e(t)$, where $m_e(t)$ is the total mass of the gases expelled from time $t = 0$ to time t. [Hint: Let $\mu(t)$ be the momentum of the exhaust gases at time t and note that $T_{\mathrm{mo}} = \frac{d}{dt} \mu(t)$. Now argue as in the appropriate discussion in Section 14.4B.]

14H. More About Gravity

Return to the homogeneous spherical shell of Section 14.5B. The assumption made there was that the particle P of mass m was outside the spherical shell. We will now suppose that it is inside. For example, place it at c as shown in Figure 14.47. What is the gravitational force of the shell on P equal to this time? Begin by reviewing Section 14.5. Then regard the shell as being composed of thin circular bands or wires as in Section 14.5B. Note that $c < R$.

35. Explain why the gravitational force exerted by the portion of the shell lying to the left of the mass m, i.e., to the left of the plane through c (see Figure 14.47) is equal to

$$\frac{GmM}{2R} \int_{-R}^{c} \frac{c - x}{(c^2 - 2cx + R^2)^{3/2}} \, dx$$

and show that the value of this integral is equal to $\frac{GmM}{2R} \frac{1}{c^2} (R - (R^2 - c^2)^{\frac{1}{2}})$.

36. Explain why the gravitational force exerted by the portion of the shell lying to the right of the mass m, i.e., to the left of the plane through c, is equal to

$$\frac{GmM}{2R} \int_{c}^{R} \frac{x - c}{(c^2 - 2cx + R^2)^{3/2}} \, dx.$$

[Hint: The strategy of Exercise 35 works with some modifications. The distance from the point P to the circle centered at x is now $x - c$.]

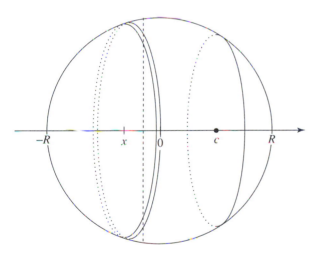

Figure 14.47

37. Show that the value of the definite integral of Exercise 36 is equal to $\frac{GmM}{2R} \frac{1}{c^2} (R - (R^2 - c^2)^{\frac{1}{2}})$. [Hint: Because $R \geq c$, note that $(c^2 - 2cR + R^2)^{\frac{1}{2}} = R - c$.]

38. The force diagram of Figure 14.48 summarizes the conclusions of Exercises 35–37. What can you conclude about the force with which the entire spherical shell acts on the point P?

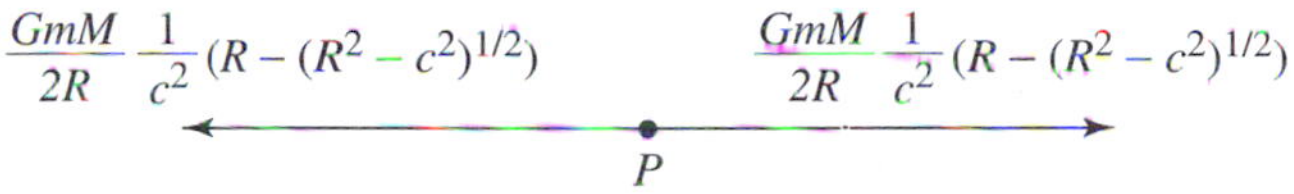

$$\frac{GmM}{2R} \frac{1}{c^2} (R - (R^2 - c^2)^{1/2}) \qquad \frac{GmM}{2R} \frac{1}{c^2} (R - (R^2 - c^2)^{1/2})$$

$$\longleftarrow \quad \bullet \quad \longrightarrow$$
$$P$$

Figure 14.48

14I.   About Kepler's Laws

39. Study the discussion of Section 14.6B and then verify the equality

$$\frac{d^2 y}{dt^2} = \sin \theta \left[\frac{d^2 r}{dt^2} - r \cdot \left(\frac{d\theta}{dt} \right)^2 \right].$$

We now turn to a matter referred to at the end of Section 14.6, namely the derivation of the laws of Kepler from those of Newton. So we will assume the following (in addition to Newton's three basic laws of motion from the introduction to Chapter 7):

i. An object P of nonzero mass m is in motion and subject to a single centripetal force, in other words, a force that acts in the direction of a fixed point S.

ii. The magnitude of this force is equal to $F = C \frac{m}{r^2}$ for some positive constant C, where r is the distance between P and S.

Taking this as given, the task is to derive the three laws of Kepler mathematically (without making additional assumptions about the realities in question). Refer to Figure 14.39, and notice that the orbit is the graph of some function $r = r(\theta)$ in polar coordinates. Also, $\theta = \theta(t)$ and hence $r = r(\theta(t))$ are functions of time $t \geq 0$. Parts of the task have already been achieved. It was shown in Section 14.6 that the single assumption (i) implies Kepler's second law of "equal areas in equal times." This single assumption also provided the equalities

$$F = m \left[r \cdot \left(\frac{d\theta}{dt} \right)^2 - \frac{d^2 r}{dt^2} \right] \quad \text{and} \quad r^2 \frac{d\theta}{dt} = 2\kappa$$

where κ is Kepler's constant.

We will now outline the verification of Kepler's first law, namely that the orbit is an ellipse, a parabola, or a hyperbola. So it is an ellipse if it is a closed orbit. The details are left as exercises. We will set $u = u(\theta) = \frac{1}{r(\theta)}$. So $r(\theta) = \frac{1}{u(\theta)} = u(\theta)^{-1}$.

40. Verify that $\frac{d^2 r}{dt^2} - \frac{4\kappa^2}{r^3} = -\frac{C}{r^2}$.

41. Make use of the chain rule (the version that allows the cancellation of differentials, see the very end of Section 8.5A) to show that $\frac{dr}{dt} = -2\kappa \frac{du}{d\theta}$. Then differentiate this equation (and use the chain rule again) to show that $\frac{d^2r}{dt^2} = -4\kappa^2 u^2 \frac{d^2u}{d\theta^2}$. [Hint: $\frac{d}{dt}(\frac{du}{d\theta}) = \frac{d^2u}{d\theta^2} \cdot \frac{d\theta}{dt}$]

42. Use a combination of the equations derived in Exercises 40 and 41 to show that

$$\frac{d^2u}{d\theta^2} + u = \frac{C}{4\kappa^2}.$$

Now let $z = u - \frac{C}{4\kappa^2}$. Observe that $\frac{dz}{d\theta} = \frac{du}{d\theta}$ and $\frac{d^2z}{d\theta^2} = \frac{d^2u}{d\theta^2}$. So by a substitution,

$$\frac{d^2z}{d\theta^2} + z = 0.$$

43. Use appropriate results from Section 13.5 to conclude that

$$z = A \sin\theta + B \cos\theta$$

for some constants A and B. It follows that $u = A\sin\theta + B\cos\theta + \frac{C}{4\kappa^2}$.

At this point, we will chose the polar axis (see Figure 14.39) in such a way that the distance r between P and S has its smallest value, say d, when $\theta = 0$. This means that u has a maximum when $\theta = 0$.

44. Review basic facts about the first and second derivative (from Sections 8.7 and 10.6) and then show that $A = 0$ and $B \geq 0$. Then verify that $B = \frac{1}{d} - \frac{C}{4\kappa^2}$ and

$$r = r(\theta) = \frac{\frac{8\kappa^2}{C}}{2(1 + (\frac{4\kappa^2}{Cd} - 1)\cos\theta)}$$

where d is the minimum distance between S and P, and $\frac{4\kappa^2}{Cd} - 1 \geq 0$.

45. Let $L = \frac{8\kappa^2}{C}$ and $\varepsilon = \frac{4\kappa^2}{Cd} - 1$. Study Section 13.4A and conclude that the polar graph of the function $r(\theta)$ is a conic section with the following properties: The conic section has S at a focus and it has latus rectum L. If $\varepsilon \leq 1$, then it is an ellipse; if $\varepsilon = 1$, it is a parabola; and if $\varepsilon > 1$, it is a hyperbola.

Kepler's second law has now been established. The verification of the third law remains. This is routine, but (as we will see) it requires an additional assumption. Suppose that the object's orbit is an ellipse and that T is the period of the orbit. By Section 4.5, we can choose Cartesian coordinates such that the ellipse has an equation of the form $\frac{x^2}{a^2} + \frac{y^2}{b^2} = 1$, where a is the semimajor axis and $a \geq b$. Because the area of this ellipse is $ab\pi$, it follows from Kepler's second law

that $\frac{ab\pi}{T} = \kappa$. By Exercise 9 of Chapter 7, the latus rectum $L = \frac{8\kappa^2}{C}$ is also equal to $\frac{2b^2}{a}$. Check the computation

$$\frac{a^3}{T^2} = a\frac{\kappa^2}{b^2\pi^2} = \frac{C}{8\kappa^2}\frac{2\kappa^2}{\pi^2} = \frac{C}{4\pi^2}.$$

The additional assumption that is needed is that the constant C is independent of the object P. If this is assumed, then the ratio $\frac{a^3}{T^2}$ is independent of the object P, as required by Kepler's third law. The universal law of gravitation tells us that $C = GM$ where M is the mass of S. So in the case of gravity, it is true that C is independent of P.

The aims of Exercises 14I have been accomplished. We have shown that all three laws of Kepler are mathematical consequences of those of Newton.

14J. Free Fall with Air Resistance

Consider an object in free fall near the Earth's surface. We know that the force of gravity acts downward with magnitude mg, where m is the mass of the object and g is the gravitational constant. The force due to the resistance of the air, also called drag, acts on the object in an upward direction. The magnitude of this force depends on the size of the object as well as its shape! For example, the air resistance on a beach ball is greater than that on a tennis ball, even if they have the same mass. The sleek, pointed bullet of a modern gun will experience less air resistance than the spherical bullet of an old musket, even if their cross-sectional areas as well as their masses are the same. With the cannonball in mind, we will now assume that the shape of the object is a sphere.

Experimental evidence has shown that the drag on a spherical object due to air resistance is given by

$$D = \frac{1}{4}\rho A v^2$$

where ρ is the density of the air, $A = \pi r^2$ is the cross-sectional area of the sphere, and v is the velocity of the object. The net force (see Figure 14.49) on the falling object is therefore

$$F = -mg + \frac{1}{4}\rho A v^2.$$

Suppose that the object starts its fall at time $t = 0$, and let $v(t)$ be its velocity at any time $t \geq 0$.

46. Set $k = \sqrt{\frac{4mg}{\rho A}}$ and show that the velocity function $v(t)$ satisfies the differential equation $\frac{k^2}{g}\frac{dv}{dt} = v^2 - k^2$. Explain why the acceleration $\frac{dv}{dt}$ is negative. Then deduce that $-k < v < k$.

$$\frac{1}{4}\rho A v^2.$$

mg

Figure 14.49

47. Use the method of separation of variables and reverse common denominators (see Sections 13.3 and 13.5) to show that

$$\frac{1}{2k}\ln(k-v) - \frac{1}{2k}\ln(k+v) = \frac{g}{k^2}t + \text{constant}.$$

Suppose from now on that the object is dropped from rest. So the initial velocity $v(0)$ is zero. Show that the constant of the equation in Exercises 47 is zero and that

$$v = v(t) = \frac{-k(e^{\frac{2g}{k}t} - 1)}{e^{\frac{2g}{k}t} + 1}$$

is the solution. Our analysis has produced an explicit formula for the velocity of the spherical object for any time $t \geq 0$ into its fall. Recall that $k = \sqrt{\frac{4mg}{\rho A}}$ and observe that the velocity function is completely determined by the constants g and ρ and the mass m and cross-sectional area A of the object. A combination of Exercises 46 and 47 shows that the speed $s(t)$ of the object (this is the absolute value of the velocity) satisfies

$$s(t) = |v(t)| = \frac{k(e^{\frac{2g}{k}t} - 1)}{e^{\frac{2g}{k}t} + 1} < k = \sqrt{\frac{4mg}{\rho A}}.$$

It follows that k is a limit on the speed of the object. (This is consistent with what we know about skydiving. The speed of a skydiver in freefall does not exceed a certain limit.)

48. Suppose that two spherical objects have the same mass/(cross-sectional area) ratio and that they are both dropped at time $t = 0$ with zero initial velocity. Explain why they have the same speed for all $t \geq 0$.

49. Show that if the two falling spheres of Exercise 48 have the same speed at a single instant $t > 0$, then they have the same speed for all $t \geq 0$. [Hint: Show that the function $f(x) = \frac{x-1}{x+1}$, where $x \geq 0$, is one-to-one. To do this, show that it is a decreasing function.]

50. Suppose that the object is allowed to fall for a long time t and study $s(t)$. Use L'Hospital's Rule (see Section 8.5B) to show that

$$\lim_{t\to\infty} s(t) = k = \sqrt{\frac{4mg}{\rho A}}.$$

(A more intuitive explanation of this equality is this: As t is pushed to infinity, $x = e^{\frac{2g}{k}t}$ becomes huge, and in the process, $e^{\frac{2g}{k}t}+1$ and $e^{\frac{2g}{k}t}-1$ get close to each other.)

The limit k is the *terminal speed* of the object. We will use the more suggestive notation s_∞ for the terminal speed. A substitution gives us the formula

$$s(t) = \frac{s_\infty(e^{\frac{2g}{s_\infty}t} - 1)}{e^{\frac{2g}{s_\infty}t} + 1}.$$

51. Use some of the graphing strategies of Section 10.8 to show that the graph of the function $s(t)$ has the general shape shown in Figure 14.50. What is the connection between the acceleration of the object and the slope at $(0, 0)$?

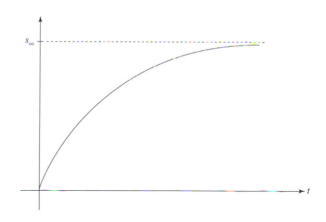

Figure 14.50

The next task is to find an explicit expression for the distance $y(t)$ of the falling spherical object above the ground at any time t.

52. Show that $-\frac{1}{s_\infty} \cdot \frac{dy}{dt} = \frac{e^{at}-1}{e^{at}+1}$, where $a = \frac{2g}{s_\infty}$.

53. Use the substitution $u = e^{at}$ (refer to Section 13.3A) to show that

$$\int \frac{e^{at} - 1}{e^{at} + 1}dt = \frac{1}{a}\int \frac{u - 1}{u(u + 1)}du.$$

54. Use "common denominators in reverse" (see Section 13.3) to verify that

$$\int \frac{u-1}{u(u+1)}\,du = -\ln u + 2\ln(u+1) + \text{constant}.$$

55. Use the conclusions of Exercises 52–54 to show that

$$y(t) = s_\infty t - \frac{(s_\infty)^2}{g}\ln(e^{\frac{2g}{s_\infty}t} + 1) + y_0 + \frac{(s_\infty)^2}{g}\ln 2.$$

where $y_0 = y(0)$ is the height of the object above the ground when it is dropped at time $t = 0$. Suppose that two spherical objects have the same mass/(cross-sectional area) ratio. Verify that if they are dropped at the same time from the same height, then they are at the same height above the ground at any time $t \geq 0$.

14K.   Returning to the Leaning Tower

Section 6.4 studied a cannonball in free fall from the Leaning Tower of Pisa. (The results will be reviewed in a moment.) Air resistance was assumed to be negligible and was ignored. We will now study the falling cannonball again, but this time we will take air resistance into account. In this study we will use the formulas of Exercises 14J. In order to be able to do so, we need certain information. First, we need to know the cross-sectional area A of the spherical cannonball. Let's take the diameter of the cannonball to be exactly $\frac{1}{2}$ foot. So the radius is $r = 0.25$ feet and the cross-sectional area is $A = \pi(\frac{1}{4})^2 = 0.1964$ ft^2. We also need to know the mass m of the cannonball. Traditionally, cannonballs were made of iron. We will take iron that has a mass of 15.20 slugs per cubic foot. Because the volume of the cannonball is $\frac{4}{3}\pi(\frac{1}{4})^3 = \frac{\pi}{48} = 0.0655$ ft^3, it has a mass of $m = (0.0655)(15.20) = 0.9956$ slugs. (As an aside, note that 1 slug of mass weighs 32.17 pounds. So the cannonball weighs $(32.17)(0.9956) = 32.03$ pounds.) Finally, we need to know that the density of the air at sea level is $\rho = 0.0023$ slugs per cubic foot.

The questions that we will be considering are numerically delicate. Even small changes in the constants and round-off errors can produce rather different results. We will therefore work with an accuracy of four decimal places. This means, for example, that r is taken with an accuracy of $r = 0.2500$, that $g = 32.1700$, and that the iron of the cannonball has a mass 15.2000 slugs per cubic foot. (Remember, round off only after a computation is complete.)

Suppose first that the cannonball is dropped from an unspecified height at time $t = 0$ with an initial velocity of zero.

56. Show that the terminal speed of the cannonball is $s_\infty = 532.5536$ feet/sec, that $\frac{2g}{s_\infty} = 0.1208$, and hence that the speed of the cannonball is

$$s(t) = 532.5536\,\frac{e^{0.1208t} - 1}{e^{0.1208t} + 1}$$

feet/sec at any time $t \geq 0$.

57. Determine the speed of the cannonball at $t = 2$, $t = 5$, and $t = 10$ seconds. How many seconds will it take for the cannonball to reach 90% of its terminal velocity? (We are assuming that the cannonball has been dropped from a great enough height for these questions to make sense.)

We now turn to the analysis of two identical cannonballs falling side by side from the 177 foot high tower of Pisa. They are released at the same time $t = 0$ and both have an initial velocity of zero. The one on the left falls in a vacuum and the one on the right is subject to air resistance. See Figure 14.51. The one on the left will hit the ground first. But by how much? The one on the left will have the greater speed at impact. Are the differences substantial?

Let's briefly recall the following facts from Section 6.4 for the situation in a vacuum. At any time $t \geq 0$ the speed of the cannonball is $gt = 32.17t$ feet/sec and its height above the ground is $\frac{g}{2}t + y_0 = -16.085t + 177$ feet. Setting this expression equal to zero and solving for t tells us that the time of impact is $\sqrt{\frac{177}{16.085}} = 3.3172$ seconds. The speed at impact is $(32.17)(3.3172) = 106.7143$ feet/sec.

For the remainder of this exercise our attention is turned to the situation with air resistance, in other words, to the cannonball falling in Figure 14.51 on the right.

58. Insert all the necessary constants into the appropriate formula to show that the cannonball on the right is

$$y(t) = 532.5536t - 8816.0813\ln(e^{0.1208t} + 1) + 6287.8419$$

feet above the ground at any time $t \geq 0$.

59. Show that the cannonball on the right is

$$y(3.3172) = 1766.5868 - 8053.0056 + 6287.8419 = 1.4231$$

feet above the ground at the instant $t = 3.3172$ that the one on the left impacts.

So the cannonball falling in air will be about 1.4 feet above the ground at the instant that the one falling in a vacuum impacts. Not very much after a fall of about 177 feet!

The cannonball falling in air will impact a split second after $t = 3.3172$. To find out precisely when this occurs, we must set

$$y(t) = 532.5536t - 8816.0813\ln(e^{0.1208t} + 1) + 6287.8419 = 0$$

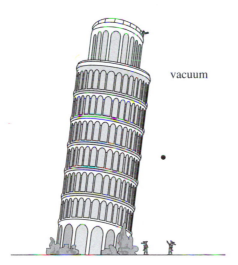
vacuum

air

Figure 14.51

and solve for t. If this is done, either by Newton's method (see Exercises 8K) or by computer (for example, *Maple* or *Mathematica*), one finds that the answer is $t = 3.3307$. Indeed, by substituting this value into the formula for $y(t)$, we can check that

$$y(3.3307) = 1773.7763 - 8061.6184 + 6287.8419$$

$$= -0.0002 \text{ feet.}$$

60. Show that the speed of the cannonball on the right at the time of impact is

$$s(3.3307) = (532.5536)\frac{1.4953 - 1}{1.4953 + 1} = 105.7119 \text{ feet/sec.}$$

As expected, this cannonball has a slightly lower speed at impact than the 106.7143 feet/sec of the one falling in a vacuum. What is perhaps unexpected is the fact that the speed at impact is far below the terminal speed of 532.5536 feet/sec. The cannonball would have to fall much longer to get close to its terminal speed. Check what its speed would be if it could fall for, say, 100 seconds.

Table 14.1 compares the speeds and heights of the two cannonballs at times $t = 0$, $t = 1$, $t = 2$, $t = 3$, $t = 3.3172$ (time of impact in a vacuum), and $t = 3.3307$ (time of impact in air).

Table 14.1				
time	speed in feet/sec		height in feet	
	vacuum	air	vacuum	air
$t = 0$	0	0	177	177
$t = 1$	32.17	32.1274	160.915	160.9908
$t = 2$	64.34	64.0245	112.66	112.9555
$t = 3$	96.51	95.4610	32.235	33.2409
$t = 3.3172$	106.7143	105.2858	0	1.4231
$t = 3.3307$	—	105.7119	—	0

14L. Dropping a Basketball from the Leaning Tower

A basketball with a radius of 0.39 feet and a weight of 1.3 pounds is regulation size (courtesy Rawlings Sporting Goods). Continue to work with an accuracy of four decimal places.

61. Show that the cross-sectional area of the basketball is $A = 0.4778 \text{ ft}^2$, that its mass is $m = 0.0404$ slugs, and that its terminal velocity is $s_\infty = 68.7795$ feet/sec.

62. Drop the basketball from the Leaning Tower of Pisa at time $t = 0$. Take air resistance into account and compute the speed $s(t)$ and the height $y(t)$ of the ball at any time $t \geq 0$.

63. Fill in the blanks of Table 14.2.

Table 14.2

elapsed time	height above the ground in feet		
	vacuum	cannonball	basketball
$t = 0$	177	177	177
$t = 1$	160.915	160.9908	
$t = 2$	112.66	112.9555	
$t = 3$	32.235	33.2409	
$t = 3.3172$	0	1.4231	
$t = 3.3307$	—	0	
$t =$	—	—	0

14M.   More About Springs

Fasten one end of a spring with spring constant k to a horizontal ceiling. Attach an object of mass m to the other end of the spring and let the system reach a state of equilibrium. The end of the spring with the weight attached is now displaced h units downward (and nothing moves).

64. Explain why $kh = mg$, where g is the gravitational constant.

   Figure 14.52 shows the suspended spring. A y-axis has been placed in such a way that the center of mass of the object is at $y = 0$ when the system is in its equilibrium state. Now pull the mass down A units. Release it at time $t = 0$, and let $y(t)$ be the position of the center of mass of the object at any time $t \geq 0$. Observe that $y(0) = -A$.

65. Analyze Figure 14.52 carefully and compute the net force $F(t)$ on the object at any time $t \geq 0$. Use your result to show that $y(t)$ satisfies the differential equation

$$y''(t) = \frac{-k}{m} y(t).$$

66. Assume that the velocity imparted to the mass when it is released is zero, i.e., assume that $y'(0) = 0$. Use appropriate results from Section 13.5 to show that

$$y(t) = -A \cos\left(\sqrt{\frac{k}{m}} t\right).$$

67. How long will it take for the object to move through one complete up-and-down cycle? [Hint: Refer to the graph of the cosine (see Figure 10.29 of Section 10.8), and observe that for one complete cycle, we need to have $0 \leq \sqrt{\frac{k}{m}} t \leq 2\pi$.]

68. Continue to suppose that the object is released from $-A$, but that it is given an initial velocity of $y'(0) = v_0$. Find a formula for $y(t)$ under this condition.

69. We will now consider a specific case. The following information is given about the spring, the mass suspended from it, and the initial state of the system:

   i. A force of 40 newtons is necessary to hold the spring extended 0.05 meters (5 centimeters) beyond its natural length.
   ii. The mass of the object that is attached is $\frac{1}{2}$ kg.
   iii. The mass is pulled down 0.1 meters from the position that it has at equilibrium and it is released at time $t = 0$ with a downward velocity of -2 meters/sec.

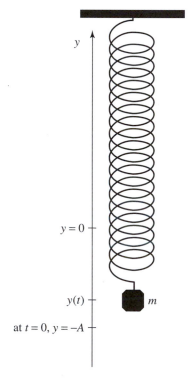

Figure 14.52

Find the function $y(t)$ explicitly, using the given data. How long will it take for the object to move through one complete up-and-down cycle?

14N.   About the Universe

70. According to Einstein's theory of Special Relativity, no physical object can move faster than the speed of light c. Incredibly, some distant galaxies have been observed to travel in excess of $0.9c$. Use the value $c = 3.0 \times 10^5$ kilometers per second and information about the constant $H_0 = H(t_0)$, to provide the estimate

$$11 \times 10^9 \text{ light years} \le R(t_0) \le 22 \times 10^9 \text{ light years}$$

for the present size of the universe.

71. Give an estimate for the mass M_u of the universe under the assumption that $\Omega = 1$.

Notes

[1]This formula seems to have first appeared in a series of articles in 1904 and 1905 in the *Revue d'Artillerie* (Théorie des affuts à Déformation by J. Challéat, Capitaine d'artillerie), where it was referred to as the semi-empirical formula of Captain Le Duc.

[2]For the details refer to A. G. Webster, On the Springfield Rifle and the Le Duc formula, *Proceedings of the National Academy of Sciences*, vol. 6 (1920), 289.

[3]See pages 380 and 381 in R. L. Wilson, *Winchester, an American Legend*, Random House, New York, 1991.

[4]Force, acceleration, velocity, and momentum are all vector quantities because they have not only a magnitude but also a direction. (See Section 9.1B.) If an object moves along a coordinatized line and a force acts on it along the line of its motion, then the dual aspect of these quantities is accommodated mathematically as follows. If f is the magnitude of a force, then the force is represented by f if it acts in the positive direction of the axis and by $-f$ if it acts in the negative direction. The understanding is the same for acceleration, velocity, and momentum. In the current discussion of forces acting on systems of particles, force, velocity, and momentum are no longer considered to "act" along a fixed line. As a consequence, a definitive version of this discussion requires the introduction and use of vector calculus. But this is beyond the purpose of this text.

[5]A third basic law of physics is the law of conservation of angular momentum. Define the *angular momentum* of a rotating object to be the product $I \cdot \omega$ of its moment of inertia I times its angular velocity ω. Refer back to Section 9.3 for the meaning of these concepts. The *law of conservation of angular momentum* says that the total angular momentum of an isolated system remains constant. The principles of conservation of energy, momentum, and angular momentum seem to govern all processes that can possibly arise in nature. They are the three most important principles of physics.

[6]All the data for the V-2 are based on the *Upper Atmosphere Research Report No.1*, Naval Research Laboratory Report R-2955, October 1, 1946.

[7]Something that is apparent immediately as one flips through the pages of these volumes is that it uses the notation of Leibniz rather than that of Newton. One hundred years after Newton's death, the superior approach of Leibniz was firmly established even at Cambridge.

[8]The word *theory* is used here to mean "a coherent group of general propositions used as principles of explanation of a class of phenomena," rather than "a proposed explanation whose status is still conjectural." Standard dictionaries offer both definitions of this word. Incidentally, this double meaning has added confusion to the debate about the validity of important areas of science such as the theory of evolution.

[9]The observed perihelion advance is 574 seconds of angle per century. Of these, 531 seconds are accounted for by the perturbing effect of the other planets on the Sun-Mercury system. Leverrier found that the largest contribution comes form Venus, 278 seconds, and next Jupiter at 153 seconds. The Earth's effect is third with 90 seconds, and the remaining planets contribute about 10 seconds. Thus the total contribution coming from Newtonian astrophysics

is about 531 seconds per century. The remaining 43 seconds is unaccounted for by the Newtonian theory.

[10]This "fudge factor" was introduced for the purpose of bringing the theory into line with observed (or in this case unobserved) facts. Einstein would later call the cosmological constant the greatest blunder of his life. Recent observations suggest that the cosmological constant will be making a comeback. See Dennis Overbye's compelling account, A Famous Einstein "Fudge" Returns to Haunt Cosmology, in the Science Times section of the New York Times of Tuesday, May 26, 1998. Incidentally, New York Times articles are obtainable (for a fee) via http://archives.nytimes.com/archives/.

[11]The special issuee Magnificent Cosmos, *Scientific American*, Spring 1998, Volume 9, Number 1, contains lots of infomation about our universe. This ranges from facts about our solar system to recent information about distant galaxies.

Index

acceleration, 164–165
    angular, 266
accelerator mass spectrometer, 342
Agnesi, Maria Gaetana, 236, 240, 331
    Instituzioni analitiche, 240, 246
Almagest, 32–39, 43–46
α particle, 336
American system of units, 200–201, 374–375
Ammann, Othmar, 258
analytic geometry, 81–86
angle, 6–8
    of departure, 167
    of elevation, 167
    of incidence, 272, 273
    of inclination, 265–266
    of reflection, 272
    of refraction, 273
angular
    acceleration, 266
    momentum, 533
    velocity, 266
annuities, 391–392
antibiotics production, 370–371, 382
antiderivative, 134–135, 312–313
Apollonius, 33, 52, 80
    Conics, 52
aphelion, 48, 74, 107
Archimedes, 9, 17–21, 250, 287–288
    and the crown, 287–288
    law of hydrostatics, 255
    law of the lever, 67
    Method, 63–65, 67, 92
    number scheme, 18–19
    On the Equilibrium of Planes, 64
    Quadrature of the Parabola, 60–62

    spiral, 455
    The Sandreckoner , 19–21
    theorem, 62
area, 56–57
    and polar coordinates, 460–462
    and the definite integral, 120–125, 153–155,
        438–441, 443–446, 460–462
    function, 149–151, 153–154, 439–441
    Leibniz's approach to, 120–125
    Newton's approach to, 148–151, 153–155
    of a circular sector, 57
    of an ellipse, 94
    of a parabolic section, 60–65
    of a parallelogram, 56
    of a rectangle, 56
    of a surface, 443–446
    of a triangle, 56–57
Aristarchus, 13, 19
    On Magnitudes and Distances, 13–17
Aristotle, 30, 76–77
astronomical eccentricity, 92
astronomical unit (AU), 22, 94–95
asymptote, 318–319
    horizontal, 318
    vertical, 318
asymptotic behavior, 322
atom, structure of, 335–336, 374–375
atomic mass unit, 336
atomic number, 336
autumnal equinox, 30
Avogadro's number, 336

ballistics, 170–173, 488–492
    interior, 488–492
    formula, 488, 490

ballistic pendulum, 171–172, 497–498
banking, 386–393
basketball, 26, 175–176, 287, 531–532
Becquerel, Henri, 335
β particle, 336
Bernoulli, Jakob and Johann, 208, 240
Big Bang, 518
Big Crunch, 519
bonds, 394–395
    and interest rates, 394–395
    coupon payments, 394
    face value, 394
    maturity, 394
    trading price, 394
Brahe, Tyco, 73–74

calculus, 110–111, 208–210
    differential, 117–120, 151–153, 217–220, 222–228
    fundamental theorem of, 125–128, 153–155
    integral, 120–125, 148–153, 436–443
    notation, 123, 145
Cambrian explosion, 349
carbon dating, 353–354
cardioid, 456
Cartesian coordinate system, 81
Cauchy, Augustin Louis, 140, 208
Cavalieri, Bonaventura, 92
    principle, 93–94
Cavendish, Henry, 196, 203
celestial
    equator, 30, 31
    navigation, 69
    sphere, 30, 31
center of force, 180
center of mass, 64, 67
centripetal force, 180, 185–191
centroid, 64
chain
    reaction, 375–377
    rule, 224, 226
characteristic triangle, 126
Chauvet caves, 352, 355
circle, 86–87
    area, 57
    circumference, 8
    sector, 57
Closed Universe, 517
Columbus, 70
comets, 105–106
components, 254

    horizontal, 254
    vertical, 254
compression, 263
concavity, 314–318
    and Newton's method, 331
    down, 314
    test, 315
    up, 314
conic sections, 52–56, 83–86, 90–92, 115–116
    polar equations for, 458–460
conjugate, 211
conservation of angular momentum, 533
conservation of energy, 485
conservation of momentum, 495–497
Consumer Price Index, 395–397, 424–425, 433
consumer surplus, 419–422
    and shifts in price, 421–422
continuous function, 212–217
    at a number, 212
    on an interval, 215
    on the domain, 215
continuity, 212–217
    criterion, 212
    theorem, 215
coordinate, 81
    axes, 81
    system, 81, 453
coordinates, 81
    polar, 453
Copernicus, 72–73
    De Revolutionibus Orbium Celestium, 72
cosine, 11–12, 87–90
cosmological principle, 518
cost,
    average total, 405
    average variable, 405, 426
    capital, 403
    fixed, 404
    functions, 403–405
        U-shaped, 405
    labor, 403
    marginal, 404–405, 426
    variable, 404
cost analysis for electric utilities, 410–412
Cournot, Augustin, 386, 428–429, 433
critical
    mass, 376–377
    number, 232
culture, 365
    medium, 365

curvature, 283
curve, length of a, 138–140

Darwin, Charles, 334–335
    The Origin of Species, 334
Dead Sea Scrolls, 352, 355
decay constant, 337
decreasing functions, 230–234
definite integrals, 135, 436–438
    and area, 120–125, 153–155, 438–441
    and consumer surplus, 420–422
    approximation of, 156–159, 466–467
    as area in polar coordinates, 460–462
    as impulse of a force, 493
    as length of a curve, 138–140
    as surface area, 443–446
    as volume of revolution, 136–138
    as work of a force, 483
    evaluation of, 156–159, 448–453
degree, 6–7
De L'Hospital, 236, 240
    Analyse, 240, 246
    pulley problem, 251
    rule, 227
demand, 397
    and supply, 397–399
    curve, 398
    function, 398
    price elastic, 400
    price inelastic, 400
derivatives, 132–134, 218–219
    as rates of change, 220–222
    computation of, 133–134, 151–153, 222–226
    of exponential functions, 302, 308
    of inverse trigonometric functions, 311–312
    of logarithm functions, 308
    of trigonometric functions, 228–230
    rules for, 153, 222–225
Descartes, Rene, 80–81
    Discourse de la méthod, 80–81
    formula for a lens, 281
    La Géométrie, 81
differentiable function, 217–220
    at a number, 218
    on an interval, 219
    on the domain, 219
differentiability criterion, 218
differential, 126
differential calculus, 117–120, 151–153, 217–220
differential equations, 462–466
    and chemical reactions, 472
    and physical reality, 463–464
    and radioactive decay, 473
    boundary conditions for, 465
    first order, 463, 472–473
    general solutions of, 465
    initial conditions for, 465
    order of, 463
    second order, 463
    separable, 465
    solutions of, 463, 466–467
differentiation formulas, 223, 224, 225, 229, 230, 302,
        306, 308, 311, 312
Diophantus, 48
    Arithmetica, 48, 297
disintegration constant, 337
distance formula, 82
distance to stars, 21–22, 533
domain of a function, 209

e, definition of, 302
Earth,
    distance to Moon, 17
    distance to Sun, 17, 94–95
    formation of, 342–344
    geologic history of, 342–347
    life on, 348–351
    mass of, 196
    orbit around Sun, 41–43, 72, 101–102, 194–195
    radius of, 9–10
eccentricity,
    astronomical, 92
    linear, 91–92
    of planetary orbits, 95
E-coli, 364, 383
    experiment, 368–370
ecliptic, 30, 31
    obliquity of, 30, 41
economies of scale, 426
Eddington, Arthur Stanley, 19, 21
Einstein, Albert, 374, 514
electron volt, 374
    MeV, 374
elasticity, 400
ellipse, 54–55, 90–92
    area of, 94
    center of, 54
    diameter of, 54
    eccentricities of, 91–92
    equation of, 91

ellipse (*cont.*),
    focal points of, 54
    major axis of, 54
    minor axis of, 54
    semimajor axis of, 74, 91
    semiminor axis of, 91
    standard equation of, 91
    standard position of, 91
energy, 482–485
    and matter, 374–376
    and work, 482–483
    binding, 374
    conservation of, 485
    kinetic, 482–483
    potential, 485
epicycles, 43–45
equation,
    graph of, 83
    in polar coordinates, 453
    linear, 115
    of a circle, 86
    of a hyperbola, 115–116
    of a line, 114–115
        point-slope form, 114
        slope-intercept form, 115
    of a parabola, 84–85
    of an ellipse, 91
    of motion, 168–169, 528–531
    quadratic, 116
equations and graphs, 83
Eratosthenes, 9–10
Escherich, Theodor, 383
Euclid, 17, 24, 250
    Elements, 24, 32, 332
Euler, Leonhard, 59, 140, 208, 286, 300, 302, 511
    Introductio in Analysin Infinitorum, 208, 300
exponential functions, 300–302
    derivatives of, 302, 308
exponential growth, 356–357, 365–366
exponents, laws of, 301
eyepiece, 279

Fermat, Pierre de, 251, 297
    last theorem, 297
    principle of least time, 273
fermentation process, 370–372
    batch, 372
    continuous, 382
fermentor, 370
Fermi, Enrico, 376

Feynman, Richard, 203
first derivative test, 230
fission, 375
    fragments, 375
force, 180, 253–254
    and impulse, 493–494
    and momentum, 495–497
    and work, 476–483
    centripetal, 180, 185–191
    components of, 254
        horizontal, 254
        vertical, 254
    compression, 263
    field, 484–485
    gravitational, 165–166, 193–197, 505–506
    in polar coordinates, 511–515
    magnitude of, 180, 253–254
    moment of, 267–268
    net, 254
    parallelogram, law for, 254
    resultant, 254
    strong nuclear, 374
    variable, 476–477
fossil record, 349–351
Fourier, Jean Baptiste, 123, 208
free fall, 477–478, 480–482, 528–531
Friedmann, Alexander, 520
    equation, 520
functions, 128–132, 140, 209
    antiderivative of, 134–135, 312–313
    concavity of, 314
    constant, 129
    continuous, 212–217
    decreasing, 230–234
    definite integral of, 135, 437–438
    derivative of, 132–133, 218–220
    differentiable, 217–220
    domain of, 209
    elementary, 466
    exponential, 300–302
    graph of, 129–130
    graphing of, 320–322
    in polar coordinates, 454
    increasing, 230–234
    inverse, 302–306
    linear, 129
    logarithm, 306–308
    maximum and minimum values of, 135–136,
        216, 231–233, 236
    one-to-one, 303

periodic, 322,
    periodicity of, 322
    quadratic, 132
    simple, 154
    symmetry for, 321
fundamental theorem,
    of arithmetic, 24
    of calculus, 125–128, 153–155, 436, 441, 442–443
Fuss, Nicolaus, 250

g, gravitational constant, 76–77, 107, 165–166, 180, 476
G, universal gravitational constant, 196–197
Galileo, 76–80, 250, 265
    experiments, 76–79, 265, 269–272, 298
    law of inertia, 77
    Starry Messenger, 78
    telescope, 279
γ ray, 336
Gauss, Karl Friedrich, 98, 107, 208, 516
    equation, 98, 106–107
geologic column, 349
geologic history, 342–346
George Washington Bridge, 258, 263–264
Gerbert, 68
gnomon, 10, 30–32
Goddard, Robert Hutchings, 499
    A Method for Reaching Extreme Altitudes, 499
Gompertz, Benjamin, 381
    population model, 381–82
graphs, 83
    asymptotic behavior of, 318–320, 322
    in polar coordinates, 454–459
    intercepts of, 83
    of equations, 83
    of exponential functions, 301
    of functions, 129–130, 320–322
    of inverse functions, 305
    of inverse trigonometric functions, 311–312
    of logarithm functions, 306
    of polar equations, 454–460
    of trigonometric functions, 323–325
    periodicity of, 322
    shifting of, 321–322
    symmetry of, 321
gravitational constant, 76–77, 107, 165–166, 180, 196–197
gravity, 165–166, 193–197, 505–506, 510–511, 516
    and shape, 506–511
    and the inverse square law, 190–191, 193–196, 505–511
Greek,
    Anthology, 23, 26, 48

geometry, 32–35, 52–56
    mathematics, 5, 72, 107
    number system, 18–19, 23
growth of micro-organisms, 364–368
    death phase, 366–367
    doubling time, 365
    exponential growth phase, 365
    lag phase, 365
    stationary phase, 366
gun barrel,
    analysis of, 488–490
    caliber, 490
    forces in, 489
    pressure in, 489–490

half-life, 341
Halley, Edmund, 173–174
    comet, 105–106
Harrison, John, 69
Herschel, William, 197, 285
Hipparchus, 32, 33, 36
Hooke, Robert, 173–174
    law for springs, 486
Hubble, Edwin, 516
    constant, 516–517
    law, 516
Hubble Space Telescope, 517, 521
human life on Earth, 351
Huygens, Christiaan, 103, 107, 111, 285
hyperbola, 55, 115–116
    axis of, 55
    center of, 55
    equation of, 115–116
    focal points of, 55
hyperbolic surface, 332

image
    real, 278
    virtual, 279
implicit differentiation, 244
impulse of a force, 493
    and momentum, 494
    as a definite integral, 493
inclined plane, 76–77, 265–266, 269–272
income stream, 391–392
increasing functions, 230–234
indefinite integral, 312–313
index
    of inertia, 267–269
    of refraction, 275
Industrial Revolution, 198

inertia,
  index of, 267–269
  law of, 77, 180
  moment of, 267
inflation, 395–397
  continuous rate of, 396
  rate of, 395
integral calculus, 120–125, 148–153, 436–443
  and area, 120–125, 148–151, 438–441
  and power series, 156–159
  methods of, 448–453
integral,
  definite, 123, 135, 436–438
  indefinite, 312–313
integral formulas, 124, 135, 138, 139, 154, 155, 446,
      448, 450, 451, 461
integrand, 448
integration,
  by parts, 451–453
  by substitution, 448–450
  by trigonometric substitution, 450–451
intercepts, 83
interest, 387–388
  compounding of, 387–388
intermediate value theorem, 216–217
internal combustion engine, 524
interval, 215
  closed, 215
  open, 215
inverse function, 302–306
  derivative of, 306
  graph of, 305
inverse square law, 187–191
  and polar coordinates, 511–515
inverse trigonometric functions, 311–312
investment plan, 389–390
ion, 337
ionization chamber, 337
irrational number, 5
isotope of an element, 336, 372

Jevons, Stanley, 386
joule, 374

κ, Kepler's constant, 185
Kepler, Johann, 73–75
  Astronomia Nova , 74, 95
  constant κ, 185
  equation, 99, 100–101
  laws of planetary motion, 74–75, 181–185,
      191, 203, 511–515, 527–528

  mathematics of orbits, 94–99, 101–102
    computer model for, 107
kilowatt, 433

L'Hospital, see De L'Hospital
Lagrange, Joseph Louis, 140, 197–198, 203, 208,
Laplace, Pierre Simon, 197–198, 203
law of,
  conservation, 485, 495–497, 533
  exponents, 301
  Hubble, 516
  hydrostatics, 255
  inertia, 77
  inverse square, 187–191
  logarithms, 306–307
  motion, 180
  planetary motion, 74–75
  refraction, 274
  the lever, 67
  universal gravitation, 193–196
latitude, 69
least squares, 408–410
Le Duc formula, 488, 533
Leibniz, Gottfried Wilhelm, 111, 117, 120, 125–126,
      128–129, 140–141, 145, 174
  differential calculus, 117–120
  fundamental theorem of calculus, 125–128
  integral calculus, 120–125
  notation, 123, 145
  Nova methodus, 117
length, 4
lens maker's equation, 281
lenses, 277–284
  as magnifying glass, 283–284
  axis of symmetry of, 277
  basic properties of, 277–280
  compound, 279
  curvature of, 283
  Descartes's formula for, 281
  diopters of, 283
  focal length of, 277
  focus of, 277
  focusing, 283
  image of, 278
    real, 278
    virtual, 279
  magnification of, 283, 285
  objective, 279
  power of, 283
  spherical, 278

symmetric, 277
thin, 278
Leonardo da Vinci, 70
Lerner Index, 429–430
lever, 67
fulcrum of, 67
law of, 67
Leverriere, 197, 515, 533
Libbey, Willard F., 353
light, 285–286
reflection of, 272–273
refraction of, 273–275, 295–297
speed of, 294–295, 298
light year (LY), 22, 294–295
limit, 7, 12, 57, 117–120, 140, 209–211
notation, 7, 12, 57, 140, 209–211
test, 209–211
"$\frac{0}{0}$" type, 210–211
linear
equation, 115
eccentricity, 91–92
lines, 111–115
equations of, 114–115
slope of, 113–115
logarithmic differentiation, 308
logarithms, 306–309
as computational tool, 309
base of, 306
derivatives of, 308
laws of, 307
natural, 307
logistics model, 358–361
longitude, 69, 70
Lucy, 351
lunes, 66, 70

M'Kendrick and Pai, 368
M.K.S. system of units, 200–201, 374–375, 433
magnification, 283, 284–285
main cable, 259–262, 263–264, 289–290
safety factor, 264
ultimate strength, 264
marginal
cost, 404–405
revenue, 414
markets, 397–399
clearing of, 398
competitive, 399
equilibrium price in, 398
forces in, 398

forecasting, 403
mechanisms of, 398
monopoly on, 399
supply and demand in, 398–399
Mars, orbit of, 74, 95–96
Marshall, Alfred, 386, 433
mass number, 336
mass
of Earth, 196
of Sun, 196
units of, 200–201, 374–375
mathematics, 522
and economics, 422–423
from the Middle Ages, 68
from the 16th century, 68
Greek, 72, 107
matter and energy, 374
maximum value, 135–136, 216, 231–233, 236
local, 231–233
max/min theorem, 232
mean value theorem, 227
measurable, 4
mega light year, 516–517
Mercury, orbit of, 95, 515–516, 533
meridian, 30
Method, 63–65, 92
method
of least squares, 408–410
of Newton, 246–247, 331
of successive approximations, 99–101
methods of integration, 448–453
integration by parts, 451–453
reverse common denominator, 360–361, 448
substitution, 448–450
trigonometric substitution, 450–451
Michelson's experiment, 294–295
micro-organisms, 364–368, 373
minimum value, 135–136, 216, 231–233, 236
local, 231–233
moment,
of force, 267–269
of inertia, 267
momentum, 492–498
and force, 493–495
and impulse, 493–494
conservation of, 495–497
Monod, Jacques, 382
equation, 382
Moon,
age and formation of, 377–378

Moon (*cont.*),
    Apollo missions to, 377–378, 505
    distance from Earth, 17
    orbit of, 191–193, 194
mortgages, 423
moving points, 160–166
muzzle velocity, 171, 490–491

navigation, 69
newton, 200–201, 372
Newton, Isaac, 141, 148–149, 173–174, 178, 180–181,
       197–198, 202–203, 285
    De Analysi, 148–151
    differential calculus, 151–153
    fundamental theorem of calculus, 153–155
    integral calculus, 156–159
    inverse square law, 187–191, 200, 203
    law of universal gravitation, 193–194, 200
    laws of motion, 180
    method, 246–247, 331
    Methodus Fluxionum, 160
    Principia, 173–174, 180–181
    System of the World, 181, 191
    telescope, 285
norm of a partition, 437
nuclear,
    bomb, 377
    clocks, 334–335, 344–347
    fission, 375
    fusion, 377
    reaction, 375–377
        controlled, 376
        subcritical, 376
        supercritical, 376
number system,
    Archimedes, 18–19
    Greek, 23
    rational, 5, 24
    real, 5–6, 24

Ω, expansion constant, 520
objective lens, 279
OPEC, 401–403, 412
    and gasoline prices, 424–425
Open Universe, 519, 520–521
optics, 250–251, 272–280
orbit
    of the Moon, 191–193
    of the planets, 39–43, 73–75, 94–99, 101–103,
        187–191, 193–195,
output, 403–404

break even, 415
    profit maximizing, 415

Pappus's theorems, 144–145
parabola, 52–54, 83–86
    axis of, 53
    directrix of, 53
    equations of, 84–86
    focus of, 53
parabolic section, 53
    vertex of, 53
parallax, 22
parallelogram law, 167, 254
partition, 437
    norm of, 437
perihelion, 48, 74, 107
period,
    of an orbit, 75
    of a function, 322
periodic, 322
Perrault's problem, 125–126, 309–311
π, 8–9
planets,
    discovery of, 197
    orbits of, 39–43, 73–74, 94–99, 101–103,
        187–195
plate tectonics, 343–344
Plato, 32
Plutarch, 17
point of inflection, 315
polar
    axis, 453
    coordinate system, 453
polar coordinates, 453
    and area, 460–462
    and conic sections, 458–460
    graphs in, 454–460
    transformation equations for, 456
pole, 453
population growth, 356–364
    exponential, 357, 365–367
    logistics model for, 358–361
    of micro-organisms, 364–368
position function, 160
potassium-argon clock, 347–348
power
    rule, 153, 223, 225
    series, 159
present value, 390–392
price, 398–401

actual, 399
cost-plus, 413
elasticity, 400–401
        of demand, 400
        of supply, 401
market equilibrium, 398
taker, 399
prime number, 24
Principia, 173–174, 180–181
        and gravity, 193–197
        and Kepler's laws, 190–191, 203, 511
        and polar coordinates, 511–515
        and the inverse square law, 187–191
        and the law of universal gravitation, 194–197
        and the orbit of the Moon, 191–193
        impact of, 197–198
principle
        Cavalieri's, 93–94
        of least time, 273
probability theory, 176–178
        and radioactivity, 342
producer surplus, 428
product rule, 223
production costs, 403–406
profit, 414–416
        maximizing, 415
        maximizing for a refinery, 416–419
projectiles, 166–173
pseudosphere, 332
Ptolemy, Claudius, 32, 46
        Almagest, 32–39, 43–46
        Geographia, 69–70
        solar model, 35–39
        theorem, 34
        theory of epicycles, 43–45
pulley problem, 251, 256
Pythagorean theorem, 4, 46
Pythagoreans, 5, 26

quadrants, 81
quadratic
        equation, 116
        formula, 23–24
quotient rule, 223

radian measure, 7–8, 87–88
radioactive elements, 336–337
        decay or disintegration constant of, 337
        half-life of, 341
        series of, 342
radioactivity, 335–342, 374

and probability, 342
        equations for, 339, 341, 473
rate
        of change, 220–222
        of decrease, 221
        of increase, 220
rational number, 5, 24
rationalizing, 211
real numbers, 5–6
rectangular coordinate system, 81
refraction, 273, 295–297
        index of, 275
        law of, 274
regions in the plane, 86
Renaissance, 52
resultant, 254
retrograde motion, 32, 44
revenue, 412–414
        functions, 413–414
        marginal, 414
        total, 413
Richter scale, 329
Riemann, Georg Bernhard, 208, 516
Riemann sum, 437
right triangle, 4, 10–12
rockets, 498–505
        equation of, 503
        propulsion of, 498–503
Roebling, John, 258
Roemer, Olaus, 298
Roentgen, Wilhelm, 335
rotation, 266–267, 290–293
roundoff error, 157, 492
rubidium-strontium clock, 346
Rutherford, Ernest, 334, 335, 341–342
        experiments, 336–340

safety factor, 264
Scientific Revolution, 198
seasons, 32
secant, 104
second derivative test, 317
semimajor axis, 74, 91
semiminor axis, 91
Shroud of Turin, 352, 355
sigma notation, 59, 437
sine, 11–13 , 87–90
slope, 113
slug, 201
Smith, William, 334

Snell, Willebrord, 251
    law of refraction, 274–276
solar system,
    stability, 197–198
    structure of, 72–76, 193–198
specific growth rate, 358
speed, 163–164
    of light, 294–295, 298
spring constant, 486
Springfield rifle, 490–492
springs, 485–488, 532–533
    Hooke's law of, 486
    natural length of, 486
stair master, 523
stellar parallax, 22
substitution method, 448–450
successive approximations, 99–101
sum rule, 223
summer solstice, 30, 41, 102
Sun,
    and nuclear fusion, 377
    distance from Earth, 17, 94–95
    mass of, 196
    position of, 41–43, 74–75
    size of, 17
supply,
    curve, 398–399
    function, 398–399
supply and demand, 397–401
surface,
    hyperbolic, 332
    of revolution, 443
surface area, 443–447
    and enzyme activity, 469–470
suspension bridges, 250, 257–265
    anchorages of, 259
    center span of, 258
    dead load of, 258
    deck of, 258
    live load of, 259
    main cable of, 259–262, 263–264, 289–290
    parameters $d, s,$ and w for, 260
    sag of, 259
    side spans of, 258
    strands of, 263–264, 289–290
    suspender cables of, 259
    towers of, 259
symmetric,
    about the origin, 321
    about the y-axis, 321

symmetry, 321
systems of units, 200–201, 374–375, 433

tangent, 11–12, 90
tangent line, 117–120
    existence of, 217–220
    slope of, 117–120
taxes, 432–433
telescope, 279–280, 284–285
    Galileo's, 78–79, 279
    Newton's, 285
    reflecting, 285
    refracting, 279
tension, 256
test points, 234
theory of relativity, 332, 516
thrust, 499–501
    momentum component of, 500
    pressure component of, 500
torque, 267–268
tractrix, 142, 309–310, 332
trigonometric functions, 87–90, 323–325
    derivatives of, 228–230
    inverses of, 311–312
trigonometric substitution, 450–451
trigonometry, 11–13, 47, 87–90
tropics of Cancer and Capricorn, 41
Tsiolkovsky, Konstantin, 498–499

ultimate strength, 264
unit circle, 87
units, 200–201, 374–375, 433
universal gravitation, 193–196
universe, 13–17, 30–32, 193–196, 197–198, 515–522
    age, 521–522
    average density of, 520
    Big Bang, 518
    Big Crunch, 519
    Closed, 519
    critical density of, 520
    expansion of, 516–519
    Open, 519
    size of, 21–22, 533
    structure of, 516–519
uranium decay, 334, 341–342, 375–377, 473

vectors, 253–254
    components of, 254
    direction of, 253–254
    magnitude of, 253–254
velocity, 162–164

angular, 266
Verhulst, Pierre Francois, 358–359
vernal equinox, 30, 101
Viète, 68
volume of revolution, 136–138

watt, 433
Weierstrass, Karl, 208
weight, 180
winter solstice, 30
work, 476–484
    and kinetic energy, 482–485
    as a definite integral, 482–483
world population, 356–364

year, 32, 101